AF816549

Raman Spectroscopy for Nanomaterials Characterization

Challa S. S. R. Kumar
Editor

Raman Spectroscopy for Nanomaterials Characterization

With 313 Figures and 11 Tables

Springer

Editor
Challa S. S. R. Kumar
Center for Advanced Microstructures and Devices
Baton Rouge, LA
USA

ISBN 978-3-642-20619-1 e-ISBN 978-3-642-20620-7
DOI 10.1007/978-3-642-20620-7
Springer Heidelberg Dordrecht London New York

Library of Congress Control Number: 2012930989

© Springer-Verlag Berlin Heidelberg 2012

This work is subject to copyright. All rights are reserved, whether the whole or part of the material is concerned, specifically the rights of translation, reprinting, reuse of illustrations, recitation, broadcasting, reproduction on microfilm or in any other way, and storage in data banks. Duplication of this publication or parts thereof is permitted only under the provisions of the German Copyright Law of September 9, 1965, in its current version, and permission for use must always be obtained from Springer. Violations are liable to prosecution under the German Copyright Law.
The use of general descriptive names, registered names, trademarks, etc. in this publication does not imply, even in the absence of a specific statement, that such names are exempt from the relevant protective laws and regulations and therefore free for general use.

Printed on acid-free paper

Springer is part of Springer Science+Business Media (www.springer.com)

Editor-in-Chief

Challa S. S. R. Kumar
Center for Advanced Microstructures and Devices
Baton Rouge, LA
USA

Contents

List of Contributors

Ramón A. Álvarez-Puebla Departamento de Química-Física and Unidad Asociada, CSIC-Universidade de Vigo, Vigo, Spain

A. Anders Technische Universität Dresden, Physikalische Chemie und Elektrochemie, Dresden, Germany

Valery N. Bliznyuk Materials Science and Engineering Department, College of Engineering and Applied Sciences, Western Michigan University, Kalamazoo, MI, USA

Colin J. Campbell EaStCHEM, School of Chemistry, University of Edinburgh, Edinburgh, UK

Carlo Carraro Department of Chemical and Biomolecular Engineering, University of California, Berkeley, CA, USA

Gabriela Chakarova Oxonica Inc, Mountain View, CA, USA

C. Y. Chan Department of Physics, The Chinese University of Hong Kong, Shatin, NT, Hong Kong

Han-Wei Chang Department of Chemical Engineering, National Chung Hsing University, Taichung, Taiwan

Hyeonsik Cheong Department of Physics, Sogang University, Seoul, South Korea

Remy Cromer Oxonica Inc, Mountain View, CA, USA

H. Dietz Technische Universität Dresden, Physikalische Chemie und Elektrochemie, Dresden, Germany

Gregory S. Doerk IBM Almaden Research Center, San Jose, CA, USA

Karen Faulds Centre for Molecular Nanometrology, Department of Pure and Applied Chemistry, WestCHEM, University of Strathclyde, Glasgow, UK

Kentaro Furusawa Nanophotonics Laboratory, RIKEN, Wako, Saitama, Japan

N. L. Garrett School of Physics, University of Exeter, Devon, UK

Cristina Gellini Dipartimento di Chimica "Ugo Schiff", Università di Firenze, Sesto Fiorentino, Italy

Yury Gogotsi Department of Materials Science and Engineering and A.J. Drexel Nanotechnology Institute, Drexel University, Philadelphia, PA, USA

Norihiko Hayazawa Nanophotonics Laboratory, RIKEN, Wako, Saitama, Japan; Near–Field NanoPhotonics Research Team, RIKEN, Wako, Saitama, Japan; CREST, Japan Science and Technology Agency, Kawaguchi, Saitama, Japan

Ed Holland Oxonica Inc, Mountain View, CA, USA

Ping-Chieh Hsu Department of Chemical Engineering, National Chung Hsing University, Taichung, Taiwan

Katsuyoshi Ikeda Division of Chemistry, Graduate School of Science, Hokkaido University, Sapporo, Japan

Massimo Innocenti Dipartimento di Chimica "Ugo Schiff", Università di Firenze, Sesto Fiorentino, Italy

Dae Hong Jeong Department of Chemistry Education, Seoul National University, Seoul, Korea

Bong-Hyun Jun School of Chemical and Biological Engineering, Seoul National University, Seoul, Korea

Ramesh Kattumenu Department of Mechanical Engineering and Materials Science, School of Engineering and Applied Sciences, Washington University, St. Louis, MO, USA

Mikael Käll Department of Applied Physics, Chalmers University of Technology, Göteborg, Sweden

Gunsung Kim Department of Chemistry Education, Seoul National University, Seoul, Korea

N. E. Korniyenko Physics Department, Taras Shevchenko National University of Kyiv, Kyiv, Ukraine

Chang H. Lee Department of Mechanical Engineering and Materials Science, School of Engineering and Applied Sciences, Washington University, St. Louis, MO, USA

Yoon-Sik Lee School of Chemical and Biological Engineering, Seoul National University, Seoul, Korea

J. Li Department of Physics, The Chinese University of Hong Kong, Shatin, NT, Hong Kong

Roya Maboudian Department of Chemical and Biomolecular Engineering, University of California, Berkeley, CA, USA

Robert C. Maher Department of Physics, Imperial College London, Blackett Laboratory, London, UK

Maurizio Muniz-Miranda Dipartimento di Chimica "Ugo Schiff", Università di Firenze, Sesto Fiorentino, Italy

Michael J. Natan Oxonica Inc, Mountain View, CA, USA

A. P. Naumenko Physics Department, Taras Shevchenko National University of Kyiv, Kyiv, Ukraine

Michael A. Ochsenkühn EaStCHEM, School of Chemistry, University of Edinburgh, Edinburgh, UK

H. C. Ong Department of Physics, The Chinese University of Hong Kong, Shatin, NT, Hong Kong

Sebastian Osswald Department of Physics, Graduate School of Engineering and Applied Science, Naval Postgraduate School, Monterey, CA, USA

Nicolas Pazos-Perez Departamento de Química-Física and Unidad Asociada, CSIC-Universidade de Vigo, Vigo, Spain

Sharron G. Penn Oxonica Inc, Mountain View, CA, USA

W. Plieth Technische Universität Dresden, Physikalische Chemie und Elektrochemie, Dresden, Germany

Fanyao Qu Instituto de Física, Universidade de Brasília, Brasília, Brazil

Kerstin Ramser Department of Computer Science and Electrical Engineering, Luleå University of Technology, Luleå, Sweden

G. Sandmann Technische Universität Dresden, Physikalische Chemie und Elektrochemie, Dresden, Germany

Michael Y. Sha Oxonica Inc, Mountain View, CA, USA

Srikanth Singamaneni Department of Mechanical Engineering and Materials Science, School of Engineering and Applied Sciences, Washington University, St. Louis, MO, USA

Maria A. G. Soler Instituto de Física, Universidade de Brasília, Brasília, Brazil

Atsushi Taguchi Nanophotonics Laboratory, RIKEN, Wako, Saitama, Japan; CREST, Japan Science and Technology Agency, Kawaguchi, Saitama, Japan

Alvarado Tarun Nanophotonics Laboratory, RIKEN, Wako, Saitama, Japan

Dmitri A. Tenne Department of Physics, Boise State University, Boise, ID, USA

Lianming Tong Department of Applied Physics, Chalmers University of Technology, Göteborg, Sweden

Yu-Chen Tsai Department of Chemical Engineering, National Chung Hsing University, Taichung, Taiwan

Kohei Uosaki Division of Chemistry, Graduate School of Science, Hokkaido University, Sapporo, Japan;International Center for Materials Nanoarchitectonics, National Institute for Materials Science, Tsukuba, Japan

Mary M. Y. Waye School of Biomedical Sciences, The Chinese University of Hong Kong, Shatin, NT, Hong Kong

Hongxia Xu Oxonica Inc, Mountain View, CA, USA

J. B. Xu Department of Electronic Engineering, The Chinese University of Hong Kong, Shatin, NT, Hong Kong

V. M. Yashchuk Physics Department, Taras Shevchenko National University of Kyiv, Kyiv, Ukraine

Duhee Yoon Department of Physics, Sogang University, Seoul, South Korea

Angle-Resolved Surface-Enhanced Raman Scattering

1

C. Y. Chan, J. Li, H. C. Ong, J. B. Xu, and Mary M. Y. Waye

1 Definition of the Topic

This chapter describes the use of angle-resolved techniques to study the Raman scattering arising from periodic metallic arrays. In fact, by matching the angle-dependent reflectivity with Raman spectroscopy, it is possible to single out a particular electromagnetic resonance and study its contribution to surface-enhanced Raman scattering. A wealth of information can be obtained.

2 Overview

Surface-enhanced Raman scattering (SERS) offers high specificity in molecular identification and is a promising technique for making biological and chemical sensors. However, a major weakness of this technique is the lack of reliable SERS substrates. Currently available substrates usually contain irregular active sites that suffer from strong spatial and temporal fluctuation in Raman intensity and thus do not produce stable Raman output. On the other hand, because of their well-defined structure, periodic metallic arrays excite more controllable local fields and can be exploited for achieving high sensitive SERS. The SERS obtained from these periodic structures exhibit strong spectral and angular dependences that are associated with the generation and decay processes of the various electromagnetic

C.Y. Chan • J. Li • H.C. Ong (✉)
Department of Physics, The Chinese University of Hong Kong, Shatin, NT, Hong Kong

J.B. Xu
Department of Electronic Engineering, The Chinese University of Hong Kong, Shatin, NT, Hong Kong

M.M.Y. Waye
School of Biomedical Sciences, The Chinese University of Hong Kong, Shatin, NT, Hong Kong

Challa S. S. R. Kumar (ed.), *Raman Spectroscopy for Nanomaterials Characterization*,
DOI 10.1007/978-3-642-20620-7_1, © Springer-Verlag Berlin Heidelberg 2012

resonance modes. In this chapter, we employ the angle-resolved reflectivity and Raman spectroscopy to study the SERS arising from periodic arrays and discuss how the basic properties of surface plasmon polaritons (SPPs) affect the SERS profiles. In fact, by correlating both reflectivity and Raman spectroscopy and their related parameters, it is possible to unravel the underlying mechanism of SERS and thereby to determine the conditions that are critical in governing the resulting enhancement factor.

3 Introduction

In general, when a monochromatic light is scattered by a molecule, in addition to the Rayleigh scattering that produces exactly the same wavelength (frequency) as the incident photons, several sharp peaks next to the Rayleigh line are also observed due to the inelastic scattering of photons [1]. Because these distinctive peaks arise from the change in vibrational, rotational, or electronic energy of the excited molecule, they provide a feasible means for "fingerprint-like" molecular identification [2]. As a result, since the discovery of the Raman effect by C.V. Raman in 1922 [3], Raman scattering spectroscopy has become a popular analytical tool in materials characterization, biological and chemical sensing, etc. [4, 5]. However, Raman scattering suffers from one major drawback, that is, it has a very small scattering cross section σ_{RS}, which is ~12–14 orders of magnitude lower than that of fluorescence σ_{F} [1, 2]. In other words, for every 10^{12} photons incident on a molecule, only one or less undergoes Raman scattering. Therefore, Raman scattering does not produce strong signal easily, particularly in situations where the concentration of target analyte is very low. However, this limitation can be partly overcome by the fact that Raman intensity is proportional to the square of the incident field. For a simple oscillator-like molecule vibrating with angular frequency ω_v, the Raman intensity I_{RS} can be expressed as [1, 4]:

$$I_{RS} \propto \left(\frac{\partial\alpha}{\partial Q}E_0[\cos(\omega_o - \omega_v)t + \cos(\omega_o + \omega_v)t]\right)^2, \tag{1.1}$$

where α is the molecular polarizability, Q is the nuclear displacement, and ω_o and E_o are the angular frequency and the amplitude of the incident field, respectively. The first and second components within the square brackets are defined as Stokes $\hbar(\omega_o - \omega_v)$ and anti-Stokes $\hbar(\omega_o + \omega_v)$ emissions, respectively [1]. Nevertheless, this equation clearly states that increasing the field strength E_o of the excitation light can possibly increase both the Stokes and anti-Stokes Raman emissions.

As a result, several schemes including cavity-enhanced Raman [6], photonic crystal–enhanced Raman [7], surface-enhanced Raman [8], etc., have been put forward in an attempt to increase the electromagnetic field strength for excitation. Among them, surface-enhanced Raman scattering (SERS), by far, is one of the most popular methods since it is simple and effective. By placing the target analytes in close proximity to a rough metal surface, Raman signal can be enhanced by 10^4–10^6 times over that from the flat surface [8]. Fleischmann and his coworkers [9] first realized SERS by discovering that rough metal surface could produce enhanced Raman signal at a level comparable to fluorescence. Since then, much work has been done in an attempt to explain this phenomenon. Eventually, after almost four decades of debate, two key mechanisms, namely, electromagnetic and chemical, are attributed to be responsible for SERS [8]. In fact, the electromagnetic enhancement arising from surface plasmon polaritons (SPPs) [10–12] is believed to play a major role in governing the overall Raman enhancement. In this scenario, SERS is proposed to involve five steps [10]. (1) In the first step, the impinging photon at ω_{exc} and incident angle θ_{exc} is coupled to a metal surface and produces SPPs for excitation. (2) Theses ingoing SPPs then polarize the molecules, creating large effective dipole moments, which results in excitation enhancement. (3) The molecular polarizations change if the molecules change vibrational state. (4) The change in molecular polarizations couples energy into SPPs at scattered frequency. (5) Finally, these outgoing SPPs decay into radiation at ω_{RS} and scattered angle θ_{RS}, giving rise to emission enhancement. Therefore, the Raman enhancement depends strongly on the generation and decay of ingoing and outgoing SPPs, which in turn determine the respective field strengths. For example, ingoing SPPs that have higher generation and lower decay rates would result in stronger field strength for excitation enhancement whereas sustaining a strong scattering field at a high radiative decay rate would be essential for the outgoing SPPs to decouple energy to the far field. Implicitly, the enhancement factor G_{SERS} for SERS is the product of the field enhancement factors for two SPPs in steps (1) and (4) and thus is given as [8]:

$$G_{\text{SERS}} \sim \left|\frac{E_{\text{loc}}(\omega_{\text{exc}}, \theta_{\text{exc}})}{E_{\text{inc}}(\omega_{\text{exc}}, \theta_{\text{exc}})}\right|^2 \left|\frac{E_{\text{loc}}(\omega_{\text{RS}}, \theta_{\text{RS}})}{E_{\text{inc}}(\omega_{\text{RS}}, \theta_{\text{RS}})}\right|^2, \tag{1.2}$$

where E_{loc} is the local field on the metal surface and E_{inc} is the incident field. $\left|\frac{E_{\text{loc}}(\omega_{\text{exc}}, \theta_{\text{exc}})}{E_{\text{inc}}(\omega_{\text{exc}}, \theta_{\text{exc}})}\right|^2$ and $\left|\frac{E_{\text{loc}}(\omega_{\text{RS}}, \theta_{\text{RS}})}{E_{\text{inc}}(\omega_{\text{RS}}, \theta_{\text{RS}})}\right|^2$ are defined as the excitation and emission enhancement factors, respectively, and they both display strong spectral and angular dependences. Maximizing both factors results in optimal Raman output [13–16].

Although the connection between SERS and SPPs is established in (1.2), how to effectively produce desirable SERS on a given metallic structure remains challenging. To address this issue, different plasmonic systems at different dimensionalities have been proposed, fabricated, and characterized with the same intention to maximize the local field strength at ω_{exc} and ω_{RS} [13–16]. A variety of structures have been found to support SERS well, including roughened metal electrodes, metal nanoparticles colloids, metal island films, metal nanorods, etc., or more specifically, substrates that contain nanoscale gaps or "hotspots" are particularly attractive since they concentrate the field in a very small mode volume, thereby leading to very high field strength [17–19]. However, even though impressive Raman enhancements are obtained from these structures, they are found to be very irreproducible since the size, shape, and density of hotspots cannot be controlled precisely. Strong fluctuation in Raman enhancement is found very often even following the same recipe in sample fabrication. Consequently, SERS suffers from the lack of trustworthy substrates. To overcome this barrier, attention has slowly been shifted toward the use of periodic metallic nanostructures since their fabrication and implementation are more reliable [5, 13, 15, 16, 20, 21]. The geometry of the lattice and individual basis can be tailored at the length scale of nanometers by using the state-of-the-art lithographic methods such as photo- and electron beam-lithography that have high throughput and reproducibility. In addition, as the generation and decay of SPPs are now more controllable because of the well-defined geometry, the local fields can be tuned to optimal by systematically varying the geometry. Nevertheless, these artificial patterns are potential in developing functional SERS substrates that on the one hand produce extremely strong Raman signal even at low concentration and on the other hand are simple, stable, and dependable.

3.1 Angular Dependence of Surface Enhanced Raman Scattering

3.1.1 Attenuated Total Internal Reflection

The Raman excitation and emission enhancements are angle dependent because they associate with the generation and decay processes of SPPs. In general, from the dispersion relation of a flat metal surface given in Fig. 1.1a, since the momentum of nonradiative SPPs are larger than that of the incident light, SPPs cannot be excited and decoupled directly [22]. As a result, several momentum matching schemes, including attenuated total internal reflection (ATR), grating coupling, etc., have been developed to overcome this limitation. In the ATR setup, for example, by impinging through a high refractive index prism, the incident light gains additional momentum to generate SPPs on the metal surface facing the air (Fig. 1.1b) [22]. By scanning the specular reflectivity of a monochromatic light as a function of incident angle, a dip is observed at angle θ_{exc} signifying the excitation of SPPs (Fig. 1.1c). Similarly, SPPs could also be scattered through the prism to far field during the decay process by following the same path but in a reverse direction to yield SPP emission [23]. As a result, as shown in Fig. 1.1a, consider in SERS that the excitation laser and the Raman Stokes emission have frequency ω_{exc} and ω_{RS}, an increase of Raman intensity would be seen when the laser in the prism is incident at angle θ_{exc} and the outgoing Raman emission is detected at θ_{RS} so that both the excitation and emission enhancements are optimized. Ushioda and Sasaki [24, 25] verified the above argument by observing strong angular dependence of SERS under ATR geometry. Both the angle dependence and the measured enhancement factor were found to agree well with the theoretical prediction deduced within the above framework. However, as the current interest in SERS is moving toward the

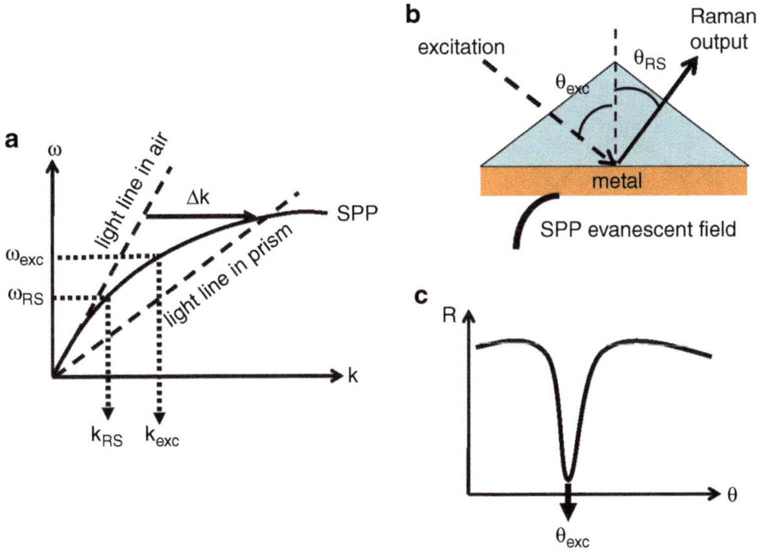

Fig. 1.1 (**a**) The (partial) dispersion relation of SPPs on a flat metal surface (*solid line*). In ATR, the light line in air shifts to the one in the prism by Δk (*dash lines*) and the ingoing and outgoing SPPs are excited if the laser and Raman emission at ω_{exc} and ω_{RS} follow the dispersion relation at k_{exc} and k_{RS} (*dot lines*). (**b**) The schematic diagram of the ATR setup. Strong Raman intensity is observed when the laser and the Raman emission are impinged at and collected from the prism at θ_{exc} and θ_{RS} given as $k = \sqrt{\varepsilon}\frac{\omega}{c}\sin\theta$, where ε is the dielectric constant of the prism. (**c**) Excitation of SPPs is signified at θ_{exc} by scanning the reflectivity R of a monochromatic light as a function of incident angle θ

development of lab-on-a-chip type biosensor that requires miniaturization of the active region, the use of a bulky prism coupler for exciting SPPs thus is not desirable.

3.1.2 Periodic Metallic Arrays

In fact, SPPs can be directly coupled to and decoupled from periodic metallic arrays. When a metal surface is indented with periodic corrugation, the surface waves scatter, leading to gaining or losing their momentum according to the phase-matching condition [22]. For example, for a one-dimensional (1D) periodic structure, light can be coupled into SPPs by following the phase-matching equation given as [22]:

$$\vec{k}_x = \vec{k}_{SPP} \pm n\vec{G}, \tag{1.3}$$

where $\vec{k}_x$ and $\vec{k}_{SPP}$ are the wavevectors of the in-plane x-component of the incident field (Fig. 1.4) and the SPPs, respectively and n is the integer. $\vec{G}$ is the reciprocal unit vector in x-direction and is equal to $2\pi/P$, where P is the period of the lattice. $\vec{k}_x$ is related to the incident angle θ by $2\pi\sin\theta/\lambda$. Therefore, in analogy to electronic

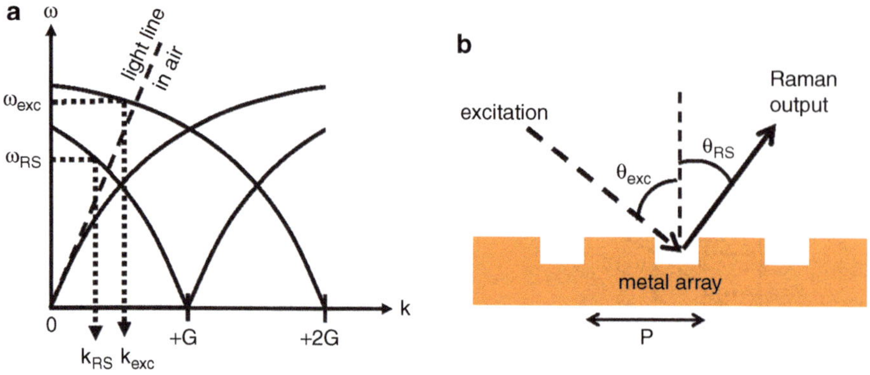

Fig. 1.2 (**a**) The (partial) dispersion relation of SPPs on 1D metallic array with period P (*solid lines*). Part of the dispersion relation now is within the light line (*dash line*) and excitation of ingoing and outgoing SPPs is possible if the laser and Raman emission at ω_{exc} and ω_{RS} follow the dispersion relation at k_{exc} and k_{RS} (*dot lines*). (**b**) Strong Raman intensity is observed from the array when the laser and the Raman emission are impinged at and collected from the array at θ_{exc} and θ_{RS} given as $k = \frac{\omega}{c}\sin\theta \pm n\frac{2\pi}{P}$

[26] and photonic [27] crystals, the dispersion relation of SPPs reproduces itself for every $\pm\vec{G}$ and yields the one being shown in Fig. 1.2a. From the figure, it is clear that some of the SPP resonance modes are now located within the light lines and they can be directly excited from or scattered to the far field without the need of prism coupler. As a result, as shown in Fig. 1.2b, we would expect similar angular dependence of SERS from the periodic arrays as that from the ATR setup, and strong excitation and emission enhancements can be found if the incident and detection angles of the laser and Raman emission coincide with the dispersion relation given in Fig. 1.2a. Tsang, Kirtley, and Bradley [28] were among the first to observe Raman enhancement from molecular monolayer attached on a Ag 1D diffraction grating when both the excitation and detection conditions were set according to the dispersion relation. Since then, several groups including Philpott [29], Knoll [30], Otto [31], etc., had demonstrated similar results, independently providing evidence in supporting the SPP-mediated SERS from metallic gratings. However, it was only recently that Baumberg et al. [32] had completely mapped out the angle-dependent reflectivity and SERS concurrently, unambiguously demonstrating the direct correlation between SPPs and SERS.

Other than the Bloch-like propagating SPP modes, different localized resonances can also be found in periodic structures [33]. Unlike the dispersive propagating SPPs, these localized modes are nondispersive with angle and their properties rely primarily on the detailed structure of the individual basis rather than on the lattice [34]. For example, localized resonance can be excited within 1D groove [33] and two-dimensional (2D) cylindrical hole when the hole size is larger than the critical diameter [35–37]. In addition, the field profile of localized modes is strongly concentrated around the bases instead of spreading all over the surface. These modes could interact with propagating SPPs, creating mixed resonances that affect the resulting field profile [33, 34]. Thus, changing the geometry of the basis can have a large impact on SERS. Nevertheless, due to the fact that the propagating and localized resonances is so geometry dependent that slight modification of the structure could cause significant change in the outcome, a complete understanding of the dependence of SERS on geometry is of great importance. In addition, knowing that a vast number of geometrical parameters including periodicity, hole shape, size, etc., are involved, a lengthy and systematic study is anticipated. Finally, the development of useful characterization tools to reveal the generation and decay processes of SPPs and their influences on SERS will certainly add value to this kind of study.

Although micro-Raman spectroscopy has been widely used for studying SERS, it cannot single out individual resonance modes one by one if more than one resonance modes are supported in the plasmonic systems since micro-Raman employs a large N.A., close to 0.9, objective that covers a wide range of angle for excitation and signal collection. All the resonance modes are excited and collected at the same time, making the resulting SERS spectrum difficult for interpretation. Therefore, an analytical tool that combines both angle- and wavelength-resolved techniques so that a particular mode in the dispersion relation can be pinpointed and its contribution to SERS can be characterized individually would be advantageous over the conventional micro-Raman setup. In this chapter, we discuss the use of angle-resolved reflectivity and Raman spectroscopy to study the optical properties of different electromagnetic modes arising from periodic arrays. In particular, by correlating the reflectivity and SERS, we are able to identify the roles SPP and other localized resonances play in SERS. Knowing that both SPPs and SERS are geometry dependent, we study the basic properties of SPPs, including decay lifetime and coupling efficiency, as a function of array geometry and relate these properties to SERS in attempt to develop a rational approach in designing a SERS substrate with good performance.

4 Experimental and Instrumental Methodology

4.1 Sample Fabrication

A variety of lithographic methods have been used for fabricating periodic arrays at the length scale of nanometers, including electron-beam lithography [15, 16, 38], nanosphere lithography [5, 13, 20, 21], focused ion beam [39], nanoimprint lithography [40, 41], interference lithography [42], etc. Although electron-beam lithography, focused ion beam, and nanoimprinting offer the best precision, stability, and reproducibility, they suffer from high instrumentation cost and moderate throughput. Here, we introduce interference lithography (IL) as our routine sample preparation method due to its low cost, high throughput, and good reproducibility. More importantly, this technique can be scaled up to wafer level without much difficulty [43].

1D and 2D periodic metallic arrays are fabricated by using IL and thin film deposition. Basically, IL crosses two identical laser beams to spatially create an

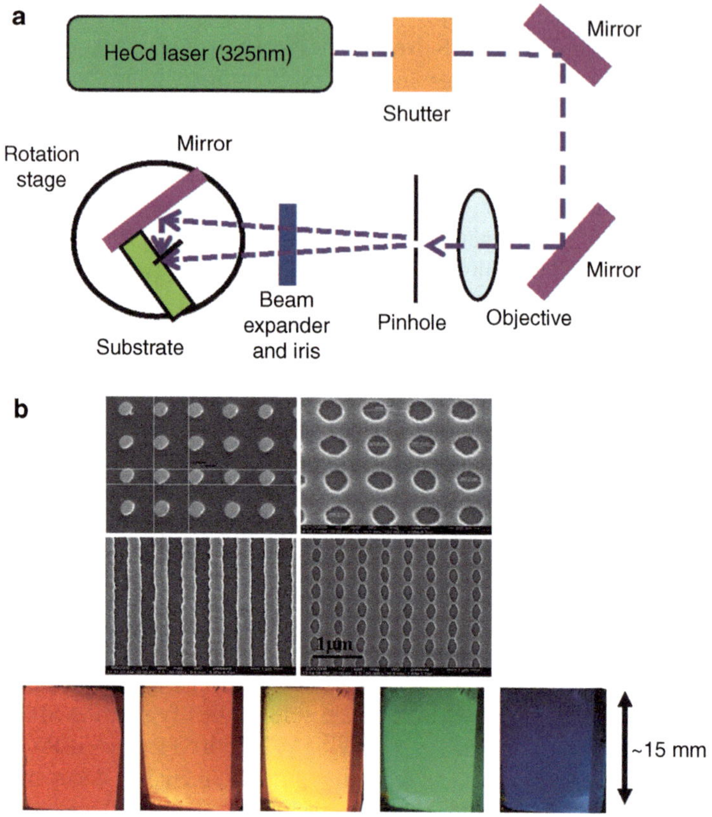

Fig. 1.3 (**a**) The schematic diagram of Lloyd's mirrior configuration for interference lithography. (**b**) The scanning electron microscopy (SEM) images of different periodic arrays fabricated by interference lithography (*top*). A typical sample is viewed at different angles and monochromatic colors are displayed at each viewing angle indicating the sample is highly uniform (*bottom*)

interference pattern to expose negative or positive photoresist [43]. The dimensionality of the arrays can be controlled by controlling the number of exposure; one and two exposures produce 1D and 2D patterns, respectively. In addition, in 2D case, negative photoresist yields well like structure whereas positive photoresist creates dot-like pattern. By changing the incident angle, the exposure time, and the thickness of photoresist, a wide range of geometrical parameters such as groove width, hole size, disk size, periodicity, etc., can be controlled independently (see Fig. 1.3b). After the exposure and development of the photoresist, periodic dielectric templates are obtained. In our study, a Lloyd's mirror configuration is used and its schematic is shown in Fig. 1.3a [44]. This setup consists of a HeCd laser at $\lambda = 325$ nm as the light source and a spatial filter for producing a uniform and clean Gaussian beam. The beam then passes a beam expander and an iris so that the central part of the Gaussian beam is collected to mimic a plane wave-like beam.

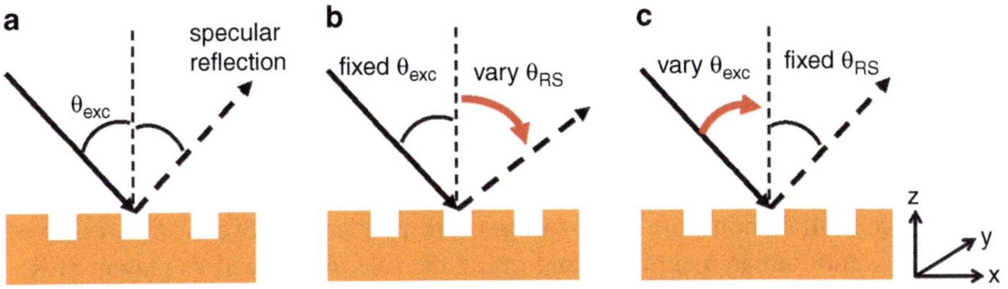

Fig. 1.4 Different configurations for (**a**) angle-dependent reflectivity measurement, (**b**) angle-dependent Raman detection scan, and (**c**) angle-dependent Raman incident scan

At the sample stage, a flat mirror is mounted perpendicular to the sample on a rotation stage. As part of the laser beam is reflected by the mirror, it interferes with the nonreflected part to form an interference pattern on the photoresist. The period P is given as: $P = \frac{\lambda}{2\sin\theta}$, where θ is the incident angle [44]. After lithography and baking, the templates are then transferred into an electron-beam or a thermal evaporation thin film deposition system for metallization. As the thickness of the metal film is larger than the skin depth of metal, the metallic arrays are considered as semi-infinitely thick.

4.2 Characterizations

For the angle-resolved measurements, a home-built computer-controlled goniometer is used and two types of measurements, namely, angle-dependent reflectivity and Raman spectroscopy, as shown in Fig. 1.4, are performed. For the reflectivity measurements, the specular reflection is collected from the sample by using a collimated white light from a quartz lamp as the excitation source (Fig. 1.4a). The reflection spectra are then obtained by normalized raw spectra with respect to that of the incident light. On the other hand, for the Raman measurements, two types of scans are conducted and they are shown in Fig. 1.4b, c. For the detection scans, the incident angle is fixed while the detection angle is varied independently. Likewise, for the incident scans, the incident angle will be scanned at a given detection angle. All the optical signals are collected by a CCD-detection system with a 0.25 m spectrograph. Polarizers are placed after the light sources and the sample for selecting the desired excitation and detection polarizations. Therefore, in the following sections, p-p is defined as both the incidence and detection are oriented in p-polarization while s-s refers to the s-polarization. Specifically for Raman measurements, a notch filer is used to remove the unwanted Rayleigh laser line from the spectra. Finally, by plotting the reflectivity and Raman spectra as a function of wavelength and angle, reflectivity and Raman mappings are obtained. The reflectivity mappings identify all the possible resonances on the metallic periodic structures, or simply map out the dispersion relations.

5 Key Research Findings

5.1 One-Dimensional (1D) Metallic Gratings

1D metallic gratings are first presented since they represent the simplest periodic plasmonic system among others. In shallow gratings where only propagating SPPs prevail, we shall demonstrate a one-to-one correspondence between angle-resolved reflectivity and Raman mappings, confirming the role resonant SPPs play in SERS. Enhanced Raman intensity is observed whenever the excitation laser and Raman emissions coincide with the ingoing and outgoing SPPs, leading to a strong angular dependence of SERS. On the other hand, in thick gratings, other than SPPs, localized resonances are also observed from the reflectivity mappings but they produce angle insensitive SERS. To account the experimental results, Rigorous Coupled-Wave Analysis (RCWA) method is used to simulate the reflectivity and field patterns, which are found to be compatible with the experimental results. Nevertheless, from these results, we find different types of resonances can give rise to SERS at different degrees of enhancement depending on the geometry of the gratings. In fact, Baltog et al. have shown that SERS enhancement factor of 10^4 can be obtained from 1-D Ag gratings [45, 46]. Kahl and Voges have numerically calculated the effects of grating geometry on SERS and Raman enhancement of 10^5 can be observed from binary and shallow gratings [47].

A scanning electron microscopy (SEM) image of a shallow Ag grating with period $P = 840$ nm, groove width $W = 140$ nm, and depth $D = 50$ nm is shown in the inset of Fig. 1.5. After grating fabrication, a thin layer of 10^{-3} M Rhodamine 6G (R6G), which serves as the Raman probe, is spin coated on the surface and then thoroughly rinsed with ethanol. The grating is then examined by a micro-Raman spectrometer and its Raman spectrum is shown in Fig. 1.5, with the peaks located at 553, 557, and 562 nm, or Raman shift at 1,372, 1,502, and 1,662 cm^{-1}, that are attributable to the symmetric modes of the in-plane C-C stretching vibration [48], confirming the presence of R6G on the surface. To reveal the possible resonances evolving from the grating, p-p and s-s angle-dependent reflectivity taken in the direction perpendicular to the grooves have been performed and two mappings are shown in Fig. 1.6a, b. From the p-p mapping, two dispersive dark (low reflectivity) loci are observed and they can be identified as the propagating SPP modes given by modifying the phase-matching condition in (1.3) as [22]:

$$\frac{\sin\theta}{\lambda} = \frac{1}{\lambda}\sqrt{\frac{\varepsilon_{Ag}\varepsilon_d}{\varepsilon_{Ag}+\varepsilon_d}} + \frac{n}{P}, \tag{1.4}$$

where λ is the incident wavelength, ε_d and ε_{Ag} are the dielectric constants of the dielectric medium and Ag, and n is an integer. We take $\varepsilon_d = 1.06$ as the effective dielectric constant for R6G/air and $\varepsilon_{Ag} = \varepsilon_H - \frac{(\varepsilon_S-\varepsilon_H)\omega_p{}^2}{\omega^2+\gamma^2}$, where ω_p is the plasma frequency $= 1.72\times10^{16}$ rad/s and ω is the angular frequency [49]. ε_H and ε_S are considered as 6.18 and 5.45, respectively, and $\gamma = 2.3\times10^{13}$ Hz

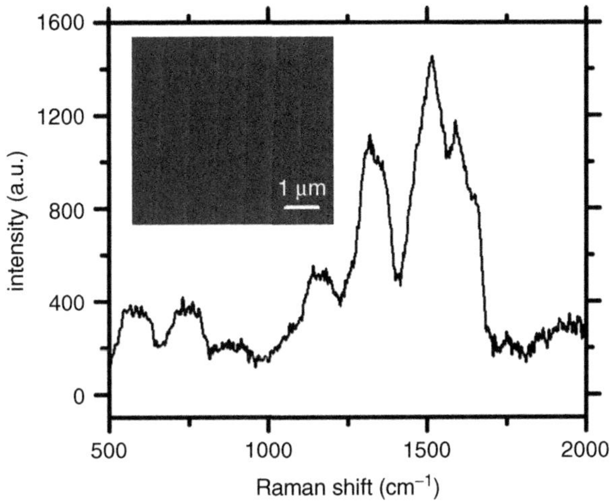

Fig. 1.5 (**a**) The plane view of SEM image of 1D Ag grating with period = 840nm, depth = 50nm, and groove width = 140nm is shown in the inset. A typical micro-Raman spectrum of the corresponding *Ag* grating coated with a thin layer of R6G. The peaks correspond to the symmetric modes of the in-plane C-C stretching vibration of R6G

[49]. The calculations are overlaid on Fig. 1.6a as the solid lines and they show the loci are due to $n = 1$ and -2 SPP modes. On the other hand, the bright (high reflectivity) loci are Wood's anomalies (WAs) and they can be deduced by replacing the $\sqrt{\varepsilon_{Ag}\varepsilon_d / \varepsilon_{Ag} + \varepsilon_d}$ in (1.4) with ε_d [50]. The red dash lines calculated from the WA equation are illustrated in Fig. 1.6a, matching well with the reflectivity maxima. No features are being seen from the s-s mapping, indicating the absence of any SPP resonances since they cannot be excited under s-polarization in 1D grating. To verify our results, RCWA is used to simulate the p-p and s-s reflectivity mappings of the grating and the results are shown in Fig. 1.6c, d [51]. The simulations reproduce the experiments well although slight discrepancies are seen that could be due to film roughness, system misalignment, etc. It is clear that other than the SPPs and WAs, no other resonances are found in this grating.

Then, the p-p and s-s Raman detection mappings of the same grating taken at the incident angle of 27° are shown in Fig. 1.7a, b. The angle of 27° is chosen so that the 514 nm Ar^+ laser coincides exactly with the $n = 1$ ingoing SPPs, giving rise to excitation enhancement. All the spectra are corrected by subtracting the background from the raw data as R6G is known to exhibit broad fluorescence peaks at ~562 nm. Apparently, from the p-p Raman mapping, clear SERS loci are visible and they correlate well with the 1 and -2 SPP modes being found in the reflectivity mapping. To examine this closely, Fig. 1.8a compares the cuts extracted from the reflectivity and Raman mappings at 562 nm, which corresponds to one of the C-C stretching bands, confirming an almost complete correlation between reflectivity and SERS intensity. The enhanced SERS at particular detection angles $\theta_{RS} = 17°$ and 23° is due to the emission enhancement when the energy emitted from the R6G molecules are coupled to the 1 and -2 outgoing SPPs before they subsequently radiate via the phase-matching condition. In contrast, from the s-s Raman mapping, weak and nondispersive SERS is found, indicating roughness-induced SERS plays

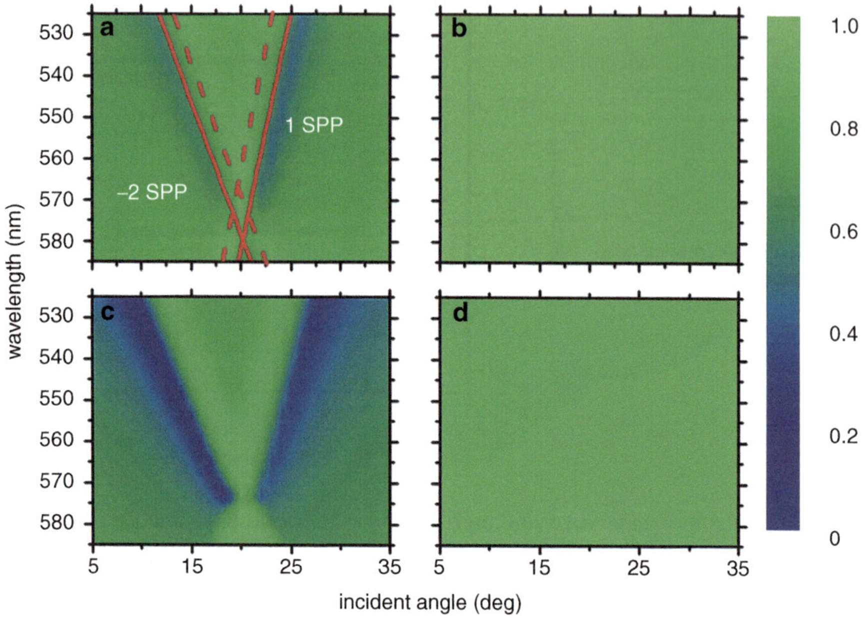

Fig. 1.6 The experimental (**a**) p-p and (**b**) s-s angle-dependent reflectivity mappings of Ag grating taken in the direction normal to the grooves. The dark loci correspond to the low reflectivity regions, indicating the excitation of SPPs. The solid lines are deduced from the 1D SPP phase-matching equation, indicating that 1 and -2 SPPs are excited. On the other hand, the dash lines are deduced from the Wood's anomalies equation. The RCWA simulated (**c**) p-p and (**d**) s-s reflectivity mappings and they reproduce the experiment well

a negligible role. We also have measured the p-p and s-s incident scans in Fig. 1.7c, d by fixing the detection angle at 23° so that the emission enhancement at 562 nm is optimized. From the p-p incident mapping, we see the increase of SERS intensity at incident angles of 9° and 27° agreeing well with the excitation enhancement at 514 nm by exciting the 1 and -2 ingoing SPPs. The corresponding reflectivity and Raman cuts taken at 514 nm and 562 nm are shown in Fig. 1.8b. As a result, the comparison between the reflectivity and Raman mappings unambiguously demonstrates the significance of SPPs on SERS. For the SERS enhancement factor, we have determined the factor at 562 nm and $\theta_{RS} = 23°$ to be ~12 in Fig. 1.8a by taking the intensity ratio of p-p to s-s detection scans. To verify this, Fig. 1.9 shows the electric field patterns calculated by RCWA at $\lambda_{exc} = 514$ nm, $\theta_{exc} = 27.6°$ and at $\lambda_{RS} = 562$nm, $\theta_{RS} = 23.6°$ given in Fig. 1.8, and the average field enhancements are estimated to be 4.1 and 4.3, yielding the total enhancement as 17.6, which is on the same order as the experimental result.

We notice from the p-p detection and incident scans that nondispersive Raman enhancements are also observed at the off-resonance regions, other than the on-resonance counterparts. Although these residual backgrounds are relatively weak, they are much stronger than the signals obtained from the s-s Raman mappings and

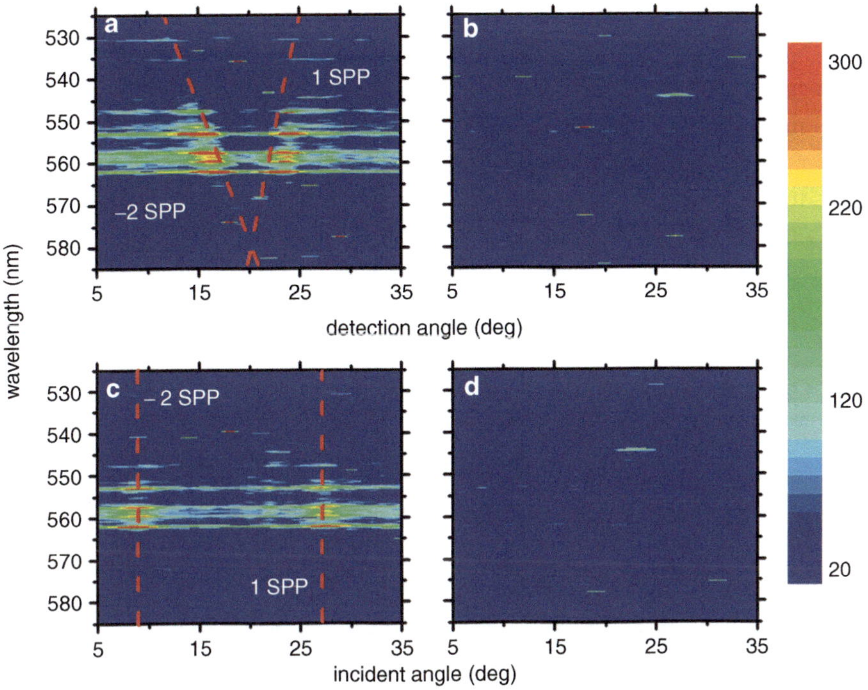

Fig. 1.7 The (**a**) p-p and (**b**) s-s angle-resolved Raman detection mappings of the grating taken at the incident angle of 27°. The strong Raman loci correlate well with the reflectivity mappings, indicating the emissions arise from 1 and -2 outgoing SPPs. The (**c**) p-p and (**d**) s-s angle-resolved Raman incident mappings of the grating taken at detection angle of 23°. The strong emissions correspond to the excitation of ingoing SPPs by laser $\lambda_{exc} = 514$ nm

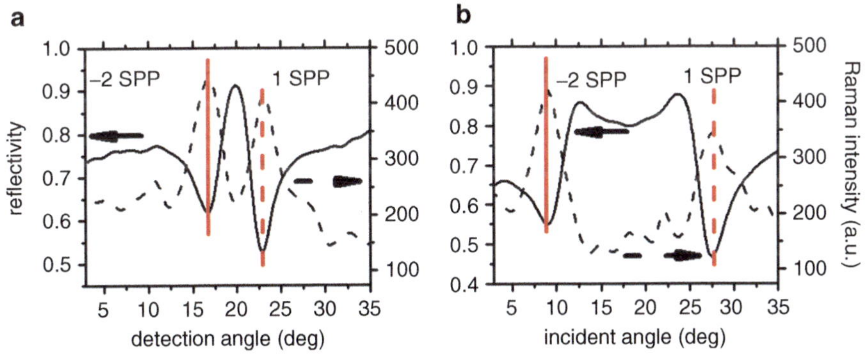

Fig. 1.8 Several cuts extracted from the reflectivity and Raman detection and incident scans. (**a**) The comparison between the reflectivity (*solid*) and Raman intensity taken from the detection scan (*dash*) at λ=562 nm and θ_{exc}=27°. (**b**) The comparison between the reflectivity (*solid*) and Raman intensity taken from the incident scan (*dash*) taken at λ=562 nm and θ_{RS}=23°. An almost complete correlation between reflectivity and Raman intensity is clearly seen by the vertical solid and dash lines

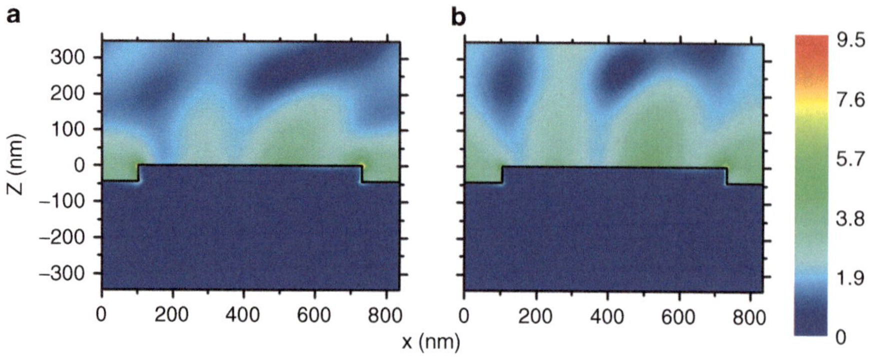

Fig. 1.9 RCWA simulations of electric field intensity at (**a**) 514 nm, 29.6° and (**b**) 562 nm, 23.6°. The intensity of the incident light is taken as unity

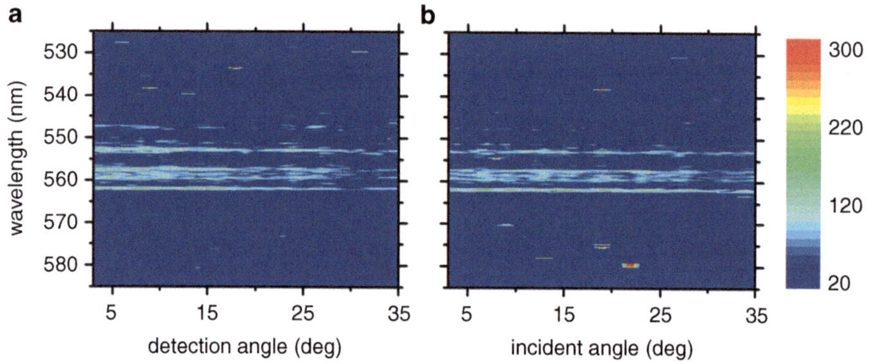

Fig. 1.10 The (**a**) p-s Raman detection mapping taken at incident angle of 27° and the (**b**) s-p Raman incident mapping taken at detection angle of 23°

thus cannot be attributed to fluorescence or roughness-induced SERS as the s-s mappings would have exhibited similar features as well. In addition, the backgrounds are not induced by other localized resonances since no such modes are revealed from the corresponding reflectivity mapping. To elucidate the origin of this background, we have performed the Raman detection scan at incident angle = 27° under p-s configuration in Fig. 1.10a in which the polarizations of the excitation and detection are orthogonal to each other. This special configuration allows us to divide the generation and decay processes separately. Comparing with the p-p detection scan in Fig. 1.7a, one clearly sees the background remains but the strong emissions at the outgoing SPP regions vanish. In fact, in the p-s scan, the p-polarized laser excites the ingoing SPPs while at the same time blocking the Raman emission arising from the outgoing SPPs. More specifically, when the molecules decay after enhanced excitation, the energy is transferred via two paths, either outgoing SPPs or direct radiation. The s-polarized detection collects

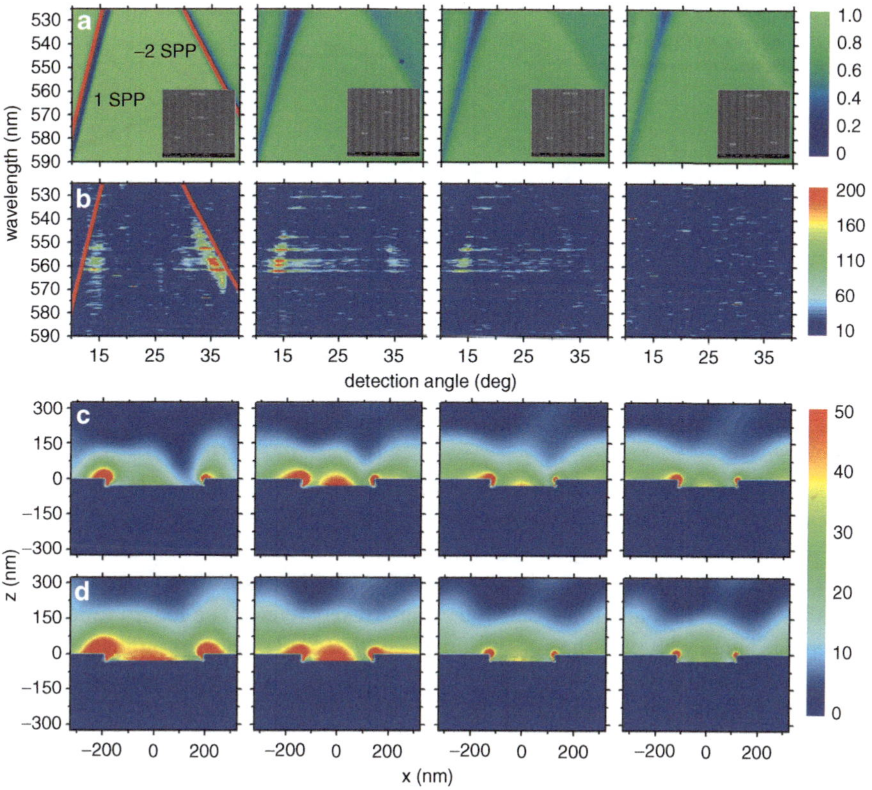

Fig. 1.11 The (**a**) p-p reflectivity mapping of Ag gratings with period = 650 nm, depth = 30 nm and groove width = 377, 277, 247, and 230 nm (from *left to right*). The dash lines are deduced from the SPP phase-matching equation revealing 1 and -2 SPPs are excited. Insets are the corresponding SEM images. (**b**) The p-p Raman detection mappings of the same gratings taken at incident angle = ~20°. RCWA simulations of electric field intensity at (**c**) 514 nm, ~20° and (**d**) 562 nm, ~13°

emissions from the second path but not the first. Since the second path does not invoke any angular dependence, it results in nondispersive backgrounds. Therefore, the SERS observed from this p-s detection scan is due solely to the excitation enhancement. On the other hand, the s-p incident mapping taken at detection angle of 23° is shown in Fig. 1.10b. In this scan, while the radiation arising from the outgoing SPPs is detected, no enhanced excitation is expected since no ingoing SPPs are generated. The molecules are excited from a direct, nonenhanced path and their emission radiating via the outgoing SPPs is being detected. As a result, the p-s and s-p scans differentiate the excitation and emission enhancement channels and the backgrounds are due to the combination of direct and enhanced channels.

To illustrate the dependence of SERS on geometry, we have studied the angle-resolved Raman of 1D Ag gratings with different groove widths. Fig. 1.11 summarizes the SEM images of various gratings, their corresponding p-p reflectivity

Fig. 1.12 The plots of Raman intensity of different SPP modes with groove width

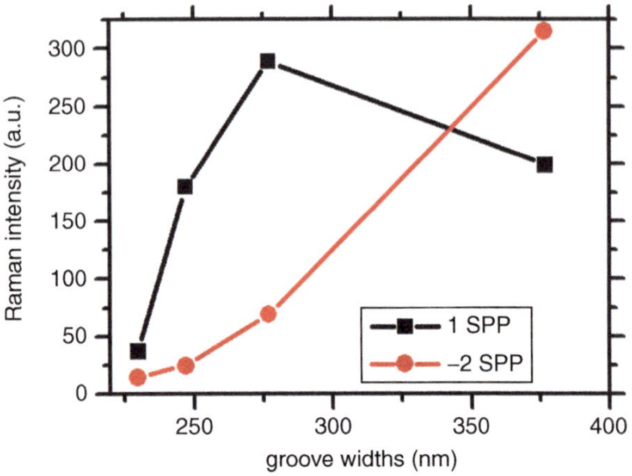

mappings, p-p Raman detection mappings taken at incident angle = 20° for maximizing the excitation enhancement, and field patterns simulated at 514 nm, $\theta_{exc} = \sim 20°$ and 562 nm, $\theta_{RS} = 13°$ by using RCWA [52]. From the SEM images, the groove widths W are determined to be 377, 277, 247, and 230 nm and the period P and depth D are found to be 650 and 30 nm, respectively. The loci in reflectivity mappings agree well with the generation of 1 and -2 SPP modes, as indicated by the solid lines. More importantly, regardless of the groove width, all the Raman mappings correlate very well with the reflectivity mappings in which SERS arises primarily from the outgoing SPPs. As shown in Fig. 1.12, we see although the Raman intensity from 1 SPPs initially increases with increasing the groove width, it reaches the maximal at $W = 277$ nm before declining. On the other hand, the Raman intensity from -2 SPPs increases almost linearly with groove width. These results demonstrate not only a close connection exists between SERS and SPPs, but also a strong dependence of SERS on grating geometry. In particular, the SERS arising from different outgoing SPP modes can have different geometry dependences. Nevertheless, within the picture of dipole scattering, wider groove facilitates larger polarizability and thus larger scattering cross section, which results in stronger SERS emission [53]. Finally, from Fig. 1.11c, d, we see the simulated field patterns of 1 SPPs reproduce the SERS profiles well in which the field strengths of both ingoing and outgoing SPPs reach the highest at $W = 277$ nm and decrease afterward. From the field patterns, considering the Raman enhancement for $W = 230$ nm grating to be unity, we estimate the relative enhancement factors for different groove widths as 1, 1.9, 4.8, and 3.5, which are consistent with the measured values as 1, 4.8, 7.8, and 5.4 although discrepancies are expected.

Other than the propagating SPPs, localized resonances also enhance Raman emission [54]. To support localized resonances in 1D grating, a wide and deep groove is being studied. As an illustration, we have fabricated a 300 nm deep Ag grating with period and groove width = 800 and 260 nm. Its p-p and s-s reflectivity mappings are shown in Fig. 1.13a, b, which exhibit more features than the shallow counterparts. In particular, broad and nondispersive bands are observed from both

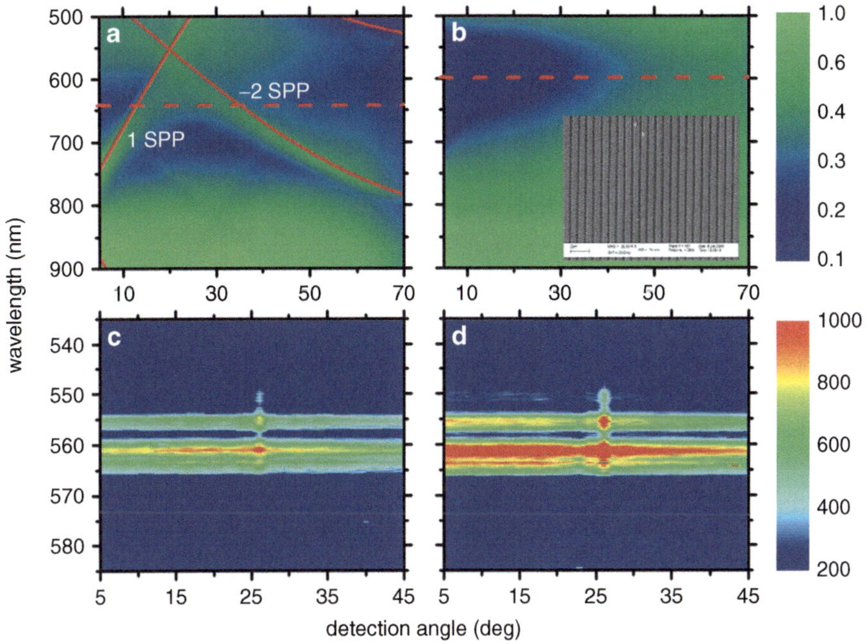

Fig. 1.13 The (**a**) p-p and (**b**) s-s reflectivity mappings of a Ag grating with period = 800 nm, depth = 340 nm, and groove width = 260 nm. The solid lines are deduced from the SPP phase-matching equation indicating 1 and -2 SPPs are excited. The horizontal dash lines at ~650 and 600 nm indicate the excitation of localized modes under p- and s-polarizations. In (**a**), the localized mode actually couple with the SPPs leading to the formation of mixed modes. Inset: the corresponding SEM image. The (**c**) p-p and (**b**) s-s Raman detection mappings of the sample taken at incident angle = 27°, showing Raman emissions that are independent of angle

p-p and s-s mappings at ~650 and ~600 nm, respectively, indicating the excitation of localized resonances. In fact, they evolve from two types of local modes in which their spectral positions can be approximately given as [55, 56]:

$$\lambda_p = \frac{2(D + d_{\text{skin}})}{N + 1/2 - 1/2\pi}\sqrt{\frac{W + 2(c/\omega_{\text{p}})}{W}}, \tag{1.5}$$

and

$$\lambda_s = \frac{2}{\sqrt{\left(\frac{M}{2(D+d_{skin})}\right)^2 + \left(\frac{1}{W+d_{skin}}\right)^2}}, \tag{1.6}$$

for the p- and s-excited modes. λ_p and λ_s are the resonant wavelengths, c is the speed of light, and N and M are the orders of modes, which are integers. A skin depth $d_{\text{skin}} = 50$ nm has been introduced for Ag in these two equations to phenomenologically take into account the field penetration into the side walls and the bottom of the grooves to mimic the case of real metal. From the above equations, we estimate the resonant wavelengths to be 620 and 576 nm for the fundamental modes and they agree qualitatively with our measurements. Nevertheless, as shown in the equations, the resonant positions depend only on the depth and width of individual grooves but not on the period. It is also noted that these two modes originate differently, thus displaying different field profiles. While the field of the p-excited modes is localized predominately at the side walls of the groove [33, 34], that of s-excited modes is more concentrated in the groove cavity [56]. From the p-p mapping, we see the localized mode actually couple with the $n = 1$ and -2 SPPs when their energy and momentum are identical, creating mixed modes that possess both propagating and localized characters [33, 34]. In addition, small plasmonic gaps are formed at the cross points [33, 34].

Once it is shown that the localized resonances are present in the thick grating, angle-dependent Raman spectroscopy is then performed and the p-p and s-s detection mappings are shown in Fig. 1.13c, d by fixing the incident angle of the Ar^+ laser at 27°. Comparing with the shallow gratings, we first notice that the p-p mapping is dominated by nondispersive SERS rather angular SERS loci. In addition, strong SERS is observed from the s-s mapping although no SPPs are expected. The Raman mappings are connected with the reflectivity mappings, demonstrating the localized resonances are also possible sources for SERS. However, it is noted that although the intensities of s-s mapping are found to be stronger than those of p-p mapping over a wide range of angles, it is not proper to claim the s-excited local mode is superior over the p-excited one or propagating SPPs in yielding SERS because we have made no attempt in optimizing the contribution of SERS by any resonances in this grating. In fact, the contribution of SPPs is very likely to be diminished due to the undesirable geometry for SPP generation and decay in this case. Nevertheless, our results are consistent with a recent study conducted by Fujimaki et al. that SERS can be observed from 1-D grating under not only p-polarization but also s-polarization [57]. We also stress that surface roughness does not play a role in our grating since, although not shown, the R6G SERS maps obtained from flat Ag exhibit typical dipole emission and much weaker intensities.

5.2 Two-Dimensional (2D) Circular Hole Arrays

After studying the 1D gratings, we then move on to discuss the 2D hole arrays. 2D arrays present a more complicated system than the 1D counterpart since they involve more geometrical parameters to determine the field profile and strength [58]. With this consideration, we shall restrict our study to only cylindrical hole in which the number of geometrical parameters can be reduced to only period P, hole radius R, and hole depth D. Furthermore, unlike particle or disk arrays where only localized resonances are present, hole arrays provide not only localized modes but also propagating SPPs [59]. In fact, hole arrays have been widely studied for the applications in extraordinary transmission [60], light extraction from light emitting diodes [61], surface plasmon resonance sensing [62, 63], and most notably, surface-enhanced Raman scattering [32, 39, 64].

As SERS depends strongly on the generation and decay processes of SPPs, it is useful to understand the dependence of decay lifetime τ_{SPP} and coupling efficiency of SPPs on array geometry. In the following, we focus on this aspect and discuss the influence of these two parameters on SERS. In principle, the SPP decay lifetime can be measured by time- or frequency-domain methods since they are connected by the Fourier transformation [65]. In time-domain, an ultrashort laser pulse first excites the SPP mode and its temporal behavior is then measured by time-resolved techniques [66]. Here, we employ frequency domain method in which the reflectivity spectrum is considered as a series of line shape functions [67]. Since the full-width-at-half-maximum (FWHM) of the function is governed by the lifetime of the resonance mode as $\tau = \hbar/\Gamma$, where τ is the lifetime and Γ is the FWHM, it is possible to deduce the SPP lifetime τ_{SPP} by extracting the linewidth from the reflectivity spectrum. As a result, we have determined τ_{SPP} and resonant wavelength λ_{res} of the SPP modes in the reflectivity mappings by fitting the reflection spectra with the Fano function given as [68]:

$$R = R_0 + \frac{A(\xi + q)^2}{1 + \xi^2}, \tag{1.7}$$

where $\xi = \frac{2(E - E_{res})}{\Gamma}$, E and E_{res} is the photon energy and the resonant energy, respectively. R_o is the offset function to take into account the reflectivity background, q is the Breit–Wigner–Fano coupling coefficient, and A is a coefficient describing the contribution of the zero-order continuum state coupled to the discrete resonant state. On the other hand, the coupling efficiency is determined as: $(R_{base} - R_{res})/R_{base}$, where R_{base} is the reflectivity taken at off-resonance wavelength and R_{res} is the reflectivity at the SPP resonance [69].

2D circular hole arrays with period = 670 nm have been fabricated by IL. As an illustration, the insets in Figs. 1.14 and 1.15 show the SEM plane-view images of two series of Ag arrays with hole depth $D = 100$ and 280 nm and radius R varying from 85 to 180 nm. From the figures, we see the holes are arranged in cubic structure so that the periods in x- and y-directions are identical. After sample preparation, the samples are then capped with a self-assembled monolayer (SAM), 6-mercaptopurine, to prevent surface oxidation as well as to serve as a Raman probe. SAM serves as a better Raman probe than R6G since it is chemically attached on the metal surface with exactly one monolayer so that comparison can be made between samples directly. In addition, 6-mercaptopurine does not produce strong fluorescence, thus background subtraction is not necessary.

Figures 1.14 and 1.15 show the p-p and s-s reflectivity mappings of the corresponding arrays taken in the Γ-X direction. At shallow hole depth $D = 100$ nm, as shown in Fig. 1.14, several propagating SPPs are clearly visible

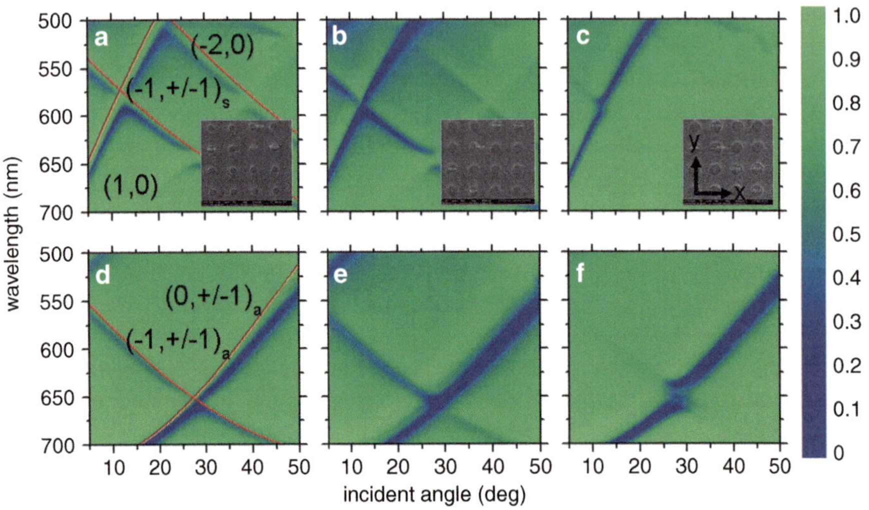

Fig. 1.14 The (**a–c**) p-p and (**d–f**) s-s reflectivity mappings of 2D Ag circular hole arrays with hole radius = 102.5, 140, and 180 nm (from *left to right*) and depth = 100 nm. The dash lines are deduced from the SPP phase-matching equation indicating (1,0), (−2,0), and $(-1,\pm1)_s$ SPPs are excited under p-polarization while $(0,\pm1)_a$ and $(-1,\pm1)_a$ SPPs are excited under s-polarization. The corresponding SEM images are shown in the insets

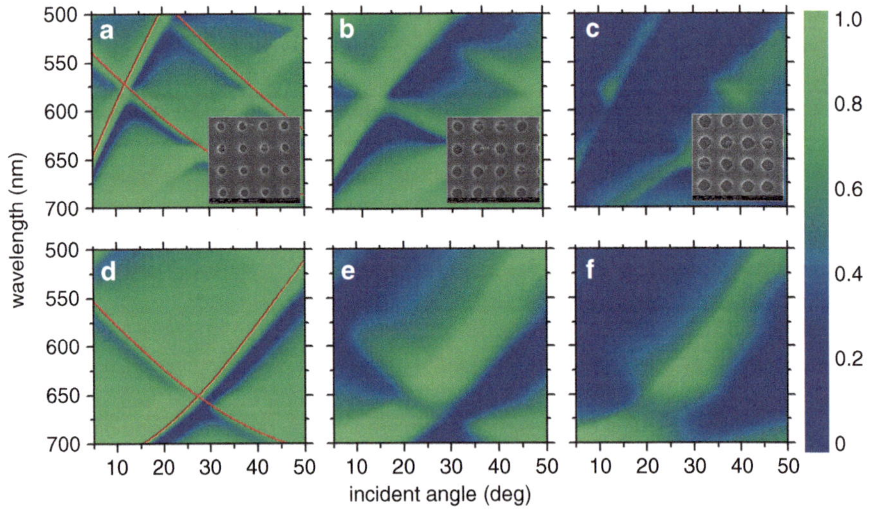

Fig. 1.15 The (**a–c**) p-p and (**d–f**) s-s reflectivity mappings of 2D Ag circular hole arrays with hole radius = 85, 140, and 180 nm (from *left to right*) and depth = 280 nm. The dash lines are deduced from the SPP phase-matching equation indicating (1,0), (−2,0), and $(-1,\pm1)_s$ SPPs are excited under p-polarization while $(0,\pm1)_a$ and $(-1,\pm1)_a$ SPPs are excited under s-polarization. The corresponding SEM images are shown in the insets. At larger hole radius, localized resonance evolves and actually couple with the SPPs making the features become blurry

and their loci are given by extending the phase-matching equation from 1D to 2D as [70, 71]:

$$\vec{k} = \frac{2\pi}{\lambda}\sqrt{\frac{\varepsilon_{Ag}\varepsilon_d}{\varepsilon_{Ag}+\varepsilon_d}}\hat{u} + \frac{2\pi}{P}\left(n_x\hat{x} + n_y\hat{y}\right), \tag{1.8}$$

where n_x and n_y are integers, $\vec{k}$ is the in-plane wavevector of the incident wave with its magnitude equal to $(2\pi/\lambda)\sin\theta$, $\hat{u}$ is the unit vector along the propagating direction of the excited SPPs and finally, $\hat{x}$ and $\hat{y}$ are the unit vectors. The solid lines deduced from (1.8) are superimposed on the mappings revealing that (1,0), (−2,0), and $(-1,\pm1)_s$ SPP modes are excited under p-polarization while $(0,\pm1)_a$ and $(-1,\pm1)_a$ SPP modes are excited under s-polarization [70–72]. $(0,\pm1)_s$ SPP mode is not seen because of its extremely low coupling efficiency compared with other modes [73]. The excitation of (0,±1) (and also (−1,±1)) modes under p- and s-polarizations is originated from the *A* and *B* eigenmodes of the array, respectively, which yields the $(0,\pm1)_s$ and $(0,\pm1)_a$ modes with different field symmetries [70–72]. The $(0,\pm1)_s$ mode is symmetric with respect to the plane of incidence, that is, the x–z plane, whereas the $(0,\pm1)_a$ mode is asymmetric. As a result, p- and s-polarized lights can be coupled to different modes depending on their symmetry; the p-polarized light is coupled to the symmetric $(0,\pm1)_s$ mode and the asymmetric $(0,\pm1)_a$ mode is excited by the s-polarized light. Examining all arrays indicate that the loci of all SPP modes are predominately governed by the period as described by (1.8) although a weak dependence on hole size is expected since the resonant positions are also dependent on the effective refractive index, n_{eff}, of the air/Ag interface [74]. Varying the hole radius and depth would change the area/volume fraction of the void in the unit cell of array as well as n_{eff}, thus leading to slight variation in the resonant position. Other than the resonant positions, it is found, for a given SPP mode, the reflection dips become broader when the hole radius and depth increase as well as the resonant wavelength λ_{res} decreases [65, 67]. As the linewidth of the reflection dip is inversely proportional to lifetime, larger hole size gives shorter SPP lifetime.

When the hole depth increases to 280 nm, the reflectivity mappings for small radius remain more or less the same as the shallow counterparts. However, for larger hole radius, a broad and low reflectivity band appears at ~500–650 nm and the SPP loci become blurry regardless of the polarization (Fig. 1.15). Similar to the case of 1D gratings, this nondispersive band is due to the excitation of localized resonance confined within the cylindrical hole [75]. In addition, this mode actually couples with all SPP modes under p- and s-polarizations. These couplings result in the formation of mixed modes and plasmonic gaps although these features are not quite distinguishable from the mappings. Nevertheless, consider that the fundamental TE_{11} mode (λ_w) supported in a cylindrical hole is roughly given as [36, 75]:

$$\lambda_w = \frac{2\pi}{\sqrt{\left(\frac{b}{R+d_{skin}}\right)^2 + \left(\frac{\pi}{2D}\right)^2}}, \tag{1.9}$$

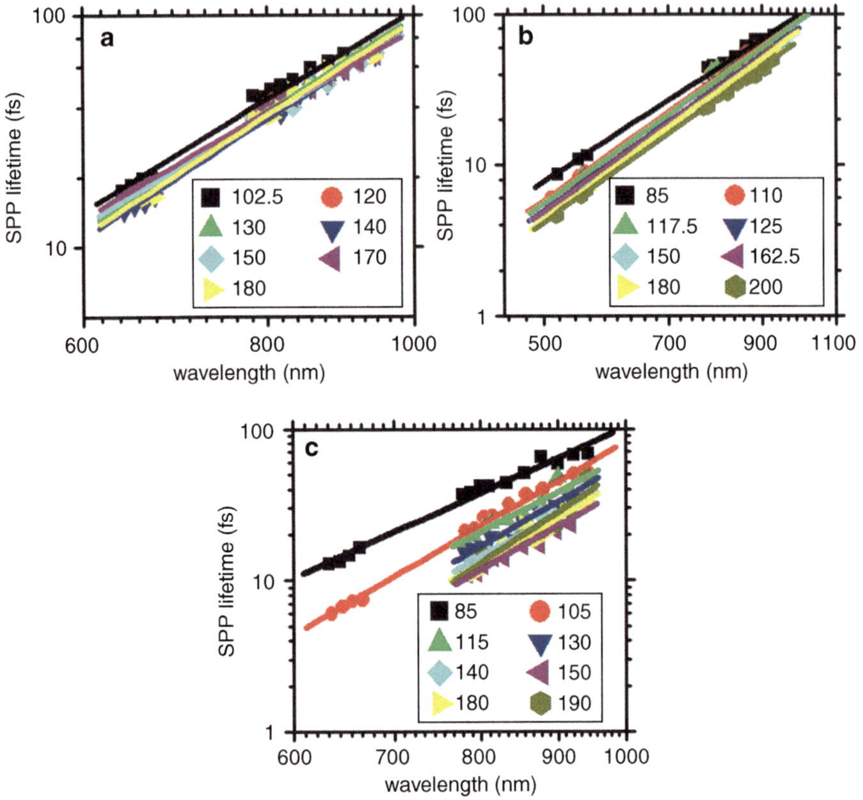

Fig. 1.16 The plots of decay lifetime τ_{SPP} of (1,0) and (−1,0) SPP modes with resonant wavelength λ_{res} at different hole radii and depths = (**a**) 100 nm, (**b**) 120 nm, and (**c**) 280 nm. All the lifetimes are determined by Fano fitting of the reflectivity spectra. The solid lines are the linearly fittings of the lifetimes to extract the n factor, indicating $\tau_{SPP} \propto \lambda_{res}{}^n$. The legends are hole radius in the unit of nm

where $d_{skin} = 50$ nm for Ag and b is the first order Bessel constant $= 1.841$ [75]. The λ_w for different R are then determined as 317, 340, and 356 nm for $D = 100$ nm and 426, 561, and 642 nm for $D = 280$ nm. As a result, one can see that only the thick arrays at $D = 280$ nm, but not the shallow ones, support the local mode within our spectrum of interest. In addition, from (1.9), we see the localized mode red shifts with increasing radius and depth, which results in different degrees of coupling with SPPs depending on the hole size.

Once all the modes have been identified, we then move on to study the effects of hole radius and depth on SPP lifetime τ_{SPP}. The lifetimes of (1,0) and (−1,0) SPP modes of all samples are determined by Fano fitting and are plotted in Fig. 1.16 with λ_{res}. Only the modes that are not coupled with the localized mode are considered here for simplification. Apparently, all the SPP modes at different hole sizes seem to follow a log-log quasi-linear relationship indicating $\tau_{SPP} \propto \lambda^n$ [67, 76, 77]. To fully illustrate

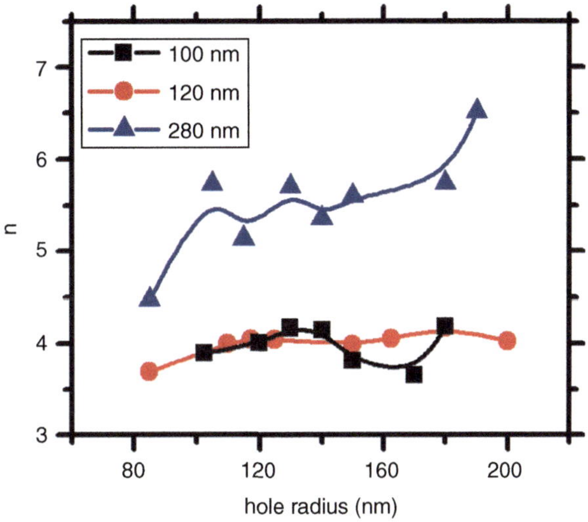

Fig. 1.17 The plots of n as a function of hole radius for different hole depths

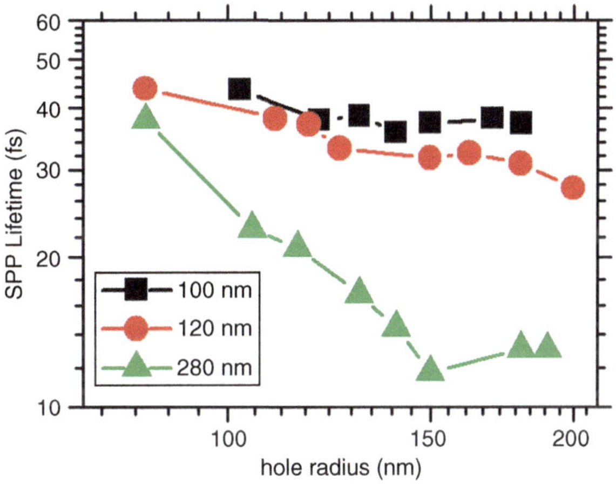

Fig. 1.18 The dependence of τ_{SPP} on hole radius for different depths taken at $\lambda_{res} = 800$ nm

the dependence of n on geometric parameters, we have extracted n from all the samples and plot them in Fig. 1.17 as a function of hole radius. From the figure, it is clear that n varies from 3.5 to 7 depending on hole radius and depth. *n* remains almost constant for different radii at shallow depths, that is, when $D = 100$ and 120 nm, but increases to ~7 gradually with radius at larger depth $D = 280$ nm. We also have plotted the dependence of τ_{SPP} of all the samples on hole radius at $\lambda_{res} = 800$ nm in Fig. 1.18. It is found that at shallow depths, τ_{SPP} does not vary much with radius but begins to follow R^{-m}, where m increases from 0 to higher value at larger hole size [78].

To account the above scenarios, we follow Ref [78] that the observed τ_{SPP} actually consists of two components, namely, absorption and radiative scattering losses, and is given as: $1/\tau_{SPP} = 1/\tau_{abs}+1/\tau_{scatt}$, where τ_{abs} and τ_{scatt} are absorption and radiative scattering lifetimes, respectively. With the fact that all the hole sizes are smaller than the wavelengths being considered, the absorption lifetime can be qualitatively expressed as:

$$\tau_{abs} \propto \lambda \frac{\mathrm{Re}(\varepsilon_{Ag})^2}{2\pi \mathrm{Im}(\varepsilon_{Ag})} \propto \lambda^2, \tag{1.10}$$

which is independent of geometry. On the other hand, within the framework of Mie scattering, the scattering loss τ_{scatt} is proportional to P/σ, where σ is the SPP scattering cross section and is given as

$$\sigma = \frac{72\pi^5}{3}\left[\frac{R^4H^2}{\lambda^4}\left(\frac{(\varepsilon_{Ag}-1)}{(\varepsilon_{Ag}+2)}\right)^2 - 5\pi^4\frac{R^{\frac{20}{3}}H^{\frac{10}{3}}}{\lambda^8}\left(\frac{(\varepsilon_{Ag}-2)(\varepsilon_{Ag}-1)}{(\varepsilon_{Ag}+2)^2}\right)^2\right], \tag{1.11}$$

after the inclusion of the quadrupolar corrections in cylindrical hole [78]. From the above equation, we see the first component indicates the dipolar, or Rayleigh, scattering with $(R/\lambda)^4$ dependence whereas the second one represents the quadrupolar counterpart. Therefore, $\tau_{scatt} \propto P/\left(\frac{R^4H^2}{\lambda^4} - \frac{R^{\sim 6.7}H^{\sim 3.3}}{\lambda^8}\right)$ and it shows different wavelength dependence depending on the hole radius and depth. We expect from the above formulations that for small radius and depth, τ_{SPP} should follow between λ^2 and λ^4, but becomes more λ^8 dominant at larger radius and depth. In fact, these deductions agree well with our observations. At small depth, n remains at $\sim$3.5 for all radii being considered indicating absorption loss is dominant over the radiative scattering loss due to the weak scattering of SPPs on the weakly corrugated surfaces [78]. However, when both radius and depth increase, the radiative loss begins to take charge and n increases gradually to 4 and then to $\sim$7 whereas m increases from 0 to 4. As a result, at a fixed period, we believe the decay process of these SPP modes in our periodic systems can be understood within the interplay between the intrinsic absorption and radiative scattering of SPPs by single isolated holes. With this in mind, these simple analytical formulations should assist us in designing desired excitation and emission enhancements once their influences on the generation and decay processes of SPPs are known.

The coupling efficiencies determined from all samples are plotted in Fig. 1.19 as a function of λ_{res}. For small hole depths as shown in Fig. 1.19a, b, one can see the coupling efficiency in general decreases with increasing wavelength and decreasing hole radius. These trends are consistent with the increase of polarizability of individual holes due to the increase of hole area, thereby leading to the increase of absorption and scattering cross sections [79]. In fact, Prikulis et al. studied the optical properties of single metallic hole by elastic scattering measurements and drew similar conclusions. However, for larger depth in Fig. 1.19c, the behavior becomes more complicated as the coupling efficiency decreases slightly with increasing hole radius. The explanations of these features are not trivial since quadrupolar contribution becomes dominant in this regime and more work is currently underway.

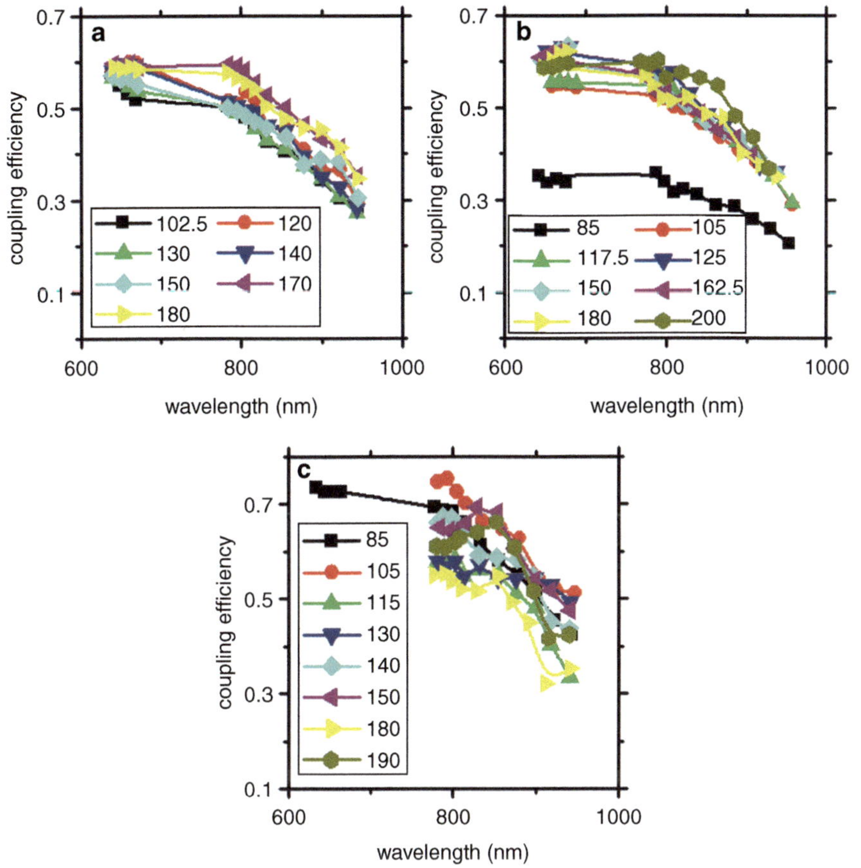

Fig. 1.19 The plots of coupling efficiency of (1,0) and (−1,0) SPP modes with resonant wavelength λ_{res} at different hole radii and depths = (**a**) 100 nm, (**b**) 120 nm, and (**c**) 280 nm. The legends are hole radius in the unit of nm

For the Raman analysis, in order to makes use of the conclusions being drawn from the studies of lifetime and coupling efficiency, we shall focus only on the case of shallow arrays (i.e., D = 100 and 120 nm) where localized mode is absent, excluding the possibility that the mixed SPP-localized modes may alter the generation and decay processes from pure SPPs. From the reflectivity mappings, we select the incident angle at 10° for D = 100 nm and 48° for D = 120 nm so that these off-resonance angles do not generate any ingoing SPPs at $\lambda_{exc} = 514$ nm, thus minimizing the excitation enhancement [64]. In this way, we focus primarily on the decay process caused by the outgoing SPPs. The p-p Raman detection mappings taken from the arrays in Figs. 1.14 and 1.15 are shown in Fig. 1.20 without the removal of background. One can see, similar to the 1D gratings, the SERS loci are connected to those of the reflectivity regardless of the hole radius. In fact, the bright lines at 538, 542, 550, 560 nm, etc., are the peaks that are attributable to the Stoke

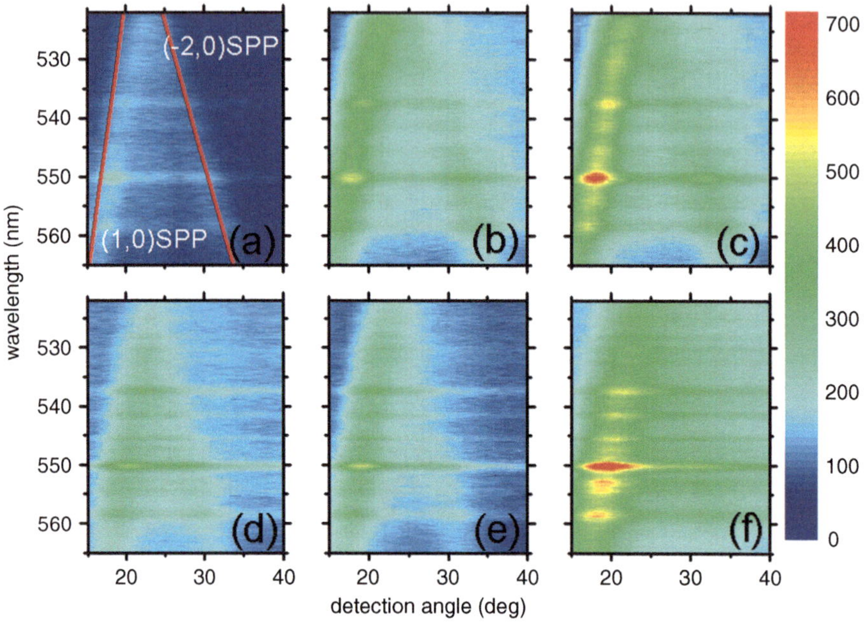

Fig. 1.20 The p-p Raman detection mappings of 2D hole arrays with depth = (**a–c**) 100 nm and (**d–f**) 120 nm taken at incident angle = 10° and 48° respectively. The hole radii are (**a**) 102.5, (**b**) 120, (**c**) 140, (**d**) 110, (**e**) 120, and (**f**) 140 nm. The solid lines are deduced from the SPP phase-matching equation, indicating the Raman emissions are due to the decay of (1,0) and (−2,0) outgoing SPPs

Raman signatures of 6-mercaptopurine [80]. No peak is found at 592 nm indicating the absence of S–H bond and a single monolayer is formed on the entire Ag surface [80]. The cuts extracted from the reflectivity and Raman mapping at $\lambda_{RS} = 550$ nm are shown in Fig. 1.21 for further examination and a one-to-one correspondence between them is clearly demonstrated for all samples. More importantly, we see that at $\theta_{RS} = \sim 18°$ and $\sim 30°$ where (1,0) and (−2,0) SPP modes prevail, the Raman peak intensities in general increase with increasing hole radius. We have plotted the Raman enhancement factors with hole radius for (1,0) and (−2,0) modes at 550 nm in Fig. 1.22. The Raman enhancement factor is estimated by using the expression given as [15, 16]: $I_{array} \times N_{flat}/I_{flat} \times N_{array}$, where I_{array} and I_{flat} are the Raman peak intensities obtained from the array and flat metal film while N_{array} and N_{flat} are the number of 6-mercaptopurine molecules attached on the array and film, respectively. In fact, assuming a uniform coverage of the monolayer on the perforated array and film, N_{flat}/N_{array} can be approximated as: $1/(1 - \pi R^2/4P^2)$. From Fig. 1.22, for two hole depths, we see that both the enhancement factors of (1,0) and (−2,0) modes increase with hole size. They all reach ~15–20 when radius increases to ~200 nm. To clarify the dependence of Raman emission on the coupling efficiency and lifetime, we have also plotted the lifetime and coupling efficiency of (1,0) SPP mode with radius at $\lambda = 550$ nm in Fig. 1.23 so that comparison can be made

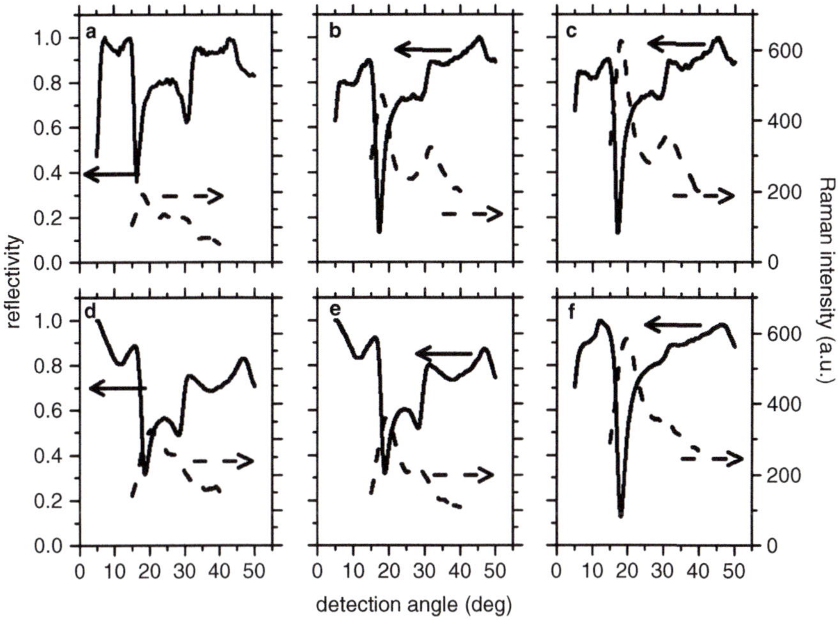

Fig. 1.21 Several cuts extracted from the corresponding reflectivity (*solid*) and Raman detection (*dash*) mappings at 550 nm. Good correlation between reflectivity and Raman intensity is clearly seen

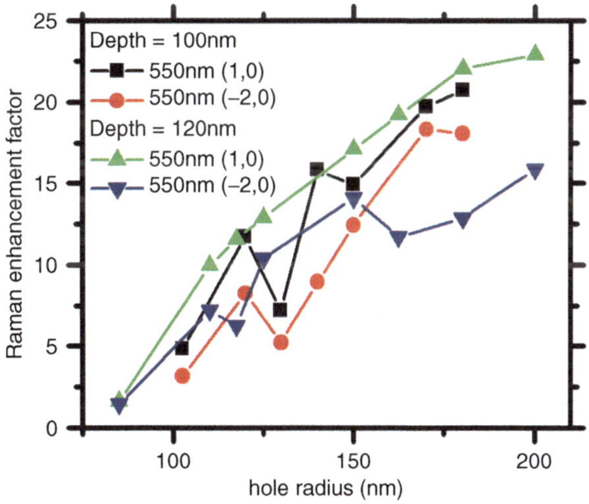

Fig. 1.22 The plots of Raman enhancement factor of (1,0) and (−2,0) SPP modes with hole radius at two hole depths

directly. Here, we only consider the (1,0) SPP mode in which the formulations given above are applicable. One can see from the figure that SPP lifetime decreases slowly with hole radius for both hole depths, indicating the increase of the radiative decay rate. On the other hand, all the coupling efficiencies increase almost

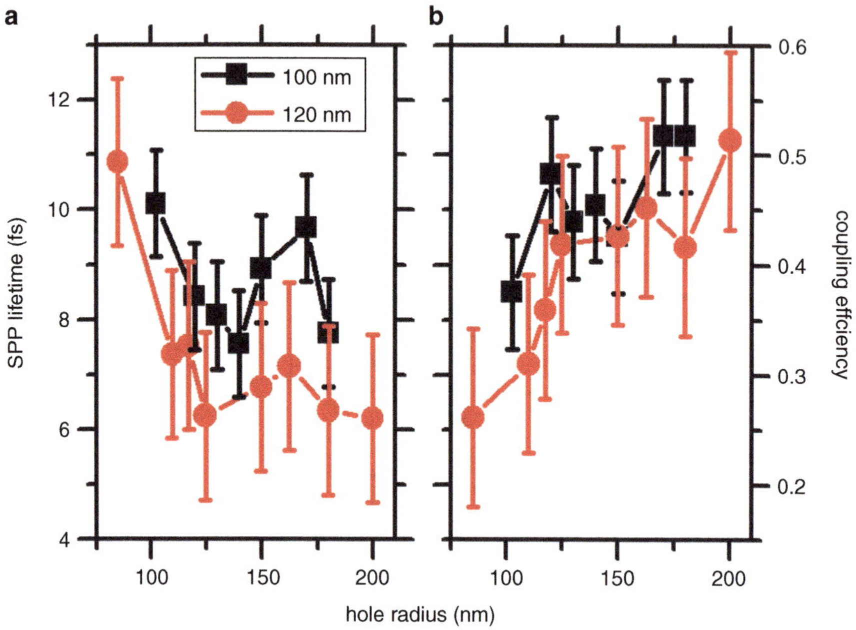

Fig. 1.23 The plots of (**a**) SPP lifetime and (**b**) coupling efficiency of (1,0) SPP mode taken at λ_{res} = 550 nm for two hole depth = 100 and 120 nm as a function of hole radius

monotonically with hole size, implying larger hole radius favors the energy coupling from the molecules to the outgoing SPPs. Therefore, by comparing the enhancement factors with the SPP lifetimes and coupling efficiencies, it is evident that they are all connected together supporting the fact that both SPP coupling and decay play a major role in the emission enhancement. For SPP-mediated SERS, larger hole size facilitates stronger coupling of Raman emission from the 6-mercaptopurine molecules to SPPs as well as faster radiative decay rate to recover the SPPs back to free space photons [61].

6 Conclusions and Future Perspective

Various 1D and 2D periodic metallic arrays have been fabricated by interference lithography and their optical properties have been studied by angle-resolved reflectivity and Raman spectroscopy. The combination of these two techniques has been proven to be an effective tool in understanding the underlying mechanism of SERS, especially the electromagnetic origin of SERS. They discern a particular electromagnetic resonance mode in the dispersion relation for detailed investigation. It is evident that the generation and decay processes of the resonance modes both govern SERS, and how to quantitatively determine their influences on SERS has been the key subject in our research. For surface plasmon–mediated SERS, we have demonstrated the importance of SPP decay lifetime and coupling efficiency in controlling the decay process of outgoing SPPs as well as the resulting Raman enhancement. Higher radiative decay rate and larger coupling efficiency favor more efficient energy transfer from excited molecules to the far field, yielding stronger Raman emission.

Many other issues also deserve attention. For example, the generation process of ingoing SPPs is also of importance since it participates equally with the decay process in maneuvering SERS. The connection between lifetime, coupling efficiency, and the excitation enhancement should be identified in order to maximize the overall Raman enhancement. In addition to the SPP mediation, localized resonances have also been shown to play a major role in enhancing Raman emission. However, its contributions to SERS have not yet been studied in detail by far. All these issues, once resolved appropriately, should allow us to lay out a guideline in rationally designing high sensitive and stable SERS substrates based on periodic arrays.

Acknowledgments This research was supported by the Chinese University of Hong Kong (CUHK) through the RGC Competitive Earmarked Research Grants (402807, 402908, and 403308) and the Shun Hing Institute of Advanced Engineering (BME-p2-08). The authors thank D.Y. Lei, A.I. Fernández-Domínguez, and S.A. Maier from Imperial College, London, and J.T.K. Wan and P.M. Hui from CUHK for valuable discussions.

References

1. Ferraro JR, Nakamoto K, Brown C (2003) Introductory Raman spectroscopy. Academic, New York
2. Smith E, Dent G (2005) Modern Raman spectroscopy, a practical approach. Wiley, New York
3. Raman CV, Krishnan KS (1928) A new radiation. Ind J Phys 2:387
4. Weber WH, Merlin R (2010) Raman scattering in materials science. Springer, Berlin
5. Anker JN, Hall WP, Lyandres O, Shah NC, Zhao J, Van Duyne RP (2008) Biosensing with plasmonic nanosensors. Nat Mater 7:442
6. Fainstein A, Jusserand B (2007) Raman scattering in resonant cavities. In: Cardona M, Merlin R (eds) Light scattering in solid IX, vol 108. Springer-Verlag, Berlin, p 17
7. Rottwitt K, Nielsen K, Povlsen JH, Emiliyanov G, Hansen TP, Jensen JB (2005) Raman spectroscopy using photonic crystal fibers. Proceedings of SPIE in photonic crystals and photonic crystal fibers for sensing applications, vol 6005, pp 387
8. Kneipp K, Moskovits M, Kneipp H (eds) (2006) Surface enhanced Raman scattering – Physics and application, vol 103. Springer, Berlin
9. Fleischmanna M, Hendraa PJ, McQuillana AJ (1974) Raman spectra of pyridine adsorbed at a silver electrode. Chem Phys Lett 26:163
10. Etchegoin P, Cohen LF, Hartigan H, Brown RJC, Milton MJT, Gallop JC (2003) Electromagnetic contribution to surface enhanced Raman scattering revisited. J Chem Phys 119:5281
11. Le Ru E, Etchegoin P (2008) Principles of surface enhanced Raman spectroscopy: and related plasmonic effects. Elsevier Science, New York
12. Moskovits M (1985) Surface enhanced spectroscopy. Rev Mod Phys 57:783
13. McFarland AD, Young MA, Dieringer JA, Van Duyne RP (2005) Wavelength-scanned surface enhanced Raman excitation spectroscopy. J Phys Chem B 109:11279
14. Qin LD, Zou SL, Xue C, Atkinson A, Schatz GC, Mirkin CA (2006) Designing, fabricating, and imaging Raman hot spots. Proc Nat Acad Sci 103:13300

15. Laurent G, Félidj N, Grand J, Aubard J, Lévi G, Hohenau A, Aussenegg FR, Krenn JR (2006) Raman scattering images and spectra of gold ring arrays. Phys Rev B 73:245417
16. Félidj N, Truong SL, Aubard J, Levi G, Krenn JR, Hohenau A, Leitner A, Aussenegg FR (2004) Gold particle interaction in regular arrays probed by surface enhanced Raman scattering. J Chem Phys 120:7141
17. Kneipp K, Wang Y, Kneipp H, Perelman LT, Itzkan I, Dasari RR, Feld MS (1997) Single molecule detection using surface enhanced Raman scattering. Phys Rev Lett 78:1667
18. Bosnick KA, Jiang J, Brus LE (2002) Fluctuations and local symmetry in single-molecule Rhodamine 6G Raman scattering on silver nanocrystal aggregates. J Phys Chem B 106:8096
19. Nie S, Emory SR (1997) Probing single molecules and single nanoparticles by surface enhanced Raman scattering. Science 275:1102
20. Camden JP, Dieringer J, Zhao J, Van Duyne RP (2008) Controlled plasmonic nanostructures for surface enhanced spectroscopy and sensing. Acc Chem Res 41:1653
21. Stiles P, Dieringer J, Shah NC, Van Duyne RP (2008) Surface enhanced Raman spectroscopy. Ann Rev Anal Chem 1:601
22. Raether H (1988) Surface plasmons on smooth and rough surfaces and on gratings, Springer Tracts in Modern Physics. Springer, Berlin
23. Gryczynski Z, Matveevan E, Calander N, Zhang J, Lakowicz JR, Gryczynski I (2007) Surface plasmon coupled emission, Surface plasmon nanophotonics. Springer, Berlin
24. Ushioda S, Sasaki Y (1983) Raman scattering mediated by surface-plasmon polariton resonance. Phys Rev B 27:1401
25. Kurosawa K, Pierce RM, Ushioda S, Hemminger JC (1986) Raman scattering and attenuated-total-reflection studies of surface-plasmon polaritons. Phys Rev B 33:789
26. Kittel C (2004) Introduction to solid state physics. Wiley, New York
27. Joannopoulos JD, Johnson SG, Winn JN, Meade RD (2008) Photonic crystals: molding the flow of light. Princeton University Press, Princeton
28. Tsang JC, Kirtley JR, Bradley JA (1979) Surface enhanced Raman spectroscopy and surface plasmons. Phys Rev Lett 43:772
29. Girlando A, Philpott MR, Heitmann D, Swalen JD, Santo R (1980) Raman spectra of thin organic films enhanced by plasmon surface polaritons on holographic metal gratings. J Chem Phys 72:5187
30. Knoll W, Philpott MR, Swalen JD, Girlando A (1982) Surface plasmon enhanced Raman spectra of monolayer assemblies. J Chem Phys 77:2254
31. Arnold M, Bussemer P, Hehl K, Grabhorn H, Otto A (1992) Enhanced Raman scattering from benzene condensed on a silver grating. J Mod Opt 39:2329
32. Baumberg JJ, Kelf TA, Sugawara Y, Cintra S, Mamdouh E, Abdelsalam E, Bartlett PN, Russell AE (2005) Angle-resolved surface enhanced Raman scattering on metallic nanostructured plasmonic crystals. Nano Lett 5:2262
33. Porto JA, García-Vidal FJ, Pendry JB (1999) Transmission resonances on metallic gratings with very narrow slits. Phys Rev Lett 83:2845
34. López-Rios T, Mendoza D, García-Vidal FJ, Sánchez-Dehesa J, Pannetier B (1998) Surface shape resonances in lamellar metallic gratings. Phys Rev Lett 81:665
35. Ruan Z, Qiu M (2006) Enhanced transmission through periodic arrays of subwavelength holes: the role of localized waveguide resonances. Phys Rev Lett 96:233901
36. Laux E, Genet C, Ebbesen TW (2009) Enhanced optical transmission at the cutoff transition. Opt Exp 17:6920
37. Gordon R, Brolo A (2005) Increased cut-off wavelength for a subwavelength hole in a real metal. Opt Express 13:1933
38. Yu Q, Guan P, Qin D, Golden G, Wallace PM (2008) Inverted size-dependence of surface enhanced Raman scattering on gold nanohole and nanodisk arrays. Nano Lett 8:1923
39. Brolo AG, Arctander E, Gordon R, Leathem B, Kavanagh KL (2004) Nanohole-enhanced Raman scattering. Nano Lett 4:2015

40. Lucas BD, Kim JS, Chin C, Guo LJ (2008) Nanoimprint lithography based approach for the fabrication of large-area, uniformly oriented plasmonic arrays. Adv Mater 20:1129
41. Alvarez-Puebla R, Cui B, Bravo-Vasquez J, Veres T, Fenniri H (2007) Nanoimprinted SERS-active substrates with tunable surface plasmon resonances. J Phys Chem C 111:6720
42. Li J, Iu H, Luk WC, Wan JTK, Ong HC (2008) Studies of the plasmonic properties of two-dimensional metallic nanobottle arrays. Appl Phys Lett 92:213106
43. Space Nanotechnology Laboratory, Massachusetts Institute of Technology, Cambridge, Boston, USA. http://snl.mit.edu/
44. O'Reilly TB, Smith HI (2008) Linewidth uniformity in Lloyd's mirror interference lithography systems. J Vac Sci Technol B 26:2131
45. Baltog I, Primeau N, Reinisch R, Coutaz JL (1995) Surface enhanced Raman scattering on silver grating: optimized antennalike gain of the Stokes signal of 10^4. Appl Phys Lett 66:1187
46. Baltog I, Primeau N, Reinisch R, Coutaz JL (1996) Observation of stimulated surface enhanced Raman scattering through grating excitation of surface plasmons. JOSA B 13:656
47. Kahl M, Voges E (2000) Analysis of plasmon resonance and surface enhanced Raman scattering on periodic silver structures. Phys Rev B 61:14078
48. Michaels AM, Nirmal M, Brus LE (1999) Surface enhanced Raman spectroscopy of individual Rhodamine 6G molecules on large Ag nanocrystals. J Am Chem Soc 121:9932
49. Penninkhof JJ (2006) Tunable plasmon resonances in anisotropic metal nanostructures, Ph.D. thesis, Utrecht University
50. Ghaemi HF, Thio T, Grupp DE, Ebbesen TW, Lezec HJ (1998) Surface plasmons enhance optical transmission through subwavelength holes. Phys Rev B 58:6779
51. RSoft Design Group, Inc., New York, USA. http://www.rsoftdesign.com/
52. Chan CY, Li J, Xu JB, Ong HC (2009) The dependence of surface enhanced Raman scattering on the groove size of one-dimensional metallic gratings. MRS Spring Meeting, San Francisco
53. Lévêque G, Martin OJF, Weiner J (2007) Transient behavior of surface plasmon polaritons scattered at a subwavelength groove. Phys Rev B 76:155418
54. Chen C, Hutchison JA, Clemente F, Kox R, Uji-I H, Hofkens J, Lagae L, Maes G, Borghs G, Van Dorpe P (2009) Direct evidence of high spatial localization of hot spots in surface enhanced Raman scattering. Angew Chem Inter Ed 48:9932
55. Lei DY, Wan JTK, Ong HC (unpublished work)
56. Skigin DC, Depine RA (1999) Resonant enhancement of the field within a single ground-plane cavity: comparison of different rectangular shapes. Phys Rev E 59:3661
57. Fujimaki H, Suzuki Y, Hatta A (1994) Raman scattering from Ag metal gratings coated with p-nitrobenzoic acid films. J Raman Spec 25:303
58. García de Abajo FJ (2007) Colloquium: light scattering by particle and hole arrays. Rev Mod Phys 79:1267
59. Martín-Moreno L, García-Vidal FJ, Lezec HJ, Pellerin KM, Thio T, Pendry JB, Ebbesen TW (2001) Theory of extraordinary optical transmission through subwavelength hole arrays. Phys Rev Lett 86:1114
60. Ebbesen TW, Lezec HJ, Ghaemi HF, Thio T, Wolff PA (1998) Extraordinary optical transmission through sub-wavelength hole arrays. Nature 391:667
61. Li J, Xu JB, Ong HC (2009) Hole size dependence of forward emission from organic dyes coated with two-dimensional metallic arrays. Appl Phys Lett 94:241114
62. Gordon R, Brolo AG, Sinton D, Kavanagh KL (2010) Resonant optical transmission through hole-arrays in metal films: physics and applications. Laser Photon Rev 4:311
63. Gordon R, Sinton D, Kavanagh KL, Brolo AG (2008) A new generation of sensors based on extraordinary optical transmission. Acc Chem Res 41:1049
64. Chan CY, Xu JB, Waye MY, Ong HC (2010) Angle resolved surface enhanced Raman scattering (SERS) on two-dimensional metallic arrays with different hole sizes. Appl Phys Lett 96:033104

65. Sönnichsen C, Franzl T, Wilk T, von Plessen G, Feldmann J, Wilson O, Mulvaney P (2002) Drastic reduction of plasmon damping in gold nanorods. Phys Rev Lett 88:077402
66. Ropers C, Park DJ, Stibenz G, Steinmeyer G, Kim J, Kim DS, Lienau C (2005) Femtosecond light transmission and subradiant damping in plasmonic crystals. Phys Rev Lett 94:113901
67. Kim DS, Hohng SC, Malyarchuk V, Yoon YC, Ahn YH, Yee KJ, Park JW, Kim J, Park QH, Lienau C (2003) Microscopic origin of surface-plasmon radiation in plasmonic band-gap nanostructures. Phys Rev Lett 91:143901
68. Genet C, van Exter MP, Woerdman JP (2003) Fano-type interpretation of red shifts and red tails in hole array transmission spectra. Opt Comm 225:331
69. Gao H, Henzie J, Lee MH, Odom TW (2008) Screening plasmonic materials using pyramidal gratings. Proc Natl Acad Sci 105:20146
70. Iu H, Li J, Ong HC, Wan JTK (2008) Surface plasmon resonance in two-dimensional nanobottle arrays. Opt Exp 16:10294
71. Li J, Iu H, Wan JTK, Ong HC (2009) The plasmonic properties of elliptical metallic hole arrays. Appl Phys Lett 94:033101
72. Sauvan C, Billaudeau C, Collin S, Bardou N, Pardo F, Pelouard J-L, Lalanne P (2008) Surface plasmon coupling on metallic film perforated by two-dimensional rectangular hole array. Appl Phys Lett 92:011125
73. Collin S, Sauvan C, Billaudeau C, Pardo F, Rodier JC, Pelouard JL, Lalanne P (2009) Surface modes on nanostructured metallic surfaces. Phys Rev B 79:165405
74. Coe JV, Rodriguez KR, Teeters-Kennedy S, Cilwa K, Heer J, Tian H, Williams SM (2007) Metal films with arrays of tiny holes: spectroscopy with infrared plasmonic scaffolding. J Phys Chem C 111:17459
75. Catrysse PB, Shin H, Fan SH (2005) Propagating modes in subwavelength cylindrical holes. J Vac Sci Technol 23:2675
76. Li J, Iu H, Lei DY, Wan JTK, Xu JB, Ho HP, Waye MY, Ong HC (2009) Dependence of surface plasmon lifetimes on the hole size in two-dimensional metallic arrays. Appl Phys Lett 94:183112
77. Muller R, Malyarchuk V, Lienau C (2003) Three-dimensional theory on light-induced near-field dynamics in a metal film with a periodic array of nanoholes. Phy Rev B 68:205415
78. Lei DY, Li J, Fernández-Domínguez AI, Ong HC, Maier SA (2010) Geometry dependence of surface plasmon polariton lifetimes in nanohole arrays. ACS Nano 4:432
79. Prikulis J, Hanarp P, Olofsson L, Sutherland D, Kall M (2004) Optical spectroscopy of nanometric holes in thin gold films. Nano Lett 4:1003
80. Yang H, Liu Y, Liu Z, Yang Y, Jiang J, Zhang Z, Shen G, Yu R (2005) Raman mapping and in-situ SERS spectroelectrochemical studies of 6-Mercaptopurine SAMs on the gold electrode. J Phys Chem B 109:2739

SERS-Encoded Particles

2

Nicolas Pazos-Perez and Ramón A. Álvarez-Puebla

1 Definition of the Topic

SERS-encoded particles show great promise for high-throughput multiplexed measurements of biologically active species (such as pathogens and/or disease markers) in complex fluids. These particles can also be applied for in vivo imaging of tissues and cells.

2 Overview

Here we show how SERS can be used for the preparation of encoded particles. These particles have been establish as a solid and reliable analytical technique for the multiplex detection in extremely low amounts of a wide variety of bioanalytes in complex matrices such as biological fluids. SERS-encoded particles for indirect detection and labeling can be implemented on chips, microfluidics or even inside living cells, tissues, and a variety of microorganisms. However, there are still open challenges, mainly related to the reproducibility of the methods for substrate fabrication, in particular when dealing with the formation of hot spots, which are responsible for the highest enhancement factors, but their efficiency is extremely sensitive toward small geometrical details within the nanostructure. Additionally, although portable Raman spectrometers are available, most of the published reports are based on very sophisticated instruments that will not find a place in routine analysis labs or hospitals. Thus, the use of SERS codification, particularly in biomedical applications, has a great potential, as demonstrated by many examples, but is open to new developments that will undoubtedly continue amazing us in the near future.

N. Pazos-Perez • R.A. Álvarez-Puebla (✉)
Departamento de Química-Física and Unidad Asociada, CSIC-Universidade de Vigo, Vigo, Spain

Challa S. S. R. Kumar (ed.), *Raman Spectroscopy for Nanomaterials Characterization*,
DOI 10.1007/978-3-642-20620-7_2, © Springer-Verlag Berlin Heidelberg 2012

3 Introduction

The ideal strategies for drug discovery or clinical diagnosis should allow users to simultaneously investigate the presence of a large number of disease markers within each sample (i.e., high-throughput screening) and draw a rapid and accurate diagnostic on the activity of a group of substances or the health condition of a patient. These assays typically include screening chemical libraries for compounds of interest and identification of particular target molecules, such as antigens, antibodies, nucleotides, and peptides, either in test or real samples. The thousands of individual reactions that are typically required for these purposes should be preferably carried out all at the same time. Promising tools for this purpose are the "multiplex technologies," which allow multiple discrete assays to be performed simultaneously within the same sample. Perhaps one of the most important challenges in developing multiplex assays is the necessity to track each reaction. For example, if only one reaction of interest occurs in a group of thousands of molecules that are screened against a certain target analyte, it must be possible to determine which molecule was responsible for that particular reaction. A promising approach requires each of the microcarriers to be labeled, so as to allow the identification of molecules bound to their surface. This method allows uniquely encoded microcarriers to be mixed and subjected to an assay simultaneously.

An assay platform that is rapidly gaining prominence for a variety of applications from diagnosis and biodetection to combinatorial chemistry and drug discovery are encoded microparticles. Such encoded particles address some of the limitations posed by conventional substrates. For instance, advantages offered by these substrates include: (a) amenability to high-throughput screening and multiplexing [1, 2], (b) larger surface area for receptor conjugation or solid phase synthesis, (c) improved accessibility of the analytes to the entire sample volume for interaction with the bead-conjugated receptors, and (d) greater versatility in sample analysis and data acquisition. Encoded particles may be grouped into various categories based on the particle size range (nano [3, 4] or micro [5, 6]) and the codifying strategy (external labeling or self-encoding). External labeling requires the addition of codifiers to the microplatform, such as dyes [7, 8], photonic crystals [9, 10], or semiconductor quantum dots [11]. Conversely, the self-encoding approach benefits from the chemical/physical properties of the materials used in the synthesis of the microparticles, such as shape (Fig. 2.1a–c) [6, 12], vapor permeability [13], spectroscopic properties [14, 15], or composition (Fig. 2.1d, e) [16].

Most of the external labeling approaches today rely on electronic spectroscopy techniques (i.e., fluorescence or surface plasmon resonance) [17]. Although such strategies are powerful and convenient, the simplicity of the obtained spectra, usually just one broad band, limits the number of different codes available for a given system. In this context, the use of surface-enhanced Raman scattering (SERS) and surface-enhanced resonance Raman scattering (SERRS) spectroscopy as codifying alternatives offers unique advantages such as: (a) ultrasensitive detection down to the single molecule level [18–20] that will decrease dramatically the

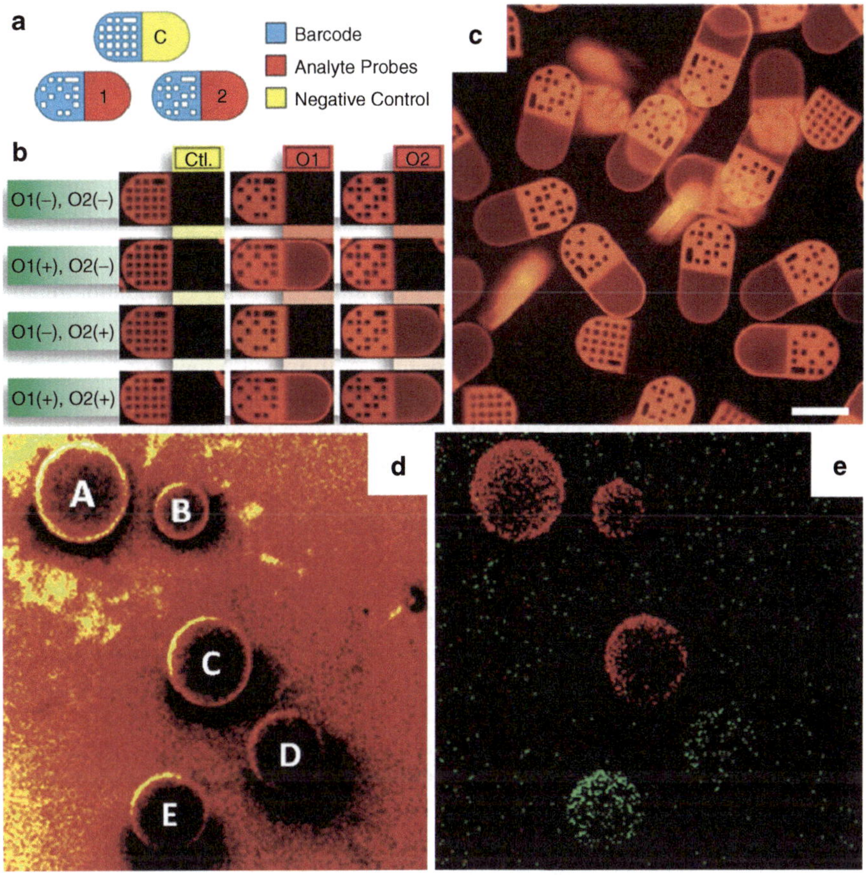

Fig. 2.1 (**a–c**) Particle encoding through lithographic marks. Particles consisting of two regions, a fluorescent labeled lithographic mark and probe area (**a–b**). Fluorescence in the probe regions indicates target detection. (**c**) Scale bar, 100 μm. (**d–e**) ToF-SIMS compositional imaging of polymer beads on a silicon substrate showing the total ion image (*left*) and the chloride (*green*)/ bromide (*red*) ion distribution (*right*) in a 1,700 × 1,700 μm^2 field. (Reproduced with permission from references [6], [**a–c**] and [16], [**d–e**]. Copyright 2007, Science Publishing Group, and 2006, American Chemical Society, respectively)

deconvolution times, allowing for real-time applications such as high-throughput screening in flow-cytometry or microfluidic systems [21, 22]; (b) an unlimited number of barcodes since SERS/SERRS spectra are essentially vibrational fingerprints, unique for each molecule, with the possibility of expanding the codified library to the infinite by combining different tags with similar SERS cross sections [23]; and (c) the use of the same hybrid systems as optical enhancing platforms for fluorescence in sandwich key-and-lock applications such as ELISA-like microarrays, with the subsequent decrease in the detection limits of pathogens or pathogen markers [24, 25].

4 Experimental and Instrumental Methodology

Although the fabrication and characteristics (material, size, biocompatibility, and others) of the desired SERS-encoded particles strongly depend on each specific application, all of them share some common requirements: (a) the presence of an optical enhancer, (b) the addition of the SERS code, and (c) protection with a silica or polymer of the SERS code from leaching while providing a suitable surface for biofunctionalization [26, 27].

The simplest way to produce SERS-encoded nanoparticles comprises the SERS codification of single plasmonic colloids and encapsulation within silica or polymer shells (Fig. 2.2) [28–30]. Regarding the plasmonic particle, which is key for the generation of the electromagnetic field necessary to increase the Raman signal of the label molecule upon excitation with the appropriate light [31, 32], a number of materials have been demonstrated to possess radiative localized surface plasmon resonances in the visible or the infrared, but gold and silver remain the most used by far due to their low toxicity, stability, and optical efficiency [33]. In recent years, the colloidal nanofabrication procedures have evolved sufficiently to allow the synthesis of nanoparticles with almost any shape and size [34–39]. On the other hand, silica coating of nanoparticles is also a mature technology [40]. Thus, the key step for the codification of single nanoparticles relies on the incorporation of the Raman label to give rise to a signal as strong as possible without compromising colloidal stability. Unfortunately, sorption isotherms are not applicable to this problem as usually both particle and Raman label concentrations are so low that the equilibrium concentration (the proportion of the initial concentration of added Raman label that was not retained onto the nanoparticles surface) cannot be determined by conventional analytical techniques. Therefore, codification of nanoparticles with the appropriate amount of label is a trial and error exercise rather than an established protocol. Another important drawback related to encoded single particles is the achievable SERS intensity. The formation of the protective shell around the single colloidal particles inhibits plasmon coupling and in turn the generation of hot spots (specific gaps between particles where the electromagnetic field is extremely high due to coupling between their plasmon resonances [18]). Even though the recent demonstration of the heterogeneous distribution of plasmon modes within anisotropic particles [41] is leading to the development of highly active nanoparticles [42, 43], still the most popular approach for the fabrication of nanometer-sized SERS-encoded particles is the use of aggregates. Particle aggregation can be easily promoted by changing the solvent, increasing the ionic strength of the suspension [44], decreasing pH [45], or just spontaneously due to the adsorption of the Raman label [46]. Notwithstanding, these uncontrollable aggregation processes are usually undesired because they lead to a random distribution of hot spots with the subsequent heterogeneous signal from particle to particle. Thus, several approaches were developed toward controlling aggregation. One of the oldest and probably most successful approaches comprises the so-called composite organic-inorganic nanoparticles (COIN) [47, 48]. The COIN approach consists of the controlled label-induced aggregation of silver particles. Then a shell of

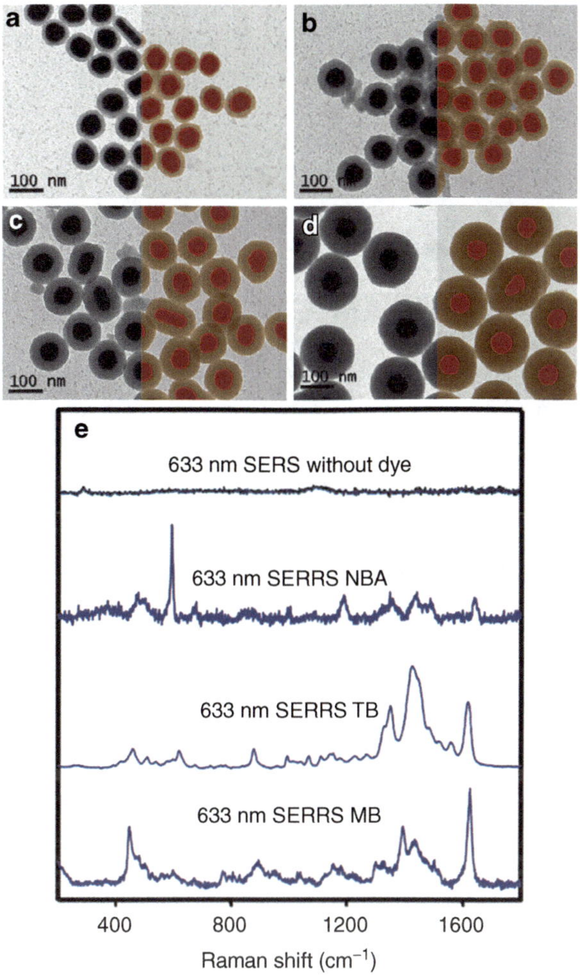

Fig. 2.2 TEM images of 60.4 nm Au spherical particles, encoded with blue dyes and coated with different silica shell thicknesses: (**a**) 11.7 ± 1.5, (**b**) 22.4 ± 2.3, (**c**) 28.7 ± 1.5, and (**d**) 43.7 ± 3.5 nm. The right half of the images has been colored for better visualization (*red*: Au; *brown*: silica). (**e**) SERRS spectra of the particles, both unlabeled and labeled with Nile blue A (NBA), toluidine blue O (TB), and methylene blue (MB NBA, TB, and MB. All spectra were obtained upon excitation with a 633 nm laser (Adapted with permission from Ref. [28], [**d–e**]. Copyright 2009, American Chemical Society)

bovine serum albumin (BSA) is spontaneously retained onto the COIN. This shell not only protects the code but also allows for further functionalization with biomolecules such us proteins (antibodies). Other strategies for the generation of aggregated encoded particles relies on inducing aggregation by increasing the ionic strength followed by silica coating of silver [49] or gold colloids [50]. In fact, these particles have recently been demonstrated to give rise to a sufficiently intense signal to be individually identified in a Raman flow-cytometer [51]. Nevertheless, both methods still yield uncontrollable aggregates with the ensuing fluctuations in their SERS intensity. For this reason, an increasing activity is observed toward the development of controlled methods for the fabrication of SERS-active dimers including spheres, cages, pyramids, or even rods (Fig. 2.3) [52–58].

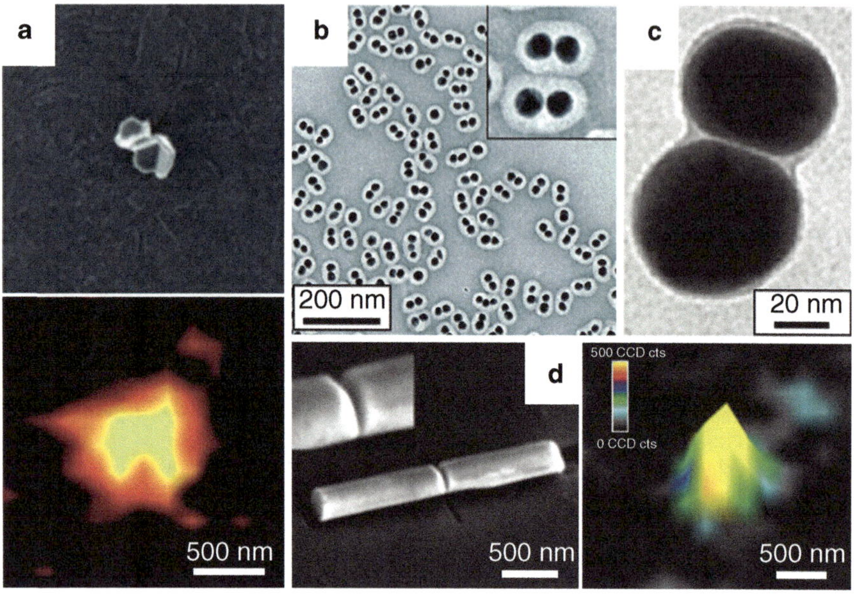

Fig. 2.3 (**a**) SEM and SERS images of a bipyramid dimer. TEM Image of a Au@Ag (**b**) and a Au (**c**) nanosphere dimer. (**d**) SEM image and nearfield map of a gold nanorod dimer (Adapted with permission from Refs. [52, 55, 57] and [58], respectively. Copyright 2009 and 2010 American Chemical Society)

For some applications such as multiplex microarrays, the possibility of observing and resolving single particles is necessary. Since state-of-the-art micro-Raman systems attached to confocal microscopes have a spatial resolution of about 500 nm, these applications require encoded particles of larger size, usually in the submicrometer or even in the micrometer scales. Although the most convenient approach for obtaining these particles may seem the use of large metal particles, radiative damping effects become increasingly significant as particle size is increased, leading to negligible SERS enhancement for particles larger than ca. 200 nm. One exception to this rule can be found in single particles produced by galvanic replacement with gold of a sacrificial particle made of a material with a lower reduction potential [59]. By choosing the appropriate thermodynamic conditions [60], this method yields particles with similar size and shape as the sacrificial templates but with a porous texture (Fig. 2.4a–c) [61]. Alternatively, it has been proven that silver or gold nanoparticles can act as polymer cross-linkers. This method is based on the high affinity of sulfur for gold and silver surfaces. By using nanoparticles with different sulfur-substituted styrene monomers [62, 63] and adding those to common styrene monomers in a typical emulsion polymerization reactor it is possible to obtain SERS-encoded beads with sizes of several micrometers (50–150 μm) where the encoding agent is the sulfur-styrene capping agent (Fig. 2.4d–f) [64]. Regarding the extrinsically labeling

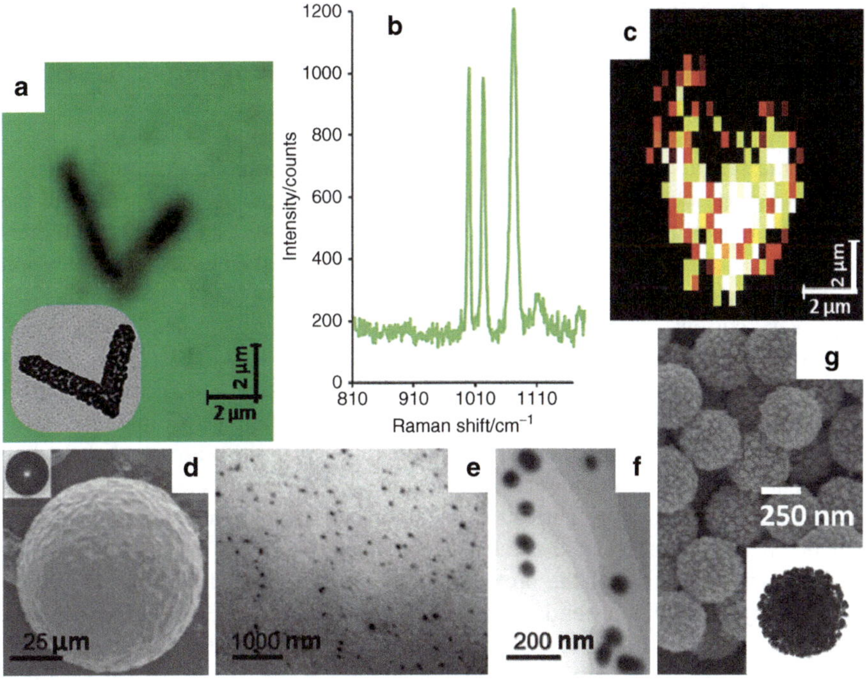

Fig. 2.4 (**a**) Optical, SEM (*inset* TEM) and (**c**) SERS mapping (1,072 cm^{-1}, ring breathing, (**b**)) of a single gold checkmark after exposure to BT in the vapor phase. (**d**) Optical and SEM micrograph of a SERS-active microsphere cross-linked with a mercaptostyrene-coated AgNP. (**d** and **f**) Cross-sectional TEM micrographs showing the distribution of AgNP inside the microspheres. (**g**) SiO_2@Ag beads prepared by in situ reduction of $AgNO_3$ onto the silica particles (Reproduced with permission from Refs. [61], [**a–c**], [64] [**d–f**], and [66] [**g**]. Copyright 2008, American Chemical Society (**a–c**, **g**) and, 2009, Elsevier Publishing Group [**d–f**])

approaches, the most common strategy is the use of a low SERS cross section material (i.e., silica, titania, polymers such as polystyrene, agarose, or others) as a solid support for the assembly of premade nanoparticles on its surface. Methods for the deposition include self-assembly of the nanoparticles directly onto surface amino or thiol groups, and the use of polyelectrolytes in a layer-by-layer technique. An alternative mode for the generation of plasmonic particles on the surface of the support, which may be less controlled but highly effective, is based on the in situ reduction of the metal salts (Fig. 2.4g) [65–67], which can be combined with the deposition of gold seeds, grown for generation of highly active hot spots [42]. One of the main characteristics of these materials is their high colloidal stability, so that codification can be achieved by directly immersing the beads in an excess of the SERS code followed by cleaning through centrifugation or sedimentation. As in the previous cases, encoded nanoparticles are then protected by coating them with silica [29, 67], polymer [27, 68], or even carbon shells [69].

5 Key Research Findings

5.1 Encoded Particles for Multiplex High-Throughput Screening and Key-and-Lock Sensors

Rapid and sensitive techniques to detect antigens and biomolecules are central to various aspects of human health, from determining the precise diagnostic and the proper therapeutic treatments to preventing and controlling pandemics and bioterrorism. Most of the existing detection assays rely on specific molecular recognition events such as antibody-antigen, DNA-DNA, or receptor-ligand interactions [70]. However, current technologies such as enzyme-linked immunosorbent assay (ELISA), radioimmunoassay (RIA), fluoroimmunoassay (FIA) or printed microarrays, and serologic testing are costly, labor intensive, and have little capacity for multiplex sampling. To overcome these limitations, simple and cost-effective analytical techniques with multiplexing capabilities are needed. In line with that, the use of bioconjugated SERS-encoded microparticles for their direct use in suspension or for fabrication of biodetection devices on micropatterned silicon chips emerges as an alternative technology [71–73].

Biofunctionalization of the protective shell of the encoded particles is usually achieved by conventional solid phase procedures (Fig. 2.5). The most extended may still be the so-called EDC chemistry (i.e., carbodiimide chemistry with 1-ethyl-3(3-dimethylaminopropyl)–carbodiimide (EDC) and *N*-Hydroxybenzotriazole (HoBt) as coupling agents) [74, 75], while other emerging methods such as click chemistry [76, 77] are gaining prominence. EDC chemistry is straightforward for particles coated with polymers containing carboxylic or amino functionalities. Otherwise, as in silica coatings, a primary self-assembly step with a carboxylic or amino silane is required to attach these groups onto the particle surface. Once the capture antigen (diagnosis) or antibody (biodetection) is covalently attached, encoded microparticles can be used as versatile platforms for sandwich key-and-lock sensors. Encoded particles can be used either in suspension assays (Fig. 2.5a) or patterned into microchips (Fig. 2.5b). In a typical suspension assay, a mixture of different antibodies, each of them previously labeled with one particular encoded particle, is added to the fluid of interest. After reaction with their specific antigens, the particles are separated by centrifugation, washed, and exposed to a detection antibody (labeled with a fluorophore), which selectively binds to particles that were complexed to the corresponding capture antibody. After fluorophore binding, the particles are centrifuged and washed again, and a small portion is cast onto a glass slide and analyzed in a micro-Raman system. Positive diagnosis of the presence of a pathogen, disease marker, etc., requires the presence of both fluorescence and SERS tags on the same particle [78]. Although in principle this method does not require homogeneity in size or shape, if automation by flow cytometry or microfluidics is to be implemented, size homogeneity is essential to avoid interference due to different Rayleigh scattering signals (which are morphology dependent) [21]. An alternative route to suspension assays is the generation of microchips containing encoded particles within well-defined spatial regions and fixed to

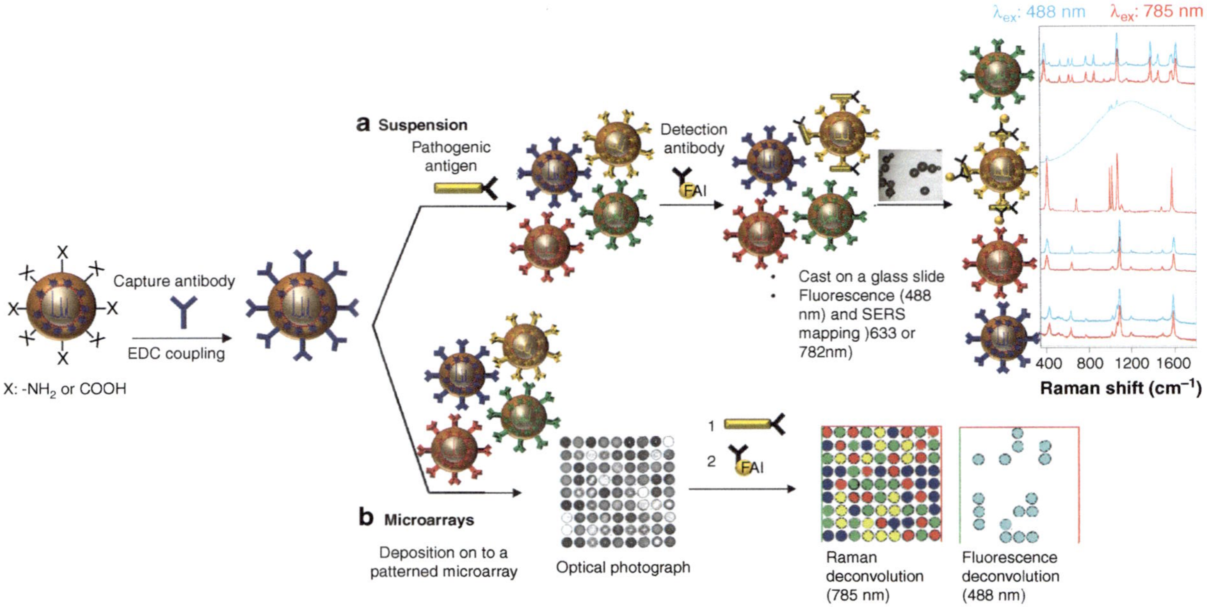

Fig. 2.5 Schematic image illustrating the biofunctionalization of SERS encoded particles; (**a**) Strategy for the detection of analytes of interest, i.e., pathogenic antigen, with suspension experiments; (**b**) Detection route through a microarray patterning. (**a**) Suspension strategy. Briefly, a pool of different barcoded beads, each carrying a single antibody, is brought into contact with the biological fluid of interest. After washing the beads, fluorescently labelled detection antibodies are added, and the pool is washed again. An aliquot is cast onto a glass slide and mapped with two laser lines. 488 nm laser line excites fluorescence on the reacted beads. Reacted antibody is identified by the SERS barcode of the bead (785 nm). (**b**) Fabrication of FIA-like microarrays. The pool of beads is deposited onto a micropatterned silicon chip with a rubber tip spatula under an optical microscope. The array is put into contact, first with the biological fluid of interest, washed and put into contact with fluorescently labelled detection antibodies. The microarray is then mapped with two laser lines: the 488 nm excitation laser line shows fluorescence on the reacted beads, while the reacted antibody is identified by the SERS barcode of the bead (785 nm)

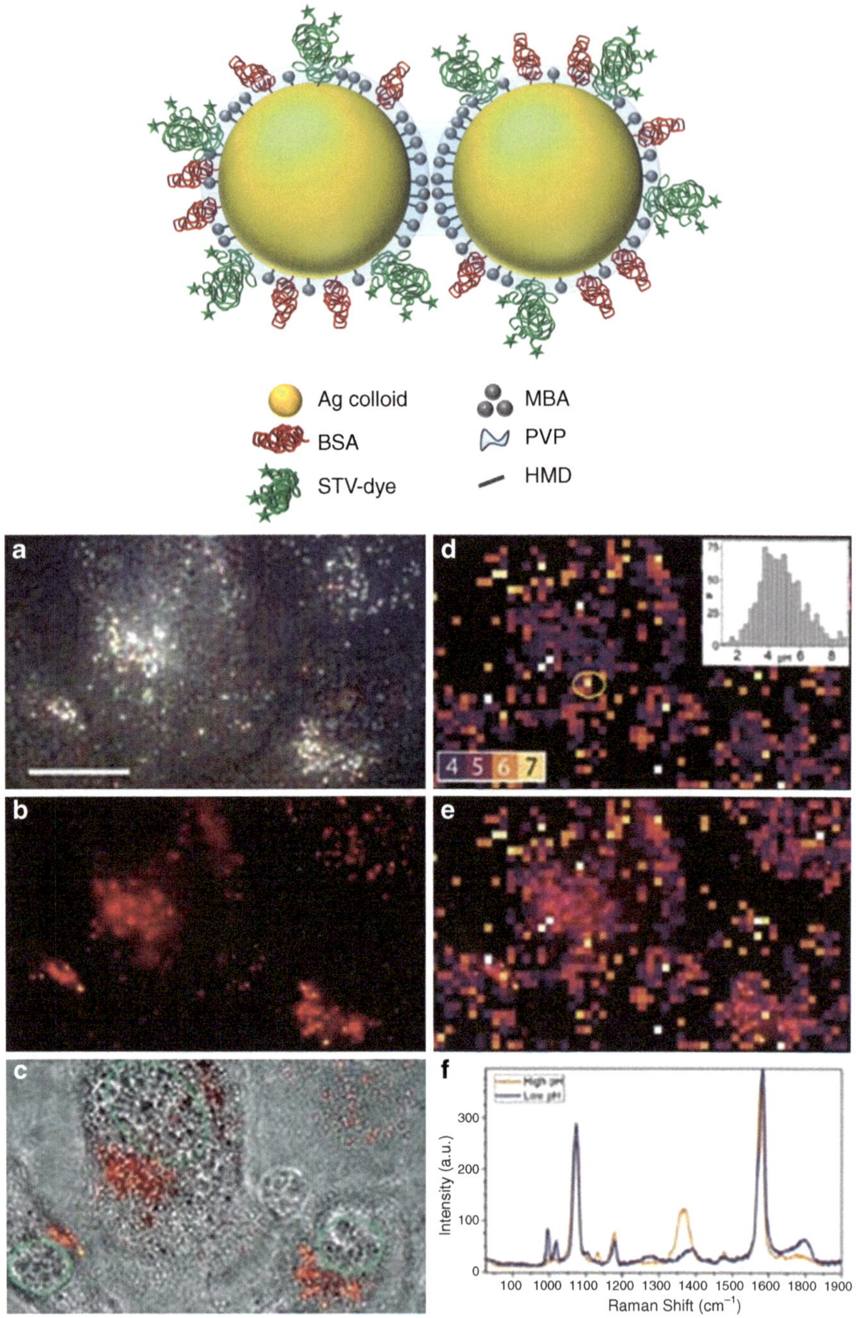

Fig. 2.6 (continued)

a surface, which allows fast recognition and avoids the need of washing steps such as centrifugation [7, 79], On the other hand, the readout from the chip can be performed automatically. Notwithstanding, the preparation of these platforms requires using particles with a high homogeneity in both size and shape [5, 14]. This approach presents many advantages with respect to the conventional ones. Intensity of SERS/SERRS spectra is huge, which enables very short detection times in the order of milliseconds. On the other hand, by taking advantage of the presence of the nanoparticles, surface-enhanced fluorescence of the labeling dye should be registered as well. This enhancement is described to be about 100-fold (as compared with regular laser fluorescence spectroscopy, around 20-fold of normal fluorescence microscopes, giving rise to a final enhancement factor of about 2,000-fold), at the optimized distance optical enhancer (nanoparticle)-molecular probe. The latter allows for a dramatic increase in the sensitivity (amount of agent/disease marker present to be detected) with detection limits down to the femtomolar regime [66]. Both suspension and microchip approaches are currently applied to a variety of health problems including biodetection of extremely infectious viruses such as dengue [80] or flu [81], cancer diagnosis [10, 65, 82], or genotyping [7, 83].

5.2 SERS Imaging of Cells and Tissues

Encoded particles do not only pave the way toward the design of new and fast advanced sensor devices, capable of monitoring multiple parameters in a single readout, but can also be applied in bioimaging. In 1991, Nabiev et al. [84] reported the SERS spectra of the antitumor drug doxorubicin, recorded from treated cancer cells incubated with citrate-reduced silver colloids. These spectra were the first feasibility data demonstrating the promising perspectives of SERS spectroscopy in bioimaging, and it has currently become a key topic of research. SERS bioimaging is based on the functionalization of encoded particles with bioligands showing affinity for specific receptors either of the cell membrane in tissues or single

Fig. 2.6 Schematic representation of the assembly of SERS-active clusters. The raw colloid is assembled using a linker (hexamethylenediamine, HMD). Polymer (PVP) addition stabilizes the aggregate and halts further aggregation. Streptavidin (STV) and bovine serum albumin (BSA), known to passively adsorb to Au and Ag surfaces, are added, and the Raman tag (4-mercaptobenzoic acid, MBA) is allowed to infuse through the coat and into the junction between the nanoparticles. The final complex is both fluorescent and Raman active. Correlated fluorescence and SERS. (**a**) Reflected white-light image of encapsulated Ag clusters within HeLa cells. Scale bar 20 μm. (**b**) Epifluorescent image obtained from the STV dye in the particle coat. (**c**) Overlay of transmitted DIC white light and fluorescence, with nuclei outlined. (**d**) SERS map colorized for pH value determined from ratios of pH-sensitive bands. Regions of low SERS are masked *black*. Inset is the histogram of pH values. (**e**) Overlay of fluorescence and pH map. (**f**) SERS spectra of the circled regions in (**d**) (Adapted with permission from Ref. [56]. Copyright 2010; Wiley-VCH)

organisms, and even for organelles inside the cell [85]. Regarding tissue imaging, this technique has been widely applied to the study of cancer tumors both in vitro [57, 65, 67, 86–88] and in vivo [89, 90].

Probably one of the more spectacular applications of SERS-encoded particles relies on the multiplex spectral imaging of the interior of living cells. Due to the specific particle internalization mechanisms by living cells [91, 92], not all the SERS-encoded particles are optimal systems for the study of the internal structure/behavior of these organisms. SERS-encoded aggregates are usually large and their size either completely inhibits the internalization or results in the inclusion of a very small number of "big particles" dramatically decreasing the spatial resolution [93]. Thus, to solve this situation, the use of single particles has been popularized for recording information from the interior of a cell [94–96]. However, because of the absence of hot spots in single nanoparticles, the internalization of a large amount of particles is necessary [94, 96], additionally requiring long acquisition times to record a high-quality signal. This also results in the absence of spatial resolution, as well as cell damage. As a possible solution to this drawback, single nanoparticles were encoded with fluorophores to increase the signal intensity as a result of electronic resonance [97]. This approach, however, requires the use of visible lasers, out of the biological transparency window, thus decreasing the penetration of the beam while increasing the damage to the cell because of the focalization of a high-energy laser, which also promotes the intrinsic fluorescence of the cell [56]. Thus, a real solution, developed by Moskovits group, is based on the controlled fabrication and codification of particle dimers (Fig. 2.6) [56]. These dimers have sizes that are small enough to be internalized by eukaryotic cells and give suitable spatial and spectral resolution [98, 99].

6 Conclusions and Future Perspective

We have shown that SERS-encoded particles have been established as a solid and reliable analytical technique for the detection in extremely low amounts of a wide variety of bioanalytes. SERS-encoded particles for indirect detection and labeling can be implemented on chip or even inside living cells, tissues, or a variety of microorganisms.

However, there are still open challenges, mainly related to the reproducibility of the methods for substrate fabrication, in particular when dealing with the formation of hot spots, which are responsible for the highest enhancement factors, but their efficiency is extremely sensitive toward small geometrical details within the nanostructure. Additionally, although portable Raman spectrometers are available, most of the published reports are based on very sophisticated instruments that will not find a place in routine analysis labs or hospitals. Thus, the use of SERS codification, particularly in biomedical applications, has a great potential, as demonstrated by many examples, but is open to new developments that will undoubtedly continue amazing us in the near future.

Acknowledgment This work has been supported by the Spanish MEC (CTQ2011-23167 and MAT2008-05755/MAT and Xunta de Galicia 08TMT008314PR).

References

1. Fenniri H, Alvarez-Puebla R (2007) High-throughput screening flows along. Nat Chem Biol 3:247
2. Bake KD, Walt DR (2008) Multiplexed spectroscopic detections. Ann Rev Anal Chem 1:515
3. Nam J-M, Park S-J, Mirkin CA (2002) Bio-barcodes based on oligonucleotide-modified nanoparticles. J Am Chem Soc 124:3820
4. Bruchez M, Moronne M, Gin P, Weiss S, Alivisatos AP (1998) Semiconductor nanocrystals as fluorescent biological labels. Science 281:2013
5. Bravo-Vasquez JP, Alvarez-Puebla RA, Fenniri H (2007) Self-encoded polymer beads for microarray technologies. Sens Actuators B 125:357
6. Pregibon DC, Toner M, Doyle PS (2007) Multifunctional encoded particles for high-throughput biomolecule analysis. Science 315:1393
7. Sun L, Yu C, Irudayaraj J (2007) Surface-enhanced Raman scattering based nonfluorescent probe for multiplex DNA detection. Anal Chem 79:3981
8. Doering WE, Nie S (2003) Spectroscopic tags using dye-embedded nanoparticles and surface-enhanced Raman scattering. Anal Chem 75:6171
9. Cunin F, Schmedake TA, Link JR, Li YY, Koh J, Bhatia SN, Sailor MJ (2002) Biomolecular screening with encoded porous-silicon photonic crystals. Nature Mater 1:39
10. Zhao Y, Zhao X, Hu J, Xu M, Zhao W, Sun L, Zhu C, Xu H, Gu Z (2009) Encoded porous beads for label-free multiplex detection of tumor markers. Adv Mater 21:569
11. Han MY, Gao XH, Su JZ, Nie S (2001) Quantum-dot-tagged microbeads for multiplexed optical coding of biomolecules. Nat Biotechnol 19:631
12. Vaino AR, Janda KD (2000) Euclidean shape-encoded combinatorial chemical libraries. Proc Natl Acad Sci USA 97:7692
13. Dickinson TA, Michael KL, Kauer JS, Walt DR (1999) Convergent, self-encoded bead sensor arrays in the design of an artificial nose. Anal Chem 71:2192
14. Raez J, Blais DR, Zhang Y, Alvarez-Puebla RA, Bravo-Vasquez JP, Pezacki JP, Fenniri H (2007) Spectroscopically encoded microspheres for antigen biosensing. Langmuir 23:6482
15. Fenniri H, Terreau O, Chun S, Oh SJ, Finney WF, Morris MD (2006) Classification of spectroscopically encoded resins by Raman mapping and infrared hyperspectral imaging. J Comb Chem 8:192
16. Chun S, Xu J, Cheng J, Ding L, Winograd N, Fenniri H (2006) Spectroscopically encoded resins for high throughput imaging time-of-flight secondary ion mass spectrometry. J Comb Chem 8:18
17. Braeckmans K, De Smedt SC, Leblans M, Roelant C, Demeester J (2002) Encoding microcarriers: present and future technologies. Nat Rev Drug Discov 1:1
18. Brus L (2008) Noble metal nanocrystals: plasmon electron transfer photochemistry and single-molecule Raman spectroscopy. Acc Chem Res 41:1742
19. Camden JP, Dieringer JA, Zhao J, Van Duyne RP (2008) Controlled plasmonic nanostructures for surface-enhanced spectroscopy and sensing. Acc Chem Res 41:1653
20. Lombardi JR, Birke RL (2009) A unified view of surface-enhanced Raman scattering. Acc Chem Res 42:734
21. Sebba DS, Watson DA, Nolan JP (2009) High throughput single nanoparticle spectroscopy. ACS Nano 3:1477
22. Watson DA, Brown LO, Gaskill DR, Naivar M, Graves SW, Doorn SK, Nolan JP (2008) A flow cytometer for the measurement of Raman spectra. Cytometry A 73A:119

23. Fenniri H, Chun S, Terreau O, Bravo-Vasquez J-P (2007) Preparation and infrared/Raman classification of 630 spectroscopically encoded styrene copolymers. J Comb Chem 10:31
24. Zhang J, Gryczynski I, Gryczynski Z, Lakowicz JR (2006) Dye-labeled silver nanoshell – bright particle. J Phys Chem B 110:8986
25. Aslan K, Wu M, Lakowicz JR, Geddes CD (2007) Fluorescent core – shell Ag@SiO_2 nanocomposites for metal-enhanced fluorescence and single nanoparticle sensing platforms. J Am Chem Soc 129:1524
26. Kirchner C, Liedl T, Kudera S, Pellegrino T, Muñoz Javier A, Gaub HE, Stölzle S, Fertig N, Parak WJ (2005) Cytotoxicity of colloidal CdSe and CdSe/ZnS nanoparticles. Nano Lett 5:331
27. Tromsdorf UI, Bruns OT, Salmen SC, Beisiegel U, Weller H (2009) A highly effective, nontoxic T1 MR contrast agent based on ultrasmall PEGylated iron oxide nanoparticles. Nano Lett 9:4434
28. Fernández-López C, Mateo-Mateo C, Álvarez-Puebla RA, Pérez-Juste J, Pastoriza-Santos I, Liz-Marzán LM (2009) Highly controlled silica coating of PEG-capped metal nanoparticles and preparation of SERS-encoded particles. Langmuir 25:13894
29. Küstner B, Gellner M, Schütz M, Schöppler F, Marx A, Ströbel P, Adam P, Schmuck C, Schlücker S (2009) SERS labels for red laser excitation: silica-encapsulated SAMs on tunable gold/silver nanoshells. Angew Chem Inter Ed 48:1950
30. Wang G, Park H-Y, Lipert RJ, Porter MD (2009) Mixed monolayers on gold nanoparticle labels for multiplexed surface-enhanced Raman scattering based immunoassays. Anal Chem 81:9643
31. Moskovits M (1985) Surface-enhanced spectroscopy. Rev Mod Phys 57:783
32. Lee SJ, Guan Z, Xu H, Moskovits M (2007) Surface-enhanced Raman spectroscopy and nanogeometry: the plasmonic origin of SERS. J Phys Chem C 111:17985
33. García de Abajo FJ (2007) Colloquium: light scattering by particle and hole arrays. Rev Mod Phys 79:1267
34. Murphy CJ, Sau TK, Gole AM, Orendorff CJ, Gao J, Gou L, Hunyadi SE, Li T (2005) Anisotropic metal nanoparticles: synthesis, assembly, and optical applications. J Phys Chem B 109:13857
35. Burda C, Chen X, Narayanan R, El-Sayed MA (2005) Chemistry and properties of nanocrystals of different shapes. Chem Rev 105:1025
36. Grzelczak M, Pérez-Juste J, Mulvaney P, Liz-Marzán LM (2008) Shape control in gold nanoparticle synthesis. Chem Soc Rev 37:1783
37. Pastorizo-Santos I, Liz-Marzán LM (2009) N, N-dimethylformamide as a reaction medium for metal nanoparticle synthesis. Adv Funct Mater 19:679
38. Cobley CM, Rycenga M, Zhou F, Li ZY, Xia Y (2009) Etching and growth: an intertwined pathway to silver nanocrystals with exotic shapes. Angew Chem Inter Ed 48:4824
39. Xia Y, Xiong Y, Lim B, Skrabalak SE (2009) Shape-controlled synthesis of metal nanocrystals: simple chemistry meets complex physics? Angew Chem Inter Ed 48:60
40. Guerrero-Martínez A, Pérez-Juste J, Liz-Marzán LM (2010) Recent progress on silica coating of nanoparticles and related nanomaterials. Adv Mater 22:1182
41. Nelayah J, Kociak M, Stéphan O, De Abajo FJG, Tencé M, Henrard L, Taverna D, Pastoriza-Santos I, Liz-Marzán LM, Colliex C (2007) Mapping surface plasmons on a single metallic nanoparticle. Nat Phys 3:348
42. Rodríguez-Lorenzo L, Álvarez-Puebla RA, Pastoriza-Santos I, Mazzucco S, Stéphan O, Kociak M, Liz-Marzán LM, De Abajo FJG (2009) Zeptomol detection through controlled ultrasensitive surface-enhanced Raman scattering. J Am Chem Soc 131:4616
43. Pazos-Pérez N, Rodríguez-González B, Hilgendorff M, Giersig M, Liz-Marzán LM (2010) Gold encapsulation of star-shaped FePt nanoparticles. J Mater Chem 20:61
44. Fresnais J, Lavelle C, Berret JF (2009) Nanoparticle aggregation controlled by desalting kinetics. J Phys Chem C 113:16371
45. Alvarez-Puebla RA, Arceo E, Goulet PJG, Garrido JJ, Aroca RF (2005) Role of nanoparticle surface charge in surface-enhanced Raman scattering. J Phys Chem B 109:3787

46. Moskovits M, Vlčkova B (2005) Adsorbate-induced silver nanoparticle aggregation kinetics. J Phys Chem B 109:14755
47. Su X, Zhang J, Sun L, Koo T-W, Chan S, Sundararajan N, Yamakawa M, Berlin AA (2004) Composite organic – inorganic nanoparticles (COINs) with chemically encoded optical signatures. Nano Lett 5:49
48. Lutz BR, Dentinger CE, Nguyen LN, Sun L, Zhang J, Allen AN, Chan S, Knudsen BS (2008) Spectral analysis of multiplex Raman probe signatures. ACS Nano 2:2306
49. Brown LO, Doorn SK (2008) A controlled and reproducible pathway to dye-tagged, encapsulated silver nanoparticles as substrates for SERS multiplexing. Langmuir 24:2277
50. Brown LO, Doorn SK (2008) Optimization of the preparation of glass-coated, dye-tagged metal nanoparticles as SERS substrates. Langmuir 24:2178
51. Goddard G, Brown LO, Habbersett R, Brady CI, Martin JC, Graves SW, Freyer JP, Doorn SK (2010) High-resolution spectral analysis of individual SERS-active nanoparticles in flow. J Am Chem Soc 132:6081
52. Stoerzinger KA, Hasan W, Lin JY, Robles A, Odom TW (2010) Screening nanopyramid assemblies to optimize surface enhanced Raman scattering. J Phys ChemLett 1:1046
53. Rycenga M, Camargo PHC, Li W, Moran CH, Xia Y (2010) Understanding the SERS effects of single silver nanoparticles and their dimers, one at a time. J Phys ChemLett 1:696
54. Jin R (2010) Nanoparticle clusters light Up in SERS. Angew Chem Inter Ed 49:2826
55. Chen G, Wang Y, Yang M, Xu J, Goh SJ, Pan M, Chen H (2010) Measuring ensemble-averaged surface-enhanced Raman scattering in the hotspots of colloidal nanoparticle dimers and trimers. J Am Chem Soc 132:3644
56. Pallaoro A, Braun GB, Reich NO, Moskovits M (2010) Mapping local pH in live cells using encapsulated fluorescent SERS nanotags. Small 6:618
57. Braun GB, Lee SJ, Laurence T, Fera N, Fabris L, Bazan GC, Moskovits M, Reich NO (2009) Generalized approach to SERS-active nanomaterials via controlled nanoparticle linking, polymer encapsulation, and small-molecule infusion. J Phys Chem C 113:13622
58. Li S, Pedano ML, Chang S-H, Mirkin CA, Schatz GC (2010) Gap structure effects on surface-enhanced Raman scattering intensities for gold gapped rods. Nano Lett 10:1722
59. Sun Y, Xia Y (2002) Shape-controlled synthesis of gold and silver nanoparticles. Science 298:2176
60. Alvarez-Puebla RA, Ross DJ, Nazri GA, Aroca RF (2005) Surface-enhanced Raman scattering on nanoshells with tunable surface plasmon resonance. Langmuir 21:10504
61. Sanchez-Iglesias A, Grzelczak M, Rodríguez-González B, Álvarez-Puebla RA, Liz-Marzán LM, Kotov NA (2009) Gold colloids with unconventional angled shapes. Langmuir 25:11431
62. Perez-Pineiro R, Dai S, Alvarez-Puebla R, Wigginton J, Al-Hourani BJ, Fenniri H (2009) Synthesis of sulfur-containing aryl and heteroaryl vinyls via Suzuki-Miyaura cross-coupling for the preparation of SERS-active polymers. Tetrahedron Lett 50:5467
63. Al-Hourani BJ, Bravo-Vasquez JP, Hermann High LR, Fenniri H (2007) Synthesis and characterization of aryl thioacetyl styrene monomers: towards a new generation of SERS-active polymers. Tetrahedron Lett 48:9144
64. Farah AA, Alvarez-Puebla RA, Fenniri H (2008) Chemically stable silver nanoparticle-crosslinked polymer microspheres. J Colloid Interf Sci 319:572
65. Chon H, Lee S, Son SW, Oh CH, Choo J (2009) Highly sensitive immunoassay of lung cancer marker carcinoembryonic antigen using surface-enhanced Raman scattering of hollow gold nanospheres. Anal Chem 81:3029
66. Kim K, Lee YM, Lee HB, Shin KS (2009) Silver-coated silica beads applicable as core materials of dual-tagging sensors operating via SERS and MEF. ACS App Mater Interface 1:2174
67. Jun BH, Noh MS, Kim J, Kim G, Kang H, Kim MS, Seo YT, Baek J, Kim JH, Park J, Kim S, Kim YK, Hyeon T, Cho MH, Jeong DH, Lee YS (2010) Multifunctional silver-embedded magnetic nanoparticles as SERS nanoprobes and their applications. Small 6:119

68. Yang MX, Chen T, Lau WS, Wang Y, Tang QH, Yang YH, Chen HY (2009) Development of polymer-encapsulated metal nanopartictes as surface-enhanced Raman scattering probes. Small 5:198
69. Shen A, Chen L, Xie W, Hu J, Zeng A, Richards R, Hu J (2010) Triplex Au-Ag-C core-shell nanoparticles as a novel Raman label. Adv Funct Mater 20:969
70. Edelstein RL, Tamanaha CR, Sheehan PE, Miller MM, Baselt DR, Whitman LJ, Colton RJ (2000) The BARC biosensor applied to the detection of biological warfare agents. Biosens Bioelectron 14:805
71. Zavaleta CL, Smith BR, Walton I, Doering W, Davis G, Shojaei B, Natan MJ, Gambhir SS (2009) Multiplexed imaging of surface enhanced Raman scattering nanotags in living mice using noninvasive Raman spectroscopy. Proc Natl Acad Sci USA 106:13511
72. Freeman RG, Raju PA, Norton SM, Walton ID, Smith PC, He L, Natan MJ, Sha MY, Penn SG (2005) Use of nanobarcodes particles in bioassays. Meth Mol Biol 303:73
73. Golightly RS, Doering WE, Natan MJ (2009) Surface-enhanced Raman spectroscopy and homeland security: a perfect match? ACS Nano 3:2859
74. Sassolas A, Leca-Bouvier BD, Blum LJ (2008) DNA biosensors and microarrays. Chem Rev 108:109
75. Wong LS, Khan F, Micklefield J (2009) Selective covalent protein immobilization: strategies and applications. Chem Rev 109:4025
76. Iha RK, Wooley KL, Nyström AM, Burke DJ, Kade MJ, Hawker CJ (2009) Applications of orthogonal "click" chemistries in the synthesis of functional soft materials. Chem Rev 109:5620
77. Louie A (2010) Multimodality imaging probes: design and challenges. Chem Rev 110:3146
78. Cao YC, Jin R, Mirkin CA (2002) Nanoparticles with Raman spectroscopic fingerprints for DNA and RNA detection. Science 297:1536
79. Qin L, Banholzer MJ, Millstone JE, Mirkin CA (2007) Nanodisk codes. Nano Lett 7:3849
80. Stoermer RL, Cederquist KB, McFarland SK, Sha MY, Penn SG, Keating CD (2006) Coupling molecular beacons to barcoded metal nanowires for multiplexed, sealed chamber DNA bioassays. J Am Chem Soc 128:16892
81. Doering WE, Piotti ME, Natan MJ, Freeman RG (2007) SERS as a foundation for nanoscale, optically detected biological labels. Adv Mater 19:3100
82. Sha MY, Xu HX, Natan MJ, Cromer R (2008) Surface-enhanced Raman scattering tags for rapid and homogeneous detection of circulating tumor cells in the presence of human whole blood. J Am Chem Soc 130:17214
83. Sha MY, Walton ID, Norton SM, Taylor M, Yamanaka M, Natan MJ, Xu C, Drmanac S, Huang S, Borcherding A, Drmanac R, Penn SG (2006) Multiplexed SNP genotyping using nanobarcode particle technology. Anal Bioanal Chem 384:658
84. Nabiev IR, Morjani H, Manfait M (1991) Selective analysis of antitumor drug interaction with living cancer cells as probed by surface-enhanced Raman spectroscopy. Eur Biophys J 19:311
85. Xiao M, Nyagilo J, Arora V, Kulkarni P, Xu DS, Sun XK, Dave DP (2010) Gold nanotags for combined multi-colored Raman spectroscopy and x-ray computed tomography. Nanotechnology 21:1–8
86. Wu LY, Ross BM, Hong S, Lee LP (2010) Bioinspired nanocorals with decoupled cellular targeting and sensing functionality. Small 6:503
87. Deutscher SL (2010) Phage display in molecular imaging and diagnosis of cancer. Chem Rev 110:3196
88. Kim JH, Kim JS, Choi H, Lee SM, Jun BH, Yu KN, Kuk E, Kim YK, Jeong DH, Cho MH, Lee YS (2006) Nanoparticle probes with surface enhanced Raman spectroscopic tags for cellular cancer targeting. Anal Chem 78:6967
89. Qian XM, Nie SM (2008) Single-molecule and single-nanoparticle SERS: from fundamental mechanisms to biomedical applications. Chem Soc Rev 37:912
90. Qian X, Peng X-H, Ansari DO, Yin-Goen Q, Chen GZ, Shin DM, Yang L, Young AN, Wang MD, Nie S (2008) In vivo tumor targeting and spectroscopic detection with surface-enhanced Raman nanoparticle tags. Nat Biotechnol 26:83

91. Nativo P, Prior IA, Brust M (2008) Uptake and intracellular fate of surface-modified gold nanoparticles. ACS Nano 2:1639
92. See V, Free P, Cesbron Y, Nativo P, Shaheen U, Rigden DJ, Spiller DG, Fernig DG, White MRH, Prior IA, Brust M, Lounis B, Levy R (2009) Cathepsin L digestion of nanobioconjugates upon endocytosis. ACS Nano 3:2461
93. Sanles-Sobrido M, Exner W, Rodríguez-Lorenzo L, Rodríguez-González B, Correa-Duarte MA, Álvarez-Puebla RA, Liz-Marzán LM (2009) Design of SERS-encoded, submicron, hollow particles through confined growth of encapsulated metal nanoparticles. J Am Chem Soc 131:2699
94. Kneipp J, Wittig B, Bohr H, Kneipp K (2010) Surface-enhanced Raman scattering: a new optical probe in molecular biophysics and biomedicine. Theor Chem Acc 125:319
95. Kneipp J, Kneipp H, Wittig B, Kneipp K (2010) Following the dynamics of pH in endosomes of live cells with SERS nanosensors. J Phys Chem C 114:7421
96. Kneipp J, Kneipp H, Wittig B, Kneipp K (2007) One- and two-photon excited optical pH probing for cells using surface-enhanced Raman and hyper-Raman nanosensors. Nano Lett 7:2819
97. Kneipp J, Kneipp H, Rice WL, Kneipp K (2005) Optical probes for biological applications based on surface-enhanced Raman scattering from indocyanine green on gold nanoparticles. Anal Chem 77:2381
98. Cho EC, Au L, Zhang Q, Xia Y (2010) The effects of size, shape, and surface functional group of gold nanostructures on their adsorption and internalization by cells. Small 6:517
99. Giljohann DA, Seferos DS, Daniel WL, Massich MD, Patel PC, Mirkin CA (2010) Gold nanoparticles for biology and medicine. Angew Chem Inter Ed 49:3280

Biomedical SERS Studies Using Nanoshells

3

Michael A. Ochsenkühn and Colin J. Campbell

1 Overview

There is a need in biomedical sciences for improved techniques for monitoring biomolecular interactions and physiological processes, especially those which can report in real time. New techniques may find use in (for example) live-cell imaging or in vitro diagnostics – one such technique is surface-enhanced Raman spectroscopy (SERS). While SERS has been demonstrated using a variety of metal surfaces this chapter focuses on applications which use nanoshells, a type of engineered nanoparticle. We focus on biological applications of SERS, illustrating the discussion with examples of investigations of biomolecular conformational changes, biomolecular interactions, and intracellular studies.

2 Introduction

Raman spectroscopy is an inelastic scattering technique that probes the vibrational energy levels within molecules and was first reported by C.V. Raman, who was awarded the Nobel Prize in Physics in 1930 for his discoveries. Nowadays, Raman spectroscopy is an established technique and is regularly used to analyze chemicals and materials, often in combination with infrared absorption spectroscopy since the two techniques are mutually complementary, that is, vibrational transitions that are Raman active are not IR active. In Raman scattering, light interacts with a molecule

M.A. Ochsenkühn • C.J. Campbell (✉)
EaStCHEM, School of Chemistry, University of Edinburgh, Edinburgh, UK

Challa S. S. R. Kumar (ed.), *Raman Spectroscopy for Nanomaterials Characterization*,
DOI 10.1007/978-3-642-20620-7_3, © Springer-Verlag Berlin Heidelberg 2012

and polarizes the electron cloud elevating the molecule to a short-lived state, called a "virtual state." This state is not stable and the photon is quickly re-radiated [1]. The difference in energy between the incident and scattered photon is characteristic of a transition between vibrational levels in the molecule and is determined by such factors as the identity of the atoms or the formation of higher order structural motifs.

The main advantages of Raman spectroscopy are: independence on excitation wavelength which allows the experimentalist to choose a wavelength suitable for the sample under study, and the ability to probe large biochemical compounds or structures, such as cells or tissues, without the need for markers. While Raman spectroscopy sounds like the perfect technique for analytical chemistry and biochemistry, its major drawback is a very low scattering cross section that results in a weak signal, which can be obscured by fluorescence or elastically scattered light [1].

This drawback can be overcome by using surface-enhanced Raman spectroscopy (SERS) which was first discovered by Fleischmann et al. as they attempted to measure Raman spectra of pyridine on an electrochemically roughened silver surface [2]. The signal was much higher than expected and the authors explained this finding on the basis of increased surface area compared to a polished electrode surface. In 1977, a surface-plasmon-enhanced mechanism was proposed by van Duyne et al. [3] and this explanation has now become widely accepted. SERS is now often employed in order to increase the sensitivity and signal strength of Raman spectra and since signal enhancements of up to 10^{14} have been reported, this has enabled extremely sensitive measurements such as single molecule studies [4]. Enhancement of the Raman signal occurs when a molecule is in close proximity to a metal nanoparticle or nanostructured metallic surface. The interaction of light with packets of electron density at the metal–dielectric interface gives rise to surface plasmon resonance (SPR) [5]. SPR results in an increase in local electromagnetic field strength which leads to an enhancement of the Raman signal from molecules at the metal surface. While silver and gold are the most commonly used metals since they are relatively resistant to corrosion and exhibit SPR with light in the visible or near IR regions, several other metals have also been used to demonstrate SERS [6].

There are two theories that attempt to explain the SERS effect. The first is the electromagnetic theory that describes an interaction between the incoming radiation and surface plasmons causing an enhancement in the local electric field and the second is the charge transfer theory where a bond is formed between the metallic surface and the molecule. It is thought that both mechanisms can contribute to the signal increase [5, 6]. For the molecule to experience an increase in its polarizability and therefore its Raman cross section it should be orientated perpendicular to the surface plasmon wave vector – as a result, not all Raman active modes of a molecule are necessarily SERS active. SERS is a powerful technique in the study of single molecules [7] and has been used in a variety of applications from biohazard detection [8] through the study of biomolecule structures, for example, proteins [9] to nucleic acid sequence detection [10]. SERS has even been used to image cells and has the advantage that it gives specific chemical information about distinct locations where nanoparticles have been delivered [11].

3 Experimental and Instrumental Methodology

This section describes the methods which we have been used for preparing substrates for SERS measurements. All of the described methods employed commercially available gold NS from Nanospectra Biosciences Inc (Houston, TX). Modification of particle surfaces was accomplished by forming a self-assembled monolayer of molecules of interest [12].

3.1 Formation of NS Aggregates on Silanized Glass Slides

NS suspended in ddH_2O at a concentration of 40×10^9 particles/mL were vortexed and ultrasonicated for 5 min before being placed in 3 μL spots on aminopropyl triethoxy silanized (APTES) slides and left until dry. APTES slides are commercially available and can be purchased from (for example) Schott or Corning. The slides were then washed five times in excess water in order to remove any free particles. Particles typically formed layered aggregates, which could be visualized using atomic force microscopy (AFM).

3.2 NS Surface Modification 4-Mercapto Benzoic Acid (4MBA)

3.2.1 Functionalization of Aggregates

In order to modify NS with 4MBA, a 10 mM solution of 4MBA in 9:1 H_2O/EtOH was prepared. This solution was then pipetted as a 10 μL drop on the aggregated spots described above and left at RT overnight. On the next day, the self-assembled monolayer was washed ten times with ddH_2O and the slides were spun dry at 1,000 rpm for 1 min in a table top Eppendorf centrifuge.

3.2.2 Functionalization of Particles for Intracellular Use

NS suspensions were sterilized by autoclaving. 300 μL of sterile NS solution of 10×10^9 NS/mL was added to 10 μL 4-MBA (1 mM) and incubated overnight at room temperature. Excess 4-MBA was washed off by careful recovery of NS from the reaction solution by centrifugation. The sample of modified NS was washed a further four times in ddH_2O.

3.2.3 Modification of Aggregates with 5′ Thiol Modified Aptamers and Oligonucleotides

Stock solutions of 5′ thiol modified aptamer(100 μM) were diluted to 100 μL in water and treated with tris (2-carboxyethyl)phosphine (TCEP) (1 mg) for 30 min at RT, gently agitated and purified using a G-25 MicroSpinTM size exclusion column purchased from Amersham Biosciences (Buckinghamshire, UK). The aptamer solution was then diluted to a concentration of 10 μM in buffer consisting of (Tris–HCl 50 mM, NaCl 140 mM, $MgCl_2$ 10 mM, KCl 10 mM, pH 7.4), heated to 95°C for 5 min, cooled down to 72°C for 15 min and then cooled to RT over 30 min.

15 μL of aptamer solution was pipetted onto the NS aggregate spots and incubated for 18 h at 30°C in a humid environment. Excess aptamer was removed by rinsing the slide five times in the same buffer used for incubation. In order to minimize nonspecific interactions, the spots were incubated with a solution of mercaptohexanol (MCH) (1 mM, 15 μL) in the above buffer for 30 min and rinsed five times afterward with the same buffer.

3.3 SERS Experimental Setup

3.3.1 Microscope

An InVia Confocal Raman Micrscope AFM system (Renishaw, UK) fitted with a 50x Leica water immersion objective (NA of 0.8) and a 785 nm diode laser delivering up to 7 mW of laser power was used.

3.3.2 SERS Measurements on Aptamer Modified NS

SERS measurements on NS bound oligonucleotides were performed using ≈30 μW of laser power at the sample and a typical acquisition time of 10 s between 800 and 1,800 cm^{-1}. The spectra shown are averages of 25 consecutive acquisitions collected from different aggregates.

4 Key Research Findings

4.1 Nanoshells and Spectroscopy

In order to collect SERS spectra, nanoscaled structures are usually either fixed onto large surfaces or free in suspension in the form of nanoparticles. Suitable particles are available in a variety of forms: simple spherical particles, rods, [13] rice-like shapes, [14] sea urchins [15], or even cages [16]. Their size can vary from 1 to 500 nm, and this factor has an important impact on their surface plasmon resonance properties.

Nanoshells are a particular class of spherical nanoparticles, invented by Halas et al., which comprise a dielectric core coated with a thin noble metal film. The core can range from solid materials like silica [17] to hollow particles, [18] and to date Au, Ag, and Cu have been used as metal layers [19, 20]. This particular approach to nanoparticle engineering allows sensitive control over both the size and the optical properties of the particles [21]. In fact, the electronic and optical properties of nanoshells can be approximated with the classic Mie-scattering theory and more accurately predicted with a theoretical method based on a time-dependent local density approximation (TDLDA) [22–24]. Nanoshells are typically made by first coating a dielectric particle with amino silane which is used as a polyelectrolyte layer on which to assemble 1–3 nm gold particles. The seeded gold particles are then chemically annealed using HAuCl2 solution and the reaction time can be varied in order to control the layer thickness [17, 25]. The process of nanoshell synthesis is well described elsewhere [26, 27]. Nanoshells have found a range of applications including SPR sensing of biomolecular interactions in blood, [28] pH measurements, [29] enhancement of fluoresence signals [30] and for investigating the interaction between ibuprofen and lipid bilayer membranes [31]. Very recently, even the measurement of SERS signals inside NS has been reported by Zhang et al. [32]. A particularly noteworthy application, the thermal ablation of cancer with antibody modified NS, exploits local heating caused by SPR, which specifically destroys soft tissue cancers with great success [33]. Depending on the chosen application, the optical properties of NS can be tuned. Since our particular aim is to demonstrate the potential of NS for biological Raman spectroscopic applications we typically use particles of 150 nm diameter, consisting of a silica core of 120 nm diameter coated in a 30 nm layer of gold, which exhibit a maximum absorbance at 780 nm where cells or tissue exhibits very low absorbance and low autofluorescence.

4.2 Biomolecular Attachment

Biomolecular detection has been demonstrated using several approaches which facilitate SERS on nanoscale roughened noble metal surfaces or by using nanoparticles in a "sandwich" format [34]. A few approaches for SERS detection of proteins have used aptamers, which are oligonucleotide probes that bind target molecules with high specificity [35]. In these cases, Raman active dyes are generally attached to one end of the aptamer and are drawn closer to the surface of the SERS substrate during binding of the target protein leading to a large enhancement of the dye spectrum [36, 37]. Perhaps the best characterized aptamer of this kind is the thrombin-binding aptamer (TBA) [38], which can be used as an inhibitor of blood coagulation. The wealth of characterization data available for this aptamer makes it an ideal candidate to use as a model-system for the development of new detection methods. A stacked G-quadruplex structure is an important motif adopted by the aptamer upon binding thrombin and its disruption has been shown to destabilize the aptamer/protein complex [39]. Having these features in mind, we speculated that by monitoring the SERS spectrum of this motif, we could detect aptamer/protein binding. The ability to measure molecular binding events in real time, without labeling, is critically important in aiding our understanding of biological processes [40–43].

A recurring problem of SERS is spectral reproducibility which is often attributed to variability in nanoparticle preparations or variability in nanoparticle aggregation. Quality controlled and commercially available APTES modified glass slides and Auroshell^{TM} NS can be used to assemble a very convenient substrate for SERS spectroscopy where NS form small aggregates with a size of ca. 0.5–5 μm diameter on the surface [44] and which often comprise NS multilayers (Fig. 3.1). NS aggregation gives rise to the formation of "hot-spots," which are areas of enhanced electromagnetic field. Similarly, other reproducible NS-based SERS platforms have been reported with controlled regular monolayers or distributions of single and paired NS [45]. In order to maximize the plasmon resonance, it is important to

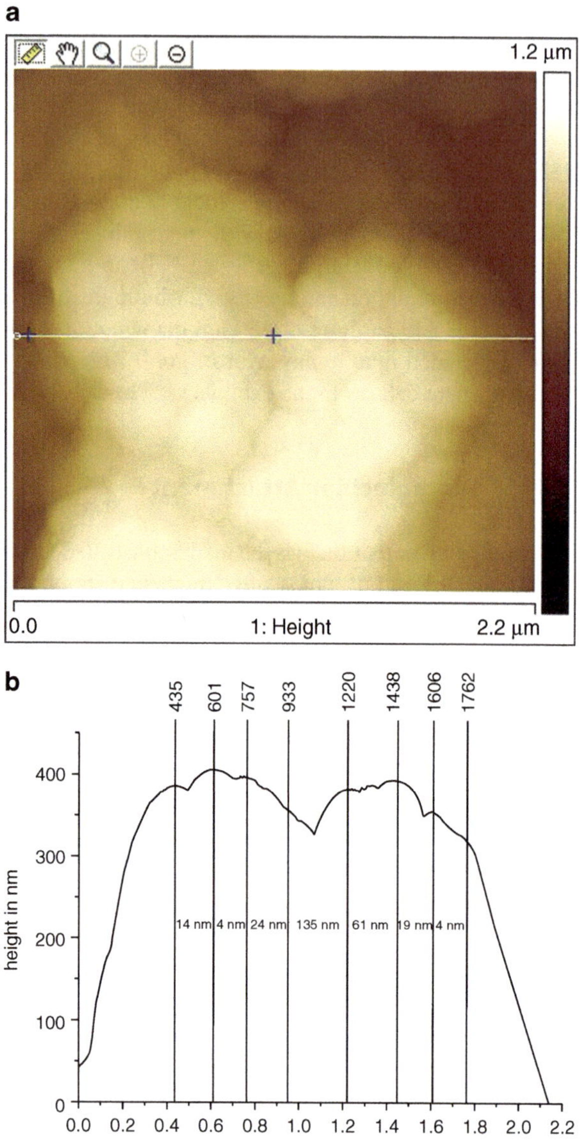

Fig. 3.1 (**a**) Representative AFM image of NS aggregate on amino-silanized glass slides functionalized with TBA. (**b**) Profile of cross section of the same NS aggregate showing distances between NS estimated by using the highest point of NS curvature as the center of the particle

choose the right laser wavelength – this allows the maximum signal strength from the lowest laser power. The field enhancement which gives rise to SERS decays rapidly within a tightly confined zone surrounding the nanoparticles, therefore the best way to obtain good signal is through the direct attachment of the molecule of interest to the metal surface. The most common way of attaching biomolecules covalently to a gold surface is through the use of a thiol-linker which can be readily attached to most molecules and the separation between the surface and the molecule

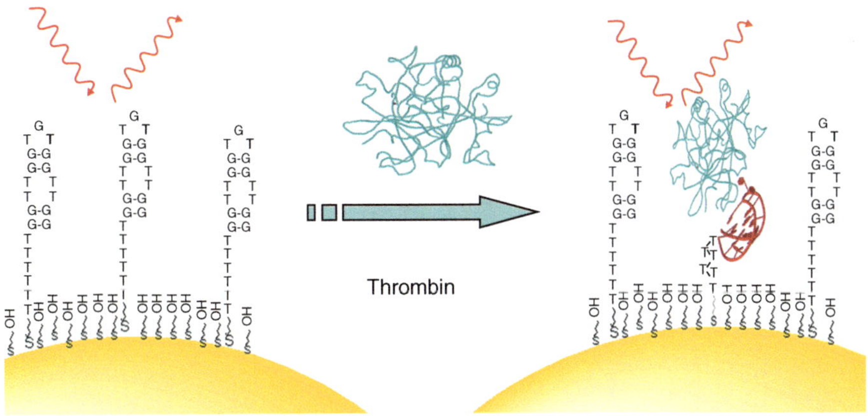

Fig. 3.2 Model illustrating the TBA aptamer modified sensor surface before and after incubation with Thrombin [49]

can be tuned through the use of a suitable spacer [46, 47]. Specifically, we attached the aptamer to the gold surface using a thiolated 6x thymine linker which allows improved secondary structure formation and aptamer–target interaction. Presentation of the aptamer was further improved by blocking unreacted surface sites on the gold surface with a layer of mercapto-hexanol, which also serves to reduce nonspecific binding (Fig. 3.2) [48]. This also has an important influence on the Raman spectrum of the surface bound oligonucleotide as can be seen from Fig. 3.3. The first spectrum shows a strong signal in the NS aromatic ring shift region at around 1,450–1,600 cm^{-1} which we attribute to nonspecifically bound aptamer. The second spectrum, after MCH incubation for 30 min and washing in buffers shows only comparatively weak signal (e.g., the ubiquitous DNA phosphate stretch at $\approx$1,080 cm^{-1}) – we attribute this difference to a rearrangement of the aptamer layer caused by surface passivation.

When measuring spectra of biological molecules and probing biomolecular interactions, it is important to ensure that the sample is not uncontrollably heated as a result of the measurement. Local plasmonic heating has been described for a variety of nanostructures and has even been used for the thermal ablation of tumor cells. Even though thermal ablation requires significantly higher laser powers than we use for spectroscopy, we sought to determine whether plasmonic excitation leads to local heating using our experimental conditions. Raman spectroscopy actually provides the means to investigate this since the ratio of anti-Stokes/Stokes intensities is proportional to the temperature. Over the course of a 1.5 h long experiment with continuous excitation using a laser power of $\approx$30 μW at RT, we acquired spectra and plotted the anti-Stokes/Stokes intensity ratio [49]. The anti-Stokes and Stokes peaks which we measure at −568 and 570 cm^{-1}, respectively, can be assigned to the CS-stretch and should be uniformly enhanced and abundant in signal in both inelastic Raman scattering modes [50]. Figure 3.4 shows the spectrum at zero minutes. One spectrum is the average of 10 acquisitions measured every 10 min. The

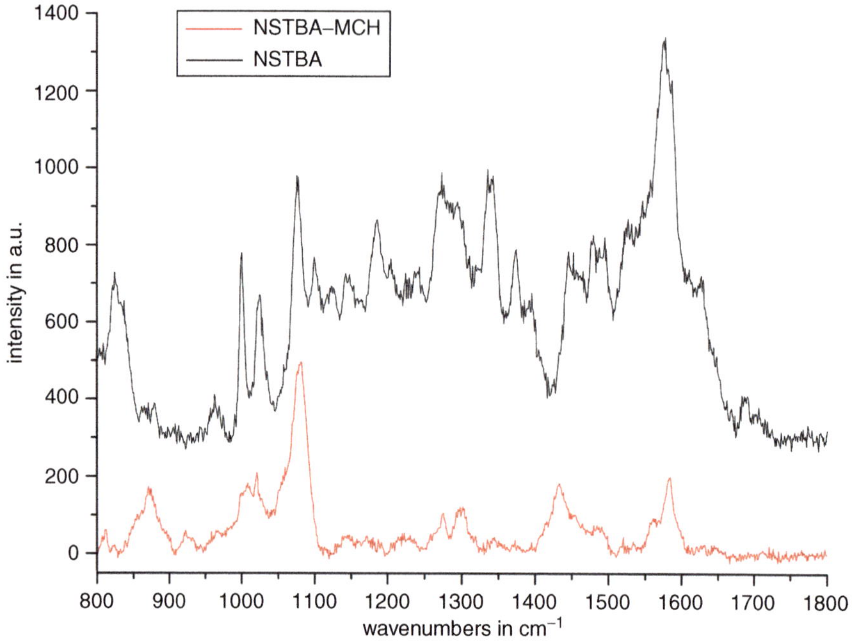

Fig. 3.3 Raman spectra acquired for NS-TBA before formation of a mixed monolayer of MCH on the NS surface

anti-Stokes/Stokes ratios do not vary significantly over the period of the measurement and this finding confirms that no significant local heating occurs.

4.3 Conformational Changes

In order to demonstrate that G-quadruplex conformational changes can be probed using Raman spectroscopy we first sought to investigate the intrinsic spectral features of this structural motif using an oligo selected by Yoshida et al. [51] to form a G-quadruplex (GQ). GQ modified nanoshells (referred to as NS-GQ) were prepared the same way as NS-TBA (described above).

Immediately following attachment of the GQ oligo to the NS, we observed a similar spectrum as for NS-TBA with the phosphate backbone stretch being the most prominent feature around 1,080 cm^{-1} (Fig. 3.5). After 20 min incubation in a buffer containing potassium and magnesium ions, which we had previously found to promote GQ formation, the spectrum is significantly different. The SERS spectrum of NS-GQ contains very distinct Raman features at 820, 1,132, 1,480 and 1,560 cm^{-1} (Fig. 3.5). We attribute these spectral changes to formation of the G-quadruplex, and suggest that formation of a GQ structure positions key nucleotides closer to the gold surface and in a better position to allow interaction between

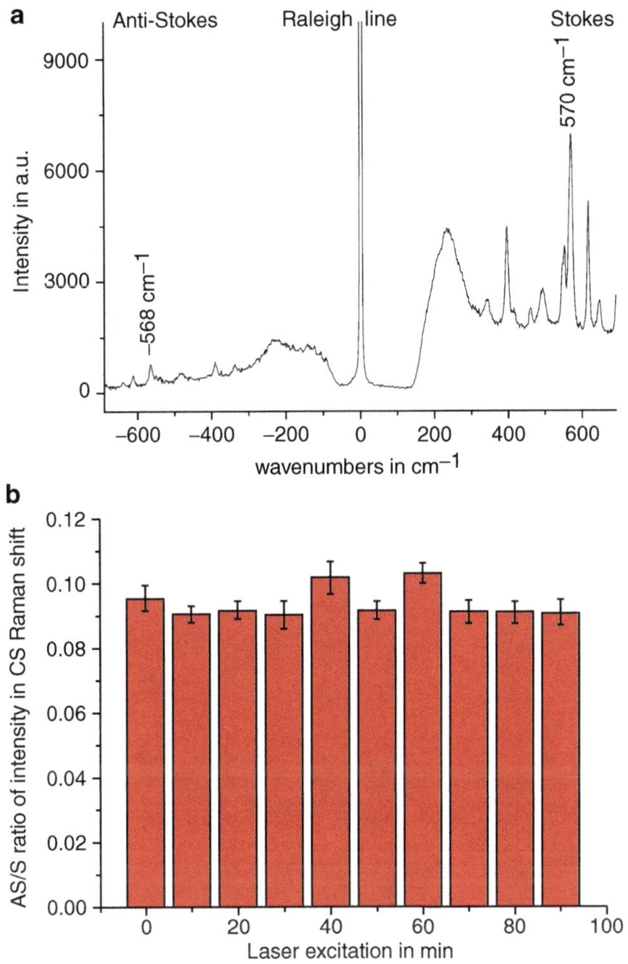

Fig. 3.4 (**a**) Representative spectrum from Anti-stokes/Stokes scattering experiment for the investigation of particle heating. (**b**) Comparison of the calculated AS/S ratio for the integrals of the peaks at −568 and 570 cm^{-1} respectively. Error bars show one single standard deviation of the integrals of 10 spectra

specific vibrational modes and the strong electromagnetic field caused by SPR. The Raman spectral features of GQ correlate well with vibrational modes of its constituent bases (guanine and thymine). In previous SERS studies by Halas et al. [52], it was shown that the hybridization of oligonucleotides leads to changes in Raman shifts. The same authors have also described the correlation between dsDNA orientation and SERS signal intensity and found signal increases depending on the angle between the molecule and the surface [53].

To investigate the temperature-dependent behavior of this oligonucleotide and to illuminate the coherence between Raman signal and oligonucleotide secondary

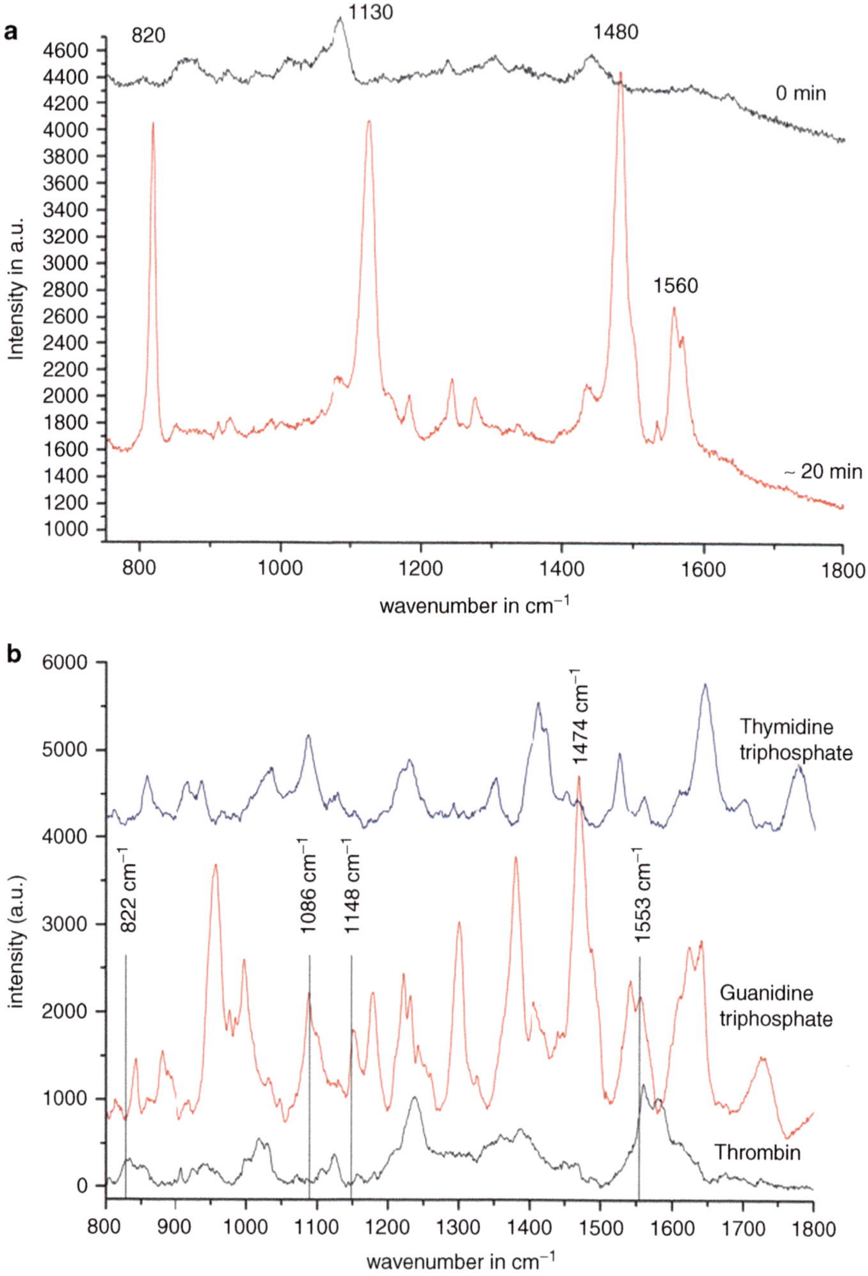

Fig. 3.5 (**a**) Spectral changes from NS-GQ before and after incubation in buffer. (**b**) Spectra of thymidine triphosphate, Guanidine triphosphate, and thrombin

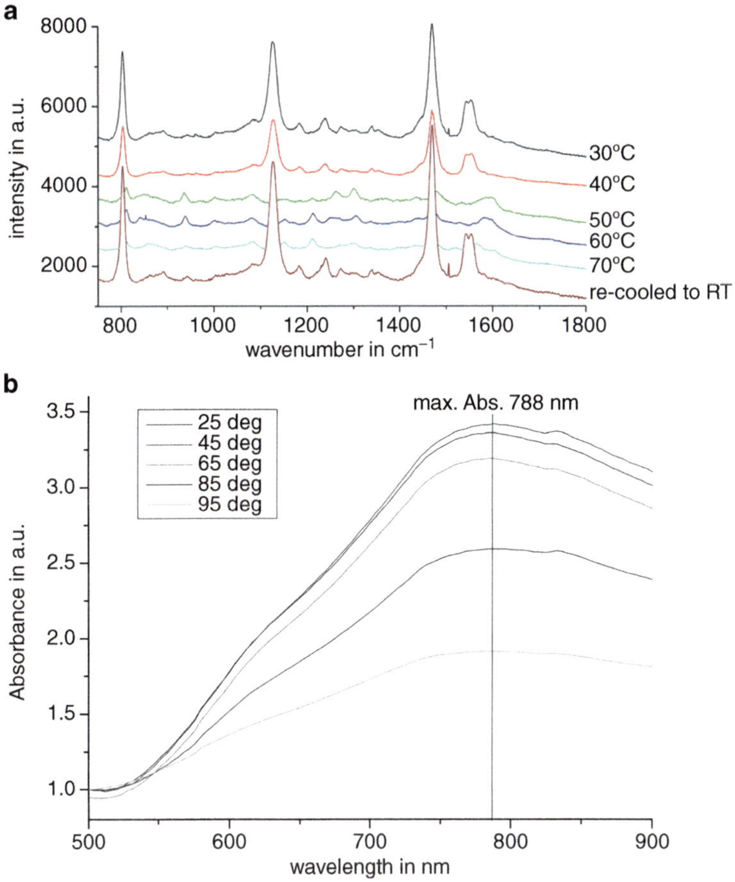

Fig. 3.6 (**a**) SERS spectra of NS-GQ over a range of temperatures. Spectra show averages of 20 spectra with acquisition times of 10 s. Temperatures were increased in 10°C steps from 30°C up to 70°C and the intensity of the marker bands decreased in intensity. Cooling back to RT after the 70°C measurement gave spectra with the original high intensity signals suggesting that the oligonucleotide refolds to its original structure at low temperature. (**b**) Influence of temperature on absorbance of NS. Change in absorbance over the temperature range observed in GFA melting experiment is negligible and not the cause of signal reduction [49]

structure, a melting experiment, shown in Fig. 3.6, was conducted by gradually heating the sensor from 30°C to 70°C with subsequent slow cooling to RT to allow secondary structure reformation. On temperature increase the signal decreases suggesting the gradual loss of secondary structure. Careful cooling back to RT leads to the same characteristic Raman shifts and intensities, and shows that the formation of the GQ motif is completely reversible. Further experiments confirmed that this apparent melting behavior was not caused by a temperature-dependent loss in NS absorbance (Fig. 3.6b). This simple experiment therefore confirmed that the changes in SERS signal can be safely attributed to changes in molecular conformation.

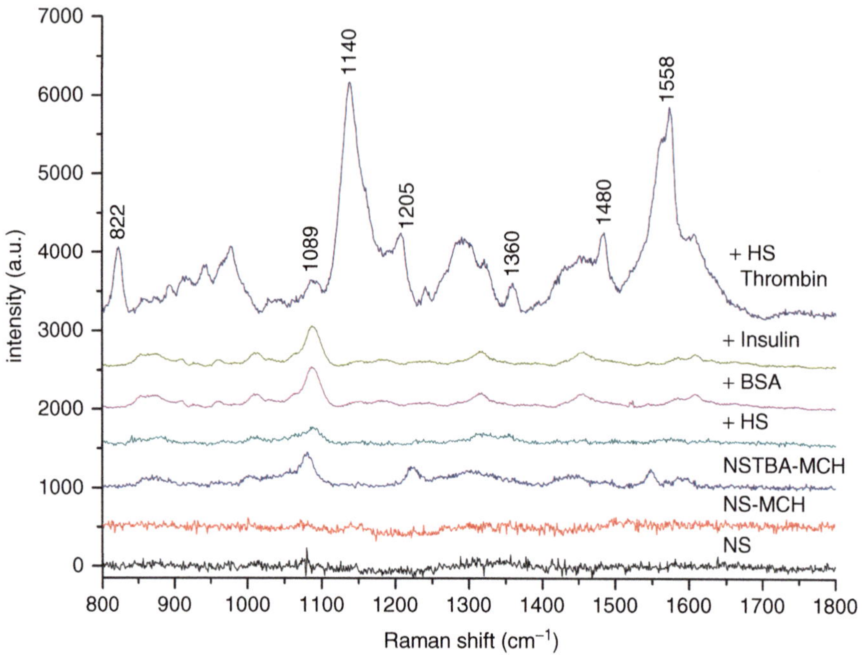

Fig. 3.7 Spectral changes caused by thrombin (10 nM) binding to NSTBA sensor. The dominant peaks can be assigned to DNA vibrational modes with a lesser contribution from protein vibrational modes. Controls with BSA (10 mg/ml), insulin 600 nM and 1% human serum show no nonspecific protein binding. Lower three traces show spectra of NS, NS plus blocking agent (mercaptohexanol) and the fully assembled NSTBA sensor. Spectra are averages of 10 acquisitions

4.4 Using SERS Spectral Changes to Detect Aptamer-Protein Binding

Having established that SERS can be used to detect changes in the secondary structure of an aptamer, we sought to detect the conformational change in a thrombin-binding aptamer (TBA) caused by binding to thrombin. As shown above the SERS spectrum of the TBA functionalized NS has only a few distinct features including the phosphate backbone stretch at 1,090 cm^{-1} aliphatic vibrational modes at 850–900 cm^{-1}, aromatic vibrational modes at 1,000 cm^{-1}, and heterocyclic ring vibrational modes between 1,200 and 1,300 cm^{-1} and between 1,550 and 1,600 cm^{-1} [49]. In these conditions, the secondary structure of the oligonucleotide does not form when not bound to thrombin. On incubation with 10 nM thrombin for 30 min at room temperature there are significant spectral changes. Figure 3.7a shows a comparison of the signal obtained after incubation with 10 nM thrombin with a series of control experiments. Most importantly, only on incubation with thrombin does the aptamer spectrum change and on incubation with different nonspecific proteins such as bovine serum albumin (BSA), insulin, and human serum (HS) the aptamer spectrum does not significantly change.

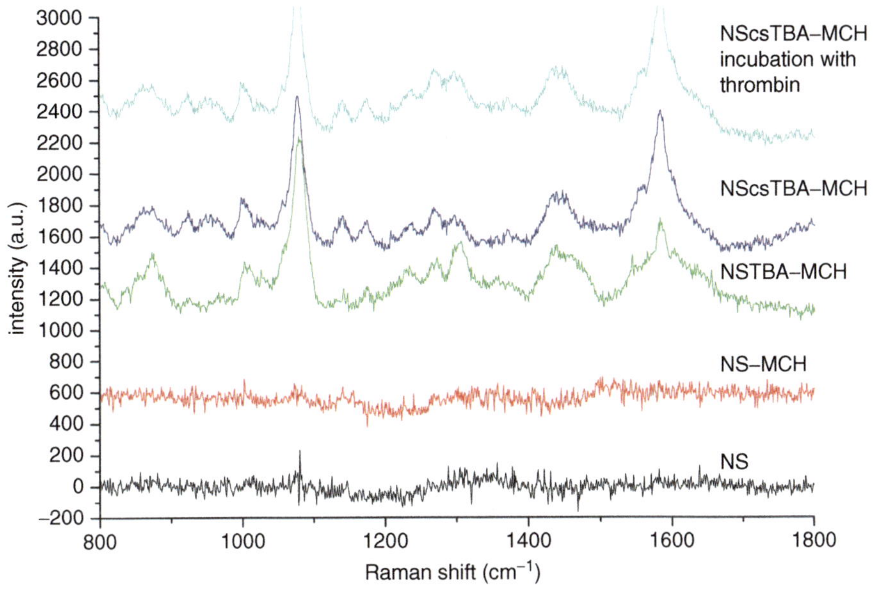

Fig. 3.8 SERS spectra of control sequence aptamer (csTBA) and csTBA incubated for 30 min with 10 nM of thrombin in binding buffer. Also shown are SERS spectra of NS with mercaptohexanol and the spectra for NS with TBA plus MCH

The spectral features of the aptamer/thrombin complex are significantly similar to those of the GQ shown above and some can be similarly attributed to vibrational modes of guanine and thymine. Broad and weak features such as those between 900 and 1,000 cm^{-1} as well as 1,250 and 1,330 cm^{-1} and a very broad feature covering the region around 1,400 cm^{-1} can be attributed to vibrational modes of the protein as can be seen in the SERS spectrum of thrombin also shown in Fig. 3.5b.

In summary, the sharp Raman features which appear after incubation with 10 nM thrombin at 822 cm^{-1}, 1,140, and 1,558 cm^{-1} can be assigned to the combined C2′-endo and C3′-endo modes of the 2′-deoxyribose sugars, the C—O—C stretch at 1,140 cm^{-1}, and guanine ring modes at 1,558 cm^{-1}. At 1,480 cm^{-1}, we observe another feature with lower intensity which can also be assigned to a guanine ring mode [49]. Further signal increases can be assigned to vibrational modes of protein, such as the amide III backbone vibration at 1,220 cm^{-1}, CH_2 stretching mode at 1,440–1,460 cm^{-1}, or tyrosine aromatic ring vibrations at 1,610 cm^{-1} [54, 55]. A control experiment using a random oligonucleotide did not show any changes on thrombin incubation (Fig. 3.8).

To study the reproducibility and the reusability of the sensor, successive washing steps were performed with solutions of 8 M Urea and 0.2 M Ca^{2+}-EDTA followed by re-incubation with 10 nM thrombin. Urea was chosen to destabilize the protein aptamer complex and Ca^{2+}/EDTA was chosen to disrupt the G-quadruplex. Spectra recorded after washing exhibit the low intensity unbound signal and signal increases repeatedly after incubation with 10 nM thrombin – these two states can be

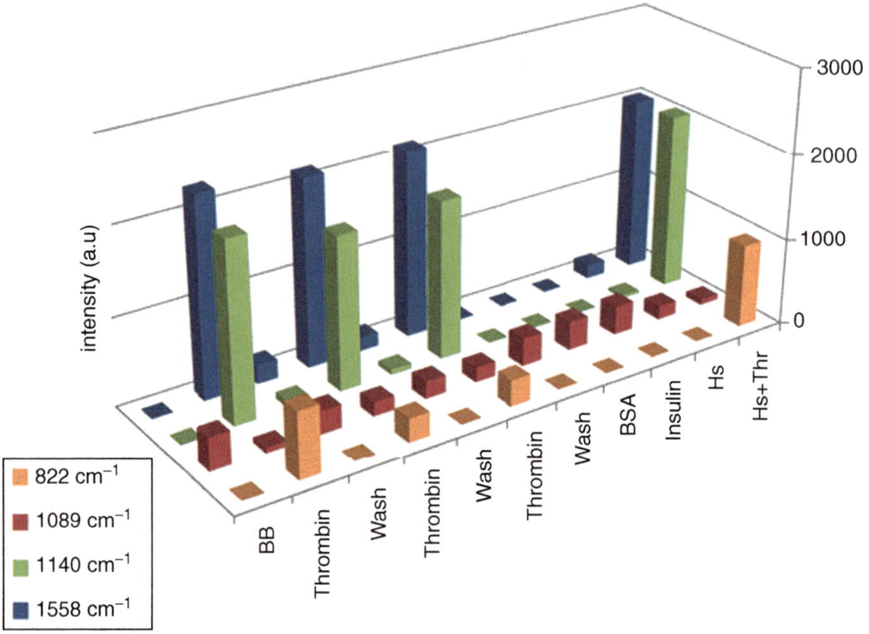

Fig. 3.9 Washing and target incubation cycles of NSTBA with 10 nM thrombin. Graph shows intensities of dominant peaks. Peak intensities are highly reproducibly at every incubation cycle. A further incubation in which 10 nM thrombin was dissolved in a 1% solution of human serum produced similar peak intensities. Control experiments incubating only with bovine serum albumin, insulin, and human serum do not show any increased signals compared to the washed sensor [49]

cycled between in a reversible fashion (Fig. 3.9). This Figure also compares the characteristic binding signals of the protein/aptamer complex with those from control measurements. Furthermore, we observe the same spectral changes when incubating the sensor with thrombin doped into human serum (Fig. 3.9 last row) confirming that we can observe this specific biomolecular interaction even in a complex biological matrix and obtain the same signal.

To determine the limit of detection (LOD) for thrombin detection, a freshly prepared NS-TBA spot was incubated with different concentrations of thrombin. The concentrations of target protein in solution ranged from 0 to 50 μM increasing in steps of 1 order of magnitude in concentrations beginning with 500 yM. A blank sample, containing no analyte, was also measured to determine the detection limit, which is defined as the average plus 3 standard deviations of the blank measurement. Different protein concentrations were incubated with NS-TBA for 30 min and spectra taken as described before, raising the protein concentration continuously (Fig. 3.10). Between every further incubation, a washing step with buffer was included to remove unbound thrombin from the sensor. The first significant data point above the limit of detection lies at 50 aM and this sensor can therefore comfortably detect in the sub-femtomolar region.

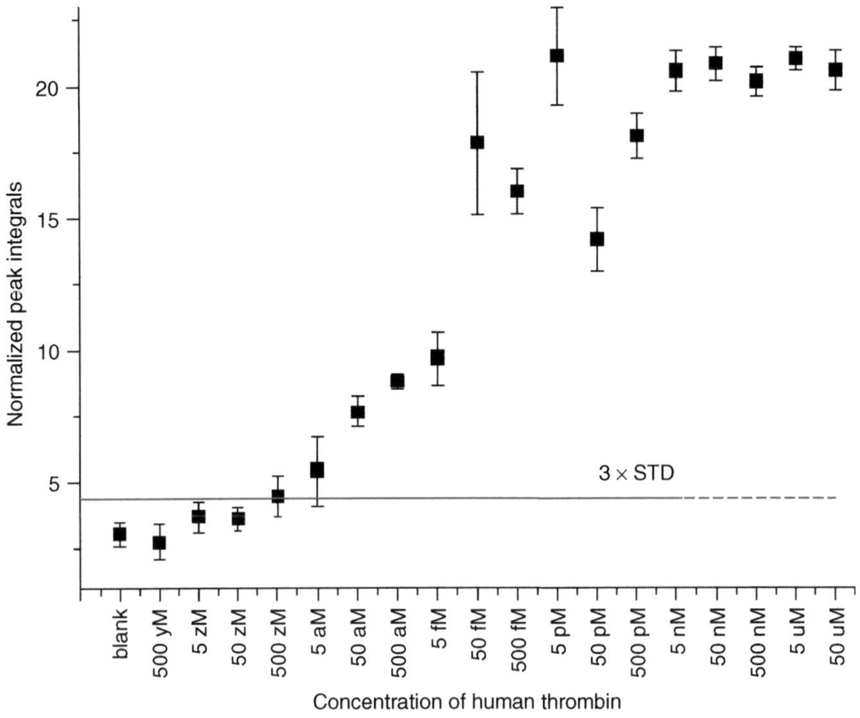

Fig. 3.10 Result of the limit of detection (LOD) determination for NS-TBA in binding buffer for the cognate target protein human α thrombin. The sensor was incubated with protein concentrations between 500 yM and 50 μM. The *red line* indicates the LOD (mean plus 3 standard deviations of the blank measurement) [49]

4.5 Intracellular SERS

The NS used in this work are designed to be excited with lasers in the NIR region, which have excellent transmittance through tissue. To date these NS have been added to cells for tumor destruction through plasmonic heating, but as demonstrated above they also have excellent potential for use in spectroscopic investigations of biomolecules [28, 49].

This section therefore demonstrates the use of NS to acquire spectra from the intracellular environment of living and viable NIH/3T3 mouse fibroblast cells. First, spectra from nonenhanced Raman measurements collected using detached living NIH/3T3 cells were compared with SERS spectra from NS inside NIH/3T3 cells on quartz coverslips (UQG Ltd., Cambridge). In order to obtain a detectable bulk Raman spectrum, 50 detached single cells were probed using 50 mW laser power with a 785 nm laser for an acquisition time of 180 s per cell. SERS spectra were collected on cells incubated for 3 h with 300 fM of NS. After locating the gold nanoparticle light scattering pattern spectra were acquired for 5 s with a laser power

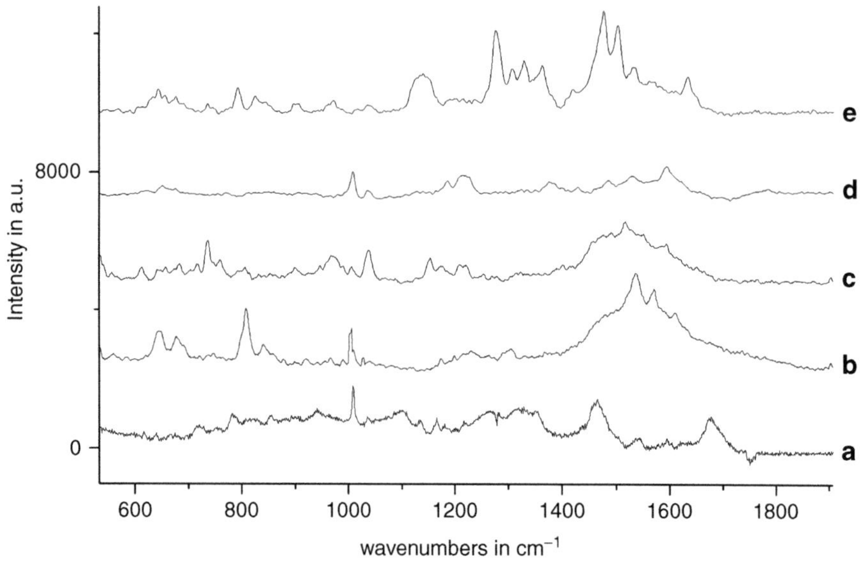

Fig. 3.11 (**a**) Averaged Raman spectrum of 50 different NIH/3T3 fibroblast cells measured at a 785 nm excitation wavelength, 50 mW laser power at the sample for 180 s. (**b**–**e**) SER spectra taken by excitation of NS in different cells at the same wavelength with 5 s acquisition time and 3 mW laser power at the sample [56]

of 3 mW at the sample. The SERS spectra were taken in several single living cells at different locations (Fig. 3.11). Comparing the two different kinds of Raman measurements some features are clearly visible in the bulk Raman as in the SERS spectra, such as the peak at around 1,005 cm^{-1} attributable to phenylalanine or the Amide-III mode and the beta sheet band at 1,215 cm^{-1}. However, we see no DNA-related peaks in the SERS spectra; this is expected if NS are only found in the cytosol. Other peak assignments can be found in Table 3.1. Equation 3.1 shows the assumptions made in order to estimate the intracellular surface enhancement. Four aspects of the measurement are compared: the probed volume, laser power, acquisition time, and signal to noise. The calculated estimate for the enhancement of the Raman signal of cytosolic compounds with NS has been determined to be 10^{10} compared to nonenhanced Raman measurements. This number is in reasonable agreement with the reported enhancement factor of 10^{12} for NS used in vitro [26, 46].

$$EF_{\mathrm{est}} = \frac{Vol_{\mathrm{bulk}}}{Vol_{\mathrm{SERS}}} \cdot \frac{T_{\mathrm{bulk}}}{\cdot T_{\mathrm{SERS}}} \cdot \frac{I_{\mathrm{bulk}}}{I_{\mathrm{SERS}}} \cdot \frac{SN_{\mathrm{SERS}}}{SN_{\mathrm{bulk}}} \tag{3.1}$$

EF_{est}: Enhancment factor estimate
Vol_{bulk}: Total cellular volume probed using bulk native Raman
Vol_{SERS}: Cytoplasmic volume probed by SERS where

Table 3.1 Comparison of Raman shifts in single cell and intracellular SERS measurements. Values in columns a to e are taken from corresponding spectra in Fig. 3.11 [56]

Cell Raman	Assignment for Raman features	SERS from NS in cells			
a)		b)	c)	d)	e)
640	C—C twist in Tyrosine	646		642	644
675	C—S stretch in Cysteine	678	684		
720	C—N stretch in lipid/adenine		736		737
783	DNA: O—P—O backbone stretch in T/C				
825	Out of plane ring breath in tyrosine				825
856	In plane ring breathing mode tyrosine	848			850
893	C—C skeletal stretch in protein		900	905	896
941	C—C skeletal stretch in protein		967		972
1,008	Sym. ring breathing mode phenylalanine	1,005	1,005	1,007	
1,036	C—H in plane bending mode phenylalanine	1,037	1,038	1,036	
1,100	DNA: O—P—O backbone stretching				
1,109	DNA: O—P—O backbone stretching				
	C—N stretch in polypeptide chains		1,133	1,138	1,139
1,166	C—C stretching in proteins	1,174	1,174		
1,217	Amide III: beta-sheet	1,216	1,209	1,214	1,214
1,266	Amide III: beta-sheet/adenine/cytosine				1,275
1,329	Guanine			1,323	1,329
1,354	Polynucleotide chain (DNA bases)				
	Possible porphyrin stretches			1,375	1,366
1,465	(v (C=C)+v (C—C)+v (C=O .-. H))			1,485	1,475
	Aromatic ring stretches		1,516		1,511
1,543	Lipid stretches	1,535		1,528	1,534
	Tyrosine stretch	1,570		1,593	1,570
1,595	Ring mode (Adenine/guanine)				
1,619	C=C bending in phenylalanine and tyrosine	1,616			
1,677	Amide I: alpha-helix				

$Vol_{\text{SERS}}=\text{NSEA}_{\text{maxvol}}-\text{NS}_{\text{vol}}$ where $\text{NSEA}_{\text{maxvol}}$ represents the volume of a sphere with radius 10 nm greater than the NS wherein SERS enhancement is expected and NS_{vol} represents the volume of a nanoshell

T_{bulk}: Acquisition time for unenhanced Raman measurements of cells

T_{SERS}: Acquisition time used for SERS of cells with NS

I_{bulk}: Laser power used for unenhanced Raman measurements

I_{SERS}: Laser power used for SERS measurements

SN_{bulk}: Signal to noise ratio for unenhanced Raman spectra

SN_{SERS}: Signal to noise ratio for SERS spectra

Although we have demonstrated a method for intracellular Raman investigations, the actual information which can be acquired is rather nonspecific. For a useful intracellular sensor high specificity to either certain cellular conditions, like pH or redox potential, or for the detection of marker molecules for different cellular processes is desirable.

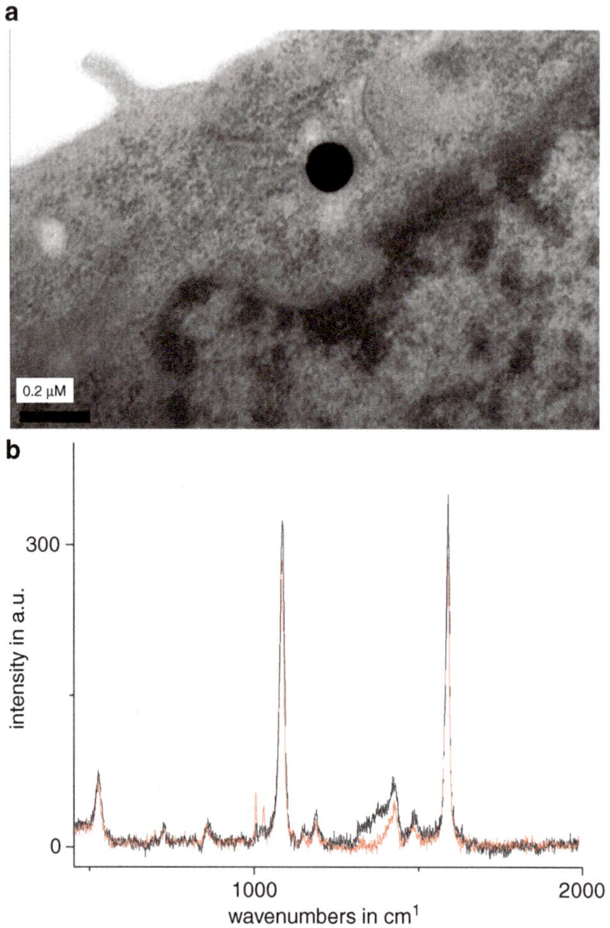

Fig. 3.12 (**a**) TEM image taken of a single NS in NIH/3T3 fibroblast cell located near the nuclear membrane. (**b**) SERS spectra of 4 mercaptobenzoic-acid functionalized NS in NIH/3T3 fibroblast. Overlying spectra depict NS-4MBA in solution (*black*) and inside NIH/3T3 cells (*red*) determining the pH to be 7.4 and 6.5, respectively [56]

In order to measure pH, we used NS functionalized with 4-mercapto benzoic acid (4MBA). Intracellular pH determination has been demonstrated by Kneipp et al. using 4MBA modified gold nanoparticles where the particles were probably located in lyzosomes [57]. For the use of a nanosensor, it is of high importance that surface functionalities remain intact whilst inside the cell and that spectra are not complicated by signals originating from foreign molecules adhering to the sensor surface. We used 4MBA functionalized NS for this simple test as 4MBA is known to easily form self-assembled monolayers on Au surfaces [12, 29]. The spectra (Fig. 3.12) of 4MBA-NS inside cells and free in media do not show any obvious differences suggesting that the monolayer is stable and that there is little nonspecific adsorption to the surface. The dominant features in the spectra are the aromatic ring breathing modes at 1,081 and 1,590 cm^{-1}. Other less intense features describe the S-C stretching mode of the thiol function at 527 cm^{-1}, different carboxyl modes at 860 cm^{-1} (d(COO—)), 1,150 cm^{-1} (C—COOH), 1,430 cm^{-1} ($-COO^-$), and at 1,700 cm^{-1} the protonated carboxylic acid group (−COOH),[29] which are all dependent on the pH of the surrounding solution. Spectra in Fig. 3.12b show that the pH can be determined in growth media and inside the cell. By comparing the intensity ratios of the peaks at 1,430 and 1,700 cm^{-1} in both spectra with those from a pH calibration curve, the pH can be determined as 7.4 in media and 6.5 in the cell. A similar approach was recently reported for intracellular imaging by using mixed monolayers of 4-MBA modified polyethyleneglycolthiols on NS as reported Bishnoi et al. [58].

In order to demonstrate a useful intracellular nanosensor, it is crucial to show that introduction of the sensor to the cell has no detrimental effects on normal cell physiology. This final section describes an investigation of nanoshell uptake, pathway of uptake, and cellular response.

4.6 Nanoshell Uptake

In order to investigate nanoshell uptake by cells we incubated fibroblast cells with gold NS under a variety of treatments. When we incubate the cells with NS suspended in standard tissue culture media (DMEM) we can see several particles inside the cell. Under these incubation conditions, we find that most NS are encapsulated in a lipid vesicle, and we assume that the NS have picked up this coating on their journey through the plasma membrane. Notably, when we incubate the cells with NS suspended in serum-free media, the majority of particles are to be found free of any coating in the cytoplasm. While we still do not fully understand the uptake mechanism, this is an important finding since it demonstrates the potential for targeted delivery of NS to specific cellular locations through control of culture conditions [56].

4.7 Cell Viability

Whenever you add material like nanoshells to a cell, especially when the purpose is to *monitor* cellular physiology, you need to ensure that the material does not adversely *affect* the cell's physiology. To this end, we carried out three important assays: one which measures cell viability and two others which look for markers of cell death. Figure 3.13a shows the result of a cell viability assay carried out using increasing concentrations of NS. Cells were seeded at relatively low coverage and their growth monitored as a function of time. A control sample of cells, grown in the absence of nanoshells, was included for comparison. It is clear from the data that even up to high concentrations of NS the cells continue to grow normally, which suggests that NS do not affect cell growth. In order to rule out the possibility that

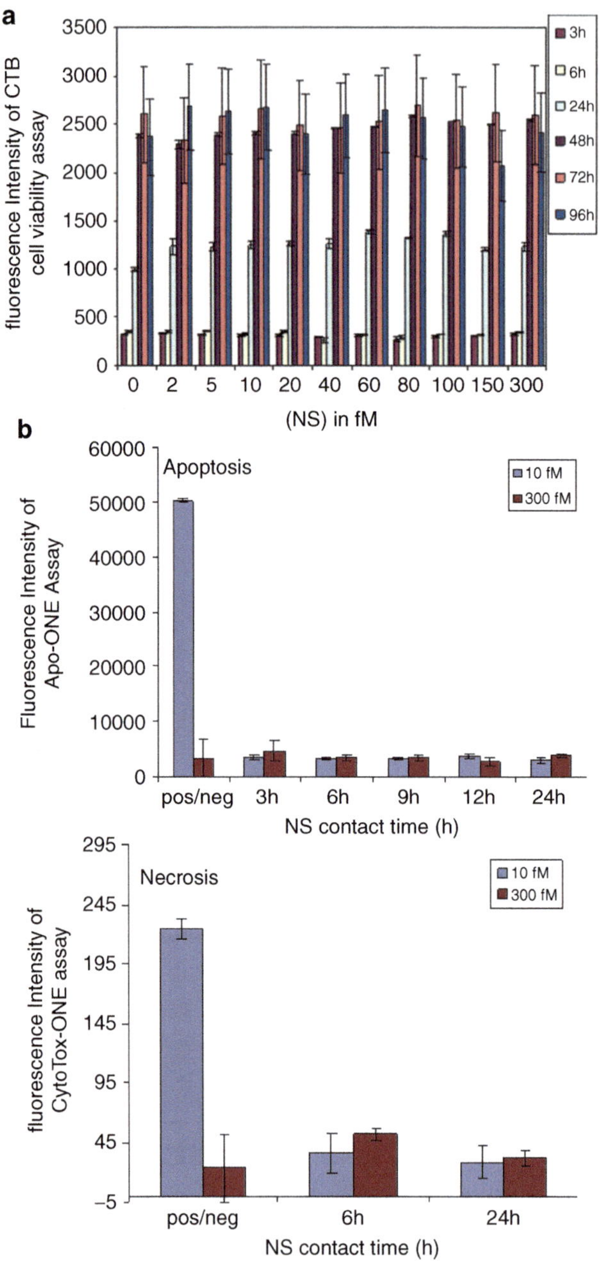

Fig. 3.13 Investigation of the response of murine fibroblast cells incubated with different concentrations of NS. (**a**) Cell viability test conducted with the Cell-Titer-Blue assay over 4 days incubation of cells with a range of NS concentrations between 0 and 300 fM (0–1,800 NS/cell). (**b**) Assays of apoptosis and necrosis markers at two selected NS concentrations (10 and 300 fM), results show no difference to untreated cell controls

some cells were growing normally while others were dying, we measured markers of apoptosis (caspase) and markers of necrosis (extracellular lactase), including negative and positive controls for comparison. Again, it is clear (Fig. 3.13b) that the level of cell-death markers is not significantly different from an untreated sample and significantly lower than a sample in which cell death had been induced. These measurements confirm that NS can be used to measure SERS spectra from distinct intracellular environments without affecting cell physiology.

5 Conclusions and Future Perspective

In the past decade SERS has become a hugely important technique for the study of biological systems. Its utility has been demonstrated in a variety of systems spanning from fundamental studies of drug–membrane interactions to applications in in vitro diagnosis. One of the challenges in measuring reliable, quantitative SERS spectra is the development of reliable, reproducible surfaces on which to measure these spectra. Nanoshells are an excellent example of such an engineered nanostructure whose absorbance can be tuned to work well with biological systems and whose strong plasmon resonance allows SERS spectra to be collected from single particles. We have used this chapter to illustrate the general utility of nanoshells for biological investigations by focusing on specific examples from our lab. The surface sensitivity of SERS allows conformational changes to be measured and we have used this aspect to probe conformational switching in a G-quadruplex oligonucleotide. The ability to probe conformational changes has led us to use SERS to measure protein concentration on the basis of changes in aptamer structure caused by protein binding. This finding opens the door to a general method of protein detection using SERS and structure-switching aptamers, although new methods of aptamer selection and design may be required in order to generate a broader suite of functional aptamers. As a probe of biomolecular structure, SERS is not limited to investigations of nucleic acid structure and we are currently using a similar approach to investigate peptide structural changes [59].

The broad utility of SERS makes it an attractive alternative to fluorescence for intracellular imaging and it can be used to measure pH in mouse fibroblast cells. Importantly, delivery of NS to cells can be controlled and shows promise for selective targeting of specific organelles. In a complex living system such as a cell, it is important to establish that adding an imaging contrast agent such as a modified nanoparticle does not have an adverse effect on physiology. Investigations using a variety of assays have determined that cells continue to grow in an unperturbed manner and do not undergo either programmed or necrotic cell death. Since several research groups have established the principal of intracellular SERS imaging, there is a need for a much greater array of modified nanoparticles which can probe, for example, a larger range of protein concentrations, biomolecular interactions, small molecule concentrations, or states such as redox potential. Furthermore, in order to ensure that measurements are made in specific locations, there is a need to improve targeting of nanoparticles to specific organelles in a predictable and reproducible manner. While this will entail a considerable interdisciplinary research effort, it is essential to ensure the translation of this technique from a niche area of physical science research to a technique of choice for life scientists.

References

1. Smith E, Dent G (2006) Modern Raman spectroscopy – A practical approach. Wiley, Hoboken
2. Fleischmann M, Hendra PJ, McQuillan AJ (1974) Raman spectra of pyridine adsorbed at a silver electrode. Chem Phys Lett 2:163–166
3. Jeanmaire DL, Van Duyne RP (1977) Surface Raman spectroelectrochemistry Part I. Heterocyclic, aromatic, and aliphatic amines adsorbed on the anodized silver electrode. J Electroanal Chem 84:1–20
4. Le Ru EC, Blackie E, Meyer M, Etchegoin Surface PG (2007) Enhanced Raman scattering enhancement factors: a comprehensive study. J Phys Chem C 111(37):13794–13803
5. Novotny L, Hecht B (2006) Prinicples of nano-optics. Cambridge University Press, Cambridge
6. Schasfoort RBM, Tudos AJ (2008) Handbook of surface plasmon resonance. RSC Publishing, Cambridge
7. Nie S, Emory SR (1997) Probing single molecules and single nanoparticles by surface-enhanced Raman scattering. Science 275:1102–1106
8. Ryu K, Haes AJ, Park HY, Nah S, Kim H, Chung J, Yoon MY, Han SH (2010) Use of peptide for selective and sensitive detection of an anthrax biomarker via peptide recognition and surface-enhanced Raman scattering. J Raman Spectrosc 41:121–124
9. Feng M, Hiroyasu Tachikawa H (2008) Surface-enhanced resonance Raman spectroscopic characterization of the protein native structure. J Am Chem Soc 130:74437448
10. MacAskill A, Crawford D, Graham D, Faulds K (2009) DNA sequence detection using surface-enhanced resonance Raman spectroscopy in a homogeneous multiplexed assay. Anal Chem 81:8134–8140
11. Kneipp J, Kneipp H, Wittig B, Kneipp K (2007) One- and two-photon excited optical pH probing for cells using surface-enhanced Raman and hyper-Raman nanosensors. Nano Lett 7:2819–2823
12. Jadzinsky PD, Calero G, Ackerson CJ, Bushnell DA, Kornberg RD (2007) Structure of a thiol monolayer protected gold nanoparticle at 1.1 Å resolution. Science 318:430
13. Liu Y-J, Zhang Z-Y, Dluhy RA, Zhao Y-P (2010) The SERS response of semiordered Ag nanorod arrays fabricated by template oblique angle deposition. J Raman Spectrosc. doi:10.1002/jrs.2567
14. Duff DG, Baiker A, Edwards PP (1993) A new hydrosol of gold clusters formation and particle size variation. Langmuir 9:2301–2309
15. Su YH, Lai WH, Chen WY, Hon MH, Chang H (2007) Surface plasmon resonance of gold nano-sea-urchin. App Phys Lett 90:1819051–1819053
16. Skrabalak SE, Chen J, Au L, Lu X, Li X, Xia Y (2007) Gold nanocages for biomedical applications. Adv Mater 19:3177–3184
17. Brinson BE, Lassiter JB, Levin CS, Bardhan R, Mirin N, Halas NJ (2008) Nanoshells made easy: Improving Au layer growth on nanoparticle surfaces. Langmuir 24:14166–14171
18. Lux F, Lerouge F, Bosson J, Lemercier G, Andraud C, Vitrant G, Baldeck PL, Chassagneux F, Parola S (2009) Gold hollow spheres obtained using an innovative emulsion process: towards multifunctional Au nanoshells. Nanotechnology 20:355603–355609
19. Alvarez-Puebla RA, Ross DJ, Abbas-Nazri G, Aroca RF (2005) Surface-enhanced Raman scattering on nanoshells with tunable surface plasmon resonance. Langmuir 21:10504–10508

20. Huang C-C, Hwu JR, Su W-C, Shieh D-B, Tzeng Y, Yeh C-S (2006) Surfactant-assisted hollowing of Cu nanoparticles involving halide-induced corrosion-oxidation processes. Chemistry 12(14):3805–3810
21. Rasch MR, Sokolov KV, Korgel BA (2009) Limitations on the optical tunability of small diameter gold nanoshells. Langmuir 25(19):11777–11785
22. Nehl CL, Grady NK, Goodrich GP, Tam F, Halas NJ, Hafner JH (2004) Scattering spectra of single gold nanoshells. Nano Lett 4(12):2355–2359
23. Prodan E, Nordlander P, Halas NJ (2003) Electronic structure and optical properties of gold nanoshells. Nano Lett 3:1411–1415
24. Kirakosyan AS, Shahbazyan TV (2008) Vibrational modes of metal nanoshells and bimetallic core-shell nanoparticles. J Chem Phys 129:34708.1–34708.7
25. Westcott SL, Jackson JB, Radloff C, Halas NJ (2002) Relative contributions to the plasmon line shape of metal nanoshells. Phys Rev B 66:155431
26. Clare SE, Halas NJ (2004) Surface-enhanced Raman scattering on tunable plasmonic nanoparticle substrates. Proc Nati Acad Sci USA 101(52):17930–17935
27. Preston TC, Signorell R (2009) Growth and optical properties of gold nanoshells prior to the formation of a continuous metallic layer. ACS Nano 3(11):3696–3706
28. Qiana W, Tana Y, Dinga S, Wanga Y (2008) A label-free biosensor based on gold nanoshell monolayers for monitoring biomolecular interactions in diluted whole blood. Biosens Bioelectron 23(7):1166–1170
29. Bishnoi SW, Rozell CJ, Levin CS, Gheith MK, Johnson BR, Johnson DH, Halas NJ (2006) All-optical nanoscale pH meter. Nano Lett 6(8):3277–3281
30. Bardhan R, Grady NK, Cole JR, Joshi A, Halas NJ (2009) Fluorescence enhancement by au nanostructures: nanoshells and nanorods. ACS Nano 3(3):744–752
31. Levin CS, Kundu J, Janesko BG, Scuseria GE, Raphael RM, Halas NJ (2008) Interactions of ibuprofen with hybrid lipid bilayers probed by complementary surface- enhanced vibrational spectroscopies. J Phys Chem B 112:14168–14175
32. Zhang P, Guo Y (2009) Surface-enhanced Raman scattering inside metal nanoshells. J Am Chem Soc 131:3808–3809
33. Lal S, Clare SE, Halas NJ (2008) Nanoshell-enabled photothermal cancer therapy: Impending clinical impact. Acc Chem Res 41(12):1842–1851
34. Lin C-C, Yang Y-M, Chen Y-F, Yang T-S, Chang H-C (2008) A new protein A assay based on Raman reporter labeled immunogold nanoparticles. Biosens Bioelectron 24:178–183
35. Chen J, Jiang J, Gao X, Liu G, Shen G, Yu R (2008) A new aptameric biosensor for cocaine based on surface-enhanced Raman scattering spectroscopy. Chemistry 14(27):8374–8382
36. Wang Y, Wei H, Li B, Ren W, Guo S, Dong S, Wang E (2007) SERS opens a new way in aptasensor for protein recognition with high sensitivity and selectivity. Chem Comm 28:5220–5222
37. Hu J, Zheng P-Ch, Jiang J-H, Shen G-L, Yu R-Q, Liu G-K (2008) Electrostatic interaction based approach to thrombin detection by surface-enhanced Raman spectroscopy. Anal Chem 81:87–93
38. Padmanabhan K, Tulinsky A (1996) An ambiguous structure of a DNA 15-mer thrombin complex. Acta Crystallogr Sect D 52(2):272–282
39. Ogasawara S, Maeda H (2009) Reversible photoswitching of a G-quadruplex. Angew Chem Int Ed 48:6671–6674
40. Cooper M (2003) Label-free screening of bio-molecular interactions. Anal Bioanal Chem 377(5):834–842
41. Rich RL, Myszka DG (2007) Higher-throughput, label-free, real-time molecular interaction analysis. Anal Biochem 361(1):1–6
42. Duguid JG, Bloomfield VA, Benevides JM, Thomas GJ (1996) DNA melting investigated by differential scanning calorimetry and Raman spectroscopy. Biophys J 71(6):3350–3360
43. Willets K (2009) Surface-enhanced Raman scattering (SERS) for probing internal cellular structure and dynamics. Anal Bioanal Chem 394(1):85–94

44. Sealy C (2006) Self-assembly leaves no stain: nanoparticles. Nanotoday 1:13
45. Levin CS, Kundu J, Barhoumi A, Halas NJ (2009) Nanoshell-based substrates for surface enhanced spectroscopic detection of biomolecules. Analyst 134:1745–1750
46. Talley CE, Jackson JB, Oubre C, Grady NK, Hollars CW, Lane SM, Huser TR, Nordlander P, Halas NJ (2005) Surface-enhanced Raman scattering from individual Au nanoparticles and nanoparticle dimer substrates. Nano Lett 5(8):1569–1574
47. Herne TM, Tarlov MJ (1997) Characterization of DNA probes immobilized on gold surfaces. J Am Chem Soc 38:2016–2020
48. Balamurugan S, Obubuafo A, McCarley RL, Soper S, Spivak D (2008) Effect of linker structure on surface density of aptamer monolayers and their corresponding protein binding efficiency. Langmuir 390(4):1009–1021
49. Ochsenkuehn MA, Campbell CJ (2010) Probing biomolecular interactions using surface enhanced raman spectroscopy: label-free protein detection using a g-quadruplex dna aptamer. Chem Comm 16:2799–2801
50. Maher RC, Cohen LF, Gallop JC, Le Ru EC, Etchegoin PG (2006) Temperature-dependent anti-stokes/stokes ratios under surface-enhanced Raman scattering conditions. J Phys Chem B 110:6797–6803
51. Yoshida W, Mochizuki E, Takase M, Hasegawa H, Morita Y, Yamazaki H, Sode K, Ikebukuro K (2009) Selection of DNA aptamers against insulin and construction of an aptameric enzyme subunit for insulin sensing. Biosens Bioelectron 24(5):1116–1120
52. Barhoumi A, Zhang D, Tam F, Halas NJ (2008) Surface-enhanced Raman spectroscopy of DNA. J Am Chem Soc 130(16):5523–5529
53. Billinghurst BE, Oladepo SA, Loppnow GR (2009) pH-dependent UV resonance Raman spectra of cytosine and uracil. J Phys Chem B 113(20):7392–7397
54. Moger J, Gribbon AP, Sewing A, Winlove MG (2007) Feasibility study using surface-enhanced Raman spectroscopy for the quantitative detection of tyrosine and serine phosphorylation. Biochim Biophys Acta 1770(6):912–918
55. Yarasi S, Billinghurst BE, Loppnow GR (2007) Vibrational properties of thymine, uracil and their isotopomers. J Raman Spectrosc 38(9):1117–1126
56. Ochsenkühn MA, Jess PRT, Stoquert H, Dholakia K, Campbell CJ (2009) Nanoshells for surface-enhanced Raman spectroscopy in eukaryotic cells: cellular response and sensor development. ACS Nano 3(11):3613–3621
57. Kneipp J, Kneipp H, McLaughlin M, Brown D, Kneipp K (2006) In vivo molecular probing of cellular compartments with gold nanoparticles and nanoaggregates. Nano Lett 6(10): 2225–2231
58. Huang Y, Swarup VP, Bishnoi SW (2009) Rapid Raman imaging of stable, functionalized nanoshells in mammalian cell cultures. Nano Lett 9:2914–2920
58. Ochsenkühn MA, Borek J, Phelps R, Campbell CJ (2011) Redox potential dependence of peptide structure studied using SERS. Nano Lett 11:2684–2688

Naturally Inspired SERS Substrates 4

N. L. Garrett

1 Definition of Topic

This topic concerns the application of naturally occurring nanostructures to surface-enhanced Raman spectroscopy.

2 Overview

Surface-enhanced Raman scattering (SERS) is a powerful spectroscopic analytical tool with a wide range of applications. Many market sectors benefit from SERS, in fields as diverse as: clinical diagnostics, pharmaceuticals, forensics, environmental evaluation, and biochemical studies. Demand for improved sensitivity, throughput, specificity, and cost efficiency in industry constantly drives advances in assay technology. This chapter outlines the recent advances in SERS substrate production inspired by naturally occurring nanostructures and their application to the problem of detecting protein-binding events.

3 Introduction

Since the turn of the twentieth century, interesting optical effects from optically small metal interfaces and structures have been the subject of much scrutiny. Molecules adsorbed onto rough metal surfaces, particles, or island films exhibit dramatically different optical properties to those of the free molecules [1]. Perhaps

N.L. Garrett
School of Physics, University of Exeter, Devon, UK

Challa S. S. R. Kumar (ed.), *Raman Spectroscopy for Nanomaterials Characterization*,
DOI 10.1007/978-3-642-20620-7_4, © Springer-Verlag Berlin Heidelberg 2012

the most widely known effect of this nature is surface-enhanced Raman scattering (SERS) where the Raman signal from molecules adsorbed onto suitable metallic substrates can be millions of times stronger than that from the free molecule. The SERS effect has found wide application in biology, medicine, electrochemistry, and materials science through such techniques as SERS of single molecules, nanostructures and transition metals, tip-enhanced Raman Scattering (TERS), surface-enhanced hyper-Raman scattering (SEHRS), ultraviolet-excited SERS (UV-SERS) [2]. Other examples of such metal enhancement exist, such as surface-enhanced infrared absorption [1] and plasmon-enhanced spectroscopy [2].

Recently, several reports have demonstrated the potential of SERS as a label-free readout for monitoring protein-binding events [3–5]. The potential for low detection limits and higher sensitivities available with SERS have been exploited in recent biological assay research [6–8] but this has been limited by the irreproducible enhancement provided by the substrates available for aqueous experiments, such as colloidal metals. Due to the requirement for conjugation of colloidal metallic particles to give substantial enhancements and the change of enhancement with aggregation, colloidal-based SERS methods are not reproducible, and metallic particles are prone to settling out of suspension during scanning.

Many available SERS substrates require the analyte to be deposited and dried prior to scanning. This is unsuitable for most biological analytes, such as proteins, since the process of desiccation alters their configuration and hence their Raman spectrum [9]. Drop-coated deposition has been claimed to produce un-enhanced Raman spectra of proteins analogous to those obtained from aqueous proteins [10] by providing a "natural" environment. However, with protein structure being strongly dependent upon the hydration of the molecule, it is essential that aqueous conditions are maintained in order to perform rigorous investigation of proteins. Protein-binding systems are complex interactions which require optimal protein structure for binding to occur. Variation in pH, hydration, and temperature can profoundly alter on protein structure, thus preventing binding events which can lead to a dramatic reduction in the detection sensitivity [11].

SERS assays using proteins have traditionally required highly complex systems with multiple interactions to produce a detectable event [3, 5], usually with colloidal noble metal particles or a SERS chemical label providing the means of detecting binding events. These assays can provide high sensitivity to low levels of analyte in solution, but the interaction of metal colloids and chemical SERS marker molecules has the potential to adversely affect the biological molecules of interest. Intricate multiple-stage assays are undesirable for complex biological systems, where assays are required to be highly robust if they are to be of use. More cutting-edge assays have used dielectric-core metal-coated nanoshells bioconjugated to antibodies [5, 12–14]. These assays produce excellent sensitivity combined with low levels of undesired biological interaction between the label and the analyte. However, with the nanoshells free in solution, SERS spectra can be difficult to reproduce unless the analyte is spun down using centrifugation or immobilized in some manner; these processes can be damaging to some biological samples.

The ideal substrate for SERS would provide highly reproducible SERS enhancement factors of the order of 10^6 or higher. Other requirements include the ability to perform rapid SERS biological assays whilst the samples are hydrated, which are sensitive over a range of concentrations that are comparable with ELISA and other highly sensitive assaying techniques. A principle challenge to providing these qualities in a SERS substrate is designing a suitable nanostructure that is not prohibitively expensive to produce or use in high through-put applications. In search of less expensive methods for SERS substrate production, in recent years scientists have taken inspiration from naturally occurring nanostructures.

Scientists in a broad variety of fields exploit various forms of nanostructures inspired by nature [15]. Naturally occurring nanostructures that have evolved for functions involving visual appearance have been the subject of a remarkable surge in research interest [16–20]. This has built upon a range of somewhat earlier studies [21, 22]. Numerous insects, birds, and plants exhibit a wide array of complex periodic and quasi-periodic ultrastructure. Their nanostructures can contribute to many biological functions: highly unsaturated color appearances for short- or long-range conspecific communication [23, 24]; angle-independent color [25] and specular or diffuse broadband appearances [26] for specialized camouflaging; linearly [24] or circularly polarized reflection [27] for high-level communication, represent a few of these.

One investigation into the optical properties of cicada wings found that the "quasiperiodic" nanostructured antireflective coating on the chitinous wing surfaces provided an excellent SERS substrate, with enhancement factors of approximately 10^6 [28]. These structures have been reproduced using a biotemplating approach for silver-coated nanoimprinted optical fibers in SERS sensing [29]. The imprinting procedure involved producing heat-cured elastomer casts of cicada wing segments, imprints of which were formed on wet polymer in contact with glass fibers under a three-step curing procedure. Although the silver-coated substrates in this study were found to be robust and inexpensive, the SERS spectra from thiophenol monolayers deposited onto these imprinted nanofibers were found to provide approximately 50% lower enhancement factors than those of the cicada wings themselves. Since silver coatings are prone to forming prohibitively thick oxide layers over time, the production of these substrates on an industrial scale would require a more chemically inert and biocompatible coating, such as gold. Gold is known to produce lower SERS enhancement factors than silver, sometimes by as much as a factor of 10^2 [30] and hence would reduce the imprinted nanofibers' enhancement factors significantly.

Another recently developed naturally inspired silver-coated SERS substrate employed the use of algae diatomic microshell monolayers which yielded maximum enhancement factors of 10^6 [31]. The substrates produced in this method were not macroscopically homogeneous, instead exhibiting a large degree of structural variability across the substrate which could prove to provide inconsistent SERS spectra. A variety of production methods for macroscopically homogeneous two-dimensional SERS substrates have been developed, including chemical etching, electron-beam lithography, chemical etching, colloid immobilization, annealing of metal ion-implanted silicon, and nanosphere lithography [32–35]. The substrates produced using these techniques provide reproducible SERS enhancement factors of the order of 10^5–10^7. Unfortunately, these production methods are generally prohibitively expensive, complex, and ultimately unsuitable to mass production. Since the substrates' nanostructures are too delicate to withstand cleaning, they are only suitable for single use. For these reasons, cheaper, faster, and more reproducible manufacturing processes are required. This chapter discusses the research undertaken which attempts to address this issue by identifying and reproducing naturally occurring nanostructures for use in SERS.

4 Experimental and Instrumental Methodology

4.1 Preparation and Characterization of SERS Substrates

Since there is such a wide variety of naturally occurring nanostructures with different sizes, spacing, and periodicities of the surface features, it is to be expected that not all will be suitable for SERS. Therefore, when selecting candidate structures for further investigation, one is posed with the question of how to determine which are most suitable. Some naturally occurring nanostructured surfaces have been found to exhibit color changes after being coated with gold, as a result of localized surface plasmon resonances (LSPRs) causing selective photon absorption [36]. Given that LSPRs are a major contributing factor to the SERS effect, it is perhaps reasonable to select and compare potential nanostructured SERS candidate materials on the basis of such color changes after metal deposition. The nanostructured surfaces may also be compared with the shape, spacing, and pitch of structures on well-established SERS substrates using electron microscopy. However, these methods are not always sufficient to determine which nanostructured surface will provide the most effective SERS substrate. For this, a quantitative approach must be taken.

In order to determine the efficacy of a SERS substrate for a given adsorbed molecule, it is necessary to quantify the degree to which the substrate increases the Raman scattered signal with respect to the non-SERS Raman scattered signal of the same molecule. This "enhancement factor" (EF) is most usually defined as:

$$\mathrm{EF} = \frac{I_{\mathrm{SERS}}/N_{\mathrm{Surf}}}{I_{\mathrm{RS}}/N_{\mathrm{Vol}}},$$

where N_{Vol} and N_{Surf} are the average number of molecules within the focal volume for the non-SERS Raman measurement, and the average number of adsorbed molecules on the SERS substrate within the focal volume, respectively [37]. The value for EF is therefore obtained by determining the average enhanced Raman signal intensity per molecule adsorbed to the substrate surface and dividing this value by the average signal obtained per molecule in bulk solution or solid.

A variety of molecules have been used for determining the enhancement factors of SERS substrates, including Rhodamine 6, crystal violet, and thiophenol. The choice of molecule in this task is nontrivial since resonance and "pre-resonance" effects may artificially skew the calculated EF, especially for dye molecules. Although the Raman cross sections for dye molecules are known to be larger than those for smaller molecules by up to 10^6 times, it is often not appreciated that pre-resonance effects mean these dyes may still exhibit large Raman cross sections even under nonresonance conditions [37]. Enhancements from pre-resonance effects can contribute as much as $10\times$ to the EF when the excitation laser used is up to as much as several hundred wavenumbers below the molecule's electronic transition wavelength. The pre-resonance effect has been shown to lead to calculated EFs that are several orders of magnitude greater when dye molecules are used in preference to smaller molecules. These findings prompted the postulation that the highest reported calculated SERS enhancement factors may, in fact, have been several orders of magnitude larger than they would have been if smaller molecules had been used.

Thiophenol, also known as benzene thiol, is a small molecule that does not exhibit the pre-resonance issues of dye molecules. Thiophenol has been shown to produce relatively stable monolayers on silver and gold surfaces which make it a convenient reference compound for SERS substrates [28, 29, 38, 39]. When the SERS substrate is immersed in an ethanolic solution of thiophenol, the sulfur groups form covalent bonds with the metal surface, forming self-assembled monolayers over time, as illustrated in Fig. 4.1.

The standard preparation technique for these monolayers in SERS experiments is immersion of the substrate for 10 min in a 10 mM ethanolic solution of thiophenol [28, 29, 40]. The resulting enhanced Raman spectra obtained from the thiophenol monolayers exhibit well-characterized peaks as shown by Garrett et al. in Fig. 4.2 [40]. For the SERS spectra of the thiophenol monolayers shown in Fig. 4.2, Garrett et al. gave peak assignments of the most prominent Raman bands based on those given in previous work [41, 42] as follows; 1,574 cm^{-1} (C–C stretching); 1,072 cm^{-1} (in plane C–C–C stretch and C–S stretching); 1,022 cm^{-1} (out of plane C–H stretching); 999 cm^{-1} (out-of-plane C–C–C stretch); 630 cm^{-1} (out-of-plane C–C–C and C–S stretching) [40].

Enhancement factors of naturally occurring nanostructures typically range from 10^6 for gold-coated structures, such as cicada and butterfly wings, and 10^7 for silver-coated structures [28, 40]. The choice of metal is very important when producing SERS substrates. Silver is expected to give a greater SERS enhancement than gold due to the greater participation of interband transitions in silver's dielectric function. However, since silver oxidizes rapidly, SERS substrates coated with silver have a short shelf life compared with those coated in gold. Additionally, silver may react with molecules being investigated with SERS. For these reasons, gold is often preferred to silver, especially for applications where the substrate needs to be inert, in spite of the lower enhancement factors achieved.

To prepare naturally occurring nanostructures for SERS, the samples must be sectioned, cleaned, mounted onto a suitable medium (such as a glass substrate), and

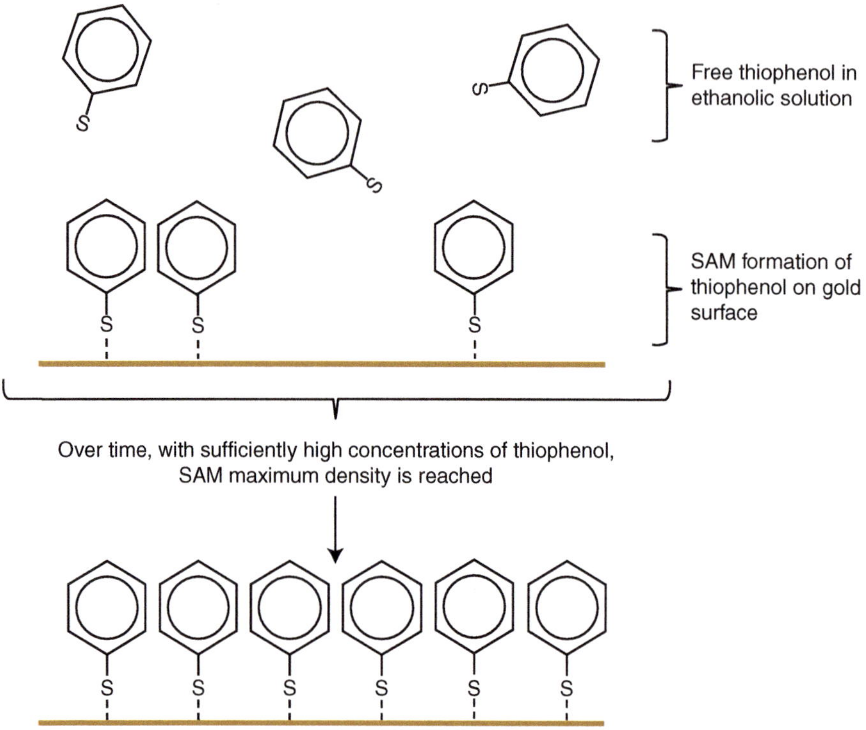

Fig. 4.1 Schematic diagram illustrating the formation of a self-assembled monolayer (SAM) of thiophenol on a gold surface

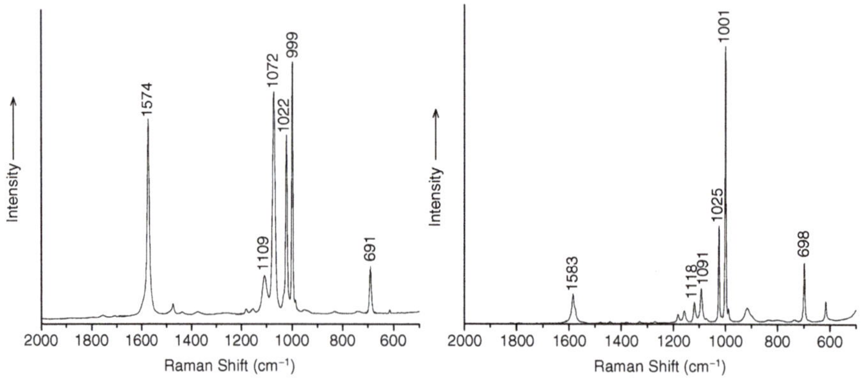

Fig. 4.2 SERS spectra of a thiophenol monolayer on a butterfly wing coated with 70 nm silver (*left*) and an un-enhanced Raman spectrum of neat thiophenol in solution (*right*), with permission from [40]

coated with the metal of choice. The method of sectioning and mounting is nontrivial and the choice of mounting method will depend on the intended experimental procedure. For example, when affixing sections of *Graphium weiskei* wings to ultraclean glass microscope slides SERS substrates, Garrett et al. used an optical adhesive (Norland Optical Adhesive, Cranbury, USA) that cured under ultraviolet light [40]. This adhesive formed a strong bond with the glass that was able to withstand immersion in acidic solutions for hours at a time, as well as maintaining its integrity during metal deposition under vacuum. Some of the most reliable methods for coating substrates with metal in a controllable fashion are physical vapor deposition techniques such as sputter coating and thermal evaporation. In these processes, metal is deposited on a surface at a controlled rate (often monitored using a quartz crystal microbalance), first forming island films, and gradually producing continuous layers of metal as the deposited thickness increases. Once coated in metal, the samples may then be characterized using electron microscopy and by using a characterization compound such as thiophenol to determine the SERS enhancement factors.

One of the main purposes of identifying naturally occurring nanostructures for use in SERS is to produce novel substrates. Many methods have been used to manufacture nanostructures for SERS substrates, including electron beam lithography, chemical etching, colloid immobilization, annealing of metal ion-implanted silicon, and nanosphere lithography [32–35]. As explained earlier, although these methods provide reproducible substrates with high enhancement factors, they are time consuming and often prohibitively cost inefficient for single-use substrates. Therefore, alternative manufacturing processes have been investigated to produce copies of nanostructures found in nature for use in SERS. The work undertaken to investigate naturally occurring nanostructures on butterfly wings by Garrett et al. and cicada wings by Stoddart and Kostovski et al. for SERS will now be described, together with their techniques for replicating the nanostructures.

4.2 Butterfly Wings as SERS Substrates for Protein-Binding Assaying

Butterfly wings have long been known to exhibit nanostructured surfaces composed of a structural protein, chitin [21, 43, 44]. These structures give rise to a wide range of startling optical effects such as vivid coloration and iridescence. The *G. weiskei* butterfly investigated for use as a SERS substrate is shown in Fig. 4.3 [40]. SEM images of the dorsal side of the wings shown in Fig. 4.4 indicated that the blue, purple, and pink regions (marked A – F on Fig. 4.3) exhibited conical nanostructures innervated with hair-like scales, which were similar to the *Sensilla trichodea* found on a different variety of *Graphium* butterfly reported elsewhere [45]. The conical nanostructures, shown in Fig. 4.4, were typically found to be ~2 μm in height and ~1 μm in diameter at their widest point.

In contrast to the brightly colored regions of the butterfly wings, the brown areas were found to be covered by shield-shaped scales as shown in Fig. 4.5, which were

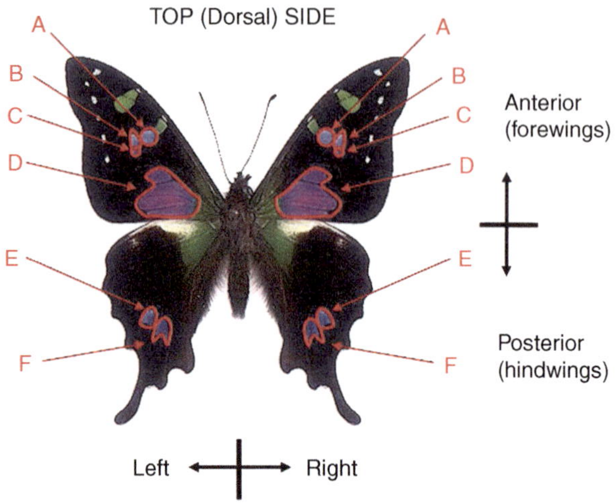

Fig. 4.3 Regions of interest on *G. weiskei* butterfly wings

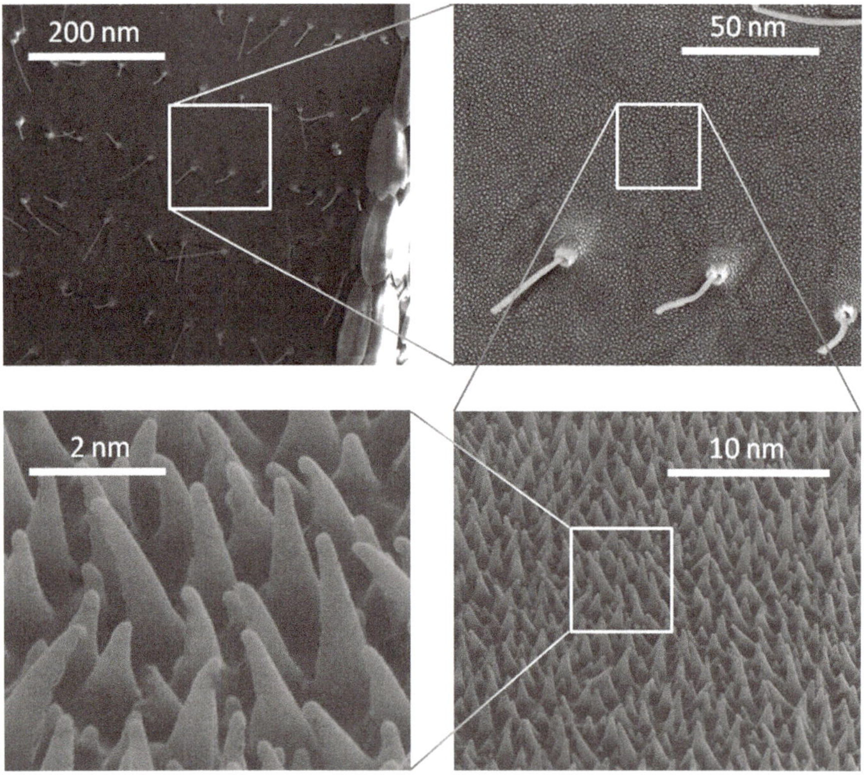

Fig. 4.4 SEM images of the brightly colored regions of *G. weiskei* typically found in regions *A–F* of Fig. 4.3, illustrating the macroscopic hair-like scales (*Sensilla trichodea*) that are characteristic of these regions and the quasi-periodic nanocone structures found between them (Reproduced with permission from [40])

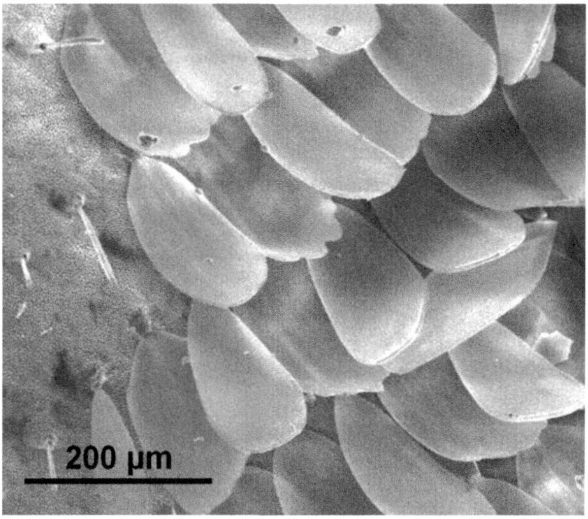

Fig. 4.5 SEM image of the shield-shaped scales covering the brown regions on *G. weiskei*

approximately ~200 nm in length and ~80 nm in width. These regions did not exhibit any nanoscale features suitable for use in SERS, and so were disregarded for the remainder of the study.

To demonstrate the application of these butterfly wing SERS substrates to the problem of protein-binding detection, Garrett et al. devised a protein-binding assay. There are a variety of methods for binding proteins to a metal surface, some of which depend on the amino acid composition of the protein, others of which do not. Direct adsorption via covalent bonds between sulfur groups in proteins and metal surfaces has been used with some success [43]; however, this method is unsuitable for use in Raman spectroscopy-based assays, as nonspecific binding may result in a wide variety of conformational orientations of the molecules with respect to the metal's surface. This can lead to Raman spectra that are difficult to reproduce. Not only that, but the analyte may bind to the metal as well as the antibody, which can lead to noisy spectra.

Other methods of protein binding to metals involve the formation of self-assembled monolayers (SAMs) of linker molecules. Self-assembled monolayers of linker molecules can be straightforward to produce [44, 45] and enable a greater degree of specificity for antibody binding, both in terms of molecular conformation on the SAM surface and in terms of which protein is bound. The model immunoassay was performed by conjugating avidin onto the metalized wing surfaces and using the SERS spectra to detect biotin binding as a function of biotin concentration. The linker molecule 3-mercaptopropanoic acid (MPA) forms covalent bonds with metals via the terminal sulfur groups and can be conjugated to proteins via their NH_2/COOH groups using carbodiimide coupling. When forming monolayers, the MPA molecules can orient themselves in different conformations with respect to the first carbon atom (that which is bound to the sulfur atom). These conformations are termed *gauche* and *trans* configurations, as shown schematically in Fig. 4.2.

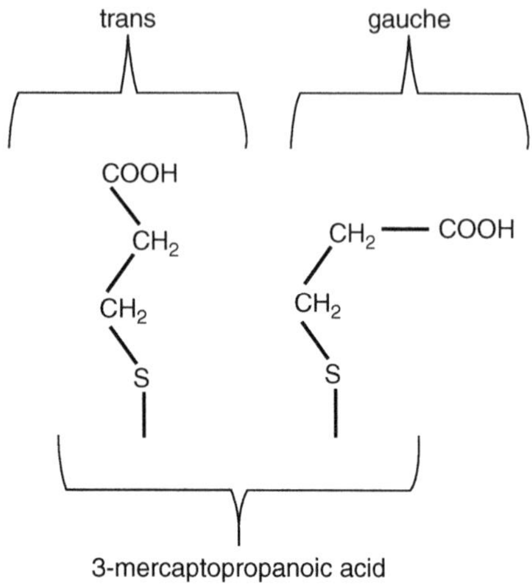

Fig. 4.6 A schematic diagram depicting the chemical conformation of *gauche* and *trans* configurations of 3-mercaptopropanoic acid

The ratio of *gauche* to *trans* configurations in the monolayers is affected by a number of factors including pH, concentration of the molecule in solution, the time over which the SAM is allowed to form, and properties of the metal surface [46].

Immediately after immersion in a solution of MPA, the monolayer packing density on the metal surface is low and the *trans* conformation predominates. Over time, as more molecules adsorb to the metal surface, the *gauche* conformation becomes more common. With silver, the metal surface has a tendency to oxidize in aqueous environments which results in desorption of molecules from the monolayer over time; thus the molecular orientation tends toward the *trans* configuration (Fig. 4.6).

The ratio of *gauche* to *trans* configured molecules may be determined by examining the Raman spectrum. For MPA, the 654 cm^{-1} band is characteristic for a *gauche* S-C-C chain and the 735 cm^{-1} band is characteristic for a *trans* conformer [48–50]. The relative peak heights for the *gauche* and *trans* Raman peaks gives an indication of the relative proportions of the number of molecules in these orientations. In the *gauche* conformation, the terminal binding groups of MPA are less able to bind with proteins during conjugation, so a monolayer of binding molecules in the *trans* conformation is preferable. In order to determine the optimum deposition time for MPA, a time study of the *gauche*: *trans* peak heights was undertaken.

In order to functionalize the metal-coated substrates with MPA, they were immersed in an ethanolic solution of 0.02 M MPA. The MPA layer was activated using a solution of 0.002 M ethyl dimethylaminopropyl carbodiimide (EDC) and 0.005 M n-hydroxy succinimide (NHS) in 2-(*N*-morpholino) ethanesulfonic acid (MES) buffer solution (20 mM MES, 0.1 M NaCl, pH5). The carboxylate groups of the MPA react with NHS in the presence of EDC to form NHS-esters which can then react with amine groups of proteins. After 15 min of activation, the metalized

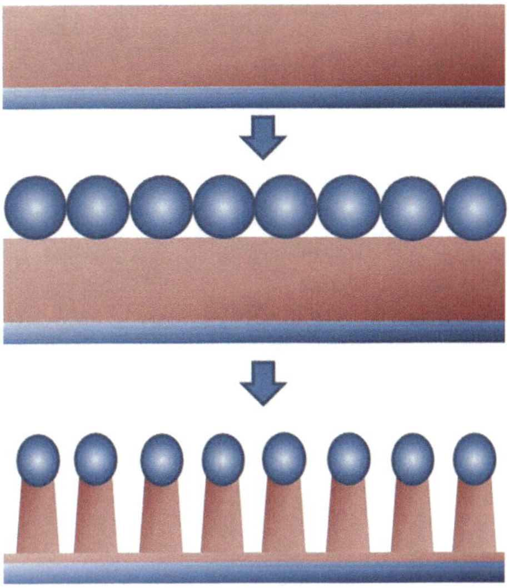

Fig. 4.7 Schematic diagram depicting the reactive ion etching (RIE) technique used to produce arrays of nanoscale pillars of polymer on a glass substrate

samples were rinsed in phosphate buffered saline (PBS) at pH 7. 50 μL aliquots of avidin solution (Sigma Aldrich) at 10 mg/mL in PBS at pH 7 were deposited onto each silvered wing section, and left to incubate for 30 min at 4°C. After incubation, the samples were rinsed in PBS pH 7 and immersed in a quenching solution of 1 M ethanolamine in PBS for 30 min at 4°C to deactivate any unreacted NHS-esters. Following immersion in the quenching solution, the silvered samples were rinsed in PBS pH 7. Aliquots of biotin solutions over a range of concentrations were then added and allowed to bind for 20 min at 4°C. The samples were then rinsed in Millipore water (PURITE, Oxford, UK) and the active area kept hydrated under a droplet of PBS to prevent the proteins from denaturing.

Normal (un-enhanced) Raman spectra of biotin and avidin were obtained using the following procedure. Aqueous solutions at working concentrations of the various chemicals used throughout these experiments were deposited onto aluminum-coated glass microscope slides. Once these drops had dried sufficiently to produce the characteristic "coffee ring" shaped deposition, normal Raman spectra were acquired for comparison with the SERS spectra.

The nanoconical arrays found on the butterfly wings were replicated using a reactive ion etching (RIE) technique illustrated schematically in Fig. 4.7, which was first devised by Weekes et al. [46]. Briefly, glass substrates were spin-coated with a 1.5 μm thick layer of PMMA, which was then annealed at 180°C. Nano-structured arrays were produced by forming hexagonally ordered monolayers of 390 nm diameter polystyrene microspheres upon the PMMA, which were then used as masks for reactive-ion etching (RIE). Reactive ion etching was applied for 8 min in oxygen plasma produced by a 15 W RF source at a base pressure of oxygen at 50 mTorr.

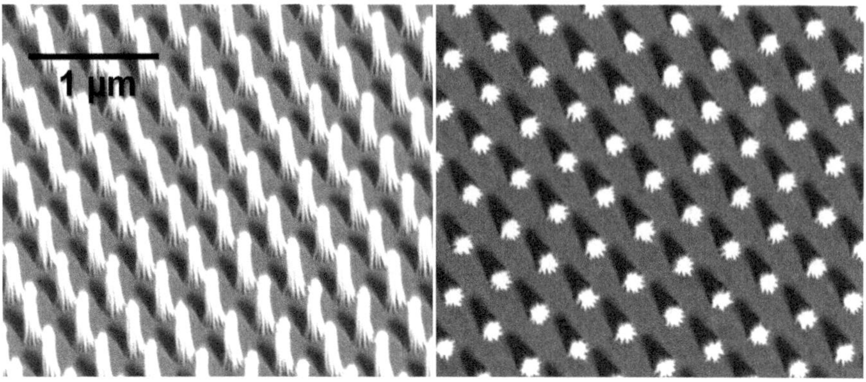

Fig. 4.8 SEM images of the biomimetic nanopillar substrate with pillar heights of 524 nm peak-to-peak distances of 390 nm

The RIE samples produced using this technique were formed over areas of ~1 cm^2. These samples could easily be broken up to create dozens of samples ready for batch-coating with metal, thus providing an efficient means of substrate production. Their SERS enhancement factors were determined in the same manner as for the butterfly wing substrates, and were found to be of similar magnitude.

The RIE biomimetic wing substrates were coated with silver and gold, and investigated in the same manner as the butterfly wings. When imaged using SEM, as shown in Fig. 4.8, they were found to exhibit highly ordered and regular, reproducible arrays of nanopillars.

4.3 Cicada Wings for SERS and a Simple Lithographic Technique for Replicating Their Nanostructure

The optical effects of the quasi-periodic tapered papillary structures found on the surfaces of cicada wings were initially investigated by Stoddart et al. [28]. As illustrated in Fig. 4.9, using scanning electron microscopy, they found that the pattern generated by these nanostructures on the wing surface was generally hexagonal within domains of between 1 and 2 μm^2. Unlike the *G. weiskei* butterfly, this structural morphology was found to be the same in the cicadas irrespective of gender, and was identical on both the upper and lower surfaces of the wings.

Sections of wing samples from *Cicadetta celis* cicadas were coated in a layer of high purity (99.99%) silver 60 nm thick using a Kurt J Lesker CMS-18HV electron beam deposition system controlled by a quartz crystal microbalance. These cicada wing substrates were found to produce a SERS enhancement factor of 10^6 when self-assembled monolayers of thiophenol were formed on their metalized surfaces. More recent work by Kostovski et al. has focused on

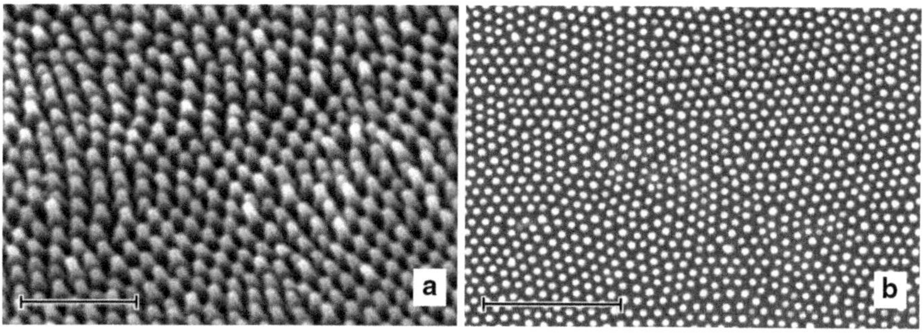

Fig. 4.9 SEM images to illustrate the nanostructures on gold-coated *C. celis* cicada wings taken (**a**) at an angle to the surface of the wing, and (**b**) from a direction normal to the upper surface. Scale bars are 1 μm and 2 μm in (**a**) and (**b**) respectively (Reproduced with permission from [28])

trying to replicate these cicada wing nanostructures for high throughput, high precision, and cost-effective SERS substrate production [29, 47].

The techniques employed by Kostovski et al. have involved nanoimprint ligthography, initially using a relatively labor-intensive technique illustrated in Fig. 4.10. In this method, moulds were made of 9 mm^2 sections of *Cyclochila australasiae* cicada wing by drip coating them with an elastomer solution (h-PDMS). Filling of the nanostructure with the elastomer was encouraged by degassing the sample under vacuum for 10 min, after which time the wing was placed coated side down onto a glass wafer.

The elastomer was cured by heating at 60°C for 12 h before the structure was pried off the wing manually with a scalpel blade. Once formed, these moulds were used to imprint nanostructures onto polymer coatings attached to the end of optical fibers, which was cured using a combination of heating and exposure to ultraviolet light. Finally, the optical fiber tips were coated in silver deposited at an angle of 60° to the fiber's long axis. The cicada wings, nanostructures, and optical fibers are illustrated in Fig. 4.11.

Although effective, this nano imprint lithography technique produced substrates with slightly lower SERS enhancements than those provided by the wings themselves. Additionally, the multistage method described is time consuming and labor intensive. However, Kostovski et al. recently improved upon this technique using a more rapid method for imprinting arrays of optical fibers, as illustrated in Fig. 4.12.

The optical fibers were loaded in an array of up to 40 fibers each placed in an individual U-shaped groove cut into an aluminum block, shown in Fig. 4.13. This technique allowed multiple fibers to be imprinted with the cicada wing mould in a simple and rapid manner that was not critically reliant upon alignment. The resolution obtained with this method was better than 15 nm, producing bidirectional optical fiber prober for SERS with enhancement factors only slightly less than those obtained from the original structures on the cicada wings.

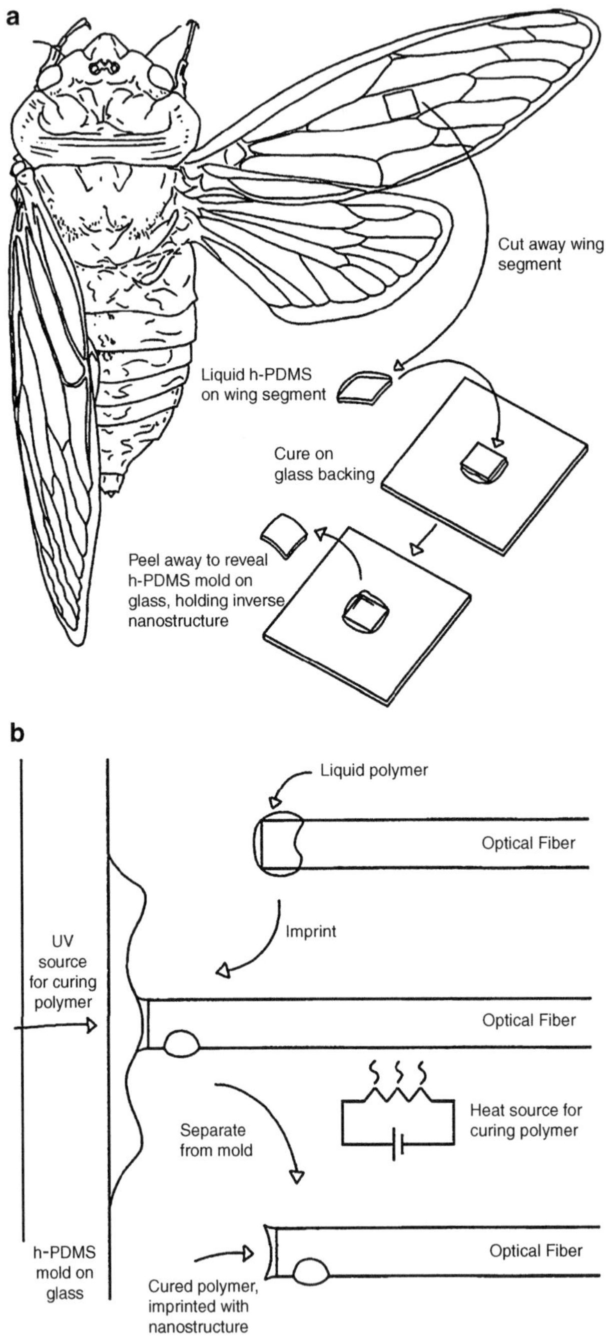

Fig. 4.10 Schematic process diagram illustrating the method used by Kostovski et al. to reproduce the nanostructures found on cicada wings, reproduced with permission [29]

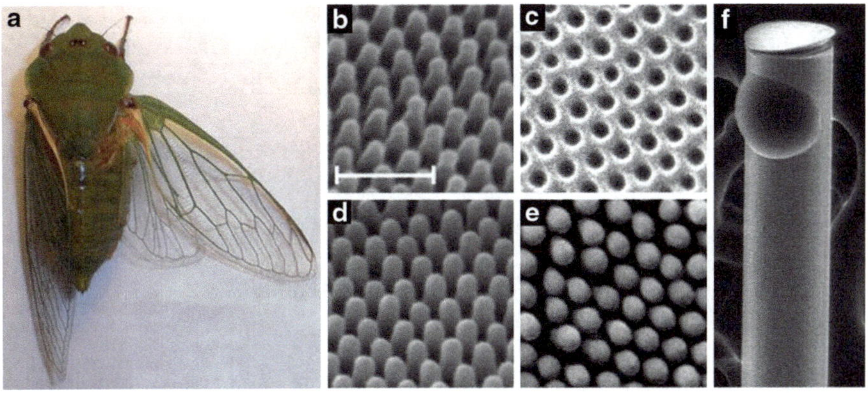

Fig. 4.11 (**a**) A photograph of the Australian greengrocer cicada. SEM images of (**b**) the nanotemplate of the cicada wing nanostructures (the scale bar is 500 nm), (**c**) the inverse h-PDMS mould, (**d**) an optical fiber exhibiting the polymer nanostructure replica, (**e**) the replica coated in silver, and (**f**) a macroscopic image of the imprinted fiber (Reproduced with permission from [29])

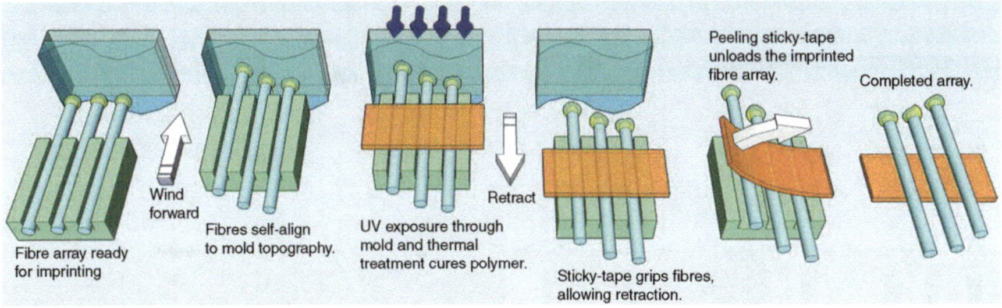

Fig. 4.12 A process schematic diagram of the rapid, self-aligned array-imprinting sequence used by Kostovski et al. to reproduce nanostructures on optical fibers using a nonplanar mould (Reproduced with permission from [47])

5 Key Research Findings

To determine the enhancement factors of *G. weiskei* wings and the biomimetic analogs made using reactive ion etching, Garrett et al. estimated the number of Raman counts per molecule in the SERS spectrum of the substrates by assuming that the thiophenol monolayer was at maximum density [40]. The maximum coverage of alkanethiols on thermally evaporated gold substrates has been shown to be 21%; 2.5×10^{14} molecules/cm^2 $\pm 0.1 \times 10^{14}$ molecules [48]. Assuming maximum metal coverage, this provides a conservative estimate for the enhancement factor, since maximum density monolayers of alkanethiols typically take 1–2 h to form, whereas SERS substrates have typically been characterized using SAMs of thiophenol formed over 10 min [49], hence in order to be able to compare the substrates investigated in this study with others reported elsewhere, it was necessary

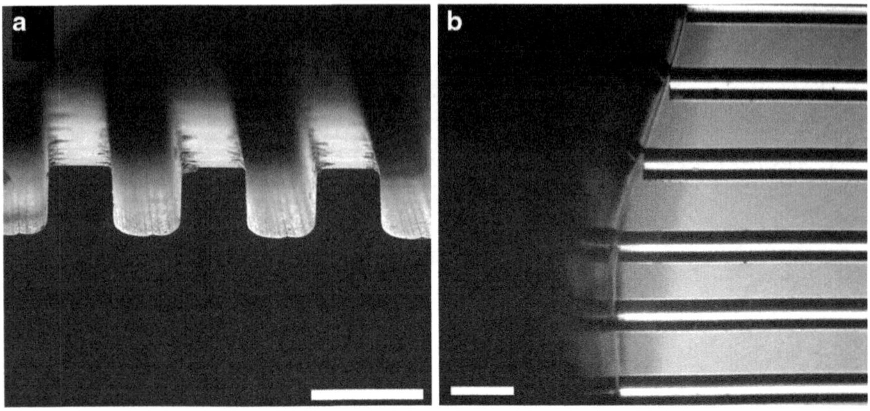

Fig. 4.13 SEM image of the diced "U-grooves" (**a**), and a photograph of the fiber-array (**b**) self-aligning against a nonplanar mould, as used by Kostovski et al. (Reproduced with permission from [47])

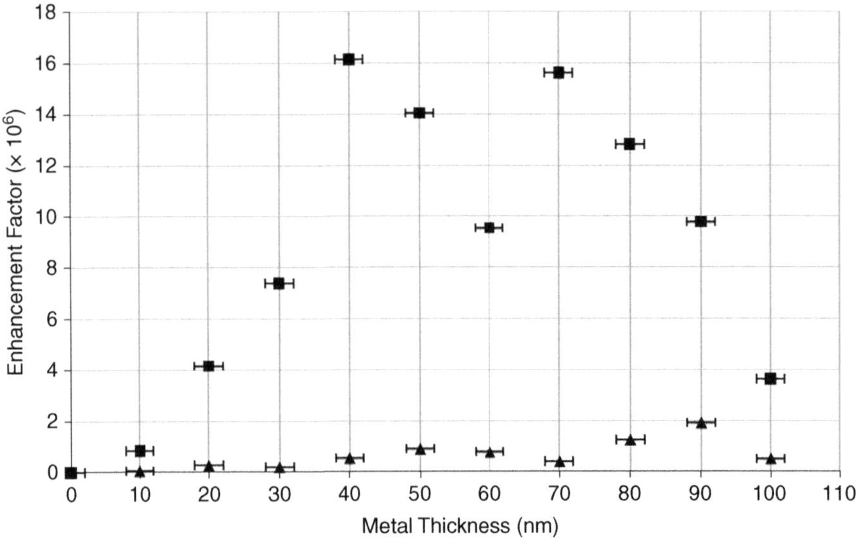

Fig. 4.14 Enhancement factor of metalized wing nanostructures as a function of metal deposition thickness: ■ data points correspond with silver, ▲ with gold

to use this reduced SAM formation time. The area of the substrate illuminated by the objective was calculated from the Airy disc diameter; hence the number of thiophenol molecules in the focal area was estimated at $6.0 \times 10^6 \pm 1.3 \times 10^4$.

A range of metal thicknesses was explored, in order to determine the optimum deposition thickness. Average enhancements over three wing sections and thirty scans were taken for each data point, the standard errors of which were too small to depict on the graph in Fig. 4.14. The enhancement factor depends on a combination of the metal thickness over each nanocone, the geometric feature sharpness (both at the tips of the nanocones and at the cavities at their bases), and the nanocone spacing. Therefore, as the deposited metal thickness increases, these features will be changed, causing both the intensity and resonant wavelength of the surface plasmons to be altered, which has led to the graph shape shown in Fig. 4.14. The difference in peak thickness and intensity for gold and silver arises from their differing dielectric constants.

When corrected for acquisition times, the maximum enhancement factors for the metalized *G. weiskei* wing surfaces were found to be $1.9 \times 10^6 \pm 5.8 \times 10^4$ for gold, and $1.6 \times 10^7 \pm 1.8 \times 10^5$ for silver. For the biomimetic substrates, the maximum enhancement factors were $1.1 \times 10^6 \pm 5.1 \times 10^4$ for gold and $1.4 \times 10^7 \pm 1.7 \times 10^5$ for silver [40].

The MPA molecule binds to metal surfaces via a covalent bond from the sulfur group. If the concentration of MPA molecules on a metal surface is high, the molecules tend to orient themselves in a *trans* configuration, whereas if the concentration is low, the MPA molecules will have a higher instance of *gauche* configuration. In the *gauche* configuration, the protein-binding site of the MPA molecule is less accessible than those in the *trans* configuration; the optimal MPA deposition time was determined by performing depositions for a range of times and comparing the ratios of the *gauche* and *trans* peaks (at 654 and 735 cm^{-1}, respectively) [50–52]. After immersing metalized wing substrates in ethanolic solutions of MPA for a range of times up to 18 h, the optimal deposition time was determined as 2 h. Upon activation of the MPA monolayer, the carboxyl band at 900 cm^{-1} shifted to 928 cm^{-1}, indicating that the carboxyl group was dissociated (Fig. 4.15).

Once avidin molecules were bound to an MPA monolayer, SERS spectra were acquired (see Fig. 4.16). The most prominent peaks in the hydrated avidin SERS spectrum were tentatively assigned to phenylalanine (1,001 cm^{-1}, 1,030 cm^{-1}), tryptophan (759 cm^{-1}, 875 cm^{-1}, 930 cm^{-1}, 960 cm^{-1}, 1,001 cm^{-1}, 1,357 cm^{-1}, 1,546 cm^{-1}, and 1,580 cm^{-1}), tyrosine (827 cm^{-1}, 852 cm^{-1}), C–C stretching (930 cm^{-1}), COO symmetric stretching (1,400 cm^{-1}). These peak assignments corresponded well with the spontaneous Raman spectrum of avidin (see Fig. 4.17). The amide I and amide III regions were located at 1,665 and 1,237 cm^{-1}, respectively. Each avidin peak lay within 6 cm^{-1} of the same peak assignments for the lyophilized avidin spectrum reported by Fagnano et al. [53]. A 6 cm^{-1} discrepancy is reasonable given the conformational changes that can occur between hydrated and lyophilized proteins, combined with the 2–3 cm^{-1} limit of resolution of the system.

It was found that upon binding to biotin, the intensity of the 960 cm^{-1} tryptophan peak increased relative to the 1,001 cm^{-1} phenylalanine peak (see Fig. 4.17), as reported by Fagnano et al. [53]. Since the avidin-binding site contains the amino acids phenylalanine, alanine, asparagine, and tryptophan [54], the 960 cm^{-1} tryptophan peak and the 1,001 cm^{-1} phenylalanine peak were chosen as binding indicator peaks for investigation in the SERS assay. The amide I and amide III bands did not change significantly upon addition of biotin.

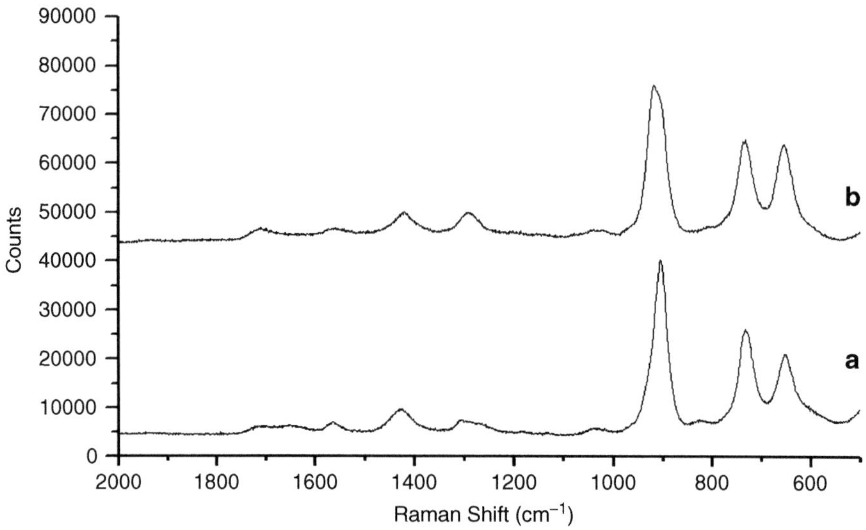

Fig. 4.15 SERS spectra of MPA (**a**) prior to activation with EDC/NHS, (**b**) after activation (Reproduced with permission from [40])

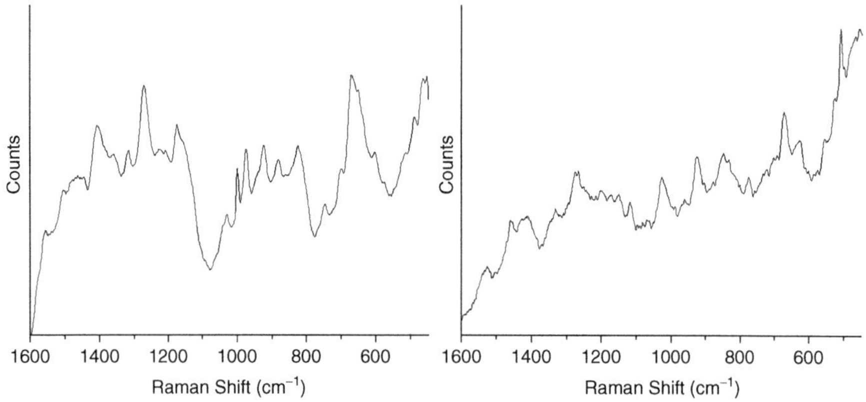

Fig. 4.16 SERS spectra of avidin bound to gold-coated wing (*right*) and the avidin-biotin complex bound to gold-coated wing (*left*) (Reproduced with permission from [40])

Upon addition of biotin, two distinct peaks were found in the avidin-biotin complex spectra which were indicative of the complex formation, 975 and 1,000 cm^{-1}, which were attributed to the avidin-binding site amino acids phenylalanine, alanine, asparagine, and tryptophan [54, 55]. It was found that the 1,000 cm^{-1} peak decreased in strength relative to the 975 cm^{-1} peak with increasing biotin concentration as shown in Fig. 4.19; an attribute exploited to perform a biotin assay. When plotted on a log scale as shown in Fig. 4.18, the assay gave a log-linear response over a range of 0.5–1,000 nM (equivalent to 0.12 ng/mL–0.24 μg/mL)

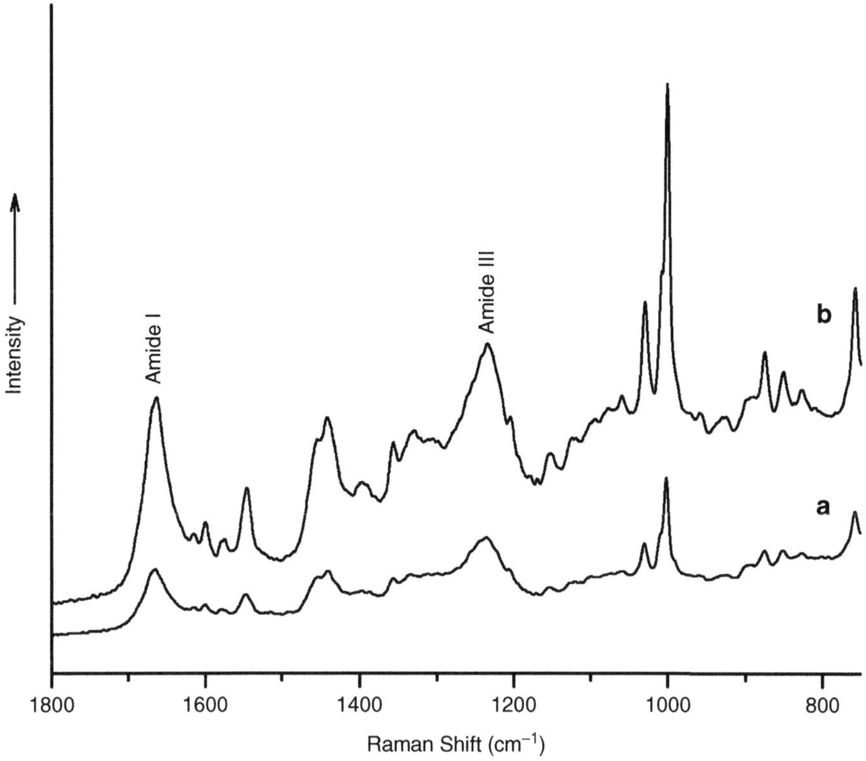

Fig. 4.17 Normal Raman spectra for avidin (**a**) and avidin-biotin (**b**) (Reproduced with permission from [40])

solutions of biotin suspended in buffer solution. The limit of detection of this assay was 0.2 nM (49 pg/mL).

Gold-coated butterfly wing nanostructures can provide excellent SERS substrates for use with protein assaying without the need for dehydrating the sample – a major advantage over other nanostructures surfaces used as SERS substrates. The entire avidin/biotin assay took no more than 4 h from surface conjugation to SERS scanning, which is comparable to some of the more rapid ELISA assays commercially available. The volume of analyte used was very small (only 30 μL), which would be an advantage for applications in areas such as forensics where it is often impractical or impossible to perform assays on large volumes of analyte.

When coated with 90 nm of gold, the substrates produced were found to provide consistent enhancement for up to 2 weeks after evaporation, when stored in a dessicator. Silver-coated substrates generally provided enhancement factors that were only consistent when used within 48 h after evaporation, due to oxide layer formation. In addition, gold is known to be markedly more biologically inert than

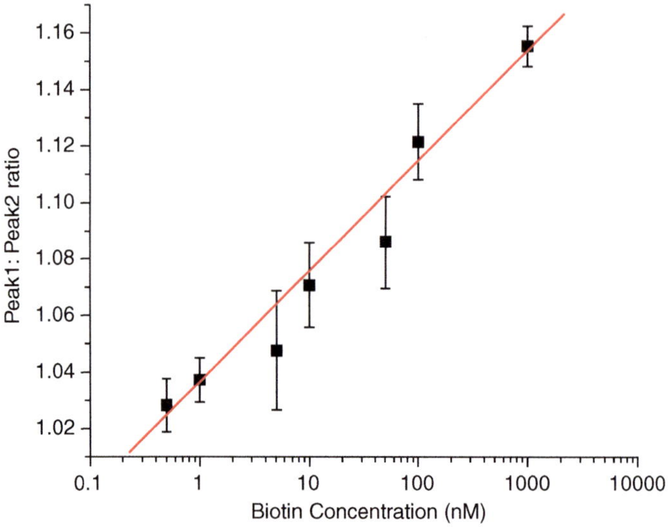

Fig. 4.18 Ratios of the heights of the peaks at 975 cm^{-1} (peak 1) and 1,000 cm^{-1} (peak 2) in the SERS spectra of the avidin-biotin complex for gold-coated butterfly wings for a range of concentrations of biotin in buffer solution (Reproduced with permission from [40])

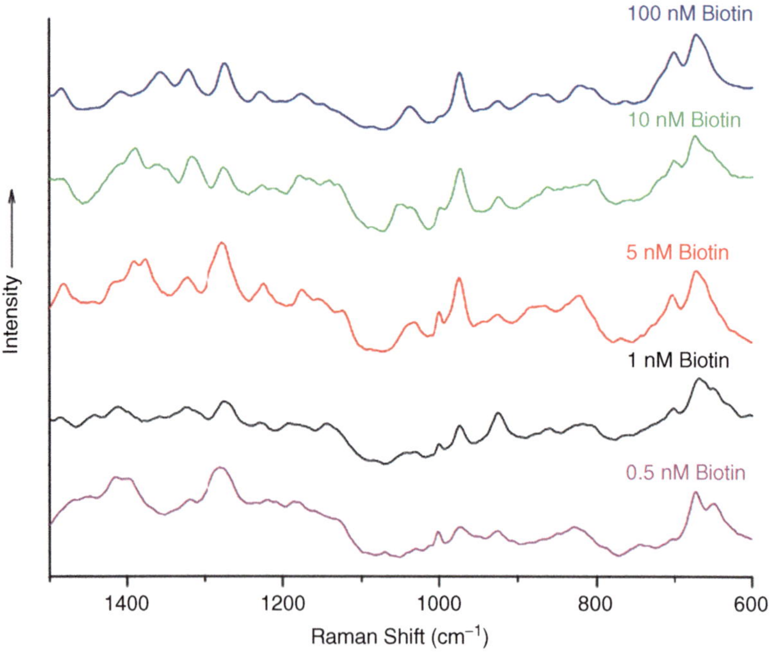

Fig. 4.19 SERS spectra of metalized wing-bound avidin exposed to a range of biotin solution concentrations (Reproduced with permission from [40])

silver, hence gold was preferred for biological applications in spite of the fact that the resulting enhancement factor was ~10× smaller than that obtained with silver.

In order to demonstrate the potential for mimicking the conical nanostructures of the butterfly wings for commercial applications, a simple reactive ion etching technique was employed. The biomimetic substrates showed excellent SERS enhancement factors (1.1×10^6 with a 90 nm coating of gold and 1.4×10^7 for a 70 nm coating of silver), and preliminary biocompatibility experiments showed that these substrates exhibit the remarkable property of enabling SERS signals to be obtained when the scanning area is submerged under a droplet of buffer solution. Future experiments will focus on combining these biomimetic substrate production techniques with nanotransfer printing methods in order to speed up production times and reduce costs, for use with biological applications.

6 Conclusions and Future Perspective

One of the main advantages of the wing-based SERS assay devised by Garrett et al. was the large range of analyte concentrations over which a log-linear response was measured (0.12 ng/mL–0.24 μg/mL) without the need for dilution of the analyte, as illustrated in Fig. 4.18. Until now, SERS assays have been effective over typical analyte concentration ranges spanning only one to two orders or magnitude [3, 4, 7, 8, 14], while ELISA assays have been made with effective concentration ranges of pg/ml to ng/mL [52]. This assay technique has bridged the gap between SERS and ELISA assays, potentially paving the way for commercially viable SERS substrates for undertaking assays of wet biological samples. Since SERS provides detailed structural and chemical information about assay analytes, with effective concentration detection ranges as large as those used in ELISA, SERS could become a much more effective immunoassay choice.

Now that the efficacy of biomimetic substrates for SERS protein assays has been proven, the next stage of research in this field should involve developing a rapid production method of these substrates. The reactive ion etching technique employed to produce the biomimetic substrates in this chapter have yielded highly reproducible nanostructured arrays over areas 1 cm^2 in size after only an 8 min exposure to the oxygen plasma, which could then be broken down to produce up to 25 substrates. Subsequent evaporation of gold took ~2.5 h for up to 12 substrates simultaneously, depending on the thickness of metal required. The total time taken to produce 25 such biomimetic substrates using the techniques outlined above is estimated at 4 h, although for industrial applications this could be greatly improved by adopting a production line technique.

Elastomer templates for nanoimprinting of the type devised by Kostovski et al. for their optical fiber nanoimprint lithography could be used to rapidly reproduce the butterfly wing nanostructures or those produced by reactive ion etching. This would further reduce production time and costs. Whatever means are employed to improve the production techniques for these butterfly and cicada wing substrates, it is clear that there is great potential for their application in biosensing.

References

1. Osawa M (2001) Surface-enhanced infrared absorption. In: Kawata S (ed) Near-field optics and surface plasmon polaritons. Springer, Berlin/Heidelberg, pp 163–187
2. Yonzon CR, Stuart DA, Zhang X et al (2005) Towards advanced chemical and biological nanosensors–an overview. Talanta 67(3):438–448
3. Dou X, Takama T, Yamaguchi Y et al (1997) Enzyme immunoassay utilizing surface-enhanced Raman scattering of the enzyme reaction product. Anal Chem 69(8):1492–1495
4. Moger J, Gribbon P, Sewing A et al (2007) Feasibility study using surface-enhanced Raman spectroscopy for the quantitative detection of tyrosine and serine phosphorylation. Biochim Biophys Acta Gen Subj 1770(6):912–918
5. Xu SP, Ji XH, Xu WQ et al (2005) Surface-enhanced Raman scattering studies on immunoassay. J Biomed Opt 10(3):031112/1–031112/12
6. Dou X, Yamaguchi Y, Yamamoto H et al (1998) NIR SERS detection of immune reaction on gold colloid particles without bound/free antigen separation. J Raman Spectrosc 29(8): 739–742
7. Xu SP, Wang LY, Xu WQ et al (2003) Immunological identification with SERS-labeled immunogold nanoparticles by silver staining. Chem J Chin Univ Chin 24(5):900–902
8. Zhang ML, Yi CQ, Fan X et al (2008) A surface-enhanced Raman spectroscopy substrate for highly sensitive label-free immunoassay. Appl Phys Lett 92(4):043116
9. Levy Y, Onuchic JN (2006) Water mediation in protein folding and molecular recognition. Annu Rev Biophys Biomol Struct 35:389–415
10. Ortiz C, Zhang D, Xie Y et al (2006) Validation of the drop coating deposition Raman method for protein analysis. Anal Biochem 353(2):157–166
11. Ronald A, Stimson WH (1998) The evolution of immunoassay technology. Parasitology 117: S13–S27
12. Hirsch LR, Jackson JB, Lee A et al (2003) A whole blood immunoassay using gold nanoshells. Anal Chem 75(10):2377–2381
13. Ji X, Xu S, Wang L et al (2005) Immunoassay using the probe-labeled Au/Ag core-shell nanoparticles based on surface-enhanced Raman scattering. Colloid Surf A Physicochem Eng Aspects 257–258:171–175
14. Rohr TE, Cotton T, Fan N et al (1989) Immunoassay employing surface-enhanced Raman-spectroscopy. Anal Biochem 182(2):388–398
15. Alexander RM (2005) Book Reviewed: The gecko's foot. Bio-inspiration: engineered from nature (Peter Forbes Hardback book, 272 p, ISBN 0007179901, Class number 620.0042) Nature 438(7065):166–166
16. Kinoshita S, Yoshioka S (2005) Structural colors in biological systems. Osaka University Press, Osaka
17. Parker AR (1999) Inverterbrate structural colour. In: Savazzi E (ed) Functional morphology of the invertebrate skeleton. Wiley, New York
18. Potyrailo RA, Ghiradella H, Vertiatchikh A et al (2007) Morpho butterfly wing scales demonstrate highly selective vapour response. Nat Photonics 1:123–128
19. Srinivasarao M (1999) Nano-optics in the biological world: beetles, butterflies, birds, and moths. Chem Rev 99(7):1935–1961
20. Vukusic P, Sambles JR (2003) Photonic structures in biology. Nature 424(6950):852–855
21. Ghiradella H, Aneshansley D, Eisner T et al (1973) Ultraviolet reflection of a male butterfly: interference color caused by thin-layer elaboration of wing scales. Science 179(4071):415
22. Land MF (1972) The physics and biology of animal reflectors. Prog Biophys Mol Biol 24: 75–106
23. Vukusic P, Wootton RJ, Sambles JR (2004) Remarkable iridescence in the hindwings of the damselfly Neurobasis chinensis chinensis (Linnaeus) (Zygoptera: Calopterygidae). Proc R Soc Lond Ser B Biol Sci 271(1539):595–601

24. Yoshioka S, Kinoshita S (2007) Polarization-sensitive color mixing in the wing of the Madagascan sunset moth. Opt Expr 15(5):2691–2701
25. Volkmer A, Cheng JX, Xie XS (2001) Vibrational imaging with high sensitivity via epidetected coherent anti-stokes Raman scattering microscopy. Phys Rev Lett 8702(2):023901
26. Vukusic P, Hallam B, Noyes J (2007) Brilliant whiteness in ultrathin beetle scales. Science 315(5810):348–351
27. Jewell SA, Vukusic P, Roberts NW (2007) Circularly polarized colour reflection from helicoidal structures in the beetle *Plusiotis boucardi*. New J Phys 9:99
28. Stoddart PR, Cadusch PJ, Boyce TM et al (2006) Optical properties of chitin: surface-enhanced Raman scattering substrates based on antireflection structures on cicada wings. Nanotechnology 17(3):680–686
29. Kostovski G, White DJ, Mitchell A et al (2009) Nanoimprinted optical fibres: biotemplated nanostructures for SERS sensing. Biosens Bioelectron 24(5):1531–1535
30. Kneipp J, Kneipp H, Kneipp K (2008) SERS – a single-molecule and nanoscale tool for bioanalytics. Chem Soc Rev 37:1052–1060
31. Pelletier MJ, Altkorn R (2001) Raman sensitivity enhancement for aqueous protein samples using a liquid-core optical-fiber cell. Anal Chem 73(6):1393–1397
32. Abdelsalam ME, Bartlett PN, Baumberg JJ et al (2005) Electrochemical SERS at a structured gold surface. Electrochem Commun 7(7):740–744
33. Baker GA, Moore DS (2005) Progress in plasmonic engineering of surface-enhanced Raman-scattering substrates toward ultra-trace analysis. Anal Bioanal Chem 382(8):1751–1770
34. Hunyadi SE, Murphy CJ (2006) Bimetallic silver-gold nanowires: fabrication and use in surface-enhanced Raman scattering. J Mater Chem 16(40):3929–3935
35. White DJ, Mazzolini AP, Stoddart PR (2007) Fabrication of a range of SERS substrates on nanostructured multicore optical fibres. J Raman Spectrosc 38(4):377–382
36. Yoshida A, Motoyama M, Kosaku A et al (1997) Antireflective nanoprotuberance array in the transparent wing of a hawkmoth, *Cephonodes hylas*. Zool Sci 14(5):737–741
37. Le Ru EC, Blackie E, Meyer M et al (2007) Surface enhanced Raman scattering enhancement factors: a comprehensive study. J Phys Chem C 111(37):13794–13803
38. Viets C, Hill W (2000) Single-fibre surface-enhanced Raman sensors with angled tips. J Raman Spectrosc 31(7):625–631
39. Haynes CL, McFarland AD, Van Duyne RP (2005) Surface-enhanced Raman spectroscopy. Anal Chem 77(17):338A–346A
40. Garrett NL, Vukusic P, Ogrin F et al (2009) Spectroscopy on the wing: naturally inspired SERS substrates for biochemical analysis. J Biophotonics 2(3):157–166
41. Bryant MA, Pemberton JE (1991) Surface Raman-scattering of self-assembled monolayers formed from 1-alkanethiols at Ag. J Am Chem Soc 113(10):3629–3637
42. Bryant MA, Pemberton JE (1991) Surface Raman-scattering of self-assembled monolayers formed from 1-alkanethiols – behavior of films at Au and comparison to films at Ag. J Am Chem Soc 113(22):8284–8293
43. Brakefield PM, French V (1999) Butterfly wings: the evolution of development of colour patterns. Bioessays 21(5):391–401
44. Vukusic P, Sambles R, Lawrence C et al (2001) Sculpted-multilayer optical effects in two species of Papilio butterfly. Appl Opt 40(7):1116–1125
45. Dey S, Hooroo RNK, Bhattacharjee CR (1998) Electron microscopy and spectroscopical studies on the coloured patches on the wing of a butterfly, *Graphium sarpedon* (Lepidoptera: Papillionidae) with reference to their photobiological and electrical properties. Pigment Cell Res 11(1):1–11
46. Weekes SM, Ogrin FY, Murray WA et al (2007) Macroscopic arrays of magnetic nanostructures from self-assembled nanosphere templates. Langmuir 23(3):1057–1060
47. Kostovski G, Chinnasamy U, Jayawardhana S et al (2011) Sub-15 nm optical fiber nanoimprint lithography: a parallel, self-aligned and portable approach. Adv Mater 23(4):531–535

48. O'Dwyer C, Gay G, Viaria de Lesegno B et al (2005) Advancing atomic nanolithography: cold atomic Cs beam exposure of alkanethiol self assembled monolayers. J Phys Conf Ser 19:109–117
49. Briand E, Salmain M, Henry JM et al (2006) Building of an immunosensor: how can the composition and structure of the thiol attachment layer affect the immunosensor efficiency? Biosens Bioelectron 22(3):440–448
50. Michota A, Kudelski A, Bukowska J (2002) Molecular structure of cysteamine monolayers on silver and gold substrates – comparative studies by surface-enhanced Raman scattering. Surf Sci 502:214–218
51. Kudelski A (2002) Raman study on the structure of 3-mercaptopropionic acid monolayers on silver. Surf Sci 502:219–223
52. Kudelski A, Hill W (1999) Raman study on the structure of cysteamine monolayers on silver. Langmuir 15(9):3162–3168
53. Fagnano C, Fini G, Torreggiani A (1995) Raman spectroscopic study of the avidin-biotin complex. J Raman Spectrosc 26(11):991–995
54. Huang TS, Delange RJ (1971) Egg white avidin. 2. Isolation, composition, and amino acid sequences of tryptic peptides. J Biol Chem 246(3):686
55. Stewart S, Fredericks PM (1999) Surface-enhanced Raman spectroscopy of amino acids adsorbed on an electrochemically prepared silver surface. Spectrochim Acta A Mol Biomol Spectrosc 55(7–8):1641–1660

Nonlinear Raman Scattering Spectroscopy for Carbon Nanomaterials

5

Katsuyoshi Ikeda and Kohei Uosaki

1 Definition of the Topic

Nonlinear Raman scattering spectroscopy is a multiphoton spectroscopy that enables access to vibrationally excited molecular levels. Through nonlinear optical processes, this technique allows us to study rich molecular information which cannot be reached by linear optical method.

2 Overview

Nonlinear vibrational spectroscopy provides accessibility to a range of vibrational information that is hardly obtainable from conventional linear spectroscopy. Recent progress in the pulsed laser technology has made the nonlinear Raman effect a widely applicable analytical method. In this chapter, two types of nonlinear Raman techniques, hyper-Raman scattering (HRS) spectroscopy and time-frequency two-dimensional broadband coherent anti-Stokes Raman scattering (2D-CARS) spectroscopy, are applied for characterizing carbon nanomaterials. The former is used as an alternative for IR spectroscopy. The latter is useful for studying dynamics of nanomaterials.

K. Ikeda (✉)
Division of Chemistry, Graduate School of Science, Hokkaido University, Sapporo, Japan

K. Uosaki
Division of Chemistry, Graduate School of Science, Hokkaido University, Sapporo, Japan
and
International Center for Materials Nanoarchitectonics, National Institute for Materials Science, Tsukuba, Japan

Challa S. S. R. Kumar (ed.), *Raman Spectroscopy for Nanomaterials Characterization*,
DOI 10.1007/978-3-642-20620-7_5, © Springer-Verlag Berlin Heidelberg 2012

3 Introduction

After the discovery of fullerenes or carbon nanotubes [1, 2], carbon nanomaterials have become of significant interest for many applications because of their unique electrical, optical, and mechanical properties [3–9]. For example, single-walled carbon nanotubes (SWNTs) can behave as semiconductors or metals, depending on their chirality [10]. Many attempts are, therefore, made to develop chirality-selective synthetic methods [11, 12]. Vibrational spectroscopy is indispensable in materials science because the analysis of vibrational spectra can tell us a great deal about the chemical identity and even the environment of the molecule being detected. Indeed, linear vibrational spectroscopy, for example, infrared absorption (IR) or Raman scattering (RS), are widely utilized in various research fields because of their convenience. However, much more detailed characterization is often required for further understanding properties of nanomaterials. In this sense, nonlinear spectroscopy should be useful for such an analysis; higher-order optical phenomena can, in principle, give us rich information that is hardly obtainable from conventional linear spectroscopy [13]. For example, sum-frequency vibrational spectroscopy (SFVS), which is one of the well-known second-order nonlinear spectroscopy, can probe *interfacial* chemical information; such selectivity is not expected in the conventional linear spectra [14].

In linear and nonlinear Raman spectroscopy, vibrational transitions are indirectly induced; incident and scattered photon energies are much larger than vibrational energy of molecules. Hence, these spectroscopic methods have several great advantages for in situ measurements or microscopic observations [15]. Moreover, electronic information is available in addition to vibrational information when incident photons are in resonance with molecular excited states. This chapter deals with two types of nonlinear Raman spectroscopy: hyper-Raman scattering (HRS) spectroscopy and broadband coherent anti-Stokes Raman scattering (CARS) spectroscopy. As an application example, these methods are applied for characterizing carbon nanomaterials.

4 Experimental and Instrumental Methodology

4.1 Selection Rules

Among various spectroscopic techniques, IR and RS are commonly utilized in experimental science. IR-active modes and RS-active modes are known to be mutually exclusive in centrosymmetric molecules because these optical processes are realized as one-photon and two-photon transitions, respectively (Fig. 5.1). It is important to obtain full vibrational information when characterizing materials. Therefore, a combined use of both methods should be helpful for understanding their chemical structures. However, this is practically difficult; due to relatively low energies of infrared photons, IR spectroscopy sometimes faces problems, e.g., in situ observation in IR-opaque aqueous solution. RS spectroscopy often suffers with

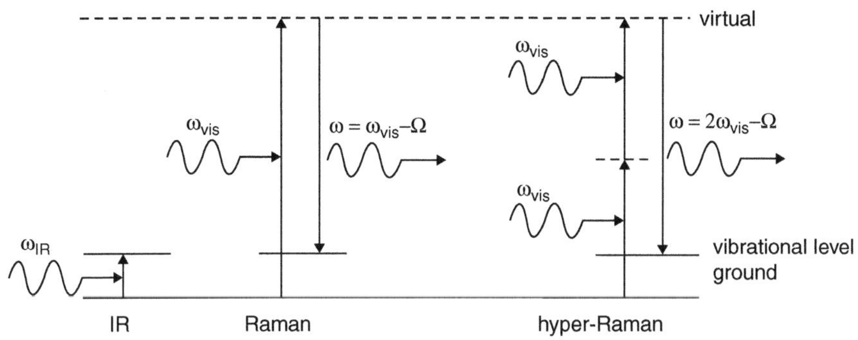

Fig. 5.1 Energy diagram of IR absorption (IR), Raman scattering (RS), and hyper-Raman scattering (HRS)

fluorescence originated from an excited sample. Nonlinear spectroscopy provides an opportunity to access many more vibrational modes under such conditions.

Linear and nonlinear optical processes are described as follows [16]. The dipole moment $\boldsymbol{\mu}$ induced in a molecule placed in an electromagnetic field with frequency of ω is given by

$$\mu_i = \mu_i^0 + \alpha_{ij} \cdot E_j(\omega) + 1/2\beta_{ijk} : E_j(\omega)E_k(\omega) + \cdots, \tag{5.1}$$

where $\boldsymbol{\mu}^0$, $\boldsymbol{\alpha}$, and $\boldsymbol{\beta}$ are the permanent dipole moment, the polarizability tensor, and the hyperpolarizability tensor, respectively, and i, j, k refer to the molecular coordinates. In this equation, the first-order expansion of the $\boldsymbol{\mu}_0$, $\boldsymbol{\alpha}$, and $\boldsymbol{\beta}$ in the normal coordinate of vibration $\boldsymbol{Q}$ is, respectively, responsible for IR, RS, and HRS. That is, in the case of HRS, hyperpolarizability modulations by nuclear vibrations of molecules are represented as the expansion in the series of the normal coordinates $\boldsymbol{Q}$ near the equilibrium $\boldsymbol{\beta}^0$:

$$\beta_{ijk} = \beta_{ijk}^0 + \sum_l \left(\partial\beta_{ijk}/\partial Q_l\right)_0 Q_l + \cdots, \tag{5.2}$$

and the second term stands for HRS. So, HRS is realized as an inelastic scattering process accompanied with optical frequency doubling. Note that HRS is allowed even in centrosymmetric media because the final state is different from the initial state. Since this second-order effect is a three-photon process with two-photon upward and one-photon downward transitions, as shown in Fig. 5.1, one can easily understand that selection rules for HRS are rather similar to those for IR according to the parity [16–18]. Moreover, HRS can access so-called silent modes, which are IR- and RS-inactive, because of less restrictive selection rules in nonlinear optical processes [18]. This feature is quite important as the analysis tool of molecules.

In a similar manner, one can also consider third-order nonlinear vibrational spectroscopy. CARS is one of well-known third-order nonlinear optical phenomena

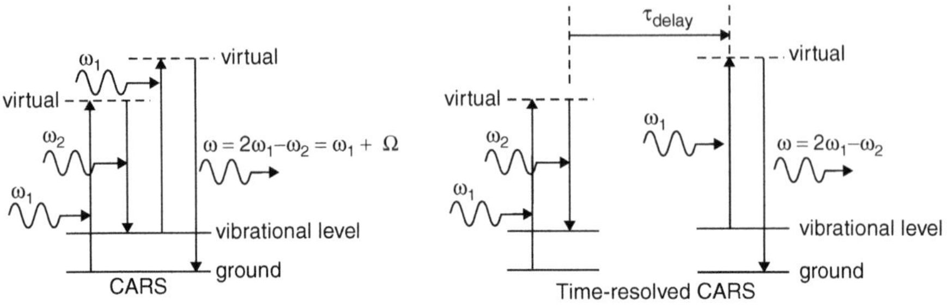

Fig. 5.2 Energy diagrams of coherent anti-Stokes Raman scattering (CARS) and time-resolved CARS

used as vibrational spectroscopic tools [13]. As shown in Fig. 5.2, the CARS process is realized as a combined process of Raman scattering events. Hence, the selection rules for CARS are same as those for conventional RS. However, there is a significant difference between CARS and other *incoherent* Raman processes (RS and HRS). In the case of CARS, the first and second incident fields, with an energy difference equal to a vibrational energy induce transition from the ground state to a vibrational state with the same energy separation. Therefore, CARS is normally utilized for selective observation of a single vibrational mode.

4.2 Hyper-Raman Scattering Spectroscopy and Microscopy

According to the selection rules, the HRS spectroscopy can in principle be used as an alternative for IR spectroscopy [19]. Indeed, this nonlinear spectroscopy has several advantages; IR-mode detection is possible even in IR-opaque media and its spatial resolution is much better than IR microscopy. Moreover, HRS signals appear in the doubled frequency region, which is far from the intense excitation laser line, and hence, low-energy vibration modes are easily observable.

In spite of such remarkable features of HRS, spectroscopic applications of this technique have been very limited so far because of the extremely small scattering cross section, typically in the order of 10^{-65} cm^4 per molecule [20]. However, the recent progress in laser technology has made the effect a widely applicable analytical method, by combining with signal enhancement techniques. HRS signals can benefit from electronic resonances to a molecular excited state and from plasmonic resonances of a substrate metal surface [19–23]. In the former case, which is called resonance HRS, the appearance of enhanced spectra is closely related to molecular symmetries. On the basis of the vibronic theory [24], hyperpolarizability can be expanded into three terms (Ziegler's A, B, and C terms) according to the order of the vibronic interactions:

$$\beta = A + B + C. \tag{5.3}$$

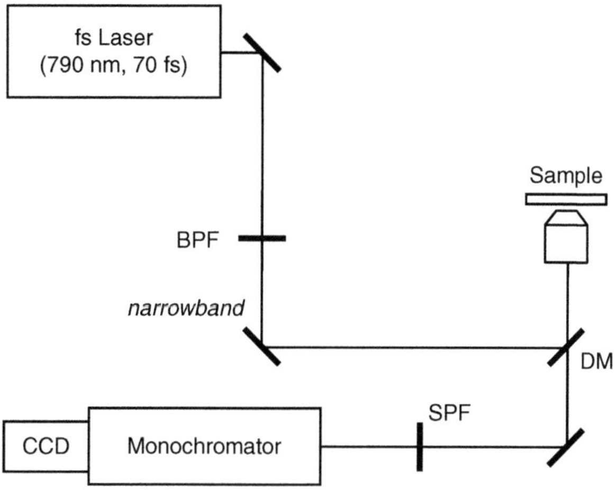

Fig. 5.3 Schematic illustration of a fs-laser-based hyper-Raman scattering microscope system. *BPF* narrow band-pass filter, *DM* dichroic mirror, *SPF* shortpass filter

In centrosymmetric molecules, HRS gains intensity via Herzberg–Teller term (the first vibronic B-term), indicating that IR-active modes and silent modes are enhanced. In the case of non-centrosymmetric molecules, however, Franck–Condon mechanism (A-term) dominantly contributes to the enhancement. Moreover, the mutual exclusive rules between HRS and RS are broken, and hence, some of RS-active modes selectively appear in the spectra. In the case of plasmonic enhancement, the spectral appearance is more sensitive to molecular orientations at the metal surface because of the surface selection rules [25].

Figure 5.3 shows a schematic illustration of our HRS measurement system [26–28]. A mode-locked Ti:sapphire laser (Spectra-Physics, Tsunami) was used to induce HRS. Pulses of 70 fs at 790 nm with a repetition rate of 82 MHz were spectrally narrowed by a custom-made laser line filter (Optical Coatings Japan). The obtained pulse width was 14 cm^{-1} FWHM with a pulse duration of ~1 ps (measured by autocorrelation). These pulses were introduced into an inverted microscope system (Nikon, TE-2000) with a 36×, 0.52 N.A. reflective microscope objective or a 100×, 1.49 N.A. oil-immersion microscope objective. Backscattered photons were collected by the same objective and filtered by dichroic mirrors (Optical Coatings Japan). Finally, HRS signals were detected by a charge-coupled camera (Princeton Instruments, PIXIS: 400B) with a polychromator (ACTON, SP2500i).

4.3 Broadband CARS and 2D-CARS Microscopy

The conventional *narrowband* CARS process probes one particular vibrational mode selectively. Conversely, so-called broadband CARS measurements, using ultrashort pulsed laser sources, can probe multiple RS-active vibrational modes simultaneously [19, 29–31]. In the case of two-beam broadband CARS method, one of the two beams has a narrow bandwidth and the other a broad bandwidth. Therefore, the technical issue is how to generate these beams from a single laser source. Typically, subpicosecond pulses from a conventional solid-state femtosecond laser

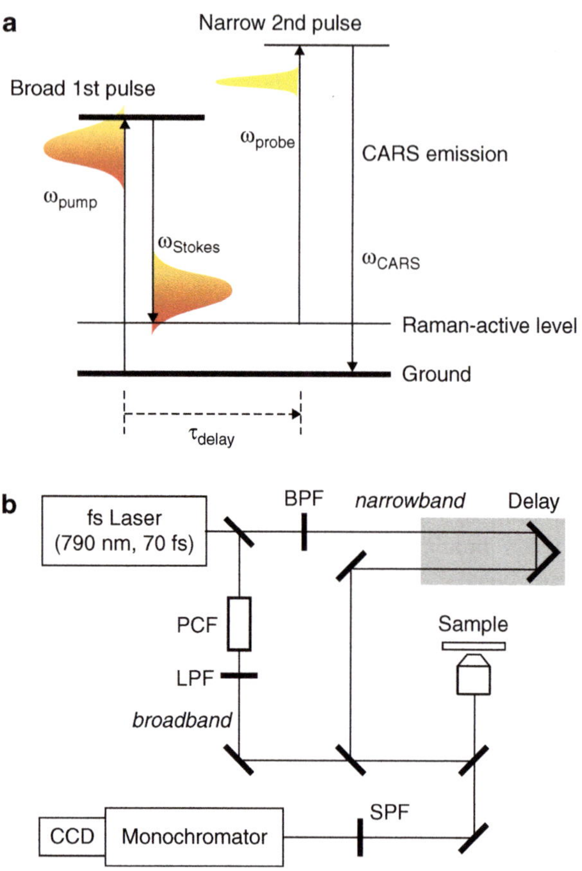

Fig. 5.4 (**a**) The energy diagram of the three-color CARS process, where the pump and Stokes photons (ω_{pump} and ω_{Stokes}) are provided from the single broadband pulse and the probe photon (ω_{probe}) from the delayed narrowband pulse. (**b**) Experimental setup of a time-frequency 2D-CARS microscope: *BPF* band-pass filter, *PCF* photonic crystal fiber, *LPF* longpass filter, *SPF* shortpass filter

system are convenient for this purpose. A photonic crystal fiber and an ultra-narrow band-pass filter are used to generate supercontinuum and narrowband pulses, respectively, from these pulses [19, 29, 30]. A sub-10-fs ultrashort laser pulse can be also utilized as a broadband light source [31]. The most important issue in this technique is the momentum conservation in a wide frequency region. Since this phase-matching condition is relaxed when the excitation beams are tightly focused, a microscope system is suitable for the broadband CARS method [19].

The experimental setup for the broadband CARS is rather simple because only two pulses are needed for three-color CARS emission, as shown in Fig. 5.4a; a *broadband* first pulse impulsively promotes molecules to vibrationally excited states through a two-photon Raman process, and a delayed *narrowband* second pulse induces anti-Stokes Raman emission from coherent superpositions to the ground state [29]. By changing the delay time for the second pulse, therefore, one can expect to probe dynamical behaviors of multiple RS-active modes. Such a two-dimensional observation in the time-frequency domains should be effective for detailed analysis of nanomaterials.

Our experimental setup of the 2D-CARS microscope is shown in Fig. 5.4b [32]. The 70-fs output from the Ti:sapphire oscillator was split into two beams. One of the beams was introduced into a photonic crystal fiber (Crystal Fibre, Femtowhite 800) to generate a coherent supercontinuum. Then, the continuum was conditioned with an 800 nm long-pass filter. The other beam was spectrally narrowed by the custom-made laser line filter (Optical Coatings Japan, $\Delta\nu = 14\ \mathrm{cm}^{-1}$ FWHM). The obtained two beams were introduced collinearly into the microscope objective lens. The chirping of the *broadband* pump beam was carefully avoided so that the CARS signals are obtained in a wide spectral region. The *broadband* CARS emission in back-reflected direction was analyzed by the CCD spectrometer.

5 Key Research Findings

5.1 Recent Progress in Hyper-Raman Spectroscopy

5.1.1 Plasmonic and Molecular Resonance Enhancement of Hyper-Raman Scattering

Since the HRS scattering cross section is so small, enhancement of the signal intensity is indispensable for a practical use of the HRS technique in the field of molecular science. As already mentioned, one of the methods for signal enhancement is to use plasmonic resonances of metal nanostructures [20, 21, 25]. This enhancement technique is called surface-enhanced hyper-Raman scattering (SEHRS) because it is similar to that of surface-enhanced Raman scattering (SERS). In SERS or SEHRS, both excitation absorption and scattered emission benefit from plasmonic field enhancement. Therefore, the plasmonic enhancement factor M is described as

$$M_{SERS} = \left|\frac{E_{loc}(\omega)}{E_0(\omega)}\right|^2 \left|\frac{E_{loc}(\omega-\Omega)}{E_0(\omega-\Omega)}\right|^2 \quad \text{and} \tag{5.4}$$

$$M_{SEHRS} = \left|\frac{E_{loc}(\omega)}{E_0(\omega)}\right|^4 \left|\frac{E_{loc}(2\omega-\Omega)}{E_0(2\omega-\Omega)}\right|^2 \tag{5.5}$$

for SERS and SEHRS, where $E_0(\omega)$, $E_0(\omega-\Omega)$, and $E_0(\omega-\Omega)$ represent incident, Raman-shifted, and hyper-Raman-shifted fields, respectively, and E_{loc} denotes the localized field [33]. According to these equations, one can understand that M_{SEHRS} can be larger than M_{SERS} when both of fundamental and harmonic resonances contribute to the enhancement. Although colloidal aggregates of metal nanoparticles are commonly used for SEHRS measurements, it is effective to design and construct metallic nanostructures artificially [26] because plasmonic properties of such structures are strongly dependent on their shape. A possibility of the single molecular detection is now discussed with the SEHRS technique in spite of the small scattering cross section [20].

The other enhancement method is to use molecular electronic resonances [19, 22, 23]. A most interesting topic in this enhancement method is so-called molecular near-field effect reported by Kano et al. [34]. This effect, explained by intermolecular vibronic coupling, makes HRS spectroscopy highly sensitive to intermolecular interactions; one can observe HRS modes of solvent molecules adjacent to the solute dye molecules via the intensity borrowing from the dye molecules. That is, solvent-solute interactions are selectively observed without impairment from signals of the bulk solvent. It is almost impossible to observe a similar effect in resonance RS spectroscopy because of the unavoidable contribution from Franck–Condon type resonance.

5.1.2 Resonance Hyper-Raman Spectra of Carbon Nanomaterials

In this section, we explain our recent results on resonance HRS spectra of fullerene and carbon nanotubes [27, 28]. Although HRS has been mainly utilized for characterizing dielectric materials in the field of physics, we believe that these examples clearly show a possibility of HRS as a spectroscopic tool in the fields of molecular science and nanomaterials science.

HRS Spectra of Fullerene

Buckminsterfullerene C_{60} has unique properties such as third-order optical nonlinearity, superconductivity, and ferromagnetism [3–6, 28]. These properties are related to the highly delocalized and strongly correlated π-electrons in the spherical cage with the icosahedral I_h symmetry. It is of great importance to study such a high degenerate electron-vibration coupling. In spite of many theoretical and experimental efforts, however, our understanding of fullerenes is still limited because of the elaborate network of the coupled electronic and vibrational structures. According to the group theory, C_{60} has 174 internal degrees of freedom with 46 distinct vibrational coordinates having the representation [35].

$$\Gamma_{\mathrm{vib}} = 2a_g + 1a_u + 3t_{1g} + 4t_{1u} + 4t_{2g} + 5t_{2u} + 6g_g + 6g_u + 8h_g + 7h_u \tag{5.6}$$

The numbers of IR- and RS-active modes are 4 ($4t_{1u}$) and 10 ($2a_g$ + $8h_g$), respectively. All other modes are IR- and RS-forbidden *silent* modes (a_u, t_{1g}, t_{2g}, t_{2u}, g_g, g_u, h_u). Conversely, the number of HRS-active modes are 22 ($4t_{1u}$ + $5t_{2u}$ + $6g_u$ + $7h_u$) including 18 silent modes. Since these HRS modes are mutually exclusive with RS-active modes, combination of resonance RS and resonance HRS should be useful for understanding the electron-vibration coupled system.

Microcrystalline films of C_{60} molecules (Strem Chemicals, Inc., 99.9% grade) were obtained on a cover slip by evaporation of C_{60}-saturated toluene solution. C_{60} molecules are known to occupy fcc lattice points in the solid phase, resulting in symmetry lowering from I_h to T_h. However, the symmetry-lowering perturbation is almost negligible at room temperature. This is because intermolecular van der Waals forces are much weaker than the intramolecular covalent forces within individual molecules and, therefore, molecules at room temperature are freely rotating in the lattice [36]. The measurement system was already described in the earlier section.

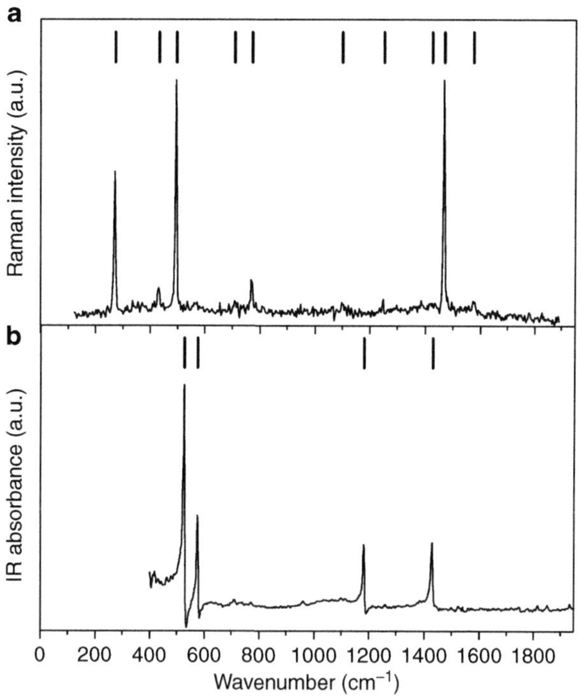

Fig. 5.5 (**a**) Raman spectrum of C_{60} microcrystals excited by 785-nm diode laser radiation. (**b**) IR absorption spectrum of C_{60} microcrystals, taken by the KBr method. Theoretically calculated Raman-active and IR-active modes are marked in the spectra [28]

Figure 5.5 shows vibrational modes of C_{60} microcrystals probed by the conventional IR and RS spectroscopy. The RS spectrum, measured with 785-nm excitation, showed all of RS-active modes, $2a_g + 8h_g$, whereas the IR spectrum clearly showed the IR-active modes, $4t_{1u}$, indicating that both spectra were consistent with the theoretical calculations of a free C_{60} molecule.

HRS spectrum of the crystalline C_{60} is shown in Fig. 5.6. Peak positions of theoretically calculated HRS-active and RS-active modes are also marked with black and gray bars in the spectrum. The most intense peak at 0 cm^{-1} is due to hyper-Rayleigh scattering, which has probably originated from magnetic dipole contributions (second harmonic generation at the glass substrate surface was negligible in the present experimental configuration) [37]. In the Stokes shift region, there appeared several peaks: intense stretching vibrational modes around 1,300–1,600 cm^{-1} and bending vibrations around 500–800 cm^{-1}. Compared with the theoretical calculation, all of HRS-active modes of the icosahedral C_{60}, that is, the infrared-active t_{1u} and the silent t_{2u}, g_u, and h_u, were seen in the spectrum. Moreover, Raman-active modes were also found in the spectrum; for example, the 270-cm^{-1} peak was obviously associated with the lowest mode of C_{60}, Raman-active $h_g(1)$ mode. A possibility of photodegradation or photopolymerization of C_{60} molecules [6] was excluded by Raman observation after the HRS measurement. Presumably, it is due to the Franck–Condon type mechanism of the magnetic dipole two-photon upward and electric dipole one-photon downward transitions.

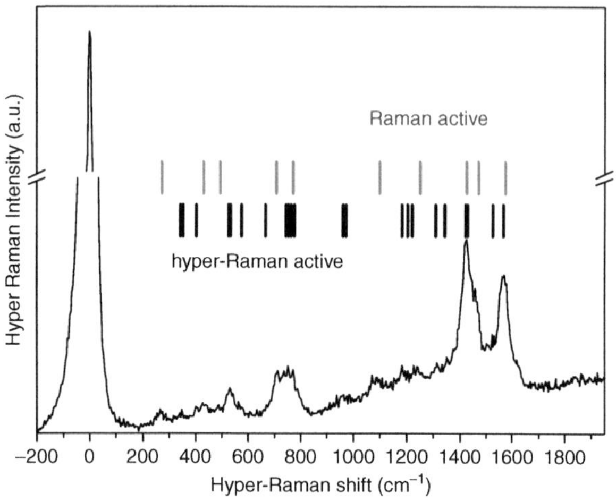

Fig. 5.6 Hyper-Raman spectrum of C_{60} microcrystals excited by 790-nm pulse radiation from the Ti-sapphire laser, with theoretically calculated hyper-Raman-active modes (*black bars*) and Raman-active modes (*gray bars*) [28]

As for the observation of the silent modes, some other techniques are available: inelastic neutron scattering, high-resolution electron energy loss, and low-temperature fluorescence spectroscopy in neon and argon matrices [38]. Compared to these techniques, HRS is more widely applicable under various experimental conditions because it does not require ultrahigh vacuum or low-temperature condition. Isotopic perturbations are also utilized for spectroscopic activation of silent modes, but electron-vibration coupling properties may be also affected. Therefore, the direct observation is meaningful for characterizing nanomaterials.

HRS Spectra of Single-Walled Carbon Nanotubes

Single-walled carbon nanotubes (SWNTs) can behave as semiconductors or metals, depending on their chirality. Therefore, vibrational spectroscopy is indispensable for characterizing chemical structures. Especially, resonance Raman spectroscopy is considered to be one of most powerful tools, which provides us rich information about nanotube diameters or defect densities [10, 27, 39]. However, much more detailed characterization is required for further development of carbon-based nanotechnology.

It has been theoretically predicted that achiral SWNTs ($m = n$ or $m = 0$) possess 8 RS-active and 3 IR-active modes, whereas chiral SWNTs ($0 < m \neq n$) have 14 RS-active and 6 IR-active modes [10]. However, there are only few experimental reports on the IR-active phonon modes because of the difficulty in detection [40, 41]. The difficulty is, in part, because SWNTs do not support a static dipole moment, and hence much weaker dynamic dipole moment has to be taken into account. Moreover, CNT samples usually contain carbon impurities that may induce similar IR absorption, making IR assignments much more difficult. Under these circumstances, resonance HRS spectroscopy is expected to be useful for characterizing SWNTs.

In the experiments, SWNTs with ~0.8 nm diameter (95 wt% grade) were purchased from Carbon Nanotechnologies, Inc. The quantity of 0.002 wt% of the nanotubes were dispersed into 2,2,3,3-tetrafluoro-1-propanol by sonication for 1 h; the solution was cast on a cover slip, and the solvent was allowed to evaporate.

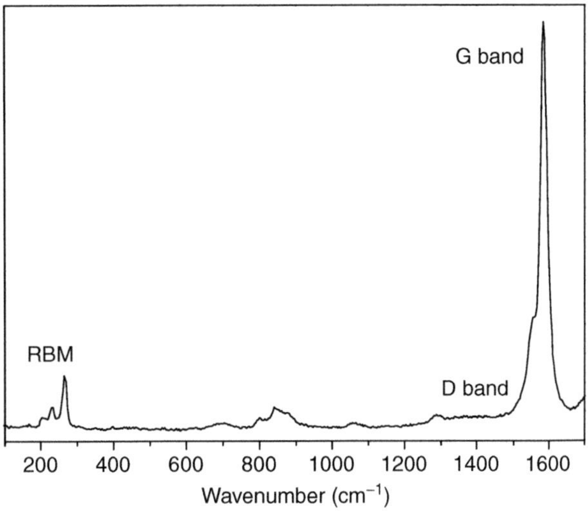

Fig. 5.7 Raman spectrum of SWNTs excited by 785-nm CW radiation. The observed peaks around 1,590, 1,300, and 260 cm^{-1} correspond to the tangential graphite-like modes (G-band), the disorder-induced modes (D-band), and the radial breathing modes (RBM), respectively [27]

The sample density was high enough to exhibit ensemble-averaged spectra of the nanotubes with diameter distribution.

Figure 5.7 shows a typical RS spectrum of the SWNTs measured by 785-nm CW laser excitation. Three peaks of radial breathing modes (RBM) were found to be in the range of 200–270 cm^{-1}, suggesting that semiconducting SWNTs with diameters of 0.9–1.2 nm were excited resonantly through the interband transitions of E_{22}^{S} [42]. The shape of tangential graphite-like modes (G-band) around 1,600 cm^{-1} also suggests that the semiconducting SWNTs were in resonance [40]. In addition to these first-order Raman bands, the disorder-induced band (D-band) was weakly observed around 1,300 cm^{-1}.

Among these three characteristic Raman bands, RBM should be HRS-inactive. Figure 5.8 shows a HRS spectrum of the SWNTs excited by the spectrally narrowed pulse radiation of 790 nm. In the frequency-doubled Stokes shift region, the SWNTs showed significant peaks around the D- and G-band regions, but no RBM in the spectrum, as expected. The highest peak at 0 cm^{-1} (395 nm) corresponds to hyper-Rayleigh scattering, which is probably due to chiral SWNTs [43].

It is known that the position of the Raman D-band shows excitation energy dependence [44]. The 790- and 395-nm excited D-bands indeed appeared around 1,300 and 1,390 cm^{-1}, respectively [27]. According to the D-band position in the HRS spectrum, both fundamental and harmonic resonances to intermediated states were thought to contribute to the enhancement of the HRS intensity. Conversely, the G-band shape, which is decomposed into G^{+} and G^{-}, suggested the contribution of harmonic resonances to the E_{11}^{M} transition. The observed HRS G^{-}-band was able to be fitted with Breit–Wigner–Fano (BWF) type line shape, which is characteristic in Raman spectra of metallic SWNTs [45].

The IR modes are theoretically expected to be around 680, 780, 870, 1,250, 1,360, and 1,500 cm^{-1} in addition to the D- and G-bands. Though Fig. 5.7 seems to display several tiny structures around the expected regions, the question is whether

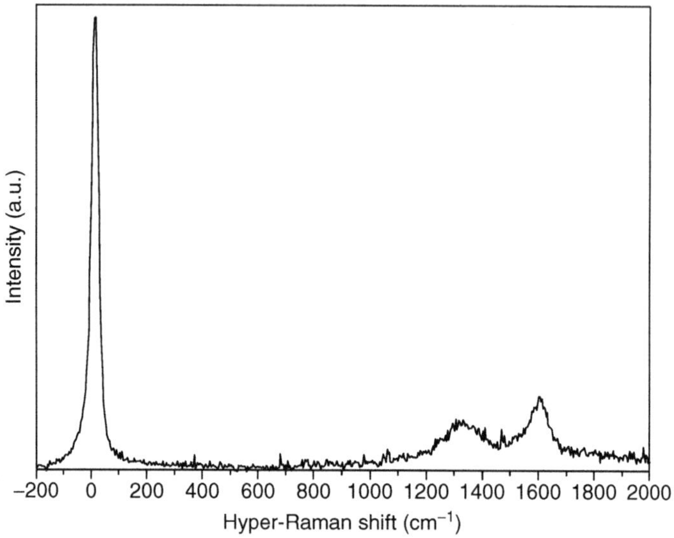

Fig. 5.8 Hyper-Raman scattering spectra of SWNTs excited by ps-pulsed radiation of 790 nm from the mode-locked Ti:sapphire laser. SWNTs exhibited the G-band and D-band peaks in addition to the hyper-Rayleigh scattered peak [27]

the observed HRS peaks are truly associated with IR modes. In this connection, we note that HRS and RS transitions are mutually exclusive only in centrosymmetric systems. In non-centrosymmetic systems such as chiral SWNTs, however, certain Raman modes can be HRS-active. In the case of resonance HRS, such Raman modes gain intensity via the Franck–Condon type mechanism (Ziegler's A-term) [24]. Nevertheless, one could assume that the present resonance HRS spectra included information on the IR-active modes of the SWNTs, because of the absence of RBM.

The previously reported IR spectrum of SWNTs exhibited many more peaks compared with the theoretical prediction, and then these peaks were tentatively assigned to second-order modes [40]. In the case of IR absorption, the existence of carbon impurities always impaired detection and assignment of intrinsic SWNT's IR-active modes so severely that careful purification of samples was essential. Conversely, resonant HRS spectroscopy can exclude such a possibility. By measuring chirality-controlled SWNTs, detailed information will be available from HRS measurements.

5.2 Recent Progress in Broadband CARS Microscopy

5.2.1 Three-Color Broadband CARS Microscopy

Conventional CARS spectroscopy is carried out using the two-color CARS process induced by continuous-wave or narrowband pulsed laser sources. On the basis of the progress in the broadband pulsed laser technology, broadband CARS methods are now drawing much attention [29–31]. One of most remarkable examples is a single-pulse CARS; a spectrally controlled broadband single pulse plays three roles: pump, Stokes, and probe beams [46]. In a practical use, however, two-pulse CARS spectroscopy is rather simple. Another advantage of the two-pulse CARS method is the capability to obtain temporal information of vibrational modes by observing delay-dependent CARS signals. Since the broadband CARS can obtain multiple vibrational modes simultaneously, the time-resolved broadband CARS can probe vibrational information in time-frequency domains. We will show in the next section our example of the two-dimensional observation of SWNT phonons by using the time-resolved broadband CARS [32].

5.2.2 2D-CARS Spectra of SWNTs

In semiconducting SWNTs, spectroscopic studies showed that the electronic excited states were excitonic in character with extraordinary high binding energy because of exciton wave function localization [47–49]. Due to such a strong electron-phonon coupling, SWNTs are expected to exhibit special electron transport properties such as superconductivity [7]. Moreover, ultrafast modulation of the electronic and optical properties could be possible via coherent control of SWNT-lattice vibration through the strong electron-phonon coupling [50]. To study interactions and dynamics of electrons and phonons in SWNTs is of great importance for understanding their unique properties.

In spontaneous resonance RS spectroscopy, which provides vibrational information in *frequency* domain, dynamical information is averaged out and convoluted in line shapes of the spectra. Hence, there is a practical limitation to extract information about phonon dynamics from frequency domain due to inhomogeneous broadening of line shapes. On the other hand, pump-probe spectroscopy is a useful technique to probe directly electron-phonon dynamics in *time* domain. In the typical pump-probe method like time-resolved electronic absorption, however, phonon dynamics appears as decaying oscillatory components; the phonon information is obtained by the Fourier transform. Therefore, the sensitivity and frequency resolution in this technique are relatively low especially for high-frequency phonons due to the limitation of the laser pulse duration. In the case of conventional time-resolved CARS [51], frequency-domain information is not available without scanning of excitation wavelength. Since decoherence processes could be accompanied with a frequency shift or modulation, there are both practical and fundamental considerations in time-frequency 2D spectroscopy for more detailed dynamical studies. In this sense, 2D *broadband* CARS spectroscopy is useful for studying dynamical behaviors. To demonstrate this technique, we focus on dynamics of G- and D-bands of SWNTs. Since the D-band intensity is related to defect density in graphitic structures, both as-produced and UV-irradiated SWNTs were utilized for the experiments.

Figure 5.9a shows a resonance anti-Stokes Raman spectrum of UV-irradiated SWNTs measured by using quasi-monochromatic excitation with the *narrowband* 790-nm beam (1.57 eV), containing information about the electron-phonon coupling in frequency domain. The characteristic Raman bands, G- and D-bands, were clearly observed. The relatively large D-band indicates the introduction of defects

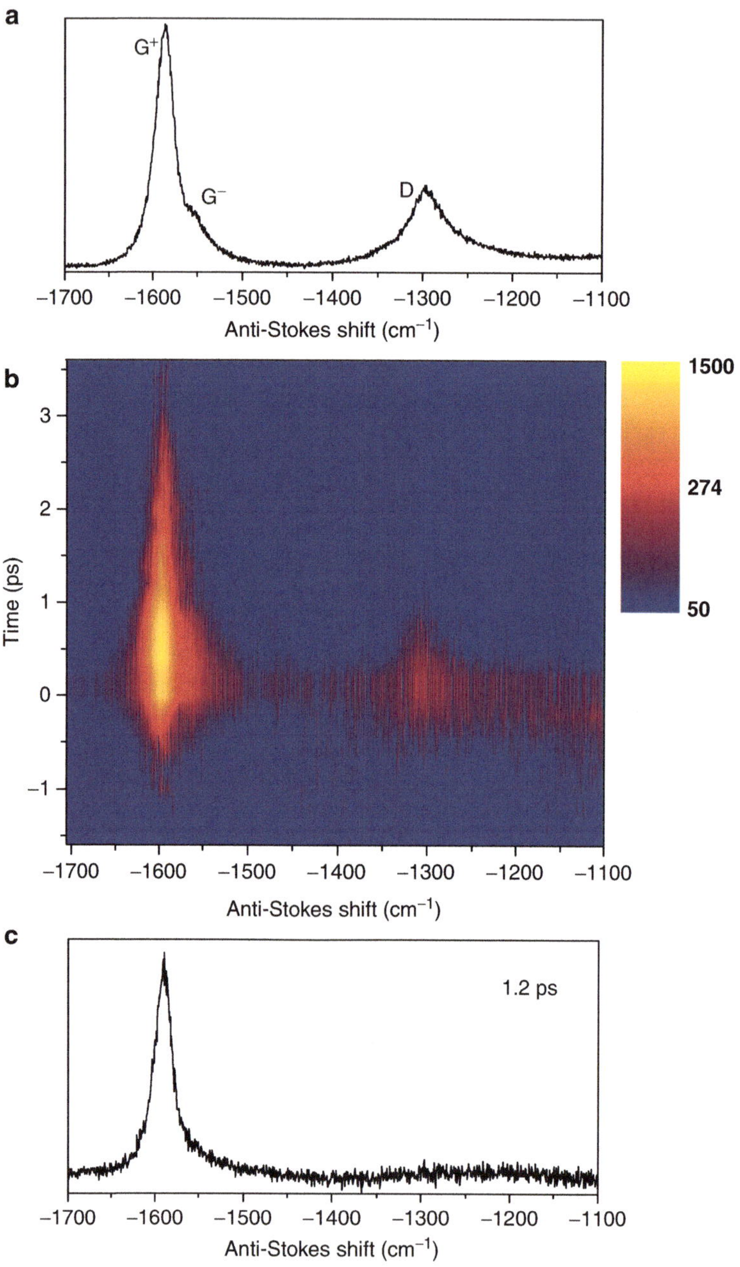

Fig. 5.9 (**a**) Anti-Stokes Raman spectrum of SWNTs excited by the 790-nm narrowband pulses. (**b**) Time-frequency 2D-CARS spectra of the SWNTs. (**c**) Time-resolved CARS spectrum of SWNTs at 0.84 ps [32]

in the SWNTs. Although the anti-Stokes Raman intensity is proportional to the population of optical phonons, it is often much larger than the value expected for thermal equilibrium in the case of SWNTs because Stokes and anti-Stokes Raman scattering can gain intensities from different nanotubes [52].

The 2D-CARS spectra for the same SWNTs are shown in Fig. 5.9b. To our knowledge, this is the first example of CARS spectra of SWNTs in a wide frequency region including G- and D-bands. (So far, there have been few reports on CARS emission from SWNTs even in conventional *narrowband* CARS at single frequency [53].) At the time delay of 0 ps, a nonresonant background signal was observed as a sum of various multiphoton processes [19]. After the background rise vanished, the induced G-band peak still remained with the picosecond-order decay, indicating coherent vibrations of the nanotube lattices. Conversely, the D-band vanished immediately after the excitation. Since CARS probes coherent vibrations generated by pairs of pump and Stokes photons, the observed D-band behavior suggests that the D-phonons has low coherency. Such incoherent D-band behavior is consistent with the double resonance mechanism [44]. In this mechanism, the zone-boundary D-phonons are allowed by the combination with elastic electron backscattering in the Brillouin zone because of the momentum conservation. In other words, the D-band resonance involves an incoherent electron scattering process mediated by a lattice defect. Conversely, the excitation of the zone-center G-phonons is not accompanied with incoherent electron-defect scattering, resulting in the coherent vibrations of the lattices. That is, the time-frequency 2D spectra directly showed the difference of the resonance mechanisms between G- and D-phonons.

Interestingly, the G-band shape in the time-resolved CARS spectrum is slightly different from those in the conventional Stokes and anti-Stokes Raman spectra; the G^- intensity was weaker in the time-resolved CARS spectrum. It is known that the G^- line shape in spontaneous Raman spectra is significantly affected by resonances to metallic SWNTs; a characteristic broadening is observed as a result of resonances between phonons and an electronic continuum (plasmons) [54]. Although the present excitation energy is mainly resonant to semiconducting SWNTs, a small contribution from metallic SWNTs is still expected [55]. By considering these circumstances, the observed difference in the G^- peak could be due to the fast relaxation of the G-phonons in metallic SWNTs. Indeed, a pump-probe study on RBM dynamics suggested that coherence was not generated in metallic SWNTs [56]. For further discussion, individual SWNT should be examined with time-resolved CARS spectroscopy.

Next, the relaxation process of G^+-phonons is analyzed by assuming that the inhomogeneous contribution is small and the response function can be modeled as an instantaneous rise at a delay time T_0 followed by an exponential decay with time constant $T_2/2$. Here, the total dephasing time T_2 is expressed by using the pure dephasing time τ_{ph} and the population lifetime T_1 as follows [51]:

$$\frac{2}{T_2} = \frac{2}{\tau_{ph}} + \frac{1}{T_1} \tag{5.7}$$

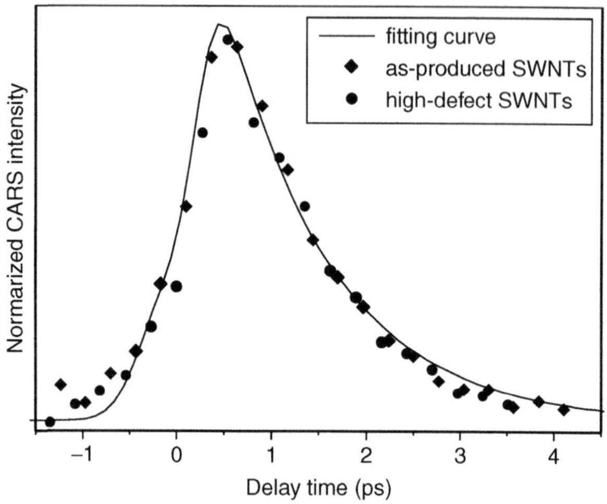

Fig. 5.10 CARS intensity change at 1,583 cm^{-1} in SWNTs as a function of delay time between the broadband pump and narrowband probe. The *solid curve* is obtained from a model that assumes instantaneous generation of hot G^+-phonons, but takes into account the finite instrumental time resolution and the exponential dephasing [32]

The experimental data was fitted, as shown in Fig. 5.10, to a convolution of this response function with the instrument response function. As the result, the decay time $T_2/2$ was estimated to be 1.1 ± 0.1 ps. Recently, the population lifetime T_1 of G-phonons was measured by *incoherent* time-resolved anti-Stokes Raman scattering and the lifetime was found to be 1.1–1.2 ps in semiconducting SWNTs [57]. Therefore, one can reasonably assume $\tau_{ph} >> T_1$ at room temperature. This result is consistent with the conventional Raman line width of semiconducting SWNTs [58]. The observed short lifetime of the G-phonons implies anharmonic mode coupling between G-phonons and RBM-phonons [59]. In fact, a frequency modulation of the G mode by the RBM has been reported, suggesting the anharmonic coupling between these vibrations [56].

6 Conclusion and Future Perspective

Nonlinear Raman spectroscopy has several unique features because of being the multiphoton optical processes. Although we focused on characterizing carbon nanomaterials in this chapter, these features must be generally useful for spectroscopy of nanomaterials. According to the selection rules, the second-order HRS can access IR-active modes and some of the silent modes. Since the scattered signals are detectable even in IR-opaque media, the HRS spectroscopy can be useful as an alternative of IR spectroscopy in biochemical imaging or electrochemical research. From the viewpoint of the optical diffraction limit, furthermore, nonlinear optical spectroscopy has much better spatial resolution than the linear one [15]. This is because the excitation efficiency of higher-order nonlinear effects is proportional to the high-order powers of the excitation light intensity, and the signals are dominantly generated from the focused spot. Indeed, CARS has become a popular tool for depth-resolved imaging in the fields of materials science and biochemical research.

The small cross sections in multiphoton processes are of course a weakness of nonlinear spectroscopy. Especially in microscopy, this problem becomes serious because of the small volume of a sample. By the use of the signal enhancement techniques, however, the disadvantage can be turned into an advantage of background-free selective measurements. For example, the combined use of HRS with the plasmonic enhancement provides us a chemical imaging with nanoscale spatial resolution; when a laser-illuminated metal tip is located adjacent to a sample surface, signal enhancement is locally induced near the tip. This spatial resolution is expected to overcome optical diffraction limit. Such tip-enhanced spectroscopy has already been reported in conventional CARS [15].

Time-resolved measurements are also one of the advantages in multiphoton spectroscopy. In the case of nanomaterials, physical and chemical properties are largely affected by interaction with surrounding environments. Therefore, the dynamical study should be useful for characterizing nanomaterials from the viewpoints of nanodevice applications.

Acknowledgments This research was partially supported by Grant-in-Aid for Scientific Research (A) (2006–2009, No. 18205016) from JSPS and that on Priority Area "Strong Photon-Molecule Coupling Fields" (No. 470) and Global COE program (Project No. B01: Catalysis as the Basis for Innovation in Materials Science) from MEXT, Japan.

References

1. Kroto HW, Heath JE, O'Brien SC, Curl RF, Smalley RE (1985) C60: buckminsterfullerene. Nature 318:1622
2. Iijima S (1991) Helical microtubules of graphitic carbon. Nature 354:56
3. Meth JS, Vanherzeele H, Wang Y (1992) Dispersion of the third-order optical nonlinearity of C60. A third-harmonic generation study. Chem Phys Lett 197:26
4. Hebard AF, Rosseinsky MJ, Haddon RC, Murphy DW, Glarum SH, Palstra TTM, Ramirez AP, Kortan AR (1991) Superconductivity at 18 K in potassium-doped C60. Nature 350:600
5. Allemand P-M, Khemani KC, Koch A, Wudl F, Holczer K, Donovan S, Grüner G, Thompson JD (1991) Organic molecular soft ferromagnetism in a FullereneC60. Science 253:301
6. Rao AM, Zhou P, Wang K-A, Hager GT, Holden JM, Wang Y, Lee W-T, Bi X-X, Eklund PC, Cornett DS, Duncan MA, Amster IJ (1993) Photo-induced polymerization of solid C60 films. Science 259:955
7. Tang ZK, Zhang L, Wang N, Zhang XX, Wen GH, Li GD, Wang JN, Chan CT, Sheng P (2001) Superconductivity in 4 Angstrom single-walled carbon nanotubes. Science 292:2462
8. Treacy MMJ, Ebbesen TW, Gibson JM (1996) Exceptionally high Young's modulus observed for individual carbon nanotubes. Nature 381:678
9. Tans SJ, Verschueren ARM, Dekker C (1998) Room-temperature transistor based on a single carbon nanotube. Nature 393:49
10. Barros EB, Jorio A, Samsonidze GG, Capaz RB, Filho AGS, Filho JM, Dresselhaus G, Dresselhaus MS (2006) Review on the symmetry-related properties of carbon nanotubes. Phys Rep 431:261
11. Maeda Y, Kimura S, Kanda M, Hirashima Y, Hasegawa T, Wakahara T, Lian Y, Nakahodo T, Tsuchiya T, Akasaka T, Lu J, Zhang X, Gao Z, Yu Y, Nagase S, Kazaoui S, Minami N, Shimizu T, Tokumoto H, Saito R (2005) Large-scale separation of metallic and semiconducting single-walled carbon nanotubes. J Am Chem Soc 127:10287

12. Zheng M, Jagota A, Strano MS, Santos AP, Barone P, Grace Chou S, Diner BA, Dresselhaus MS, Mclean RS, Bibiana Onoa G, Samsonidze GG, Semke ED, Usrey M, Walls DJ (2003) Structure-based carbon nanotube sorting by sequence-dependent DNA assembly. Science 302:1545
13. Shen YR (1984) The principles of nonlinear optics. Wiley, New York
14. Shen YR (2000) Surface nonlinear optics: a historical perspective. IEEE J Sel Top Quant 6:1375
15. Hayazawa N, Ichimura T, Ikeda K, Kawata S (2009) In: Yakovlev V (ed) Biochemical applications of nonlinear optical spectroscopy. CRC Press, Boca Raton, p 239
16. Denisov VN, Mavrin BN, Podobedov VB (1987) Hyper-Raman scattering by vibrational excitations in crystals, glasses and liquids. Phys Rep 151:1
17. Cyvin SJ, Rauch JE, Decius JC (1965) Theory of hyper-Raman effects (nonlinear inelastic scattering): selection rules and depolarization ratios for the second-order polarizability. J Chem Phys 43:4083
18. Christie JH, Lockwood DJ (1971) Selection rules for three- and four-photon Raman interactions. J Chem Phys 54:1141
19. Shimada R, Kano H, Hamaguchi H (2006) Hyper-Raman microspectroscopy: a new approach to completing vibrational spectral and imaging information under a microscope. Opt Lett 31:320
20. Kneipp J, Kneip H, Kneipp K (2006) Two-photon vibrational spectroscopy for biosciences based on surface-enhanced hyper-Raman scattering. Proc Natl Acad Sci USA 103:17149
21. Leng W, Woo HY, Vak D, Bazan GC, Kelly AM (2006) Surface-enhanced resonance Raman and hyper-Raman spectroscopy of water-soluble substituted stilbene and distyrylbenzene chromophores. J Raman Spectrosc 37:132
22. Shoute LCT, Bartholomewa GP, Bazan GC, Kelley AM (2005) Resonance hyper-Raman excitation profiles of a donor-acceptor substituted distyrylbenzene: one-photon and two-photon states. J Chem Phys 122:184508
23. Mizuno M, Hamaguchi H, Tahara T (2002) Observation of resonance hyper-Raman scattering of all-trans-retinal. J Phys Chem A 106:3599
24. Chung YC, Ziegler LD (1988) The vibronic theory of resonance hyper-Raman scattering. J Chem Phys 88:7287
25. Golab JT, Sprague JR, Carron KT, Schatz GC, Van Duyne RP (1988) A surface enhanced hyper-Raman scattering study of pyridine adsorbed onto silver: experiment and theory. J Chem Phys 88:7942
26. Ikeda K, Takase M, Sawai Y, Nabika H, Murakoshi K, Uosaki K (2007) Hyper-Raman scattering enhanced by anisotropic dimer-plasmons on artificial nanostructures. J Chem Phys 127:111103
27. Ikeda K, Saito Y, Hayazawa N, Kawata S, Uosaki K (2007) Resonant hyper-Raman scattering from carbon nanotubes. Chem Phys Lett 438:109
28. Ikeda K, Uosaki K (2008) Resonance hyper-Raman scattering of fullerene C60 microcrystals. J Phys Chem A 112:790
29. Lee YJ, Liu Y, Cicerone MT (2007) Characterization of three-color CARS in a two-pulse broadband CARS spectrum. Opt Lett 32:3370
30. Paulsen HN, Hilligsoe KM, Thogersen J, Keiding SR, Larsen JJ (2003) Coherent anti-stokes Raman scattering microscopy with a photonic crystal fiber based light source. Opt Lett 28:1123
31. Furusawa K, Hayazawa N, Kawata S (2010) Two-beam multiplexed CARS based on a broadband oscillator. J Raman Spectrosc 41:840
32. Ikeda K, Uosaki K (2009) Coherent phonon dynamics in single-walled carbon nanotubes studied by time-frequency two-dimensional coherent anti-stokes Raman scattering spectroscopy. Nano Lett 9:1378
33. Itoh T, Yoshikawa H, Yoshida K, Biju Y, Ishikawa M (2009) Evaluation of electromagnetic enhancement of surface enhanced hyper Raman scattering using plasmonic properties of binary active sites in single Ag nanoaggregates. J Chem Phys 130:214706

34. Shimada R, Kano H, Hamaguchi H (2008) Intensity enhancement and selective detection of proximate solvent molecules by molecular near-field effect in resonance hyper-Raman scattering. J Chem Phys 129:024505
35. Weeks DE, Harter WG (1989) Rotation-vibration spectra of icosahedral molecules. II. Icosahedral symmetry, vibrational eigenfrequencies, and normal modes of buckminsterfullerene. J Chem Phys 90:4744
36. Yannoni CS, Johnson RD, Meijer G, Bethune DS, Salem JR (1991) Carbon-13 NMR study of the C60 cluster in the solid state: molecular motion and carbon chemical shift anisotropy. J Phys Chem 95:9
37. Mene'ndez J, Page JB (2000) In: Cardona M, Güntherodt G (eds) Light scattering in solids VIII. Springer, Berlin
38. Banfi GP, Fortusini D, Bellini M, Milani P (1997) Wave-dispersed two-photon absorption of C60. Phys Rev B 56:R10075
39. Rao AM, Richter E, Bandow S, Chase B, Eklund PC, Williams KW, Menon M, Subbaswarry KR, Thess A, Smalley RE, Dresselhaus G, Dresselhaus MS (1997) Diameter-selective Raman scattering from vibrational modes in carbon nanotubes. Science 275:187
40. Kim UJ, Liu XM, Furtado CA, Chen G, Saito R, Jiang J, Dresselhaus MS, Eklund PC (2005) Infrared-active vibrational modes of single-walled carbon nanotubes. Phys Rev Lett 95:157402
41. Kuhlmann U, Jantoljak H, Pfänder N, Bernier P, Journet C, Thomsen C (1998) Infrared active phonons in single-walled carbon nanotubes. Chem Phys Lett 294:237
42. Kataura H, Kumazawa Y, Maniwa Y, Umezu I, Suzuki S, Ohtsuka Y, Achiba Y (1999) Optical properties of single-wall carbon nanotubes. Synth Met 103:2555
43. De Dominies L, Fantoni R (2006) Effects of electrons statistic on carbon nanotubes hyperpolarizability frequency dependence determined with sum over states method. J Raman Spectrosc 37:669
44. Saito R, Jorio A, Souza Filho AG, Dresselhaus G, Dresselhaus MS, Pimenta MA (2002) Probing phonon dispersion relations of graphite by double resonance Raman scattering. Phys Rev Lett 88:027401
45. Brown SDM, Jorio A, Corio P, Dresselhaus MS, Dresselhaus G, Saito R, Kneipp K (2001) Origin of the Breit-Wigner-Fano lineshape of the tangential G-band feature of metallic carbon nanotubes. Phys Rev B 63:155414
46. Dudovich N, Oron D, Silberberg Y (2002) Single-pulse coherently controlled nonlinear Raman spectroscopy and microscopy. Nature 418:512
47. Wang F, Dukovic G, Brus LE, Heinz TF (2005) The optical resonances in carbon nanotubes arise from excitons. Science 308:838
48. Spataru CD, Ismail-Beigi S, Benedict LX, Louie SG (2004) Excitonic effects and optical spectra of single-walled carbon nanotubes. Phys Rev Lett 92:077402
49. Lüer L, Hoseinkhani S, Polli D, Crochet J, Hertel T, Lanzani G (2009) Size and mobility of excitons in (6, 5) carbon nanotubes. Nat Phys 5:54
50. Lim Y-S, Yee K-J, Kim J-H, Ha'roz EH, Shaver J, Kono J, Doom SK, Hauge RH, Smalley RE (2006) Coherent lattice vibrations in single-walled carbon nanotubes. Nano Lett 6:2696
51. Fayer MD (2001) Ultrafast infrared and Raman spectroscopy. Marcel Dekker, New York
52. Brown SDM, Corio P, Marucci A, Dresselhaus MS (2000) Anti-stokes Raman spectra of single-walled carbon nanotubes. Phys Rev B 61:R5137
53. Baltog I, Baibarac M, Lefrant S (2005) Coherent anti-stokes Raman scattering on single-walled carbon nanotubes and copper phthalocyanine thin films excited through surface plasmons. J Opt A Pure Appl Opt 7:632
54. Pimenta MA, Marucci A, Empedocles SA, Bawendi MG, Hanlon EB, Rao AM, Eklund PC, Smalley RE, Dresselhaus G, Dresselhaus MS (1998) Raman modes of metallic carbon nanotubes. Phys Rev B 58:R16016
55. Souza Filho AG, Jorio A, Samsonidze GG, Dresselhaus G, Saito R, Dresselhaus MS (2003) Raman spectroscopy for probing chemically/physically induced phenomena in carbon nanotubes. Nanotechnology 14:1130

56. Gambetta A, Manzoni C, Menna E, Meneghetti M, Cerullo G, Lanzani G, Tretiak S, Piryatinski A, Saxena A, Martin RL, Bishop AR (2006) Nat Phys 2:515
57. Kang K, Ozel T, Cahill DG, Shim M (2008) Nano Lett 8:4642
58. Jorio A, Fantini C, Dantas MSS, Pimenta MA, Souza Filho AG, Samsonidze GG, Brar VW, Dresselhaus G, Dresselhaus MS, Swan AK, Ünlü MS, Goldberg BB, Saito R (2002) Phys Rev B 66:115411
59. Song D, Wang F, Dukovic G, Zheng M, Semke ED, Brus LE, Heinz TF (2008) Phys Rev Lett 100:225503

Ag/Carbon Nanotubes for Surface-Enhanced Raman Scattering

6

Han-Wei Chang, Ping-Chieh Hsu, and Yu-Chen Tsai

1 Definition of the Topic

The surface-enhanced Raman scattering (SERS)-active substrates were prepared by electrodeposition of Ag nanoparticles in multiwalled carbon nanotubes (MWCNTs)-based nanocomposites for SERS sensor application.

2 Overview

The novel SERS-active substrates were prepared by electrodeposition of Ag nanoparticles in the MWCNTs-based nanocomposites. The formation of Ag-MWCNTs nanocomposite was characterized by scanning electron microscopy and energy dispersive X-ray spectroscopy. The application of the Ag-MWCNTs nanocomposite in SERS was investigated by using rhodamine 6G (R6G). The present methodology demonstrates that the Ag-MWCNTs nanocomposite is suitable for SERS sensor.

3 Introduction

For the past two decades, SERS has been widely studied due to its high sensitivity and specificity [1–5]. It is a powerful tool for determining vibrational spectra of probe molecules on roughened metal substrates [6, 7]. In general, there are two simultaneously operative mechanisms to describe the predominantly SERS effect:

H.-W. Chang • P.-C. Hsu • Y.-C. Tsai (✉)
Department of Chemical Engineering, National Chung Hsing University, 250 Kuo Kuang Road, Taichung, Taiwan

Challa S. S. R. Kumar (ed.), *Raman Spectroscopy for Nanomaterials Characterization*,
DOI 10.1007/978-3-642-20620-7_6, © Springer-Verlag Berlin Heidelberg 2012

electromagnetic enhancement mechanism (EM) and chemical enhancement mechanism (CHEM) [8]. The EM mechanism is based on the amplified electromagnetic fields generated upon excitation of the localized surface plasmon resonance of roughness nanostructures surface [9–12] and the CHEM enhancement is based on the polarizability of the molecule induced from electronic resonance/charge transfer between probe molecules and roughened metal substrates. Various metallic materials such as Ag [13], Au [14], and Cu [15] have been developed and investigated as SERS-active roughness substrates. Accordingly, it is of special interest to fabricate the SERS substrates from rough surfaces to nanostructures of metals materials. Recently, many reports have shown the deposition of various metal nanoparticles onto the surface of carbon nanotubes (CNTs) for SERS active substrates [16–18]. The combination of metal nanoparticles and CNTs was effectively utilized to enhance the SERS effect [11, 17, 18]. Currently, many techniques have been developed to obtain SERS-active substrates, including Langmuir–Blodgett (LB) [19, 20], self-assembly [19, 21, 22], lithography [9, 23], and electrodeposition [17, 18, 24].

3.1 SERS

Raman spectroscopy is commonly used as a detection method for noninvasive identification and molecule sensing. It offers many advantages, such as fluorescent counterparts, monitoring molecule excitement at any wavelength, and narrow peak width, and it can represent a chemical fingerprint between the molecular composition and structure [25]. However, conventional Raman scattering have some limits that include extremely small cross section and the interference of fluorescence [26, 27], which leads to the need for intense laser excitation sources and a large number of molecules to give adequate signals.

In 1974, Fleischman et al. found that Raman scattering of adsorbed pyridine species on Ag electrodes could be enhanced by several orders of magnitude [28]. This phenomenon is called SERS. The SERS effect is observed by analyte molecules adsorbed upon roughened metallic surfaces, and Raman scattering intensity can be enhanced for several orders of magnitude greater than in solution or on an ordinary smooth metal surface. The magnitude of the enhancement in Raman scattering depend on the chemical nature of the adsorbed molecules and the roughness of the surface [7]. Furthermore, the SERS spectroscopy shows the possibility of overcoming the relatively low sensitivity of Raman spectroscopy and the suppression of fluorescence. SERS has continued to attract much attention for its potential applications in electrochemistry [29], optics [7], catalysis [30], and biological sensing devices [5, 11, 35].

There are two types of SERS mechanisms, which are responsible for the observation of the SERS enhancement [8]; one type is the long-range EM effect and the other is the short-range CHEM effect. The EM effect is believed to be the result of localized surface plasmon resonance electric fields (hot spot) set up onto the roughened metallic surfaces [9, 10, 31]. The probe molecules residing within these hot spots will be strongly excited and subsequently emit amplified Raman signals. The Raman enhancement factor attained in EM effect was typically in the range of 10^6–10^{10}. The CHEM results from the charge-transfer excitation of analyte molecules [32]. This normally brings about an additional SERS enhancement factor in the order of 10^2. The two mechanisms operating simultaneously can be observed in SERS, but it is widely believed that the EM effect plays a greater role in SERS enhancement than the CHEM effect.

Further enhancement of the SERS can be achieved through precise control over the parameters at the metal particle size scale [10] Most SERS-active substrates were made from pure metallic nanostructures such as metal nanoparticles [33–35], metal particle arrays [5], roughened metal surfaces [36], or a combination with metal nanostructures and other nanomaterials [17, 18, 29, 37–39]. Recently, many strategies have shown the adsorbation of molecules on the surface of Ag and Au substrates for SERS applications [40]. SERS-active Ag nanostructures substrates are required to satisfy certain conditions with good reproducibility and stability [39]. For this reason, it is indispensable to develop and optimize the methods to prepare the SERS-active Ag substrates [41].

3.2 Methods to Prepare SERS-Active Ag Substrates

In recent years, there are many effective methods for the preparation of SERS-active Ag substrates. These include Langmuir–Blodgett (LB), self-assembly, lithography, and electrodeposition.

3.2.1 Langmuir–Blodgett (LB)

The LB techniques are a powerful tool used to assemble large-scale monolayers of hydrophobic metallic nanoparticles on a water surface [42]. The advantages of the LB techniques are the good reproducibility of the metallic nanoparticles substrates and the ability to disperse the particles on the surface of water [43]. Recently, the LB methods have received interest as a process for the fabrication of thin films SERS substrates.

Lu's group has developed a novel technique for synthesizing high-density nanoparticle thin film by LB techniques on a water surface and transferring the particle monolayer to a polymer membrane as SERS substrates. They found that the optical response of the Ag nanoparticles on polymer film are a dependence of temperature and demonstrated the capability to create highly ordered particle film by heating the film to temperature above the polymer's low critical solution temperature [44]. The high-density particle film prepared via LB techniques can be easily obtained, and the particle distance can be optimized to approach the strongest coupling between adjacent particles and match the plasmon resonance wavelength to the laser excitation wavelength [45]. These properties make the high-density particle film for use in SERS and for developing a highly specific label-free biomolecular sensing. The high-density Ag particle film has a strong affinity to R6G, and it is able to enhance the scattering signal and demonstrates the tunable plasmon coupling between high-density nanoparticles [46].

Tao's group demonstrated a good SERS substrate using aligned monolayers of Ag nanowires by LB techniques and found that a unique feature can arise when synthesizing the Ag nanowires to sharp vertices, noncircular pentagonal cross sections, and interwire coupling. By using aligned monolayers of Ag nanowires as SERS substrates, the detections of thiol, 2, 4-dinitrotoluene, and R6G on Ag nanowires substrates have a high electromagnetic field enhancement factors (2×10^5 for thiol and 2, 4-dinitrotoluene, 2×10^9 for R6G). The enhancement factors change is caused by the specific interaction of the nanowires (sharp vertices, noncircular pentagonal cross sections, and interwire coupling) that result in larger field enhancement factors, offering higher sensitivity and enabling SERS experiments with a broad selection of excitation sources [47].

Chowdhury et al. demonstrated Rhodamine 123 (Rh 123) monolayer organized on the Ag island film by the LB techniques. Results showed that the Rh 123 molecules in a monolayer on the Ag island film offered a high SERS activity. The huge SERS enhancement intensity confirms the charge transfer interaction between the Rh 123 and the Ag island film by physisorption of Rh 123 on the Ag island film [48].

3.2.2 Self-assembly

The self-assembly technique has attracted much attention since they were observed by Decher in 1991 [49]. Self-assembly is the fundamental principle that provides the precise control of the resulting assemblies and the thickness of an individual layer on the nanometer scale by variation in the bulk concentration of the metal colloids suspension, deposition time, pH, and transport conditions [50]. Recently, the functionalization of metal nanoparticles has opened up new opportunities for the construction of nanostructured self-assembly films to fabricate novel SERS-active Ag substrates.

Peng et al. have applied λ-DNA networks-Ag nanoparticles hybrid films as highly rough and stable SERS-active substrates to develop sensors. The λ-DNA networks-Ag nanoparticles composite were self-assembled from λ-DNA networks onto cetyltrimethylammonium bromide-capped Ag nanoparticles coated on the charged mica surface until the desired number of bilayers was reached. The self-assembled techniques have the capability to create a large amount of Ag particles and the higher surface roughness of hybrid films with the increase in the number of bilayers. The rough metallic hybrid films have been used for the sensitive analytical tool. A strong SERS effect could be easily explored when methylene blue adsorbed on these hybrid films [50].

3.2.3 Lithography

Lithography techniques such as nanosphere lithography and electron-beam lithography are ideal methods to fabricate the reproducible SERS substrates [51]. They have been shown to improve the substrates performance by controlling the size and shape of colloidal nanoparticles and interparticle spacing [52]. Lithography techniques are simple to implement and of low cost. These advantages make them suitable platforms for the fabrication of SERS-active substrate.

Jesus et al. indicated that the nanolithography fabrication of metal–polymer substrates can be applied to detect R6G and 1, 10-phenanthroline with SERS. The results confirm that the nanolithography technique not only reduces cost, but also creates SERS substrates with very high sensitivity that largely improve the quantitative detection of trace amounts of an analyte [53].

It has been shown that [54] a pattern of gold nanograin aggregate structures with diameters ranging between 80 and 100 nm with interstitial gap of 10–30 nm were fabricated by an e-beam lithography. The results demonstrated that reproducible Raman signals can be achieved when the concentration of myoglobin protein is attomole (10^{-18}). The use of e-beam lithographically for the preparation of gold nanograin aggregate structures provides a novel approach for the optimization of reproducible SERS substrates.

3.2.4 Electrodeposition

The electrodeposition technology has proven to be the least expensive, effective, and readily adoptable method to deposit Ag substrates for reliable SERS substrates with good reproducibility. It allows the preparation of nanostructure patterns by controlling the amount of composition, deposition time, temperature, and applied potential. The SERS substrates prepared by electrodeposition were a good candidate for the fabrication of a reproducible substrate. In principle, most of the metals including Au [55], Cu [56], and Ag [57–59] can be electrodeposited from aqueous solutions.

Recently, Sun et al. [60] proposed to fabricate Ag nanostructures by electrochemical deposition of Ag in anodic aluminum oxide templates with each pore diameter of 100 nm. The morphology of Ag substrates was characterized by FESEM. The length of the Ag nanowires is estimated to be about 2 μm from the FESEM images. In addition, the SERS enhancement factor calculated for Ag nanowires SERS substrates is approximately 10^6 in magnitude. The Ag nanowire arrays are expected to have important applications in the development of high sensitivity SERS-based substrates.

Tourwe's group developed the SERS-active electrodes by electrodeposition. During the electrodeposition process, the Ag electrode has been grown on the diamond paste electrode through a simple galvanic reaction between an aqueous solution which contains 0.5 M $LiNO_3$ and 5×10^{-4} M $AgNO_3$. The experimental results revealed that the reproducibility of SERS-active electrodes allows performing quantitative measurements at various probe molecule concentrations.

Luo et al. [61] demonstrated a simple method for the preparation of SERS active Ag nanostructures substrates by deposition of Ag nanoparticles into the designed Si holes. The morphologies of the Ag nanostructures were observed with SEM. The diameters of the Ag nanoparticles were found to be 40–60 nm. With increasing deposition time, flower-like Ag nanostructure commenced crystallization to form near the edge of the bottom surface of the Si microholes. These Ag nanostructures exhibited strong SERS enhancement, which provided an excellent platform for monitoring the R6G molecules by SERS technology [62].

3.3 CNTs-Metal Nanoparticles Composites

CNTs have attracted much attention due to their high chemical stability, high mechanical strength, large surface area, high electrical conductivity, and thermal stability [63]. Recently, many reports have been shown the deposition of various metal nanoparticles onto the surface of CNTs, such as platinum [64], plumbum [65], gold [16, 66], and Ag [16, 67]. Metal nanomaterials have received considerable interest because of their particular optical, electronic, and magnetic properties. The unique characteristics of nanoparticles are usually prepared with the control of size and shape. The introduction of metal nanoparticles to CNTs is mainly stimulated by their characteristic interfacial interactions to produce a novel nanocomposite that combines the properties of two nanoscale materials [16]. The CNTs-metal nanoparticles composites have been used in a wide range of applications, such as sensors [70] and SERS [16, 66]. In SERS, the localized surface plasmon resonance achieved by metal nanoparticles provides a strong and highly confined electromagnetic field. Thus, several research groups have successfully combined metal nanoparticles on the surface of CNTs and have attracted attention in SERS. Kumar et al. [68] compared the effect of SERS from before and after depositing Ag nanoparticles on CNTs. The substrate of CNTs-Ag nanoparticles showed a strong SERS enhancement [16]. The average intensity increased almost 134,000-fold when Ag nanoparticles were deposited on CNTs. Guo et al. reported on a simple method for preparing effective CNTs/TiO_2/Ag SERS substrates. The SERS effect was investigated using the molecular probe 4-aminothiophenol adsorbed on the surface of a CNTs/TiO_2/Ag. It was found that stronger SERS activity was obtained for CNTs/TiO_2/Ag substrates than the normal Raman spectrum of solid 4-ATP [69].

In order to prepare CNTs-metal nanoparticles composites, the CNTs need to be well dispersed. The poor dispersion of CNTs in liquid media has limited the research progress in many fields. Recently, polymers are potential choices to disperse CNTs into aqueous media for the noncovalent method. Several groups have reported that the use of noncovalent method provided a useful and effective method to improve the dispersion of CNTs in a polymer liquid media. Zhang's groups [70] demonstrated a simple method for the dispersion of single-walled carbon nanotubes (SWCNTs) in Nafion solution. The method could significantly improve the dispersion of SWCNTs. The Nafion aqueous solution–dispersed SWCNTs provide many promising and important applications. Zhang et al. [71] focused on the modification of MWCNTs by noncovalent method. They showed a new method to enhance the dispersion of MWCNTs in chitosan (CHIT) by the effect of pH. Chitosan provided a good biocompatibility and maintained the property of MWCNTs. A good biocompatibility makes the CHIT/MWCNTs system useful for various biomedical applications to improve the physical and chemical properties of materials.

This work is to introduce the SERS-active surface prepared by electrodeposition of Ag nanoparticles in MWCNTs for SERS chemical sensor application. R6G, a dye molecule, is chosen as a probe molecule because it has been well investigated by SERS.

4 Experimental and Instrumental Methodology

4.1 Preparation of Ag-MWCNTs-Nafion Nanocomposite

The ITO substrates were rinsed thoroughly with the aid of ultrasonic agitation in isopropyl alcohol, acetone, and distilled water for 15 min, respectively. Then, they were dried with a stream of high purity nitrogen. The MWCNTs-Nafion solution was made by adding 10 mg MWCNTs in the 1 mL 0.5 wt% Nafion solution and ultrasonic agitation for 1 h to form a homogeneous solution. The prepared solution was spin-coated on a sheet of ITO substrate. The solvent was allowed to evaporate at room temperature in the air. Then, the MWCNTs-Nafion-coated ITO substrate was immersed in 0.1 M KNO_3 solution containing 10 mM $AgNO_3$. The electrochemical deposition of the Ag nanoparticles was conducted at −0.3 V versus Ag/AgCl (3 M KCl) at desired charges. Finally, the Ag-MWCNTs-Nafion-modified ITO substrates were obtained.

4.2 Preparation of Ag-MWCNTs-Alumina-Coated-Silica (ACS) Nanocomposite

A MWCNTs-ACS solution was made by dispersing 15 mg MWCNTs in a 1 mL of 1 wt% ACS (DuPont) aqueous solution (pH 2) with the aid of ultrasonic agitation for 2 h. The prepared MWCNTs-ACS homogeneous solution was spin-coated on a sheet of ITO substrate. The solvent was allowed to evaporate at room temperature in the air. Electrodeposition of Ag nanoparticles on MWCNTs-ACS-coated ITO substrate was carried out in an electrochemical cell that consisted of 0.1 M KNO_3 solution containing 10 mM $AgNO_3$ and a constant potential of −0.3 V versus Ag/AgCl (3 M KCl) was applied under stirring condition.

5 Key Research Findings

5.1 Ag-MWCNTs-Nafion Nanocomposite for SERS

SERS-active substrates were prepared by attaching Ag nanoparticles in the MWCNTs-Nafion for SERS chemical sensor application. The preparation of Ag nanoparticles onto the surface of CNTs by one-step electrochemical method is considered as an effective procedure when compared with others which are time consuming and the impurity might be involved during the preparation. The FESEM images of MWCNTs-Nafion nanocomposite before and after electrodeposition of Ag nanoparticles are shown in Fig. 6.1a, b, respectively. Figure 6.1b reveals that the Ag nanoparticles, with an average diameter of approximately 100 nm in the MWCNTs-Nafion nanocomposite and some nanoparticles, aggregated to form large Ag particles. The Ag particle size greater than 100 nm is efficient for SERS at 633 nm excitation [72]. The EDS was used to confirm the formation of Ag in

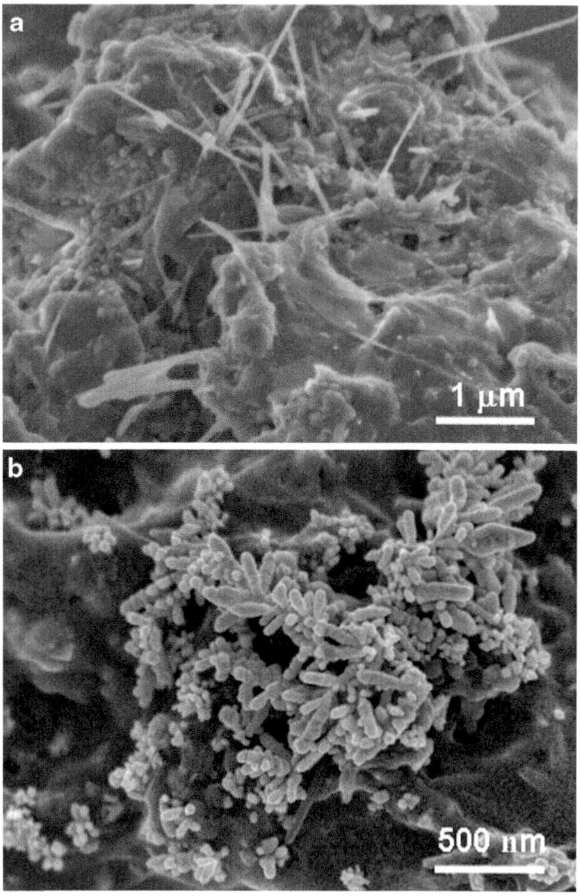

Fig. 6.1 FESEM images of MWCNTs-Nafion nanocomposite before (**a**) and after (**b**) electrochemical deposition of Ag nanoparticles

MWCNTs-Nafion nanocomposite. The EDS patterns of MWCNTs-Nafion nanocomposite before and after electrochemical deposition of Ag nanoparticles are shown in Fig. 6.2a, b, respectively. It indicates that C is the major element which is derived from MWCNTs and Nafion. The Ag element can be identified in Fig. 6.2b after the electrochemical deposition of Ag nanoparticles. The prepared Ag-MWCNTs-Nafion nanocomposite was used to detect the typical SERS active probe molecule such as R6G. For the determination, the Ag-MWCNTs-Nafion-nanocomposite-modified ITO substrates were placed on the bottom of a liquid cell with 35 mL aqueous solution. Two control experiments at the Ag-coated and Ag-carbon black (CB)-Nafion-coated ITO substrates were performed for comparison. The Raman spectra of 10^{-4} M R6G in aqueous solution at the surfaces of MWCNTs-Nafion, Ag, Ag-CB-Nafion, and Ag-MWCNTs-Nafion-modified ITO substrates are shown in Fig. 6.3a–d, respectively. The peaks at 1,186, 1,310, 1,362, 1,509, and 1,650 cm^{-1} are observed at all the surfaces. The strong peaks at 1,362, 1,509, and 1,650 cm^{-1} in the SERS spectra can be assigned to aromatic C–C

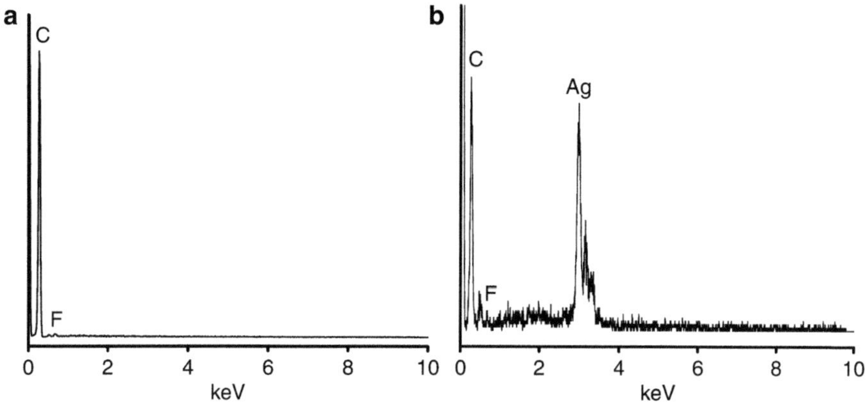

Fig. 6.2 EDS patterns of MWCNTs-Nafion nanocomposite before (**a**) and after (**b**) electrochemical deposition of Ag nanoparticles (Reprinted with permission from Ref. [18]. © 2009 Elsevier)

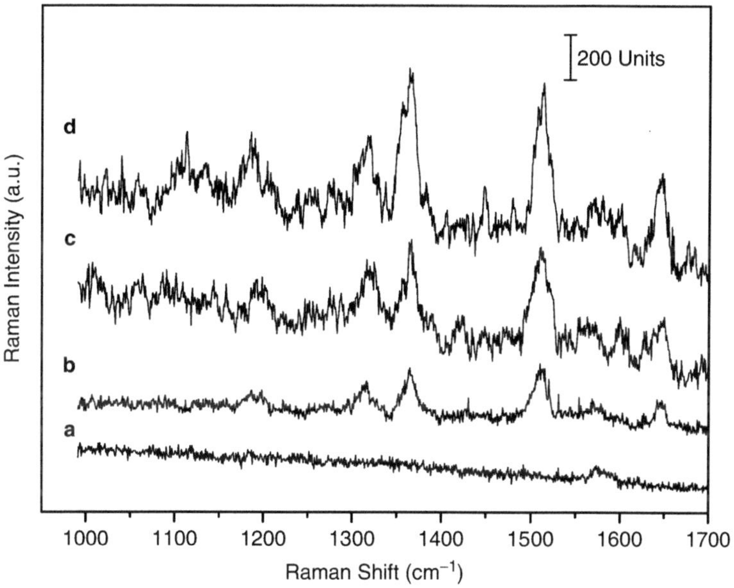

Fig. 6.3 Raman spectra of 10^{-4} M R6G in aqueous solution on the surfaces of MWCNTs-Nafion (**a**), Ag (**b**), Ag-CB-Nafion (**c**), and Ag-MWCNTs-Nafion (**d**) modified ITO substrates (Reprinted with permission from Ref. [18]. © 2009 Elsevier)

stretching vibrations and indicated that the R6G were adsorbed on the surface of Ag-MWCNTs-Nafion. The characteristic Raman peaks in Fig. 6.3 were in agreement with other works [47, 72–74]. The SERS intensity of R6G obtained at the surface of Ag-MWCNTs-Nafion is higher than those at the surfaces of Ag and Ag-CB-Nafion. The results show that the three-dimensional nanostructure of

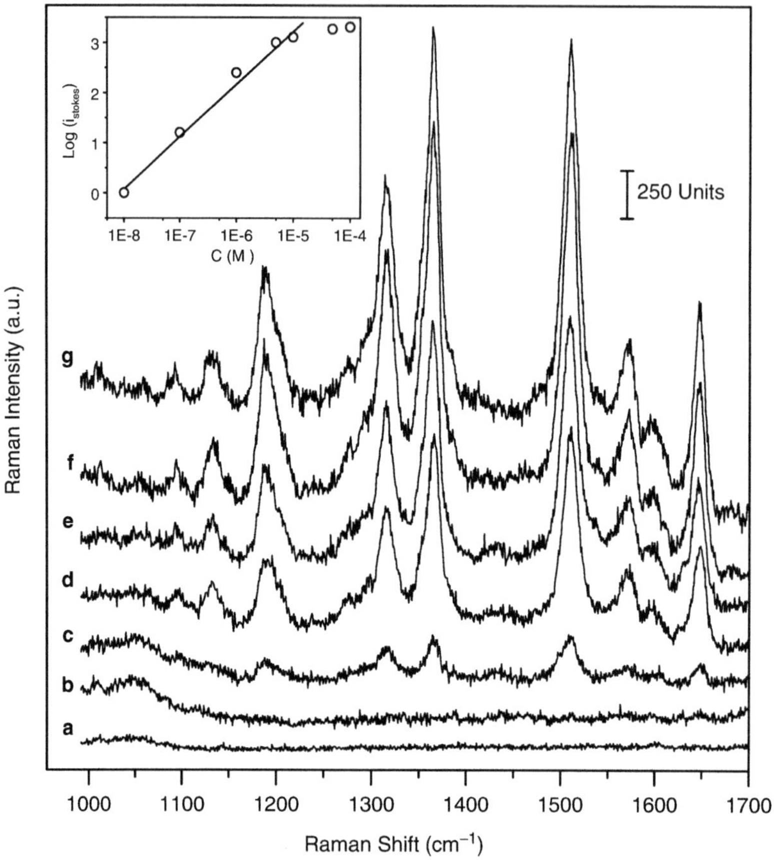

Fig. 6.4 Raman spectra obtained at Ag-MWCNTs-Nafion-modified ITO substrate for successive additions of R6G concentrations of 10^{-8} (**a**), 10^{-7} (**b**), 10^{-6} (**c**), 5×10^{-6} (**d**), 10^{-5} (**e**), 5×10^{-5} (**f**), 10^{-4} (**g**) M in aqueous solution. The inset shows the calibration curve of Raman peak intensity (at 1,509 cm^{-1}) of R6G obtained at Ag-MWCNTs-Nafion-modified ITO substrate (Reprinted with permission from Ref. [18]. © 2009 Elsevier)

MWCNTs-Nafion is a suitable matrix for Ag nanoparticles loading for SERS sensor applications. Different concentrations of R6G were added into the liquid cell in order to evaluate the potential applications of the Ag-MWCNTs-Nafion-nanocomposite-film-modified ITO substrates. Raman spectra of Ag-MWCNTs-Nafion to the additions of varying concentrations of R6G in aqueous solution are shown in Fig. 6.4. It can be seen that the Raman intensities increased with successive addition of R6G. These results demonstrate that R6G can be detected with the Ag-MWCNTs-Nafion substrates. The calibration curve of Raman peak intensity (at 1,509 cm^{-1}) at the Ag-MWCNTs-Nafion-modified ITO substrates to the addition of varying concentrations of R6G in aqueous solution is shown in the inset in Fig. 6.4. The linear range is 10^{-8}–10^{-5} M with a correlation factor of 0.994.

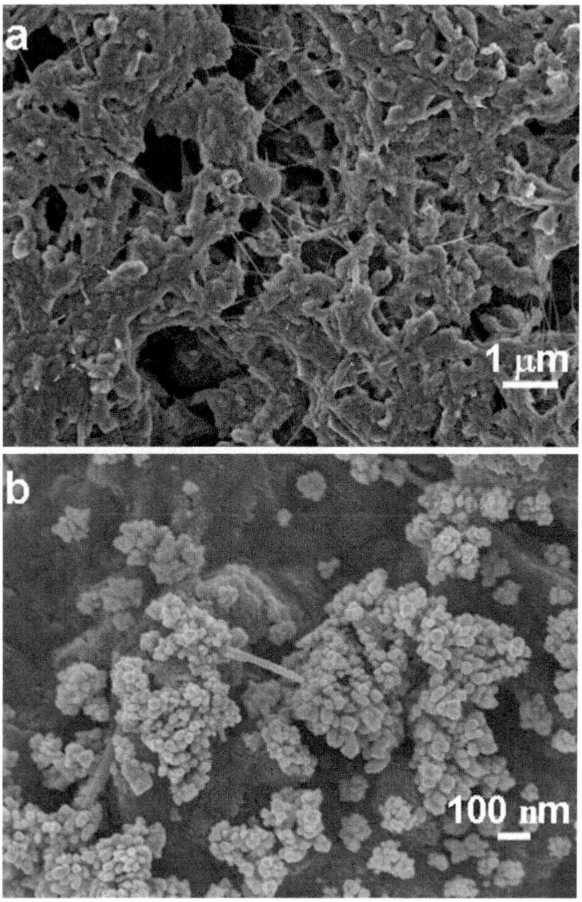

Fig. 6.5 FESEM images of MWCNTs-ACS (**a**) and Ag-MWCNTs-ACS (**b**)

5.2 Ag-MWCNTs-ACS Nanocomposite for SERS

It has been reported that ACS provides an effective way for dispersion of CNTs in aqueous solution in electroanalytical applications [75, 76]. The FESEM images of MWCNTs-ACS-coated ITO, and Ag-MWCNTs-ACS-coated ITO are shown in Fig. 6.5a, b, respectively. It can be seen in Fig. 6.5a that the MWCNTs are well dispersed in the MWCNTs-ACS nanocomposite. The formation of Ag in MWCNTs-ACS after electrochemical deposition can be seen in Fig. 6.5b. Ag particles were with diameters about 100 nm in the MWCNTs-ACS nanocomposite and exhibited some spherical nanostructures. The Ag nanoparticles are in electrical contact with the ITO substrates throughout the MWCNTs-ACS nanocomposite. The Ag particle size prepared by this method is suitable and provided a high number of good hot spots for SERS.

In order to compare the SERS performance between the Ag-MWCNTs-ACS-coated ITO and Ag-coated ITO substrates, the Raman intensity for R6G at the

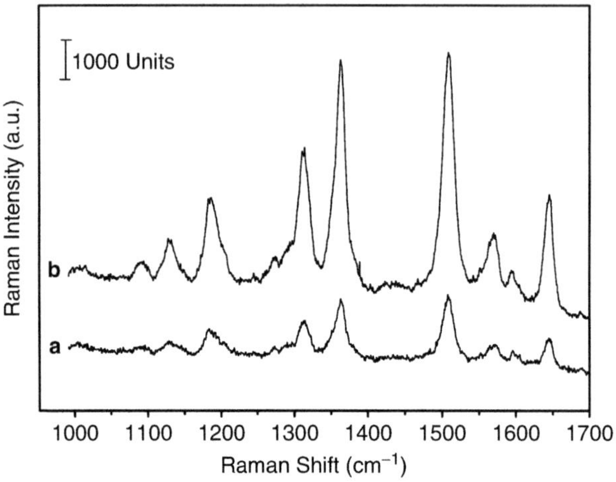

Fig. 6.6 Raman spectra obtained at the surfaces of Ag-coated ITO (**a**) and Ag-MWCNTs-ACS-coated ITO for 10^{-4} M R6G in aqueous solution (Reprinted with permission from Ref. [17]. © 2009 Elsevier)

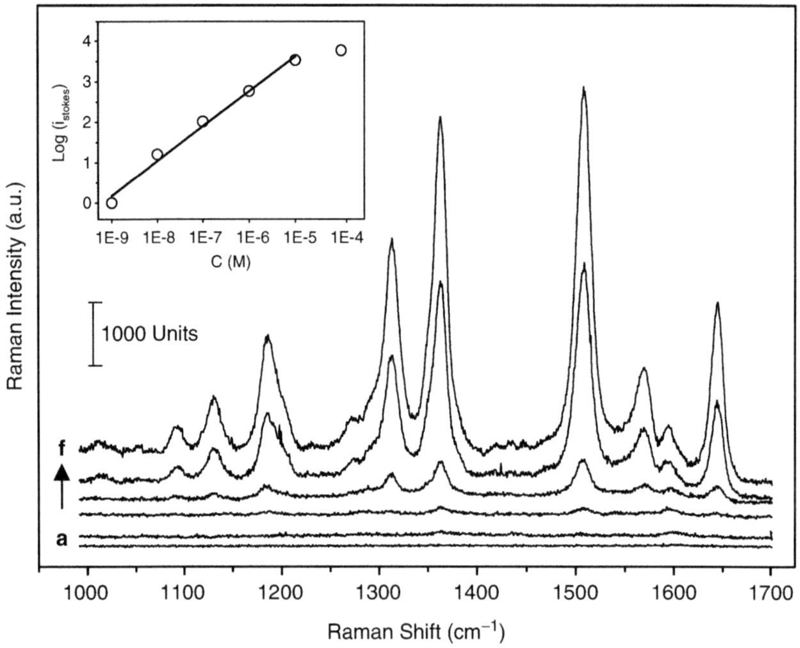

Fig. 6.7 Raman spectra obtained at the Ag-MWCNTs-ACS-modified ITO substrate for successive additions of R6G concentrations of 10^{-9} (**a**), 10^{-8} (**b**), 10^{-7} (**c**), 10^{-6} (**d**), 10^{-5} (**e**), and 10^{-4} (**f**) M in aqueous solution. The inset shows the Raman peak intensity at 1,509 cm^{-1} of R6G drawn as a function of the R6G concentration in a logarithmic scale at the Ag-MWCNTs-ACS-modified ITO substrate (Reprinted with permission from Ref. [17]. © 2009 Elsevier)

Ag-MWCNTs-ACS-coated ITO and Ag-coated ITO substrates were studied. The Raman spectra of the R6G in aqueous solution at the surfaces of Ag-modified ITO and Ag-MWCNTs-ACS-modified ITO substrates are shown in Fig. 6.6a, b, respectively. The Raman intensity of R6G obtained at the Ag-MWCNTs-ACS-coated ITO is greater than that of Ag-coated ITO. It can be seen that the Ag- MWCNTs-ACS-coated ITO has a considerable effect on the Raman spectra with improvements of more than four times of magnitude as compared with the Ag-coated ITO. This may be attributed to the large surface area of MWCNTs. These results indicate that the Ag-MWCNTs-ACS nanocomposite is a highly SERS-active substrate. Raman spectra of Ag-MWCNTs-ACS nanocomposite to the additions of varying concentrations of R6G in aqueous solution are shown in Fig. 6.7 and the Raman peak intensity of R6G at 1,509 cm^{-1} obtained at the Ag-MWCNTs-ACS-coated ITO versus different R6G concentration in a logarithmic scale is shown in the inset in Fig. 6.7. The Raman intensities of R6G at 1,509 cm^{-1} vary linearly with the concentration of R6G in the range of 10^{-9}–10^{-5} M. These results indicate that the Ag-MWCNTs-ACS nanocomposite is suitable for SERS-active substrates.

6 Conclusion and Future Perspective

In conclusion, it has been shown that the SERS techniques offer a means of sensitive detection of probe molecules. An efficient and simple SERS-active substrate prepared by electrodeposition of Ag on MWCNTs has been developed. The prepared Ag-MWCNT nanocomposites exhibited good SERS performance and also featured a simple application process. The technique may have a potential use for in situ determination of analytes. Therefore, such a work will lead to a very promising future for applications in SERS chemical sensors.

References

1. Dejesus MA, Giesfeldt KS, Oran JM, Abu-Hatab NA, Lavrik NV, Sepaniak MJ, Nogueira HIS, Soares-Santos RCR, Cruz SMG, Trindade T (2002) Adsorption of 2,2′-dithiodipyridine as a tool for the assembly of silver nanoparticles. J Mater Chem 12:2339–2342
2. Bae SJ, Lee C, Choi IS, Hwang CS, Gong M, Kim K, Joo SW (2002) Adsorption of 4-biphenylisocyanide on gold and silver nanoparticle surfaces: surface-enhanced Raman scattering study. J Phys Chem B 106:7076–7080
3. Suh YD, Schenter GK, Zhu L, Lu HP (2003) Probing nanoscale surface enhanced Raman-scattering fluctuation dynamics using correlated AFM and confocal ultramicroscopy. Ultramicroscopy 97:89–102
4. Seitz O, Chehimi MM, Cabet-Deliry E, Truong S, Felidj N, Perruchot C, Greaves SJ, Watts JF (2003) Preparation and characterisation of gold nanoparticle assemblies on silanised glass plates. Colloids Surf A 218:225–239
5. Zhang X, Shah NC, Duyne RPV (2006) Sensitive and selective chem/bio sensing based on surface-enhanced Raman spectroscopy (SERS). Vib Spectrosc 42:2–8

6. Bjerneld EJ, Foldes-Papp Z, Kal M, Rigler R (2002) Single-molecule surface-enhanced Raman and fluorescence correlation spectroscopy of horseradish peroxidase. J Phys Chem B 106:1213–1218
7. Kwon K, Lee KY, Kim M, Lee YW, Heo J, Ahn SJ, Han SW (2006) High-yield synthesis of monodisperse polyhedral gold nanoparticles with controllable size and their surface-enhanced Raman scattering activity. Chem Phys Lett 432:209–212
8. Doering WE, Nie S (2002) Single-molecule and single-nanoparticle SERS: examining the roles of surface active sites and chemical enhancement. J Phys Chem B 106:311–317
9. Haynes CL, Duyne RPV (2003) Plasmon-sampled surface-enhanced Raman excitation spectroscopy. J Phys Chem B 107:7426–7433
10. Felidj N, Aubard J, Levi G, Krenn JR, Hohenau A, Schider G, Leitner A, Aussenegg FR (2003) Optimized surface-enhanced Raman scattering on gold nanoparticle arrays. Appl Phys Lett 82:3095–3097
11. Lim JK, Kim IH, Kim KH, Shin KS, Kang W, Choo J, Joo SW (2006) Adsorption of dimethyl sulfide and methanethiolate on Ag and Au surfaces: surface-enhanced Raman scattering and density functional theory calculation study. Chem Phys 330:245–252
12. Vlckova B, Pavel I, Sladkova M, Siskova K, Slouf M (2007) Single molecule SERS: perspectives of analytical applications. J Mol Struct 834–836:42–47
13. Barber ZH, Clyne TW (2002) Ag nanoparticle induced surface enhanced Raman spectroscopy of chemical vapor deposition diamond thin films prepared by hot filament chemical vapor deposition. J Appl Phys 91:6085–6088
14. Joo SW, Chung TD, Jang WC, Gong M, Geum N, Kim K (2002) Surface-enhanced Raman scattering of 4-cyanobiphenyl on gold and silver nanoparticle surfaces. Langmuir 18:8813–8816
15. Cejkova J, Prokopec V, Brazdova S, Kokaislova A, Matejka P, Stepanek F (2009) Characterization of copper SERS-active substrates prepared by electrochemical deposition. Appl Surf Sci 255:7864–7870
16. Niu Z, Fang Y (2006) Surface-enhanced Raman scattering of single-walled carbon nanotubes on silver-coated and gold-coated filter paper. J Colloid Interface Sci 303:224–228
17. Tsai YC, Hsu PC, Lin YW, Wu TM (2009) Electrochemical deposition of silver nanoparticles in multiwalled carbon nanotube-alumina-coated silica for surface-enhanced Raman scattering-active substrates. Electrochem Commun 11:542–545
18. Tsai YC, Hsu PC, Lin YW, Wu TM (2009) Silver nanoparticles in multiwalled carbon nanotube–Nafion for surface-enhanced Raman scattering chemical sensor. Sensor Actuator B 138:5–8
19. Hou X, Wu L, Xu W, Qin L, Wang C, Zhang X, Shen J (2002) Self-assembly and Langmuir–Blodgett (LB) film of a novel hydrogen-bonded complex: a surface enhanced Raman scattering (SERS) study. Colloids Surf A 198–200:135–140
20. Chowdhury J, Pa P, Ghosh M, Misra TN (2001) Surface-enhanced Raman scattering of Rhodamine 123 in silver hydrosols and in Langmuir–Blodgett films on silver Islands. J Colloid Interface Sci 235:317–324
21. Nishiyama K, Tahara S, Uchida Y, Tanoue S, Taniguchi I (1999) Structural differences in self-assembled monolayers of anthraquinone derivatives on silver and gold electrodes studied by cyclic voltammetry and in situ SERS spectroscopy. J Electroanal Chem 478: 83–91
22. Conoci S, Valli L, Rella R, Compagnini G, Cataliotti RS (2002) A SERS study of self-assembled (4-methylmercapto) benzaldehyde thin films. Mater Sci Eng C 22:183–186
23. Coluccio ML, Das G, Mecarini F, Gentile F, Pujia A, Bava L, Tallerico R, Candeloro P, Liberale C, Angelis FD, Fabrizio ED (2009) Silver-based surface enhanced Raman scattering (SERS) substrate fabrication using nanolithography and site selective electroless deposition. Microelectron Eng 86:1085–1088
24. Vosgrone T, Meixner AJ, Anders A, Dietz H, Sandmann G, Plieth W (2005) Electrochemically deposited silver particles for surface enhanced Raman spectroscopy. Surf Sci 597:102–109

25. Shibata N, Horigudhi M, Edahiro T (1981) Raman spectra of binary high-silica glasses and fibers containing GeO_2, P_2O_5 and B_2O_3. J Non-Cryst Solids 45:115–126
26. Cai WB, Ren B, Li XQ, She CX, Liu FM, Cai XW, Tian ZQ (1998) Investigation of surface-enhanced Raman scattering from platinum electrodes using a confocal Raman microscope: dependence of surface roughening pretreatment. Surf Sci 406:9–22
27. Bolboaca M, Iliescu T, Kiefer W (2004) Infrared absorption, Raman, and SERS investigations in conjunction with theoretical simulations on a phenothiazine derivative. Chem Phys 298:87–95
28. Fleischmann M, Hendra PJ, McQuillan AJ (2004) Raman spectra of pyridine adsorbed at a silver electrode. Chem Phys Lett 26:163–166
29. Jacintho GVM, Corio P, Rubim JC (2007) Surface-enhanced Raman spectra of magnetic nanoparticles adsorbed on a silver electrode. J Electroanal Chem 603:27–34
30. Mao CF, Vannice MA (1995) Formaldehyde oxidation over Ag catalysts. J Catal 154:230–244
31. Hossain MK, Shibamoto K, Ishioka K, Kitajima M, Mitani T, Nakashima S (2007) 2D nanostructure of gold nanoparticles: an approach to SERS-active substrate. J Lumin 122–123:792–795
32. Shibamoto K, Katayama K, Sawada T (2007) Ultrafast charge transfer in surface-enhanced Raman scattering (SERS) processes using transient reflecting grating (TRG) spectroscopy. Chem Phys Lett 433:385–389
33. Mulvaney SP, He L, Natan MJ, Keating CD (2003) Three-layer substrates for surface-enhanced Raman scattering: preparation and preliminary evaluation. J Raman Spectrosc 34:163–171
34. Smejkal P, Siskova K, Vlckova B, Pfleger J, Sloufova I, Slouf M, Mojzes P (2003) Characterization and surface-enhanced Raman spectral probing of silver hydrosols prepared by two-wavelength laser ablation and fragmentation. Spectrochim Acta A 59:2321–2329
35. Wu ZS, Zhoua GZ, Jiang JH, Shen GL, Yu RQ (2006) Gold colloid-bienzyme conjugates for glucose detection utilizing surface-enhanced Raman scattering. Talanta 70:533–539
36. Weibenbacher N, Gobel R, Kellner R (1996) Ag-layers on non-ferrous metals and alloys. A new substrate for Surface Enhanced Raman scattering (SERS). Vib Spectrosc 12:189–195
37. Li ZM, Zhang N, Li FT (2006) Studies of the adsorption state of activated carbon by surface-enhanced Raman scattering. Appl Surf Sci 253:2870–2874
38. Yu DG, Lin WC, Lin CH, Chang LM, Yang MC (2007) An in situ reduction method for preparing silver/poly(vinyl alcohol) nanocomposite as surface-enhanced Raman scattering (SERS)-active substrates. Mater Chem Phys 101:93–98
39. Gong JL, Liang Y, Huang Y, Chen JW, Jiang JH, Shen GL, Yu RQ (2007) Ag/SiO_2 core-shell nanoparticle-based surface-enhanced Raman probes for immunoassay of cancer marker using silica-coated magnetic nanoparticles as separation tools. Biosens Bioelectron 22:1501–1507
40. Muniz-Miranda M, Neto N, Sbrana G (2003) SERS intensities of benzodiazines adsorbed on silver nanoparticles. J Mol Struct 651–653:85–90
41. Zheng J, Li X, Gu R, Lu T (2002) Comparison of the surface properties of the assembled silver nanoparticle electrode and roughened silver electrode. J Phys Chem B 106:1019–1023
42. Swami A, Kumar A, Selvakannan PR, Mandal S, Sastry M (2003) Langmuir–Blodgett films of laurylamine-modified hydrophobic gold nanoparticles organized at the air–water interface. J Colloid Interface Sci 260:367–373
43. Luccio TD, Antolini F, Aversa P, Scalia G, Tapfer L (2004) Structural and morphological investigation of Langmuir–Blodgett SWCNT/behenic acid multilayers. Cairbon 42: 1119–1122
44. Obeid R, Park JY, Advincula RC, Winnik FM (2009) Temperature-dependent interfacial properties of hydrophobically end-modified poly(2-isopropyl-2-oxazoline)s assemblies at the air/water interface and on solid substrates. J Colloid Interface Sci 340:142–152
45. Fleger Y, Mastai Y, Rosenbluh M, Dressler DH (2009) Surface enhanced Raman spectroscopy of aromatic compounds on silver nanoclusters. Surf Sci 603:788–793

46. Lu Y, Liu GJ, Lee LP (2008) High density self assembled nanoparticle film with temperature-controllable interparticle spacing for deep sub-wavelength nanolithography using localized surface plasmon modes on planar silver nanoparticle tunable grating. Microelectron Eng 85:486–491
47. Tao A, Kim F, Hess C, Goldberger J, He R, Sun Y, Xia Y, Yang P (2003) Langmuir-blodgett silver nanowire monolayers for molecular sensing using surface-enhanced Raman spectroscopy. Nano Lett 3:1229–1233
48. Chowdhury J, Pal P, Ghosh M, Misra TN (2001) Surface-enhanced Raman scattering of Rhodamine 123 in silver hydrosols and in Langmuir–Blodgett films on silver islands. J Colloid Interface Sci 235:317–324
49. Decher G, Hong JD (1991) Buildup of utrathin multilayer films by a self-assembly process.1. consecutive adsorption of anionic and cationic bipolar amphiphiles on charged surfaces. Makromol Chem Macromol Symp 46:321–327
50. Decher G, Hong JD (1991) Buildup of utrathin multilayer films by a self-assembly process.1. consecutive adsorption of anionic and cationic bipolar amphiphiles on charged surfaces. Makromol Chem Macromol Symp 46:321–327
51. Kahl M, Voges E, Kostrewa S, Viets C, Hill W (1998) Periodically structured metallic substrates for SERS. Sensor Actuator B 51:285–291
52. Vo-Dinh T (1998) Surface-enhanced Raman spectroscopy using metallic nanostructures. Trend Anal Chem 17:557–582
53. Dejesus MA, Giesfeldt KS, Oran JM, Abu-Hatab NA, Lavrik NV, Sepaniak MJ (2005) Nanofabrication of densely packed metal–polymer arrays for surface-enhanced Raman spectrometry. Appl Spectrosc 59:1501–1508
54. Das G, Mecarini F, Deangelis F, Prasciolu M, Liberale C, Patrini M, Difabrizio E (2008) Attomole (amol) myoglobin Raman detection from plasmonic nanostructures. Microelectron Eng 85:1282–1285
55. Bozzini B, Romanello V, Mele C (2007) A SERS investigation of the electrodeposition of Au in a phosphate solution. Surf Coat Tech 201:6267–6272
56. Hope GA, Woods R (2004) Transient adsorption of sulfate ions during copper electrodeposition. J Electrochem Soc 151:C550–C553
57. Viga A, Radoib A, Mumoz-Berbel X, Gyemantc G, Marty JL (2009) Impedimetric aflatoxin M1 immunosensor based on colloidal gold and silver electrodeposition. Sensor Actuator B 138:214–220
58. Fu C, Zhou H, Peng W, Chen J, Kuang Y (2008) Comparison of electrodeposition of silver in ionic liquid microemulsions. Electrochem Commun 10:806–809
59. Koo HC, Cho SK, Lee CH, Kim SK, Kwon OJ, Kim JJ (2008) Silver direct electrodeposition on Ru thin films. J Electrochem Soc 155:D389–D394
60. Sun B, Jiang X, Dai S, Du Z (2009) Single-crystal silver nanowires: preparation and Surface-enhanced Raman Scattering (SERS) property. Mater Lett 63:2570–2573
61. Luo LB, Chen LM, Zhang ML, He ZB, Zhang WF, Yuan GD, Zhang WJ, Lee ST (2009) Surface-enhanced Raman scattering from uniform gold and silver nanoparticle-coated substrates. J Phys Chem C 113:9191–9196
62. Lucotti A, Zerbi G (2007) Fiber-optic SERS sensor with optimized geometry. Sensor Actuator B 121:356–364
63. Ouyang Y, Fang Y (2005) A new surface-enhanced Raman scattering system for carbon nanotubes. Spectrochim Acta A 61:2211–2213
64. Hu X, Wang T, Wang L, Guo S, Dong S (2007) A general route to prepare one- and three-dimensional carbon nanotube/metal nanoparticle composite nanostructures. Langmuir 23:6352–6357
65. Sullivan JA, Flanagan KA, Hain H (2009) SUZUKI coupling activity of an aqueous phase Pd nanoparticle dispersion and a carbon nanotube/Pd nanoparticle composite. Catal Today 145:108–113

66. Chu H, Wang J, Ding L, Yuan D, Zhang Y, Liu J, Li Y (2009) Decoration of gold nanoparticles on surface-grown single-walled carbon nanotubes for detection of every nanotube by surface-enhanced Raman spectroscopy. J Am Chem Soc 131:14310–14316
67. Sun Y, Liu K, Miao J, Wang Z, Tian B, Zhang L, Li Q, Fan S, Jiang K (2010) Highly sensitive surface-enhanced Raman scattering substrate made from superaligned carbon nanotubes. Nano Lett 10:1747–1753
68. Kumar R, Zhou HB, Cronin SB (2007) Surface-enhanced Raman spectroscopy and correlated scanning electron microscopy of individual carbon nanotubes. Appl Phys Lett 91:223105
69. Guo S, Dong S, Wang E (2008) Constructing carbon-nanotube/metal hybrid nanostructures using homogeneous TiO_2 as a spacer. Small 4:1133–1138
70. Zhang J, Gao L, Sun J, Liu Y, Wang Y, Wang J, Kajiura H, Li Y, Noda K (2008) Dispersion of single-walled carbon nanotubes by nafion in water/ethanol for preparing transparent conducting films. J Phys Chem C 112:16370–16376
71. Zhang J, Wang Q, Wang L, Wang A (2007) Manipulated dispersion of carbon nanotubes with derivatives of chitosan. Carbon 45:1911–1920
72. Maxwell DJ, Emory SR, Nie S (2001) Nanostructured thin-film materials with surface-enhanced optical properties. Chem Mater 13:1082–1088
73. Hildebrandt P, Stockburger M (1984) Surface-enhanced resonance Raman spectroscopy of Rhodamine 6 G adsorbed on colloidal silver. J Phys Chem 88:5935–5944
74. Chan S, Kwon S, Koo TW, Lee LP, Berlin AA (2003) Surface-enhanced resonance Raman of small molecules from silver-coated silicon nanopores. Adv Mater 15:1595–1598
75. Tao A, Kim F, Hess C, Goldberger J, He R, Sun Y, Xia Y, Yang P (2003) Langmuir-blodgett silver nanowire monolayers for molecular sensing using surface-enhanced Raman spectroscopy. Nano Lett 3:1229–1233
76. Tsai YC, Chiu CC, Tsai MC, Wu JY, Tseng TF, Wu TM, Hsu SF (2007) Dispersion of carbon nanotubes in low pH aqueous solutions by means of alumina-coated silica nanoparticles. Carbon 45:2823–2827

Raman Spectroscopy of Carbon Nanostructures: Nonlinear Effects and Anharmonicity

7

A. P. Naumenko, N. E. Korniyenko, V. M. Yashchuk, Srikanth Singamaneni, and Valery N. Bliznyuk

1 Definition of the Topic

We have analyzed the influence of the annealing temperature, structural disorder, and the frequency of a continuous excitation laser radiation ν_L on the first- and the second-order Raman spectra of several nanostructured carbon materials including single-wall carbon nanotubes (SWCNT), SWCNT-polymer composites, and nanostructured single-crystalline graphites. Consideration of the high-order nonlinear effects in Raman spectra and anharmonicity of characteristic Raman bands (such as G, G′, and D modes) provides important information on the vibration modes and collective (phonon-like) excitations in such 1D or 2D confined systems

2 Introduction

Raman spectra of various carbon nanostructures, which include fullerenes (0D), nanotubes (1D), and graphene (2D), have been a subject of intensive studies over the last two decades [1–8]. The Raman spectra of these carbon nanostructures have

A.P. Naumenko (✉) • N.E. Korniyenko • V.M. Yashchuk
Physics Department, Taras Shevchenko National University of Kyiv, Kyiv, Ukraine

S. Singamaneni
Department of Mechanical Engineering and Materials Science, School of Engineering and Applied Sciences, Washington University in St. Louis, St. Louis, MO, USA

V.N. Bliznyuk (✉)
Materials Science and Engineering Department, College of Engineering and Applied Sciences, Western Michigan University, Kalamazoo, MI, USA

Challa S. S. R. Kumar (ed.), *Raman Spectroscopy for Nanomaterials Characterization*, DOI 10.1007/978-3-642-20620-7_7, © Springer-Verlag Berlin Heidelberg 2012

been employed to identify their atomic and electronic structure. For example, so-called radial breathing mode (RBM) and graphitic modes of carbon nanotubes are sensitive to the diameter and electronic nature (metallic and semiconducting) of carbon nanotubes. Even more recently, it has been demonstrated that the second harmonic of the defect mode in Raman spectra can be used to precisely differentiate between single and bilayer graphene structure [9]. Furthermore, the vibrational signature of these nanostructures is important for advanced applications such as carbon nanoelectronics (determination of electronic properties), nanocomposites (level of dispersion and orientation), and nanosensors (detection of strains) [10–15].

While the fraction of the surface atoms in any nanoparticluate system is high (typically up to 50–90%), the same reaches 100% in the case of single-walled carbon nanotubes (SWCNT). SWCNT and fullerenes exhibit high nonlinear susceptibilities in infrared and visible range of light due to their enhanced vibration anharmonicity, high electron polarizability, and electron delocalization [16]. These peculiarities can possibly result in highly nonequilibrium vibrational modes and electronic states due to nonlinear interaction of thermally excited vibration modes. The latter effects, in our view, are responsible for the high functional activity of nanoparticles and other activated media. Therefore, detailed studies of physical properties of SWCNT are very important for both basic research in the field of materials science and practical applications. In this context a comparative study of the vibrational spectra of SWCNT and various types of graphitic materials seems to be crucially important.

Consideration of high-order nonlinear effects in Raman spectra should give important information on anharmonicity of vibration modes, which in turn provides a deeper insight into physical properties and structure of carbon materials and their response to a continuous laser radiation. In this chapter, in addition to the above-stated data, we report on the vibrational bands of SWCNT in the range of overtones of $2\nu_{D,G}$ (D and G modes) in the range of 2,680–3,180 cm^{-1} as well as their sum harmonics $\nu_D+\nu_G$ and $\nu_G+\nu_{RBM}$ including contribution provided by a low frequency breathing mode $\nu_{RBM}\sim 160$ cm^{-1}. We performed a comparative study of the first- and the second-order Raman spectra of SWCNT with their flat analogs – several graphitic materials including single crystals of highly oriented pyrolytic graphite (HOPG). We have analyzed the influence of the annealing temperature, structural disorder, and the frequency of a continuous laser excitation ν_L on the Raman spectra of such carbon nanomaterials. We demonstrate that the doublet G-band of SWCNT is analogous to a corresponding Raman band of graphite and corresponds to a disordered state of the latter. The study of the harmonic Raman bands allowed characterization of anharmonicity of the vibration modes. We have found a correlation between the anharmonicity of G and D Raman modes of SWCNTs and the wavelength of the Raman excitation λ_L. The anharmonicity of the D mode is shown to increase while those of the G mode to decrease with the laser frequency increase. Simultaneously, a significant enhancement of the harmonics' intensities with the frequency of the laser radiation has been observed. These effects together with the observed anomalous behavior of the vibration anharmonicity highlight a requirement of addressing the question of the influence of a coherent laser radiation on the physical properties of nanotubes. Finally, we also discuss and compare the fine structure of the Raman spectra observed for SWCNT and graphitic materials as those having similarity in their carbon hybridization.

3 Experimental and Instrumental Methodology

Raman spectroscopy of completely black materials like graphite or carbon black became possible only after lasers began to be applied as a source. In the first publication by F. Tuinstra and J. L. Koenig [17] devoted to Raman scattering in graphite a 488 nm line (~300 mW) of an Ar-ion laser was used to induce the Raman spectrum; the 514.5 nm line was used occasionally. The scattered light was observed at 90° angle from the incident beam. To resolve Raman scattering peaks from the Rayleigh scattering, a double monochromator (Spex 1400) was used. The single crystals of graphite were mounted on a goniometer head to observe the Raman scattering at different orientations of the crystal with respect to the incident beam. The pellets of the microcrystalline samples were mounted in such a way that the incident beam made an angle of 30° with the flat front surface. The illuminated area of the samples (~10 μm in diameter) was examined with the aid of an optical microscope, which was mounted on top of the sample holder.

Since the time of the first detection of the Raman spectrum of graphite, the Raman equipment has progressed to further advancement. Nowadays Raman and resonance Raman spectra of various materials, especially SWCNTs, can be acquired using standard commercial micro-Raman spectrometers and lasers [18, 19]. Typical measurements use a back-scattering geometry and 50×, 80×, and 100× objective lenses. The Raman spectra from SWCNTs are excited by different discrete lines from Ar-Kr and He-Ne lasers[1] and lines from continuous range of Ti:Saphire and Dye lasers. Laser powers ~10 mW are typically used. Excitation with several laser lines gives more realistic characterization of the diameter distribution in the sample due to resonance conditions achieved for various wavelength-nanotube diameter combinations [20].

In the present work, because of strong absorption in graphite and carbon materials at optical frequencies (the absorption coefficient is 2.4–2.6 cm^{-1} and the skin depth at $\lambda_L = 514.5$ nm is about 400 Å), the Brewster backscattering geometry was used in Raman experiments. Particularly several lines of Ar^+ laser with the wavelength of 514.5, 488, and 476.5 nm and power of ~50 mW were selected with

[1]Argon ion visible lines (8 UV lines and 2 IR lines are ignored): 454.6, 457.9, 465.8, 472.6, 476.5, 488.0, 496.5, 501.7, 514.5, 528.7 nm; Krypton ion visible lines (4 UV and 8 IR lines are ignored) 406.7, 413.1, 415.4, 468.0, 476.2, 482.5, 520.8, 530.9, 568.2, 647.1, 676.4 nm. The most common HeNe lasers by far produce light at a wavelength of 632.8 nm in the red portion of the visible spectrum. Green (543.5 nm), yellow (594.1 nm), and orange (604.6 and 611.9 nm) HeNe lasers are also available but are not nearly as "efficient" as the common red type ones since the spectral lines that need to be amplified are much weaker at these wavelengths (http://www.repairfaq.org)

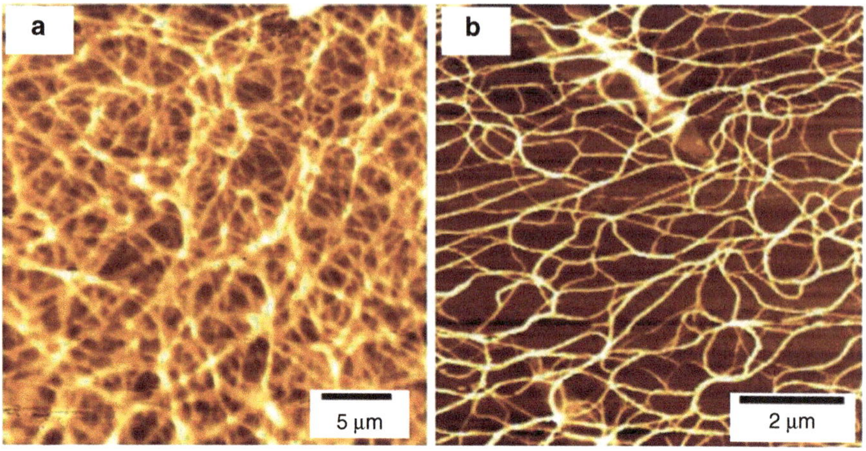

Fig. 7.1 (**a**) Large scale AFM (using Autoprobe CP microscope in tapping mode) image showing the SWCNT mat with high purity confirming the complete removal of the metal catalyst residue and (**b**) Higher magnification image showing the uniform diameter of SWCNT cast on a silicon substrate

a prism located outside of laser resonator and a cylindrical lens was used to focus light in a 10×0.1 mm^2 spot. It is worth mentioning that the larger illuminated area can efficiently eliminate the local heterogeneity (more reliable data) and minimize the probability of radiation damage (heating) of the sample due to resonance absorption. The spectra were detected by an automated double spectrometer DFS-24 (LOMO, Russia), equipped with a cooled photo multiplier and registration system working in a photon counting mode. In connection to numerical analysis, the spectra were digitalized in wide frequency range with a fixed increment (from 1 up to 5 cm^{-1}). As the scattering cross sections of the carbon materials under study are relatively low, additional measures were taken for noise minimization. Particularly, for relatively broad ~3–5 cm^{-1} Raman bands, long signal acquisition time and optimal numerical flattening with a variable spectral window were applied for enhancing the signal to noise ratio.

Untreated SWCNTs were purchased from Carbolex Inc. The AP Grade carbon nanotubes were 50–70% pure with nickel and yttrium catalysts making up the remainder. The nanotubes dispersed in the selected solvent (chloroform) were filtered through Anatop filters with a pore size of 200 nm (Whatman). The filtration process resulted in high purity of SWCNT as depicted in the AFM images (Fig. 7.1). Graphite samples ("NII Grafit," Moscow, Russia) were used as reference samples and have been previously studied by X-ray diffraction and Raman scattering techniques. For structural characterization of the samples, DRON-1.5 X-ray diffractometer was employed either in powder diffraction mode (polycrystalline samples) or in surface reflection mode (single crystals).

The careful analysis of observed Raman bands included several steps: determination of line positions and line shapes; deconvolution of multiple peaks; finding relative intensity and spectral position of individual lines in composite Raman bands upon laser energy variation. Following mathematical treatment procedures included fittings of multiple peaks subtraction of the normalized spectra performing the second derivative spectra analysis; and analysis of the asymmetry and fine band structure. Detailed study of the individual components of Raman doublets has been applied based on consideration of the mathematical procedures of the normalized spectra subtraction and the second derivative of the original Raman spectra. In the latter case, the spectral position of the individual components could be revealed even in cases of strong overlap of the bands. We demonstrate below how interpretation of the bands is possible based on the existence of sharp edges of the bands and their one-side "smearing" under conditions of reduction of the crystallite size. Useful information about the existing fine structure of the bands was also extracted from the analysis of the second derivatives of corresponding Raman spectra.

4 Key Research and Finding

4.1 Overview

Raman spectroscopy has been used as one of the most important techniques suitable for investigation and characterization of carbon and graphitic materials including carbon nanotubes. For the first time, carbon nanotubes were studied with this technique in 1991 (multiwalled carbon nanotubes MWCNTs [21]). Raman spectra of SWCNTs were reported in 1993 in two papers submitted independently [22, 23]. First Raman spectroscopy was used as a tool for characterizing the purification and synthesis process, as well as to study physical properties. Nowadays, Raman spectroscopy has developed to a powerful tool for structural characterization of SWCNTs. In 1997, Rao et al. [7] have shown a dependence of the Raman spectra of SWCNTs bundles on the excitation laser energy E_L, due to a strong resonance effect between E_L and van Hove singularities in the joint density of electronic states in SWCNTs. Next year, it was shown by M.A. Pimenta et al. that Raman spectra could be used to distinguish between metallic and semiconducting SWNTs [24]. The next important event in the history of Raman studies of carbon nanotubes happened in 2001 when Jorio et al. [5] demonstrated the possibility of detecting Raman spectra from a single isolated single-wall carbon nanotube. The suggested procedure has enabled direct determination of the structural parameters of SWCNTs from their Raman spectra. Appropriate reviews on the structure and properties of carbon nanotubes can be found in several recent publications [2, 25].

Raman spectra of SWCNTs as well as the most common experimental techniques of their characterization have been also thoroughly discussed in literature [3, 26]. The strongest Raman bands of SWCNTs are the RBM band (radial breathing mode in the range 100–300 cm^{-1}), and the G-band (tangential mode at around 1600 cm^{-1}) as shown in Fig. 7.2 [27]. Two more characteristic, but weaker bands are the D-band (disorder-induced band in the range 1300–1400 cm^{-1}) and the G′-band (sometimes called D*-band) at around 2600–2800 cm^{-1}.

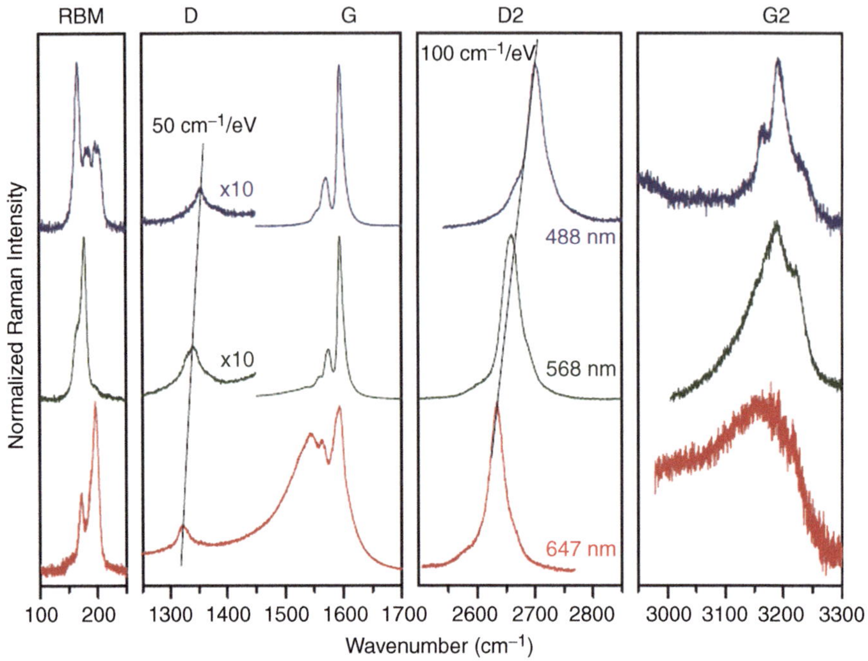

Fig. 7.2 Most important Raman lines of single-wall carbon nanotubes as excited with three different laser lines. *RBM* radial breathing mode, *D* defect-induced line, *G* graphitic line, *D2* overtone of D-line, *G2* overtone of G-line. The thin straight lines indicate the dispersion of the modes. All spectra in one slot were normalized to unit height (Reprinted with permission from Kuzmany H, Plank W, Schaman CH, Pfeifer R, Hasi F, Simon F, Rotas G, Pagona G, Tagmatarchis N (2007) Raman scattering from nanomaterials encapsulated into single-wall carbon nanotubes. Journal of Raman Spectroscopy 38 (6): 704–713, John Wiley & Sons, Ltd.)

The RBM is an important phonon mode in Raman spectra, which provides information on the nanotubes geometrical parameters. The RBM frequency is inversely proportional to the tube diameter $\omega_{RBM} = 248/d_t(\mathrm{cm}^{-1})$ (see Ref. [5]) for isolated SWNTs on a SiO_2 substrate and $\omega_{RBM} = 234/d_t + 10(\mathrm{cm}^{-1})$ for nanotube bundles [28, 29].

The existence of the D-band in Raman spectrum of disordered and ion-implanted graphite, different carbon structures (amorphous and diamond-like carbon) was known for a long time [30, 31] but its origin was not understood for several decades. The energy of the D-mode in Raman spectra depends on the wavelength of exciting laser line [32, 33]. Thus, while increase of the energy of excitation by 1 eV causes 48 cm^{-1} upshift (to higher frequency) of the D-mode line, its overtone shifts at 96 cm^{-1}/eV. In the case of single-wall carbon nanotubes, a resonance effect similar to that in sp^2 hybridization-type carbons is observed. Raman studies of SWCNT bundles, [34–37] and isolated SWNTs, [38–44] revealed a dispersive behavior – that is, changing of the characteristic bands position depending on the excitation energy. Frequency shifts of the D-mode of 38 cm^{-1}/eV (overtone

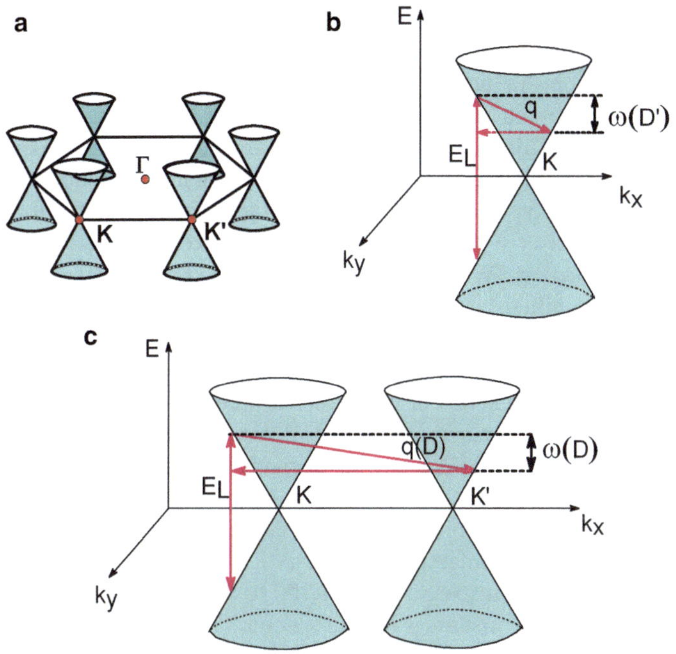

Fig. 7.3 (**a**) 2D Brillouin zone of graphene showing characteristic points K and Γ and "Dirac cones" located at the six corners (K points). (**b**) Second-order double resonance scheme for the D′ peak (close to Γ); (**c**) Raman spectral process for the D peak (involving two neighboring K points of the Brillouin zone K and K′). E_L is the incident laser energy. (After Ref. [46, 48])

90 cm^{-1}/eV), 53 cm^{-1}/eV (overtone 106 cm^{-1}/eV) for SWNT, and 43 cm^{-1}/eV for MWNT have been reported [45]. Thomsen and Reich first calculated the Raman spectra of the D-mode using the concept of double resonances (DR) [46]. The DR Raman theory works well for explaining the dispersive character of this band and relevant to these features of second-order spectra [47]. The DR also describes the strong intensity of G′-band and overtones in SWNTs Raman spectra. Within the DR concept, Raman scattering is considered as a fourth order process (see Fig. 7.3) [46, 48]. The activation process responsible for the appearance of the D peak and involving two neighboring K points of the Brillouin zone (Fig. 7.3a) is shown in Fig. 7.3c. This Figure demonstrates a laser induced excitation of an electron/hole pair and electron–phonon scattering with an exchanged momentum q~K as well as possible scattering events due to defects and electron/hole recombination. The DR condition is reached when the energy is conserved in all the above transitions. A similar process to Fig. 7.3c is possible "intra-valley" as shown in Fig. 7.2b. This process activates phonons with a small q, resulting in the so-called D′ peak, which can be seen around ~1620 cm^{-1} in a defect graphite [49]. In Ref. [50] the disorder-induced D-band and some other non-center Raman modes of graphite and single-wall carbon nanotubes have been assigned to phonon modes in the corresponding Brillouin zones.

The G-mode in SWCNTs has correspondence to G-band in graphite. However, in contrast to the latter, which describes as the anti-phase displacements of the two dissimilar carbon atoms in the same unit cell, multiplicity (splitting) of this band in SWCNTs is caused by curvature effects arising from "wrapping" of a graphene sheet into a tube [51]. This curvature dependence generates a diameter dependence, thus making the G-band a probe also for the tube diameter, while the G-band frequency dependence on externally induced strains is very rich but still controversial [52]. The G peak of SWCNTs can be fit with two components: G^+ and G^- [53–57]. Semiconducting SWCNTs have sharp G^+ and G^- peaks, the metallic ones have a broad downshifted G^-. The G^+ and G^- peaks are commonly assigned to longitudinal optic LO (axial) and transversal optic TO (circumferential) modes, respectively. Authors of Ref. [58] presented a set of simple formulas which quantitatively explained a series of experiments, ranging from the slopes and the *G*-peak full width at half maximum (FWHM) parameters in graphite to the G^- peak's position and G^- FWHM diameter dependence in SWCNTs. For metallic nanotubes, G^+ and G^- peaks are traditionally assigned to TO (circumferential) and LO (axial) modes, while for semiconducting SWNTs G^+ and G^- peaks are assigned to LO (axial) and TO (circumferential) modes.

The second-order Raman spectra[2] from 1,700 to 3,300 cm^{-1} are relatively weak and broad compared to the first order Raman spectra. There are many papers devoted to the second-order Raman processes in graphite [49, 59–62] and carbon nanotubes [8, 34, 63]. The interpretation of the spectral behavior of the second-order bands in terms of DR have been done [3, 64, 65]. The corresponding phonon is involved in an inter-valley scattering process of "in-plane" modes (LO or TO) around the K-points in the two-dimensional (2D) Brillouine zone (BZ, Fig. 7.4) [66, 67].

The phonon dispersion relation around the K-point can be obtained from the G′-band of the corresponding Raman spectra [57]. Some other bands of the second-order spectrum, namely, M, iTOLA, G′, 2iTO, and 2G bands, have been assigned in References: [37, 68, 69] (see definitions in the text below).

The observed M feature near 1,750 cm^{-1} for SWCNT bundles and for several laser lines is shown on Fig. 7.5 [68]. This feature have been analyzed in terms of two components with frequencies ω_M^- and ω_m^+, where the lower frequency mode ω_M^- exhibits a weakly dispersive behavior ($\Delta\omega_M^- \sim 30\ cm^{-1}$ as E_L is varied from 1.58 to 2.71 eV), while the frequency ω_m^+ is basically independent of E_L. These two features near 1,750 cm^{-1} are attributed to overtones of the out-of-plane (oTO), infrared-active mode at 867 cm^{-1} in graphite. Here, the M^+ feature is identified with an intra-valley ($q = 0$) scattering process, and the M^- with an inter-valley ($q = 2k$) process (see Fig. 7.5b). A downshift of about 20 cm^{-1} occurs for both the M^+ and M^--band features

---

[2]If the two-phonon state is formed from phonons belonging to the same branch of the phonon dispersion curves and have equal energy, the corresponding state is called an overtone. If the two-phonon state represents the sum or difference of two phonons with unequal energy the state, it is termed a combination.

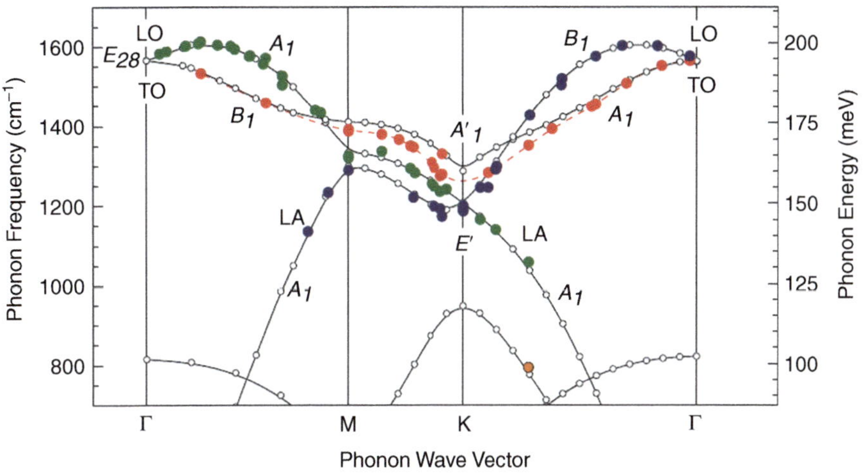

Fig. 7.4 Phonon dispersion of graphite from inelastic X-ray scattering along the in-plane high symmetry directions (*full circles*). The *dashed line* is a cubic-spline interpolation to the TO-derived branch. The open circles show an ab-initio calculation for graphene, downscaled by 1% (Reprinted with permission from Maultzsch J, Reich S, Thomsen C, Requardt H, Ordejón P (2004) Phonon dispersion in graphite. Physical Review Letters 92 (7): 075501-1-4. Copyright (2004) by the American Physical Society)

in SWCNTs relative to 2D graphite because of diameter-dependent curvature effects. The M-band modes are further enhanced in SWCNTs by van Hove singularity effects, and by symmetry-breaking effects associated with SWCNT curvature.

The higher-frequency iTO+LA combination mode[3] is highly dispersive and up shifts from 1864 to 2000 cm^{-1} as E_L varies from 1.58 to 2.71 eV [68]. This feature can be assigned as a combination mode not seen in graphite, but enhanced by the $q = 2k$ double-resonance process. The experimental dispersion of this feature is shown in Fig. 7.5a. Figure 7.5b demonstrates the relation of this mode dispersion to phonon branches in 2D graphite [3].

Weak features have been recognized around 2450 cm^{-1} [8]. Since this band has no or very small dispersion irrespective of E_L the authors assigned it as "q = 0" branch of DR Raman scattering as discussed in graphite [70].

So-called intermediate frequency modes (IFM) lie in the spectral region between ω_{RBM} and ω_G (600–1100 cm^{-1}) [3]. Some of IFMs, which have the dispersive dependence on laser excitation energy, are attributed to combination modes described by DR theory.

[3]This band called iTOLA because it is attributed to a combination of two intra-valley phonons: the first from the in-plane transverse optical branch (iTO) and the second phonon from the longitudinal acoustic (LA) branch, iTO+LA, where the acoustic LA phonon is responsible for the large dispersion that is observed experimentally [69].

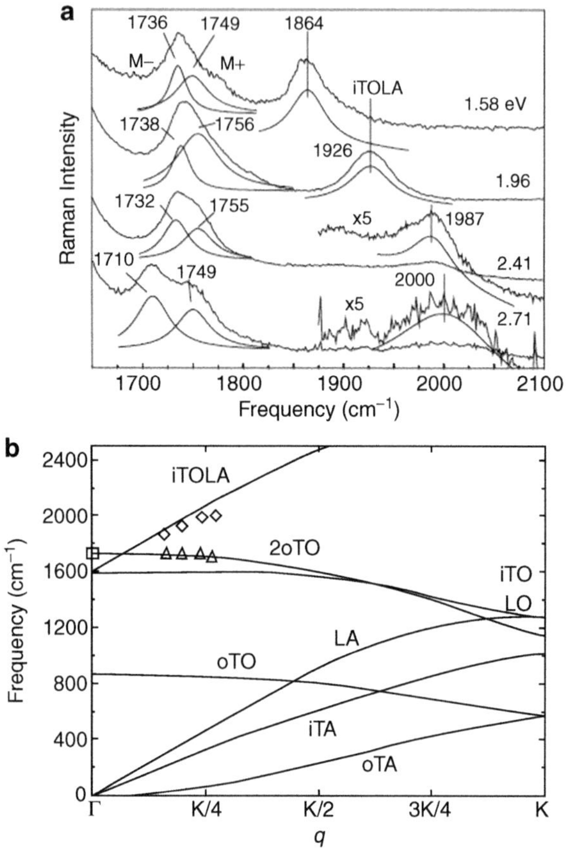

Fig. 7.5 (**a**) Lorentzian fits of the Raman spectra taken at several E_L values for the M-bands near 1750 cm^{-1} and the highly dispersive iTOLA feature observed at 1950 cm^{-1} in SWCNT bundles. Peak frequencies (cm^{-1}) and E_L values (eV) are displayed. (**b**) Phonon dispersion for 2D graphite along the Γ–K direction, including the dispersion for 2oTO and iTO+LA. Squares, up-triangles and diamonds denote experimental values for the M ($q = 0$ and $q = 2k$) and iTO+LA ($q = 2k$) bands, which are here placed on the phonon dispersion curves in accordance with double resonance theory [68] (Reprinted with permission from Brar VW, Samsonidze GG, Dresselhaus MS, Dresselhaus G, Saito R, Swan AK, Ünlü MS, Goldberg BB, Souza Filho AG, Jorio A (2002) Second-order harmonic and combination modes in graphite, single-wall carbon nanotube bundles, and isolated single-wall carbon nanotubes. Physical Review B 66 (15): 027403-1-4. Copyright (2002) by the American Physical Society)

4.2 Comparative Analysis of Raman Spectra of Carbon Nanotubes and Graphite

In addition to many other publications dealing with the so-called G-band of the Raman spectra of SWCNT, we consider the structure of the fundamental bands together with the structure of their harmonics $2\nu_G$ and $2\nu_D$, and their sums $\nu_{G+}\ \nu_D$ and $\nu_G + \nu_{RBM}$. Figure 7.6a shows the review Raman spectra of SWCNT, single

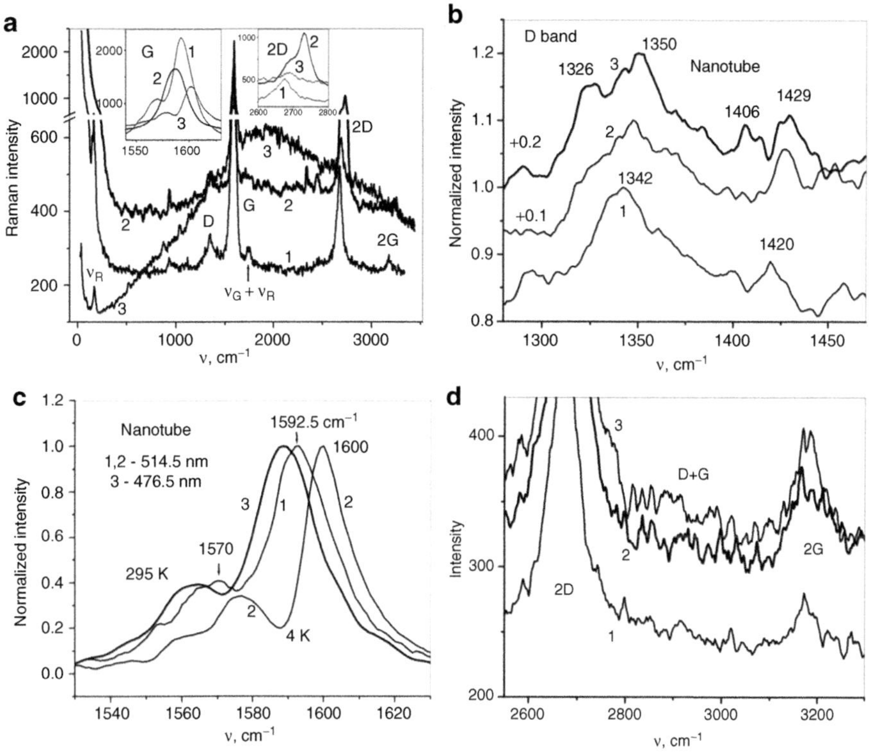

Fig. 7.6 Review spectra of SWCNT (1), HOPG single crystal (2) and polyurethane/SWCNT (2%) composite (3) under excitation with 514.5 nm (**a**), and fragments of Raman spectra of SWCNT in the spectral region corresponding to D-band (**b**), G-band (**c**) and a region of the second-order harmonics $2\nu_G$, $2\nu_D$ and $\nu_{G}\ _{+}\ \nu_D$ (**d**) under excitation with 514.5 (1), 488 (2) and 476.5 nm (3). Insets in Fig. 7.6a show variation of the intensity in the region of the most intensive bands ν_G and $2\nu_D$. The spectra shown in Fig. 7.6b for different λ_L values are displaced in *vertical* direction

crystal of HOPG and of polyurethane-based nanocomposite containing 2% SWCNTs under excitation with 514.5 nm.

One can see that G-band is characterized with the highest intensity in the spectra (Fig. 7.6a). It should be noted as well that the SWCNT $2\nu_D$ band is noticeably higher in intensity in comparison with the main D band, which can be seen from the lower curve 1. Broadband noise is typically not strong in Raman spectra of SWCNT but became more noticeable for the bulk systems (graphite and polymer-SWCNT composite, curves 2 and 3 in the Fig. 7.6a). SWCNT possess also a low-frequency vibration mode at $\nu_{RBM} \approx 160\ \text{cm}^{-1}$ corresponding to radial oscillations of carbon atoms in plane of the cross section, which exhibits a strong variation with nanotube diameter [7]. The SWCNT vibration spectra show a sum harmonic signal ν_G+ν_{RBM} with a spectral shape, which will be discussed below.

Raman spectra of SWCNT in the range of D- and G-bands and the second-order bands under excitation with various wavelengths are shown in Fig. 7.6b–d, correspondingly. The D band is characterized with a fine structure of the spectrum containing at least five spectral components (Fig. 7.6b). Reduction of the wavelength of the excitation λ_L causes shift of the D band maximum to higher frequency. In addition, the side components of this peak, for instance, those at 1326 and $\sim$ 1425 cm^{-1}, are enhanced in their intensities. The effect is more pronounced for lower frequency component, which may be due to strengthening of the contribution from sp^3 hybridization. This can be explained as condition close to a resonance as well as by appearance of vibration states of carbon materials in the region of high wavevectors.

Natural graphite has the maximum of D band near 1375 cm^{-1}, which corresponds to limiting frequencies of the longitudinal acoustic branch near the edge of the Brillouin zone. This maximum is shifted to 1350 cm^{-1} position for highly disordered graphite samples (e.g., for polycrystalline graphite with a crystallite size reduced to $\sim$ 200 Å). The intensity of the D-band in Raman spectra of the disordered graphite samples often can exceed that of the G-band. As an example, the graphite sample with the size of crystallites $\sim 10^3$ Å has the D-band approximately twice the intensity of G-band. However, SWCNT's structure is characterized with smaller structural distortions and therefore the intensity of D-band in their Raman spectra is typically weak.

The G-band of nanotubes in Raman spectra is shifting to lower frequencies with the increase of the energy of excitation photons (Fig. 7.6c), which is contrary to the effect of higher frequency shift of the D-band position and behavior of the graphitic materials, which are characterized with a hypsochromic shift. We believe the behavior of the G-band of SWCNT Raman spectra is the result of two mutually exclusive trends. On one hand, due distortions in the crystalline structure of the graphitic layers, the G-band maximum shifts to higher frequency range similar to the behavior observed in disordered graphite samples. On the other hand, due to cylindrical symmetry of the nanotubes (i.e., bending of the grapheme layer) the electron states of the latter are characterized with higher contribution from sp^3 hybridization, which leads to decrease of the frequency of C-C skeleton vibrations of the carbon frame. Competition of these two trends has different effect in different parts of the Brillouin zone, which can be visualized by variation of the excitation frequency and observation of phonon excitation corresponding to different wavevectors. One should expect that variation in the hybridization of carbon atoms corresponding to different structures will appear first of all in the region of higher wavevectors. In fact, this region corresponds also to the case where such variations are easier to detect. Particularly, for $\lambda_L = 514.5$ nm, the frequency of the SWCNT G-band maximum $\nu_{SWCNT} = 1592.5$ cm^{-1} is higher compared to the frequency of the G-band of isotropic polycrystalline graphite $\nu_G(200\ \text{Å}) = 1587$ cm^{-1} having the size of crystallites of 200Å, while the position of these bands is reversed ($\nu_{SWCNT} = 1588$ cm^{-1}, ν_G (200 Å) = 1590 cm^{-1}) for $\lambda_L = 488$ nm. The effect can be explained by the appearance of two opposite trends in different regions of the Brillouin zone. Generally, by lowering the λ_L, G-band in Raman spectra of SWCNTs are displaced bathochromically. However, lowering the

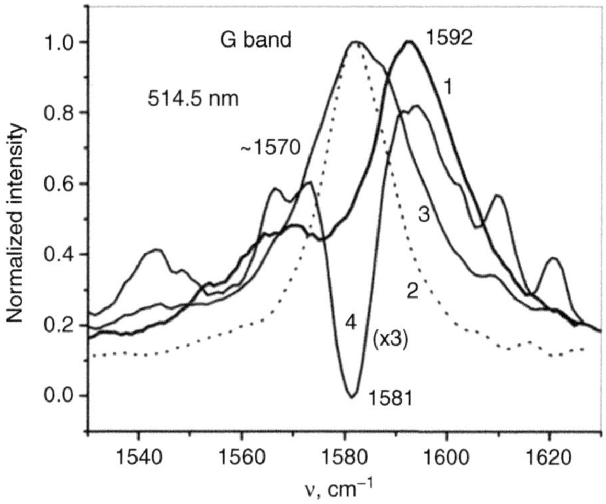

Fig. 7.7 Comparison of Raman spectra for carbon nanotubes (1) and normalized G-bands of a single crystalline (2) polycrystalline anisotropic graphite sample with oriented grains and L≈300 Å (3) and the difference spectrum corresponding to the curves 2 and 3 (4). All spectra correspond to room temperature and excitation with λ_L=514.5 nm and are normalized to the maximum of the intensity of the corresponding Raman band. The difference spectrum (4) is multiplied by a factor of 3 for better comparison with the other spectra

temperature to 4 K causes high frequency shift of ν_G and gives an evidence for strengthening of the C–C bonds in nanotubes, which is in contrary to the behavior of other crystalline forms of carbon.

In the region of the second-order bands, a weak sum harmonic (i.e., ν_D+ν_G band) is observed (Fig. 7.6d) in addition to 2ν_D and 2ν_G harmonics. As the frequency of the laser excitation radiation ν_L increases, the intensity of the harmonic bands as well as the intensity of the broadband noise also increases but not the intensities of the fundamental bands, which apparently remain the same. In order to separate useful vibration bands, the broadband noise was approximated with a linear or a quadratic power function and was subtracted from the corresponding Raman spectra. As discussed above, the intensity of the allowed transition of 2ν_D harmonic was significantly higher than the intensity of the fundamental D band (ν_D). It was also higher than the intensity of the 2ν_G harmonic despite the fact that in the region of the main bands the intensity of the fundamental G-band was dominating.

Comparison of the position and shape of G-band of carbon nanotubes and various graphite samples is shown in Fig. 7.7. The G-band of SWCNTs has a doublet structure 1570–1592 cm^{-1} (curve 1 in Fig. 7.7). Due to the fact that distortion of the "graphene layer" along the nanotube main axis is minimal, the comparison of nanotube Raman spectra with the variation of the G-band of HOPG (possessing a low level of distortions) is reasonable. In view of this similarity, we have compared the Raman spectra of a single crystal and a polycrystalline samples (with the oriented grains) possessing the crystallites of the size 300 Å (curves 2 and 3 in the Fig. 7.7). The increase of the disorder in graphite (in the samples of natural and polycrystalline isotropic graphite) results in broadening and shift of the G-band maximum to higher frequencies while the lower wavelength edge of the band being practically unaffected. Simultaneously, enhancement of the broadband noise has been observed.

Figure 7.7 shows Raman spectra corresponding to ν_G bands of the graphite samples under study as well as their difference spectrum and gives a comparison with corresponding spectrum of the nanotubes recorded under excitation with the wavelength $\lambda_L = 514.5$ nm. As can be seen, the main maxima of the difference spectrum are in good agreement with the maxima of G-band position for SWCNT. The difference spectrum also demonstrates that disordering of the crystalline structure of anisotropic polycrystalline graphite (with oriented grains of the size ~300Å) is stronger in comparison to the nanotubes, which is evident from relatively high intensity of the other Raman bands. In particular, low frequency 1543 cm^{-1} maximum in the subtraction spectrum corresponds to an additional band with the intensity strongly depending on the wavelength of excitation. Its intensity considerably increases when the Raman spectrum is excited with shorter wavelength ($\lambda_L = 488$ nm). Furthermore, the two high frequency bands (1608 and 1622 cm^{-1}) exhibit similar behavior, that is, their intensities also increase under conditions of smaller wavelength excitation in natural and isotropic graphite samples. Therefore, the doublet structure of the G-band can be related to a distortion of the graphene layer, which is supported by the increase of the intensity for this band in comparison to the graphitic G-band (see inset in Fig. 7.10a).

4.3 Harmonic Bands and D+G Sum Mode

In this section, we aim to compare the shape of fundamental Raman bands and their harmonics. Due to high variation of the corresponding intensities, it is natural to compare the bands after normalization to the corresponding maxima. It is worth noting that the positions of maximum of the $2\nu_D$ harmonic maximum and the D-band shift to higher frequencies with ν_L increase while the maxima of the corresponding G-bands shift to lower frequencies. Comparison of the harmonic bands $2\nu_G$, $2\nu_D$ for SWCNT and graphite single crystals is illustrated in Fig. 7.8a, b.

In comparison to D and G harmonics in graphite single crystal, the corresponding SWCNT bands are significantly shifted to a lower frequency range. Despite the fact that ν_G-band for nanotubes has higher frequency than corresponding band in graphite (1592 cm^{-1} vs 1582 cm^{-1}) giant frequency shifts are observed for $2\nu_D$ and $2\nu_G$ harmonic bands (56 and 63 cm^{-1} correspondingly) as can be seen from comparison of Fig. 7.8a, b. Considerable broadening of the G harmonic vibrational band of SWCNT in comparison to that of the graphite single crystal sample can be also seen.

Comparison of the shape of the first-order peaks (ν_D and ν_G) and corresponding second-order harmonics ($2\nu_D$, $2\nu_G$) for the case of $\lambda_L = 476.5$ nm excitation is shown in Fig. 7.9a, b. The frequencies of the first-order spectra have been doubled for such comparison which gives a fair assessment between the experimentally

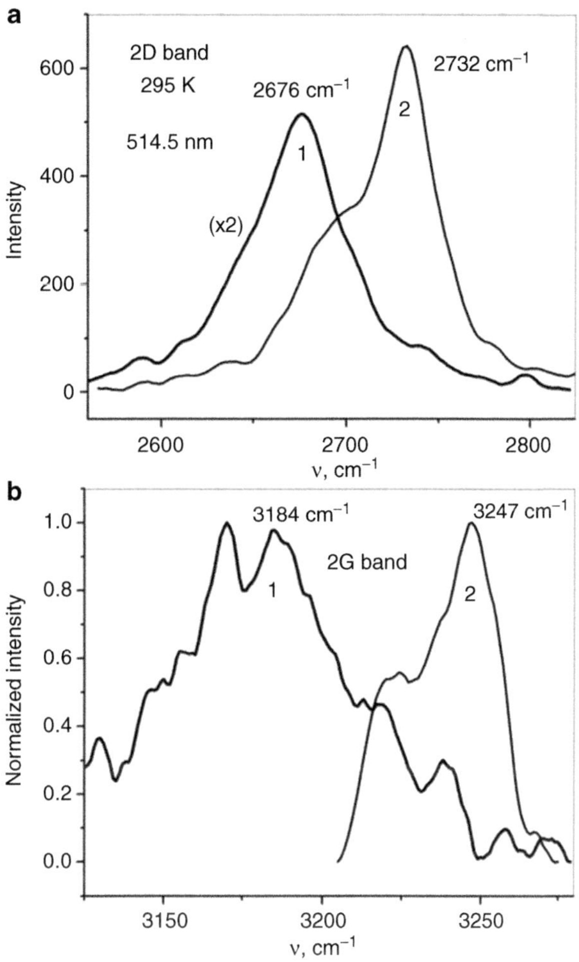

Fig. 7.8 Comparison of harmonic bands $2\nu_D$ (**a**) and $2\nu_G$ (**b**) highlighted in the Raman spectra for SWCNT (curves 1) and graphite single crystal (curves 2)

observed and the calculated values based on the fundamental bands harmonic spectra of $2\nu_D$ and $2\nu_G$. Comparison of the first-order and the second-order bands is interesting because of the possibility of observation of all available vibrational states. In the latter case, even the states corresponding to the border of the Brillouin zone and the opposite directions of the corresponding wavevectors of the interacting phonon waves can be observed. On the contrary, only the states with low wavevectors (k∼0) can be detected for the first-order spectra. We estimate the lowest limit of possible $2\nu_D$ and $2\nu_G$ harmonics broadening in our case. Under conditions of phonon dispersion, the real bands of SWCNT harmonic oscillations may be significantly broader. Therefore, the applied procedure for the comparison between the first- and the second-order Raman bands gives deeper insight into cooperative vibration modes of the nanotubes.

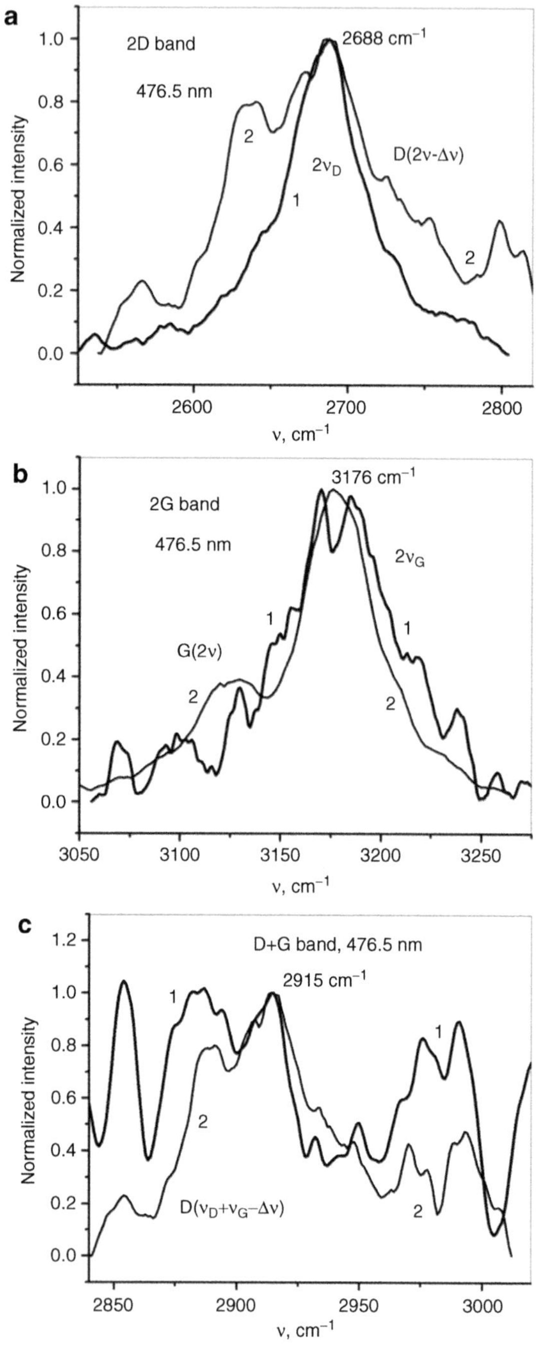

Fig. 7.9 Comparison of the experimentally observed SWCNT Raman spectra (curves 1) in the frequency range of harmonic bands $2\nu_D$ (**a**), $2\nu_G$ (**b**), and of the sum tone $\nu_D+\nu_G$ (**c**) (laser excitation with the wavelength $\lambda_L=476.5$ nm) with theoretically calculated bands based on the main bands ν_D and ν_G spectra (curves 2). All bands are normalized to the corresponding maxima and the calculated bands are shifted to the lower frequencies on values $\Delta\nu$ of 14 cm^{-1} (**a**) and 24 cm^{-1} (**c**) to achieve coincidence of the theoretical bands maxima positions with the experimental ones

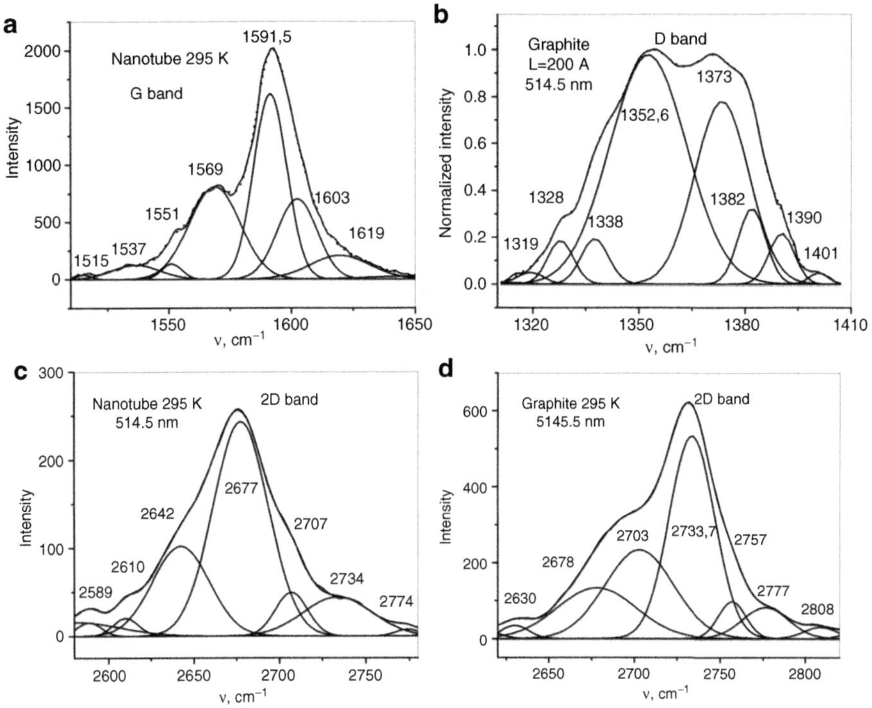

Fig. 7.10 Results of numerical deconvolution of the main bands and harmonic tones in Raman spectra into composing lines: ν_G band of SWCNT (**a**), ν_D band of an isotropic polycrystalline graphite (**b**), $2\nu_D$ tone spectrum of SWCNT (**c**), and $2\nu_D$ tone spectrum of the graphite single crystals (**d**)

Due to a pronounced anharmonic behavior, the maximum position of the experimentally observed harmonic band $2\nu_D$ is shifted to lower frequencies in comparison to the theoretically calculated value. Therefore, the calculated band of the harmonic vibration $I_D(2\nu)$ in Fig. 7.9a has been shifted 14 cm^{-1} to lower frequencies to achieve a coincidence with the experimentally observed band. On the other hand, in the case of $2\nu_G$ harmonic, theoretical band position matches well the experimentally observed maximum due much smaller anharmonicity of the G mode for the nanotubes. Therefore, G- and D-bands show very different anharmonic behavior as will be considered in more detail below.

First of all, the observed $2\nu_D$ harmonic band is anomalously sharp in comparison to the one theoretically calculated from the main D-band and does not have such a pronounced internal structure as the latter (Fig. 7.9a). Due to a presence of two maxima (around 1570 cm^{-1} and around 1592 cm^{-1}) in the main G-band, one should expect three peaks in the $2\nu_G$ harmonic band (two peaks corresponding to a doubled frequency of these two peaks and one peak corresponding to their sum) with a spectral interval of ~22 cm^{-1}. However, this is not the case even with consideration of a possible fine structure for the constituting bands. Relatively narrow width of the $2\nu_D$ and $2\nu_G$ harmonic bands and absence of the expected fine structure in them (obviously present in the corresponding first order spectra) can be due to a spectral compression effect, often observed in the nonlinear optics of short pulses. As far as we know, such anomalous narrowness of harmonic vibration bands was not observed before for condensed molecular systems. It can be probably related to a typical signature of a nonlinear behavior of nanostructures, explained by the higher contribution from the surface atoms in comparison to the bulk ones. One can assume that due to high nonlinearity of SWCNT, nonlinear resonance effects are observed even under conditions of their thermal excitation. Therefore, we discuss in more detail the D- and G-bands' anharmonicity in the next section.

The half-widths of the second harmonic bands $2\nu_D$ and $2\nu_G$ are 56 and 64 cm^{-1}, respectively (Fig. 7.9a, b). This is practically the same frequency shift as observed for the maxima of these bands relative to similar bands observed in the single crystalline graphite sample (Fig. 7.8a, b). The half-widths observed for the first-order vibration bands ν_D and ν_G are 56 and 21 cm^{-1} correspondingly. The too narrow width of the observed tone bands does not allow for a more definite conclusion as concerning the appearance of cooperative effects in the vibration modes of the SWCNT. Nevertheless, a significant role of nonlinear interactions of their excitations shows the importance of their wave properties for the Raman spectra interpretation.

G-band of SWCNT's Raman spectra is a relatively narrow peak as discussed before. Therefore, the main contribution to the sum tone $\nu_D+\nu_G$ is expected from the more complex and broader D component. It should also be shifted to a higher frequency by $\nu_G = 1{,}588\text{–}1{,}592.5\ cm^{-1}$ for correct comparison. Such comparison of the fine structure of the sum tone $\nu_D+\nu_G$ with the frequency shifted D-band of the nanotubes is given in Fig. 7.9c. In order to achieve a good match of all the peculiarities of the D-band possessing at least six spectral components, the mathematically calculated second harmonic band had to be additionally shifted to lower frequencies by $\Delta\nu = 24\ cm^{-1}$ to match the experimentally observed spectral position. Therefore, the value of 24 cm^{-1} should be considered as the value of Raman anharmonicity. A good agreement of the all spectral components of the fine structure for the first- and the second-order harmonics (ν_D and $\nu_D+\nu_G$) registered for several different excitation wavelengths and at different temperatures makes unambiguous interpretation of the observed fine structure in the spectra. This confirms also good quality of the obtained spectra even for the weakest $\nu_D+\nu_G$ bands (see Fig. 7.6d). The intensity enhancement of the lowest and the highest frequency components in the considered sum tone is observed. This is a typical signature of a resonance Raman scattering and may be connected to a contribution from the electronic states of carbon atoms with sp^3 hybridization.

4.4 Deconvolution of the Vibration Bands into Elementary Components

It is convenient to make a comparison of the different vibration bands of SWCNT and graphite after deconvolution of the corresponding spectra into elementary

Table 7.1 Spectral positions and half-widths of the main spectral components of the most intensive bands in Raman spectra of SWCNT and graphitic materials under study

Raman band	ν_G, SWCNT		ν_D, isotropic graphite		$2\nu_D$, SWCNT		$2\nu_D$, crystalline graphite	
ν, cm^{-1}	1569	1591.5	1352.6	1373.4	2642	2677	2703	2733.7
$\delta\nu$, cm^{-1}	20.8	13.5	21.7	15.3	33.2	32.5	40.1	26.4

spectral components. We applied Lorentzian shape of the constituting lines and used the mean square procedure for spectra deconvolution with a variable spectral position, intensity, and half-width of the individual lines. Subtraction of the broadband noise has been done prior to such mathematical procedure. The number of reliable spectral component was determined from the corresponding second derivative spectra by the number of the negative minima in accordance to a standard procedure. The deconvolution results for nanotube and graphite Raman bands are shown in Fig. 7.10. The positions of the corresponding individual line maxima are marked and the corresponding wave numbers are displayed in the Figures. The half-widths of the main spectral components are presented in Table 7.1.

It is worth mentioning, that the half-widths of the main components of the ν_D band of disordered graphite are approximately equal to those of the doublet components of the ν_G band of SWCNT, which affirms the above conclusion about considerable contribution of the graphene layer deformation to the formation of the G-band in Raman spectra of such compounds. Figure 7.10b shows that distortions of the graphitic structure lead to the intensity enhancement of the low-frequency component (1352.6 cm^{-1}) in the spectra, which is opposite to the trend observed for the D-band. The latter band is shifting to higher frequencies under increasing energy of the excitation photons (see Fig. 7.10b). Similarly, the distortion of the structure and higher excitation frequencies are opposing each other in their effect on the G-band. In accordance to Fig. 7.6c the ν_G band spectral component positions are shifting to lower frequencies with the ν_L increase while an opposite effect (i.e., shift to higher frequencies) is observed for this band if structural distortions are enhanced in the system. It is worth noting that the shift observed for ν_D and ν_G bands of SWCNT Raman spectra is analogous to the shifts corresponding to deformation ν_δ and valence ν_{OH} modes of water under conditions of strengthening of the hydrogen bonding. Therefore, the vibration properties of the nanotubes are affected by their exposure to the short wavelength radiation excitation. Particularly, the D mode is more affected by excitation at higher frequencies, which is confirmed by a notable increase of the ν_D and $2\nu_D$ band spectral positions while the G-band remains practically unchanged.

We have detected a large number of spectral components for all SWCNT Raman bands under consideration. For instance, under excitation with 514.5 nm wavelength, the G-band in our spectrum can be decomposed into eight constituting lines: 1516, 1534, 1551, 1569, 1592.5, 1605, 1615, and 1624 cm^{-1}. The reliability of this fine structure is proved by its exact reproduction under excitation with different λ_L or at different temperatures as well as by simultaneous observation of the same fine structure for G- and D-bands in the Raman spectra.

To conclude this section, we consider in short the connection between individual components of the nanotube and graphite samples vibration bands. Figure 7.10c, d demonstrates that the spectral components of the SWCNT's Raman spectrum $2\nu_D$: 2642, 2677, 2707, and 2734 cm^{-1} make a good match with the components 2630, 2678, 2703, and 2733,7 cm^{-1} of the corresponding band in graphite. The only difference is that the SWCNT's band has two lower frequency component lines with higher intensity, while the corresponding graphite's band has the two higher frequency lines of the composite band more intensive. This explains a higher shift of the harmonic lines observed in nanotubes in comparison to a single crystal of graphite. Some discrepancy of the spectral positions (especially of weak peaks near the "wings" of the bands) may be ascribed to the accuracy of the applied numerical deconvolution procedure in splitting of the bands into constituting fine structure components. The observed pronounced changes in the intensity of the individual components of the Raman spectra of SWCNT and graphite are in good agreement with the suggested concept of the variation of electronic states of carbon materials under the influence of strong high-frequency light irradiation.

4.5 Anomalous Changes of Vibration Modes' Intensities and Anharmonicity in Raman Spectra of SWCNTs

We have compared the intensity of harmonic bands with one of G-bands, which has weak dependence on the wavelength of excitation. As can be seen from Fig. 7.11a, the intensity of $2\nu_D$ harmonic band increases linearly under the excitation frequency ν_L increase, and significantly exceeds the increase of the $2\nu_G$ harmonic vibration. When λ_L changes from 514.5 to 476.5 nm the intensity of the $2\nu_D$ tone increases 1.69 times. This is significantly higher than the theoretical value for the Raman spectrum enhancement (1.36 in accordance to ω^4 law). These results cannot be explained only by the resonance Raman spectrum enhancement mechanism due to the fact that the multiplication factor is different for G- and D-band harmonics. We explain the anomalous enhancement of the $2\nu_D$ and $2\nu_G$ tones by an opposite frequency displacement of the main bands ν_D, ν_G and by a different variation of the anharmonicity of D and G modes under excitation frequency ν_L increase.

The value of the vibration mode anharmonicity can be characterized by the difference between the calculated values of the first-order tones and the experimentally observed positions for the second-order maxima: $\Delta\nu_{D,G} = 2\nu_{D,G} - \nu_{(2D,2G)}$ and $\Delta\nu_{DG} = \nu_D + \nu_G - \nu_{(D+G)}$. Dependence of $\Delta\nu_{D,G}$ and $\Delta\nu_{DG}$ values on ν_L is shown in Fig. 7.11b. One can easily see that the highest anharmonicity is observed for the sum tone $\nu_D + \nu_G$. In the case of a long wavelength radiation λ_L used in our experiments, the anharmonicity of the graphite G mode is higher than that of the D mode. However, as ν_L increases, the anharmonicity of the D mode and of the sum tone $\nu_D + \nu_G$ increases while that of the G mode decreases. In other words, the deviation of the G mode position from its theoretical frequency predicted by the harmonic approximation is increasing for the nanotubes while the G mode of

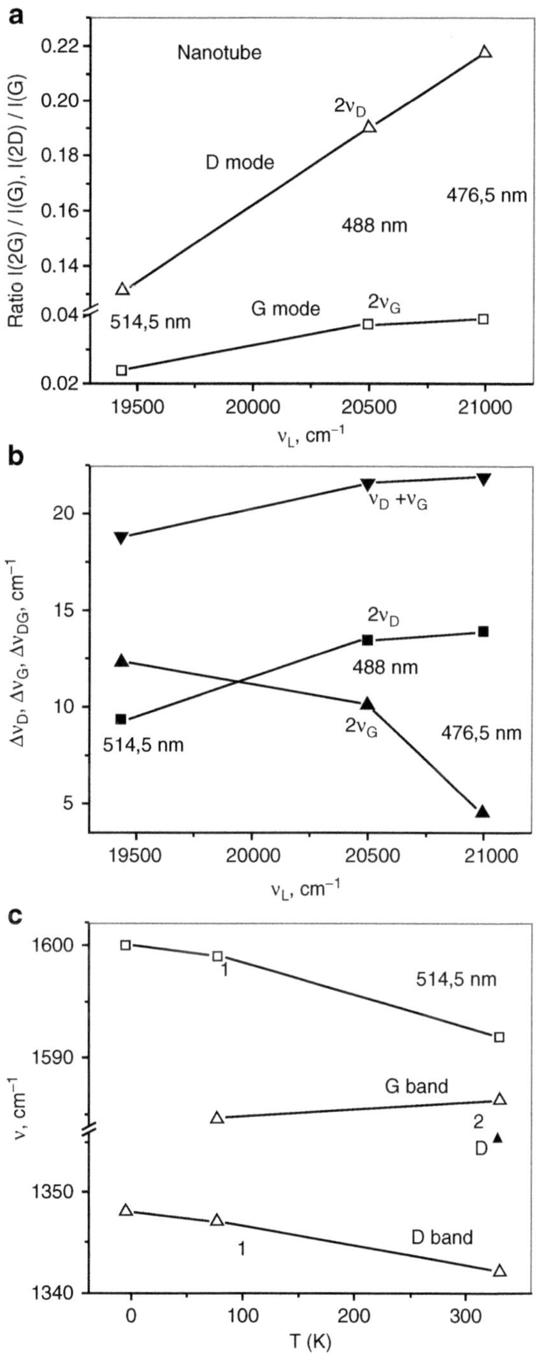

Fig. 7.11 Dependence of the ratios of the second-order to the first-order Raman band intensities (**a**) and of the experimentally observed harmonic bands (**b**) on the excitation frequency ν_L (**b**). (**c**) Temperature dependencies of the G and D Raman bands of SWCNTs (1) and graphite crystals (2)

graphite shows an opposite behavior. Anomalous behavior is also manifested in a strong dependence of the SWCNT material Raman bands on the wavelength of the excitation λ_L, as well as in a difference in anharmonicity for different vibration bands.

Such unexpected variations of the anharmonicity can be explained by considerable variations of the main electronic state under influence of high frequency excitation. Under these conditions the electronic states in the nanotubes vary in such a way that deviation from the harmonic approximation is decreasing for G mode but increasing for the D mode and the sum D+G tone.

The observed anharmonic behavior of the D mode is in good agreement with the behavior of the second-order Raman bands despite the fact that such dependence would be more natural for IR absorption spectra. It is known that the IR absorbance of the tones is determined by the value of their anharmonicity. In Raman spectra, the intensity of the bands depends on the second derivative of polarizability with respect to coordinates: $(d\alpha/dq)^2$. Therefore, in the Raman spectrum, the D mode shows enhancement of the polarizability with the anharmonicity, which follows from the variation of the electronic states in the system. For G mode, weakening of the anharmonicity with ν_L increase is accompanied by a small increase in $(d\alpha/dq)^2$. Variation of the electronic state of carbon nanotubes might be related to the effect of strong electron-phonon coupling. This effect is similar to electron–phonon interaction enhancement observed near the melting transition temperature [71]. Figure 7.11c displays opposite temperature dependencies of D and G modes in SWCNT and graphitic single crystals, which can be also related to the effects of the electron–lattice interactions but requires additional studies.

4.6 The Sum Frequency Harmonic of the SWCNT Low Frequency Mode

The increase in the frequency of excitation radiation causes the intensity of the Raman ν_{RBM} mode to increase significantly. Such effect is in agreement with a suggested concept of photo-induced variation of the electronic states. One more argument to support this concept is the increase of Rayleigh shoulder intensity with the increase of the excitation laser frequency. The main low-frequency ν_{RBM} mode appears also in a region of the sum and difference frequencies – that is, $\nu_G \pm \nu_{RBM}$ near intensive G-band as can be seen in Fig. 7.10a. In this work, we consider only the sum tone band as it is much more intensive than the difference mode. It is interesting to compare the normalized shape of the low-frequency band $\nu_{RBM} \sim 160\ \text{cm}^{-1}$ and that of the sum tone $\nu_G + \nu_{RBM}$ near 1750 cm^{-1}. For this comparison, the $\nu_G \sim 1588$–$1592\ \text{cm}^{-1}$ has been subtracted from the sum band similar to the situation when ν_D was shifted to the sum tone $\nu_D + \nu_G$ region (Fig. 7.9c). The results of such comparison at two different wavelengths of excitation radiation, with normalization to the corresponding maxima are shown in Fig. 7.12a, b. First of all, one should note that the frequency position of the low frequency ν_R mode is decreasing similar to the G mode behavior under excitation frequency ν_L increase. Remarkably, under $\lambda_L = 514.5$ nm the maxima of these bands practically coincide, which justify a low value of the corresponding anharmonicity.

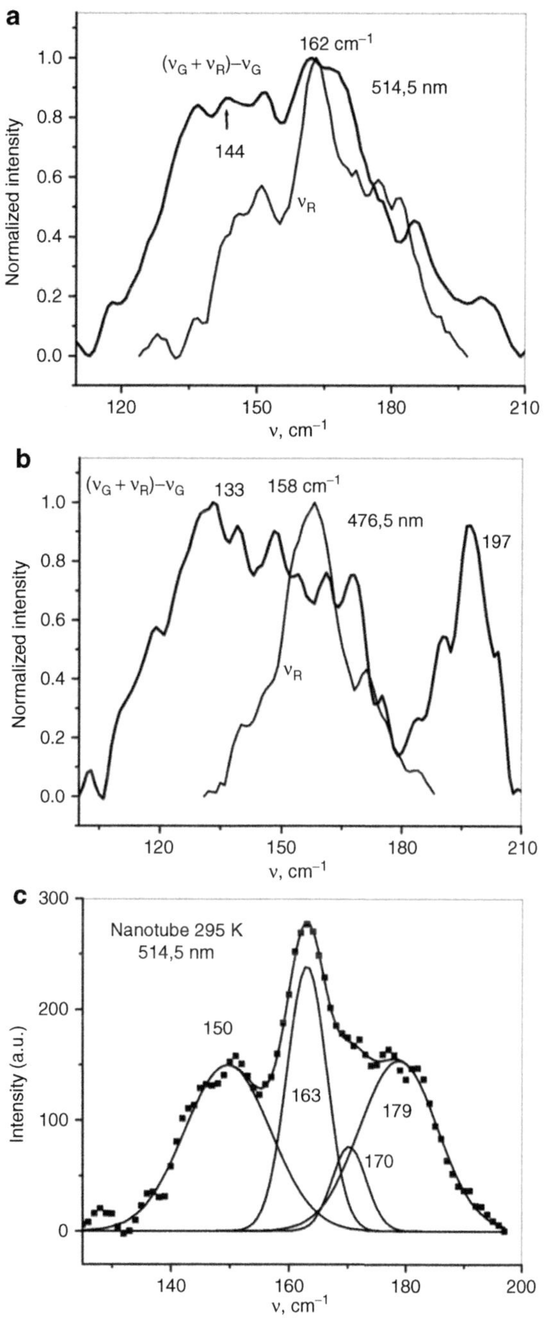

Fig. 7.12 Comparison of normalized bands of low-frequency mode and sum tone $\nu_G + \nu_{RBM}$ under excitation with $\lambda_L = 514.5$ (**a**) and 476.5 nm (**b**); decomposition of low-frequency "breathing" mode of SWCNT ν_{RBM} to individual components (**c**)

Contrary to the above considered harmonic bands, the band of the sum tone $\nu_G+\nu_{RBM}$ is much broader than the low frequency band. We explain this fact by a well-pronounced doublet structure of G-band, which has appearance in the considered sum tone. This is also confirmed by the same spectral range between the observed maxima and by similarity in the intensities of the corresponding constituting components of the doublet bands. More detailed analysis of the structure of all the observed bands is required in the future.

In case of Raman spectra excitation with shorter wavelength radiation (476.5 nm), more complex structure of the sum tone $\nu_G+\nu_{RBM}$ is observed (see Fig. 7.12b) similar to the D-band and sum tone $\nu_D+\nu_G$ behavior represented earlier in Fig. 7.10b. Importantly, the intensity of the low-frequency component of the sum tone near 1720 cm^{-1} (corresponding to the 133 cm^{-1} component in the shifted spectrum of comparison) significantly increased in intensity and a high-frequency component appeared at 197 cm^{-1}. Similar to high-frequency D- and G-bands, we have made a quantitative study of the fine structure of the SWCNT low-frequency mode ν_{RBM}. The results corresponding to excitation with 514.5 nm laser are shown in Fig. 7.12c. This band can be reliably decomposed into four spectral components. Remarkably, the half-widths of the most intensive component at 163 cm^{-1} is only 6.8 cm^{-1}, which is less than a half of the narrowest component of the ν_G band in nanotubes and confirms high quality factor of the low-frequency mode.

5 Summary and Future Perspective

Carbon nanotubes and graphite are excellent model systems to address fundamental issues related to physical materials science. This is due to relative simplicity of these materials containing just one type of atoms and small number of vibration bands as well as possibility of variation of the contributions from sp^2 and sp^3 hybrid states of carbon atoms within the same system. Raman spectroscopy is very important and powerful tool for the study and characterization of graphitic materials and carbon nanotubes especially. In this article, we give a short review of the achievements of the Raman spectroscopy in the study of the physical properties of carbon nanotubes during the last decade.

The detailed analysis of the first- and the second-order Raman spectra of SWNTs has been done. It is emphasized particularly that the second-order Raman features are rich in information about electron and phonon structure of SWNTs, which cannot be obtained from the first-order Raman spectra.

We report on the influence of a short-wave laser radiation on the properties of SWCNTs. This effect is confirmed by considerable increase of the intensities of $2\nu_G$ and $2\nu_D$ harmonics and of low-frequency radial mode ν_R in their Raman spectra under increase of the energy of exciting photons. While the G-band intensity remains almost the same, the intensity of the $2\nu_D$ band increases $\sim$1.7 times under excitation with 476.5 nm in comparison with 514.5 nm. Excitation with the 476.5 nm laser beam also leads to increase of the side bands in the fine structure of the ν_D mode as well as $\nu_D+\nu_G$ and $\nu_G+\nu_{RBM}$ harmonics. We demonstrate that hypsochromic shift of ν_D and bathochromic shift of ν_G with ν_L increase are in opposite direction to the spectral shift observed in graphitic materials due to their structural distortions. The reported peculiarities are related to photoinduced electronic state variations.

We report on anomalous behavior of the observed anharmonicity for different vibrational modes of SWCNT. This peculiarity appears in a strong dependence of the anharmonicity on the wavelength of excitation as well as in the opposite trends observed for different vibration bands. While the anharmonicity of the D mode increases with the frequency of the excitation radiation (ν_L), the same of the G mode decreases. The sum harmonic band $\nu_D+\nu_G$ is characterized with the highest anharmonicity while the same for a composed tone $\nu_G+\nu_{RBM}$ is negligible. The frequency dependence of the anharmonicity also supports the concept of variation of the electronic states under the light excitation.

The structure of vibration bands of the first and the second order in SWCNT Raman spectra has also been studied for ordered and disordered forms of graphite. This was accomplished by decomposition of the complex spectral bands into constituting components. We found proximity of spectral positions in most of spectral components of the nanotubes and graphite and considerable variation of their intensities. This also demonstrates variation of the electronic polarizabilities and can explain anomalous shifts of the harmonic bands $2\nu_G$ and $2\nu_D$ for nanotubes in comparison to corresponding bands of a single crystalline graphite. Narrow width of the low frequency mode $\nu_{RBM}{\sim}160\ cm^{-1}$ leads to reproduction of the G-band structure in the sum harmonic band $\nu_G+\nu_{RBM}{\sim}1750\ cm^{-1}$ while the complex structure of the broad ν_D band is remarkably reproduced in the $\nu_D+\nu_G$ sum tone. The narrow width of SWCNT's $2\nu_D$ and $2\nu_G$ harmonics in the Raman spectra may be related to group synchronism effects [72].

References

1. Dresselhaus MS, Dresselhaus G, Eclund PC (1996) Science of fullerenes and carbon nanotubes. Academic, New York/San Diego
2. Saito R, Dresselhaus G, Dresselhaus MS (1998) Physical properties of carbon nanotubes. Imperial College Press, London
3. Dresselhaus MS, Dresselhaus G, Saito R, Jorio A (2005) Raman spectroscopy of carbon nanotubes. Phys Rep 409:47–99
4. Dresselhaus MS, Dresselhaus G, Jorio A, Souza Filho AG, Pimenta MA, Saito R (2002) Single nanotube Raman spectroscopy. Acc Chem Res 35:1070–1078
5. Jorio A, Saito R, Hafner JH, Lieber CM, Hunter M, McClure T, Dresselhaus G, Dresselhaus MS (2001) Structural (*n*, *m*) determination of isolated single-wall carbon nanotubes by resonant Raman scattering. Phys Rev Lett 86(6):1118–1121
6. Thomsen C (2000) Second-order Raman spectra of single and multiwalled carbon nanotubes. Phys Rev B 61(7):4542–4544
7. Rao AM, Richter E, Bandow S, Chase B, Eklund PC, Williams KA, Fang S, Subbaswamy KR, Menon M, Thess A, Smalley RE, Dresselhaus G, Dresselhaus MS (1997) Diameter-selective Raman scattering from vibrational modes in carbon nanotubes. Science 275:187–191
8. Shimada T, Sugai T, Fantini C, Souza M, Cançado LG, Jorio A, Pimenta MA, Saito R, Grüneis A, Dresselhaus G, Dresselhaus MS, Ohno Y, Mizutani T, Shinohara H (2005)

Origin of the 2450 cm^{-1} Raman bands in HOPG, single-wall and double-wall carbon nanotubes. Carbon 43(5):1049–1054
9. Ferrari AC, Meyer JC, Scardaci V, Casiraghi C, Lazzeri M, Mauri F, Piscanec S, Jiang D, Novoselov KS, Roth S, Geim AK (2006) Raman spectrum of graphene and graphene layers. Phys Rev Lett 97(18):187401–187404
10. Reich S, Thomsen C, Ordejón P (2002) Electronic band structure of isolated and bundled carbon nanotubes. Phys Rev B 65(15):155411
11. Schadler LS, Giannaris SC, Ajayan PM (1998) Load transfer in carbon nanotube epoxy composites. Appl Phys Lett 73(26):3842–3844
12. Haggenmueller R, Gommans HH, Rinzler AG, Fischer JE, Winey KI (2000) Aligned single-wall carbon nanotubes in composites by melt processing methods. Chem Phys Lett 330:219–225
13. Bhattacharyya AR, Sreekumar TV, Liu T, Kumar S, Ericson LM, Hauge RH, Smalley RE (2003) Crystallization and orientation studies in polypropylene/single wall carbon nanotube composite. Polymer 44(8):2373–2377
14. Bliznyuk VN, Singamaneni S, Sanford RL, Chiappetta D, Crooker B, Shibaev PV (2006) Matrix mediated alignment of single wall carbon nanotubes in polymer composite films. Polymer 47(11):3915–3921
15. Sandler J, Shaffer MSP, Windle AH, Halsall MP, Montes-Moran MA, Cooper CA, Young RJ (2003) Variations in the Raman peak shift as a function of hydrostatic pressure for various carbon nanostructures: a simple geometric effect. Phys Rev B 67(3): 035417–035418
16. Ganeev RA, Ryasnyansky AI, Kodirov MK, Usmanov T (2001) Generation of the third harmonic in fullerene-containing polyimide films by picosecond radiation of the Nd: YAG laser. Techn Phys 46:1270–1273
17. Tuinstra F, Koenig JL (1970) Raman spectrum of graphite. J Chem Phys 53(3):1126–1130
18. Jimenez-Sandoval S (2000) Micro-Raman spectroscopy: a powerful technique for materials research. Microelectron J 31(6):419–427
19. Horiba Scientific, Raman Spectroscopy Applications. http://www.horiba.com/us/en/scientific/products/raman-spectroscopy/. Accessed 25 Nov 2011
20. Milnera M, Kürti J, Hulman M, Kuzmany H (2000) Periodic resonance excitation and intertube interaction from quasicontinuous distributed helicities in single-wall carbon nanotubes. Phys Rev Lett 84:1324–1327
21. Iijima S (1991) Helical microtubules of graphitic carbon. Nature 354:56–58
22. Iijima S, Ichihashi T (1993) Single-shell carbon nanotubes of 1-nm diameter. Nature 363: 603–605
23. Bethune DS, Kiang CH, de Vries MS, Gorman G, Savoy R, Vazquez J, Beyers R (1993) Cobalt-catalysed growth of carbon nanotubes with single-atomic-layer walls. Nature 363:605–607
24. Pimenta MA, Marucci A, Empedocles S, Bawendi M, Hanlon EB, Rao AM, Eklund PC, Smalley RE, Dresselhaus G, Dresselhaus MS (1998) Raman modes of metallic carbon nanotubes. Phys Rev B 58(24):R16016–R16019
25. Reich S, Thomsen C, Maultzsch J (2004) Carbon nanotubes. Basic concepts and physical properties. Wiley-VCH, Weinheim
26. Dresselhaus MS, Dresselhaus G, Hofmann M (2007) The big picture picture of Raman scattering in carbon nanotubes. Vib Spectrosc 45(2):71–81
27. Kuzmany H, Plank W, Schaman Ch, Pfeifer R, Hasi F, Simon F, Rotas G, Pagona G, Tagmatarchis N (2007) Raman scattering from nanomaterials encapsulated into single wall carbon nanotubes. J Raman Spectrosc 38(6):704–713
28. Kuzmany H, Plank W, Hulman M, Kramberger Ch, Grüneis A, Pichler Th, Peterlik H, Kataura H, Achiba Y (2001) Determination of SWCNT diameters from the Raman response of the radial breathing mode. Eur Phys J B 22(3):307–320

29. Milnera M, Kürti J, Hulman M, Kuzmany H (2000) Periodic resonance excitation and intertube interaction from quasicontinuous distributed helicities in single-wall carbon nanotubes. Phys Rev Lett 84:1324–1327
30. Elman BS, Dresselhaus MS, Dresselhaus G, Maby EW, Mazurek H (1981) Raman scsttering from ion-implanted graphite. Phys Rev B 24(2):1027–1034
31. Ferrari AC, Robertson J (2004) Raman spectroscopy of amorphous, nanostructured, diamond–like carbon, and nanodiamond. Philos Trans R Soc Lond A 362:2477–2512
32. Vidano RP, Fischbach DB, Willis LJ, Loehr TM (1981) Observation of Raman band shifting with excitation wavelength of carbons and graphites. Solid State Commun 39(2):341–344
33. Kastner J, Pichler T, Kuzmany H, Curan S, Blau W, Weldon DN, Delamisiere M, Draper S, Zandbergen H (1994) Resonance Raman and infrared spectroscopy of carbon nanotubes. Chem Phys Lett 221(1–2):53–58
34. Pimenta MA, Hanlon EB, Marucci A, Corio P, Brown SDM, Empedocles SA, Bawendi MG, Dresselhaus G, Dresselhaus MS (2000) The anomalous dispersion of the disorder-induced and the second-order Raman bands in carbon nanotubes. Braz J Phys 30(2):423–427
35. Jorio A, Dresselhaus G, Dresselhaus MS, Souza M, Dantas MSS, Pimenta MA, Rao AM, Saito R, Liu C, Cheng HM (2000) Polarized Raman study of single-wall semiconducting carbon nanotubes. Phys Rev Lett 85(12):2617–2620
36. Fantini C, Pimenta MA, Dantas MSS, Ugarte D, Rao AM, Jorio A, Dresselhaus G, Dresselhaus MS (2001) Micro-Raman investigation of aligned single-wall carbon nanotubes. Phys Rev B 63(16):161405–161408
37. Brown SDM, Corio P, Marucci A, Pimenta MA, Dresselhaus MS, Dresselhaus G (2000) Second-order resonant Raman spectra of single-walled carbon nanotubes. Phys Rev B 61(11):7734–7742
38. Jorio A, Souza Filho AG, Dresselhaus G, Dresselhaus MS, Saito R, Hafner JH, Lieber CM, Matinaga FM, Dantas MSS, Pimenta MA (2001) Joint density of electronic states for one isolated single-wall carbon nanotube studied by resonant Raman scattering. Phys Rev B 63(24):245416 [4]
39. Pimenta MA, Jorio A, Brown SDM, Souza Filho AG, Dresselhaus G, Hafner JH, Lieber CM, Saito R, Dresselhaus MS (2001) Diameter dependence of the Raman D-band in isolated single-wall carbon nanotubes. Phys Rev B 64(4):041401(R)
40. Souza Filho AG, Jorio A, Samsonidze Ge G, Dresselhaus G, Dresselhaus MS, Swan AK, Ünlü MS, Goldberg BB, Saito R, Hafner JH, Lieber CM, Pimenta MA (2002) Probing the electronic trigonal warping effect in individual single-wall carbon nanotubes using phonon spectra. Chem Phys Lett 354(1–2):62–68
41. Souza Filho AG, Jorio A, Swan AK, Ünlü MS, Goldberg BB, Saito R, Hafner JH, Lieber CM, Pimenta MA, Dresselhaus G, Dresselhaus MS (2002) Anomalous two-peak G′-band Raman effect in one isolated single-wall carbon nanotube. Phys Rev B 65(8):085417
42. Duesberg GS, Loa I, Burghard M, Syassen K, Roth S (2000) Polarized Raman spectroscopy on isolated single-wall carbon nanotubes. Phys Rev Lett 85(25):5436–5439
43. Hwang J, Gommans HH, Ugawa A, Tashiro H, Haggenmueller R, Winey KI, Fischer JE, Tanner DB, Rinzler AG (2000) Polarized spectroscopy of aligned single-wall carbon nanotubes. Phys Rev B 62(20):R13310–R13313
44. Yu Z, Brus LE (2001) Rayleigh and Raman scattering from individual carbon nanotube bundles. J Phys Chem B 105(6):1123–1134
45. Grűneis A, Hulman M, Kramberger C, Pichler T, Peterik H, Kuzmany H, Kataura H, Ashiba Y (2001) Oscillatory behavior of Raman modes in SWCNT. In: Proceedings of the XV international winterschool of the electronic properties of the novel materials. AIP-Conference-Proceedings, American Institute of Physics, New-York, USA
46. Thomsen C, Reich S (2000) Double-resonant Raman scattering in graphite. Phys Rev Lett 85(24):5214–5217
47. Kürti J, Zólyomy V, Grüneis A, Kuzmany H (2002) Double resonant Raman phenomena enhanced by van Hove singularities in single-wall carbon nanotubes. Phys Rev B 65(16):165433

48. Ferrari AC (2007) Raman spectroscopy of graphene and graphite: disorder, electron–phonon coupling, doping and nonadiabatic effects. Solid State Commun 143(1–2):47–57
49. Nemanich RJ, Solin SA (1979) First- and second order Raman scattering from finite-size crystals of graphite. Phys Rev B 20(2):392–401
50. Saito R, Jorio A, Souza Filho AG, Grüneis A, Pimenta MA, Dresselhaus G, Dresselhaus MS (2002) Dispersive Raman spectra observed in graphite and single wall carbon nanotubes. Physica B 323(1–4):100–106
51. Sasaki K, Saito R, Dresselhaus G, Dresselhaus MS, Farhat H, Kong J (2008) Curvature-induced optical phonon frequency shift in metallic carbon nanotubes. Phys Rev B 77(24):245441
52. Cronin SB, Swan AK, Ünlü MS, Goldberg BB, Dresselhaus MS, Tinkham M (2004) Measuring the uniaxial strain of individual single-wall carbon nanotubes: resonance Raman spectra of atomic-force-microscope modified single-wall nanotubes. Phys Rev Lett 93(16):167401
53. Jorio A, Suiza Filho AG, Dresselhaus G, Dresselhaus MS, Swan AK, Ünlü MS, Goldberg BB, Pimenta MA, Hafner JH, Lieber CM, Saito R (2002) G-band resonant Raman study of 62 isolated single-wall carbon nanotubes. Phys Rev B 65(15):155412
54. Jorio A, Fantini C, Dantas SS, Pimenta MA, Suiza Filho AG, Samsonidze G, Brar VW, Dresselhaus G, Dresselhaus MS, Swan AK, Ünlü MS, Goldberg BB, Saito R (2002) Linewidth of the Raman features of individual single-wall carbon nanotubes. Phys Rev B 66(11):115411
55. Dresselhaus MS, Jorio A, Suiza Filho AG, Dresselhaus G, Saito R (2002) Raman spectroscopy on one isolated carbon nanotube. Physica B 323(1–4):15–20
56. Telg H, Fouquet M, Maultzsch J, Wu Y, Chandra B, Hone J, Heinz TF, Thomsen C (2008) G^- and G^+ in the Raman spectrum of isolated nanotube: a study on resonance conditions and lineshape. Phys Status Solidi B 245(10):2189–2192
57. Doorn SK, O'Conneell MJ, Zheng L, Zhu YT, Huang S, Liu J (2005) Raman spectral imaging of a carbon nanotube intramolecular junction. Phys Rev Lett 94(1):016802
58. Lazzeri M, Piscanec S, Mauri F, Ferrari AC, Robertson J (2006) Phonon linewidths and electron-phonon coupling in graphite and nanotubes. Phys Rev B 73(15):155426
59. Nemanich RJ, Solin SA (1977) Observation of an anomalously sharp feature in the 2-nd order Raman spectrum of graphite. Solid State Commun 23(7):417–420
60. Vidano RP, Fishbach R (1978) New lines in the Raman spectra of carbon and graphite. J Am Ceramic Soc 61(1–2):13–17
61. Vidano RP, Fishbach R, Willis LJ, Loehrv TM (1981) Observation of Raman band shifting with excitation wavelength for carbons and graphites. Solid State Commun 39(2):341–344
62. Al-Jishi R, Dresselhaus G (1994) Lattice-denamical model for alkali-metal-graphite interaction compounds. Phys Rev B 26(8):4523–4538
63. Kastner J, Pichler T, Kuzmany H, Curran S, Blau W, Wedlon DN, Delamesiere M, Draper S, Zandbergen H (1994) Resonance Raman and infrared spectroscopy of carbon nanotubes. Chem Phys Lett 221(1–2):53–58
64. Saito R, Jorio A, Souza Filho AG, Dresselhaus G, Dresselhaus MS, Pimenta MA (2002) Probing phonon dispersion relations of graphite by double resonance Raman scattering. Phys Rev Lett 88(2):027401
65. Grüneis A, Saito R, Kimura T, Cançado LG, Pimenta MA, Jorio A, Souza Filho AG, Dresselhaus G, Dresselhaus MS (2002) Determination of two-dimensional phonon dispersion relation of graphite by Raman spectroscopy. Phys Rev B 65(15):155405
66. Maultzsch J, Reich S, Thomsen C, Requardt H, Ordejón P (2004) Phonon dispersion in graphite. Phys Rev Lett 92(7):075501
67. Samsonidze GG, Saito R, Jorio A, Souza Filho AG, Grüneis A, Pimenta MA (2000) Phonon trigonal warping effect in graphite and carbon nanotubes. Phys Rev Lett 90(2):027403
68. Brar VW, Samsonidze GG, Dresselhaus MS, Dresselhaus G, Saito R, Swan AK, Ünlü MS, Goldberg BB, Souza Filho AG, Jorio A (2002) Second-order harmonic and combination

modes in graphite, single-wall carbon nanotube bundles, and isolated single-wall carbon nanotubes. Phys Rev B 66(15):155418
69. Saito R, Grüneis A, Samsonidze GG, Brar VW, Dresselhaus G, Dresselhaus MS, Jorio A, Cançado LG, Fantini C, Pimenta MA, Souza Filho AG (2003) Double resonance Raman spectroscopy of single-wall carbon nanotubes. New J Phys 5:157
70. Cançado LG, Pimenta MA, Saito R, Jorio A, Ladeira LO, Grüneis A, Souza Filho AG, Dresselhaus G, Dresselhaus MS (2002) Stokes- and anti-stokes double resonance Raman scattering in two-dimensional graphite. Phys Rev B 66(3):035415
71. Kornienko M (2004) On a relationship between the heat of melting of molecular crystals and their optical phonons energies. Bull Kyiv University Ser Phys Math 4:466–476 (in Ukrainian)
72. Dolinchuk SG, Kornienko NE, Zadorozhnii VI (1994) Noncritical vectorial phase matchings in nonlinear optics of crystals and infrared up-conversion. Infrared Phys Techn 35:881–895

Confocal Surface-Enhanced Raman Microscopy at the Surface of Noble Metals

8

H. Dietz, G. Sandmann, A. Anders, and W. Plieth

1 Definition of the Topic

The optical and spectroscopic properties of nanoparticles are of interest for a wide variety of methodic and technical fields of applications. The investigation of nanoparticles requires convincing characterization methods with high spatial resolution. The surface-enhanced Raman spectroscopy is a sensitive tool for characterizing the chemical structure of metal nanoparticles like gold, silver, or copper. If combined with local confocal microscopy it becomes a method for delivering optical and geometrical information of nanosize metallic structures. This method is called confocal surface-enhanced Raman spectroscopic (SERS) microscopy. The fundamentals and significance of confocal SERS microscopy are described based on literature data and recent results of own studies.

2 Overview

Various routes for the preparation of silver and gold nanoparticles were introduced.

Controlled electrodeposition of silver and gold nanoparticles by the electrochemical double-pulse technique delivers samples with varying particle size from 10 to 500 nm and varying particle density.

Confocal SERS microscopy combined with subsequent local scanning electron microscopy (SEM) is able to deliver valuable optical and chemical information on the location of SERS active spots. The method as to how to retrieve in the electron micrograph the active spots seen in the confocal images is explained. The spots were analyzed with respect to band position and spectral intensity.

H. Dietz • G. Sandmann • A. Anders • W. Plieth (✉)
Technische Universität Dresden, Physikalische Chemie und Elektrochemie, Dresden, Germany

Challa S. S. R. Kumar (ed.), *Raman Spectroscopy for Nanomaterials Characterization*,
DOI 10.1007/978-3-642-20620-7_8, © Springer-Verlag Berlin Heidelberg 2012

The optical and spectroscopic data of the local nanoparticle structures investigated showed that SERS is a local and time-dependent phenomenon, because (1) only few particles are Raman active particles, (2) strongest enhancements in SERS are obtained from particle agglomerates, (3) typically the Raman radiation is emitted from irregular structures like the necks between two or more particles agglomerated, (4) a time-dependent behavior characterized by intensity fluctuations was observed.

In the investigated range from 10 to 500 nm, no significant influence of the particle size was observed.

3 Introduction

The surface of nanoparticles of the coin metal group is known to show, after structural modification, substantial enhancement of the Raman signal of adsorbates. This effect is the basis of surface-enhanced Raman scattering (SERS). For more than 30 years, SERS has been developed to a sensitive spectroscopic method for detecting molecules adsorbed at gold, silver, and copper surfaces. Though the phenomenon of surface-enhanced Raman scattering is still under discussion, it is widely accepted that the origin of SERS is closely correlated to the enhancement of (1) the local electromagnetic field at the surface of small metallic nanoparticles and of (2) the charge transfer between the adsorbed molecules and the metal particles [1–4]. An additional chemical enhancement (3) called resonance Raman scattering (RRS) is observed if the energy of the laser excitation is resonant with a molecular transition [4]. To increase the detection sensitivity, SERS and RRS can be combined to surface-enhanced resonance Raman scattering called SERRS.

Theoretical and experimental studies demonstrated that the electromagnetic field enhancement is substantially increased at surface sites having local geometric singularities. In particular, enormous field enhancement occurs at the junction between two or more almost touching nanoparticles [5–7] due to plasmon coupling. For spacings of 1–2 nm, maximum enhancement factors of about 10^{10}–10^{11} were calculated [6, 8].

In combination with strong chemical enhancement, huge factor values of the order 10^{14} were observed [9] and single molecule SERS became possible [10, 11]. The critical magnitude of the electromagnetic enhancement needed for single molecule detection, however, is still under discussion [12]. In general, due to the high structural sensitivity and the complex interaction of molecule and electromagnetic environment, quantification of SERS has been challenging. Nevertheless, surface-enhanced Raman scattering is a sensitive analytical technique for the investigation of surface species.

The SERS method is therefore often used for a variety of applications in medicine, biophysics, and microelectronics. As the SERS effect seems to be a local phenomenon, SERS as a method delivering chemical or molecular information should be combined with spatial microscopy (Fig. 8.1).

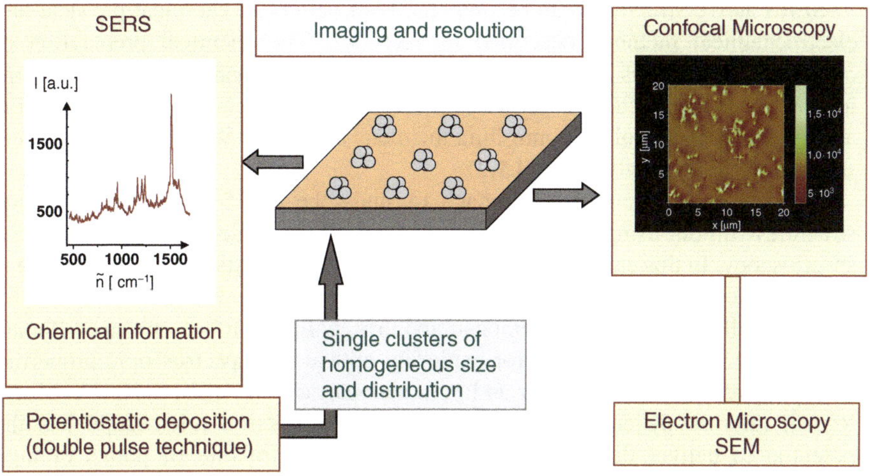

Fig. 8.1 Raman microspectroscopy as combination of surface-enhanced Raman spectroscopy and confocal microscopy

With SERS spectroscopy, the initial nanoparticle formation at the beginning of the electrocrystallization process can be studied, providing new insight into the fundamentals of electrodeposition. Surprising phenomena like band splitting, sometimes observed when monitoring the SERS activity of an adsorbed molecule in the process of electrocrystallization, indicates interaction between particle growth and SERS activity [13]. The question whether SERS sites of high activity and growth centers are identical can only be clarified by focusing on spatially resolved Raman microscopy.

It is evident that the properties of nanoparticle structures depend not only on the individual particle size, particle shape, or the degree of size dispersion but also on their spatial distribution and the degree of aggregation or film formation. Therefore, before investigating the optical and electronic properties of nanoparticles, it is important to gain knowledge of how to control the particle size, particle density, and spatial distribution.

Research has to be focused on two topics:

1. Searching for appropriate preparation techniques of nanoparticles leading to SERS-active particle arrays with preferably tailored structure
2. Optical and spectroscopic characterization of the SERS-active structures formed by means of spatially resolved Raman microscopy

The spatial resolution of the Raman signals can be achieved by using scanning confocal microscopy combined with spatial SERS imaging. The confocal laser microscope (Sect. 4.2) provides optical images of high resolution combined with improved depth of sharpness [14]. To improve the attribution of SERS signals to local structures the confocal images were compared with local SEM images (Sect. 5).

SERS active structures can be prepared by a variety of chemical physical and electrochemical methods described in Sect. 4.1. The chemical preparation of colloidal nanoparticles is frequently used (Sect. 4.1.1). An interesting electrochemical preparation procedure is the so-called double-pulse technique. This method is an electrochemical tool for controlling the metal deposition with respect to particle size and particle density (Sect. 4.1.2).

An interesting alternative approach to study the SERS activity of substrate structures without using noble metal particles is the so-called tip-enhanced SERS spectroscopy. In this case, a scanning gold or silver tip of a tunneling microscope is used to generate local SERS signals (Sect. 4.2).

In the following, it will be demonstrated that confocal surface-enhanced Raman microscopy is a powerful tool for exploring optical and spectroscopic properties of molecules adsorbed on silver and gold nanoparticles. It is shown that confocal Raman microscopy coupled with high-resolution electron microscopy has the potential to deliver detailed information about the structure of points of high Raman intensity the so-called hot spots.

4 Experimental and Instrumental Methodology

4.1 Preparation of SERS Active Surfaces

4.1.1 Common Preparation Methods

Common methods for the fabrication of metallic nanoparticle arrays are electron beam lithography, photolithography, laser ablation, colloidal synthesis, electrodeposition and, in recent time, nanosphere lithography for which a monodisperse nanosphere template acts as deposition mask. A review on advances in preparation of nanomaterials with localized plasmon resonance is given in [15].

Lithographic methods are able to generate very uniform nanoparticle arrays and can be used to monitor the preparation process. In combination with additional techniques, even various three-dimensionally shaped structures can be formed (see, e.g., [16]). However, lithographic methods suffer from the limitation that the small interparticle spacings required for huge electromagnetic enhancement are not technically feasible.

High levels of SERS activity can be expected if nanoparticle substrates were prepared by colloidal chemical methods because colloidal particles tend to aggregation, thus forming the narrow interparticle spacings needed for plasmon coupling. A traditional technique for the preparation of silver colloids is the sol formation according to Lee et al. [17].

This method is based on the chemical reduction of the metal salts in citrate solution. Metal-dye adsorbates will be formed if a dye in extremely diluted solution is added. The surface enhancement depends on the precipitation conditions, type and charge of the adsorbing dye molecule, the degree of colloidal aggregation, and many other factors [18–20], so that a full understanding of the chemistry at colloidal surfaces is difficult and the aggregation process can be hardly controlled. There is also the disadvantage that extraneous ions adsorbed within the preparation process can affect the SERS spectrum.

Besides these chemical methods, electrochemical techniques are of interest. This is because the electrodeposition is a convenient and fast method for the preparation of metallic nanoparticles on large areas of conductive substrates. However, for precise and systematic investigation of the nanoparticle properties control of the particle size, form and distribution is necessary. From this point of view, the classical electrodeposition technique from solution is not so successful, as the homogeneity in particle size and spatial particle distribution is presumably disappointing in comparison to the invasive tip-directed SPM routes [21] or deposition techniques into nanotemplates.

In order to decrease dispersion in particle size during electrodeposition, two important principles should be taken into account. Firstly, the crystal seed formation has to occur spontaneously, thus preventing progressive nucleation. Secondly, the crystal growth has to be conducted at a slow rate, that is, at low overpotential. Penner et al. [22–26] has elucidated the importance and coherence of these fundamental principles.

His research group successfully started extensive investigations on the electrodeposition of silver [24–26] and gold [26] nanoparticles on graphite surfaces. Combined with Brownian dynamic simulations for the growth of metal nanoparticle ensembles [22, 23], the work focused on the development of nontemplate, electrochemical routes to dimensionally uniform metal structures.

The simulation results clearly showed that particle-size dispersion is not only caused by a progressive type of nucleation but also, and in particular, by the spatially random distribution of nuclei initially produced in the nucleation process [22]. Even in the case of instantaneous nucleation, polydispersity occurred increasingly, depending on the extent of inter particle diffusion coupling, a consequence of the growth of the individual diffusion zones around the particles. Size heterogeneity inevitably increases as a function of growth-time, if inter particle diffusion coupling occurs. To prevent diffusion controlled growth, Penner proposed decreasing the overpotential of the electrodeposition process to below a threshold value of about −75 mV [23]. The experiments carried out under these conditions (where metal particles grow independently of their neighbors), revealed that extremely narrow particle distributions are attainable [26].

4.1.2 Double-Pulse Technique as an Electrochemical Tool for Controlling Particle Structure

The ideal method for transforming both principles, which favor monodispersity into an experimental procedure, can best be attained by applying the potentiostatic double-pulse technique. This method, introduced by Scheludko, Todorova, Kaischev, and Milchev [27, 28], is based on an extremely short nucleation pulse of high cathodic polarization, followed by a much longer growth pulse at

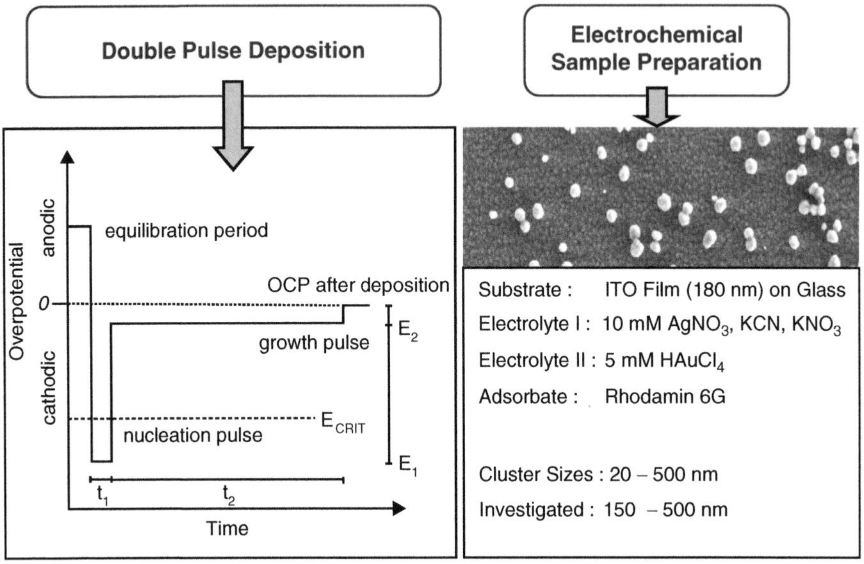

Fig. 8.2 Schematic representation of the double-pulse method utilized for the preparation of silver and gold clusters

low cathodic overpotential [29, 30]. Thus, crystal seed formation can be exclusively transferred into the first pulse and crystal growth can be conducted at a slow cathodic overpotential of 20 mV within the second pulse without new nucleation. Course and fundamentals of the double-pulse method are depicted in Fig. 8.2.

Based on a model on the features of the double-pulse technique, various structures of silver nanoparticles grown onto a thin ITO film covered glass plate were generated and characterized [30]. With this method, the conflict between both optimal conditions for nucleation and growth is partially defused. This is due to the amount of small seed additionally nucleated at the higher polarization and resolved as soon as the potential is switched over to the lower polarization of the growth pulse. The interaction of the pulse parameters was modeled, thus forming the basis for how the electrodeposition process of noble metal clusters can be variably controlled.

Silver

The electrodeposition of silver nanoparticles on 1.32 cm^2 ITO substrates was performed in separate Teflon cells of 12 ml volume using a standard three electrode setup, as previously described for silver particles [29, 30]. The electrolyte used for the silver deposition contained 0.1 M KNO_3, 0.1 M KCN, and 0.01 M $AgNO_3$ per liter.

A HEKA PG284/IEC potentiostat/galvanostat was used for the double-pulse deposition. In a first pulse of high overpotential, nucleation was forced and in a second pulse of small overpotential of about −20 mV referred to the open circuit potential (OCP), the nuclei were directed to grow to a predetermined size.

In the case of silver deposition, the OCP was −680 mV versus SCE after deposition, the nucleation pulse potentials E_1 and the deposition time t_1 and t_2 were varied. The growth pulse potential E_2 was –700 mV versus SCE. The averaged diameter of the silver cluster deposited was about 200 nm. The samples with particles were taken from the preparation bath and carefully washed and dried in a desiccator.

In the case of silver nanoparticles, carbonaceous impurities remained on the particle surface or were absorbed from the ambient atmosphere after the transfer to the Raman microscope. As revealed by the measurements (Sect. 5.2.1, Fig. 8.10), the Raman active species were graphite and nanocrystalline carbon, a typical experience in surface analysis. In the literature, the origin of carbon contamination is discussed in detail [31, 32]. One general source of contamination is the metal preparation procedure. The carbon dioxide of the laboratory air might act as another source if reacting with the alkaline electrolyte on the cluster surface after removing the sample from the cluster preparation apparatus. Ions from the preparation electrolyte adsorbed on the substrate surface (e.g., CN^-) might also be a source for carbon, despite a careful rinsing after the deposition process.

Gold

The electrodeposition of gold was made in an acidic electrolyte of 0.005 M $HAuCl_4$ per liter. The open circuit potential (OCP) after deposition was +820 mV versus SCE. The pulse potentials were varied between +800 and +600 mV versus SCE. Particles between 10 and 500 nm diameter were prepared.

In the case of gold nanoparticles, no signals of nanocrystalline carbon could be found. Therefore, in the following experiments, the dye Rhodamin 6 G was used as SERS active molecule. It shows an additional resonance effect (surface-enhanced resonance spectroscopy, SERRS). The samples were dipped into an aqueous solution of 10^{-7} M Rhodamin 6 G per liter for 12 h and subsequently dried in the desiccator.

4.2 Confocal Raman Microscopy

For investigating nanoparticle structures by means of a Raman microscope, it is necessary to increase the spatial resolution and depth of sharpness of the Raman signal.

While commercial Raman spectrometers collect the scattered light from a laser spot of size 1 μm at best and the Raman spectra also include information of sample regions above and below the focal plane, the implementation of confocal microscopy can improve the optical resolution and the depth of sharpness. The method leads to better image definition. Confocal microscopy means point illumination and that a small aperture is placed in the optical path so that only the scattered light located in a thin focus plane can reach the detector. The aperture-dependent spatial resolution is improved toward the Abbe refraction limit of about half of the wavelength used or approximately 300 nm. As a result of scanning confocal microscopy, a real three-dimensional picture of the nanoparticle structure can be generated (Sect. 5.2).

For overcoming the limit of light microscopy and further improvement in spatial resolution, the implementation of scanning near-field microscopy (SNOM) by means of a local illumination probe is an interesting approach [33–35]. The method is based on the field enhancement in the cavity between a sharp metal tip and the sample. In combination with Raman spectroscopy, this scanning probe technique is called tip-enhanced Raman spectroscopy (TERS) and enables high-resolution spatial microscopy with a lateral resolution of 50 nm [35]. Bouhelier [36] has reviewed advances in this field.

The combination of SERS and optical near field microscopy (SNOM) can be therefore an ideal technique to carry out real locally resolved Raman microscopy delivering both chemical and optical information. A similar alternative could be the combination of Raman spectroscopy and AFM.

Instead of SNOM, in many cases, particularly if the sample is to be screened for Raman active spots and their spatial distance is more than half of the wavelength of the laser, it is also confocal Raman microscopy that delivers enough morphological and spectral information on both nanoparticle structure and SERS activity, respectively. Thus, confocal Raman microscopy is interesting for a wide variety of applications in biology, medicine, and technological materials research.

Combining confocal Raman microscopy with scanning electron microscopy, the SERS active spots can be retrieved in the SEM image, if the active sample region has been labeled before investigating.

A Raman microscope typically consists of an inverse microscope equipped with a monochromator and a photon counting system. Typical equipment (Figs. 8.3, 8.4) consisted of a modified inverse microscope Zeiss Axiovert combined with one or two Avalanche photodiodes (SPCM-AQR-14) to measure the luminosity of a sample spot. The sample could be scanned using a scan table (Physik Instrumente GmbH) with a resolution of 10 nm. The microscope was equipped with a HeNe-laser ($\lambda = 633$ nm, used for Au particles) and a frequency doubled Nd:YAG-laser ($\lambda = 532$ nm), used for Ag particles. Excitation light and scattered light were separated via a dichroitic mirror (beam splitter), as can be seen from Fig. 8.3. In the beam path also a holographic notch filter (Kaiser Optical Systems, 6.0 O.D.) were installed for rejecting back-scattered excitation light of the excitation wavelength. Two modes could be applied: (1) the topography mode without using the notch filter, and (2) the Raman/fluorescence mode with use of the notch filter. For Raman measurements, the Stokes shifted photons were collected at $\lambda > 540$ nm (Ag particles) and $\lambda > 642$ nm (for Au particles). Spectra could be recorded with a 300 mm spectrograph (SpectraPro 300i, Acton, grating 1,200 lines/mm) equipped with a liquid-nitrogen cooled CCD detector. Scanning electron microscopy (SEM) images of the same sample regions were made with a Gemini 982 (Zeiss), subsequently to the optical measurements.

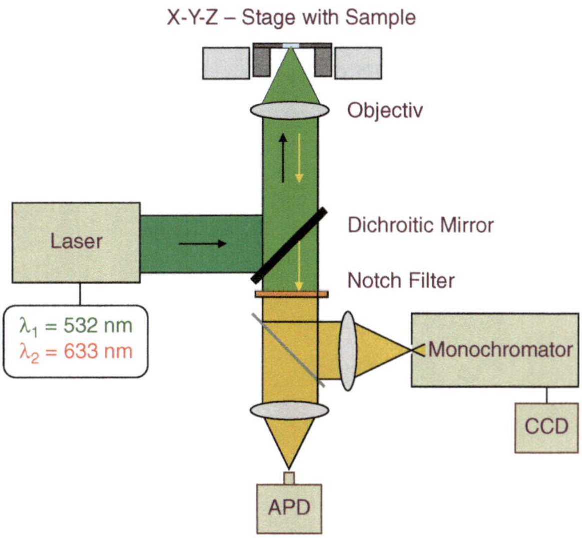

Fig. 8.3 Schematic representation of the confocal Raman microscope inclusive light path

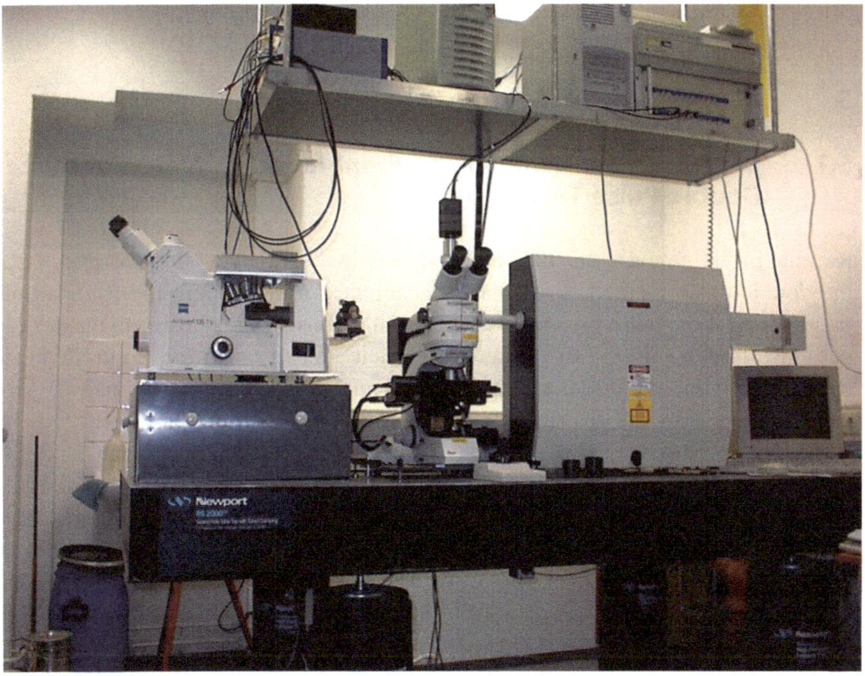

Fig. 8.4 Confocal scanning microscope used for Raman microscopy

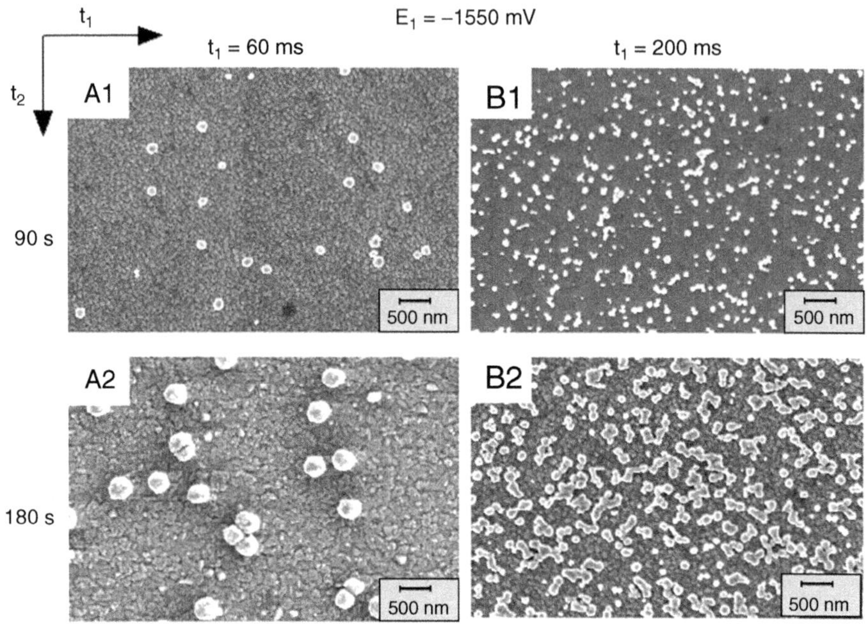

Fig. 8.5 SEM images of silver clusters on ITO substrates deposited at E1 = −1,550 mV in dependence of t_1 and t_2; $E_2 = -700$ mV; (**A1**) $t_1 = 60$ ms; $t_2 = 90$ s; (**A2**) $t_1 = 60$ ms; $t_2 = 180$ s; (**B1**) $t_1 = 200$ ms; $t_2 = 90$ s; (**B2**) $t_1 = 200$ ms; $t_2 = 180$ s; [30]

5 Key Research Findings

5.1 Control of the Nanoparticle Preparation by Means of the Double-Pulse Technique

Recent investigations on the electrodeposition of silver demonstrated that the double-pulse method is a suitable technique for controlling the nanoparticle preparation, if the pulse parameters are carefully chosen and adjusted to the desired particle structure [29, 30]. Whereas particle density can be controlled via the overpotential and duration time t_1 of the nucleation pulse E_1, the particle size can be enlarged by the growth time t_2 (Fig. 8.5, [30]).

Exclusive particle growth without additional nucleation formation is found for low particle densities until $1/\mu m^2$, thus minimizing dispersion in particle size. Responsible for this is the transition moment between nucleation pulse and growth pulse. During the nucleation pulse, a Gaussian distribution of N nuclei with radii $r > r_{crit}$ (η_1) are formed as a result of progressive nucleation and growth (Fig. 8.6a) [29]. With the transition to the extremely low overpotential η_2 of the growth pulse, the critical radius increases to the value r_{crit} (η_2).

As a consequence, dissolution of under critical $r < r_{crit}$ (η_2) occurs (Fig. 8.6b). Only the larger particles $r > r_{crit}$ (η_2) are stable enough to survive. The particle size

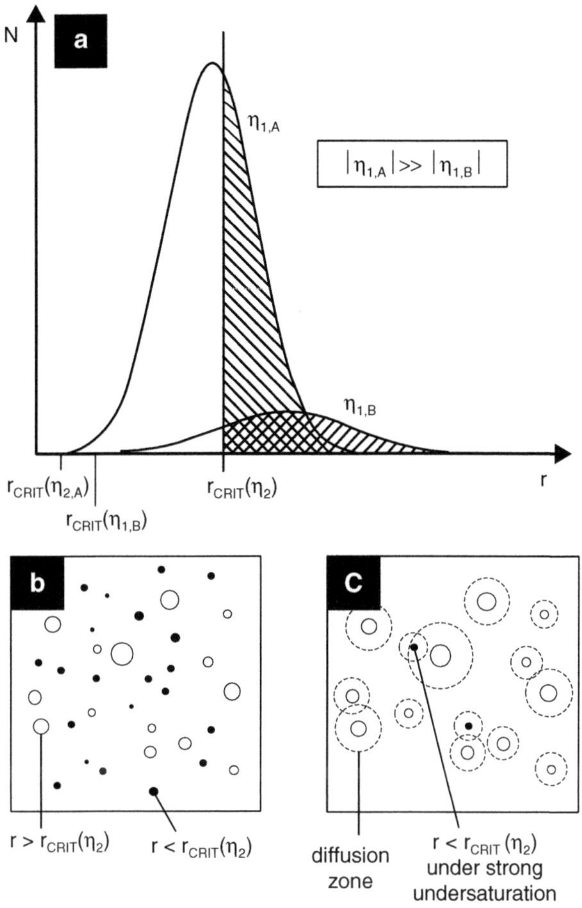

Fig. 8.6 Features of the double-pulse technique; Model on the influence of the transition moment between nucleation pulse and growth pulse in the course of the double-pulse deposition on the Gaussian particle distribution formed after the nucleation pulse [29]; (**a**) Gaussian particle distribution of N nuclei with radii $r > r_{Cr}$ (η_1) for different over potentials of the first pulse $(|\eta_{1B}| << |\eta_{1A}|)$. The hatched area of the Gaussian distribution corresponds to the number of stable particles with radii $r > r_{Cr}$ (η_2), whereas the *white* area of particles of under critical size is amputated as these particles dissolve. (**b**) Representation of the result of the particle cut off, small (*dark*) particles dissolve but larger particles (*white*) survive under the lower overvoltage of the growth pulse.(**c**) If a small particle lies in the diffusion zone of a larger particle the under saturation can favor the dissolution of the smaller ones

distribution is then cut off with respect to under critical particles, leading to narrower particle size distribution at marginal polydispersity. Only at high particle densities ($>10/\mu m^2$) nanoparticle aggregation occurs accompanied by slight decrease in mean particle density as a result of diffusion coupling effects (Fig. 8.6c).

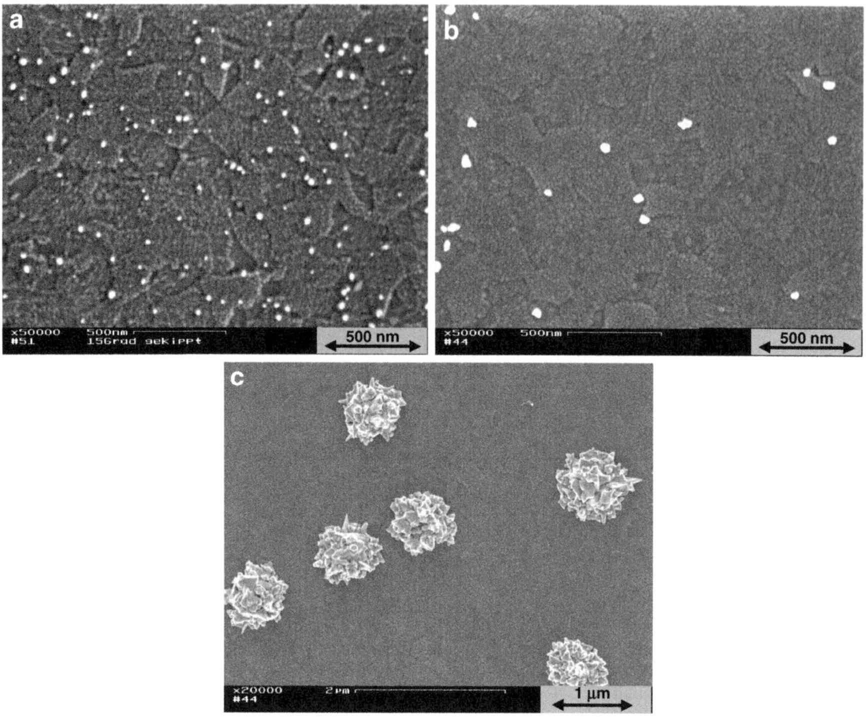

Fig. 8.7 SEM images of isolated gold clusters electrodeposited on ITO covered glass [37]. (**a**) Single pulse ($E_1 = -100$ mV/ = 0.1 s), (**b**) double-pulse ($E_1 = -100$ mV/0.1 s; $E_2 = +800$ mV/ 90 s), (**c**) double-pulse ($E_1 = -100$ mV/0.1 s; $E_2 = +600$ mV/90s)

The preparation regime and experimental experience was transferred to the generation of gold nanoparticles onto ITO and glassy carbon substrates. It was shown that various nanoparticle structures with particle sizes from 10 to 500 nm could be prepared [37].

Figure 8.7 shows that, analogous to the preparation of silver clusters, the particle size of gold can be controlled in electrodeposition. Figure 8.7a shows the particle ensemble formed after the nucleation pulse. Figure 8.7b shows the particles after the second pulse. Normally, particles formed during the first pulse (Fig. 8.7a), are growing in the second pulse. However, the electrode has anodic character at the beginning of the growth pulse, if the growth potential (of +800 mV in Fig. 8.7b) is only few mV below the Nernst equilibrium potential (+820 mV). This is due to the negative shift of the reversible potential of an electrode covered by smaller nanoparticles [38, 39]. Thus, dissolution of small nanocrystallites is initially observed, as seen by comparing Fig. 8.7a and b. This is an example of Ostwald ripening.

If the overpotential of the second pulse is substantially increased, the large particles grow under diffusion control, while the smaller nanoparticles dissolve (diffusion coupling). Characteristic surface structures are developed (Fig. 8.7c).

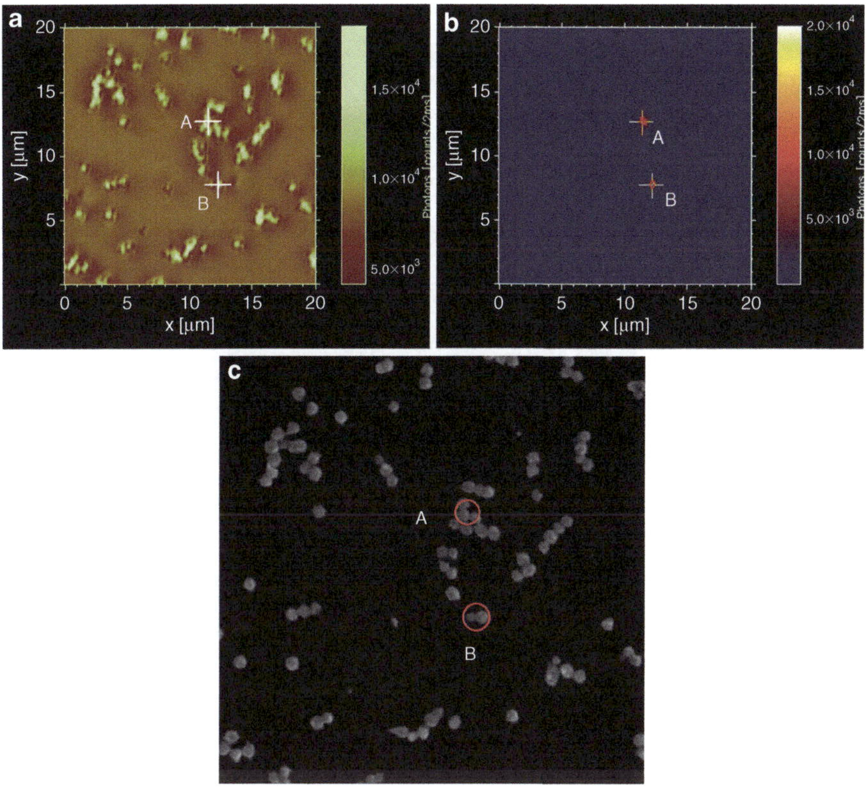

Fig. 8.8 Optical images (at an active zone I) of isolated silver clusters electrodeposited onto ITO by means of the double-pulse method ($E_1 = -1{,}550$ mV; $E_2 = -700$ mV, 25 s) [37]: (**a**) Scanning confocal microscopy image (topography mode), (**b**) Raman/fluorescence image of the same sample area, (**c**) SEM image corresponding to (**a**) and (**b**)

5.2 Confocal Microscopy and Microspectroscopy

5.2.1 Silver

Based on the model on the features of the double pulse parameters [29, 30], samples with isolated clusters and low particle density were electrochemically prepared from cyanide solution, as explained in Sect. 4.1. The particles, having a diameter of about 200 nm, were partially aggregated (Fig. 8.8a, b, c).

In Fig. 8.8a, confocal images of an active sample zone I in the topography mode are shown. Subsequently, the notch filter is brought into the light beam, eliminating the laser light. Fluorescence and Raman scattered light can pass through the filter. The image of fluorescence and Raman scattering is shown in Fig. 8.8b. Only two

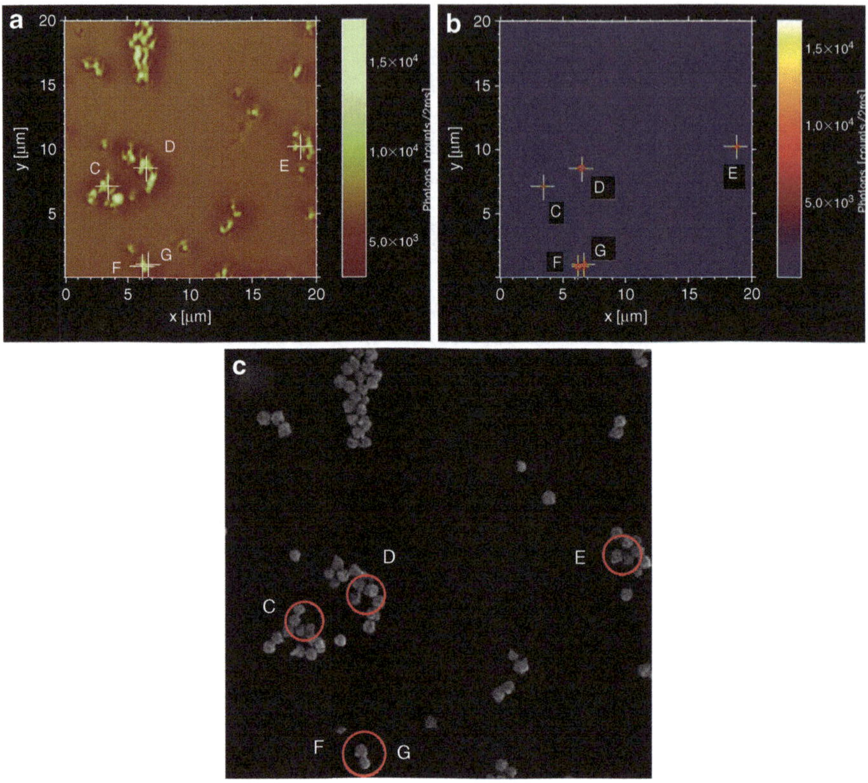

Fig. 8.9 Optical images (at an active zone II) of isolated silver clusters electrodeposited onto ITO by means of the double-pulse method ($E_1 = -1{,}550$ mV; $E_2 = -700$ mV, 25 s) [37]: (**a**) Scanning confocal microscopy image (topography mode), (**b**) Raman/fluorescence image of the same sample area, (**c**) SEM image corresponding to (**a**) and (**b**)

distinct light spots appear in the image, which are marked with the cursor and are then compared with the topography image (Fig. 8.8a). An apparent enhancement factor can be obtained from the comparison of the technical data of topographic and molecular imaging. A factor of 10^{10} was found for silver.

The result of this comparison is that the location of the Raman/fluorescence spots and the center of the clusters are not identical. The Raman/fluorescence spots are located close to surface structures, for example, the necks of the cluster agglomerations. To get even more insight into the surface structure around the light spots of Fig. 8.8b, the corresponding SEM image of the same structure could be found and is presented in Fig. 8.8c. There is clear evidence that the center of the Raman spot is a structure with two or more clusters in contact.

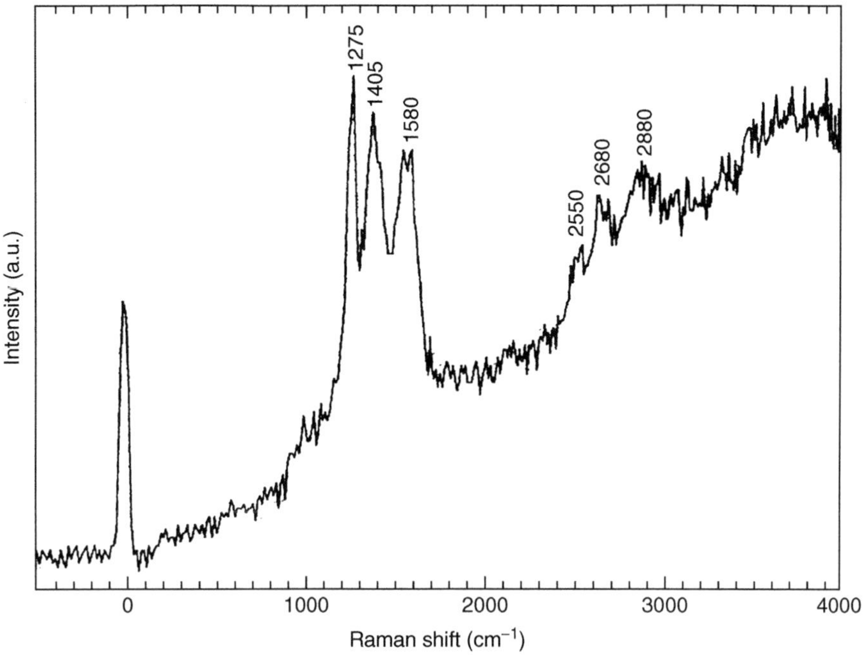

Fig. 8.10 Raman spectrum of a few carbon nanocrystallites adsorbed on a silver particle [41]

The investigation was repeated for another active area II of the same sample. In this area, besides single isolated particles also loosely aggregated clusters are to be seen (Fig. 8.9a, c). The five distinct light spots found in the Raman image (Fig. 8.9b) were again marked with the cursor and compared with the topography image (Fig. 8.9a). Again, the Raman/fluorescence spots are located in the neck zones of the cluster aggregates that could be retrieved in the SEM image (Fig. 8.9c).

It is also important to note that substantial SERS-activity found only originated from a few nanoparticles forming particle aggregates. Typically, single isolated particles were less active [40], as also confirmed by other groups [5].

The Raman spots are caused by nanocrystalline carbon adsorbed on isolated silver particles [41]. An example for such a Raman spectrum is shown in Fig. 8.10. The broad structure between 2,500 and 5,000 cm^{-1} is caused by fluorescence. The Raman bands in the region at 1,350 cm^{-1} (D-band) and 1,590 cm^{-1} (G-band) as well as the secondary Raman bands between 2,500 and 3,000 cm^{-1} were observed in the experiments. The carbon lines mainly agree with results reported in the literature [42]. Only the splitting of the D-band is different and may be indicative for probing a small number of carbon nanocrystallits. The domain size of the carbon can be estimated from the intensity ratio of the D- and G-bands according to Tuinstra and Koenig [42]. Similar experiments of

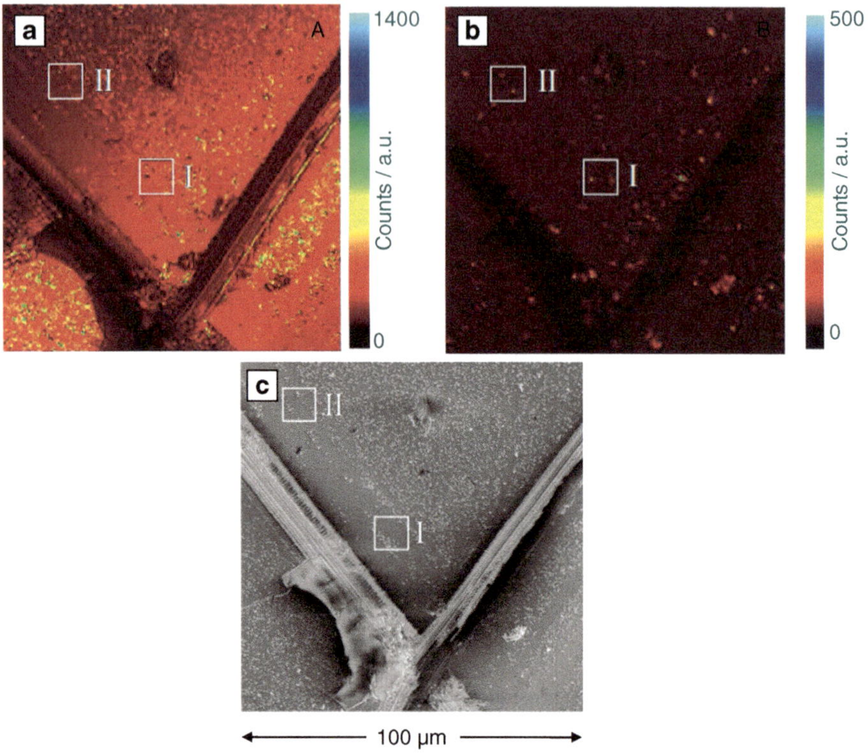

Fig. 8.11 Optical images of R6G covered isolated gold clusters electrodeposited onto ITO by means of the double-pulse method [$E_1 = -100$ mV (30 ms); $E_2 = 600$ mV (90 s)]; (**a**) Scanning confocal microscopy image (topography mode), (**b**) Raman/fluorescence image of the same sample area, (**c**) SEM image corresponding to (**a**) and (**b**). Two active zones I and II were found. The confocal image could be retrieved within the SEM image by means of cross marking

Moyer et al. [41] demonstrated that the two dimensionally domain size of nanocrystalline carbon was found to be in the order of 10 nm. Additional spectroscopic Raman measurements on nanocrystalline carbon adsorbed at separate isolated silver nanoparticles revealed temporal fluctuations in the peak intensities of the characteristic carbon bands and intermittent on/off behavior [41]. The SERS literature (e.g., [5, 10, 11]) frequently reports this phenomenon, called "blinking." Most authors interpreted the "blinking" that only few nanocrystallites were involved.

Vosgröne and Meixner [18, 19] clearly demonstrated and explored in samples with R6G at very low concentration several inhomogeneous spectroscopic features like inhomogeneous line broadening, line splitting, spectral diffusion and wandering which appeared in addition to intermittent intensity fluctuation. These features attributed to single molecule properties usually disappear in the ensemble average.

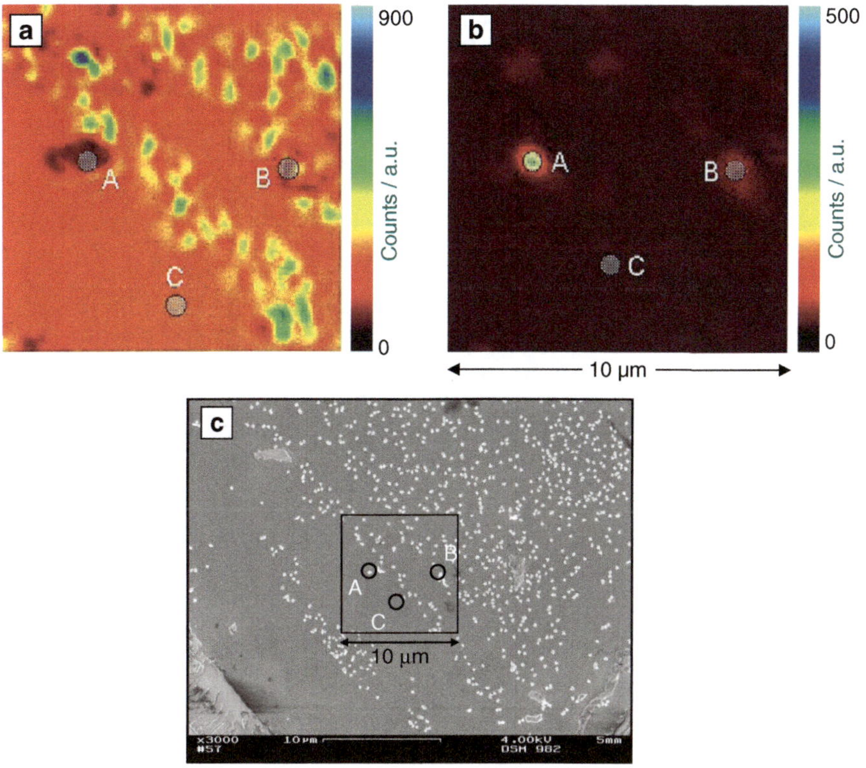

Fig. 8.12 Optical images of R6G covered gold clusters within the active zone I marked in Fig. 8.11 (active zone I at higher resolution); (**a**) Scanning confocal microscopy image (topography mode), (**b**) Raman/fluorescence image of the same sample area, (**c**) SEM image corresponding to (**a**) and (**b**). Two active points A and B could be found. The confocal image could be retrieved within the SEM image by means of cross marking

5.2.2 Gold

Results on confocal Raman microscopy of Rhodamin 6 G adsorbed on gold particles are presented in this chapter. The local microscopic information on SERS active spots can be supported by SEM. This approach is demonstrated in Fig. 8.11. Before starting the investigation on gold nanoparticles electrodeposited on ITO glasses, the sample surface was scratched by means of a sharp scalpel, thus forming a cross as marker. The cross was discovered in the white-light image and in the confocal topographical image (Fig. 8.11a). It served as point of reference for searching and retrieving SERS active areas in the confocal topography image and subsequently in the SEM image. The confocal images (Fig. 8.11a, b) and subsequent the SEM image (Fig. 8.11c) shows two active zones I and II located above the cross.

Figure 8.12 shows the active zone I of Fig. 8.11 enlarged. The active spot A, the medium-active spot B and, for comparison, the nonactive point C are marked in

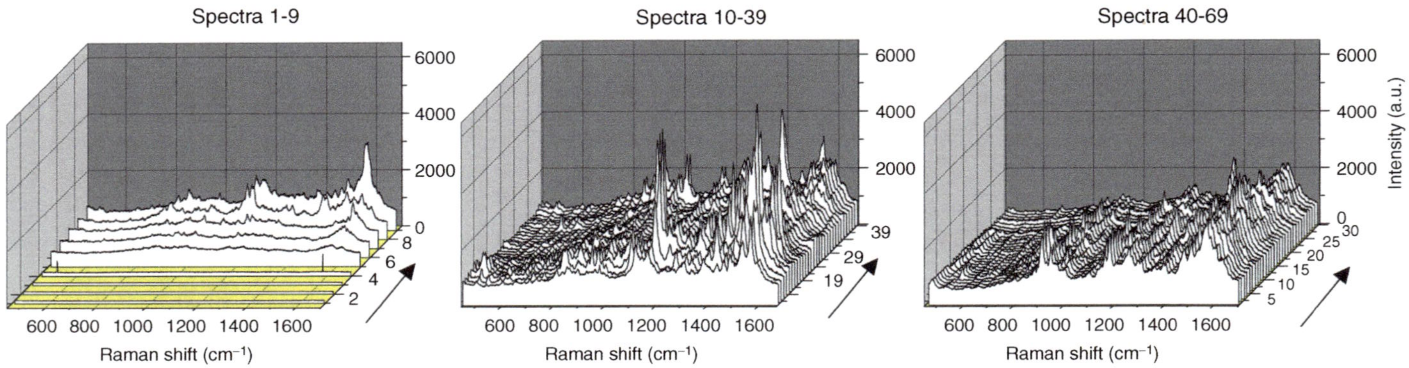

Fig. 8.13 Time dependence of subsequent SERS spectra (1–69) of R6G covered gold clusters at the active point A marked in Fig. 8.12 (active zone I)

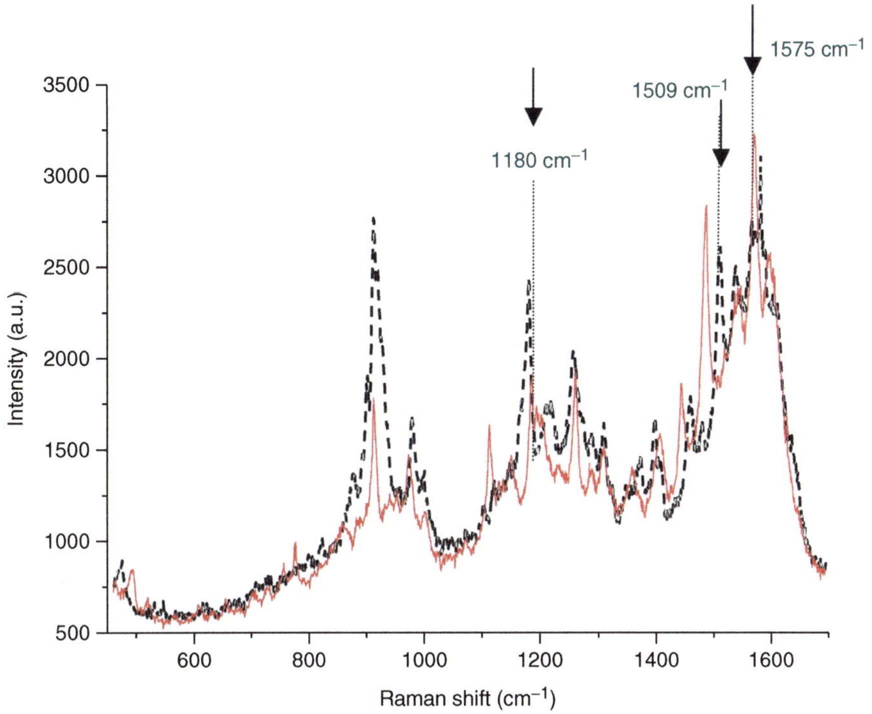

Fig. 8.14 Two SERS spectra within the sequences 10–39 of Fig. 8.13 (point A)

the confocal images (Fig. 8.12a, b). By means of the white-light images, these points could be retrieved within in the SEM image and subsequently marked (Fig. 8.12c). Thus, better local information on the active spot area of aggregated nanoparticles is obtained. It is seen that the active spot A is located at the surface of an aggregated gold particle.

Spectral SERS information of the active spot A can be derived from stationary measurements. The basis of this type of experiment is to center the spot A and to hold it in the excitation focus of the confocal microscope.

Figure 8.13 shows a sequence of subsequent SERS spectra 1–69 taken at point A. The SERS activity is time dependent and fluctuating. At the very beginning of the stationary measurements, substantial SERS activity of the strong Rhodamin 6 G peaks at 1,510 and 1,270 cm^{-1} is not visible. With time, SERS activity increasingly develops accompanied by smaller fluctuations. After longer times, the SERS activity again decreases.

To demonstrate the time-dependent spectral behavior, two Raman spectra within the sequences 10–39 of Fig. 8.13 are shown in Fig. 8.14. The strongest R6G peaks at 1,180 cm^{-1} (C–H in plane bend), 1,510, and 1,575 cm^{-1}

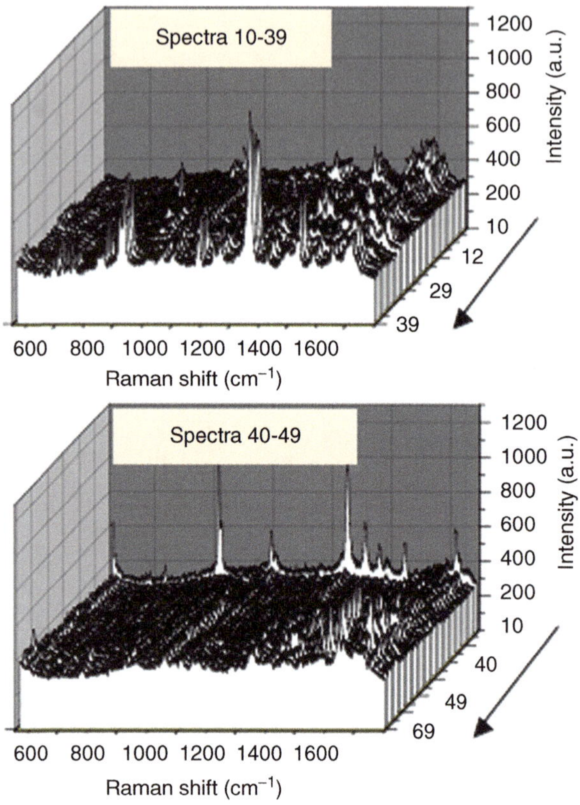

Fig. 8.15 Time dependence of subsequent SERS spectra (10–69) of R6G covered gold clusters at the active point B marked in Fig. 8.12 (active zone I)

(both aromatic C-C stretch vibrations, based on literature data) are marked with arrows. There are small deviations in intensity and peak position from sequence to sequence and the peak positions are fluctuating around the average value of literature data.

Figure 8.15 shows a sequence of subsequent SER spectra 10–69 from the point B in zone I having lower luminosity compared to point A. As expected, the lower luminosity also correlates to smaller SERS intensity, as seen from the strongest peaks at 1,510 and 1,270 cm^{-1}. Again time dependence of the peak intensities is observed. The strongest enhancement does not concern the aromatic stretch vibration but is found for the Raman peak near 1,300 cm^{-1} (1,270 cm^{-1}), thus supporting that the enhancement is dependent on the adsorption position and the distance to the metal.

Figure 8.16 shows the active zone II of Fig. 8.11 enlarged. Again, the most active points in the confocal images were marked and could be retrieved in the SEM image. The results demonstrate that only a few particles show SERS-activity.

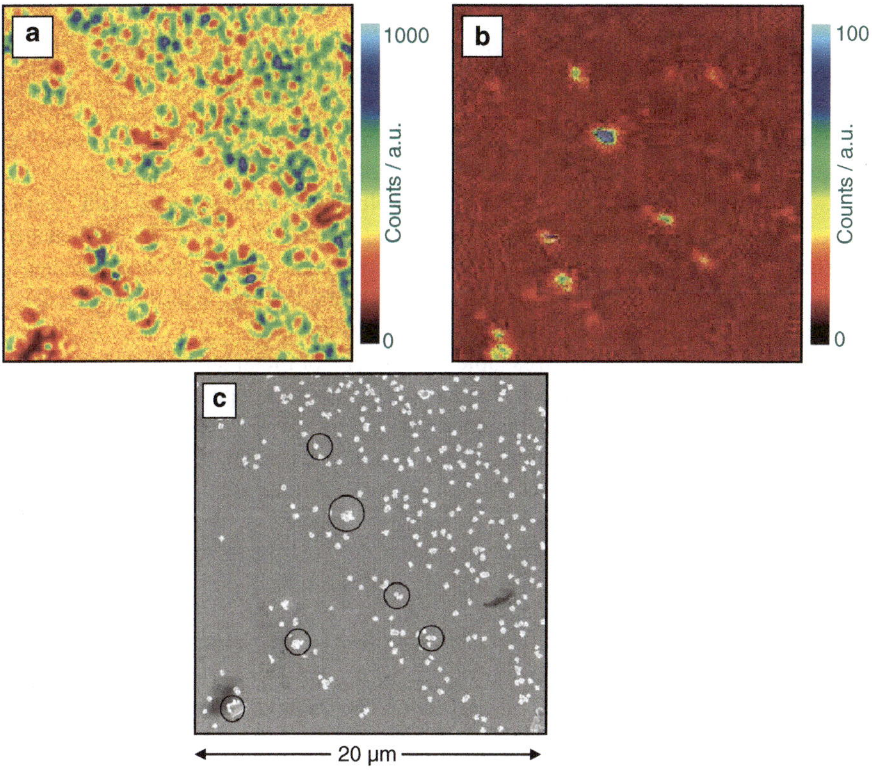

Fig. 8.16 Optical images of R6G covered gold clusters within the active zone II marked in Fig. 8.11; (**a**) Scanning confocal microscopy image (topography mode), (**b**) Raman/fluorescence image of the same sample area, (**c**) SEM image corresponding to (**a**) and (**b**). The confocal image could be retrieved within the SEM image by means of cross marking

6 Conclusion and Future Perspective

Silver and gold nanoparticles were generated electrochemically and investigated with confocal Raman microscopy. The combination of surface-enhanced Raman spectroscopy with confocal microscopy accompanied by subsequent scanning electron microscopy provided an image of the geometrical structure of the Raman spots.

The results confirm that the extreme enhancement factors of SERS are local effects. The "hot spots" were connected with surface irregularities and the strongest effects were seen on points of cluster agglomeration. An explanation might be that the polarizability tensor in this region has strong components perpendicular to the surface-enhancing charge transfer and Raman intensity. The results clearly demonstrate that the origin of the hot spots is not just an interference pattern of the electromagnetic plasmon fields [43]. Possibly, due to the huge enhancement in the junctions of particle aggregates, no significant effect of the particle size was observed contrary to reports in the literature [11, 44].

The results obtained from R6G molecules adsorbed on gold particles also revealed various dynamic spectroscopic features like fluctuations in intensity and even sporadically slight changes in spectral position. As the R6G concentration is relatively high compared to the amount of gold nanoparticles present and thermal effects can be excluded, the only possible explanation for the dynamic behavior seems to be fluctuations of the enhanced electromagnetic field [43]. Additionally, spatial and time-dependent changes in the orientation of the adsorbed molecule in the field may be also responsible for the fluctuations.

The understanding of spectroscopic phenomena of metal clusters is the basis for applications of surface-enhanced Raman spectroscopy to deposition studies of SER active metals. From the different methods, SERS is the most promising one because it makes possible the in situ monitoring of additives in the growth process. The experiments have shown that the Raman active structures are not identical with the growth centers of the deposition process. Nevertheless, the special structures showing Raman activity are irregularities, which will have adsorption properties very similar to the growth centers making still the investigation of additives with SERS very attractive.

Acknowledgments We acknowledge the financial support by the Deutsche Forschungsgemeinschaft and by the Fonds of the Chemical Industrie. We also acknowledge valuable discussions and experimental support given by Prof. A. J. Meixner (Universität Tübingen) and his former research fellows H. Kneppe, M. Weber, J. Schmidt, and T. Vosgröne.

References

1. Moskovits M (1985) Surface-enhanced spectroscopy. Rev Mod Phys 57:783
2. Otto A (1991) Surface-enhanced Raman scattering of adsorbates. J Raman Spectrosc 22:743
3. Campion A, Kambhampati P (1998) Surface-enhanced Raman scattering. Chem Soc Rev 27:241
4. Jensen L, Aikens CM, Schatz GC (2008) Electronic structure methods for studying surface-enhanced Raman scattering. Chem Soc Rev 37:1061
5. Michaels AM, Jiang J, Brus L (2000) Ag nanocrystal junctions as the site for surface-enhanced Raman scattering of single Rhodamin 6 G molecules. J Phys Chem B 104:11965
6. Xu HX, Aizpurua J, Käll M, Apell P (2000) Electromagnetic contributions to single molecule sensivity in surface-enhanced Raman scattering. Phys Rev E 62(3):4318
7. Jiang J, Bosnick K, Maillard M, Brus L (2003) Single molecule Raman spectroscopy at the junctions of large Ag nanocrystals. J Phys Chem B 107:9964
8. Hao E, Schatz GC (2004) Electromagnetic fields around silver nanoparticles and dimers. J Chem Phys 120(1):357
9. Kneipp J, Kneipp H, Kneipp K (2008) SERS-a single-molecule and nanoscale tool for bioanalytics. Chem Soc Rev 37:1052
10. Kneipp K, Wang Y, Kneipp H, Perelman LT, Itzkan I, Dasari RR, Feld MS (1997) Single molecule detection using surface-enhanced Raman scattering (SERS). Phys Rev Lett 78(9):1667

11. Nie S, Emory SR (1997) Probing single molecules and single nanoparticles by surface-enhanced Raman scattering. Science 275:1102
12. Otto A (2006) On the significance of Shalaev's "hot spots" in ensemble and single-molecule SERS by adsorbates on metallic films at the percolation threshold. J Raman Spectrosc 37:937
13. Reents B, Lacconi G, Plieth W (1994) Nucleation in the electrocrystallization process studied by surface-enhanced raman spectroscopy. J Electroanal Chem 376:185
14. Pawley JB (ed) (2006) Handbook of biological confocal microscopy, 3rd edn. Springer, New York
15. Hutter E, Fendler JH (2004) Exploitation of localized surface plasmon resonance. Adv Mater 16(19):1685
16. Fleischer M, Zhang D, Braun K, Jäger S, Ehrlich R, Häffner M, Stanciu C, Hörber JKH, Meixner AJ, Kern DP (2010) Tailoring gold nanostuctures for near-field optical applications. Nanotechnology 21:065301
17. Lee PC, Meisel DJ (1982) Adsorption and surface enhanced Raman of dyes on silver and gold sols. J Phys Chem 86:3391
18. Vosgröne T, Meixner AJ (2005) Surface-and resonance-enhanced micro-Raman spectroscopy of xanthene dyes: from the ensemble to single molecules. ChemPhysChem 6:154
19. Vosgröne T, Meixner AJ, Anders A, Dietz H, Sandmann G, Plieth W (2005) Electrochemically deposited silver particles for surface enhanced Raman spectroscopy. Surf Sci 597:102
20. Hildebrand P, Epding A, Vanhecke F, Keller S, Schrader B (1995) Molecule-specificity of surface enhanced Raman scattering. J Mol Struct 349:137
21. Ullmann R, Will T, Kolb DM (1993) Nanoscale decoration of Au(111) electrodes with Cu clusters by an STM. Chem Phys Lett 209:238
22. Fransaer J, Penner RM (1999) Brownian dynamics simulation of the growth of metal nanocrystal ensembles on electrode surfaces from solution. I. Instantaneous nucleation and diffusion-controlled growth. Phys Chem B 103:7643
23. Penner RM (2001) Brownian dynamics simulations of the growth of metal nanocrystal ensembles on electrode surfaces in solution: 2. The effect of deposition rate on particle size dispersion. Phys Chem B 105:8672
24. Zoval JV, Stiger RM, Biernacki P, Penner RM (1996) Electrochemical deposition of silver nanocrystallites on the atomically smooth Graphite basal plane. J Phys Chem 100:837
25. Zoval JV, Biernacki P, Penner RM (1996) Implementation of electrochemically synthesized silver nanocrystallites for the preferential SERS enhancement of defect modes on thermally etched Graphite surfaces. Anal Chem 68:1585
26. Liu H, Penner RM (2000) Size-selective electrodeposition of mesoscale metal particles in the uncoupled limit. J Phys Chem B 104:9131
27. Scheludko A, Todorova M (1952) Bull Acad Bulg Sci Phys 3:61
28. Milchev E, Vassileva E, Kertov V (1980) Electrolytic nucleation of silver on a glassy carbon electrode. J Electroanal Chem 107:323–336
29. Sandmann G, Dietz H, Plieth W (2000) Preparation of silver nanoparticles on ITO surfaces by a double-pulse method. J Electroanal Chem 491:78
30. Ueda M, Dietz H, Anders A, Kneppe H, Meixner A, Plieth W (2002) Double-pulse technique as an electrochemical tool for controlling the preparation of metallic nanoparticles. Electrochim Acta 48:377
31. Taylor CE, Garvey SD, Pemberton JE (1996) Carbon contamination at silver surfaces: surface preparation procedures evaluated by Raman spectroscopy and X-ray photoelectron spectroscopy. Anal Chem 68:2401
32. Norrod KL, Rowlen KL (1998) Removal of carbonaceous contamination from SERS-active silver by self-assembly of decanethiol. Anal Chem 70:4218
33. Wessel J (1985) Surface-enhanced optical microscopy. J Opt Soc Am B 2(9):1538
34. Stöckle RM, Sou YD, Deckert V, Zanobi R (2000) Nanoscale chemical analysis by tip-enhanced Raman spectroscopy. Chem Phys Lett 318:131

35. Hayazawa N, Inouye Y, Sekkat Z, Kawata S (2000) Metallized tip amplification of near field Raman scattering. Opt Commun 183:333
36. Bouhelier A (2006) Field-enhanced scanning near-field optical microscopy. Microsc Res Tech 69:563
37. Plieth W, Dietz H, Anders A, Sandmann G, Meixner A, Weber M, Kneppe H (2005) Electrochemical preparation of silver and gold nanoparticles: characterization by confocal and surface enhanced Raman microscopy. Surf Sci 597:119
38. Henglein A (1977) The reactivity of silver atoms in aqueous solutions (a γ-radiolysis study). Bunsenges Phys Chem 81:556
39. Plieth W (1982) Direct observation of size-dependent optical enhancement in single metal nanoparticles. J Phys Chem 86:3166
40. Plieth W, Dietz H, Sandmann G, Dietz H, Meixner AJ, Kneppe H, Weber M (1999) Optical phenomena of single silver clusters. In: Landolt D, Matlosz M, Sato Y (eds) Fundamentals in electrochemical deposition and dissolution. Proceedings, vol 99–33. The Electrochemical Society, Pennington, New Jersey
41. Moyer PJ, Schmidt J, Eng LM, Meixner AJ, Sandmann G, Dietz H, Plieth W (2000) Surface-enhanced Raman scattering spectroscopy of single carbon domains on individual Ag nanoparticles on a 25 ms time scale. J Am Soc 122:5409
42. Tuinstra F, Koenig JL (1970) Raman spectrum of graphite. J Chem Phys 53:1126
43. Markel VA, Shalaev VM, Zhang P, Huynh W, Tay L, Haslet TL, Moskovits M (1999) Near-field optical spectroscopy of individual surface-plasmon modes in colloid clusters. Phys Rev B 59:10903
44. Emory SR, Haskins WE, Nie S (1998) Direct observation of size-dependent optical enhancement in single metal nanoparticles. J Am Chem Soc 120:8009

Raman Spectroscopy for Characterization of Graphene

9

Duhee Yoon and Hyeonsik Cheong

1 Definition of the Topic

Raman spectroscopy is a spectroscopic tool to probe scattering of light by phonons in graphene and used to study the structural and electronic properties.

2 Overview

Raman spectroscopy is one of the most widely used characterization tool in the study of graphene, a two-dimensional hexagonal crystal of carbon atoms. Owing to the resonant nature of the light scattering process, fairly strong signals are detected in spite of the fact that only one monolayer of atoms is probed. In graphene research, Raman spectroscopy is used not only as an indispensible characterization tool to identify the number of layers, but also as a probe to investigate the mechanical, electrical, and optical properties. Here, important recent progress in Raman spectroscopic characterization of graphene is reviewed.

3 Introduction

3.1 Graphene

Graphene is a two-dimensional lattice of carbon atoms arranged in a honeycomb structure. It is the basic building block of all graphitic materials based on sp^2

D. Yoon • H. Cheong (✉)
Department of Physics, Sogang University, Seoul, South Korea

Challa S. S. R. Kumar (ed.), *Raman Spectroscopy for Nanomaterials Characterization*,
DOI 10.1007/978-3-642-20620-7_9, © Springer-Verlag Berlin Heidelberg 2012

Fig. 9.1 Optical microscope image of graphene on SiO_2/Si. The lightest-colored area corresponds to single-layer graphene

carbon-carbon bonding, including graphite, carbon nanotubes, and fullerenes. Graphite is a periodic stack of graphene sheets; a carbon nanotube is a result of rolling up a sheet of graphene into a cylinder; and fullerenes result when graphene is wrapped into a soccer-ball-like structure. Graphene has attracted much attention due to its remarkable electronic [1–3], optical [4], and mechanical properties [5] resulting from its peculiar electronic band structure. Although graphene's band structure has been studied theoretically for many years [6], experimental work on single graphene started in earnest only after Novoselov et al. [7] succeeded in isolating single layer graphene by mechanical exfoliation. So far, most of the cutting-edge studies on graphene's physical properties have used mechanically exfoliated graphene samples which usually exhibit superior electrical properties, although the quality of large-area graphene samples grown by chemical vapor deposition or by thermal treatment of SiC wafers is improving in many aspects. Mechanically exfoliated graphene on 300-nm SiO_2 covered Si substrates can be easily identified through ordinary optical microscopes as shown in Fig. 9.1. This makes Raman spectroscopy a particularly convenient tool in graphene study since one can easily find graphene and measure a Raman spectrum from a particular location of the sample using a standard micro-Raman spectroscopy system. In this chapter, the use of Raman spectroscopy for characterization of graphene is reviewed. A general review of graphene's electronic properties and other physical properties is beyond the scope of this chapter and the readers are encouraged to refer to more comprehensive reviews available elsewhere.

3.2 Crystal Structure

Figure 9.2 shows the crystal structure of graphene and its reciprocal lattice. Each unit cell, indicated in shade, contains two carbon atoms, A and B. The two unit vectors, $\mathbf{a}_1$ and $\mathbf{a}_2$, are also shown. For multilayer graphene or graphite, the Bernal

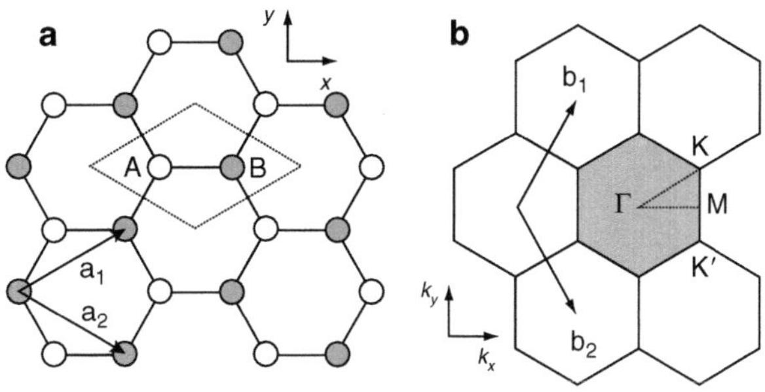

Fig. 9.2 (**a**) Graphene's lattice structure and (**b**) its reciprocal lattice

stacking (or AB stacking) is the most common configuration. The Bernal stacking is formed by putting a B carbon atom of the second layer on top of an A carbon of the first layer or vice versa. In graphite, the interlayer spacing is 0.335 nm, which is often taken as the "thickness" of a graphene single layer. The Brillouin zone of graphene is shown in Fig. 9.2b. It has a hexagonal shape, and the high symmetry points are the Γ point at the zone center, the M point in the middle of the hexagonal sides, and the two inequivalent K and K' points at the hexagonal corners.

3.3 Band Structure

Three of the four valence electrons of a carbon atom in graphene are involved in sp^2 bonding and contribute little to the electronic properties. Practically, one needs to consider only the remaining electron in the π state. Since each unit cell has two carbon atoms, two π electrons need to be considered. The electronic band structure can be described with a reasonable accuracy in the framework of the tight-binding approximation [6]. If only the nearest-neighbor interaction is considered, one gets

$$E^{\pm}(k_x, k_y) = \pm\gamma_0\sqrt{1 + 4\cos\frac{\sqrt{3}k_x a}{2}\cos\frac{k_y a}{2} + 4\cos^2\frac{k_y a}{2}}, \tag{9.1}$$

where γ_0 is the nearest-neighbor hopping energy, $a = \sqrt{3}a_0$ is the magnitude of the principal unit vector, and $a_0 = 0.142$ nm is the distance between nearest neighbors. Figure 9.3 illustrates the two energy bands E^+ and E^-. Of particular interest is the dispersion of the bands near K or K' points. One can write $\boldsymbol{k} = \boldsymbol{K} + \boldsymbol{\kappa}$, where $\boldsymbol{\kappa}$ is the crystal momentum vector relative to the K point. Then, it is straightforward to show that

$$E^{\pm}(\boldsymbol{k}) = \varepsilon^{\pm}(\boldsymbol{\kappa}) = \pm\hbar v_F|\boldsymbol{\kappa}|, \tag{9.2}$$

where v_F is the electronic group velocity and $v_F = \sqrt{3}\gamma_0 a/2\hbar \approx 10^6\text{m/s}$. Although this is a very simple approximation, it faithfully reproduces most important features of the band structure calculated with the more sophisticated ab initio numerical many-body GW calculation [8]. The most significant implications are: (1) graphene

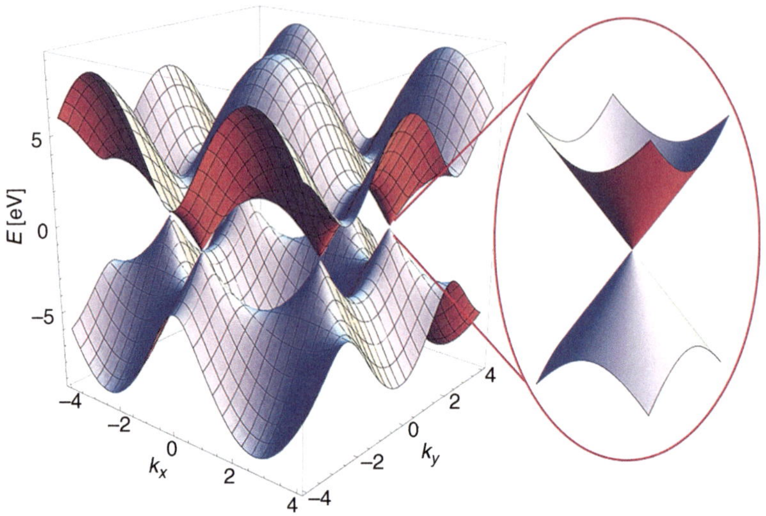

Fig. 9.3 Energy band structure of graphene calculated with tight-binding approximation

is a zero-gap semiconductor and (2) the energy dispersion near the band extrema is linear. Since each unit cell has two π electrons, the lower (negative) energy branch is completely filled and the upper (positive) energy branch is completely empty at zero temperature. The linear dispersion implies that the effective mass is zero. Therefore, one cannot use the Schrödinger equation with an effective mass in describing the behaviors of electrons in graphene. One should rather use a Dirac equation for massless relativistic fermions, and K and K' points are hence called "Dirac points." Many interesting properties of graphene are direct results of such Dirac nature of the electrons. Figure 9.4 shows the band structures of graphene calculated using the density functional theory and the more sophisticated *GW* approximation. The linear dispersion of the bands near the K point is evident in either calculation.

3.4 Raman Scattering in Graphene

Raman spectroscopy has been established as one of the most important characterization tools in graphene research. Graphene's peculiar band structure plays an important role in Raman scattering as well. Figure 9.5 is the phonon dispersion of graphene. The highest energy branches have linear dispersions near the Γ and K points due to Kohn anomalies [9] resulting from strong electron-phonon interaction [10].

Figure 9.6 shows typical Raman spectra of graphene and graphite. The Raman spectrum of graphene is dominated by two prominent features. The G band near 1,580 cm^{-1} is due to the degenerate zone-center (Γ) optical phonon mode with E_{2g} symmetry, which is found in most graphitic materials [13]. Because of the energy

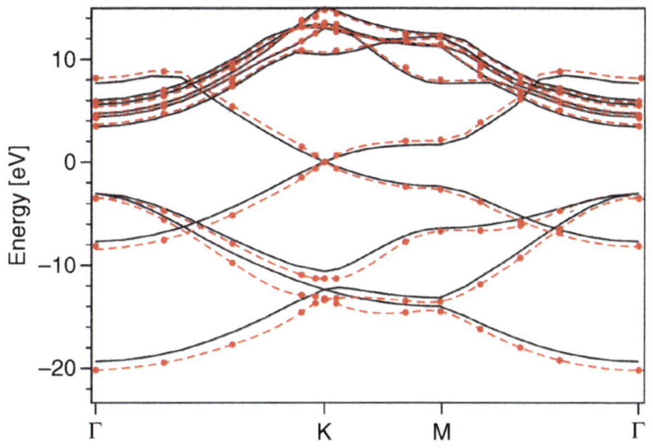

Fig. 9.4 Energy band structure of graphene. *Solid thick lines*: density functional theory with local density approximation; *circles* and *dashed lines*: *GW* approximation [8]

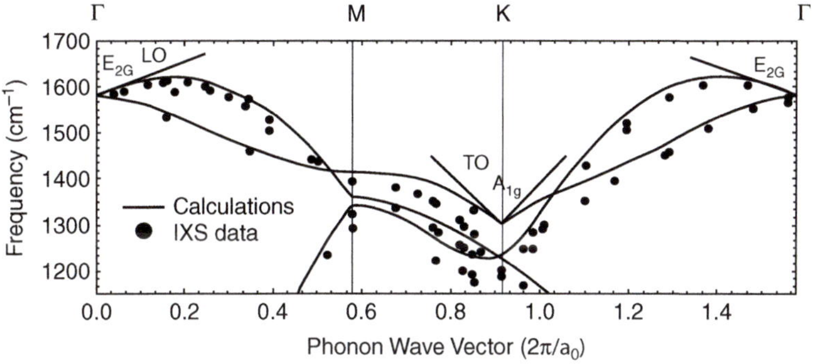

Fig. 9.5 Phonon dispersion of graphene calculated by using the density functional perturbation theory [10] compared with the inelastic X-ray scattering data [11] from graphite [12]

band structure, the Raman scattering process for $q = 0$ phonons is resonant for any excitation laser energy in the infrared and visible range (see Fig. 9.7.). This resonance explains a rather strong intensity of this Raman band despite the fact that the material being probed is only one atomic layer thick. The strong band at $\sim$2,680 cm^{-1} is due to a two-phonon scattering involving A_1' phonons near the K point and is called the 2D band, whereas some people prefer to call it a G' band. A photoexcited electron (or hole) near a K point can be scattered to a nearby K' point by emitting a phonon. In order to conserve the total momentum, the emitted phonon must have a momentum of magnitude $\sim |\boldsymbol{K} - \boldsymbol{K}'| \sim |\boldsymbol{K}|$. Therefore, a phonon near the K point of the Brillouin zone is involved. Then, another phonon of the same momentum should be emitted before the electron and the hole recombine to emit Raman scattered light. Figure 9.8 illustrates this process. This is a doubly resonant process, which explains why the 2D band is usually stronger

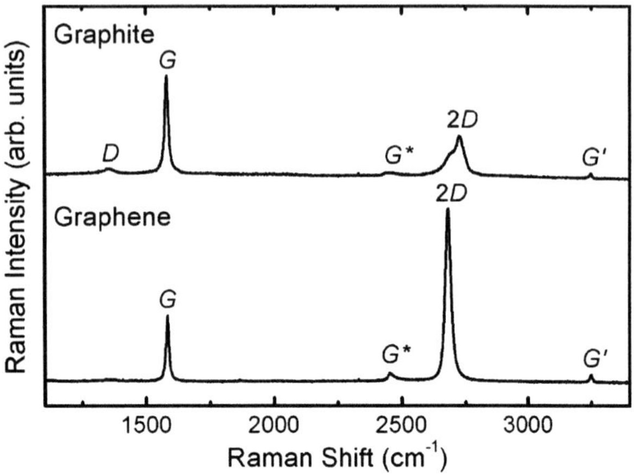

Fig. 9.6 Raman spectra of graphene and graphite

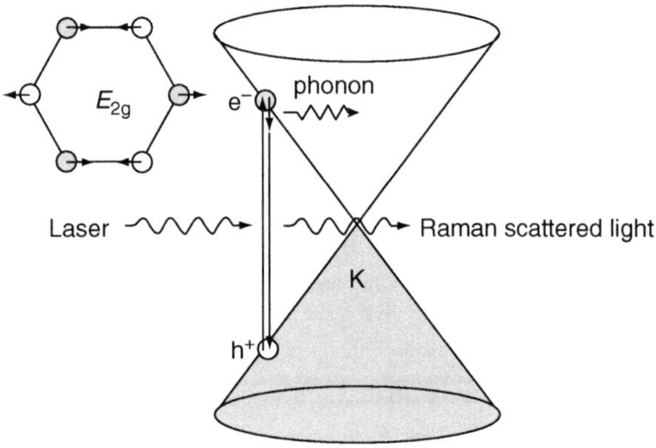

Fig. 9.7 Vibration of E_{2g} phonon mode and the resonant Raman scattering process for the G band

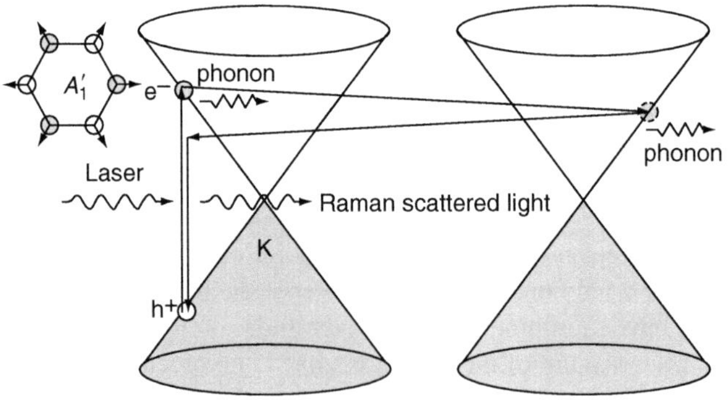

Fig. 9.8 Vibration of $A_1{}'$ phonon mode and the double resonance Raman scattering process for the $2D$ band

than the *G* band although it is due to a two-phonon scattering process. The triple resonance process [14], where the electron and the hole scatter simultaneously, can also explain the two-phonon scattering.

There are also some weak features in the Raman spectrum. The most important is the *D* band near 1,355 cm^{-1} due to a one-phonon scattering involving a phonon near the *K* point and scattering due to defects [15]. The scattering process for this band is explained in terms of a double resonance scattering between the *K* and *K′* points [14] similar to the case of the 2*D* band. It is absent or extremely weak in a spectrum from pristine graphene but becomes visible when there are significant amount of defects in the graphene sample, its intensity correlating with the defect density [16].

4 Experimental and Instrumental Methodology

4.1 Micro-Raman Spectroscopy

Since mechanically exfoliated graphene samples are very small (typically $<100 \times 100\ \mu m^2$), confocal micro-Raman systems are used. For routine measurements of Raman spectra, commercial micro-Raman spectrometer systems are good enough. For more flexible configuration of measurements, however, custom-built systems are preferred. Figure 9.9 is a schematic of a typical confocal micro-Raman spectroscopy system. Since the Raman scattering process in graphene is resonant regardless of the excitation laser energy, one may use any laser wavelength from infrared to near ultraviolet. The excitation laser light is sent to the sample through a microscope objective. The focusing of the laser spot on the sample is monitored by a digital video camera. Scattered light is collected by the same objective and sent to a spectrometer through a focusing lens. A holographic notch filter or a dielectric edge filter is used to block the laser light from entering the spectrometer. A CCD multichannel detector collects dispersed photons and records the spectrum. The spatial resolution is determined by the laser wavelength and the numerical aperture (N.A.) of the objective. For an Ar-ion laser line of $\lambda = 514.5$ nm and an objective with N.A. = 0.6, the theoretical laser spot diameter is $1.22\lambda/2\text{N.A.} = 500$ nm. The spectral resolution is determined by the focal length of the spectrometer, the groove density in the grating, and the pixel size in the CCD detector. A typical resolution value is 1–2 cm^{-1}. In order to remove unwanted wavelength components from the laser, for example, plasma discharge lines, a spike filter or a prism monochromator can be used. Also, a spatial filter can be used to remove unwanted spatial mode components from the laser beam, and a beam expander may be used to expand the beam to the optimal size for the input aperture of the objective. These will ensure that the smallest possible laser spot size is achieved. In order to move the sample laterally, a two-axis xy translation stage is used. If one uses a computer-controlled motorized stage, one can raster scan the sample to obtain Raman images. Alternatively, the beam steering mirror can be controlled with servo motors to scan the laser spot on the sample.

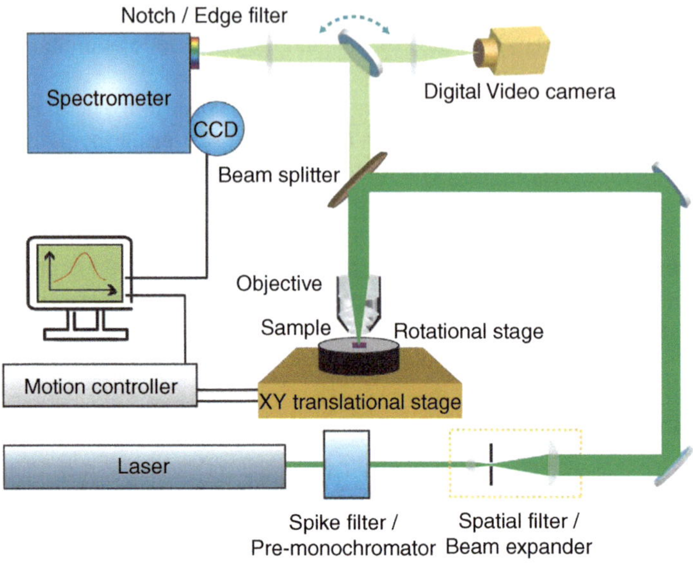

Fig. 9.9 Schematic of a confocal micro-Raman system

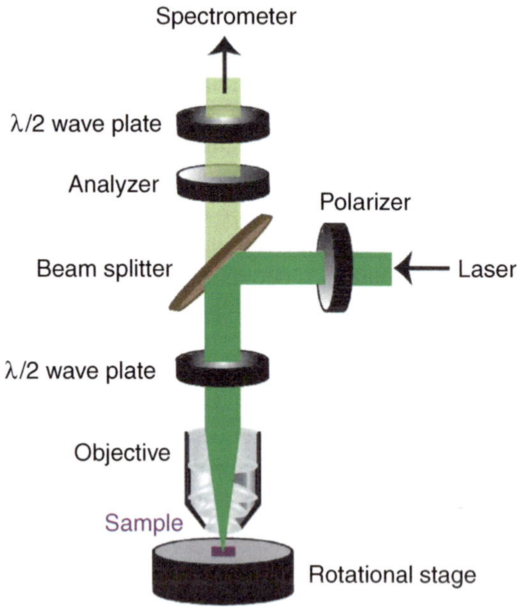

Fig. 9.10 Schematic of polarized micro-Raman setup

4.2 Polarization

Polarized Raman spectroscopy has proven to be a powerful tool in the study of the properties of graphene. The polarization analysis of the Raman scattered light provides valuable information on the symmetry-related properties. For complete polarization analysis, the polarizations of the excitation laser beam as well as the Raman scattered light entering the spectrometer must be controlled. This can be achieved with combinations of linear polarizers and wave plates. Care must be taken in order not to introduce instrument-induced polarization anisotropy. Since the throughput of a spectrometer is usually polarization dependent, the polarization direction of the light entering should be kept the same by using a half-wave plate just before the light enters the spectrometer. Another factor is the beam splitter. The reflectivity and the transmittance of the *p* wave are significantly different from those of the *s* waves for the incident angle of $\sim$45°, which is close to Brewster's angle. Therefore, the *p* wave component of the scattered light has a higher transmittance at the beam splitter than the *s* wave component. As a result, even when the signal from the sample is isotropic, it may appear anisotropic if the polarization-dependent transmittance of the beam splitter is not properly taken into account. A separate calibration of the polarization dependence of the transmittance is often necessary. Also, the wavelength-dependence of the half-wave plates should be considered. Since the phonon frequencies in graphene are fairly large, the retardation of a wave plate could be significantly different between the G band and the 2*D* band, for example. Achromatic wave plates, which produce fairly uniform retardation within $\pm$15% of the center wavelength, produce far better result than ordinary zero-order or multiple-order wave plates. Figure 9.10 illustrates a typical polarized micro-Raman measurement setup [17].

4.3 Strain

Uniaxial strain breaks the in-plane isotropy and produces many interesting effects. In order to apply strain, a piece of graphene is usually put on a flexible substrate. When the substrate is bent, the top surface is stretched whereas the bottom surface is contracted. One can use either the two-point bending or the four-point bending method (Fig. 9.11) [18]. Since the graphene sample is much smaller than the length of the substrate, the strain is essentially uniform across the sample in either case. As a result, tension is applied if the graphene sample is on the top surface, and compression is applied if the sample is on the bottom surface. Since graphene is attached to the substrate by the van der Waals force, there is a limitation on the maximum applied tension, typically less than 3%. In the case of compression, buckling can happen as well. In order to prevent slippage of the sample, some researchers deposit metal anchors on top of the graphene sample [19] or embed the graphene sample in a polymer layer [20].

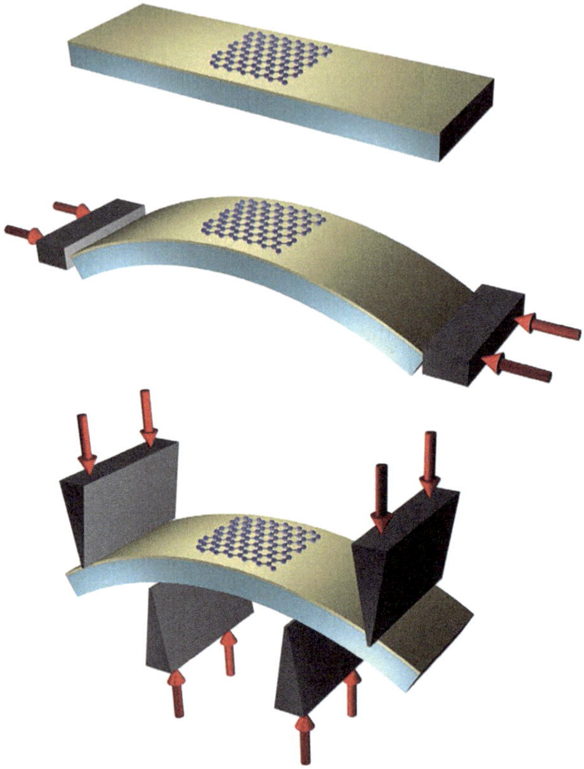

Fig. 9.11 (**a**) A graphene sample is placed in the middle part of a flexible substrate; (**b**) two-point bending; and (**c**) four-point bending [18]

5 Key Research Findings

5.1 Determination of Number of Graphene Layers

The most important application of Raman spectroscopy in graphene research is to estimate the number of graphene layers [21–24]. It is by far the most straightforward method to identify single-layer graphene. Figure 9.12 shows the evolution of the G, G^*, and $2D$ bands as the number of graphene layers is increased [24]. The intensity of the Raman G band increases with the number of layers up to seven layers and then decreases for thicker samples (~23, ~40 layers and graphite). The shape of the G band does not vary, but the variation in the intensity is sensitive up to seven layers. Therefore, the variation in the G band Raman intensity can be used to approximately identify the number of graphene layers. The Raman G^* band has a small intensity, which does not vary much, as shown in Fig. 9.2b. This band originates from a combination of the zone boundary in-plane longitudinal acoustic (iLA) phonon and the in-plane transverse optical (iTO) phonon modes [25, 26]. The shapes of the G^* bands are similar to each other, except for single-layer

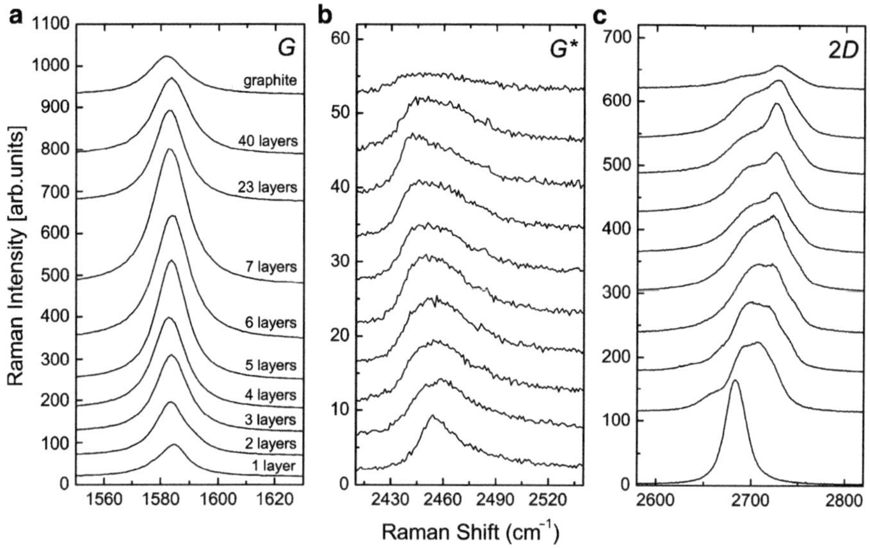

Fig. 9.12 Evolution of the (**a**) *G* band, (**b**) *G** band, and (**c**) *2D* band in the Raman spectrum as functions of the number of graphene layers [24]

graphene. The *G** band of single-layer graphene is relatively sharp compared to that for other thicknesses. Although the sharp shape of the *G** band could be evidence for single-layer graphene, the *G** band has a low signal-to-noise ratio.

The line shape of the *2D* band of single-layer graphene is unique compared to the others. It has a single Lorentzian line shape and a high intensity. The line shape of the *2D* band reflects the electronic band structure of graphene. In bilayer graphene, the inter-layer interaction splits the conduction and valence bands into two each. There are four possible pathways for the double resonance Raman scattering process as shown in Fig. 9.13 [21, 25]. As a result, the *2D* band is a convolution of four Lorentzian lines. As the number of graphene layers increases beyond two layers, the electronic band structure varies and approaches that of graphite [28, 29], and as a result, the line shape of the *2D* band approaches that of graphite. From Fig. 9.12c, it is evident that the *2D* band can be used as an exact fingerprint of graphene layers for one to four layers.

By using the Raman imaging technique, one can visualize areas of different thickness. Figure 9.14a is an optical microscope image of a graphene sample that has areas with different thickness. Areas with different shades are observed. Figure 9.14b is an atomic force microscope (AFM) topography image obtained from the area indicated by the dotted square in Fig. 9.14a. The number of graphene layers is estimated by using cross-section profiling of the AFM image. The Raman scanning was performed on a $30 \times 40\ \mu m^2$ region. Figures 9.14c–e are images for the Raman *G* band intensity, the *2D* band intensity, and the ratio of the intensity of

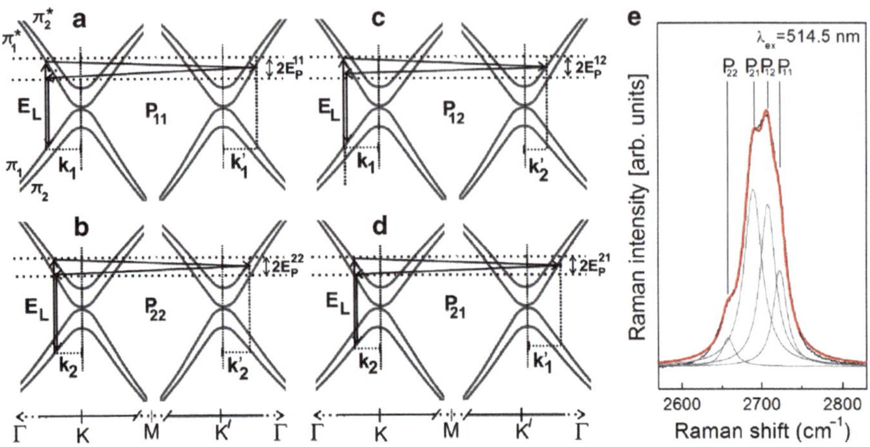

Fig. 9.13 (**a–d**) Schematic view of the electron dispersion of bilayer graphene near the K and K' points showing both π_1 and π_2 bands. The four double resonance Raman scattering processes are indicated. The wave vectors of the electrons (k_1, k_2, $k_1{}'$, and $k_2{}'$) involved in each of these four processes are also indicated [27]. (**e**) Typical $2D$ band spectrum with the four components is indicated

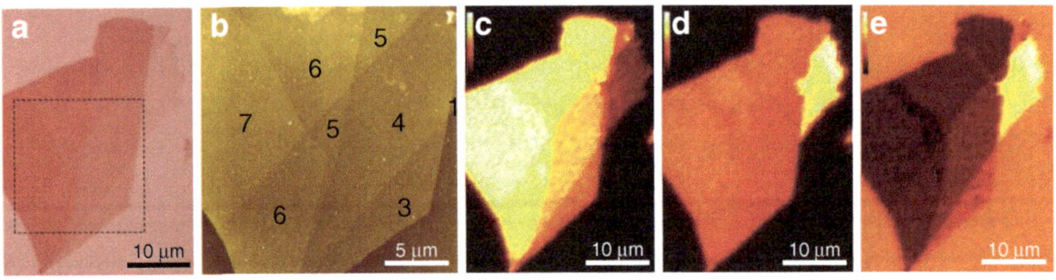

Fig. 9.14 (**a**) Optical microscope image and (**b**) AFM (contact mode) image of a graphene sample. Images of (**c**) the G band intensity, (**d**) the $2D$ band intensity, and (**e**) the ratio of the intensity of the $2D$ band to that of the G band [24]

the $2D$ band to that of the G band, respectively. The boundaries between areas with different numbers of layers are well distinguished in Fig. 9.14c because the G band intensity is sensitive to the number of layers. On the other hand, the boundaries are not well distinguished in Fig. 9.14d, except for the single-layer area because the $2D$ band intensity of single-layer graphene is much larger than the others. Because the G band intensity increases and then decreases as the number of graphene layers increases, there may be a case where the G band intensities are the same for areas with different numbers of graphene layers. Since the $2D$ band intensity tends to decrease as the number of graphene layers increases, one can better estimate the thickness in such cases by comparing the intensities of the G and the $2D$ bands. Figure 9.14e is the image of the ratio of the intensity of the $2D$ band to that of the G band.

As expected, areas of different thicknesses are more clearly resolved. Therefore, by combining the intensity images of the *G* and the *2D* bands, one can reliably estimate the number of graphene layers.

5.2 Interference Effect

The fact that graphene, which is only single atomic layer thick, can be observed through an optical microscope is a major factor in making experiments with graphene accessible to many researchers. The visibility of a graphene layer on a substrate is a result of an optical contrast enhancement due to interference effect in the dielectric layer of the substrate. Most graphene samples are made on Si substrates with an oxide layer on the top. Multiple reflections of the illuminating light in the SiO_2 layer between Si and graphene result in an interference which enhances or suppresses the optical contrast. As seen in Fig. 9.15, the optical contrast is a sensitive function of both the wavelength of the illuminating light and the thickness of the SiO_2 layer [30]. A similar interference effect plays a role in Raman spectroscopy. Figure 9.16 illustrates the effect of the interference on the Raman intensities [31]. The Raman intensities of the *G* and *2D* bands are strong functions of the thickness of SiO_2 layer (Fig. 9.16a). As the thickness is increased from 240 nm, the observed intensities increase first and show the highest values at the thickness of ~280 nm for the *G* band and ~290 nm for the *2D* bands, respectively. Since the two maxima occur at different wavelengths, the resulting observed I_{2D}/I_G ratio varies greatly; the maximum of the intensity ratio is around 9.3 and 6 times higher than the minimum (Fig. 9.16b).

The interference effect can be quantitatively analyzed using a multireflection model (MRM) of the Raman scattered light. In this model, the absorption and scattering processes are treated separately [31]. Since the difference in the wavelengths of the excitation laser and the scattered light is significant, these differences and concomitant differences in the index of refraction should be taken into account. The interference either enhances or suppresses the Raman intensity, and the enhancement factors relative to the case of a free-standing graphene film can be calculated as functions or the excitation laser wavelength and the thickness of the SiO_2 layer (Fig. 9.17). Figure 9.17c shows that the ratio of the enhancement factor of the *2D* band to that of the *G* band is a strong function of both the laser wavelength and the SiO_2 thickness. The intensity ratio of the *2D* and the *G* band is often used in the analysis of Raman data in an effort to eliminate the effects of external factors such as equipment alignment that may vary in each measurement. However, Fig. 9.17c demonstrates that care must be taken when comparing such ratios obtained from different measurements. It should also be noted that this kind of interference effect is not unique to Raman measurements but applies to any spectroscopic measurements of thin samples on dielectric layers. Appropriate choice of the dielectric layer thickness therefore can significantly enhance the measured signal in such cases.

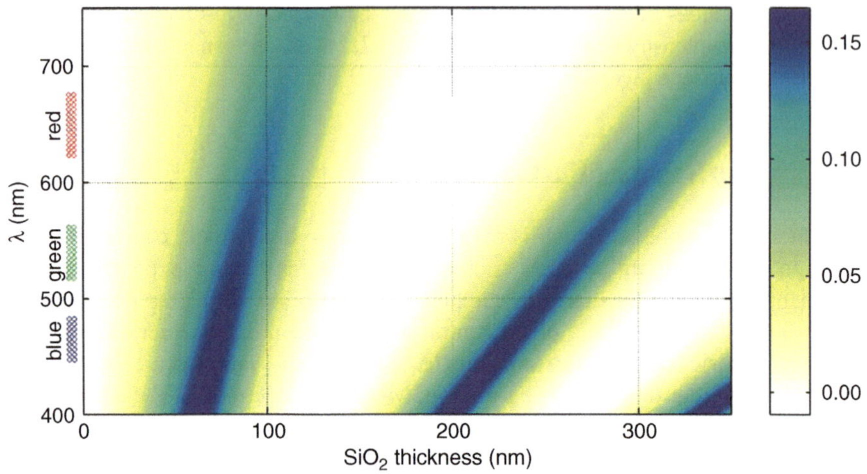

Fig. 9.15 Optical contrast as a function of wavelength and SiO_2 thickness [30]

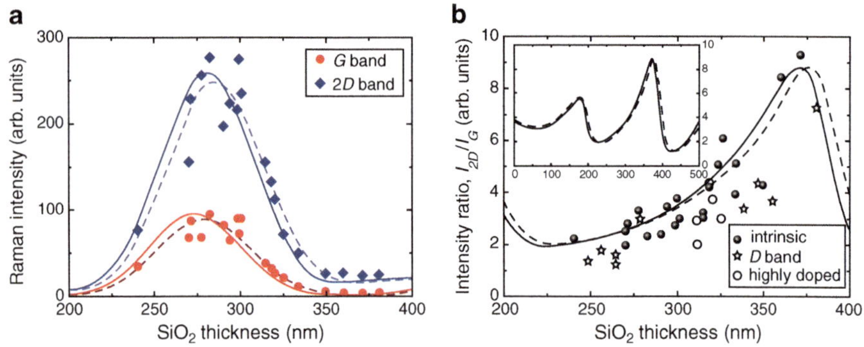

Fig. 9.16 (**a**) *G* (*circle dots*) and 2*D* (*square dots*) band Raman intensities as functions of the thickness of the SiO_2 layer. (**b**) Raman intensity ratio I_{2D}/I_G as a function of the thickness of the SiO_2 layer [31]

5.3 Effect of Charge Carriers

Controlling the charge carrier type and density is a key to electronic device application of graphene. Most of the studies so far have used either a back gate (through the substrate) or a top gate to induce charge in the graphene layer. Chemical doping of graphene with metal adatoms [32], ammonia [33], or other gases [34] has been tried, and both electron and hole doping have been reported. The effect of doping can be studied with Raman spectroscopy, too. It is well known that the frequency and the linewidth of the Raman *G* band are sensitive functions of the doping density [35, 36]. The dependence of the graphene's Raman spectrum on

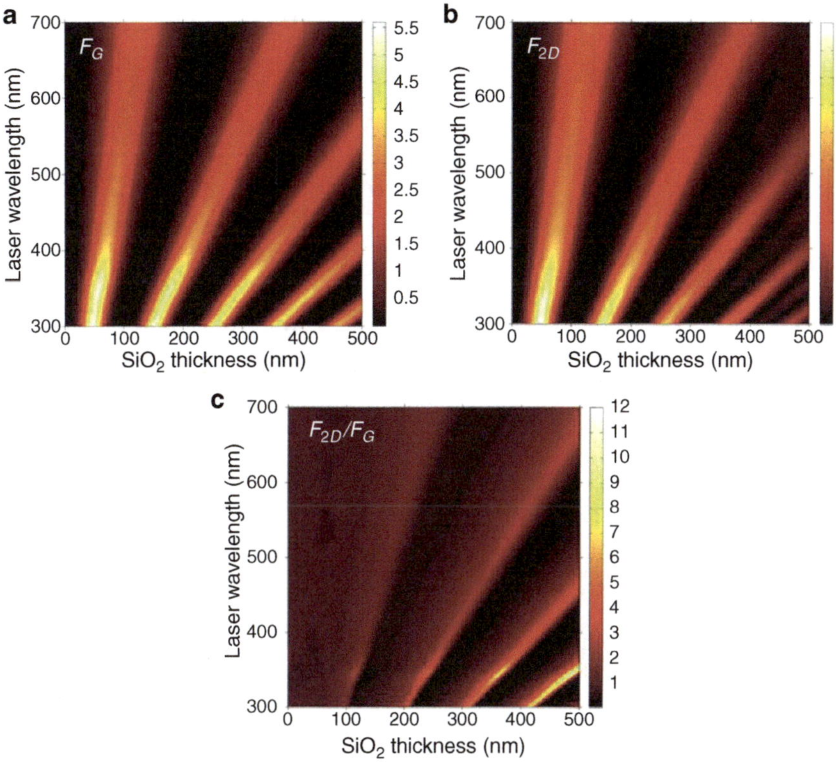

Fig. 9.17 Plots of calculated Raman enhancement factors of (**a**) G band and (**b**) $2D$ band as functions of the thickness of SiO_2 layer and the wavelength of the laser. (**c**) Ratio of the enhancement factor for the $2D$ band to that of the G band, F_{2D}/F_G [31]

the charge carrier density is due to suppression of Kohn anomaly [9, 10] by charge carriers. The physical origin of the Kohn anomaly in graphene can be explained in terms of a virtual process in which a phonon excites an electron from the valence band to the conduction band creating an electron-hole pair, and the phonon is re-created when the electron and hole recombine (Fig. 9.18). When the charge density increases such that the Fermi energy E_F is larger than $\frac{1}{2}\hbar\omega_G$, a half of the phonon energy, such electron-hole pair creation becomes impossible due to the Pauli principle. This suppresses the Kohn anomaly, and as a result, the G band blueshifts. The linewidth also becomes smaller because the phonon lifetime increases due to the blockage of the decay channel. The $2D$ band also depends on the charge density [37], although it is not as well understood.

The charge density dependence of the Raman spectrum can be used to estimate chemical doping of graphene samples [38]. The frequency of the G band, in combination of its linewidth, can be used as an indicator of the charge density of a graphene sample. However, it should be noted that the G band frequency can be

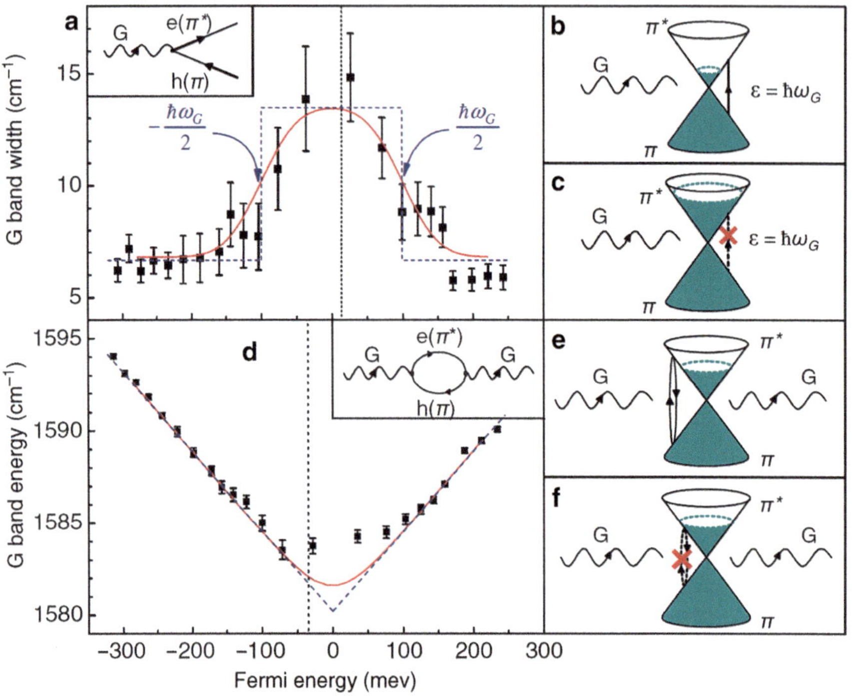

Fig. 9.18 Dependence of (**a**) the linewidth and (**d**) frequency of the *G* band. (**b**) Broadening of the *G* phonon due to decay into electron-hole pairs. (**c**) The *G*-phonon decay into the electron-hole pair is forbidden by the Pauli principle at high charge densities. (**e**) Renormalization of the *G*-phonon energy by interaction with virtual electron-hole pairs. (**f**) Virtual electron-hole pair transitions with energy ranging from 0 to 2 E_F are forbidden [35]

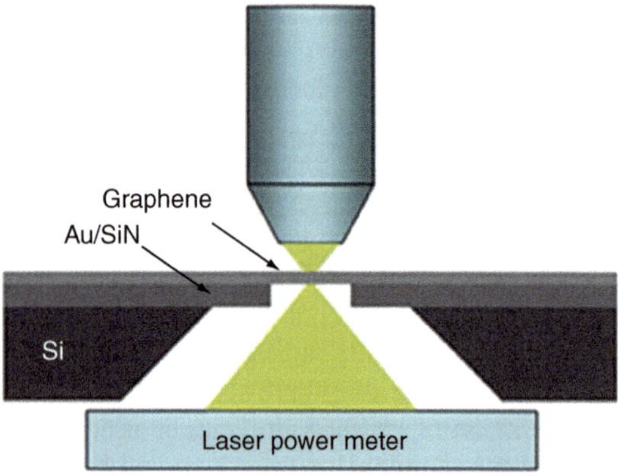

Fig. 9.19 Schematic of the experimental setup for graphene thermal conductivity measurement. The laser power meter measures how much of the excitation power is absorbed by the graphene sample [46]

shifted due to other external factors such as residual strain. The linewidth may also increase when there are structural defects in the sample. Therefore, the estimation of the charge density from a Raman spectrum is only approximate.

5.4 Thermal Effect

Unlike most solids, graphene has a negative thermal expansion coefficient near the room temperature. A theoretical calculation [39] predicts a negative thermal expansion coefficient value for temperatures up to 2,300 K and beyond, whereas an indirect estimation obtained by analyzing the ripples formed on graphene membranes suspended across trenches on SiO_2/Si substrates [40] gives a negative thermal expansion coefficient only up to ~350 K and a positive value at higher temperatures. The negative thermal expansion would cause a *blueshift* of the Raman band as the temperature is increased. However, an analysis of anharmonic effects predicts a *redshift* instead [41]. There are a few measurements of the temperature dependence of the Raman spectrum of graphene. Calizo et al. reported a *linear* coefficient of $\partial\omega_G/\partial T = -0.016\text{cm}^{-1}/°\text{C}$ for the single layer and $-0.015\ \text{cm}^{-1}/°\text{C}$ for the bilayer graphene between 83 K and 373 K [42]. On the other hand, Allen et al. reported a larger value of $-0.038\ \text{cm}^{-1}/°\text{C}$ for chemically derived graphene in the temperature range 22–400°C [43]. Chae et al. monitored the local temperature of the probed spot by monitoring the ratio of the Stokes and anti-Stokes signals and obtained a strongly *nonlinear* dependence of the *G* band frequency between 300 K and 1,600 K [44].

The thermal conductivity of graphene has been estimated by using Raman spectroscopy. In a typical experiment, a graphene sample is heated with a focused laser beam which doubles as the excitation source for Raman scattering (Fig. 9.19). An initial measurement [45] on a graphene sample suspended over a trench gave a very large value of $\sim 5 \times 10^3$ W/m·K. Similar measurements on samples suspended across a circular hole give smaller values of $2.5 \pm 1.1 \times 10^3$ W/m·K [46] and $\sim 6 \times 10^2$ W/m·K [47]. The thermal conductivity becomes smaller if the sample is supported on a substrate [46, 48].

5.5 Strain Effect

Uniaxial or biaxial strain applied to a sheet of graphene alters the interatomic distance and modifies the crystal structure, two of the fundamental parameters that determine the properties of graphene. Strain engineering, modifying the structural and electronic properties of graphene with applied strain, is therefore attracting much attention. When uniaxial strain is applied, the in-plane symmetry is broken, and as a result, the degenerate Raman *G* band splits into two peaks, G^+ and G^- (Fig. 9.20). Since the splitting is proportional to the applied strain [18, 19], one can use the splitting to estimate the built-in uniaxial strain.

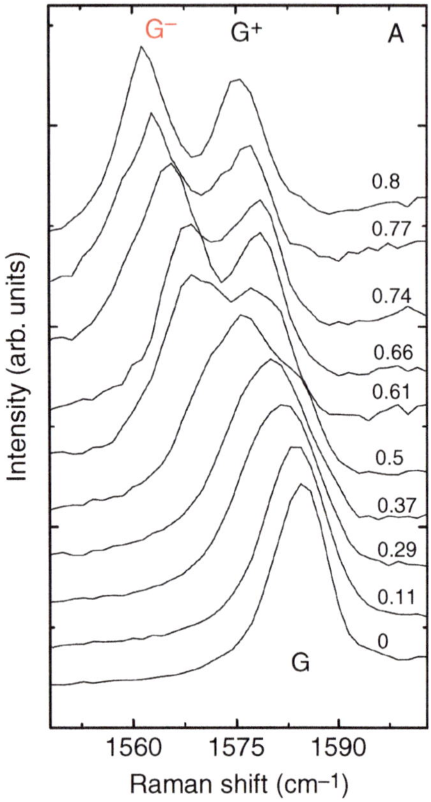

Fig. 9.20 Evolution of the G band as a function of uniaxial strain. The spectra are measured with incident light polarized along the strain direction, collecting the scattered light with no analyzer. The magnitudes of the strain, ranging from 0% to ~0.8%, are indicated on the right side of the spectra [18]

Furthermore, the polarization dependence of the two peaks can be used to determine the crystallographic orientation of a graphene sample. The intensities of the G^+ and G^-peaks follow the simple expressions [18, 19]:

$$\begin{aligned} I_{G^+} &\propto \cos^2(\theta_{in} + \theta_{out} + 3\varphi_s) \\ I_{G^-} &\propto \sin^2(\theta_{in} + \theta_{out} + 3\varphi_s), \end{aligned} \tag{9.3}$$

where θ_{in} and θ_{out} are the polarization directions of the excitation laser and the Raman scattered photons with respect to the strain direction, respectively, and φ_s is the relative angle between the strain axis and the zigzag direction of the graphene crystal (Fig. 9.21). If one sets the polarization of the incident laser parallel with the strain direction so that $\theta_{in} = 0$, the above expressions take a simple form:

$$\begin{aligned} I_{G^+} &\propto \cos^2(\theta_{out} + 3\varphi_s) \\ I_{G^-} &\propto \sin^2(\theta_{out} + 3\varphi_s) \end{aligned} \tag{9.4}$$

Therefore, by measuring the intensities of the strain-split peaks as functions of the analyzer direction, one can extract φ_s. This would unambiguously determine the

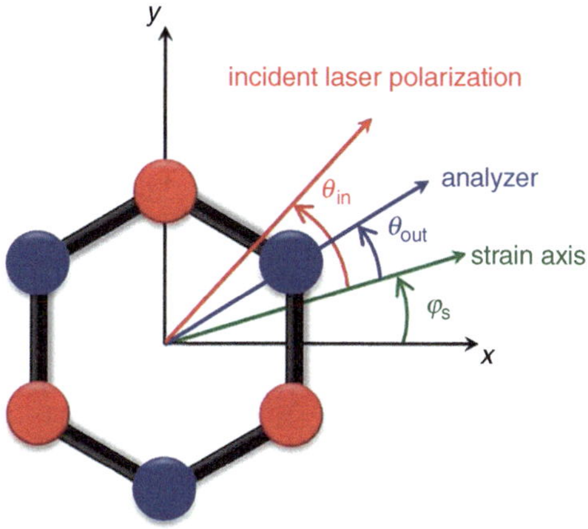

Fig. 9.21 Coordinate system in the analysis of polarization dependence of Raman spectrum from strained graphene

crystallographic direction of the graphene sample. Depending on the experimental setup, one may alternatively set $\theta_{out} = 0$ and measure the intensities as functions of θ_{in} as in Fig. 9.22. The 2*D* band also splits under strain and shows polarization dependence [49], although its interpretation is much more complicated because the double resonance two-phonon scattering reflects both the phonon dispersion and the electronic band structure.

5.6 Edges

Since an edge of a graphene sample is a kind of a line defect, it induces defect-mediated Raman scattering. The Raman *D* band near 1,355 cm^{-1}, which becomes significant only when there are significant amount of defects, becomes visible near edges. Figure 9.3a illustrates the double resonance intervalley scattering process that gives rise to the D band [50]. An electron near a *K* point with a wave vector $\vec{k}$ is excited by absorbing an incident photon. The set of all electron wave vectors that can participate in the absorption forms a circle with a radius $k = |\vec{k}|$ about the *K* point. The electron is then scattered by emitting a phonon of wave vector $\vec{q}$ to a point on a circle with a radius $k' = |\vec{k'}|$about a nearby K' point. The electron is then scattered back to $\vec{k}$ by a defect with wave vector $\vec{d} = -\vec{q}$, followed by the electron-hole recombination that emits a scattered photon. Figure 9.3b shows (1) the armchair and (2) the zigzag edges. An edge destroys the translational symmetry in the direction perpendicular to it, and as a result, the momentum conservation requirement in that direction is relaxed. Therefore, the armchair edge can contribute to a process in Fig. 9.23a with $\vec{d} = \vec{d}_a$ and the zigzag edge with $\vec{d} = \vec{d}_z$. In Fig. 9.3c, it is evident that whereas $\vec{d}_a$ can connect two states on the circles about the *K* and K' points, $\vec{d}_z$ cannot. Therefore, a perfect zigzag edge would not contribute to the *D* band. Further theoretical analysis predicts that the *D* band signal from a perfect armchair edge would be maximum if the polarizations of the excitation and the collection of scattered photons are both parallel to the armchair edge direction [50, 51].

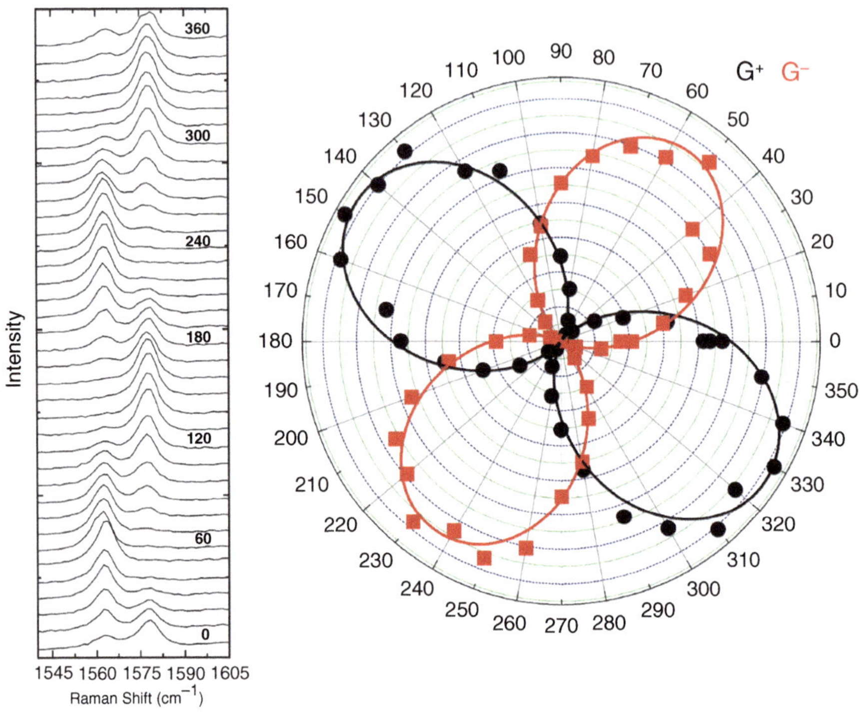

Fig. 9.22 (*Left*) Raman spectra and (*right*) polar plot of the intensities of the G^{+} and G^{-} peaks as functions of θ_{in} measured with an analyzer selecting scattered polarization along the strain axis $\theta_{out} = 0$. The polar data are fitted to $I_{G^-} \propto \sin^2(\theta_{in} + 34°)$ and $I_{G^+} \propto \cos^2(\theta_{in} + 34°)$, which gives $\varphi_s = 11.3°$ [18]

In real samples, however, the edges are never perfectly armchair or zigzag. There are many imperfections and usually armchair and zigzag edges are mixed. Therefore, Raman spectra from such edges would contain contributions from both armchair and zigzag components. One can expect that an "armchair-like" edge with predominantly armchair structure would show higher *D* band intensity than a "zigzag-like" edge. Several authors [50, 52] have imaged the *D* band intensity of graphene samples and correlated the polarization dependence and intensity with the edge types. Edges showing stronger polarization dependence are identified as being "armchair-like." However, due to the "roughness" of the edges, the identification of the edge types from the *D* band Raman scattering is not very conclusive [51, 53].

5.7 Stacking

The discussion in Sect. 4 of the Raman spectrum of bilayer and multilayer graphene is valid only for Bernal (AB) stacked graphene. Mechanically exfoliated graphene from crystalline graphite usually exhibits such characteristics. However, multilayer

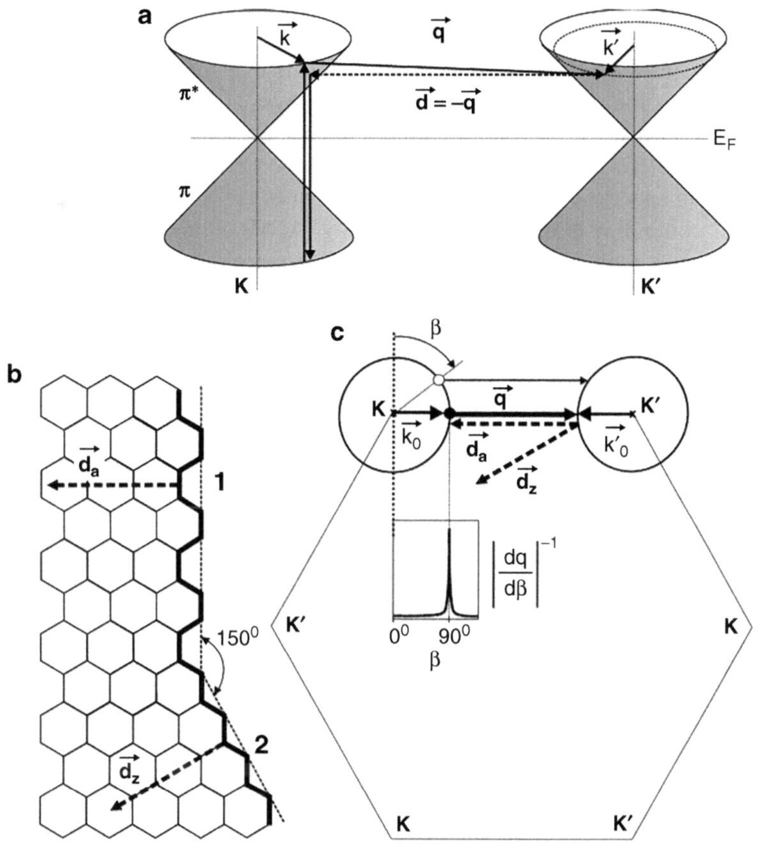

Fig. 9.23 (**a**) Representation of a double resonance Raman process that gives rise to the *D* band. (**b**) Schematic illustration of the atomic structure of the edges. (**c**) Double resonance mechanism for an armchair edge [50]

graphene produced by thermal treatment of SiC or by chemical vapor deposition often show incommensurate stacking as evidenced by Moiré patterns in scanning tunneling microscope images [54, 55]. The splitting of the conduction and valence bands in Bernal stacked bilayer or multilayer graphene, and the resultant broadening of the Raman 2*D* band is a result of interlayer interaction. However, when subsequent layers are not Bernal stacked, the interlayer interaction, which is already weak in Bernal stacked graphene, becomes even weaker, and the electronic and vibrational structures of such multilayer graphene resemble those of a collection of independent single-layer graphene pieces. The shape of the Raman 2*D* band in such cases resembles that of single-layer graphene, showing a single Lorentzian lineshape [56, 57]. This has an important implication on the scheme for determination of the number of graphene layers as discussed in Sect. 4. The lineshape of the 2*D* band by itself should not be taken as a definitive indicator of the number of layers in situations where non-Bernal stacked multilayer formation is possible.

6 Conclusions and Future Perspective

Raman spectroscopy has proven to be an invaluable tool in graphene research owing to the unique features of graphene's physical properties. In addition to characterization of structural parameters, which is usually the typical realm of Raman characterizations, the electronic and transport properties can be investigated with Raman spectroscopy. Several new techniques are also being developed. For example, combining atomic force microscopy (AFM) with Raman imaging allows investigators a unique opportunity to study the correlation between microscopic structures and the Raman characteristics. Surface-enhanced Raman scattering (SERS) or tip-enhanced Raman scattering (TERS) have been attempted with mixed results [58, 59]. It is certain that Raman spectroscopy will be an integral part of graphene research for the foreseeable future.

References

1. Geim AK, Novoselov KS (2007) The rise of graphene. Nat Mater 6:183–191
2. Katsnelson MI (2007) Grpahene: carbon in two dimensions. Mater Today 10:20–27
3. Castro Neto AH, Guinea F, Peres NMR, Novoselov KS, Geim AK (2009) The electronic properties of graphene. Rev Mod Phys 81:109–162
4. Nair RR, Blake P, Grigorenko AN, Novoselov KS, Booth TJ, Stauber T, Peres NMR, Geim AK (2008) Fine structure constant defines visual transparency of graphene. Science 320:1308
5. Lee C, Wei X, Kysar JW, Hone J (2008) Measurement of the elastic properties and intrinsic strength of monolayer graphene. Science 321:385–388
6. Wallace PR (1947) The band theory of graphite. Phys Rev 71:622–634
7. Novoselov KS, Geim AK, Morozov SV, Jiang D, Dubonos SV, Girgorieva IV, Firsov AA (2004) Electric field effect in atomically thin carbon films. Science 306:666–669
8. Trevisanutto PE, Giorgetti C, Reining L, Ladisa M, Olevano V (2008) *Ab Initio GW* many–body effects in graphene. Phys Rev Lett 101:226405
9. Kohn W (1956) Image of the Fermi surface in the vibration spectrum of a metal. Phys Rev Lett 2:393–394
10. Piscanec S, Lazzeri M, Mauri F, Ferrari AC, Robertson J (2004) Kohn anomalies and electron–phonon interactions in graphite. Phys Rev Lett 93:185503
11. Maultzsch J, Reich S, Thomsen C, Requardt H, Ordejón P (2004) Phonon dispersion in graphite. Phys Rev Lett 92:075501
12. Charlier J-C, Eklund PC, Zhu J, Ferrari AC (2008) Electron and phonon properties of graphene: their relationship with carbon nanotubes. In: Jorio A, Dresselhaus G, Dresselhaus MS (eds) Carbon nanotubes, vol 111, Topics in applied physics. Springer, Berlin/Heidelberg, pp 673–709
13. Dresselhaus MS, Jorio A, Hofmann M, Dresselhaus G, Saito R (2010) Perspectives on carbon nanotubes and graphene Raman spectroscopy. Nano Lett 10:751–758
14. Malard LM, Pimenta MA, Dresselhaus G, Dresselhaus MS (2009) Raman spectroscopy in graphene. Phys Rep 473:51–87
15. Pimenta MA, Dresselhaus G, Dresselhaus MS, Cançado LG, Jorio A, Saito R (2007) Studying disorder in graphite-based systems by Raman spectroscopy. Phys Chem Chem Phys 9:1276–1291
16. Lucchese MM, Stavale F, Martins Ferreira EH, Vilani C, Moutinho MVO, Capaz RB, Achete CA, Jorio A (2010) Quantifying ion–induced defects and Raman relaxation length in graphene. Carbon 48:1592–1597

17. Yoon D, Moon H, Son Y–W, Samsonidze G, Park BH, Kim JB, Lee Y, Cheong H (2008) Strong polarization dependence of double–resonant Raman intensities in graphene. Nano Lett 8:4270–4274
18. Mohiuddin TMG, Lombardo A, Nair RR, Bonetti A, Savini G, Jalil R, Bonini N, Basko DM, Galiotis C, Marzari N, Novoselov KS, Geim AK, Ferrari AC (2009) Uniaxial strain in graphene by Raman spectroscopy: G peak splitting, Grüneisen parameters, and sample orientation. Phys Rev B 79:205433
19. Huang M, Yan H, Chen C, Song D, Heinz TF, Hone J (2009) Phonon softening and crystallographic orientation of strained graphene studied by Raman spectroscopy. Proc Natl Acad Sci USA 106:7304–7308
20. Gong L, Kinloch IA, Young RJ, Riaz I, Jalil R, Novoselov KS (2010) Interfacial stress transfer in a graphene monolayer nanocomposite. Adv Mater 22:2694-2697
21. Ferrari AC, Meyer JC, Scardaci V, Casiraghi C, Lazzeri M, Mauri F, Piscanec S, Jiang D, Novoselov KS, Roth S, Geim AK (2006) Raman spectrum of graphene and graphene layers. Phys Rev Lett 97:187401
22. Gupta A, Chen G, Joshi P, Tadigadapa S, Eklund PC (2006) Raman scattering from high-frequency phonons in supported n-graphene layer films. Nano Lett 6:2667–2673
23. Graf D, Molitor F, Ensslin K, Stampfer C, Jungen A, Hierold C, Wirtz L (2007) Spatially resolved Raman spectroscopy of single– and few–layer graphene. Nano Lett 7:238–242
24. Yoon D, Moon H, Cheong H, Choi JS, Choi JA, Park BH (2009) Variations in the Raman spectrum as a function of the number of graphene layers. J Korean Phys Soc 55: 1299–1303
25. Mafra DL, Samsonidze G, Malard LM, Elias DC, Brant JC, Plentz F, Alves ES, Pimenta MA (2007) Deterination of LA and TO phonon dispersion relations of grapehene near the Dirac point by double resonance Raman scattering. Phys Rev B 76:233407
26. Shimada T, Sugai T, Fantini C, Souza M, Cançado LG, Jorio A, Pimenta MA, Saito R, Grüneis A, Dresselhaus G, Dresselhaus MS, Ohno Y, Mizutani T, Shinohara H (2005) Origin of the 2450 cm^{-1} Raman bands in HOPG, single–wall and double–wall carbon nanotubes. Carbon 43:1049–1054
27. Malard LM, Nilsson J, Elias DC, Brant JC, Plentz F, Alves ES, Castro Neto AH, Pimenta MA (2007) Probing the electronic structure of bilayer graphene by Raman scattering. Phys Rev B 76:201401(R)
28. Latil S, Henrard L (2006) Charge carrier in few-layer graphene films. Phys Rev Lett 97:036803
29. Saito R, Dresselhaus G, Dresselhaus MS (1993) Electronic structure of double–layer graphene tubules. J Appl Phys 73:494–500
30. Blake P, Hill EW, Castro Neto AH, Novoselov KS, Jiang D, Yang R, Booth TJ, Geim AK (2007) Making graphene visible. Appl Phys Lett 91:063124
31. Yoon D, Moon H, Son Y-W, Choi JS, Park BH, Cha YH, Kim YD, Cheong H (2009) Interference effect on Raman spectrum of graphene on SiO_2/Si. Phys Rev B 80:125422
32. Gierz I, Riedl C, Starke U, Ast CR, Kern K (2008) Atomic hole doping graphene. Nano Lett 8:4603–4607
33. Wang X, Li X, Zhang L, Yoon Y, Weber PK, Wang H, Guo J, Dai H (2009) N–doping of graphene through electrothermal reactions with ammonia. Science 324:768–771
34. Farmer DB, Golizadeh-Mojarad R, Perebeinos V, Lin Y-M, Tulevski GS, Tsang JC, Avouris Ph (2009) Chemical doping and electron–hole asymmetry in graphene devices. Nano Lett 9:388–392
35. Yan J, Zhang Y, Kim P, Pinczuk A (2007) Electric field effect tuning of electron–phonon coupling in graphene. Phys Rev Lett 98:166802
36. Pisana S, Lazzeri M, Casiraghi C, Novoselov KS, Geim AK, Ferrari AC, Mauri F (2007) Breakdown of the adiabatic Born–Oppenheimer approximation in graphene. Nat Mater 6:198–201

37. Das A, Pisana S, Chakraborty B, Piscanec S, Saha SK, Waghmare UV, Novoselov KS, Krishnamurthy HR, Geim AK, Ferrari AC, Sood AK (2008) Monitoring dopants by Raman scattering in an electrochemically top–gated graphene transistor. Nat Nanotechnol 3:210–215
38. Casiraghi C, Pisana S, Novoselov KS, Geim AK, Ferrari AC (2007) Raman fingerprint of charged impurities in graphene. Appl Phys Lett 91:233108
39. Mounet N, Marzari N (2005) First-principles determination of the structural, vibrational and thermodynamic properties of diamond, graphite, and derivatives. Phys Rev B 71:205214
40. Bao W, Miao F, Chen Z, Zhang H, Jang W, Dames C, Lau CN (2009) Controlled ripple texturing of suspended graphene and ultrathin graphite membranes. Nat Nanotechnol 4:562–566
41. Bonini N, Lazzeri M, Marzari N, Mauri F (2007) Phonon anharmonicities in graphite and graphene. Phys Rev Lett 99:176802
42. Calizo I, Balandin AA, Bao W, Miao F, Lau CN (2007) Temperature dependence of the Raman spectra of graphene and graphene multilayers. Nano Lett 7:2645–2649
43. Allen MJ, Fowler JD, Tung VC, Yang Y, Weiller BH, Kaner RB (2008) Temperature dependent Raman spectroscopy of chemically derived graphene. Appl Phys Lett 93:193119
44. Chae D–H, Krauss B, von Klitzing K, Smet JH (2010) Hot phonons in an electrically biased graphene constriction. Nano Lett 10:466–471
45. Balandin AA, Ghosh S, Bao W, Calizo I, Teweldebrhan D, Miao F, Lau CN (2008) Superior thermal conductivity of single–layer graphene. Nano Lett 8:902–907
46. Cai W, Moore AL, Zhu Y, Li X, Chen S, Shi L, Ruoff RS (2010) Thermal transport in suspended and supported monolayer graphene grown by chemical vapor deposition. Nano Lett 10:1645–1651
47. Faugeras C, Faugeras B, Orlita M, Potemski M, Nair RR, Geim AK (2010) Thermal conductivity of graphene in Corbino membrane geometry. ACS Nano 4:1889–1892
48. Seol JH, Jo I, Moore AL, Lindsay L, Aitken ZH, Pettes MT, Li X, Yao Z, Huang R, Broido D, Mingo N, Ruoff RS, Shi L (2010) Two-dimensional phonon transport in supported graphene. Science 328:213–216
49. Yoon D, Son YW, Cheong H (2011) Strain-dependent splitting of the double-resonance raman scattering band in graphene. Phys Rev Lett 106:155502
50. Cançado LG, Pimenta MA, Neves BRA, Dantas MSS, Jorio A (2004) Influence of the atomic structure on the Raman spectra of graphite edges. Phys Rev Lett 93:247401
51. Casiraghi C, Hartschuh A, Qian H, Piscanec S, Georgi C, Fasoli A, Novoselov KS, Basko DM, Ferrari AC (2009) Raman spectroscopy of graphene edge. Nano Lett 9:1433–1441
52. You Y, Ni Z, Yu T, Shen Z (2008) Edge chirality determination of graphene by Raman spectroscopy. Appl Phys Lett 93:163112
53. Gupta AK, Russin TJ, Gutiérrez HR, Eklund PC (2009) Probing graphene edges via Raman scattering. ACS Nano 3:45–52
54. Varchon F, Mallet P, Magaud L, Veuillen J–Y (2008) Rotational disorder in few–layer graphene films on 6H–SiC(000–1): a scanning tunneling microscopy study. Phys Rev B 77:165415
55. Li G, Luican A, Lopes dos Santos JMB, Castro Neto AH, Reina A, Kong J, Andrei EY (2009) Observation of Van Hove singularities in twisted graphene layers. Nat Phys 6:109–113
56. Poncharal P, Ayari A, Michel T, Sauvajol J–L (2008) Raman spectra of misoriented bilayer graphene. Phys Rev B 78:113407
57. Ni Z, Wang Y, Yu T, You Y, Shen Z (2008) Reduction of Fermi velocity in folded graphene observed by resonance Raman spectroscopy. Phys Rev B 77:235403
58. Schedin F, Lidorikis E, Lombardo A, Kravets VG, Geim AK, Grigorenko AN, Novoselov KS, Ferrari AC (2010) Surface enhanced Raman spectroscopy of graphene. ACS Nano 4:5617–5626
59. Saito Y, Verma P, Masui K, Inouye Y, Kawata S (2009) Nano–scale analysis of graphene layers by tip–enhanced near–field Raman spectroscopy. J Raman Spectrosc 40:1434–1440

SERS Hot Spots

10

Robert C. Maher

1 Overview

Hot spots are highly localized regions of intense local field enhancement believed to be caused by local surface plasmon resonances (LSPR). Formed within the interstitial crevices present in metallic nanostructures [1–4], such hot spots have been claimed to provide extraordinary enhancements of up to 10^{15} orders of magnitude to the surface-enhanced Raman scattering (SERS) signal (proportional to $/E/^4$) [5] in areas of subwavelength localization [6, 7]. As a result, hot spots are critically important for SERS and, if in sufficient density, can dominate the properties of any SERS active substrate within which they reside. Given their characteristics, hot spots are now widely acknowledged to be a prerequisite for the observation of single-molecule SERS and set the limit for the achievable spatial resolution for SERS as applied to scanning probe microscopy with high chemical specificity also known as tip-enhanced Raman spectroscopy (TERS). As a result of their importance to SERS, hot spots have generated a great deal of interest in the last 5 years. This review summarizes the key results from recent theoretical and experimental investigations focused on improving the understanding of the characteristics of SERS hot spots and their control.

2 Introduction

When surface-enhanced Raman spectroscopy (SERS) was discovered in 1974 by Fleischman, Hendra, and McQuillan, it was initially attributed to a surface concentration effect rather than to a surface enhancement effect [8]. The foundations of the

R.C. Maher
Department of Physics, Imperial College London, Blackett Laboratory, London, UK

Challa S. S. R. Kumar (ed.), *Raman Spectroscopy for Nanomaterials Characterization*, DOI 10.1007/978-3-642-20620-7_10, © Springer-Verlag Berlin Heidelberg 2012

theory and understanding of SERS were developed by Van Duyne [9], Moskovits [10], and others [11] in the years immediately following with great progress in the understanding being achieved [12]. As a result of this work, average enhancement factors for typical SERS substrates were found to be 10^6–10^8 and two enhancement mechanisms were proposed to explain the SERS phenomenon, the electromagnetic (EM) and the chemical or charge transfer (CT) mechanisms [13, 14]. The EM mechanism arises due to the presence of enhanced electromagnetic fields localized to within a few nanometers of a nanostructured metallic surface formed by surface plasmon resonances. The enhanced electromagnetic fields decay exponentially from the surface and thus result in enhanced optical processes within those molecules in close proximity to the surface. The EM enhancement is believed to provide the majority of the normal SERS enhancement (10^4–10^5) [12]. The CT enhancement results from resonant charge transfer effects between a molecule that is strongly chemically adsorbed onto the surface and the metal itself and is believed to contribute 10^2–10^3 to the enhancement. Thus, the EM mechanism is believed to be the dominant enhancement mechanism, providing the majority of the enhancement while being less molecularly dependent than the CT mechanism which should only be active for molecules that strongly interact with the surface.

At this point, it is important to clarify what is meant by the enhancement factor before discussing hot spots further. Typically, the enhancement factor is calculated as the ratio of the detected Raman signal under SERS conditions compared to the signal obtained under normal conditions, for equal numbers of active molecules and surface area exposed to the laser beam. This value is proportional to the intensity of the local electromagnetic field (E) to the fourth power, i.e., $/E/^4$ which results from the enhancement of both the incident and emitted photons [15]. Throughout the rest of the chapter, the enhancement factor that is quoted refers to this: the enhancement of SERS intensity, rather than that associated with the local electromagnetic field unless otherwise stated.

It was suggested early in the development of SERS that the majority of the signal originated from a small fraction of the total number of adsorbed molecules residing within highly localized areas of strong enhancement, or hot spots [16]. At the time, these hot spots were primarily believed to be due to highly advantageous binding sites where the contribution to the enhancement from the chemical enhancement was maximized. The enhancement factor for a molecule adsorbed at such a site was postulated to be at least one or two orders of magnitude larger than average estimates which considered all adsorbed molecules. Such hot spots were shown to be an essential feature in SERS, for instance, it was verified that covering the SERS active silver electrode with only 3% of electrodeposited Pt or Ti completely quenched the SERS effect which was attributed to the preferential deposition on the hot spots [17, 18].

However, interest in SERS waned due to the irreproducibility, the molecule specific nature, and relatively low enhancements inherently associated with SERS measurements which prohibited the development of the effect for possible applications. SERS was consigned to being characterized as an experimental curiosity.

This changed with the observation of single-molecule SERS in 1997 by two independent studies by Nie and Emory [22] and Kneipp et al. [23], which sparked

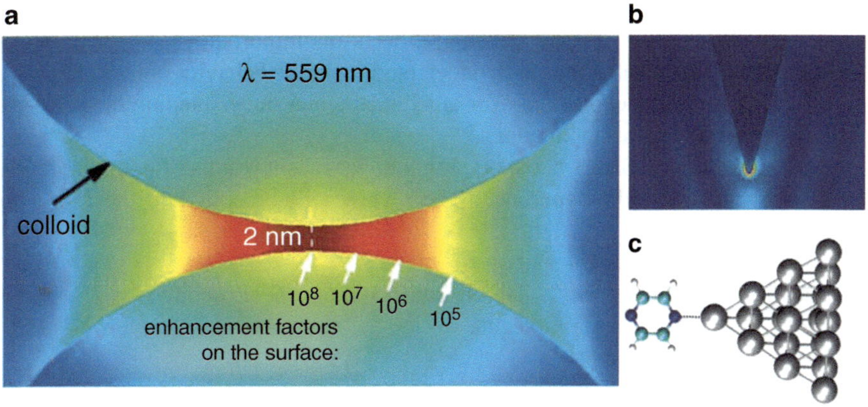

Fig. 10.1 Examples of different types of generic hot spot: (**a**) The hot spot formed between two closely placed nanoparticles resulting from a coupled plasmon resonance, (**b**) The hot spot formed at a sharp geometrical feature and (**c**) surface complex formed when a molecule adsorbs strongly onto a metal surface (Figure reproduced from Refs. [19, 20] and [21])

a renaissance in SERS. Hot spots were proposed as the source of the high signal enhancement required to allow for the detection of single molecules. While average enhancement factors for ensemble SERS measurements had typically been estimated to be 10^3–10^6, the enhancement factors for hot spots thought to be responsible for the single-molecule SERS were estimated to be as high as 10^{12}–10^{15}. However, differences in the experimental conditions led to different conclusions. For example, Kneipp et al. reported the observation of Raman spectra from single molecules of crystal violet in an aqueous solution of aggregated silver nanoparticles. This was broadly in line with the prevailing theory of SERS in that clusters will contain extremely intense localized electromagnetic fields in the interstitial spaces due to local surface plasmon resonance interactions. However, Nie and Emory reported single-molecule SERS for rhodamine 6G (Rh6G)–adsorbed to silver nanoparticles, not only for clusters but also for isolated nanoparticles where the electromagnetic enhancement is relatively weak. This suggested that the majority of the enhancement originated from molecules adsorbed to preferential binding sites and raised questions about whether hot spots in aggregated nanostructures were a feature of the aggregate geometry or the inclusion of hot particles.

Figure 10.1 summarizes the different types of generic hot spots that have been proposed in the literature. These include (a) those formed between two closely spaced nanoparticles by the coupled plasmon, (b) hot spots formed at sharp nanoscale geometrical features through the "lightening rod" effect, and (c) highly enhancing chemically active binding sites. The ambiguity in the exact enhancement mechanism at work within the observed hot spots and the number of active molecules, the magnitude of hot spot enhancement, and the lack of a robust method to define the single molecular signature have led to much debate within the literature [24, 25]. However, it is clear that the properties of any SERS active substrate where hot spots contribute to the observed signal will be completely dominated by them in the low concentration limits important for many proposed applications. As a result a thorough understanding of the behavior of these hot spots is of critical importance not only for our fundamental understanding of SERS but for its development for applications.

In this chapter on SERS hot spots, I will cover recent results of both theoretical and experimental studies approach. The review will be split into the following sections:

- The fundamentals of hot spots
- Idealized hot spot control – Tip-enhanced Raman spectroscopy
- Hot spots and the single molecule
- Imaging SERS hot spots
- Fabricating SERS hot spots
- Applications of SERS hot spots

It should be noted that while this chapter is focused on hot spots for SERS, hot spots are of significant interest for other surface-enhanced spectroscopies such as surface-enhanced fluorescence [26, 27] and surface-enhanced infrared absorption [28, 29] as well as the related field of plasmonics [30–32]. The natural link between SERS, the local surface plasmon resonance, and plasmonics has attracted a great deal of interest from the field of plasmonics in recent years. Such hot spots are potentially important for a variety of applications including plasmonic waveguides, plasmonic lithography [33–35], nano-optical waveguides [36], molecular traps [37], and plasmonic lasers [38, 39].

3 Basic Methodology

Before the results of recent investigations into the nature of SERS hot spots are reviewed, it is useful to very briefly cover the techniques and methodologies that are commonly applied to their study. The exploration of SERS hot spots has come to involve a diverse array of characterization and fabrication techniques including:

Raman spectroscopy has matured into a standard experimental technique over the years with sophisticated systems readily available from several manufacturers including Jobin Von and Renishaw. Such systems are easy to set up and use and, typically, include some form of data analysis present in the control software. More recently, there has been a drive to combine Raman systems with high-resolution topographic mapping using atomic force microscopy (AFM) stages [40] and advanced data analysis techniques such as principle component analysis in order to obtain high-resolution chemical and topographic surface characterization over large areas. Costs have come down drastically in recent years due to integration, miniaturization, and standardization of components, and several handheld Raman spectrometers are now available at reasonable cost, opening up new possibilities for applications.

Computational simulations have provided a theoretical backdrop for many experimental measurements focusing on the electromagnetic enhancement mechanism. In principle, the interaction of the incident electromagnetic radiation and the nanostructured metallic SERS substrates involves a solution of Maxwell's equations for a given geometry. In general, this is not a trivial task and exact solutions can only be obtained for simplified geometries such as isolated spheres using Mie theory. More complex arrangements of spheres and other simplified nanoparticle geometries can be simulated analytically, for example, with generalized Mie theory [41]. However, although such simplified geometries can provide a great deal of insight into the characteristics of SERS hot spots, they fail to capture the full details of the majority of experimental situations. A range of numerical solutions to Maxwell's have been developed in response to this challenge which have been recently reviewed by Zhao et al. [42].

High-resolution imaging has vastly improved since the discovery of SERS allowing for the in-depth characterization of SERS active substrates. It is now possible to investigate areas of the substrate on a scale appropriate to the subwavelength confinement of electromagnetic energy which is known to occur within hot spots. The techniques of scanning electron microscopy (SEM) [43], transmission electron microscopy (TEM) [44], and atomic force microscopy (AFM) [45] have all been applied to the topographical characterization of SERS active substrates which exhibit hot spots at high resolution. Such information can be correlated to high-resolution Raman mapping allowing hot spot geometries to be investigated in detail. This structural information can then be directly used in computational simulations to gain a more detailed understanding of hot spot characteristics. As well as this, near-field scanning optical microscopy (NSOM) [46] has been applied to image the plasmonic response of SERS active substrates below the diffraction limit allowing for a direct comparison with theory.

Fabrication of hot spots has become more easily realizable as techniques such as electron-beam lithography (ebeam) [47] and focused ion beam (FIB) [48] are readily available and capable of defining features reproducibly with resolution reaching a few nanometers. Such high-resolution control allows for the creation of metallic structures on a scale appropriate for hot spots. These techniques have allowed for the creation of isolated hot spots of varying geometries which can then be perfectly defined and modeled in computational simulations. As a result, these substrates allow for an unprecedented comparison between theoretical and experimental investigations. However, while these techniques allow for careful control of hot spot parameters, they are inherently low throughput and prohibitively expensive making them unsuitable for anything other than research purposes. In order to harness the power of hot spots for applications, several "bottom-up" schemes have been developed to allow for the large-scale production of SERS substrates with high concentrations of hot spots with well-defined average properties for applications. These techniques usually rely on sophisticated chemical preparation of colloids in solution and will be reviewed in Sect. 4.4.

These techniques have been instrumental in advancing our understanding of SERS hot spots. The more interesting results from recent years involving these techniques are reviewed in the following section.

4 Key Research Findings

Results will be split into various sections the first of which will be fundamentals of hot spots. This will include a summary of the most important developments in the theory of SERS hot spots for both the EM and CT enhancement mechanisms. The second section will cover developments in tip-enhanced Raman spectroscopy (TERS) which represents the idealized hot spot. Then some issues regarding hot spots and the single molecule will be tackled such as the magnitude of enhancement required for single-molecule detection, the effects of molecular orientation with respect to the hot spot as well as the possible influence of optical forces. Sections 4.4 and 4.5 will cover developments in the imaging and fabrication of SERS hot spots, respectively, which have important implications for theoretical modeling as well control of SERS hot spots. The chapter will conclude by summarizing some of the applications of SERS hot spots that have been recently reported.

4.1 Hot Spot Fundamentals

In recent years, it has become accepted within the SERS community that hot spots primarily arise from intense and highly localized electromagnetic fields caused by local surface plasmon resonances (LSPR). However, questions over the precise magnitude of the enhancement present at such locations have led some to suggest that there must be a "chemical" aspect to hot spots. In this section, the key results from recent work focusing on the fundamentals of hot spots with regard to these two components are summarized.

4.1.1 Electromagnetic Hot Spots

Many theoretical studies have been performed over the years investigating hot spots formed by the interaction of light with metallic nanostructures. The majority of these studies have investigated the interstitial spatial location of enhanced field using a two sphere model, schematically shown in Fig. 10.2a. This has the advantage of being relatively simple and thus readily solved while providing a good physical insight. These early studies showed that the EM enhancement for such a dimer is strongly dependent on the particle size, shape, separation as well as the dielectric properties of the surrounding environment and the metal itself and the arrangement with respect to the polarization direction of the incident light. Several studies suggested that such junctions may lead to strong highly localized enhancements which may possibly reach a factor of 10^{11} [2, 49–54]. Other systems of hot spots that have been investigated previously include linear particle chains [53, 55] and self-affine fractal silver clusters [56, 57].

More recently, Le Ru and Etchegoin have been particularly active in developing the fundamental understanding of SERS – particularly with regard to electromagnetic hot spots. It was pointed out that most of the previous studies have concentrated on simply determining the maximum enhancement and have ignored questions related to the spatial distribution of enhancements. Le Ru and Etchegoin

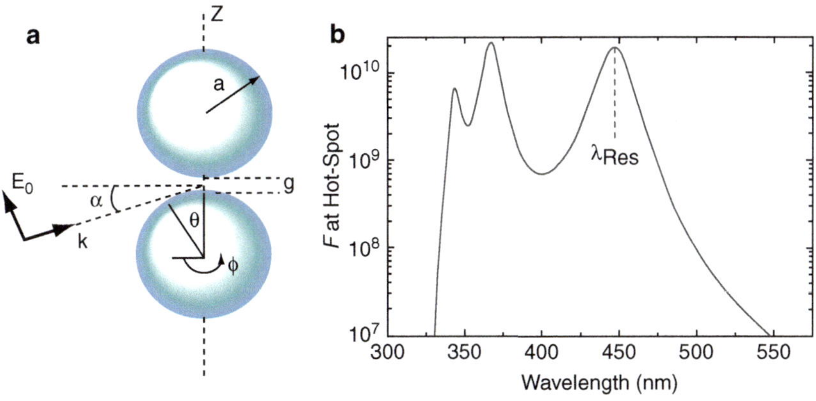

Fig. 10.2 (**a**) Schematic of simplified dimer geometry formed by two spheres of radius a separated by g for electromagnetic calculation. (**b**) Enhancement factor at the hot spot formed between two silver 25-nm spheres separated by 2 nm. The dipolar plasmon resonance occurs at $\lambda_{Res} = 448$ nm with other peaks due to higher order interactions (Figure reproduced from Ref. [58])

pointed out that while this approach can be important for determining if a certain geometry might provide sufficient enhancement to observe single molecules, it is only relevant for systems where a molecule can be specially placed within the hot spot. For example, the enhancement factor precisely at the hot spot formed between two 25-nm silver nanoparticles separated by 2 nm as a function of incident wavelength is shown in Fig. 10.2b. Typically, it is very difficult to achieve the precise molecular control required in order to position a molecule within the hot spot experimentally and thus, a more general approach to calculations can be instructive. Their studies have primarily focused on the simplified two sphere model of a hot spot. They began by confirming the results of previous studies in that the maximum enhancement is critically dependent on the separation between the particles, the radius of the particle, the dielectric material of the particle, and the dimer's orientation with respect to the polarization of the incident light. For example, the maximum enhancement of the hot spot increases significantly as the separation was decreased and for spheres of smaller diameter where the curvature of the surface is increased consistent with previous studies. As well as this the maximum enhancement was found to be increased for silver particles compared to gold through the visible range. This is due to the interband transitions present in gold in this range due to the differing optical properties of the two metals. The highly uniaxial nature of the hot spot formed by the two particles was also confirmed with optimal enhancements obtained when the incident field polarization matched the hot spot axis [59]. However, it was shown that although the maximum enhancement was found to vary significantly with sphere radius, separation, and material, this variation actually reflected a global scaling factor which applied to the entire distribution.

More relevant were their findings related to the spatial distribution of enhancements surrounding a hot spot. It is well known that hot spots are intrinsically highly

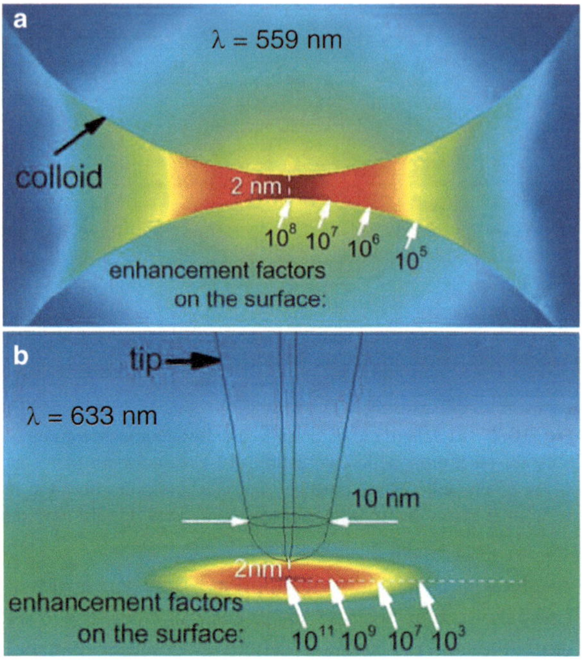

Fig. 10.3 (**a**) Enhancement factor distribution in a hot spot formed between two gold nanoparticles separated by 2 nm plotted on a logarithmic scale. (**b**) TERS enhancement factor distribution for a gold tip over a gold surface separated by 2 nm in air (Figure reproduced from Ref. [19])

localized with the enhancement dropping by several orders of magnitude over only a few nanometers [54]. However, no study had previously tackled the extent of the spatial distribution of enhancement for a hot spot in detail and considered the implications for SERS measurements. To investigate this, Le Ru et al. studied the enhancement distribution at the surface of two spherical silver nanoparticles with radii of 25 nm separated by 2 nm. For this system, the maximum enhancement due to the dipolar interaction between the spheres on resonance at 448 nm was found to be 1.9×10^{10} at the hot spot while the minimum enhancement was only 1.5×10^3. The average enhancement in the electromagnetic field at the surface of the particles at 6.7×10^7 was found to be significantly smaller than the maximum. Similar results were found for closely spaced gold nanoparticles as well as sharp gold tips in close proximity to a gold surface as shown in Fig. 10.3. Clearly, there is a large variation in the enhancement across the surface of the particles – over seven orders of magnitude for this case of silver nanoparticles. Although these characteristics of the spatial distribution of the enhancement for the hot spot were investigated on resonance, the distribution was shown to be essentially identical for all wavelengths in the visible range.

Such large variations in enhancement have significant implications for SERS experiments. For example, it was shown that if we consider a uniform distribution of molecules on the nanoparticle surface, then the enhancement for a molecule located directly at the hot spot will be 285 times larger than the average. Thus, a molecule located directly at the hot spot will contribute as much signal as 285 other molecules, dominating the detected signal. Such skewed distributions are

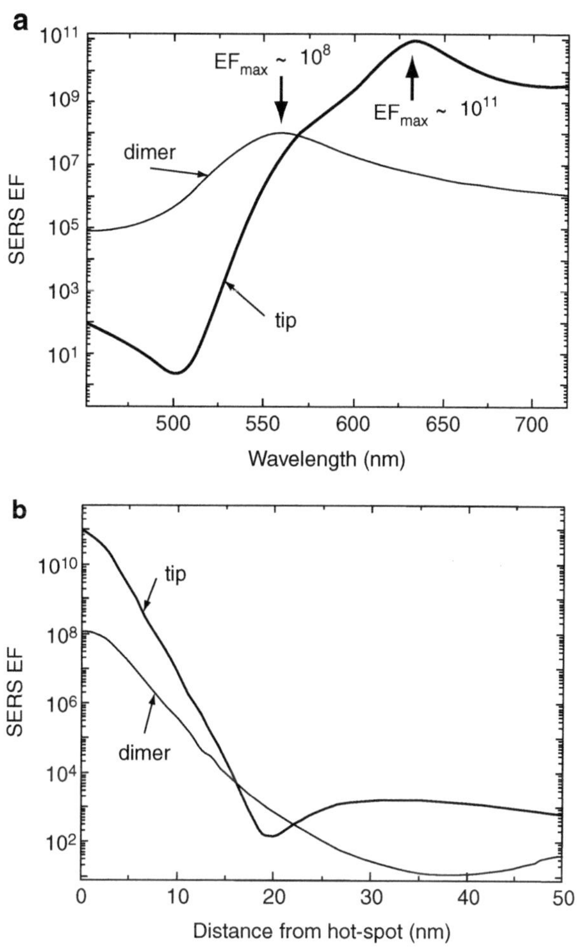

Fig. 10.4 (**a**) Enhancement factor (EF) distribution as a function of wavelength for the hot spots formed between the gold dimer and gold tip shown in Fig. 10.3. (**b**) Spatial distribution of enhancement factors for the dimer and tip. EF decreases strongly with distance and can be more than an order of magnitude smaller than the maximum only a few nanometers away (Figure reproduced from Ref. [19])

well described by the Pareto distribution [60] and in this case, 80% of the SERS signal will originate from only 0.64% of the population. This coupled with the extremely small size of the analyte and highly varying enhancement can already explain, to a large degree, the fluctuations characteristic of many SERS measurements when surface diffusion of the molecules is taken into account, as will be discussed in Sect. 4.3.3.

Additional enhancement of the local electromagnetic field may arise from the confinement of electromagnetic energy at sharp corners or edges due to the increasing confinement of the surface charge density. Figure 10.4a and b show the enhancement factor as a function of wavelength and distance-dependent SERS enhancement factor for hot spots formed between the gold dimer and sharp gold tip in close proximity to a gold substrate shown in Fig. 10.3. There are clear differences in the wavelength and intensity of the enhancement maxima as well as in the distance dependence of the enhancement. This shape effect, commonly referred to as the "lightening rod" effect, was recognized early as a possible contribution to the SERS enhancement and although it is not absent from Mie theory, it is not particularly well demonstrated in simulations involving two spherical particles. Although the effect of such structures has been investigated previously [49, 52], Boyack and Le Ru used exact T-matrix electrodynamic calculations to further investigate both the size and shape effects on the SERS enhancement for silver and gold metallic nanoparticles [61]. They showed the localized surface plasmon resonances are redshifted for elongated nanoparticles for polarization along the long axis and that the hot spot localization for the dipolar surface plasmon resonance is primarily controlled by the shape rather than the size of nanoparticles involved [61]. The spatial distribution of the enhancement on the particle surface was also found to be significantly different in comparison to that obtained on a sphere with the enhancement being magnified at the tip of the elongated nanoparticle, particularly when the incident light is polarized along the long axis. As with their previous simulations of hot spots located between spherical metallic nanoparticles, the spatial distribution of local field enhancements for elongated nanoparticles was found to be largely independent of incident wavelength and well described as a Pareto distribution. This was shown to be true for both silver and gold nanoparticles, albeit with a lower overall enhancement for the gold.

It was shown that the particles have to be significantly elongated in order to obtain enhancement factors similar to those for a hot spot formed between two spherical particles. The enhancement increased dramatically as the protrusion was "sharpened" with enhancements of up to 10^{11} for a protrusion apex of 30°. Interestingly, it was found that the enhancement decreases away from the hot spot before increasing again. This clearly shows that the presence of sharp protrusions on a nanoparticle could lead to enhancements sufficient for single-molecule detection from single particles, particularly if effects such as resonant excitation are also contributing. The existence of such features on single nanoparticles is quite likely in some cases given that the majority of nanoparticles in a colloidal suspension are highly irregular due to their nano-crystallite nature rather than perfectly spherical as assumed for many computational simulations. This result has significant implications for the single-molecule detection from single nanoparticles as will be discussed in Sect. 4.3.

Other hot spots have been considered theoretically such as those formed within grooves [62] or within particle chains [63, 64]. Such hot spots will show the same Pareto-like distribution of enhancements in the spatial sense although other parameters may change compared to the simple gap hot spot considered above. For example, Wang et al. investigated the effects of the number particles within a strongly coupled chain of particles on the generated hot spots [64]. They demonstrated that the resonance peaks of the particles are increasingly redshifted as the number of particles is increased due to the near-field coupling between the particles. They also showed that the field enhancement of the hot spots formed between the particles may be maximal for chains of more than two particles depending on the exact parameters of the near-field coupling between the particles [64].

There have also been significant developments regarding the theoretical investigation of electromagnetic field distribution. Perassi et al. have proposed a novel approach based on the computation of the variation in the volume trapped between a constant surface surrounding the nanoparticle and the nanoparticle surface boundary as a function of the enhancement itself [65]. The technique not only has several computational advantages, such as rigorous convergence criteria, but also allows for the average enhancement, localization, and distribution of the hot spot to be visualized more easily.

In recent years, many of the aspects of the electromagnetic enhancement mechanism have been confirmed through experimentation driven by improvements in the reproducibility and resolution of fabrication techniques. Nanofabrication of well-defined nano-antennae of different geometries has brought the correlation of experimental and theoretical investigations to a new level of accuracy. Mirkin and coworkers have been particularly active in the development of on-wire lithography (OWL) [67] which allows for nanodisk arrays to be fabricated with precisely controlled geometries. These efforts have shown that parameters, such as disk thickness and separation, play a crucial role in SERS. Figure 10.5 shows the correlation of experimental measurements and theoretical simulations of the SERS intensity as a function of segment length for smooth and rough rods formed using OWL. These and other studies have confirmed features of the electromagnetic enhancement mechanism such as the effect of the interparticle separation on the near-field coupling including the surprising result that the largest enhancements do not occur for the smallest gaps. This and other fabrication methods will be reviewed in Sect. 4.5.

4.1.2 Chemically Enhanced Hot Spots

SERS active hot spots were first suggested relatively early in the history of SERS by Moerl [16] and Pettinger [68]. However, these hot spots were primarily attributed to localized regions of highly optimized chemical enhancement – i.e., specific binding sites, rather than localized surface plasmon resonances. The chemical enhancement mechanism can be viewed as originating from the resonant enhancement of a charge transfer process within the metal-adsorbate complex [12]. Two distinct effects can occur within these complexes which can then contribute to the detected Raman signal. Firstly, electrons from the metal can be excited to unoccupied molecular orbitals of the adsorbed molecule and back to the metal or electrons from the occupied molecular orbitals can be excited into the Fermi level of the metal and back to the adsorbed molecule. These processes can leave the molecule in a vibrationally excited state, enhancing the Raman signal. The magnitude of the enhancement arising from this process is about two orders of magnitude [69] which is relatively small in comparison to the enhancement provided by the electromagnetic contribution. As well as this it is only applicable to molecules adsorbed directly on to the metal. As a result, chemically active hot spots are much less influential overall, nevertheless, there have been some interesting developments regarding their contribution.

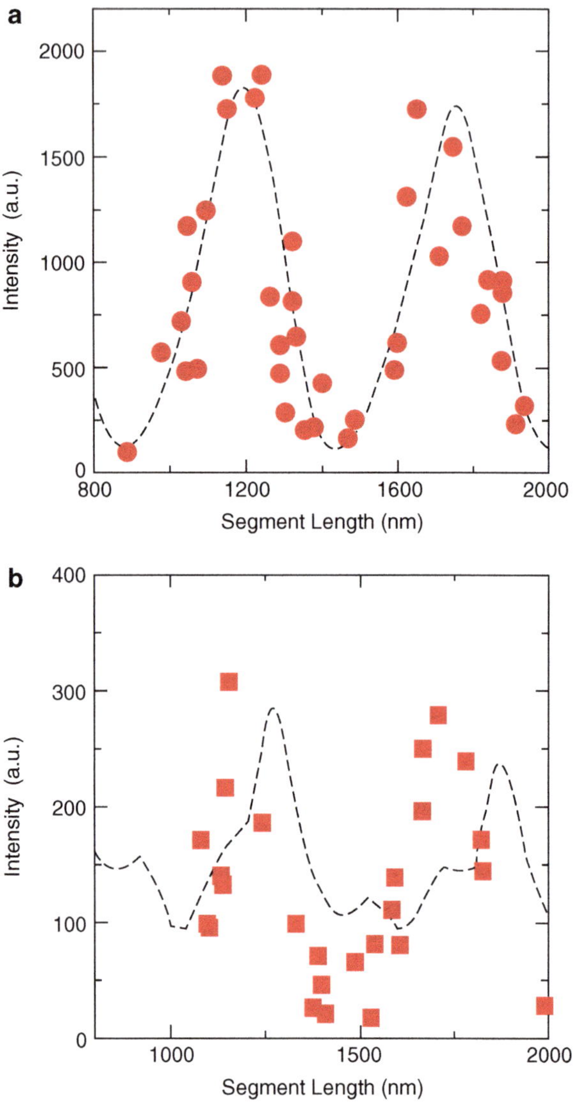

Fig. 10.5 Experimental (*symbols*) and calculated (*dashed lines*) SERS intensity as a function of segment length for smooth and rough rods produced using on-wire lithography. Theoretical values were calculated using the discrete dipole approximation and scaled to maximum experimental value (Figure reproduced from Ref. [66])

The observation of single-molecule SERS by Nie from single particles in 1997 revived interest not simply in hot spots but specifically the contribution due to the chemical enhancement [22]. At the time it was believed that the electromagnetic contribution to the enhancement for a single particle did not provide the enhancement of 10–15 orders of magnitude that was believed to be necessary for single-molecule detection. Otto, in particular, has suggested that it is unlikely that the single-molecule signals can be explained exclusively by electromagnetic hot spots and must therefore contain a significant contribution from the chemical enhancement [70]. Indeed, Kall et al. have pointed out that relatively large molecules such as globular proteins cannot physically fit into the small gaps between metallic nanostructures normally associated with LSPR hot spots. Thus, the enhancement experienced by the molecule is greatly reduced below the threshold for SM-SERS. Despite this there have been several reports of the observation of SM-SERS from proteins such as hemoglobin. Clearly, in these cases, there is room for contributions to the enhancement from additional modes of enhancement such as the CT resonance enhancement described above to allow for the detection of single molecules.

Peyser-Capadona and coworkers have gone further, suggesting that the CT enhancement mechanism may, in some cases, be strong enough to allow single-molecule observation – even in the absence of an EM contribution [71]. Their claims were based on experimental measurements on clusters of only a few Ag atoms encapsulated within peptide scaffolding which are unable to support plasmon resonances due to their small size.

Such large enhancement factors for localized and isolated hot spots from few atom Ag clusters arising from only the chemical enhancement under certain conditions are supported by calculations. Zhao working with Jensen and Schatz used time-dependent density functional theory (TDDFT) to investigate the adsorption and Raman response of pyrazine molecules [21]. Figure 10.6 shows the Raman response of (a) isolated pyrazine compared to that of pyrazine complexed to the vertex of a (b) one and (c) two 20 Ag atom clusters with enhancements of 10^2 and 10^5 predicted, respectively. Small clusters of Ag atoms have little or no plasmon response, suggesting that the chemical enhancement can be quite significant and certainly may allow for enhancement hot spots.

However, there are few studies which predict such large chemical enhancements while several similar TDDFT studies on different molecules predict much lower enhancements [72–74]. For example, a study of pyridine complexed to a 20 atom Ag cluster by the same group predicted much lower enhancement factors. Factors on the order of 10 for the static chemical enhancement, 10^3 for charge transfer effects, and 10^5 for the electromagnetic contribution emphasized that the chemical enhancement is a much smaller contribution in general [72].

More generally, many experimental studies have found the contribution from the chemical enhancement to be on this order. For example, the role of the chemical enhancement in the case of SERS hot spots has been recently investigated by Shegai and coworkers using spectroelectrochemistry [75]. They estimated the contribution from the chemical enhancement to have a lower limit of three orders of magnitude through the comparison of the intensity of 4-mercapto-pyridine (4MPy) to that of rhodamine 6 G obtained on the same type of SERS surface. This relatively large chemical enhancement was attributed to a resonance with the transfer of an electron from the metal to the adsorbed molecule. Such spectroelectrochemical investigation of individual hot spots not only allows the characterization of the interplay between the chemical and electromagnetic enhancement mechanisms but also will allow for the detailed study of the redox properties of single molecules.

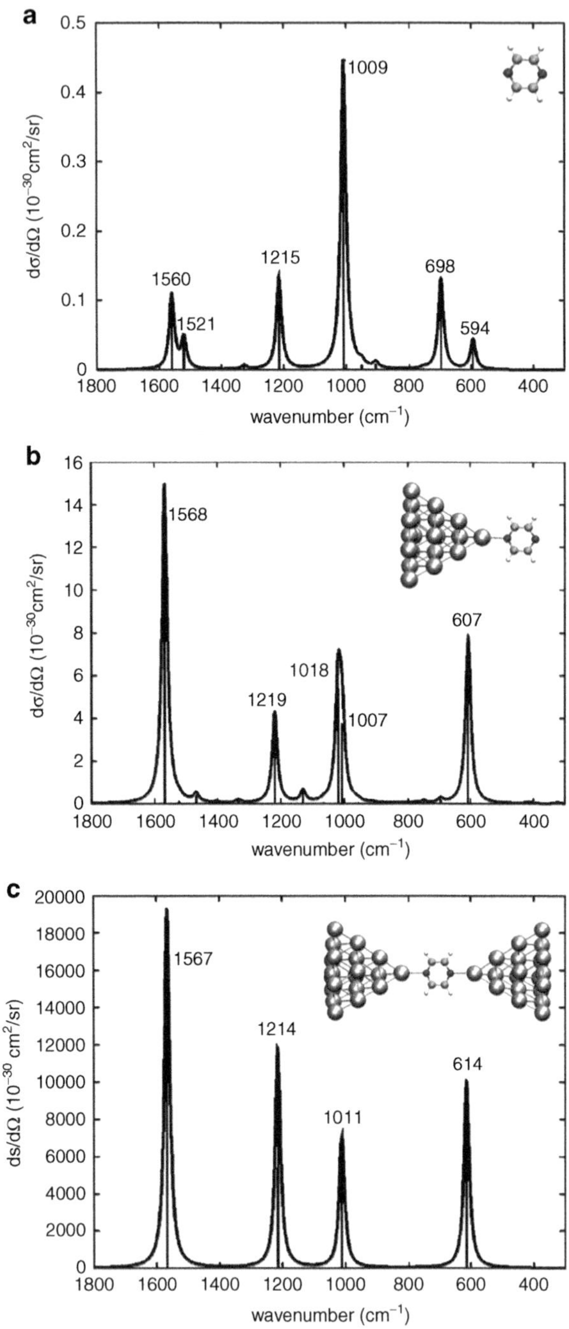

Fig. 10.6 Raman spectra of (**a**) isolated pyrazine, (**b**) complex between pyrazine and 20-Ag atom cluster, and (**c**) complex of pyrazine and two 20-Ag atom clusters calculated using time-dependent density functional theory for 514.5 nm excitation based on static polarizability derivatives. Large increases up to 5 orders of magnitude in Raman intensity are predicted from systems where little or no plasmon response exists (Figure reproduced from Ref. [21])

4.2 Idealized Hot Spot Control: Tip-Enhanced Raman Spectroscopy (TERS)

Tip-enhanced Raman spectroscopy (TERS) offers the best possible control of the characteristics a single hot spot [76, 77]. Controlling the characteristics and spatial positioning of a SERS active hot spot with respect to molecules of interest is a challenge fundamental to the development of SERS both for understanding and application. TERS involves the combination of an extremely sharp metalized tip with the spatial control of a scanning probe microscope (SPM) such as an AFM. When under illumination, the electromagnetic field is enhanced at the apex of the tip due to the LSPR and results in enhanced Raman scattering from molecules in close proximity [78, 79]. Thus, as with other SPM techniques, the resolution is determined by the size of the tip which can be extremely sharp as shown in Fig. 10.7a. The SPM feedback control allows for high-resolution control of both the vertical separation between the tip and a surface as well as high lateral resolution. This control allows precise positioning of the enhanced fields and, in principle, allows the statistical properties of a single hot spot to be probed and compared to theoretical predictions [80, 81]. Figure 10.7b shows the Raman signal obtained from a submonolayer of brilliant cresyl blue dye when a metalized tip is either in close proximity or retracted, along with the signal intensity detected as a function of tip to surface distance.

However, the experimental implementation and interpretation of TERS measurements is a significant challenge [83]. For instance, not only are all the normal problems associated with SPM techniques present, they are combined with controlling the molecular binding to the tip itself. The enhanced fields can lead to enhanced photodegradation requiring careful control of the power of the incident illumination. Care must be taken in the interpretation of TERS spectra as both the tip and the sample must be exposed to the diffraction-limited laser spot. As a result, the detected signal is a combination of an unenhanced Raman signal from a relatively large surface area and a SERS signal from the enhanced near-field area of the tip. The separation of these represents a significant challenge [79, 84]. Biological samples can be particularly challenging given that they present soft interfaces and require aqueous environments where nonspecific binding of analyte to tip can overwhelm molecular signals of interest.

Despite these difficulties, considerable progress on TERS has been achieved in the last 5 years and single-molecule TERS spectra have been measured from an increasing number of analytes. The detection of single molecules by TERS usually requires a contribution from not only the enhanced electromagnetic fields of the tip but also resonant Raman effects as well as an additional contribution from a metallic substrate.

Single-molecule TERS (SM-TERS) was first reported for malachite green adsorbed on planer metal surfaces by Raschke's group in 2006. Single-molecule detection was inferred from spectral diffusion and discretization of the Raman peak intensities. They estimated that the coupling between the tip and the sample gave rise to a Raman enhancement of 5×10^9 with the 633-nm laser being in resonance

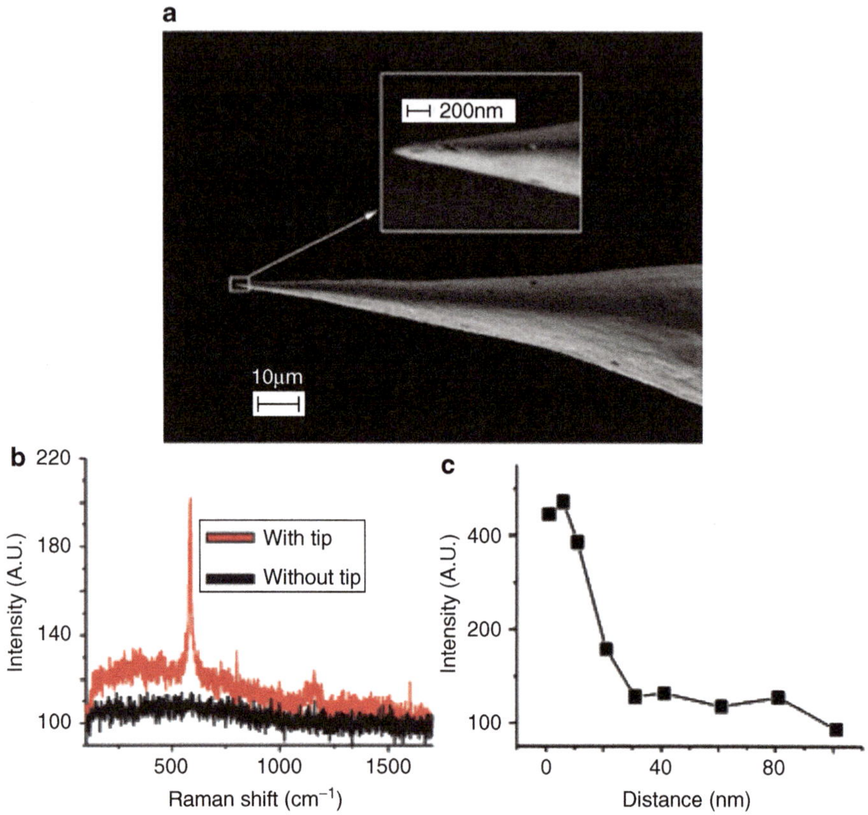

Fig. 10.7 (**a**) SEM image of an Ag tip with an apex sharper than 50 nm. (**b**) Raman spectra with and without tip of BCB at submonolayer concentration as well as (**c**) a intensity distance dependence showing significant enhancement only when the tip is within 30 nm of the surface (Figure reproduced from Ref. [82])

with both the dye and plasmon and a spatial resolution below 10 nm [85]. However, the initial claim based on fluctuating signals was met with some skepticism. Domke and Pettinger argued that the high-field enhancements would lead to a rapid photodecomposition of the adsorbed molecules which would lead to carbon contamination which is well known to cause spectral fluctuation [86]. Further work by Raschke and coworkers utilizing careful control measurements ruled these contributions out [87] and similar reports for single-molecule detection of brilliant cresyl blue [82, 88] have confirmed single-molecule detection using TERS. The high sensitivity and spatial control of TERS are now being utilized to study weak phenomena such as hydrogen bonding in DNA bases [89].

Several studies have been carried out investigating the local field enhancement of a metallic tip under illumination both for isolated metallic tips [20] and for metallic tips when in close proximity to a metallic film [19, 90]. The fine control afforded by TERS allows for a unique correlation and of theoretical and experimental aspects of SERS hot spots such as the distance dependence [91]. TERS has also been used to investigate the spatial distribution of enhancement for hot spots of nanometer dimensions present on smooth Au films with high-resolution tip-enhanced luminescence and Raman from benzotriazole molecules adsorbed onto the surface [92].

Symmetry considerations under TERS conditions have recently been considered as a method for obtaining molecular orientations [93, 94]. Berweger and Raschke considered using the polar phonon mode selection rules for TERS to determine nanocrystallographic information from solids [93]. TERS enables control of experimental parameters required to extract information from the Raman tensor such as the polarization- and k-vector-dependent field enhancement relative to the surface. The selection rules were demonstrated for both near-field TERS and far-field scattering from bulk and nanocrystalline $LiNbO_3$.

4.3 Hot Spots and the Single Molecule

Hot spots are now widely acknowledged within the SERS community to be prerequisite for the observation of single molecules due to the strong, highly localized enhancements they provide. As a result of this, the specific nature of SERS hot spots has significant implications for single-molecule SERS. Recent results regarding the enhancement factor required for single-molecule detection, molecular orientation within hot spots, statistics and fluctuations, laser forces in hot spots, vibrational pumping and molecular control of molecules within hot spots are reviewed here on their relation to SERS hot spots.

4.3.1 Enhancement Factors for Single-Molecule Detection

There has been much discussion within the literature regarding the enhancement needed to observe single-molecule SERS since the initial reports of Nie and Emory [22] and Kneipp [23]. These studies suggested huge SERS enhancements of the order of 10^{14} to 10^{15} were not only responsible for single-molecule detection but also necessary. Much discussion into the order of magnitude required for single-molecule detection and the volume to which it extends from the hot spot has followed with many conflicting claims [24, 95, 96]. More recently, there has been a growing consensus emerging that such figures are not only erroneously large but will actually lead to the physical denaturation of any analyte. As a result, recent estimates have been dramatically revised down to orders of 10^7–10^8.

Le Ru et al. demonstrated that a careful consideration of the minimum signal required for detection, the Raman cross section of the molecules involved, and the characteristics of the hot spot were required to determine the minimum enhancement factor necessary for single-molecule detection using SERS [95]. Through this process they concluded that a minimum enhancement factor of 2×10^8 was required for detection of single molecules with bare Raman cross sections of 10^{-27} cm^2 sr^{-1}. This is 10^6 times smaller than the 10^{14} initially claimed to be essential for single-molecule SERS. The bare Raman cross section of many molecules in preresonance or full resonance with the incident light either meets or exceeds this condition, suggesting that single-molecule SERS is far more common than has been previously assumed.

But what about the mechanisms involved in single-molecule SERS? The electromagnetic mechanism has been shown to theoretically provide enhancement factors up to 10^{10}–10^{11} for LSPR hot spots. This would enable detection of both resonant and nonresonant single molecules, and Dieringer et al. recently confirmed that the electromagnetic mechanism is all that is required in order to explain single-molecule sensitivity [97].

4.3.2 Molecule Orientation

SERS selection rules have been extensively studied for molecules adsorbed onto flat planar metallic surfaces [98–100]. The fixed molecular orientation of molecules that are chemically adsorbed to the surface with respect to the local field polarization at the surface can be expected to result in a strong modification of the relative intensities of the Raman active modes compared to the response from a randomly orientated population. Thus, the comparison of the Raman response from a randomly oriented population to the highly orientated surface population can be used to infer details regarding the adsorption geometry. Thus, the combination of surface selection rules with hot spots raises the intriguing possibility of determining the orientation of a single molecule and several attempts have been made [1, 101]. These studies have been complicated by the effect of the local field polarization which overwhelmingly dominates any measurement. In order for such an analysis to be valid, the molecule must adopt a fixed position with respect to the local enhancement field which is only true for certain situations, for example, when the molecule–surface interaction is particularly strong. Molecules with a preferential adsorption geometry will also have a preferential orientation with respect to the local field at the hot spot which tends to be perpendicular to the surface.

Molecules that interact less strongly with the surface will adopt a more random orientation and the relative enhancement factors for the individual modes may be different under SERS conditions compared to normal Raman conditions. Etchegoin et al. have considered surface selection rules for hot spots utilizing the depolarization ratios of the different Raman active modes of benzene thiol under SERS and normal Raman scattering conditions using both measurements and density functional calculations (DFT) [95]. Under normal conditions the Raman depolarization ratio for different modes will be dependent on their symmetry [102]. Figure 10.8a shows Raman spectra of benzenethiol for parallel and perpendicular polarization and the corresponding depolarization ratios compared to DFT. Spectra and depolarization ratios from benzenethiol under SERS conditions are shown Fig. 10.8b. The depolarization ratio is observed to be one third for all modes under SERS conditions, and the breakdown of these depolarization ratios for modes of known symmetry has been suggested as an indication of the molecular orientation [1, 4, 101, 103]. However, the local field enhancement in the hot spot is strongly dependent on the coupling between the hot spot and the external excitation. This is controlled by the angle between the polarization of the incident excitation

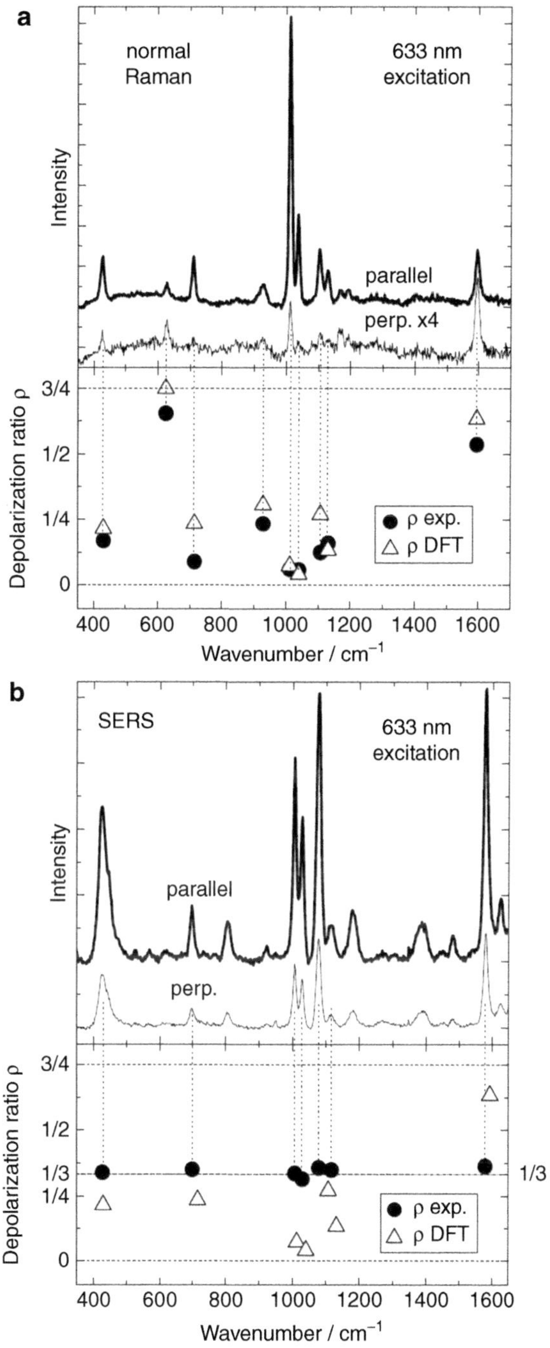

Fig. 10.8 Polarized Raman spectra and depolarization ratios of benzenethiol under (**a**) normal Raman and (**b**) SERS conditions. Depolarization ratios are shown together with predictions from DFT calculations. All modes have a depolarization ratio of 1/3 under SERS conditions due to the highly uniaxial local field polarization of the hot spots (Figure reproduced from Ref. [95])

and the axis of the hot spot. As a result, changes to the depolarization ratio for molecules located within a hot spot as a function of the polarization of the incident electromagnetic field result from the highly uniaxial nature of the hot spot rather than a specific adsorption geometry [58, 104].

Recently, Berweger and Raschke considered the polar phonon mode selection rules for TERS for the determination of nanocrystallographic information from solids using TERS as discussed in Sect. 4.2 [93]. This is a very different situation given that the location and orientation of the hot spot relative to the sample and excitation are well characterized and could be extended to the study of molecular adsorption on metals and other surfaces in the future.

4.3.3 Statistics and Fluctuations

Temporal fluctuation, or blinking, of the detected single-molecule signal has often been cited in the literature as being part of the SERS "signature" of single molecules [22, 23, 105, 106]. Several mechanisms have been proposed to explain these fluctuations such as the Brownian motion of colloidal particles or molecular surface diffusion [107]. For example, the strong localization of the hot spots limits the number of molecules residing within a hot spot at any one time increasing the effects of surface diffusion of adsorbed species [108]. Such fluctuations may even be an inevitable consequence of the enhanced photochemical effects such as photobleaching that will be expected to occur for molecules adsorbed to such nanostructured surfaces where enhanced fields exist [109, 110]. These effects have often been generalized under the "blinking" term with little discrimination. However, it is important to note that signal fluctuations with magnitudes much greater than the mean signal are a natural consequence of the unusual statistical properties of these skewed enhancement distributions. Several efforts have been made to experimentally measure the long tail distribution of enhancement in SERS [111, 112].

Etchegoin et al. investigated the influence of such long tailed enhancement distributions on SERS measurements involving any substrate dominated by the influence of hot spots [58, 113]. In many previous studies, the presence of a Poisson distribution of intensities has been presented as evidence of single-molecule detection. However, this analysis relies on the contribution of each and every molecule being identical. This is a highly unlikely scenario given the strong spatial dependence on the SERS enhancement within a hot spot. We are left asking if the single is dominated by a single highly enhanced molecule or the combination of many molecules with low enhancements.

Typically, this is difficult to investigate experimentally as measurements involving ultralow concentrations where there is less than a single molecule in the scattering volume at any particular moment suffer from issues of statistical significance. These effects have been simulated for low (100 randomly adsorbed molecules) concentration up to high concentration (7,850 randomly adsorbed molecules) corresponding to complete monolayer formation [58]. A strongly skewed distribution in the simulated detected SERS intensities was found to persist with up to 500 molecules adsorbed to the nanoparticle surface. Large intensity fluctuations were predicted even at this relatively high concentration and may even persist for monolayer coverage due to the strong first layer effect of SERS. These fluctuations are a direct manifestation of the strong localization of the hot spot and its intrinsic enhancement distribution.

Given that fluctuations in signal intensity are expected even for relatively high concentrations, it is difficult to use them as a convincing indication of single molecular signatures. Etchegoin and Le Ru went on to propose a bi-analyte method as a proof of single-molecule detection which has been used and further developed by other groups [114–116]. Goulet and Aroca have investigated the nature of the analyte concentration and hot spot localization through the fine concentration control allowed by Langmuir Blodgett monolayers coupled with the bi-analyte technique for single-molecule SERS identification [117]. Single-molecule detection was observed at a concentration that resulted in 1–10 molecules of each dye occupying the area illuminated by the laser (approximately 1 μm [2]). Clearly, this is much larger than the ultralow concentrations used in many previous studies. The appearance of single-molecule SERS at these concentrations was attributed to the spatial coincidence of the target analyte with the highly localized electromagnetic hot spot.

Signal fluctuations can be controlled and used as a characterization tool in some cases. Dos Santos and coworkers electrochemically controlled the time dependent intensity fluctuations of the SERS response of Rh6G and congo red (CR) adsorbed onto roughened silver electrodes to investigate the nature of SERS hotspots [118]. The fluctuations were attributed to a small number of SERS active molecules probing the highly enhanced fields of hot spots in real time. The cationic and anionic natures of Rh6G and CR, respectively, allowed the intensity distribution of the dyes to be controlled in a mixed solution. Such potential controlled SERS dynamics suggest that spectroelectrochemical investigation of single-molecule reduction and oxidation should be possible under optimal conditions.

4.3.4 Molecular Control in Hot Spots and Laser Forces

Experimental investigation of SERS hot spots is particularly challenging as a result of their highly localized nature. In recent years, sophisticated strategies have been developed by several groups to reproducibly and precisely place molecules within the hot spot in a controlled way. Controlled placement of molecules directly within the hot spot allows for the enhancement factor distribution to be studied in detail. These studies have allowed the long tail enhancement distribution to be probed and demonstrated the dominant contribution from the small number of molecules present directly within the hot spot.

Camargo et al. were able to isolate molecules within the hot spot formed between two closely spaced silver nanocubes using an oxygen plasma etch [119]. Fang et al. utilized the photoinduced burning of benzenethiolate molecules on a film of 330-nm nanospheres coated with silver to investigate the distribution of enhancement from substrate containing SERS hot spots [111]. The benzenethiolate molecules form a close packed self-assembled monolayer on the surface ensuring the detected signal is a sum of the full enhancement distribution, for which the

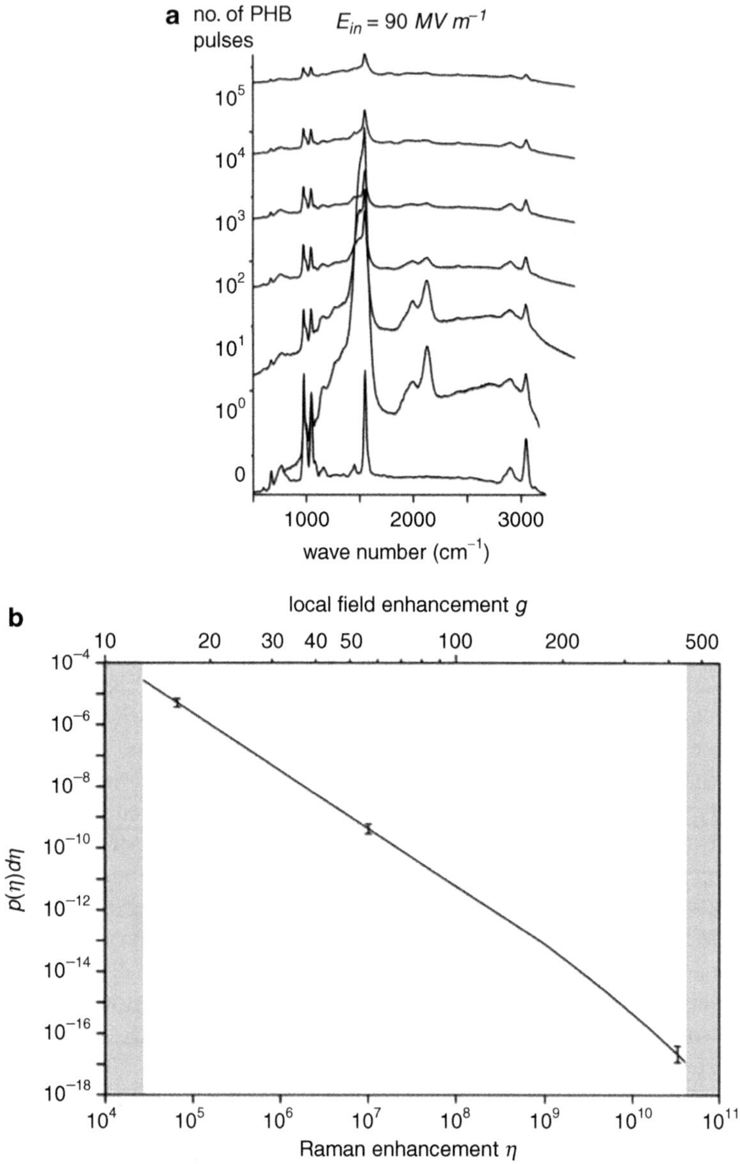

Fig. 10.9 (**a**) Raman spectra of benzenethiolate on a SERS substrate as a function of number of photobleaching pulses. (**b**) Measured distribution of SERS enhancement factors for benzenethiolate monolayer (Figure reproduced from Ref. [111])

average enhancement was approximately 10^6. Molecules were removed from the surface through a series of increasingly intense laser pulses with the molecules progressively burned away with lower enhancements. Figure 10.9a shows the Raman spectra obtained from the substrate after exposure to an increasing number of laser pulses. A clear decrease in Raman intensity is observed with increasing numbers of pulses indicating an increasing degradation in film coverage. This allowed the full distribution of enhancements to be probed, as shown in Fig. 10.9b, with the minimum and maximum being 10^4 and 10^{10} respectively. From this, it was deduced that 24% of the detected SERS signal is provided by only 63 out of 10^6 molecules.

Rather than investigating hot spots by removing molecules from a surface, Chen et al. were able to deposit amorphous carbonaceous nanoparticles into a localized hot spot using a modified electron-beam-induced deposition method (EBID) [120]. The carbon nanoparticles were deposited within a nanoslit cavity structure which was coated with gold to create a SERS active surface where hot spots exist [121].

In recent years, interest in the influence of laser forces in SERS, and in regards to single-molecule SERS in particular, has grown [122]. Laser forces become relevant for highly focused laser beams where the strong electric field gradient will attract dielectric particles and are the basis for optical tweezers. Such forces may become relevant for the interaction of dielectric particles and molecules with electromagnetic hot spots where the strong electric field might be expected to exert a significant trapping force. Optical forces under SERS conditions will have important implications for the statistical interpretation of the single-molecule SERS signals where molecules are normally assumed to be randomly positioned [122a]. The importance of such effects will depend strongly on the type of binding of the analyte to the surface and whether molecular surface diffusion might be active. The ever popular SERS analyte, rhodamine 6 G, has been observed to move toward a tight center of a tightly focused laser beam in a solution due to optical forces [123]. It is reasonable to assume that such effects would be greatly enhanced within a hot spot where strong gradients will exist. Such forces should scale with the intensity of the incident laser and we might expect that a simple investigation of the optical signals as a function of incident intensity would allow the existence of such forces to be demonstrated. However, the additional complications of enhanced photoinduced effects such as enhanced photochemistry have prevented a credible demonstration of laser forces under SERS conditions. Laser trapping of metallic particles is typically more difficult although Svedberg et al. were able to utilize optical forces in order to fabricate hot spots for SM-SERS [124]. Issues arise when working with nanometer scale objects using optical forces because the thermal energy associated with the Brownian motion of the particle can be significant.

4.3.5 SERS Pumping

Kneipp and coworkers suggested in 1996 that the intense electromagnetic fields present within SERS hot spots may lead to vibrational pumping of resident molecules [125, 126]. The effect is of interest not only fundamentally but also due to its potential as a metrological tool capable of estimating SERS cross sections of molecules in hot spots. They inferred the influence of vibrational pumping based on experimental observations of the power dependence of the anti-Stokes to Stokes (aS/S) ratio (ρ) of rhodamine 6 G (Rh6G) and crystal violet (CV) absorbed onto silver nanoparticles using off-resonant 830 nm excitation at room temperature.

However, the observations generated a great deal of discussion within the literature due to the relatively large cross sections implied by the measurements. Alternative explanations such as resonance or heating effects were proposed by various authors [127, 128].

Vibrational pumping under SERS conditions was conclusively demonstrated through temperature measurements of the aS/S ratio of Rh6G adsorbed onto dried silver colloids [129]. Under normal conditions, the vibrational population will simply be proportional to the temperature and given by a Boltzmann distribution. Under SERS conditions, each Stokes Raman process leaves an excited vibration within the molecule resulting in an increase in the vibrational population. The vibrational population may be significantly increased above the thermal level if the Raman scattering cross section becomes large enough, i.e., for a molecule within a hot spot. Under such conditions, the aS/S ratio is given by:

$$\rho = I_{aS}/I_S = \frac{\sigma_{aS}}{\sigma_S} n = A\left[\frac{\tau\sigma_S I_L}{\hbar\omega_L} + e^{-\hbar\omega_V/k_B T}\right] \qquad (10.1)$$

Equation 1 Anti-Stokes/Stokes ratio

where I_{aS} (I_S) is the anti-Stokes (Stokes) scattered intensity, σ_{aS} (σ_S) is the anti-Stokes (Stokes) cross section, n is the average phonon population of the mode, A is the asymmetry parameter which accounts for the general case where $\sigma_{aS} \neq \sigma_S$, τ is the vibrational lifetime, I_L and ω_L are the laser intensity and wavelength, respectively, T is the temperature, and ω_v is the vibrational frequency. $\hbar$ and k_B are the usual constants.

It is clear from this expression that there are two major contributions to the vibrational population during a SERS measurement, the first due to vibrational pumping through Stokes Raman scattering and second due to the normal thermal population. These are given by the first and second terms on the right hand side of (10.1) respectively. The exponential dependence of the Boltzmann distribution for the thermal population implies that the thermal population will decrease quickly with temperature leaving only the contribution from Stokes pumping.

Measurements were performed over large spatial areas encompassing many SERS active sites with extremely low power densities using a 676-nm laser between room temperature and 10 K. This avoided sample degradation through enhanced photochemical effects for the duration of the measurements. If the vibrational population were purely thermal in nature, then no anti-Stokes scattering would be expected at such low temperatures. However, clear nonthermal anti-Stokes populations were observed at low temperatures as shown in Fig. 10.10a. A full measurement of the aS/S ratio as a function of temperature, shown in Fig. 10.10b, demonstrated clear evidence for vibrational pumping, with two regimes observed where the vibrational population was dominated by the thermal and pumping contributions. Extremely large cross sections on the order of 10^{-15}–10^{-16} cm^2 for the Stokes Raman scattering of the various vibrational modes were extracted from the experimental data using (10.1) once the lifetime had been estimated from the peak widths.

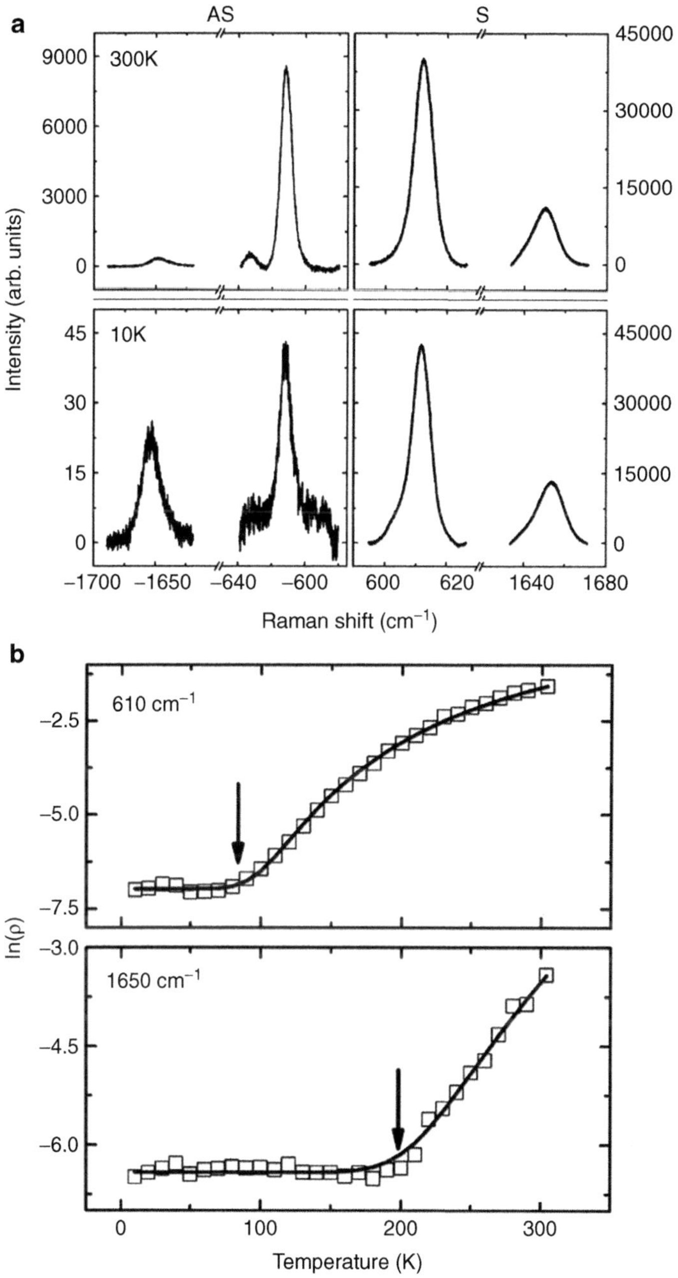

Fig. 10.10 (**a**) Anti-Stokes (*left*) and Stokes (*right*) Raman spectra from the 610 and 1650 cm^{-1} modes of Rh6G at 300 and 10 K. (**b**) Temperature dependence of the aS/S ratio of the modes fitted using (10.1). Nonthermal populations are observed for both modes at low temperatures (Figure reproduced from Ref. [129])

Such large cross sections tend to be significantly larger than estimates obtained from the detected Stokes signal. The limitations of cross sections obtained through vibrational pumping under SERS conditions were emphasized in a study of single-molecule vibrational pumping under SERS conditions [130]. This was demonstrated through the application of the bi-analyte SERS method extended to low temperature regime to measure the SERS cross section using both vibrational pumping and the normally detected Stokes intensity. Cross sections obtained from vibrational pumping were found to be consistently larger than those obtained by the Stokes intensity. Several explanations for this difference were suggested including uncertainty in the A and τ parameters used to extract the cross section from pumping or the influence of non-radiative SERS processes.

A more recent study by Kozich and Werncke has cast doubt on the precise mechanism of vibrational pumping under SERS conditions [131]. Sub-picosecond time-resolved Raman spectroscopy was used to probe the populations of the vibrational modes of Rh6G in a colloidal silver nanoparticle solution. Pronounced nonequilibrium population distributions were observed for all the modes which were slightly delayed with respect to the excitation pulse with the highest energy mode being the most strongly affected. The vibrational population then decayed with a common time constant of approximately 1 ps for all modes. Kozich and Werncke ruled out Stokes Raman scattering and attributed the nonequilibrium vibrational population to electronic excitations involving the metal-analyte surface complex which then increase vibrational populations for all modes by fluorescence and ultrafast internal conversion.

4.4 Visualizing SERS Hot Spots

The strong nanoscale localization of SERS hot spots presents a challenge for the experimental visualization of real hot spots. It is relatively common to identify hot spots through Raman mapping of samples which are then correlated to SEM of TEM microscopy [6]. Hot spots can also be directly observed and characterized using near-field visualization. This allows for the electromagnetic field distribution to be directly probed with a simultaneous measurement of the surface topography although at relatively high resolution compared to that of the hot spot. The observations obtained from such microstructural analysis can be used to generate simplified templates for fabrication of SERS active substrates with high hot spot densities. Recent developments on the investigation of the structural properties of SERS hot spots using high-resolution microscopy and near-field scanning optical microscopy (NSOM) are summarized here.

4.4.1 Near-Field Visualization

The highly localized electromagnetic fields of SERS active substrates have been studied for a number of years by various groups utilizing near-field visualization techniques. The groups of Martin Moskovits and Vladimir Shalaev investigated the electromagnetic field distribution of laser-excited optical modes of fractal clusters of silver colloids as early as 1994 using photon scanning tunneling microscopy correlated to electromagnetic simulations [132]. Highly localized hot spots where the intensity enhancement was found to be at least two times larger than the lowest intensity were observed [132] while Bozhevolnyi et al. reported fluctuations in the local intensity on the order of a factor of seven in near-field optical images of rough gold surfaces [133]. Frequency dependent, highly localized resonances were imaged in colloidal aggregates known to enable single-molecule SERS detection using near-field measurements which at the time helped to explain some of the characteristics of single-molecule signals [57]. Such intensities appear to be small given the enhancements known to exist for SERS hot spots. However, as with all near-field measurements, these investigations were limited by the finite tip resolution leading to a smoothing of sharp peaks and dips which results in a decrease or averaging of the intensity observed. Despite this, near-field visualization techniques can provide detailed information on the hot spot characteristics.

Zhang et al. directly imaged colloidal clusters of silver nanoparticles treated with rhodamine 6 G using scanning near-field microscopy [135]. They found that the SERS intensity is greatest in localized areas of the clusters – smaller than 200 nm. However, the locations of the hot spots were not necessarily observed to coincide with topographical features to which hot spots have normally been associated with such as interstices between nanoparticles. This suggests the presence of highly enhancing absorption sites on the surfaces or the field localization of surface plasmons being focused away from such features for particular excitation wavelengths.

More recently, Lin et al. have investigated the localized electromagnetic field distribution between two nanoparticles as a function of interparticle distance and polarization of the incident light using NSOM [134]. Figure 10.11a shows the optical response of gold nanoparticle pairs with separations of 0.8 and 2 radii as a function of polarization. A clear dependence in the response is observed with angle consistent with prediction from theory. Figure 10.11b shows NSOM images of a gold nanoparticle pair for light polarized parallel and perpendicular to the hot spot, emphasizing the importance of the coupling of the incident light to the hot spot axis. Although the simplified two particle system represents a model system, the results demonstrate the capability of NSOM in characterizing SERS substrates and hot spots on a fundamental level for comparison with theory.

4.4.2 Correlation of High-Resolution Microscopy to SERS

High-resolution AFM, SEM, or TEM microscopy has been used in several studies to investigate the microstructural properties of SERS active surfaces in correlation to confocal Raman imaging. Nie and Emory used tapping mode AFM to investigate the structures responsible for single-molecule SERS in their 1997 paper [22]. Single-molecule SERS was found to originate from a variety of structures including few particle aggregates, single elongated particles, and spherical faceted particles. Michaels et al. also used AFM to characterize small colloidal aggregate structures shown to allow single-molecule detection [1]. These early studies suggested that SERS hot spots resulted from faceted particles or aggregates of two or more nanocrystals.

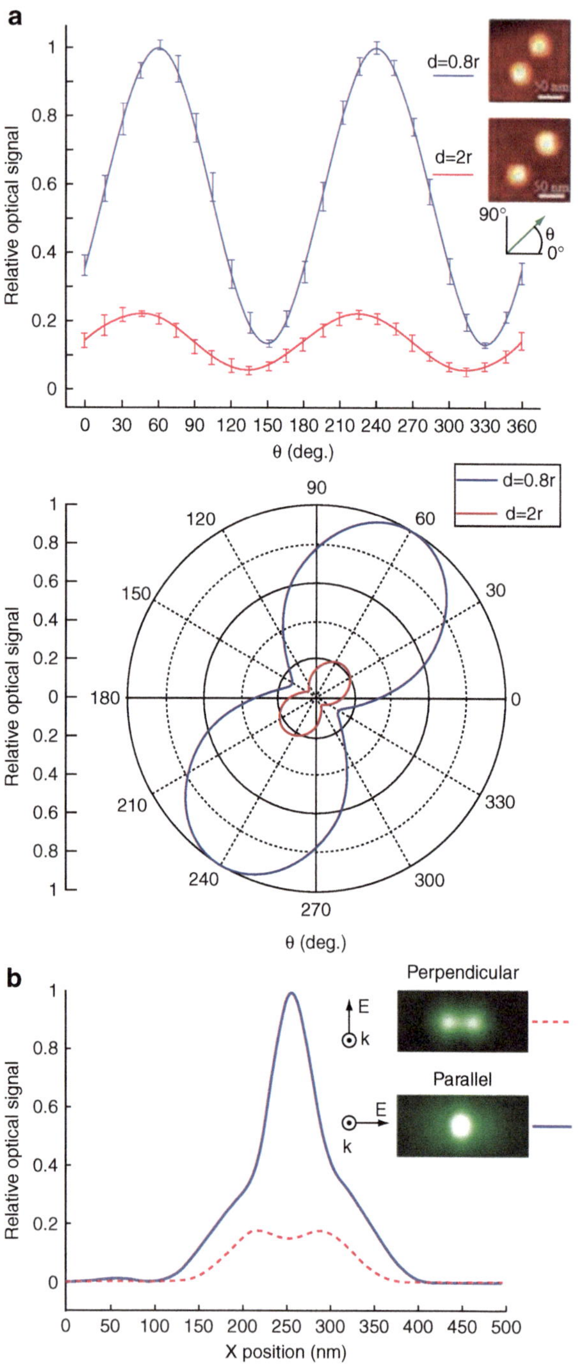

Fig. 10.11 (**a**) Optical responses of gold nanoparticle pairs as a function of polarization angle for particles separated by 0.8 and 2 radii. (**b**) NSOM images and cross sections for a gold nanoparticle pair at 532 nm for light polarized perpendicularly (**a**) and parallel (**b**) to the interparticle axis with cross-sectional view (Figure reproduced from Ref. [134])

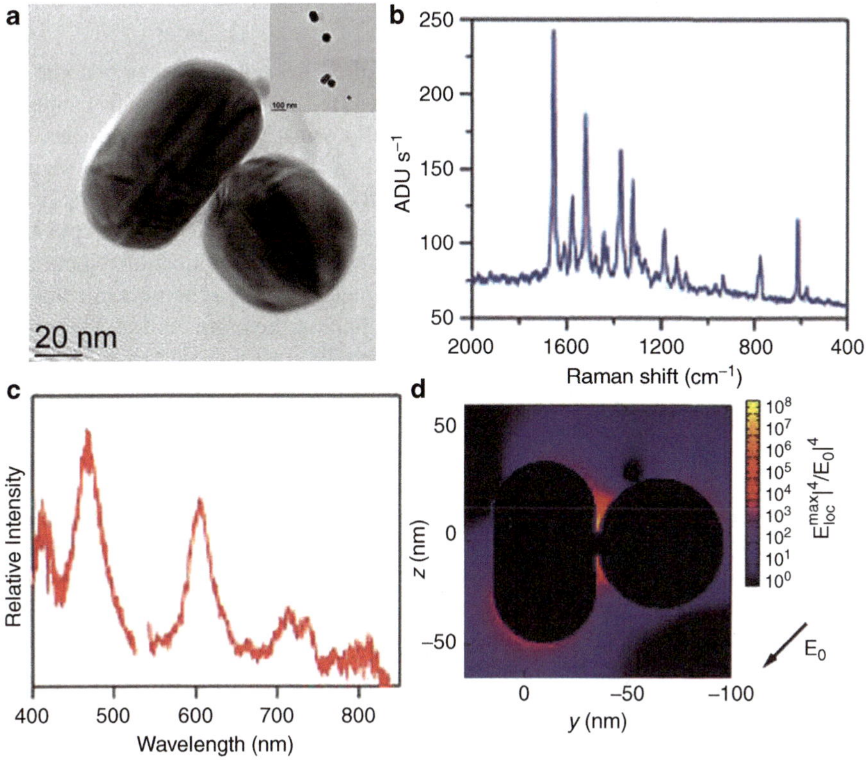

Fig. 10.12 (**a**) Correlated TEM, Raman scattering, and dark scattering spectrum of a single-molecule active site. (**b**) SERS enhancement distribution at 532 nm calculated using 3D DDA calculations. The enhancement maximum was found to be 3.9×10^8 (Figure reproduced from Ref. [136])

More recent studies have reinforced the importance of hot spots generated within the interstitial spaces between particles. Camden et al. used high-resolution TEM imaging of colloidal nanostructures correlated with confocal Raman mapping to investigate the structural characteristics of single-molecule SERS active sites [136]. The experiments were carried out using Rh6G with 532 nm resonant excitation to maximize the Raman cross section and provide the best possible conditions for single-molecule detection. Figure 10.12a shows high-resolution imaging of a dimer shown to be single-molecule SERS active. This is correlated to both (b) Raman and (c) dark field scattering spectra originating from the same structure. The high-resolution imaging allows for the structure to be digitized and the electromagnetic field distribution to be accurately calculated. The presence of interstitial gaps and crevices was found to be critically important for single-molecule detection with the simplest structures from which single-molecule SERS signals were detected being dimers. In no case was isolated colloidal particles found to be single-molecule SERS active.

4.5 Fabricating SERS Hot Spots

The reproducible and controlled fabrication of SERS hot spots has become an increasingly important area of research [137, 138]. The large-scale enhancements of 10^8–10^{10} provided by hot spots are critically sensitive to the microstructural geometry presenting a significant challenge both in terms of their rational design as well as their fabrication. Controlled fabrication of hot spots is not only important for developing the fundamental understanding of electromagnetic hot spots but also imperative if SERS is ever going to fulfill its potential as an application. Numerous methods have been developed ranging from sophisticated wet chemical methodologies which are pseudorandom and high throughput to the more controlled such as electron-beam lithography. We first review chemical methods and then highlight some of the nanolithographically produced structures frequently referred to as nano-antennae.

4.5.1 Wet Chemical Protocols

The uncontrolled aggregation of silver or gold nanoparticles through the addition of salts is typically the most common and successful method for preparation of SERS substrates with a high density of hot spots [139]. Although such aggregates provide strong SERS signals and are capable of single-molecule detection [23], they are difficult to control and as a result highly irreproducible. This irreproducibility is a combination of a lack of precise control in the aggregation conditions such as temperature, concentration, etc., and a typically broad distribution of nanoparticle sizes and shapes in the source colloidal solution. Such issues present many challenges for correlating scattering enhancements to specific hot spot features.

Great progress has been made in controlling both the size distribution and shape of gold and silver nanoparticles in solution through the use of surfactants [140, 141]. Nonspherical particles with sharp corners and edges are particularly attractive given that their high local curvatures give rise to highly concentrated local electromagnetic fields due to the lightening rod effect [54]. As a result, colloidal solutions of a diverse range of nanoparticles with sharp edges for SERS applications have been developed including cubes [142], nanorods [143], nanoshells [29, 144], prisms [145, 146], nanowires [147, 148], nanostars [149], mesoflowers [150], and octopods [151]. High-resolution SEM images of some of the nanoparticle shapes are shown in Fig. 10.13.

Several methods have been developed to enable the growth of high aspect ratio nanowires of gold and silver. Nanowire arrays have been used to demonstrate the complex interaction of factors such as excitation incident angle [152] and polarization [153] coupled with array properties such as the interstitial distances [147], layer absorbance [154, 155], and dielectric environment. Carmargo used Ag nanowires prepared by a modified protocol of $AgNO_3$ reduction with poly(vinyl pyrrolidone) and hydrochloric acid to investigate the activity of SERS hot spots formed when an Ag nanocube was placed in close proximity orientated either face nearly touching or one edge nearly touching [156].

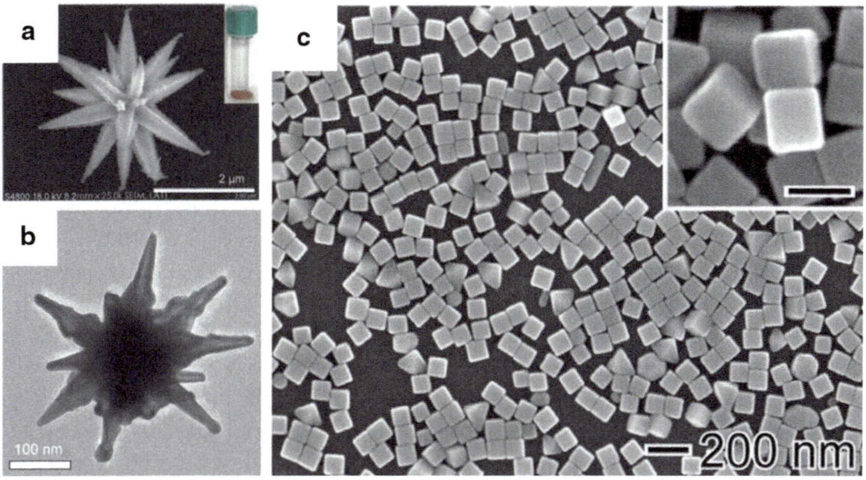

Fig. 10.13 (**a**, **b**) SEM images of gold mesoflowers produced by Sajanlal et al. (**c**) SEM images of silver nanocubes. Scale bar on the *inset* corresponds to 100 nm (Figure reproduced from Refs. [150] and [142])

Improved control of the particle size and shape is only one aspect of the problem. Great efforts have been made to control their relative position, even on a pseudorandom level, in order to maximize the hot spot intensity and density and to move away from the irreproducible and random aggregation by salts. Sophisticated chemical functionalization of the nanoparticles has been developed in order to control the degree of nanoparticle aggregation not only for SERS but other optically active systems [157]. Most of these strategies exploit the strong covalent binding between thiolated ligands and gold atoms in order to form highly stable self-assembled monolayers [158, 159]. Such methods are becoming particularly important for biological sensing applications as will be discussed in the following section.

Braun et al. demonstrated a variation on the chemically driven SERS active system that enabled the creation of hot spots and the localization onto a substrate using a short dithiol linker molecule as a model analyte in conjunction with a microsphere patterning technique [160]. This strategy allowed hot spots to be found easily and reliably using routine optical microscopy, reducing the time spent mapping large areas locating the hot spots, and enabled the correlated investigation through Raman spectroscopy combined with TEM.

Vlckova et al. demonstrated single-molecule SERS in self-assembled dimers of silver nanoparticles [161]. The self-assembly process was driven by nanoparticle functionalization with a bifunctional molecular linker (4,4′-diaminoazobenzene) which forms a molecular bridge between the particles. Molecules forming the molecular bridge are then precisely positioned within the hot spot with the molecular population being controlled by careful optimization of the linker to nanoparticle concentration ratio. Single-molecule time-dependent SERS signal fluctuations and molecular decomposition were studied in this and following studies in small aggregates of silver nanoparticles functionalized by 4,4″-diaminoterphenyl molecular bridges [162].

The self-assembly process in these studies was driven by strongly interacting groups which bind covalently to the metal surface [159]. However, several protocols for the dynamic formation of SERS hot spots using self-assembly have been developed based on DNA hybridization or π–π interactions [159, 163, 164]. As a result, a wide range of options for the controlled aggregation of silver and gold nanoparticles are now available.

Langmuir Blodgett (LB) monolayers have been increasingly used to control and optimize nanoparticle distributions and particle aggregation. Goulet working with Aroca used LB monolayers as a controlled approach to the investigation hot spots [117, 165]. Single-molecule SERS was achieved from a resonant dye mixed into fatty acid LB monolayers deposited onto nanostructure Ag island films. Detailed measurements as a function of analyte concentration showed that hot spots were highly localized, encompassing only a few molecular sites and emphasized the importance of the rare coincidence between hot spot and a single molecule involved in single-molecule measurements. Several other groups have used LB to show that the LB surface pressure is critical in controlling the SERS intensity through controlling the aggregation state of the nanoparticles involved. Mahmoud and El-Sayed have used LB to create monolayers of silver nanocubes [166] and gold nanocages [167] as well as gold nanoframes [168].

4.5.2 Lithographic Techniques

While such wet chemistry methods enable the large-scale production of hot spots which might be useful for applications, the pseudorandom control of hot spot parameters limits the detailed study of fundamental hot spot properties. Advanced lithographic techniques, such as electron-beam lithography [169] or nanosphere lithography [170, 171], allow for greater control over particle shape and size as well as interparticle separation [172]. However, there is some limitation of these techniques given the local field enhancement and localization is so critically dependent on geometrical variations below the resolution currently achievable. Despite this a number of studies have utilized these techniques to investigate the properties of hot spots formed in unusual geometries. Fromm et al. investigated the gap-dependent optical coupling of two Au triangles orientated tip to tip in a bowtie formation [173], as well as the chemical enhancement for SERS [174].

Numerous studies investigating the optical properties of precisely designed and fabricated nano-antennae have harnessed many of the aspects of the theoretical understanding of the electromagnetic enhancement mechanism. Figure 10.14a shows UV–vis spectra correlated to SEM images of arrays of Au nanowells of different aspect ratios [175]. The SERS response of the 610 cm^{-1} mode of Rh6G adsorbed to Au nanowells as a function of the number of nanowells per unit area is shown in Fig. 10.14b. Such studies have been extended to long-range arrays of metallic nanocrescents [176] and Au/SiO_2 nanodisks [177].

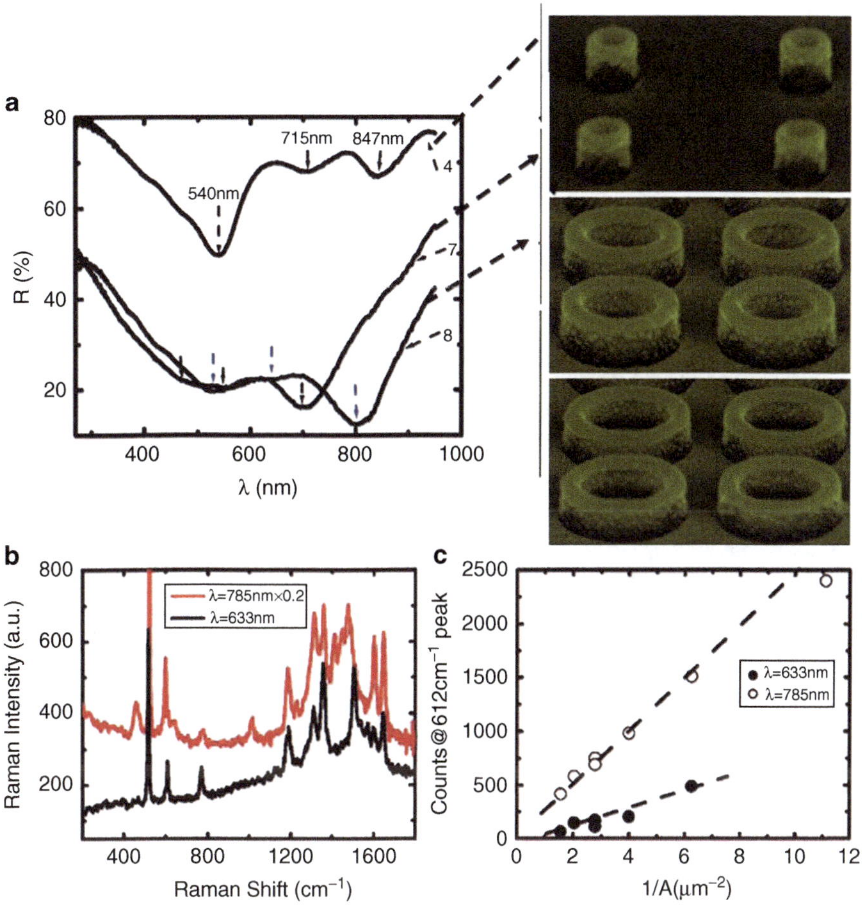

Fig. 10.14 (**a**) UV–vis spectra and SEMs of Au nanowells with different aspect ratios and pitches. (**b**) SERS spectra of Rh6G from Au nanowells and (**c**) intensity of the 612 cm^{-1} mode as a function of the number of nanowells per unit area (Figure reproduced from Ref. [175])

Mirkin and coworkers have pioneered the development of on-wire lithography (OWL) [67, 178]. OWL involves the electrochemical deposition of materials of interest into porous alumina templates with controllable pore diameter. A schematic of the scheme is shown in Fig. 10.15. The wires are released after deposition by dissolution of the template. The deposition and template dissolution processes are standard processes that are well established in the literature. OWL allows the fabrication of metal nanodisk arrays with precise control of the disk thickness, surface roughness [179], composite metals [180], and inter-disk separation [66]. Figure 10.15 shows high-resolution SEM images of the resulting rods demonstrating the control of segment separation and surface roughness. Such control enables close comparison to theoretical calculations demonstrating the importance of the electromagnetic mechanism [66].

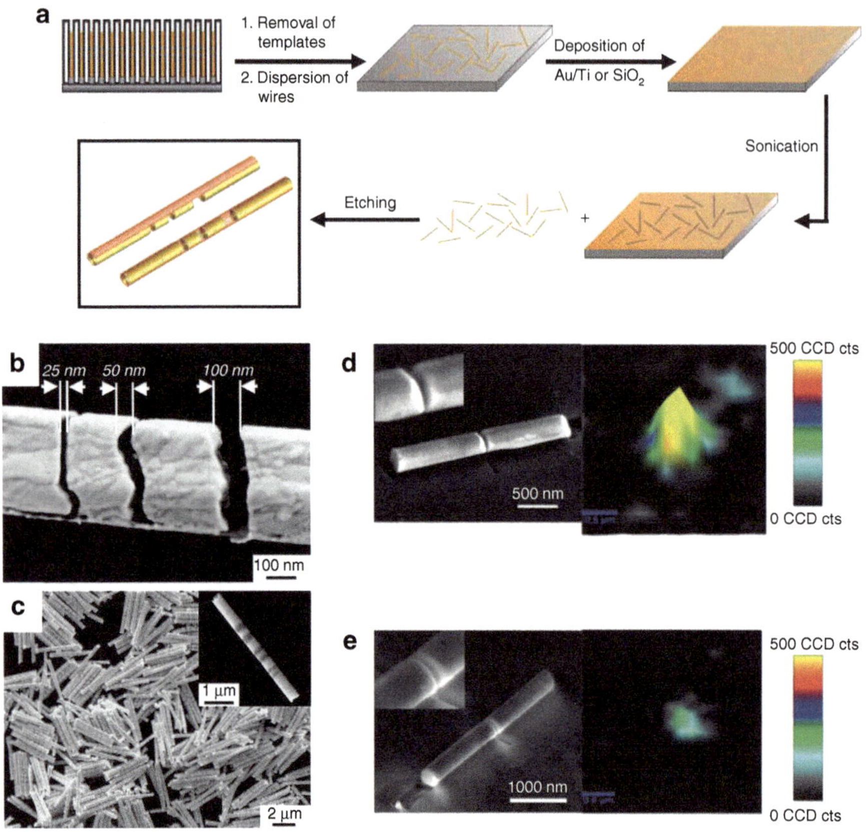

Fig. 10.15 (**a**) OWL methodology. (**b**, **c**) SEM images of multi-segment metallic Au–Ag nanowires with gaps of 25, 50, and 100 nm. (**d**, **e**) SEM and Raman microscopy of smooth and rough gapped rods produced by OWL (Figure reproduced from Refs. [67] and [178])

4.6 Applications of SERS Hot Spots

SERS holds great potential for a myriad of applications as a technique capable of both single-molecule detection and simultaneous chemical identification. It is now widely accepted that reproducible and reliable substrates containing a high density of SERS active hot spots are required to achieve such sensitive trace detection. A number of applications have been developed in the last 5–10 years including biomedical [181, 182], security [183], and drug detection [184] among others. This and progress in the development of inexpensive handheld Raman spectrometers, and one shot disposable SERS substrates, have made the application of SERS to real-world applications much more viable. A prominent example is that of the application of SERS to glucose detection [185, 186]. However, many of these applications have involved clever strategies employing SERS tags loaded with

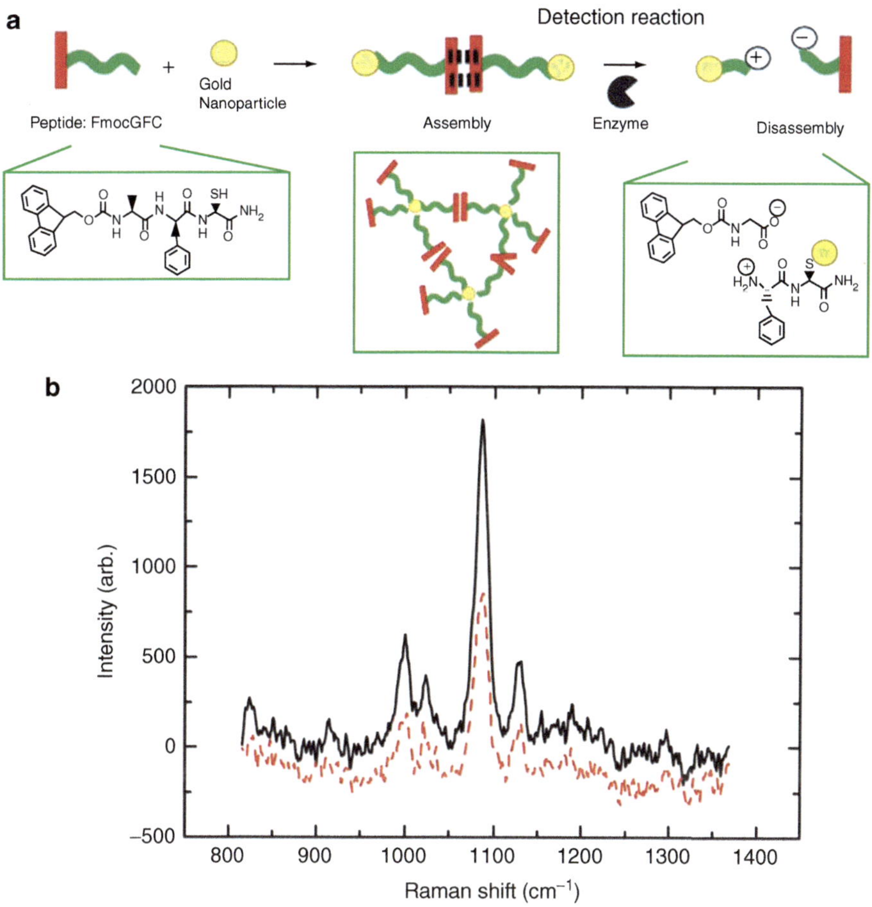

Fig. 10.16 (**a**) Scheme for the self-assembly of the peptide-functionalized 40-nm gold nanoparticles and deaggregation process. (**b**) Background corrected Raman spectra before (*black line*) and 4 h after (*dashed red line*) the addition of 7.2 nM enzyme solution

high molecular concentrations which are targeted to specific molecules of interest through surface functionalization. Relatively few, if any, applications have been developed which specifically take advantage of the strong SERS enhancements of hot spots. This is partly because, in spite of a great deal of effort, enhancement is typically gained at the expense of reproducibility.

We recently demonstrated a method that reproducibly exploits hot spots in metal nanoparticle aggregates for the trace detection of disease-specific enzymes by design [187]. The scheme, shown in Fig. 10.16a, involves the functionalization of gold nanoparticle with thiolated short peptide sequences terminated by an N-(fluorenyl-9-methoxycarbonyl) (FMOC) group. Nanoparticle aggregation is then driven by π–π interactions between the terminating FMOC groups [157]. Hot spots

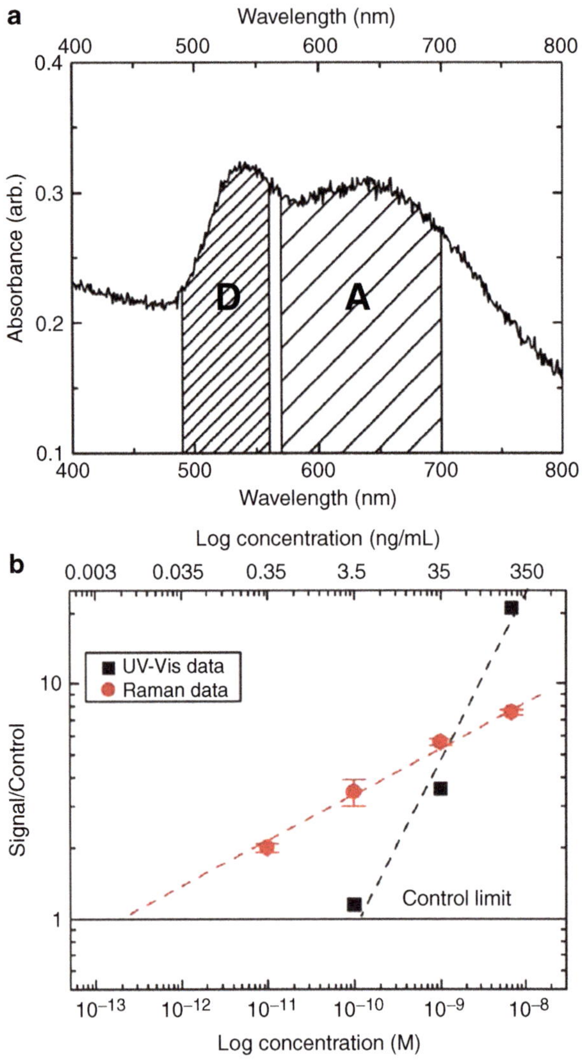

Fig. 10.17 (**a**) UV–vis extinction spectrum of an aggregated colloidal solution prior to enzyme addition. Ratio of region D and A gives an indication of aggregation state of the nanoparticles. (**b**) Detection limits for Raman and UV–vis extinction spectra normalized to control values. A possible improvement in the detection limit down to 10^{-10} M is suggested from the SERS measurements (Figure reproduced from Ref. [187])

are formed directly in between the nanoparticle with characteristics such as resonance and enhancement intensity dependent on the peptide length and nanoparticle size. As a result these properties can be precisely controlled. FMOC, with a strong characteristic SERS response, also acts as a reporter molecule which is placed precisely at the optimal hot spot position during the aggregation process as a result of its role in the nanoparticle aggregation. The detected SERS intensity was then a direct indication of the degree of nanoparticle aggregation.

The short peptide linker ligand is specifically designed to be cleaved by the target enzyme in order to enable enzyme detection. Dissociation of the aggregates follows the addition of the enzyme to the solution as the peptide linker is cleaved, aided by the electrostatic repulsion of the positively charged nanoparticles. The average number of hot spots present within the scattering volume decreases as the degree of nanoparticle aggregation reduces with the detected SERS signal decreasing in response. Figure 10.16b shows the Raman spectra obtained from a solution of aggregated nanoparticles before and 4 h after exposure to the enzyme at 7.2 nM concentration. A clear decrease in the intensity of the characteristic peak of FMOC is observed as a function of time.

The decrease in the SERS intensity can be monitored as a function of time for different concentrations in order to determine the detection limits. Figure 10.17a shows the UV–vis extinction spectrum of an aggregated colloidal solution. The ratio of the D and A regions can be used as an accurate indication of the aggregation state of the nanoparticles. This was used in order to compare the sensitivity of the SERS detection scheme. Figure 10.16a shows the detection sensitivity for both methods as a function of enzyme concentration. Extrapolation of the detection trends suggests potential detection limits on the order of 0.003 ng/mL (10^{-13} M) using SERS detection. This compares favorably to the current detection limit of normal immunoassay detection of enzymes which is on the order of 10 ng/mL [188]. Importantly, the detection of enzymes using such hybrid peptide-gold nanoparticle bioactive materials is typically faster with fewer intervening chemical steps than current technology. They also offer flexibility as the functionalizing chemistry can be tailored for the detection of many disease-specific enzymes.

Conclusions

Hot spots, or highly localized regions of intense local field enhancements, formed by localized surface plasmon resonances, are now recognized as being the dominant contributor to the detected SERS signals in the majority of experiments. There has been rapid progress in the understanding of hot spot characteristics in recent years and the geometrical characteristics of the nanostructures in which they form are well understood due to excellent correlation between theory and experiment which has been enabled by improvements in nanofabrication techniques and theoretical simulations. Hot spots are still considered to be a prerequisite for single-molecule detection although the extraordinary enhancements of up to 10^{15} have now been moderated to more reasonable estimates approaching 10^8–10^9, while new understanding of fluctuation phenomena and statistical nature of single molecular SERS has been gained through a consideration of hot spot enhancement distribution profiles. The combination of SERS and atomic force microscopy (AFM) has led to the combination of the chemical specificity of Raman and spatial mapping of AFM in the form of tip-enhanced Raman spectroscopy (TERS). However, a SERS substrate combining a high density of SERS hot spots with molecular generality is still lacking. Future research involving hot spots will most certainly be directed toward applications, particularly in the areas of biomedical diagnostics and homeland security.

References

1. Michaels AM, Jiang J, Brus L (2000) Ag nanocrystal junctions as the site for surface-enhanced Raman scattering of single rhodamine 6 G molecules. J Phys Chem B 104(50):11965–11971
2. Xu HX, Aizpurua J, Kall M, Apell P (2000) Electromagnetic contributions to single-molecule sensitivity in surface-enhanced Raman scattering. Phys Rev E 62(3):4318–4324
3. Futamata M (2006) Single molecule sensitivity in SERS: importance of junction of adjacent Ag nanoparticles. Faraday Discuss 132:45–61
4. Jiang J, Bosnick K, Maillard M, Brus L (2003) Single molecule Raman spectroscopy at the junctions of large Ag nanocrystals. J Phys Chem B 107(37):9964–9972
5. Anderson DJ, Moskovits M (2006) A SERS-active system based on silver nanoparticles tethered to a deposited silver film. J Phys Chem B 110(28):13722–13727
6. Le Ru EC, Etchegoin PG (2004) Sub-wavelength localization of hot-spots in SERS. Chem Phys Lett 396(4–6):393–397
7. Kneipp K, Kneipp H, Itzkan I, Dasari RR, Feld MS (2002) Surface-enhanced Raman scattering and biophysics. J Phys-Condens Mat 14(18):R597–R624
8. Fleischm M, Hendra PJ, Mcquilla AJ (1974) Raman-spectra of pyridine adsorbed at a silver electrode. Chem Phys Lett 26(2):163–166
9. Jeanmaire DL, Vanduyne RP (1977) Surface Raman spectroelectrochemistry. 1. Heterocyclic, aromatic, and aliphatic-amines adsorbed on anodized silver electrode. J Electroanal Chem 84(1):1–20
10. Moskovits M (1978) Surface-roughness and enhanced intensity of Raman-scattering by molecules adsorbed on metals. J Chem Phys 69(9):4159–4161
11. Albrecht MG, Creighton JA (1977) Anomalously intense Raman-spectra of pyridine at a silver electrode. J Am Chem Soc 99(15):5215–5217
12. Moskovits M (1985) Surface-enhanced spectroscopy. Rev Mod Phys 57(3):783–826
13. Aroca R (2006) Surface-enhanced vibrational spectroscopy. Wiley, Hoboken
14. Le Ru EC, Etchegoin P (2009) Principles of surface-enhanced Raman spectroscopy, 1st edn. Elsevier, Amsterdam
15. Le Ru EC, Etchegoin PG (2006) Rigorous justification of the [E](4) enhancement factor in surface enhanced Raman spectroscopy. Chem Phys Lett 423(1–3):63–66
16. Moerl L, Pettinger B (1982) The role of Cu atoms on silver electrodes in surface enhanced Raman-scattering from pyridine – giant enhancement by a minority of adsorbed molecules. Solid State Commun 43(5):315–320
17. Watanabe T, Yanagihara N, Honda K, Pettinger B, Moerl L (1983) Effects of underpotentially deposited Tl and Pb submonolayers on the surface-enhanced Raman-scattering (SERS) from pyridine at Ag electrodes. Chem Phys Lett 96(6):649–655
18. Furtak TE, Roy D (1983) Nature of the active-site in surface-enhanced Raman-scattering. Phys Rev Lett 50(17):1301–1304
19. Etchegoin PG, Le Ru EC (2008) A perspective on single molecule SERS: current status and future challenges. Phys Chem Chem Phys 10(40):6079–6089
20. Zhang WH, Cui XD, Martin OJF (2009) Local field enhancement of an infinite conical metal tip illuminated by a focused beam. J Raman Spectrosc 40(10):1338–1342
21. Zhao LL, Jensen L, Schatz GC (2006) Surface-enhanced Raman scattering of pyrazine at the junction between two Ag-20 nanoclusters. Nano Lett 6(6):1229–1234
22. Nie SM, Emery SR (1997) Probing single molecules and single nanoparticles by surface-enhanced Raman scattering. Science 275(5303):1102–1106
23. Kneipp K, Wang Y, Kneipp H, Perelman LT, Itzkan I, Dasari R, Feld MS (1997) Single molecule detection using surface-enhanced Raman scattering (SERS). Phys Rev Lett 78(9):1667–1670

24. Otto A (2006) On the significance of Shalaev's "hot spots" in ensemble and single-molecule SERS by adsorbates on metallic films at the percolation threshold. J Raman Spectrosc 37(9):937–947
25. Zhang P, Haslett TL, Douketis C, Moskovits M (1998) Mode localization in self-affine fractal interfaces observed by near-field microscopy. Phys Rev B 57(24):15513–15518
26. Barnes WL (1998) Fluorescence near interfaces: the role of photonic mode density. J Mod Opt 45(4):661–699
27. Aslan K, Geddes CD (2009) Metal-enhanced chemiluminescence: advanced chemiluminescence concepts for the 21st century. Chem Soc Rev 38(9):2556–2564
28. Osawa M (2001) Surface-enhanced infrared absorption. Near-Field Opt Surf Plasmon Polaritons 81:163–187
29. Lal S, Grady NK, Kundu J, Levin CS, Lassiter JB, Halas NJ (2008) Tailoring plasmonic substrates for surface enhanced spectroscopies. Chem Soc Rev 37(5):898–911
30. Baker GA, Moore DS (2005) Progress in plasmonic engineering of surface-enhanced Raman-scattering substrates toward ultra-trace analysis. Anal Bioanal Chem 382(8): 1751–1770
31. Ozbay E (2006) Plasmonics: merging photonics and electronics at nanoscale dimensions. Science 311(5758):189–193
32. Maier S (2007) Plasmonics: fundamentals and applications. Springer, Berlin
33. Srituravanich W, Fang N, Durant S, Ambati M, Sun C, Zhang X (2004) Sub-100 nm lithography using ultrashort wavelength of surface plasmons. J Vac Sci Technol B 22(6):3475–3478
34. Liu ZW, Steele JM, Srituravanich W, Pikus Y, Sun C, Zhang X (2005) Focusing surface plasmons with a plasmonic lens. Nano Lett 5(9):1726–1729
35. Shao DB, Chen SC (2005) Surface-plasmon-assisted nanoscale photolithography by polarized light. Appl Phys Lett 86(25):253107
36. Maier SA, Kik PG, Atwater HA, Meltzer S, Harel E, Koel BE, Requicha AAG (2003) Local detection of electromagnetic energy transport below the diffraction limit in metal nanoparticle plasmon waveguides. Nat Mater 2(4):229–232
37. Grigorenko AN, Roberts NW, Dickinson MR, Zhang Y (2008) Nanometric optical tweezers based on nanostructured substrates. Nat Photonics 2(6):365–370
38. Cubukcu E, Yu NF, Smythe EJ, Diehl L, Crozier KB, Capasso F (2008) Plasmonic laser antennas and related devices. Ieee J Sel Top Quant 14(6):1448–1461
39. Oulton RF, Sorger VJ, Zentgraf T, Ma RM, Gladden C, Dai L, Bartal G, Zhang X (2009) Plasmon lasers at deep subwavelength scale. Nature 461(7264):629–632
40. Flores SM, Toca-Herrera JL (2009) The new future of scanning probe microscopy: combining atomic force microscopy with other surface-sensitive techniques, optical and microscopy fluorescence techniques. Nanoscale 1(1):40–49
41. Gerardy JM, Ausloos M (1982) Absorption-spectrum of clusters of spheres from the general-solution of Maxwell equations. 2. Optical-properties of aggregated metal spheres. Phys Rev B 25(6):4204–4229
42. Zhao J, Pinchuk AO, Mcmahon JM, Li SZ, Ausman LK, Atkinson AL, Schatz GC (2008) Methods for describing the electromagnetic properties of silver and gold nanoparticles. Acc Chem Res 41(12):1710–1720
43. Goldstein J, Newbury DE, Joy DC, Lyman CE, Echlin P, Lifshin E, Sawyer L, Michael JR (2003) Scanning electron microscopy and x-ray microanalysis. Springer, New York
44. Williams DB, Barry Carter CB (2009) Transmission electron microscopy: a textbook for materials science. Springer, New York
45. Eaton P, West P (2010) Atomic force microscopy. Oxford University Press, Oxford
46. Abashin M (2009) Near-field characterization of photonic nanodevices: near-field scanning optical microscopy (NSOM) characterization of photonic nanodevices and nanoscale optical phenomena. Saarbrücken, Germany
47. Franssila S (2004) Introduction to microfabrication. Wiley-Blackwell, Chichester

48. Orloff J, Swanson L, Utlaut M (2002) High resolution focused Ion beams: FIB and applications. Springer, Berlin
49. Gersten JI (1980) The effect of surface-roughness on surface enhanced Raman-scattering. J Chem Phys 72(10):5779–5780
50. Gersten J, Nitzan A (1980) Electromagnetic theory of enhanced Raman-scattering by molecules adsorbed on rough surfaces. J Chem Phys 73(7):3023–3037
51. Gersten JI, Nitzan A (1985) Photophysics and photochemistry near surfaces and small particles. Surf Sci 158(1–3):165–189
52. Wang DS, Kerker M (1981) Enhanced Raman-scattering by molecules adsorbed at the surface of colloidal spheroids. Phys Rev B 24(4):1777–1790
53. Li KR, Stockman MI, Bergman DJ (2003) Self-similar chain of metal nanospheres as an efficient nanolens. Phys Rev Lett 91(22):227402
54. Hao E, Schatz GC (2004) Electromagnetic fields around silver nanoparticles and dimers. J Chem Phys 120(1):357–366
55. GarciaVidal FJ, Pendry JB (1996) Collective theory for surface enhanced Raman scattering. Phys Rev Lett 77(6):1163–1166
56. Shalaev VM, Sarychev AK (1998) Nonlinear optics of random metal-dielectric films. Phys Rev B 57(20):13265–13288
57. Markel VA, Shalaev VM, Zhang P, Huynh W, Tay L, Haslett TL, Moskovits M (1999) Near-field optical spectroscopy of individual surface-plasmon modes in colloid clusters. Phys Rev B 59(16):10903–10909
58. Le Ru EC, Etchegoin PG, Meyer M (2006) Enhancement factor distribution around a single surface-enhanced Raman scattering hot spot and its relation to single molecule detection. J Chem Phys 125(20):204701
59. Etchegoin PG, Galloway C, Le Ru EC (2006) Polarization-dependent effects in surface-enhanced Raman scattering (SERS). Phys Chem Chem Phys 8(22):2624–2628
60. Weisstein EW. The Pareto distribution. http://mathworld.wolfram.com/ParetoDistribution.html
61. Boyack R, Le Ru EC (2009) Investigation of particle shape and size effects in SERS using T-matrix calculations. Phys Chem Chem Phys 11(34):7398–7405
62. Kahl M, Voges E (2000) Analysis of plasmon resonance and surface-enhanced Raman scattering on periodic silver structures. Phys Rev B 61(20):14078–14088
63. Norton SJ, Vo-Dinh T (2008) Optical response of linear chains of metal nanospheres and nanospheroids. J Opt Soc Am A 25(11):2767–2775
64. Wang ZB, Luk'yanchuk BS, Guo W, Edwardson SP, Whitehead DJ, Li L, Liu Z, Watkins KG (2008) The influences of particle number on hot spots in strongly coupled metal nanoparticles chain. J Chem Phys 128(9):094705
65. Perassi EA, Canali LR, Coronado EA (2009) Enhancement and confinement analysis of the electromagnetic fields inside Hot spots. J Phys Chem C 113(16):6315–6319
66. Li SZ, Pedano ML, Chang SH, Mirkin CA, Schatz GC (2010) Gap structure effects on surface-enhanced Raman scattering intensities for gold gapped rods. Nano Lett 10(5):1722–1727
67. Qin LD, Zou SL, Xue C, Atkinson A, Schatz GC, Mirkin CA (2006) Designing, fabricating, and imaging Raman hot spots. P Natl Acad Sci USA 103(36):13300–13303
68. Pettinger B, Moerl L (1983) Influence of foreign metal atoms deposited at electrodes on local and nonlocal processes in surface enhanced Raman-scattering. J Electron Spectrosc 29:383–395
69. Persson BNJ (1981) On the theory of surface-enhanced Raman-scattering. Chem Phys Lett 82(3):561–565
70. Otto A (2002) What is observed in single molecule SERS, and why? J Raman Spectrosc 33(8):593–598
71. Peyser-Capadona L, Zheng J, Gonzalez JI, Lee TH, Patel SA, Dickson RM (2005) Nanoparticle-free single molecule anti-stokes Raman spectroscopy. Phys Rev Lett 94(5):058301

72. Zhao LL, Jensen L, Schatz GC (2006) Pyridine-Ag-20 cluster: a model system for studying surface-enhanced Raman scattering. J Am Chem Soc 128(9):2911–2919
73. Aikens CM, Schatz GC (2006) TDDFT studies of absorption and SERS spectra of pyridine interacting with Au-20. J Phys Chem A 110(49):13317–13324
74. Brewer KE, Aikens CM (2010) TDDFT investigation of surface-enhanced Raman scattering of HCN and CN- on Ag-20. J Phys Chem A 114(33):8858–8863
75. Shegai T, Vaskevich A, Rubinstein I, Haran G (2009) Raman spectroelectrochemistry of molecules within individual electromagnetic Hot spots. J Am Chem Soc 131(40): 14390–14398
76. Elfick APD, Downes AR, Mouras R (2010) Development of tip-enhanced optical spectroscopy for biological applications: a review. Anal Bioanal Chem 396(1):45–52
77. Domke KF, Pettinger B (2010) Studying surface chemistry beyond the diffraction limit: 10 years of TERS. Chemphyschem 11(7):1365–1373
78. Stockle RM, Suh YD, Deckert V, Zenobi R (2000) Nanoscale chemical analysis by tip-enhanced Raman spectroscopy. Chem Phys Lett 318(1–3):131–136
79. Bailo E, Deckert V (2008) Tip-enhanced Raman scattering. Chem Soc Rev 37(5):921–930
80. Hartschuh A, Sanchez EJ, Xie XS, Novotny L (2003) High-resolution near-field Raman microscopy of single-walled carbon nanotubes. Phys Rev Lett 90(9):095503
81. Pettinger B, Ren B, Picardi G, Schuster R, Ertl G (2004) Nanoscale probing of adsorbed species by tip-enhanced Raman spectroscopy. Phys Rev Lett 92(9):096101
82. Zhang WH, Yeo BS, Schmid T, Zenobi R (2007) Single molecule tip-enhanced Raman spectroscopy with silver tips. J Phys Chem C 111(4):1733–1738
83. Berweger S, Raschke MB (2010) Signal limitations in tip-enhanced Raman scattering: the challenge to become a routine analytical technique. Anal Bioanal Chem 396(1):115–123
84. Motohashi M, Hayazawa N, Tarun A, Kawata S (2008) Depolarization effect in reflection-mode tip-enhanced Raman scattering for Raman active crystals. J Appl Phys 103(3):034309
85. Neacsu CC, Dreyer J, Behr N, Raschke MB (2006) Scanning-probe Raman spectroscopy with single-molecule sensitivity. Phys Rev B 73(19):193406
86. Domke KF, Pettinger B (2007) Comment on "scanning-probe Raman spectroscopy with single-molecule sensitivity". Phys Rev B 75(23):236401
87. Neacsu CC, Dreyer J, Behr N, Raschke MB (2007) Reply to "comment on 'scanning-probe Raman spectroscopy with single-molecule sensitivity'". Phys Rev B 75(23):236402
88. Steidtner J, Pettinger B (2008) Tip-enhanced Raman spectroscopy and microscopy on single dye molecules with 15 nm resolution. Phys Rev Lett 100(23):236101
89. Zhang D, Domke KF, Pettinger B (2010) Tip-enhanced Raman spectroscopic studies of the hydrogen bonding between adenine and thymine adsorbed on Au (111). Chemphyschem 11(8):1662–1665
90. Geshev PI, Fischer U, Fuchs H (2010) Calculation of tip enhanced Raman scattering caused by nanoparticle plasmons acting on a molecule placed near a metallic film. Phys Rev B 81(12):125441
91. Pettinger B, Domke KF, Zhang D, Picardi G, Schuster R (2009) Tip-enhanced Raman scattering: influence of the tip-surface geometry on optical resonance and enhancement. Surf Sci 603(10–12):1335–1341
92. Sackrow M, Stanciu C, Lieb MA, Meixner AJ (2008) Imaging nanometre-sized hot spots on smooth Au films with high-resolution tip-enhanced luminescence and Raman near-field optical microscopy. Chemphyschem 9(2):316–320
93. Berweger S, Raschke MB (2009) Polar phonon mode selection rules in tip-enhanced Raman scattering. J Raman Spectrosc 40(10):1413–1419
94. Domke KF, Pettinger B (2009) Tip-enhanced Raman spectroscopy of 6 H-SiC with graphene adlayers: selective suppression of E1 modes. J Raman Spectrosc 40(10): 1427–1433
95. Le Ru EC, Meyer M, Blackie E, Etchegoin PG (2008) Advanced aspects of electromagnetic SERS enhancement factors at a hot spot. J Raman Spectrosc 39(9):1127–1134

96. Kneipp K, Kneipp H (2006) Surface-enhanced Raman scattering on silver nanoparticles in different aggregation stages. Isr J Chem 46(3):299–305
97. Dieringer JA, Wustholz KL, Masiello DJ, Camden JP, Kleinman SL, Schatz GC, Van Duyne RP (2009) Surface-enhanced Raman excitation spectroscopy of a single rhodamine 6 G molecule. J Am Chem Soc 131(2):849–854
98. Moskovits M (1982) Surface selection-rules. J Chem Phys 77(9):4408–4416
99. Moskovits M, Suh JS (1984) Surface selection-rules for surface-enhanced Raman-spectroscopy – calculations and application to the surface-enhanced Raman-spectrum of phthalazine on silver. J Phys Chem-Us 88(23):5526–5530
100. Moskovits M, Dilella DP, Maynard KJ (1988) Surface Raman-spectroscopy of a number of cyclic aromatic-molecules adsorbed on silver – selection-rules and molecular-reorientation. Langmuir 4(1):67–76
101. Shegai TO, Haran G (2006) Probing the Raman scattering tensors of individual molecules. J Phys Chem B 110(6):2459–2461
102. Hayes W, Loudon R (1978) Scattering of light by crystals. Wiley-Interscience, New York
103. Bosnick KA, Jiang J, Brus LE (2002) Fluctuations and local symmetry in single-molecule rhodamine 6 G Raman scattering on silver nanocrystal aggregates. J Phys Chem B 106(33):8096–8099
104. Xu HX, Kall M (2003) Polarization-dependent surface-enhanced Raman spectroscopy of isolated silver nanoaggregates. Chemphyschem 4(9):1001–1005
105. Pieczonka NPW, Aroca RF (2008) Single molecule analysis by surfaced-enhanced Raman scattering. Chem Soc Rev 37(5):946–954
106. Zhang ZL, Yin YF, Jiang JW, Mo YJ (2009) Single molecule detection of 4-dimethylaminoazobenzene by surface-enhanced Raman spectroscopy. J Mol Struct 920(1–3):297–300
107. Schleifenbaum F, Blum C, Subramaniam V, Meixner AJ (2009) Single-molecule spectral dynamics at room temperature. Mol Phys 107(18):1923–1942
108. Emory SR, Jensen RA, Wenda T, Han MY, Nie SM (2006) Re-examining the origins of spectral blinking in single-molecule and single-nanoparticle SERS. Faraday Discuss 132:249–259
109. Nitzan A, Brus LE (1981) Theoretical-model for enhanced photochemistry on rough surfaces. J Chem Phys 75(5):2205–2214
110. Goncher GM, Parsons CA, Harris CB (1984) Photochemistry on rough metal-surfaces. J Phys Chem 88(19):4200–4209
111. Fang Y, Seong NH, Dlott DD (2008) Measurement of the distribution of site enhancements in surface-enhanced Raman scattering. Science 321(5887):388–392
112. Maher RC, Zhang T, Cohen LF, Gallop JC, Liu FM, Green M (2009) Towards a metrological determination of the performance of SERS media. Phys Chem Chem Phys 11(34): 7463–7468
113. Etchegoin PG, Meyer M, Le Ru EC (2007) Statistics of single molecule SERS signals: is there a Poisson distribution of intensities? Phys Chem Chem Phys 9(23):3006–3010
114. Le Ru EC, Meyer M, Etchegoin PG (2006) Proof of single-mokeule sensitivity in surface enhanced Raman scattering (SERS) by means of a two-analyte technique. J Phys Chem B 110(4):1944–1948
115. Blackie E, Le Ru EC, Meyer M, Timmer M, Burkett B, Northcote P, Etchegoin PG (2008) Bi-analyte SERS with isotopically edited dyes. Phys Chem Chem Phys 10(28):4147–4153
116. Dieringer JA, Lettan RB, Scheidt KA, Van Duyne RP (2007) A frequency domain existence proof of single-molecule surface-enhanced Raman spectroscopy. J Am Chem Soc 129(51):16249–16256
117. Goulet PJG, Aroca RF (2007) Distinguishing individual vibrational fingerprints: single-molecule surface-enhanced resonance Raman scattering from one-to-one binary mixtures in Langmuir-Blodgett monolayers. Anal Chem 79(7):2728–2734

118. dos Santos DP, Andrade GFS, Temperini MLA, Brolo AG (2009) Electrochemical control of the time-dependent intensity fluctuations in surface-enhanced Raman scattering (SERS). J Phys Chem C 113(41):17737–17744
119. Camargo PHC, Rycenga M, Au L, Xia YN (2009) Isolating and probing the hot spot formed between two silver nanocubes. Angew Chem Int Edit 48(12):2180–2184
120. Chen C, Hutchison JA, Clemente F, Kox R, Uji-I H, Hofkens J, Lagae L, Maes G, Borghs G, Van Dorpe P (2009) Direct evidence of high spatial localization of hot spots in surface-enhanced Raman scattering. Angew Chem Int Edit 48(52):9932–9935
121. Chen C, Hutchison JA, Van Dorpe P, Kox R, De Vlaminck I, Uji-i H, Hofkens J, Lagae L, Maes G, Borghs G (2009) Focusing plasmons in nanoslits for surface-enhanced Raman scattering. Small 5(24):2876–2882
122. (a) Xu HX, Kall M (2002) Surface-plasmon-enhanced optical forces in silver nanoaggregates. Phys Rev Lett 89(24); (b) Svedberg F, Kall M (2006) On the importance of optical forces in surface-enhanced Raman scattering (SERS). Faraday Discuss 132:35–44
123. Osborne MA, Balasubramanian S, Furey WS, Klenerman D (1998) Optically biased diffusion of single molecules studied by confocal fluorescence microscopy. J Phys Chem B 102(17):3160–3167
124. Svedberg F, Li ZP, Xu HX, Kall M (2006) Creating hot nanoparticle pairs for surface-enhanced Raman spectroscopy through optical manipulation. Nano Lett 6(12):2639–2641
125. Kneipp K, Wang Y, Kneipp H, Itzkan I, Dasari RR, Feld MS (1996) Population pumping of excited vibrational states by spontaneous surface-enhanced Raman scattering. Phys Rev Lett 76(14):2444–2447
126. Maher RC, Galloway CM, Le Ru EC, Cohena LF, Etchegoin PG (2008) Vibrational pumping in surface enhanced Raman scattering (SERS). Chem Soc Rev 37(5):965–979
127. Haslett TL, Tay L, Moskovits M (2000) Can surface-enhanced Raman scattering serve as a channel for strong optical pumping? J Chem Phys 113(4):1641–1646
128. (a) Brolo AG, Sanderson AC, Smith AP (2004) Ratio of the surface-enhanced anti-Stokes scattering to the surface-enhanced Stokes-Raman scattering for molecules adsorbed on a silver electrode. Physical Review B 69(4); (b) Maher RC, Dalley M, Le Ru EC, Cohen LF, Etchegoin PG, Hartigan H, Brown RJC, Milton MJT (2004) Physics of single molecule fluctuations in surface enhanced Raman spectroscopy active liquids. J Chem Phys 121(18):8901–8910
129. Maher RC, Etchegoin PG, Le Ru EC, Cohen LF (2006) A conclusive demonstration of vibrational pumping under surface enhanced Raman scattering conditions. J Phys Chem B 110(24):11757–11760
130. Galloway CM, Le Ru EC, Etchegoin PG (2009) Single-molecule vibrational pumping in SERS. Phys Chem Chem Phys 11(34):7372–7380
131. Kozich V, Werncke W (2010) The vibrational pumping mechanism in surface-enhanced Raman scattering: a subpicosecond time-resolved study. J Phys Chem C 114(23): 10484–10488
132. Tsai DP, Kovacs J, Wang Z, Moskovits M, Shalaev VM, Suh JS, Botet R (1994) Photon scanning tunneling microscopy images of optical excitations of fractal metal colloid clusters. Phys Rev Lett 72(26):4149
133. Bozhevolnyi SI, Vohnsen B, Smolyaninov II, Zayats AV (1995) Direct observation of surface polariton localization caused by surface-roughness. Opt Commun 117(5–6):417–423
134. Lin HY, Huang CH, Chang CH, Lan YC, Chui HC (2010) Direct near-field optical imaging of plasmonic resonances in metal nanoparticle pairs. Opt Express 18(1):165–172
135. Zhang P, Smith S, Rumbles G, Himmel ME (2005) Direct imaging of surface-enhanced Raman scattering in the near field. Langmuir 21(2):520–523
136. Camden JP, Dieringer JA, Wang YM, Masiello DJ, Marks LD, Schatz GC, Van Duyne RP (2008) Probing the structure of single-molecule surface-enhanced Raman scattering hot spots. J Am Chem Soc 130(38):12616–12617

137. Banholzer MJ, Millstone JE, Qin LD, Mirkin CA (2008) Rationally designed nanostructures for surface-enhanced Raman spectroscopy. Chem Soc Rev 37(5):885–897
138. Lin XM, Cui Y, Xu YH, Ren B, Tian ZQ (2009) Surface-enhanced Raman spectroscopy: substrate-related issues. Anal Bioanal Chem 394(7):1729–1745
139. Kneipp K, Kneipp H, Itzkan I, Dasari RR, Feld MS (1999) Ultrasensitive chemical analysis by Raman spectroscopy. Chem Rev 99(10):2957–2975
140. Wiley B, Sun Y, Xia Y (2007) Synthesis of silver nanostructures with controlled shapes and properties. Accounts Chem Res 40(10):1067–1076
141. Sardar R, Funston AM, Mulvaney P, Murray RW (2009) Gold nanoparticles: past, present, and future. Langmuir 25(24):13840–13851
142. Camargo PHC, Au L, Rycenga M, Li WY, Xia YN (2010) Measuring the SERS enhancement factors of dimers with different structures constructed from silver nanocubes. Chem Phys Lett 484(4–6):304–308
143. Orendorff CJ, Gearheart L, Jana NR, Murphy CJ (2006) Aspect ratio dependence on surface enhanced Raman scattering using silver and gold nanorod substrates. Phys Chem Chem Phys 8(1):165–170
144. Li WB, Guo YY, Zhang P (2010) General strategy to prepare TiO_2-core gold-shell nanoparticles as SERS-tags. J Phys Chem C 114(16):7263–7268
145. Cui B, Clime L, Li K, Veres T (2008) Fabrication of large area nanoprism arrays and their application for surface enhanced Raman spectroscopy. Nanotechnology 19(14):145302
146. Hsu MS, Cao YW, Wang HW, Pan YS, Lee BH, Huang CL (2010) Time-dependent surface plasmon resonance spectroscopy of silver nanoprisms in the presence of halide ions. Chemphyschem 11(8):1742–1748
147. Lee SJ, Morrill AR, Moskovits M (2006) Hot spots in silver nanowire bundles for surface-enhanced Raman spectroscopy. J Am Chem Soc 128(7):2200–2201
148. Lee SJ, Guan ZQ, Xu HX, Moskovits M (2007) Surface-enhanced Raman spectroscopy and nanogeometry: the plasmonic origin of SERS. J Phys Chem C 111(49):17985–17988
149. Barbosa S, Agrawal A, Rodriguez-Lorenzo L, Pastoriza-Santos I, Alvarez-Puebla RA, Kornowski A, Weller H, Liz-Marzan LM (2010) Tuning size and sensing properties in colloidal gold nanostars. Langmuir 26(18):14943–14950
150. Sajanlal PR, Pradeep T (2009) Mesoflowers: a new class of highly efficient surface-enhanced Raman active and infrared-absorbing materials. Nano Res 2(4):306–320
151. Naumov II, Li ZY, Bratkovsky AM (2010) Plasmonic resonances and hot spots in Ag octopods. Appl Phys Lett 96(3):033105
152. Chaney SB, Zhang ZY, Zhao YP (2006) Anomalous polarized absorbance spectra of aligned Ag nanorod arrays. Appl Phys Lett 89(5):053117
153. Liu YJ, Zhao YP (2008) Simple model for surface-enhanced Raman scattering from tilted silver nanorod array substrates. Phys Rev B 78(7):075436
154. Liu YJ, Zhang ZY, Zhao Q, Dluhy RA, Zhao YP (2009) Surface enhanced Raman scattering from an Ag nanorod array substrate: the site dependent enhancement and layer absorbance effect. J Phys Chem C 113(22):9664–9669
155. Zhao YP, Chaney SB, Shanmukh S, Dluhy RA (2006) Polarized surface enhanced Raman and absorbance spectra of aligned silver nanorod arrays. J Phys Chem B 110(7):3153–3157
156. Camargo PHC, Cobley CM, Rycenga M, Xia YN (2009) Measuring the surface-enhanced Raman scattering enhancement factors of hot spots formed between an individual Ag nanowire and a single Ag nanocube. Nanotechnology 20(43):434020
157. Laromaine A, Koh LL, Murugesan M, Ulijn RV, Stevens MM (2007) Protease-triggered dispersion of nanoparticle assemblies. J Am Chem Soc 129(14):4156–4157
158. Freeman RG, Grabar KC, Allison KJ, Bright RM, Davis JA, Guthrie AP, Hommer MB, Jackson MA, Smith PC, Walter DG, Natan MJ (1995) Self-assembled metal colloid monolayers – an approach to sers substrates. Science 267(5204):1629–1632

159. Daniel MC, Astruc D (2004) Gold nanoparticles: assembly, supramolecular chemistry, quantum-size-related properties, and applications toward biology, catalysis, and nanotechnology. Chem Rev 104(1):293–346
160. Braun G, Pavel I, Morrill AR, Seferos DS, Bazan GC, Reich NO, Moskovits M (2007) Chemically patterned microspheres for controlled nanoparticle assembly in the construction of SERS hot spots. J Am Chem Soc 129(25):7760–7761
161. Vlckova B, Moskovits M, Pavel I, Siskova K, Sladkova M, Slouf M (2008) Single-molecule surface-enhanced Raman spectroscopy from a molecularly-bridged silver nanoparticle dimer. Chem Phys Lett 455(4–6):131–134
162. Sladkova M, Vlckova B, Pavel I, Siskova K, Slouf M (2009) Surface-enhanced Raman scattering from a single molecularly bridged silver nanoparticle aggregate. J Mol Struct 924–26:567–570
163. Ghadiali JE, Stevens MM (2008) Enzyme-responsive nanoparticle systems. Adv Mater 20(22):4359–4363
164. DuFort CC, Dragnea B (2010) Bio-enabled synthesis of metamaterials. Ann Rev Phys Chem 61:323–344. Annual Reviews: Palo Alto
165. Goulet PJG, Pieczonka NPW, Aroca RF (2005) Mapping single-molecule SERRS from Langmuir-Blodgett monolayers, on nanostructured silver island films. J Raman Spectrosc 36(6–7):574–580
166. Mahmoud MA, Tabor CE, El-Sayed MA (2009) Surface-enhanced Raman scattering enhancement by aggregated silver nanocube monolayers assembled by the Langmuir-Blodgett technique at different surface pressures. J Phys Chem C 113(14):5493–5501
167. Mahmoud MA, El-Sayed MA (2008) Comparative study of the assemblies and the resulting Plasmon fields of Langmuir–Blodgett assembled monolayers of silver nanocubes and gold nanocages. J Phys Chem C 112(37):14618–14625
168. Mahmoud MA, El-Sayed MA (2009) Aggregation of gold nanoframes reduces, rather than enhances, SERS efficiency due to the trade-off of the inter- and intraparticle plasmonic fields. Nano Lett 9(8):3025–3031
169. Gopinath A, Boriskina SV, Reinhard BM, Dal Negro L (2009) Deterministic aperiodic arrays of metal nanoparticles for surface-enhanced Raman scattering (SERS). Opt Express 17(5):3741–3753
170. Haynes CL, Van Duyne RP (2001) Nanosphere lithography: a versatile nanofabrication tool for studies of size-dependent nanoparticle optics. J Phys Chem B 105(24):5599–5611
171. Zhang XY, Whitney AV, Zhao J, Hicks EM, Van Duyne RP (2006) Advances in contemporary nanosphere lithographic techniques. J Nanosci Nanotechno 6(7):1920–1934
172. Haynes CL, McFarland AD, Smith MT, Hulteen JC, Van Duyne RP (2002) Angle-resolved nanosphere lithography: manipulation of nanoparticle size, shape, and interparticle spacing. J Phys Chem B 106(8):1898–1902
173. Fromm DP, Sundaramurthy A, Schuck PJ, Kino G, Moerner WE (2004) Gap-dependent optical coupling of single "Bowtie" nanoantennas resonant in the visible. Nano Lett 4(5):957–961
174. Fromm DP, Sundaramurthy A, Kinkhabwala A, Schuck PJ, Kino GS, Moerner WE (2006) Exploring the chemical enhancement for surface-enhanced Raman scattering with Au bowtie nanoantennas. J Chem Phys 124(6):061101
175. Li K, Clime L, Tay L, Cui B, Geissler M, Veres T (2008) Multiple surface plasmon resonances and near-infrared field enhancement of gold nanowells. Anal Chem 80(13):4945–4950
176. Abernathy HW, Koep E, Compson C, Cheng Z, Liu ML (2008) Monitoring Ag-Cr interactions in SOFC cathodes using Raman spectroscopy. J Phys Chem C 112(34):13299–13303
177. Su KH, Wei QH, Zhang X (2006) Tunable and augmented plasmon resonances of Au/SiO_2/Au nanodisks. Appl Phys Lett 88(6):063118
178. Qin LD, Park S, Huang L, Mirkin CA (2005) On-wire lithography. Science 309(5731):113–115

179. Banholzer MJ, Li SZ, Ketter JB, Rozkiewicz DI, Schatz GC, Mirkin CA (2008) Electrochemical approach to and the physical consequences of preparing nanostructures from gold nanorods with smooth ends. J Phys Chem C 112(40):15729–15734
180. Wei W, Li SZ, Millstone JE, Banholzer MJ, Chen XD, Xu XY, Schatz GC, Mirkin CA (2009) Surprisingly long-range surface-enhanced Raman scattering (SERS) on Au-Ni multisegmented nanowires. Angew Chem Int Edit 48(23):4210–4212
181. Qian XM, Nie SM (2008) Single-molecule and single-nanoparticle SERS: from fundamental mechanisms to biomedical applications. Chem Soc Rev 37(5):912–920
182. Hu JA, Zhang CY (2010) Surface-enhanced Raman scattering technology and its application to gene analysis. Prog Chem 22(8):1641–1647
183. Golightly RS, Doering WE, Natan MJ (2009) Surface-enhanced Raman spectroscopy and homeland security: a perfect match? ACS Nano 3(10):2859–2869
184. Patel BD, Mehta PJ (2010) An overview: application of Raman spectroscopy in pharmaceutical field. Curr Pharm Anal 6(2):131–141
185. Shafer-Peltier KE, Haynes CL, Glucksberg MR, Van Duyne RP (2003) Toward a glucose biosensor based on surface-enhanced Raman scattering. J Am Chem Soc 125(2):588–593
186. Lyandres O, Van Duyne RP, Walsh JT, Glucksberg MR, Mehrotra S (2010) Prediction range estimation from noisy Raman spectra with robust optimization. Analyst 135(8):2111–2118
187. Maher RC, Maier SA, Cohen LF, Koh L, Laromaine A, Dick JAG, Stevens MM (2010) Exploiting SERS hot spots for disease-specific enzyme detection. J Phys Chem C 114(16):7231–7235
188. Mazhar D, Ngan S, Waxman J (2006) Improving outcomes in early prostate cancer: part I – adjuvant treatment. BJU Int 98(4):725–730

Immunoassays and Imaging Based on Surface-Enhanced Raman Spectroscopy

11

Dae Hong Jeong, Gunsung Kim, Yoon-Sik Lee, and Bong-Hyun Jun

1 Definition of the Topic

This topic covers recent developments in technologies related to biomolecular detection, including bioassay and imaging based on surface-enhanced Raman spectroscopy (SERS). Bioassay includes molecular detection technologies that monitor antigen-antibody association (immunoassay) and sequence-specific interactions between two and more complementary strands of nucleic acids (hybridization assay) as well as other molecules. Simultaneous multiple target detection and imaging (multiplexing) of samples such as cells and tissues is discussed as one of the unique features of SERS-based detection technology.

2 Overview

Enzyme-linked immunosorbent assay (ELISA) is currently the most popular and common method for immunoassays. Enzymatic amplification, fluorescent antibody assays, evanescent wave interference methods, and electrochemical methods have been developed as alternatives during recent decades. Current successes in related nanotechnologies and instrumentation have led to the development of biomolecular detection systems based on SERS with higher sensitivity and more clear visualization of bio-analytes. In an intrinsic configuration, the spectra or images of the targets on cellular surfaces or in cells are directly measured and analyzed by Raman spectroscopic methods. In an extrinsic configuration, binding, monitoring, and quantification of the targets are performed by indirect measurement

D.H. Jeong (✉) • G. Kim
Department of Chemistry Education, Seoul National University, Seoul, Korea

Y.-S. Lee • B.-H. Jun
School of Chemical and Biological Engineering, Seoul National University, Seoul, Korea

Challa S. S. R. Kumar (ed.), *Raman Spectroscopy for Nanomaterials Characterization*,
DOI 10.1007/978-3-642-20620-7_11, © Springer-Verlag Berlin Heidelberg 2012

technologies utilizing SERS signature-encoded metal nanoparticles. Especially, Raman scattering-based technologies have been recognized as very strong tools for multiplexed detection and imaging owing to their very sharp and narrow bandwidth that allows broad multiplexing within a narrow optical range as well as ultrahigh sensitivity for detection due to surface-enhancement of Raman scattering with properly engineered metal nanostructures.

Biomolecular detection based on SERS can be categorized into two parts. One is for measurement and quantification of trace bio-targets such as proteins, DNAs, peptides, and pH in cells. The other is for imaging of cells, tissues, and microorganisms in cells. Since targets of interest in biological systems are multiple in many cases, multiplexing can reduce the time and costs of characterization and analysis. In addition, applicability of SERS-based detection is expanded by incorporating multifunctionality to SERS probe particles.

3 Introduction

Raman spectroscopy, based on vibrational transitions, provides information about the molecular structure, conformation, dynamics and interactions and so on. These features give the opportunity for the recognition of biological molecules and imaging of cells and tissues. However, Raman spectroscopy was not an attractive bio-analytical methodology in the early days because the Raman scattering cross section is approximately 10 orders of magnitude smaller than the absorption and fluorescence cross section of molecules. The sensitivity of Raman spectroscopy is insufficient for real applications even when high laser power is used.

However, the low sensitivity problem of Raman spectroscopy can be overcome by the optical phenomenon called surface-enhanced Raman scattering (SERS), which originates from the conjugation of molecules with the novel metal nanomaterials such as gold or silver. After the first report of such a phenomenon in 1974 [16], several techniques were developed for a wide variety of applications. Current development in nanotechnology has accelerated the researches about SERS to realize accurate, sensitive, selective, and practical sensing platform in several areas including biomolecular detection.

Nowadays SERS has become a valuable tool in the biomedical field. Because the SERS technique offers some superior advantages such as label-free, high sensitivity, and multiplexing capability, a lot of new analytical techniques based on SERS have been developed for high-throughput screening of biological molecules as well as immunoassay and bio-imaging or direct detection of trace chemicals and molecules.

Here, we describe the bio-application of the SERS techniques focused on immunoassay and imaging. It covers the developments in nanostructures for substrates, configurations of measurements, imaging methods, and bio-conjugation technologies.

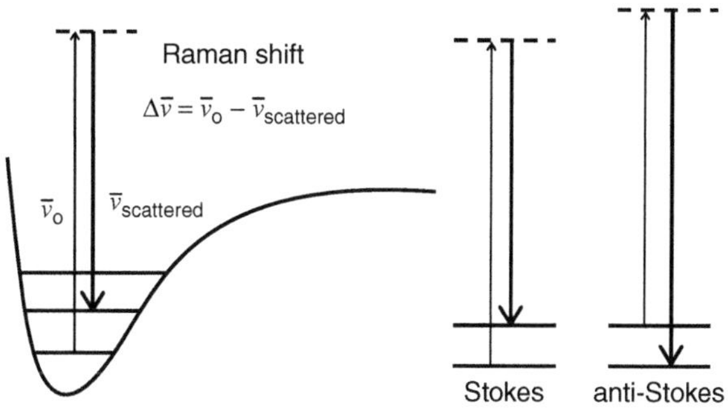

Fig. 11.1 Schematics of vibrational Raman process. Stokes process gives scattered lights that are lower in energy than the incident light, and anti-Stokes process gives scattered lights that are higher in energy than the incident light

4 Experimental and Instrumental Methodology

4.1 Raman Spectroscopy

Raman spectroscopy is a method for measuring the spectrum of inelastically scattered lights by molecules and materials. The "Stokes" shifted lights have lower energy than incident lights, and the "anti-Stokes" shifted have higher energy. The amount of the energy shift corresponds to the energy gap between the initial and final states of molecules (or materials) by the Raman process. Hence, depending on the energy of incidence, electronic, vibrational, and rotational states can be involved. In most cases of biological application, Raman spectroscopy refers to vibrational Raman spectroscopy, and the "Stokes" shifted spectrum is utilized [1] (Fig. 11.1).

Raman peaks in the spectrum are displayed as frequency shifts from the incident laser-line, or $\Delta\bar{\nu} = \bar{\nu}_0 - \bar{\nu}$. Each peak corresponds to the energy of a vibrational normal, which depends on molecular structure as well as the characteristics of chemical bonds comprising each normal mode. Hence, Raman spectrum is called the "molecular fingerprint" of the molecules and materials. Raman spectra of DNA and proteins, for example, contain rich information on their chemical bonds and structures. The Raman spectrum not only provides information about the structure, conformation, and identity of the sample but also the dynamics and interactions between biomolecules such as protein folding and DNA-protein interactions.

Raman spectroscopy can be used to detect normal modes of target molecules and also to monitor spectra of Raman labels that are used for one of the spectroscopic bar-codes. Raman bands in the vibrational Raman spectrum have intrinsically narrow bandwidths of ca. 10 cm^{-1}, which, for example, correspond to less than 0.5 nm width in the visible region below 800 nm. The fluorescence of dye molecules has a broad bandwidth of 100 nm more or less. Hence, spectral overlap between fluorescence bands is inevitable and limits their use for multiplexed analysis. Quantum dots (QDs) have narrower bandwidth than dye-based fluorescence but still have broad bands that are several tens of nanometers. Light scattering of noble metal nanoparticles caused by surface plasmon resonance is also

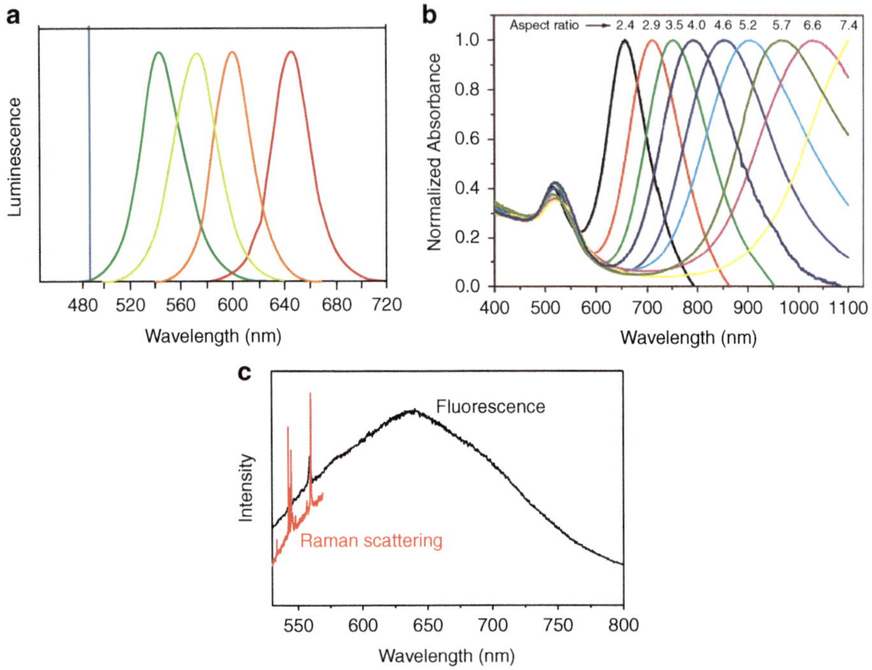

Fig. 11.2 Optical spectra of various optical probes. (**a**) Photo-luminescence spectra of Q-dots (from the work by Michalet et al. [2]), (**b**) scattering spectra of gold nanoparticles with different aspect ratio (from the work by Jain et al. [3]), and (**c**) fluorescence spectrum of FITC (*black line*) and SERS spectrum (*red line*) of benzene thiol with 514.5 nm photo-excitation (from the work by Jun et al. [4]). Raman band has much narrower bandwidth than the others

considered as a good tool for tunable and non-bleaching labeling. However, the resulting bandwidths are also close to 100 nm. This comparison exhibits that Raman spectroscopy has more than 100 times the capacity for multiplex labeling than other optical technologies in the visible range (Fig. 11.2).

Photo-excitation of molecules to their excited electronic states can lead to photochemical modification into other molecules, which then causes decay of fluorescence intensity with illumination time. Further, fluorescence signals are subject to environmental changes since the lifetimes of excited molecules can be quenched by surrounding molecules. Vibrational Raman scattering, however, involves transitions between vibrational energy levels during the Raman process, unless the energy of incident laser lights is tuned to a resonance condition with an electronic transition. Hence, molecules have little chance to undergo photochemical modification or intensity fluctuations due to the quenching of vibrational excited states. Non-photo-bleaching and less signal sensitivity on environmental change make Raman spectroscopy a unique and advantageous tool for quantitative bio-labeling compared to luminescence- and isotope-based labeling tools, which suffer intensity fluctuations or extrinsic and intrinsic decays.

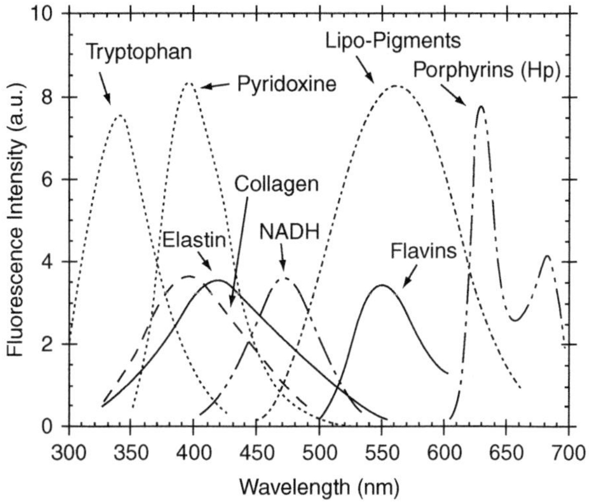

Fig. 11.3 Emission spectra of principal endogenous fluorophores (From the work by Wagnieres et al. [5])

One important issue regarding bio-detections using optical tools is the choice of spectral regions. Better sensitivity in detection as well as imaging with higher contrast depend on the signal-to-noise ratio. Fluorescence emission, which arises from endogenous fluorophores, is an intrinsic property of cells and is called "auto-fluorescence," appearing in the entire range of the visible region [5, 6]. The majority of cell auto-fluorescences originate from mitochondria and lysosomes. In tissues, the extracellular matrix, such as collagen and elastin, shows a relatively high quantum yield [7, 8]. Imaging techniques based on the monitoring of auto-fluorescence have been successfully utilized in order to obtain information about the morphological and physiological state of cells and tissues. However, escape from auto-fluorescence is necessary for applications such as the monitoring of cellular composition following malignancy or classification of cells. Unless resonance Raman techniques are employed, the laser-line for Raman excitation can be almost freely chosen from the UV to near-infrared spectrum in order to achieve the lowest background level from auto-fluorescence of the samples. The near-infrared region currently draws interest for Raman analysis of cells and tissues benefiting of minimal auto-fluorescence as well as less laser-induced damage and deeper penetration into the tissue. Additionally, Raman spectrum using laser-lines of UV to the near-infrared region suffers little spectral interference from water, which allows Raman spectroscopy to be utilized for bio-analysis (Fig. 11.3).

4.2 Surface-Enhanced Raman Scattering (SERS)

Despite all the merits mentioned above, the Raman process is so inefficient that only a small fraction of photons are inelastically scattered, resulting in the Raman scattering cross section to be lower than 10^{-10} of the typical fluorescence cross section [9]. This feature limits the application of Raman spectroscopy

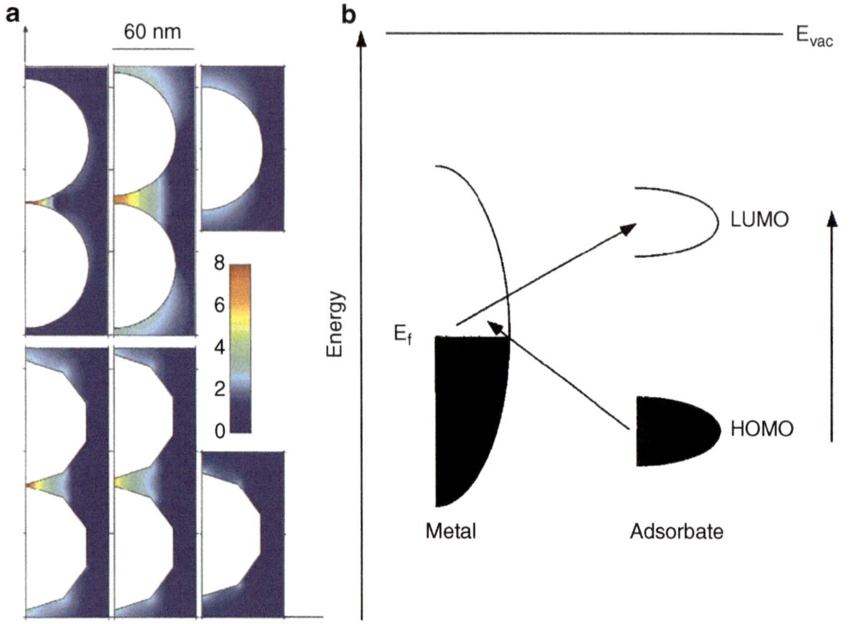

Fig. 11.4 Schemes demonstrating two SERS mechanisms. (**a**) Local field concentration near the gap between two particles [19]. The *white spheres* and *polygons* are silver, and the color scale bar indicates the electro-magnetic field enhancement. The separation distance is 1 and 5.5 nm for the first and second columns, respectively. The wavelength of the incident field is 514.5 nm with vertical polarization, and (**b**) Possible charge transfer excitations between molecule and metal [22]. The occupied and unoccupied molecular orbitals are broadened by the molecule-metal interactions. Metal-to-molecule and molecule-to-metal charge transfers are indicated by *arrows*

for bio-detection. However, this disadvantage can be overcome by using an optical phenomenon called SERS, which can observe even a single molecule adsorbed on a specially engineered nanostructure of noble metals such as gold and silver [10–13]. This phenomenon was discovered some 35 years ago [14–17] and has been continuously studied in order to understand its nature and origin and more recently to engineer optimal nanostructures for SERS [18]. The extraordinary enhancement is attributed to two mechanisms: a highly localized effect of the electromagnetic field near the metal surface [19, 20] and a chemical resonance effect [21–23] (Fig. 11.4). The former explains 10–11 orders of magnitude of signal enhancement while the latter explains 2–4 orders of magnitude.

This huge enhancement of Raman scattering ensures that Raman spectroscopy is effective for ultrasensitivity and multiplexing in the field of bio-detections. Another interesting point for SERS is the linear dependence of SERS intensity on the power of incident light despite the extraordinary nonlinear effect of signal enhancement [18]. This suggests that SERS technology can be used potentially for the quantitative measurement of analytes with ultrahigh sensitivity. SERS also depends on the wavelength of the incident laser lights. However, the resonance wavelength is determined by the surface plasmon of the metal nanostructure. The resonance conditions of the surface plasmon can be controlled and tuned for the purpose of application by choosing a metal such as gold or silver and by controlling their size and shape [24–26]. By considering application preferences such as avoidance of auto-fluorescence and sensitivity, metal species and nanostructures can be chosen and controlled for tuning of the resonance wavelength.

4.3 Nanostructures for SERS Substrates

An unexpected, large increase in Raman scattering was discovered while measuring the Raman signals of chemicals adsorbed on a silver substrate that was electrochemically roughened in order to increase surface area and consequently the Raman intensity [16]. Ever since this phenomenon was recognized as surface-enhanced Raman scattering of molecules adsorbed on roughened coinage metal surfaces, there have been a huge number of reports on increasing signal enhancement of SERS substrates that utilized many different metals such as gold, silver, copper, aluminum, etc., as well as various structures including electrochemically roughened structures, lithographically fabricated periodic structures, colloidal nanoparticles, chemically etched films and cold-deposited islands [27, 28]. Due to the pioneering works of S. Nie and K. Kneipp, independently reporting that the Raman enhancement factor of molecules adsorbed on isolated individual particles or small aggregates can be increased up to 10^{14} times of normal Raman scattering [10, 11], intensive studies have followed to explain this enormous enhancement.

The local electric field near the coinage metal nanoparticle is enhanced by a factor, g, as in $E_{loc} = g \times E_{inc}$, when the energy of the incident lights is resonant with the surface plasmon of the metal nanoparticle. The electric field of scattered light is expressed as the product of the molecular polarizability and electric field of the incident lights that the molecules experience ($E_{sca} = \alpha \times E_{loc}$). A Raman-scattered field is enhanced similarly. Therefore, if the wavelength of Raman-scattered light is close to that of the incident light, I_{SERS} will be enhanced by a factor proportional to the forth power of the local field enhancement described as $I_{SERS} \propto |\alpha|^2|g|^4 I_{inc}$ [18]. Theoretical studies by Xu and Käll [19, 20] and experimental studies [12, 29] have supplied visual guides for nanostructure fabrication of SERS hot spots, illustrating that electromagnetic fields are highly concentrated at certain positions such as the gap between nanoparticles and sharp tips of triangular nanoparticles and that the field concentration also strongly depends on polarization of incident lights.

Based on these considerations, a great deal of current research on SERS has focused on the controlled and reproducible fabrication of metallic nanostructures that produce "hot spots" in which the molecules are appropriately and predictably located for large Raman enhancement. Several strategies have been proposed for engineering such hot structures for chemical sensing applications in a reproducible and controllable manner. These include structures such as triangular nanoparticle array [30], silver nanowire bundles [31, 32], Ag nanoparticle–assembled silica

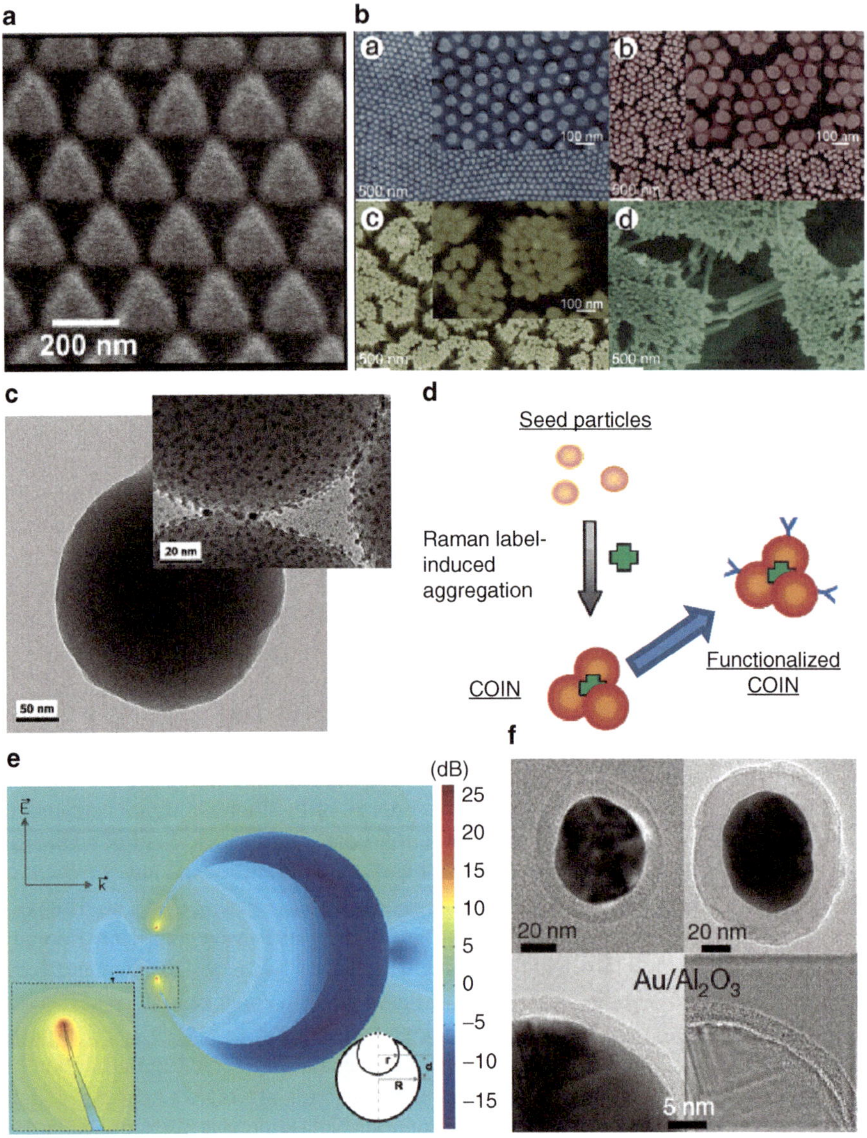

Fig. 11.5 (continued)

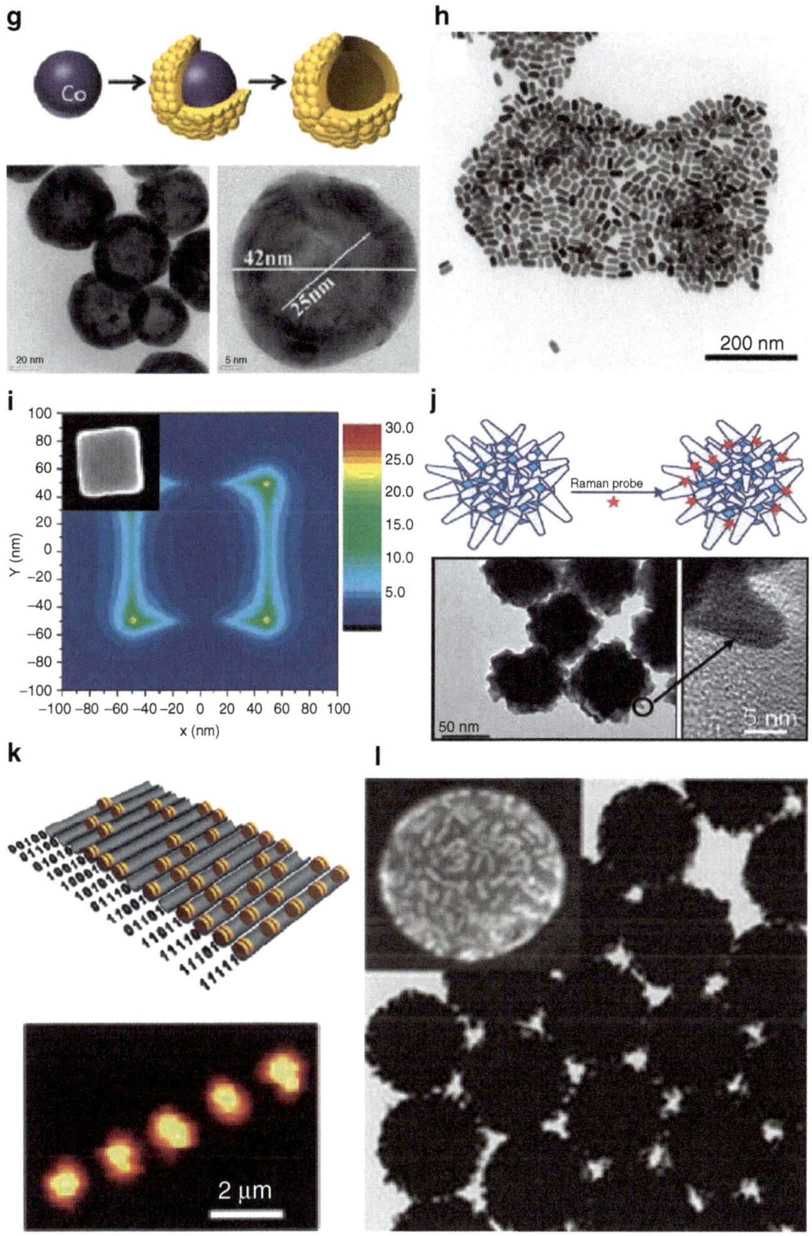

Fig. 11.5 Rationally fabricated SERS substrates. From (**a**) to (**l**), triangular nanoparticle array [30], silver nanowire bundles [32], Ag nanoparticle–assembled silica nanoparticle (SERS dots) [33], metal nanoparticle aggregates (COINs) [34], gold nanocrescent [49], gold nanoparticles with thin oxide shells [50], hollow-type gold nanoparticles [35–37], gold nanorods [38–40], nanocubes [41, 42], flower-like gold nanoparticles [43], nanodisks [46], and gold nanorods immobilized on silica nanoparticles [48]

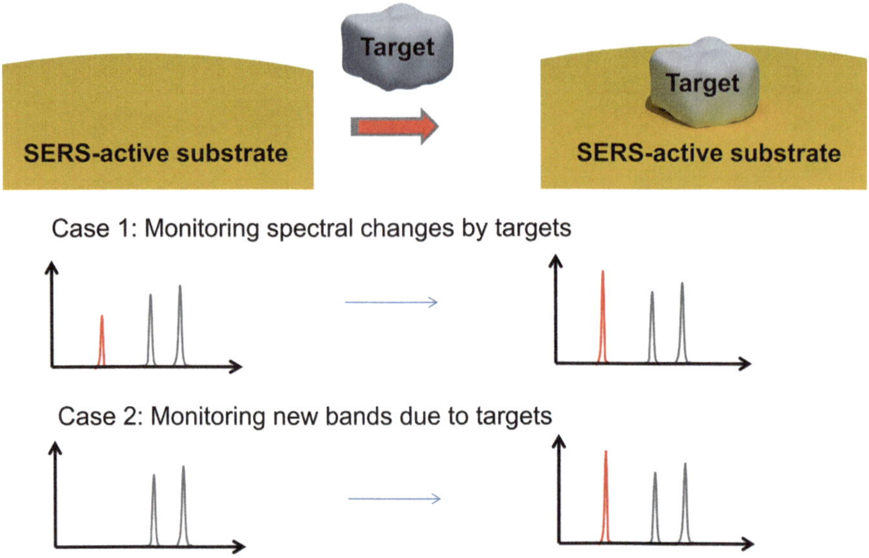

Fig. 11.6 In intrinsic configuration, SERS signals from the targets are measured. The targets can be monitored by relative intensity changes in the spectrum caused by targets (Case 1) and also by appearance of new SERS bands characteristic of the targets (Case 2)

nanoparticle (SERS dots) [33], small aggregates of Ag nanoparticles (COINs) [34], hollow-type gold nanoparticles [35–37], gold nanorods [38–40], polydimensional structure-like nanocubes [41, 42], flower-like gold nanoparticles [43], star-shaped gold nanoparticles [44], dimeric nanospheres [45], nanodisks [46], encapsulated silver nanoparticle aggregates [47], gold nanorods immobilized on silica nanoparticles [48], gold nanocrescent [49], and gold nanoparticles with thin oxide shells [50] (Fig. 11.5).

The selection of specific nanostructures relies on various aspects, and the choice also influences the wavelength of a laser-line to be employed. The surface plasmon resonance of spherical silver nanoparticles lies around 400 nm while that of their dimers and aggregates shifts to red, covering the whole visible region and showing maximum SERS enhancement around 500 nm. Gold nanoparticles have their surface plasmon resonance around 500 nm while their dimers and aggregates shift to red from 500 nm to near-infrared. Therefore, in many studies, SERS measurements were performed using a green laser-line such as a 514.5 nm laser-line for silver substrates and a red laser-line such as a 632.8 nm laser-line for gold substrates, respectively. The choice of a laser-line can be made even broader over the visible range by controlling the shapes of the silver or gold nanostructures [51–54]. With the help of current nanotechnologies, we can choose or develop nanostructures for SERS substrates for optimal sensitivity during detection, reduction of background noise under given experimental conditions, and designed functions such as multiplexing.

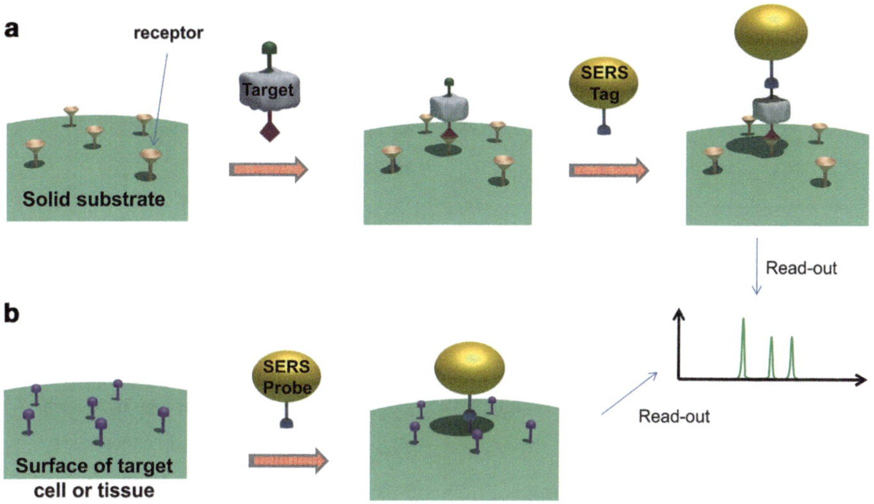

Fig. 11.7 In an extrinsic configuration, SERS signals from SERS probes indicate the presence of targets. A typical schematic of sandwich assay (**a**) and targeting on cells and tissues (**b**)

4.4 Configuration of Measurements (Intrinsic and Extrinsic Approaches)

The application of SERS for bio-detection can be classified into two approaches: intrinsic and extrinsic configurations. In the intrinsic configuration, the SERS spectra of targets themselves are measured as shown in Fig. 11.6. The target can be captured by receptors on a SERS-active substrate such as complimentary DNA strands to target DNA. In this case, the SERS spectrum of the complimentary DNA strands is observed before capturing the targets, and after the targets are bound the SERS spectral changes are compared. Targets can bind to the bare surfaces of the SERS-active substrates with affinities for metal surfaces. In this case, new SERS bands originating from the targets appear once the targets are bound to the SERS substrate. However, generally bare surfaces lack specificity for target binding.

In the extrinsic configuration, targets are monitored indirectly by measuring the signals of SERS-encoded particles, which are composed of SERS-active metal nanostructures, Raman reporter molecules linked thereon, and receptor molecules for specific targets. The particles are employed to assay target biomolecules that exist, for example, on the surfaces of cells and tissues, captured on biochips, or floating in aqueous medium such as DNA in medium. This approach is useful for targets that cannot be monitored successfully with intrinsic approaches. For example, large molecules such as proteins and lipids have spectral features that are not distinctive from one another due to similarities in chemical composition. They are not effectively placed at SERS hot spots due to their large size, resulting in loss of SERS detection sensitivity.

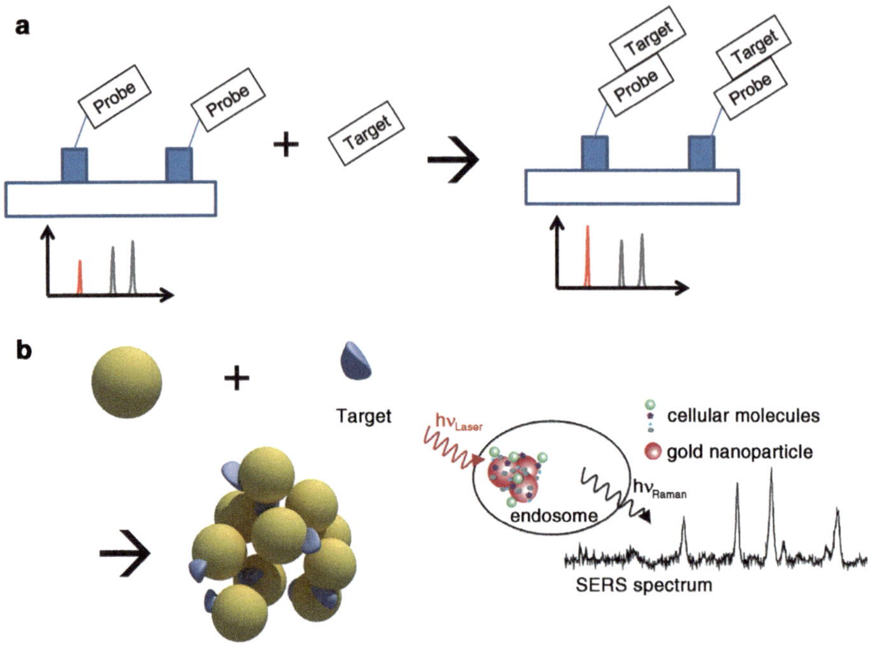

Fig. 11.8 Examples of direct measurements based on SERS. Conceptual schematics in (**a**) is re-drawn based on the work of Green et al. [55]. Sensitive SERS substrate gives different SERS signals for the probe only and probe/target hybrid. For example, Green et al. used DNA strands as the model compound. The example in (**b**) is from the work of Kneipp et al. [59]. This approach uses the hot spot generation from the colloidal aggregation induced by specific components such as salt, endosome, etc.

Typical utilization of SERS-encoded particles is illustrated in Fig. 11.7. SERS-encoded particles can be applied to sandwich assays based on solid substrates that have receptors such as antibodies. Once targets are captured by the substrate, the SERS-encoded particles are engaged as SERS labels for specific binding to the targets followed by readout of the SERS spectrum. SERS-encoded particles that contain receptor molecules such as antibodies or biomarkers specific to certain targets can also bind targets on tissue and cell surfaces. This strategy also applies to floating cells in aqueous medium.

4.5 Direct Measurements

Intrinsic approaches focus on the detection of trace of chemicals and molecules with unique Raman spectra that are small enough to bind at SERS hot spots such as tips and crevices of metal nanoparticles. Direct measurement of small chemicals has a long history in SERS. The SERS phenomenon itself was discovered by the measurement of pyridine molecules adsorbed to a silver electrode that had been electrochemically roughened to increase the surface area and therefore the signal [16].

There have been numerous studies reporting biomolecular detection using direct SERS measurements.

One type of intrinsic measurement is based on a solid plate substrate as shown in Fig. 11.8a. An example of this approach is the label-free detection of oligonucleotide molecules by measuring their SERS spectra using a silver-island substrate [55]. This approach could distinguish differences in the SERS spectrum of probe DNA strands and that of probe/target DNA strands after hybridization of target DNA strands. Another type of intrinsic measurement is based on non-immobilized nanoparticles as shown in Fig. 11.8b. For example, sub-micromolar detection of DNA/RNA mononucleotides was reported by Bell and Sirimuthu [56]. Briefly, they mixed a target solution with silver colloidal nanoparticles and aggregation-inducing salts, expecting exchange reactions between the targets and pre-adsorbed citrate ions as well as aggregation of silver nanoparticles for activation of SERS. Their choice of $MgSO_4$ salt as the aggregation-inducing agent caused colloidal aggregation for SERS hot spots generation and prevented deactivation of local field enhancement, which was commonly observed for salt-induced aggregation using chloride ions. Another group monitored enzymatic activity by measuring the surface-enhanced resonance Raman spectroscopy (SERRS) spectrum of 8-hydroxylquinolinylazo dyes, which was transformed from non-SERRS-active structures to SERRS-active structures by lipase enzymes [57]. Catecholamine secretion by rat phaeochromocytoma (PC12) cells is spatially resolved by measuring the SERS spectra of dopamine and noradrenaline with silver colloidal nanoparticles [58]. The direct insertion of metal nanoparticles into cells can be performed as a different application of non-immobilized SERS probes. Kneipp et al. injected gold nanoparticles into cultured eukaryotic cells and successfully monitored SERS spectra of nanoparticle aggregates caused by subcellular components such as endosomes [59].

4.6 Immunoassays Using Raman-Labels

As a conventional immunoassay technology, SERS-based immunoassay consists of three steps as shown in Fig. 11.7a. First, a substrate with capture antibodies is prepared. Second, targets are captured by dipping the substrate into sample solution. Third, treatment of SERS labels is performed, followed by reading of SERS signals. The key steps in the assay are the preparation of good Raman labels and maintenance of the biological activity of the capture antibodies. Various systems of substrate-based immunoassays using Raman spectroscopy have been developed in a similar manner.

Figure 11.9 illustrates a representative system of SERS-based immunoassays. Rohr et al. first proposed an SERS-based immunoassay using dye-labeled antibodies and silver-island films coated with a capture antibody [60]. In their system, a silver-island film is used for the base substrate linked to the capture antibody as well as the SERS-active substrate that enhances Raman signals of reporter molecules when the reporter dye–linked antibody conjugate is bound to the captured

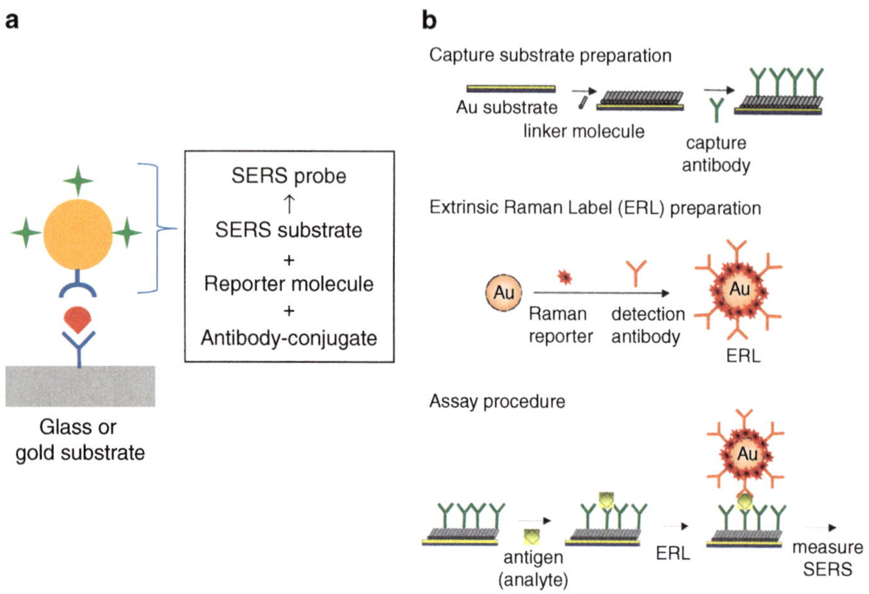

Fig. 11.9 A representative immunoassay scheme using SERS. (**a**) A general concept scheme. SERS probe is composed of SERS substrate, reporter molecule, and antibody-conjugate, and (**b**) An example of immunoassay procedure by Porter et al. [62]. This configuration uses plasmonic coupling between gold nanoparticles and gold film to enhance probe signal

antigen and the reporter dye gets close to the roughened silver surface. Dou et al. demonstrated a new system involving antibody immobilized on a solid substrate and antibody conjugate with azo compounds as reporter molecules adsorbed on silver nanoparticles [61]. Except for the fact that they utilized an enzymatic reaction to introduce reporter molecules, their system is a prototype of the most current chip-based immunoassays using SERS technology (Fig. 11.9a). More recently, Porter et al. proposed a very successful readout method for SERS-based immunoassay employing a sandwich-type assay approach [62]. This method exploits gold film as a base substrate which contains the capture antibody and a gold nanoparticle as an SERS probe which contains capture antibody and reporter molecules. Plasmonic coupling between gold nanoparticles and the gold film was utilized as signal-enhancement strategy [63].

4.7 SERS-Based Imaging

Technical requirements for SERS-based imaging is almost the same as the detection technologies discussed above, but imaging technology rather focuses on visualization of targeted area in cells and tissues. UV-resonance Raman spectroscopy utilizes resonance-enhanced Raman signal from certain chromophores for imaging of targets but is only applicable for limited species with a proper optical resonance. Near-infrared Raman spectroscopy minimizes background signals by using spectral region that is free from auto-fluorescence and uses spectral changes of certain functional groups such as amide and C–H functional groups. SERS has been successfully used as an imaging technology by directly measuring targets such as the case of monitoring neurotransmitter release in cells with an aid of silver nanoparticles on top of cell membrane [58, 64–66], as well as by indirectly measuring SERS tags [33, 67, 68]. Intriguing advantages using SERS-based imaging include the broad applicability of SERS tags that basically uses optical signature of tags and biomolecular reactions that are utilized for binding and capturing targets and the possibility of multiplexed imaging for various targets. The following discussion is focused on SERS tags–based imaging.

Various SERS tags can be used for imaging. Each SERS tag is recognized by unique SERS bands which are characteristic and distinctive between reporter molecules. SERS-active nanostructures for SERS imaging can be categorized as a single particle system and an assembled particle one. For a single particle system, gold nanoparticle is mostly used. Currently, encapsulation of nanoparticle is preferred to protect SERS signal source and adsorbed reporter molecule for better biocompatibility, and to allow easy bio-conjugation at the particle surface. In many cases, spherical gold nanoparticle was mostly utilized as well as gold nanorods [69] in order to increase signal sensitivity and tune optical region of less auto-fluorescence and background signals. Hollow-type gold nanoparticles were also applied to the SERS label-based imaging [36].

In order to get higher signal sensitivity, strategies for assembling nanoparticles were utilized to make SERS "hot spots." As assembled nanoparticle systems, two approaches are worthy of note. Well-controlled silver nanoparticle aggregates with reporter molecules named COINs can give high SERS sensitivity attributed to local field concentration at or near the gaps between assembled nanoparticles [34, 70]. Nanoparticle assembly on silica core gives ensemble-averaged SERS signature as well as large signal intensity due to large surface area that is SERS active [33]. Silica core and shell leave room for additional functionalities such as incorporation of fluorescence dyes in shell [71, 72] or magnetic particle in core [73–75].

Binding between SERS tags and targets relies on biochemical reactions, such as antibody–antigen association, and DNA hybridization as in the case of sandwich immunoassay discussed in the previous section. However, while target capture to substrate and successive tagging by SERS labels are required in immunoassay detection (Fig. 11.7a), only one step of target capture by SERS tags is necessary in imaging (Fig. 11.7b). A more detailed scheme and an example of SERS-based imaging are illustrated in Fig. 11.10. The number of spectrally distinguishable SERS tags can be numerous, up to hundreds, by simply comparing Raman bandwidth within visible optical range from 400 to 700 nm. Three different SERS tags are chosen as examples in Fig. 11.10. These tags are encoded with SERS reporter molecules and bio-conjugated for specific biomarkers expressed on the surfaces of cells or tissues. Once the SERS tags bind to target molecules on cellular or tissue surfaces, SERS spectra are obtained point-by-point in a two-dimensional area of

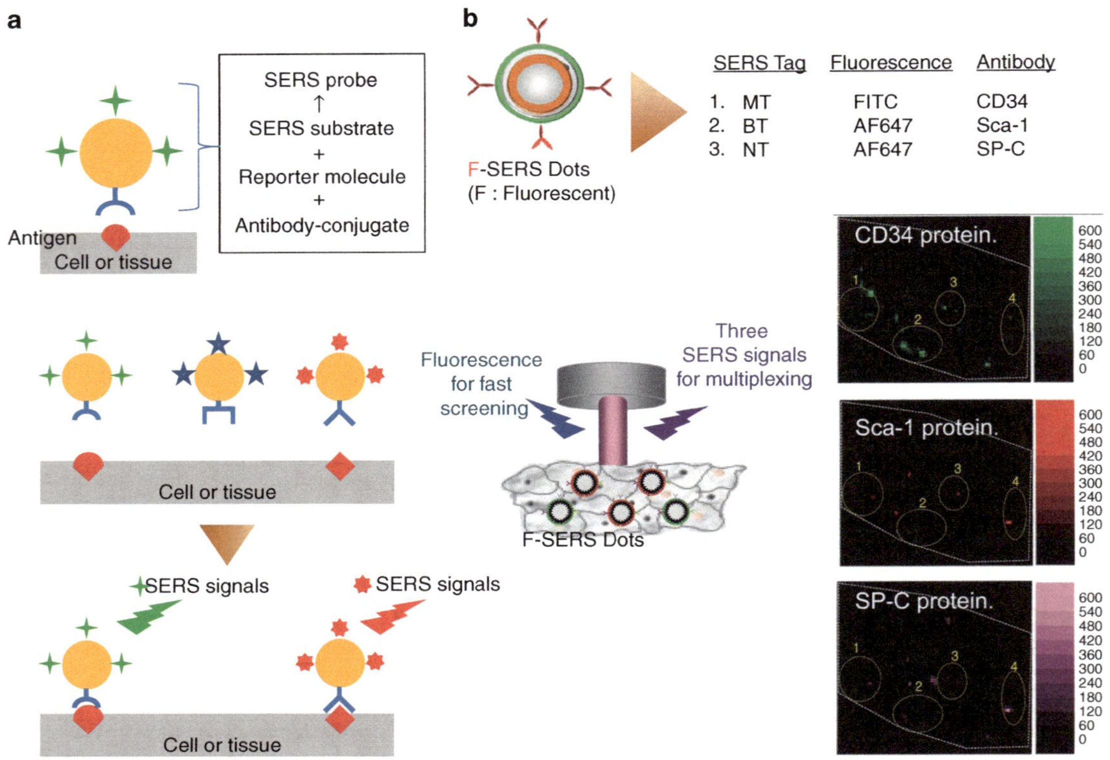

Fig. 11.10 A representative imaging scheme using SERS multiplexing. (**a**) A general concept scheme. Distribution of several targets can be imaged from a single scan by Raman intensity mapping of specific SERS band of each SERS tag, and (**b**) an example by Woo et al. [71]. The three different SERS intensity maps correspond to distributions of CD34, Sca-1, and SP-C proteins on tissue sample. They were obtained by single scan at the pre-selected area

interest. Since each SERS tag has distinguishable SERS bands, three different SERS intensity maps for each tag can be obtained simultaneously from a single set of Raman scans.

4.8 Bio-conjugation and Surface-Modification Technologies for Biological Applications

SERS probes have SERS-active nanostructures that can be bio-conjugated. They are either single nanoparticles or assembled nanoparticles of gold or silver metal, and reporter molecules are located close to or on the surfaces. They are bio-functionalized by adsorption of biomolecules such as proteins [76], DNAs [77, 78], and the other biomarkers [79] on the metal surface. In many cases, a metal affinity group like thiol-containing linkers is used to immobilize biomolecules to the metal surface as shown in Fig. 11.11. In the cases of encapsulated probes, biomolecules are linked to shell surfaces while reporter molecules are adsorbed on metal surfaces. Coupling biomolecules to the shell surface can be accomplished by several methods using carboxyl, amine, thiol, and hydroxyl group of biomolecules, which were well reviewed by Rusmini et al. [76]. Among many approaches, the utilization of active ester that is to be coupled with the amine group [80] and maleimide for coupling with thiols in biomolecule [33] are most widely used. These immobilization methods can be easily applied to carboxyl or amine group containing materials regardless of their core composition as shown in Fig. 11.11 (d $\sim$ e).

In the active ester method, the shell surface is functionalized with carboxyl group with or without spacer, and then activated by carbodiimides such as *N,N′*-dicyclohexylcarbodiimide (DCC), *N,N′*-diisopropylcarbodiimide (DIC), and 1-ethyl-3-(3-dimethylaminopropyl) carbodiimides (EDC)[70, 80]. Additives, such as *N*-hydroxybenzotriazole (HOBt) or *N*-hydroxysuccinimide (NHS), are often added to increase yields and decrease side reactions. Activated carboxyl group reacts easily with the amine group of biomolecules. When the shell is functionalized with the amine group, it can be directly converted to active ester using *N,N′*-disuccinimidyl carbonate (DSC) and 4-(dimethylamino) pyridine (DMAP) [80]. In the maleimide method, the shell surface is functionalized with the amine group, coupled with a bifunctional linker which contains maleimide group, and then reacted further with the thiol group of biomolecules forming sulfide bonding [33]. Since cysteines are not as abundant as lysines in biomolecules, random immobilization is less likely to occur.

In immunoassay and in vitro imaging, bovine serum albumin (BSA) is most widely used as a blocking agent for preventing nonspecific binding. Simply, BSA is treated to target substrate together with SERS tagging materials [33, 71, 74, 83]. However, this approach is hard to be applied in in vivo assay because of the weak affinity between SERS tagging materials and BSA, which can be detached from the tagging materials in in vivo condition. The surfaces of SERS tagging materials may also be modified with bio-inert, hydrophilic molecules such as polyethylene glycol

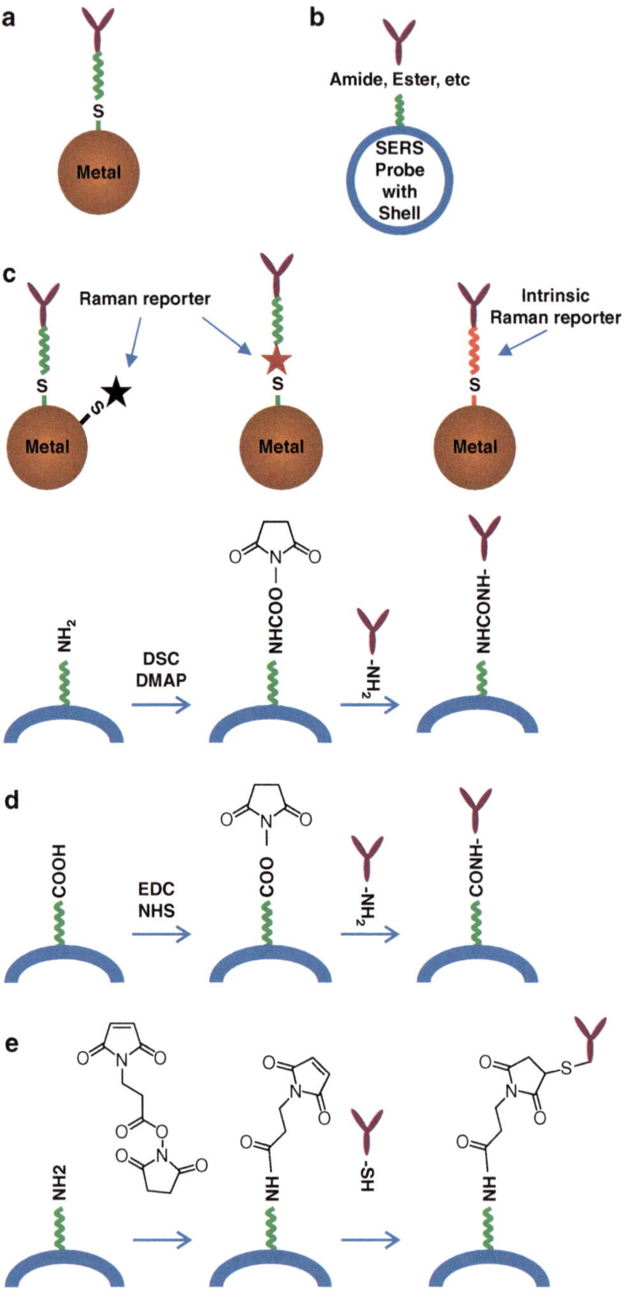

Fig. 11.11 Various methods of bio-conjugation. (**a**) Direct immobilization of the linker molecule to the metal surface, (**b**) composite formation of the metal/biocompatible shell hybrids, (**c**) examples of direct immobilization to metal surface [77, 81, 82], (**d**) examples of active ester utilization [70, 80], and (**e**) an example of maleimide-based bio-conjugation [33]

(PEG) to reduce nonspecific binding by a mechanism known as the "steric repulsion effect" [84]. Some groups have immobilized nanoparticles and BSA stably by cross-linking BSA using glutaraldehyde [70] or by using reduced BSA [43]; and other groups have used casein instead of BSA [80].

5 Key Research Findings

SERS-based bio-detection relies on the quality of SERS probes in respect of signal sensitivity, reproducibility, feasibility of quantitative analysis, and multiplexing ability. Recently, Lim et al. reported a synthetic method for SERS-active gold-silver core-shell nano-dumbbells (GSNDs) [85]. The nano gaps of GSNDs could be controlled to detect a single DNA molecule and they could be used as highly sensitive bio-labeling probes enabling quantification and multiplexing. Another innovative technique, so-called shell-isolated nanoparticle enhanced Raman spectroscopy (SHINERS), was reported by Li et al. [50]. Thin silica or alumina shell of 2 nm maintained a highly intensified local field to photo-excite target molecules on the shell surface and also prevented deactivation of surface plasmon of gold nanocore by target molecules that could adsorb on metal surface. This group successfully identified minute amounts of hydrogen on a single-crystal silicon wafer, probed the surfaces of yeast cells, and detected the insecticide parathion on an orange peel. These new techniques have the potential to be used for in situ analysis with a much greater level of sensitivity as a substrate for direct measurements and as an SERS tag for indirect measurements (Fig. 11.12).

Immuno-assay technology based on SERS has been developed by many research groups. Among them, the Porter group established a good strategy for immunoassay utilizing single gold nanoparticle with reporter molecules as SERS label, gold film as target capture substrate as well as SERS signal enhancing platform using plasmonic coupling between gold nanoparticles and the gold film [63], and readout method [62]. They could detect femto-molar prostate-specific antigen (PSA) [81].

The quality of the SERS probe is essential for the success of SERS-based immunoassays. In order to improve the quality of the SERS probe, there have been many attempts to optimize the choice of metal and the nanostructure. The SERS probe that is conjugated with antibody requires adsorption of reporter molecules. Co-adsorption methods for antibody and the Raman reporter with mixed monolayer have been widely used as in the case of Porter et al. [86]. While successful in simultaneous detection of multiple targets, Porter noticed a partial desorption of weakly adsorbed antibody as well as re-adsorption of the other Raman label. They proposed a bifunctional coating method to overcome this "cross talk" problem, using a bifunctional reporter molecule that had a reactive moiety for covalent coupling to an antibody [81]. These SERS tags are not shielded from the external environment and therefore are prone to spectral changes by molecular contact. Glass-coated, analyte-tagged gold nanoparticles (GANs) have been demonstrated by Mulvaney et al. [87], whereas encapsulated gold nanoparticles with PEG groups were demonstrated by Qian et al. [88] in order to

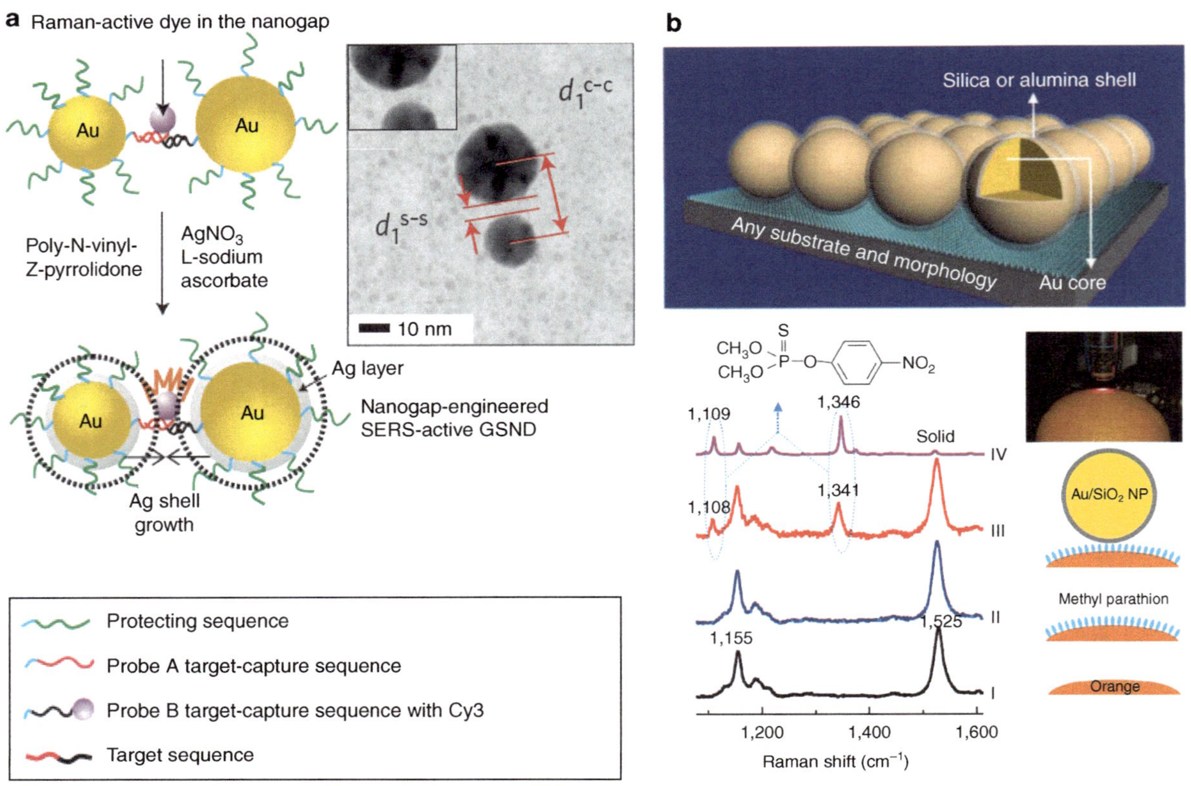

Fig. 11.12 Examples of highly sensitive SERS substrate. (**a**) SERS-active gold-silver core-shell nano-dumbbells (GSNDs) [85] and (**b**) shell-isolated-nanoparticle-enhanced Raman spectroscopy (SHINERS) [50]

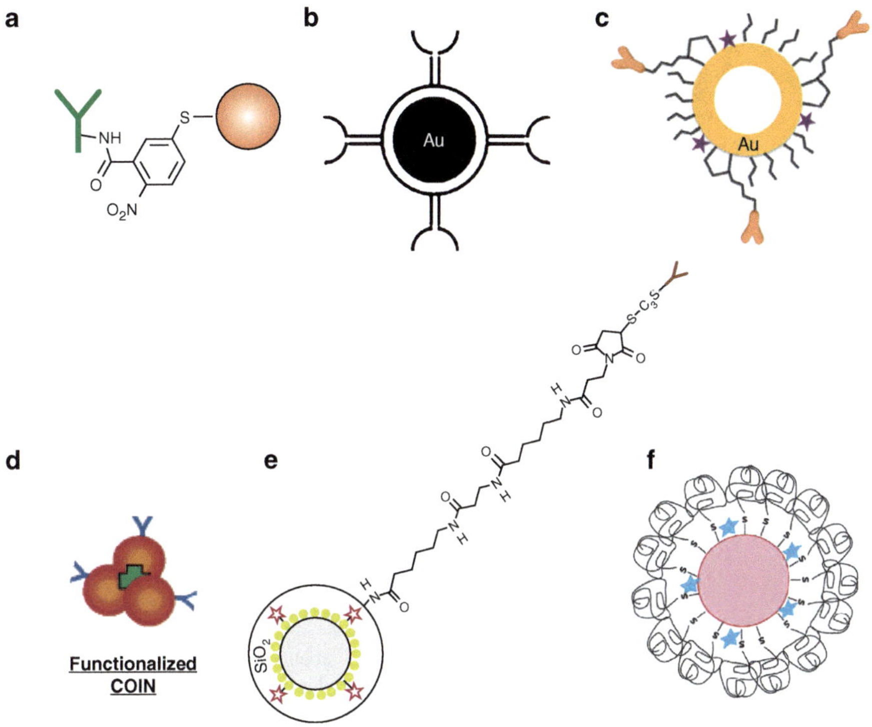

Fig. 11.13 Various types of SERS tags. (**a**) Utilization of bifunctional reporter [81], (**b**) glass-coated, analyte-tagged gold nanoparticles (GANs) [87], (**c**) hollow-type gold nanoparticle [36], (**d**) aggregates of a few silver nanoparticles (COINs) [34], (**e**) encapsulated SERS tag with silver nanoparticle-embedded silica core (SERS Dot) [33], and (**f**) pegylated gold nanoparticle [88]

overcome nonspecificity of silica encapsulation to proteins and cell surfaces [89]. Since hollow-type gold nanoparticles (HGNs) have hot spots located on a pinhole in the hollow particle structure, they result in strong enhancement of about ninefold higher than that of single nanoparticles [90]; HGNs have been demonstrated by Choo et al. [36].

While many of these SERS-tags are based on a single nanoparticle, there have been different approaches employing multiple nanoparticles within a single SERS tag. Clustering of a few silver nanoparticles with reporter molecules termed as composite organic-inorganic nanoparticles (COINs) confers higher signal strength by SERS "hot spots" between silver nanoparticles [34]. A silver nanoparticle-embedded silica core system was developed to enjoy a large ensemble-averaged SERS effect due to a large surface area with many SERS "hot spots" [33]. The surface was further encapsulated with a silica shell, and a bifunctional linker group (3-maleimidopropionic acid *N*-hydroxysuccinimide ester) was introduced for bio-conjugation (Fig. 11.13).

Several SERS tags have been successfully applied to imaging such as cells and tissues. Wabuyele et al. demonstrated the potential of SERS-based imaging

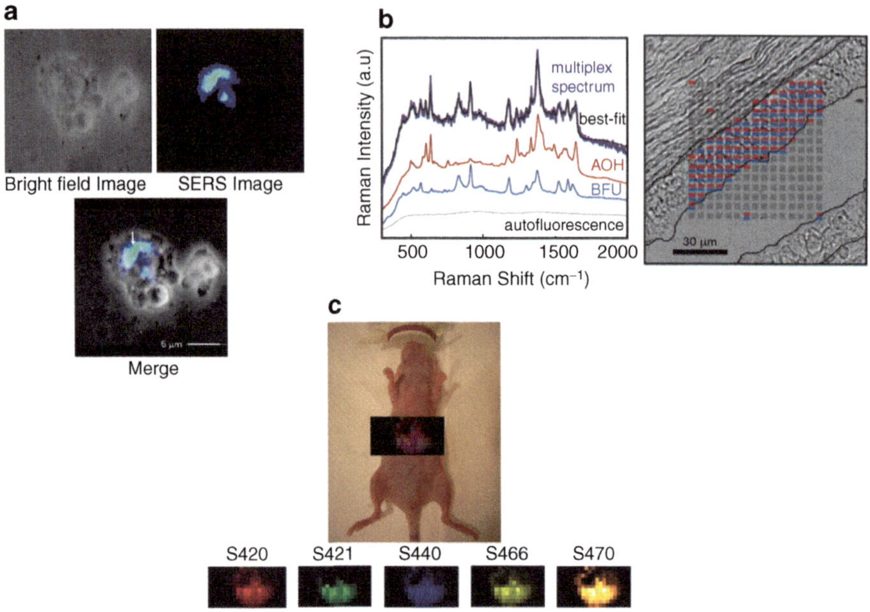

Fig. 11.14 Representative imaging studies using SERS labels. (**a**) An example of cells after passive uptake of silver nanoparticles linked with cresyl violet [91], (**b**) An example of simultaneous imaging of two biomarkers on tissue surfaces using COINs [70], and (**c**) An example of in vivo multiplex imaging [92, 93]

technology using cells that had passively uptaken silver nanoparticles linked with cresyl violet [91]. Schlücker et al. reported in situ detection of antigens on tissue surfaces by measuring SERS signals from gold nanoparticles with aromatic Raman labels covalently linked to the corresponding antibodies [67]. Kim et al. prepared silica-encapsulated silver nanoparticle assemblies on silica core and obtained specifically targeted cell images for HER2 and CD10 biomarkers [33]. Sun et al. detected PSA and CK18 on tissue surfaces simultaneously using two different SERS labels composed of silver nano-aggregates with reporter molecules [70]. Hu et al. reported imaging of membrane proteins on cells using CN SERS band of a cyano-labeled SERS probe [64]. In order to reduce measurement time in large area imaging such as for tissue samples, spectrally separated fluorescence from Raman bands was introduced to the silica-encapsulated silver nanoparticle assemblies and they were successfully applied for fast locating target area with fluorescence followed by multiplex SERS imaging of BAX and BAD antibodies for apoptosis in tissue sample [72]. The same group applied fluorescent SERS labels to find cancer stem cells in lung by tracking three biomarkers simultaneously as well as screening of the targeted area by fluorescence staining [71]. The Choo group also presented SERS imaging for HEK293 cells expressing PLCγ1 using Au/Ag core-shell nanoparticles [83]. The Nie group reported biocompatible and nontoxic nanoparticles for in vivo tumor targeting and detection based on pegylated gold

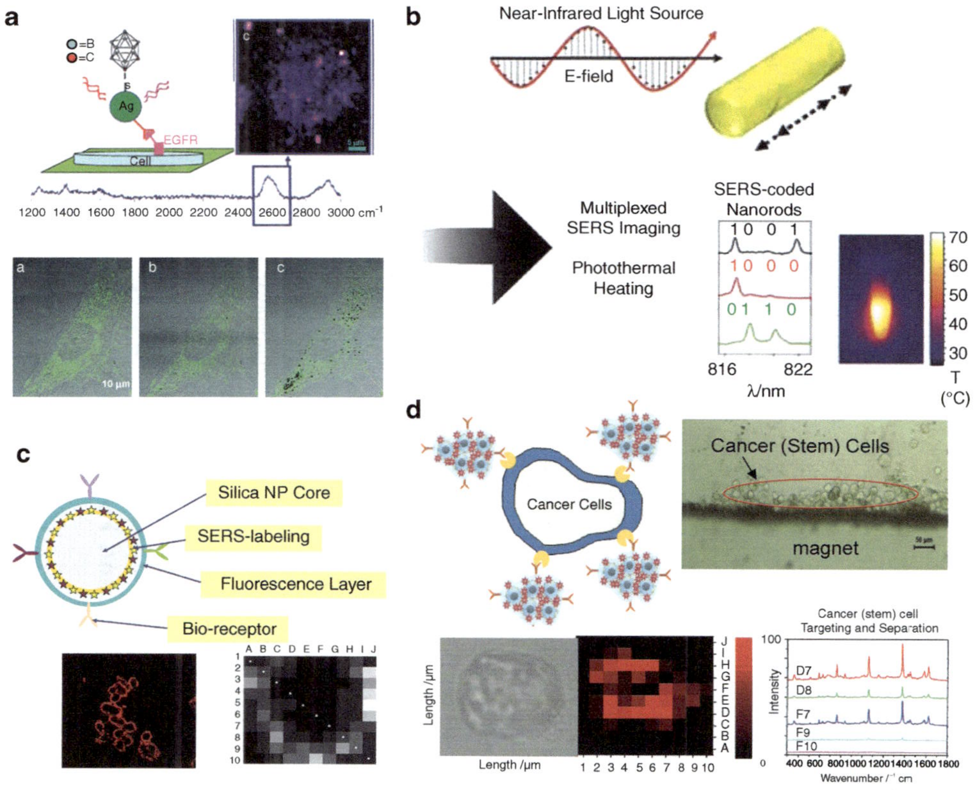

Fig. 11.15 Examples of multi-functional SERS probes. (**a**) Carborane-functionalized nanoparticles [94], (**b**) SERS-coded gold nanorods [95], (**c**) fluorescent SERS probe [71], and (**d**) magnetic SERS probe [74]

nanoparticles and surface-enhanced Raman scattering [88]. This group could overcome nonspecificity of silica encapsulation to proteins and cell surfaces. The Gambhir group applied SERS-based multiplex imaging for in vivo live animal study [92, 93]. Briefly, they injected SERS nanoprobes into the tail vein of a mouse and obtained SERS images of their natural accumulation in the liver (Fig. 11.14).

SERS probe has evolved into multifunctional ones with additional functions such as heating, fluorescence coloring, and magnetization. Several groups have reported that SERS-based multifunctional materials could be used for broad applications such as therapy, cell separation, or SERS-based imaging. Kennedy et al. prepared water-soluble carborane-functionalized nanoparticles [94]. Since carboranes are boron-rich species and the B-H Raman stretch signal falls in a spectroscopically silent region of a cell (1,800–2,700 cm^{-1}), they can be used for boron neutron capture therapy (BNCT) and SERS monitoring. Maltzahn et al. demonstrated SERS-coded gold nanorods as a multifunctional platform for multiplexed near-infrared imaging and photothermal heating [95]. The gold nanoparticles were designed to have an excitation wavelength of 785 nm for near-infrared SERS and an absorption wavelength of 810 nm for photothermal heating. Fluorescent SERS probes were introduced for fast tracking targeted area and multiplex imaging of targets. They were successfully applied for tissue analysis of large surface areas such as targeting BAX and BAD antibodies for the detection of apoptosis in tissues [72] and tracking bronchioalveolar stem cells (BASCs) with multiple biomarkers [71]. Magnetic property was combined with SERS label and used for separation of targets. Gong et al. developed sandwich assay method for cancer antigen detection by combining the Ag/SiO_2 core-shell nanoparticles embedded with rhodamine B isothiocyanate dye molecules as Raman tags and the amino group–modified silica-coated magnetic nanoparticle as immobilization matrix and separation tool [96]. They could avoid complicated pretreatment and washing steps. Jun et al. incorporated magnetic nanoparticle into SERS labels, which were composed of silica core and silver nanoparticle assemblies, and applied them for detection of biomarkers of breast-cancer cells and separation of targeted cells [74] (Fig. 11.15).

6 Conclusions and Future Perspectives

Nowadays, surface-enhanced Raman scattering (SERS) has became a valuable detection tool in various research fields. This is attributed to several advantages of SERS technology. SERS-based technology is regarded as one of the most powerful tools for multiplexing due to its narrow band width. Optically bleachless characteristics of Raman scattering and surface-modification technology of SERS labels open an optimistic route for their potential utilization as a quantitative analysis method. There is still enough room for optimization of sensitivity of SERS-based detection methods since SERS sensitivity can rise up to the single molecule detection level. Optimization of both sensitivity and reproducibility is one of the difficult problems. Along with optimization of various parameters in SERS-based detection methods, successful application of multiplexing in biological studies and medical fields can reduce repetitive and demanding processes and consequently will serve as a powerful detection method.

Multiplexed in vivo imaging is one of the important successful achievements of SERS-based assay technology in bio-related fields. Its medical application can serve as a simple and cheap diagnostic tool supporting traditional medical services such as diagnosis of cancers.

Technical improvement in instrumentation such as laser and optics will decrease the instrumentation cost and cause Raman instruments more accessible. The development of table-top SERS equipment and their application to bio-related fields would make SERS nanomaterials more popular in the medical and diagnostic fields. This can be used not only for low-priced research instrument but also for fast screening tools of hazard materials and biomaterials in the street.

References

1. Ferraro JR, Nakamoto K, Brown CW (eds) (2003) Introductory Raman spectroscopy, 2nd edn. Academic Press, San Diego
2. Michalet X, Pinaud FF, Bentolila LA, Tsay JM, Doose S, Li JJ, Sundaresan G, Wu AM, Gambhir SS, Weiss S (2005) Quantum dots for live cells, in vivo imaging, and diagnostics. Science 307:538–544
3. Jain PK, Huang X, El-Sayed IH, El-Sayed MA (2008) Noble metals on the nanoscale: optical and photothermal properties and some applications in imaging, sensing, biology, and medicine. Acc Chem Res 41:1578–1586
4. Jun BH, Kim JH, Park H, Kim JS, Yu KN, Lee SM, Choi H, Kwak SY, Kim YK, Jeong DH, Cho MH, Lee YS (2007) Surface-enhanced Raman spectroscopic-encoded beads for multiplex immunoassay. J Comb Chem 9:237–244
5. Wagnieres GA, Star WM, Wilson BC (1998) In vivo fluorescence spectroscopy and imaging for oncological applications. Photochem Photobiol 68:603–632
6. Monici M (2005) Cell and tissue autofluorescence research and diagnostic applications. Biotechnol Annu Rev 11:227–256
7. Blomfiel J, Farrar JF (1969) Fluorescent properties of maturing arterial elastin. Cardiovasc Res 3:161–170
8. Fujimoto D, Akiba KY, Nakamura N (1977) Isolation and characterization of a fluorescent material in bovine achilles-tendon collagen. Biochem Bioph Res Co 76:1124–1129
9. Long DA (ed) (1977) Raman spectroscopy. McGraw-Hill, London
10. Kneipp K, Wang Y, Kneipp H, Perelman LT, Itzkan I, Dasari R, Feld MS (1997) Single molecule detection using surface-enhanced Raman scattering (SERS). Phys Rev Lett 78:1667–1670
11. Nie SM, Emery SR (1997) Probing single molecules and single nanoparticles by surface-enhanced Raman scattering. Science 275:1102–1106
12. Bosnick KA, Jiang J, Brus LE (2002) Fluctuations and local symmetry in single-molecule rhodamine 6 G Raman scattering on silver nanocrystal aggregates. J Phys Chem B 106: 8096–8099
13. Xu H, Bjerneld EJ, Käll M, Börjesson L (1999) Spectroscopy of single hemoglobin molecules by surface enhanced Raman scattering. Phys Rev Lett 83:4357–4360
14. Albrecht MG, Creighton JA (1977) Anomalously intense Raman-spectra of pyridine at a silver electrode. J Am Chem Soc 99:5215–5217

15. Creighton JA, Blatchford CG, Albrecht MG (1979) Plasma resonance enhancement of Raman-scattering by pyridine adsorbed on silver or gold sol particles of size comparable to the excitation wavelength. J Chem Soc Farad T 2 75:790–798
16. Fleischm M, Hendra PJ, McQuilla AJ (1974) Raman-spectra of pyridine adsorbed at a silver electrode. Chem Phys Lett 26:163–166
17. Jeanmaire DL, Van Duyne RP (1977) Surface Raman spectroelectrochemistry 1. Heterocyclic, aromatic, and aliphatic-amines adsorbed on anodized silver electrode. J Electroanal Chem 84:1–20
18. Moskovits M (2005) Surface-enhanced Raman spectroscopy: a brief retrospective. J Raman Spectrosc 36:485–496
19. Xu H, Aizpurua J, Käll M, Apell P (2000) Electromagnetic contributions to single-molecule sensitivity in surface-enhanced Raman scattering. Phys Rev E 62:4318–4324
20. Kottmann JP, Martin OJF (2001) Retardation-induced plasmon resonances in coupled nanoparticles. Opt Lett 26:1096–1098
21. Moskovits M (1985) Surface-enhanced spectroscopy. Rev Mod Phys 57:783–826
22. Campion A, Kambhampati P (1998) Surface-enhanced Raman scattering. Chem Soc Rev 27:241–250
23. Vo-Dinh T (1998) Surface-enhanced Raman spectroscopy using metallic nanostructures. Trac-Trend Anal Chem 17:557–582
24. Haynes CL, Van Duyne RP (2003) Plasmon-sampled surface-enhanced Raman excitation spectroscopy. J Phys Chem B 107:7426–7433
25. McFarland AD, Young MA, Dieringer JA, Van Duyne RP (2005) Wavelength-scanned surface-enhanced Raman excitation spectroscopy. J Phys Chem B 109:11279–11285
26. Dieringer JA, McFarland AD, Shah NC, Stuart DA, Whitney AV, Yonzon CR, Young MA, Zhang XY, Van Duyne RP (2006) Surface enhanced Raman spectroscopy: new materials, concepts, characterization tools, and applications. Faraday Discuss 132:9–26
27. Chang RK, Furtak TE (eds) (1982) Surface-enhanced raman scattering. Plenum, New York
28. Banholzer MJ, Millstone JE, Qin LD, Mirkin CA (2008) Rationally designed nanostructures for surface-enhanced Raman spectroscopy. Chem Soc Rev 37:885–897
29. Okamoto H, Imura K (2008) Near-field optical imaging of nanoscale optical fields and plasmon waves. Jpn J Appl Phys 47:6055–6062
30. Haynes CL, McFarland AD, Zhao LL, Van Duyne RP, Schatz GC, Gunnarsson L, Prikulis J, Kasemo B, Käll M (2003) Nanoparticle optics: the importance of radiative dipole coupling in two-dimensional nanoparticle arrays. J Phys Chem B 107:7337–7342
31. Jeong DH, Zhang YX, Moskovits M (2004) Polarized surface enhanced Raman scattering from aligned silver nanowire rafts. J Phys Chem B 108:12724–12728
32. Lee SJ, Morrill AR, Moskovits M (2006) Hot spots in silver nanowire bundles for surface-enhanced Raman spectroscopy. J Am Chem Soc 128:2200–2201
33. Kim JH, Kim JS, Choi H, Lee SM, Jun BH, Yu KN, Kuk E, Kim YK, Jeong DH, Cho MH, Lee YS (2006) Nanoparticle probes with surface enhanced Raman spectroscopic tags for cellular cancer targeting. Anal Chem 78:6967–6973
34. Su X, Zhang J, Sun L, Koo TW, Chan S, Sundararajan N, Yamakawa M, Berlin AA (2005) Composite organic-inorganic nanoparticles (COINs) with chemically encoded optical signatures. Nano Lett 5:49–54
35. Chon H, Lee S, Son SW, Oh CH, Choo J (2009) Highly sensitive immunoassay of lung cancer marker carcinoembryonic antigen using surface-enhanced Raman scattering of hallow gold nanospheres. Anal Chem 81:3029–3034
36. Lee S, Chon H, Lee M, Choo J, Shin SY, Lee YH, Rhyu IJ, Son SW, Oh CH (2009) Surface-enhanced Raman scattering imaging of HER2 cancer markers overexpressed in single MCF7 cells using antibody conjugated hollow gold nanospheres. Biosens Bioelectron 24:2260–2263
37. Sanles-Sobrido M, Exner W, Rodriguez-Lorenzo L, Rodriguez-Gonzalez B, Correa-Duarte MA, Alvarez-Puebla RA, Liz-Marzan LM (2009) Design of SERS-encoded, submicron, hollow particles through confined growth of encapsulated metal nanoparticles. J Am Chem Soc 131:2699–2705

38. Nikoobakht B, Wang JP, El-Sayed MA (2002) Surface-enhanced Raman scattering of molecules adsorbed on gold nanorods: off-surface plasmon resonance condition. Chem Phys Lett 366:17–23
39. Wang YL, Lee K, Irudayaraj J (2010) SERS aptasensor from nanorod-nanoparticle junction for protein detection. Chem Commun 46:613–615
40. Guo HY, Lu LH, Wu C, Pan JG, Hu JW (2009) SERS tagged gold nanorod probes for immunoassay application. Acta Chim Sin 67:1603–1608
41. Kwon K, Lee KY, Lee YW, Kim M, Heo J, Ahn SJ, Han SW (2007) Controlled synthesis of icosahedral gold nanoparticles and their surface-enhanced Raman scattering property. J Phys Chem C 111:1161–1165
42. McLellan JM, Li ZY, Siekkinen AR, Xia YN (2007) The SERS activity of a supported ag nanocube strongly depends on its orientation relative to laser polarization. Nano Lett 7:1013–1017
43. Xie JP, Zhang QB, Lee JY, Wang DIC (2008) The synthesis of SERS-active gold nanoflower tags for in vivo applications. ACS Nano 2:2473–2480
44. Khoury CG, Vo-Dinh T (2008) Gold nanostars for surface-enhanced Raman scattering: synthesis, characterization and optimization. J Phys Chem C 112:18849–18859
45. Li WY, Camargo PHC, Au L, Zhang Q, Rycenga M, Xia YN (2010) Etching and dimerization: a simple and versatile route to dimers of silver nanospheres with a range of sizes. Angew Chem-Int Edit 49:164–168
46. Qin L, Banholzer MJ, Millstone JE, Mirkin CA (2007) Nanodisk codes. Nano Lett 7:3849–3853
47. McCabe AF, Eliasson C, Prasath RA, Hernandez-Santana A, Stevenson L, Apple I, Cormack PAG, Graham D, Smith WE, Corish P, Lipscomb SJ, Holland ER, Prince PD (2006) SERRS labelled beads for multiplex detection. Faraday Discuss 132:303–308
48. Wang CG, Chen Y, Wang TT, Ma ZF, Su ZM (2008) Monodispersed gold nanorod-embedded silica particles as novel Raman labels for biosensing. Adv Funct Mater 18:355–361
49. Lu Y, Liu GL, Kim J, Mejia YX, Lee LP (2005) Nanophotonic crescent moon structures with sharp edge for ultrasensitive biomolecular detection by local electromagnetic field enhancement effect. Nano Lett 5:119–124
50. Li JF, Huang YF, Ding Y, Yang ZL, Li SB, Zhou XS, Fan FR, Zhang W, Zhou ZY, Wu DY, Ren B, Wang ZL, Tian ZQ (2010) Shell-isolated nanoparticle-enhanced Raman spectroscopy. Nature 464:392–395
51. Felidj N, Aubard J, Levi G, Krenn JR, Salerno M, Schider G, Lamprecht B, Leitner A, Aussenegg FR (2002) Controlling the optical response of regular arrays of gold particles for surface-enhanced Raman scattering. Phys Rev B 65:075419(1–9)
52. Hirsch LR, Gobin AM, Lowery AR, Tam F, Drezek RA, Halas NJ, West JL (2006) Metal nanoshells. Ann Biomed Eng 34:15–22
53. Jain PK, Lee KS, El-Sayed IH, El-Sayed MA (2006) Calculated absorption and scattering properties of gold nanoparticles of different size, shape, and composition: applications in biological imaging and biomedicine. J Phys Chem B 110:7238–7248
54. Wiley BJ, Chen YC, McLellan JM, Xiong YJ, Li ZY, Ginger D, Xia YN (2007) Synthesis and optical properties of silver nanobars and nanorice. Nano Lett 7:1032–1036
55. Green M, Liu FM, Cohen L, Kollensperger P, Cass T (2006) SERS platforms for high density DNA arrays. Faraday Discuss 132:269–280
56. Bell SEJ, Sirimuthu NMS (2006) Surface-enhanced Raman spectroscopy (SERS) for sub-micromolar detection of DNA/RNA mononucleotides. J Am Chem Soc 128:15580–15581
57. Ingram A, Stokes RJ, Redden J, Gibson K, Moore B, Faulds K, Graham D (2007) 8-hydroxyquinolinyl azo dyes: a class of surface-enhanced resonance Raman scattering-based probes for ultrasensitive monitoring of enzymatic activity. Anal Chem 79:8578–8583
58. Dijkstra RJ, Scheenen W, Dam N, Roubos EW, ter Meulen JJ (2007) Monitoring neurotransmitter release using surface-enhanced Raman spectroscopy. J Neurosci Meth 159:43–50

59. Kneipp J, Kneipp H, McLaughlin M, Brown D, Kneipp K (2006) In vivo molecular probing of cellular compartments with gold nanoparticles and nanoaggregates. Nano Lett 6:2225–2231
60. Rohr TE, Cotton T, Fan N, Tarcha PJ (1989) Immunoassay employing surface-enhanced Raman-spectroscopy. Anal Biochem 182:388–398
61. Dou X, Takama T, Yamaguchi Y, Yamamoto H, Ozaki Y (1997) Enzyme immunoassay utilizing surface-enhanced Raman scattering of the enzyme reaction product. Anal Chem 69:1492–1495
62. Porter MD, Lipert RJ, Siperko LM, Wang G, Narayanana R (2008) SERS as a bioassay platform: fundamentals, design, and applications. Chem Soc Rev 37:1001–1011
63. Driskell JD, Lipert RJ, Porter MD (2006) Labeled gold nanoparticles immobilized at smooth metallic substrates: systematic investigation of surface plasmon resonance and surface-enhanced Raman scattering. J Phys Chem B 110:17444–17451
64. Hu QY, Tay LL, Noestheden M, Pezacki JP (2007) Mammalian cell surface imaging with nitrile-functionalized nanoprobes: biophysical characterization of aggregation and polarization anisotropy in SERS imaging. J Am Chem Soc 129:14–15
65. Willets KA (2009) Surface-enhanced Raman scattering (SERS) for probing internal cellular structure and dynamics. Anal Bioanal Chem 394:85–94
66. Chourpa I, Lei FH, Dubois P, Manfait M, Sockalingum GD (2008) Intracellular applications of analytical SERS spectroscopy and multispectral imaging. Chem Soc Rev 37:993–1000
67. Schlücker S (2009) SERS microscopy: nanoparticle probes and biomedical applications. ChemPhysChem 10:1344–1354
68. Lutz BR, Dentinger CE, Nguyen LN, Sun L, Zhang JW, Allen AN, Chan S, Knudsen BS (2008) Spectral analysis of multiplex Raman probe signatures. Acs Nano 2:2306–2314
69. Park H, Lee S, Chen L, Lee EK, Shin SY, Lee YH, Son SW, Oh CH, Song JM, Kang SH, Choo J (2009) SERS imaging of HER2-overexpressed MCF7 cells using antibody-conjugated gold nanorods. Phys Chem Chem Phys 11:7444–7449
70. Sun L, Sung KB, Dentinger C, Lutz B, Nguyen L, Zhang JW, Qin HY, Yamakawa M, Cao MQ, Lu Y, Chmura AJ, Zhu J, Su X, Berlin AA, Chan S, Knudsen B (2007) Composite organic-inorganic nanoparticles as Raman labels for tissue analysis. Nano Lett 7:351–356
71. Woo MA, Lee SM, Kim G, Baek J, Noh MS, Kim JE, Park SJ, Minai-Tehrani A, Park SC, Seo YT, Kim YK, Lee YS, Jeong DH, Cho MH (2009) Multiplex immunoassay using fluorescent-surface enhanced Raman spectroscopic dots for the detection of bronchioalveolar stem cells in murine lung. Anal Chem 81:1008–1015
72. Yu KN, Lee SM, Han JY, Park H, Woo MA, Noh MS, Hwang SK, Kwon JT, Jin H, Kim YK, Hergenrother PJ, Jeong DH, Lee YS, Cho MH (2007) Multiplex targeting, tracking, and imaging of apoptosis by fluorescent surface enhanced raman spectroscopic dots. Bioconjugate Chem 18:1155–1162
73. Jun BH, Noh MS, Kim G, Kang H, Kim JH, Chung WJ, Kim MS, Kim YK, Cho MH, Jeong DH, Lee YS (2009) Protein separation and identification using magnetic beads encoded with surface-enhanced Raman spectroscopy. Anal Biochem 391:24–30
74. Jun BH, Noh MS, Kim J, Kim G, Kang H, Kim MS, Seo YT, Baek J, Kim JH, Park J, Kim S, Kim YK, Hyeon T, Cho MH, Jeong DH, Lee YS (2010) Multifunctional silver-embedded magnetic nanoparticles as SERS nanoprobes and their applications. Small 6:119–125
75. Noh MS, Jun BH, Kim S, Kang H, Woo MA, Minai-Tehrani A, Kim JE, Kim J, Park J, Lim HT, Park SC, Hyeon T, Kim YK, Jeong DH, Lee YS, Cho MH (2009) Magnetic surface-enhanced Raman spectroscopic (M-SERS) dots for the identification of bronchioalveolar stem cells in normal and lung cancer mice. Biomaterials 30:3915–3925
76. Rusmini F, Zhong ZY, Feijen J (2007) Protein immobilization strategies for protein biochips. Biomacromolecules 8:1775–1789
77. Sun L, Yu CX, Irudayaraj J (2007) Surface-enhanced raman scattering based nonfluorescent probe for multiplex DNA detection. Anal Chem 79:3981–3988

78. Qian XM, Zhou X, Nie SM (2008) Surface-enhanced raman nanoparticle beacons based on bioconjugated gold nanocrystals and long range plasmonic coupling. J Am Chem Soc 130:14934–14935
79. McKenzie F, Faulds K, Graham D (2007) Sequence-specific DNA detection using high-affinity. LNA-functionalized gold nanoparticles. Small 3:1866–1868
80. Liu XJ, Knauer M, Ivleva NP, Niessner R, Haisch C (2010) Synthesis of core-shell surface-enhanced Raman tags for bioimaging. Anal Chem 82:441–446
81. Grubisha DS, Lipert RJ, Park HY, Driskell J, Porter MD (2003) Femtomolar detection of prostate-specific antigen: an immunoassay based on surface-enhanced Raman scattering and immunogold labels. Anal Chem 75:5936–5943
82. Lyandres O, Shah NC, Yonzon CR, Walsh JT, Glucksberg MR, Van Duyne RP (2005) Real-time glucose sensing by surface-enhanced Raman spectroscopy in bovine plasma facilitated by a mixed decanethiol/mercaptohexanol partition layer. Anal Chem 77:6134–6139
83. Lee S, Kim S, Choo J, Shin SY, Lee YH, Choi HY, Ha SH, Kang KH, Oh CH (2007) Biological imaging of HEK293 cells expressing PLC gamma 1 using surface-enhanced Raman microscopy. Anal Chem 79:916–922
84. Otsuka H, Nagasaki Y, Kataoka K (2003) PEGylated nanoparticles for biological and pharmaceutical applications. Adv Drug Deliv Rev 55:403–419
85. Lim DK, Jeon KS, Kim HM, Nam JM, Suh YD (2010) Nanogap-engineerable Raman-active nanodumbbells for single-molecule detection. Nat Mater 9:60–67
86. Ni J, Lipert RJ, Dawson GB, Porter MD (1999) Immunoassay readout method using extrinsic Raman labels adsorbed on immunogold colloids. Anal Chem 71:4903–4908
87. Mulvaney SP, Musick MD, Keating CD, Natan MJ (2003) Glass-coated, analyte-tagged nanoparticles: a new tagging system based on detection with surface-enhanced Raman scattering. Langmuir 19:4784–4790
88. Qian XM, Peng XH, Ansari DO, Yin-Goen Q, Chen GZ, Shin DM, Yang L, Young AN, Wang MD, Nie SM (2008) In vivo tumor targeting and spectroscopic detection with surface-enhanced Raman nanoparticle tags. Nat Biotechnol 26:83–90
89. Kaufmann S, Tanaka M (2003) Cell adhesion onto highly curved surfaces: one-step immobilization of human erythrocyte membranes on silica beads. Chem Phys Chem 4:699–704
90. Küstner B, Gellner M, Schutz M, Schöppler F, Marx A, Ströbel P, Adam P, Schmuck C, Schlücker S (2009) SERS labels for red laser excitation: silica-encapsulated SAMs on tunable gold/silver nanoshells. Angew Chem-Int Ed 48:1950–1953
91. Wabuyele MB, Yan F, Griffin GD, Vo-Dinh T (2005) Hyperspectral surface-enhanced Raman imaging of labeled silver nanoparticles in single cells. Rev Sci Instrum 76:063710(1–7)
92. Keren S, Zavaleta C, Cheng Z, de la Zerda A, Gheysens O, Gambhir SS (2008) Noninvasive molecular imaging of small living subjects using Raman spectroscopy. Proc Natl Acad Sci USA 105:5844–5849
93. Zavaleta CL, Smith BR, Walton I, Doering W, Davis G, Shojaei B, Natan MJ, Gambhir SS (2009) Multiplexed imaging of surface enhanced Raman scattering nanotags in living mice using noninvasive Raman spectroscopy. Proc Natl Acad Sci USA 106:13511–13516
94. Kennedy DC, Duguay DR, Tay LL, Richeson DS, Pezacki JP (2009) SERS detection and boron delivery to cancer cells using carborane labelled nanoparticles. Chem Commun 6750–6752
95. von Maltzahn G, Centrone A, Park JH, Ramanathan R, Sailor MJ, Hatton TA, Bhatia SN (2009) SERS-coded gold nanorods as a multifunctional platform for densely multiplexed near-infrared imaging and photothermal heating. Adv Mater 21:3175–3180
96. Gong JL, Liang Y, Huang Y, Chen JW, Jiang JH, Shen GL, Yu RQ (2007) Ag/SiO_2 core-shell nanoparticle-based surface-enhanced Raman probes for immunoassay of cancer marker using silica-coated magnetic nanoparticles as separation tools. Biosens Bioelectron 22:1501–1507

In Situ Raman Spectroscopy of Oxidation of Carbon Nanomaterials

12

Sebastian Osswald and Yury Gogotsi

1 Definition of the Topic

In situ Raman spectroscopy during heating in a controlled environment allows for a time-resolved investigation of the oxidation kinetics of carbon nanomaterials and can identify changes in material structure and composition during oxidation. In this chapter, we describe the application of in situ Raman spectroscopy to determine conditions for selective oxidation and purification of carbon nanotubes (CNT) and nanodiamond (ND).

2 Overview

The ability to synthesize carbon nanostructures, such as fullerenes, carbon nanotubes, nanodiamond, and mesoporous carbon; functionalize their surface; or assemble them into three-dimensional networks has opened new avenues for material design. Carbon nanostructures possess tunable optical, electrical, or mechanical properties, making them ideal candidates for numerous applications ranging from composite structures and chemical sensors to electronic devices and medical implants.

Unfortunately, current synthesis techniques typically lead to a mixture of different types of carbon rather than a particular nanostructure with defined size and

S. Osswald (✉)
Department of Physics, Graduate School of Engineering and Applied Science, Naval Postgraduate School, Monterey, CA, USA

Y. Gogotsi
Department of Materials Science and Engineering and A.J. Drexel Nanotechnology Institute, Drexel University, Philadelphia, PA, USA

Challa S. S. R. Kumar (ed.), *Raman Spectroscopy for Nanomaterials Characterization*,
DOI 10.1007/978-3-642-20620-7_12, © Springer-Verlag Berlin Heidelberg 2012

properties. In order to fully exploit the great potential of carbon nanostructures, one needs to provide purification procedures that allow a selective separation of carbon nanostructures, and methods which enable a control of size and surface functionalization. Oxidation is a frequently used method for purification of carbon materials, but it can also damage or destroy the sample.

Here we describe how to employ in situ Raman spectroscopy to determine the conditions for selective oxidation and purification of carbon nanotubes (CNT) and nanodiamond (ND); measure and control their crystal size; and improve the fundamental understanding of effects of temperature, quantum confinement, and surface chemistry on Raman spectra of nanocrystalline materials.

3 Introduction

Carbon is the basis of all life on earth, and without a doubt, one of the most versatile elements known to man. More than ten million carbon compounds are known today, many times more than that of any other element. Carbon itself exists in several allotropes. Its flexible electron configuration allows carbon to form three hybridization states which lead to different types of covalent bonding. The most representative macroscopic forms of carbon are graphite and diamond. In 1985, Kroto et al. discovered a third carbon allotrope, the fullerenes. While their experiments aimed at understanding the mechanisms by which long chained carbon molecules are formed in interstellar space, their results opened a new era in science – the beginning of nanotechnology.

Today nanotechnology includes the synthesis, characterization, and application of a variety of nanostructured materials. Different carbon nanostructures exist simultaneously at the nanoscale, including carbon nanotubes, carbon onions, nanodiamond (ND), and diamondoids, all showing unique and novel properties [1].

The physical and chemical properties of nanostructures differ significantly from that of bulk materials of the same composition. For example, a decrease in the crystal size results in an increase in the surface/volume ratio, which is becoming exceptionally high when entering the lower nanometer range. As a result, properties are primarily determined by the shape of the crystals and their surfaces. Therefore, controlling both the size of nanostructures as well as their surface chemistry appears to be crucial.

Unfortunately, current synthesis techniques, such as chemical vapor deposition, arc discharge, laser ablation, or detonation, typically lead to a mixture of various nanostructures, amorphous carbon, and catalyst particles rather than a particular nanostructure with defined properties, thus limiting the number of potential applications [1]. Even if pure materials were available, the size-dependence of most nanomaterial properties requires a high structural selectivity. In order to fully exploit the great potential of carbon nanostructures, one needs to provide purification procedures that allow for a selective separation of carbon nanostructures, and methods which enable size control and surface functionalization.

Several purification techniques have been developed and are either based on treatments in acids and bases, or involve the processing of other toxic substances.

While these techniques are widely used, they do not provide sufficient purity. Liquid phase purification is not an environmentally friendly process and requires corrosion-resistant equipment, as well as costly waste disposal processes. Alternative dry chemistry approaches, such as catalyst-assisted oxidation or ozone-enriched air oxidation, also require the use of aggressive substances or supplementary catalysts, which result in an additional contamination. Moreover, in many previous studies "trial and error" rather than insight and theory approaches have been applied. As a result, a lack of understanding and limited process control often lead to extensive sample losses of up to 90%. Because oxidation in air would be a controllable and environmentally friendly process, selective purification of carbon nanomaterials, such as CNT and ND, in air is very attractive. In contrast to current purification techniques, air oxidation does not require the use of toxic or aggressive chemicals, catalysts, or inhibitors and opens avenues for numerous new applications of carbon nanomaterials.

Until recently, selective oxidation of carbon nanomaterials has not been optimized or was considered not feasible as in the case of ND [2]. However, for the production and application of nanostructures on an industrial scale, it is very important to develop a simple and efficient route to selectively remove sp^2-bonded carbon from nanodiamond and amorphous carbon from nanotubes with minimal or no loss of diamond or nanotubes.

In situ Raman spectroscopy allows for a detailed and time-resolved investigation of the kinetics of complex physical/chemical processes such as oxidation. Using in situ Raman spectroscopy, one is able to monitor the oxidation and related structural changes of carbon nanostructures in real time, in order to identify the optimum purification conditions.

While oxidation is mainly used for purification of carbon nanostructures, it may also be an efficient tool for size control and surface modification. By selectively oxidizing, for example, smaller carbon nanotubes or diamond crystals, it can provide a simple technique for narrowing size distributions in carbon nanomaterials. Finally, modification of the porosity (activation) of carbon is another example of controlled oxidation and may allow optimization of the pore structure and surface area of nanoporous carbide-derived carbon (CDC) for various applications.

4 Experimental Methodology

4.1 Oxidation of Carbon Nanomaterials

In a typical in situ Raman study, CNT samples are heated in a heating stage, operated in air or controlled environment between 20°C and 600°C. Powders must be dispersed in a solvent, such as ethanol, to produce a thin film of CNTs on a glass slide. The samples are then kept in the heating stage and placed under the microscope of the Raman spectrometer. The oxidation process described in this chapter followed two different heating procedures. The first procedure (nonisothermal) includes heating from room temperature up to 600°C at a rate of

5°C/min while holding each measurement temperature for about 4 min. Spectra were taken in steps of 50°C in the range from 50°C to 350°C, followed by 25°C steps from 350°C to 400°C and 10°C steps from 400°C to 600°C. After reaching the desired temperature, the sample was cooled down at 20°C/min until reaching 80°C and then cooled at 10°C/min down to room temperature. In the second oxidation procedure (isothermal), samples were heated at 50°C/min and exposed to the desired temperature. In that case, Raman spectra were acquired every 15 min after reaching the dwell temperature. After 330 min, samples were cooled down with 50°C/min rate until reaching 80°C followed by cooling at ~10°C/min to room temperature. Sample size and conditions in the hot stage also influence the oxidation reaction. A potential problem might be an insufficient air circulation and oxidation in the oxygen-lean O_2–CO–CO_2 atmosphere. However, experiments with the opened stage showed no significant changes in the results, suggesting that the influence of the small sample volume is negligible.

Oxidative purification of ND powders was conducted under isothermal conditions using the heating stage and a tube furnace. Isothermal experiments included two steps: (1) rapid heating at 50°C/min to the selected temperature and (2) isothermal oxidation for 5 h in ambient air at atmospheric pressure. ND powders used for crystal size characterization were oxidized for 2, 6, 17, 26, and 42 h at 430°C, in a closed tube furnace in static air at atmospheric pressure.

4.2 Raman Spectroscopy and In Situ Studies

Raman analysis is performed using Raman microspectrometers in backscattering geometry. Analysis of nanodiamond requires the use of ultraviolet (UV) excitation and spectra were recorded using a Raman spectrometer with a 325-nm HeCd laser. Visible (VIS) Raman spectra were acquired using a 514-nm Ar-ion laser, a 633-nm HeNe laser, and a 785-nm semiconductor diode laser.

In situ Raman studies were performed using a heating/cooling stage placed under the Raman microscope. Long focus objectives are required to protect the equipment against the heat produced by the stage. To minimize the influence of sample inhomogeneities, every measurement should be done at the same spot during in situ experiments. All Raman measurements must be repeated multiple times to produce statistically reliable results.

5 Key Research Findings

5.1 Oxidation and Purification of Carbon Nanomaterials

The oxidation of carbon materials such as coal, charcoal, or graphite has been investigated for decades, and is well known as a powerful route to modify the physical and chemical properties of carbon materials. While the reaction of oxygen with carbon surfaces is one of the simplest reactions involving elemental carbon, it is also one of the most important reactions for a wide range of technological applications, such as heterogeneous oxidation catalysis, carbon activation, spacecraft engineering, fuel-efficiency, emission control, and surface functionalization, to mention a few. Until recently, carbon-oxygen reactions had three main functions: (1) burning carbon to produce energy, (2) increasing surface area and porosity of a material by removing carbon atoms from the surface (activation), and (3) formation of organic functional groups on the carbon surface (functionalization). However, during the recent years, researchers have recognized the great potential of carbon-oxygen reactions in carbon nanotechnology and took advantage of another important function: the selectivity of the oxidation process toward different forms of carbon.

The structural diversity of carbon at the nanoscale exceeds that of all other materials [1]. Detailed information on the nature of the material and the structure-dependency of the oxidation kinetics is thus crucial for providing the required selectivity. While some nanomaterials, such as carbon nanotubes, have been studied extensively and are generally well understood, other nanostructures such as nanodiamond (ND) have received much less attention. However, in order to study their properties and open avenues for new applications, one has to provide a material of high purity and defined composition.

Although much progress has been made in both synthesis and purification of carbon nanomaterials, commercial samples still contain nanostructures of different size, shape, and composition. As-produced carbon nanomaterials are frequently composed of mixtures of CNTs, fullerenes, carbon onions, amorphous carbon and graphite, which are structurally different and possess different reactivity. Since the oxidation kinetics are closely related to structural features, reaction rates and activation energies are expected to differ for the distinct carbon forms, which is an important issue for oxidation-based purification or surface functionalization. In analogy to graphite [3–6], oxidation of a carbon nanostructure [7–9] can be described by a first-order reaction, with respect to the carbon component.

Therefore, depending on the temperature, the oxidation of carbon nanomaterials can be understood as an overlap of several first-order oxidation reactions resulting from different nanostructures present in the sample. The resistance against oxidation increases with the level of graphitization and is lowest for amorphous carbon. However, while CNTs and other carbon nanomaterials are highly graphitized, curvature effects lower the activation energy, compared to that of graphite or graphene.

A large number of existing oxidation methods can be classified into dry chemistry (gas-based) and wet chemistry (liquid-based) approaches. Liquid phase oxidation involves treatments in concentrated or diluted acids (e.g., HNO_3, H_2SO_4, H_3PO_4) and/or bases (e.g., NaOH, KOH) [10–15]. One advantage of liquid phase oxidation is the ability to control the purification process and reaction rates by adjusting the concentrations of reactants. Simultaneous dissolution of metal impurities in parallel to oxidation of carbon phases is also possible. At the same time, liquid phase oxidation techniques suffer from many disadvantages. Despite being effective oxidation reactants, acids and bases are environmentally harmful and require corrosion-resistant equipment in order to run the process and to store and transport aggressive oxidizers. In some cases, oxidation with acids, bases, or oxides introduces foreign elements and leads to additional impurities, such as nitrogen- and sulfur-containing compounds, chlorine, chromium, etc., which alter the intrinsic characteristics of the carbon material and need to be removed in additional purification steps. Especially heavy metal compounds cannot be ignored and may result in a costly waste disposal process. To overcome the disadvantages of liquid oxidizers, some alternative purification techniques have been elaborated. Dry chemistry approaches are based on high-temperature reactions in gases such as air, oxygen, carbon dioxide, or ozone [10–12, 15–18].

While oxidation has been used extensively to modify and activate macroscopic forms of carbon on an industrial scale, most of the studies reported in literature provide only a qualitative description of the oxidation kinetics, and lack a fundamental understanding of the physical and chemical processes occurring at the nanoscale. The various types of carbon used in these studies were often simply referred to as "coke" or "carbon soot," even though they are characterized by distinct differences in crystal structure and surface terminations. Therefore, a comparison of available data is often difficult and interpretation of the results may lead to ambiguous conclusions.

In many experiments, the energy required for the oxidation of carbon materials is significantly lower than the activation energy predicted by theoretical calculations [19]. The process of lowering the activation energy by providing alternative reaction pathways is called catalysis [20]. Catalysts enable reactions that would be impossible in their absence, or serve to accelerate oxidation reactions. The catalyst itself is not consumed during the reaction and may participate in multiple catalytic cycles [20]. Therefore, even small amounts of catalysts in the sample can significantly affect the oxidation reaction.

A large number of different inorganic materials, ranging from noble metals to salts and oxides of alkali and alkaline earth metals, were found to catalyze the reaction of carbon with gaseous reactants such as O_2 or CO_2 at elevated temperatures [19]. The common property of these materials is that they are able to form several oxidation states in the temperature regime of the oxidation process. Although catalytic reactions are not fully understood and the subject of ongoing research, it is generally agreed that catalytic processes result from oxidation/reduction cycles on the carbon surface.

The effects of catalysts on the oxidation temperature can be significant. For example, lead (Pb), copper (Cu), silver (Ag), iron (Fe), platinum (Pt), and nickel (Ni) were found to lower the ignition temperature of graphite powder from 740°C to 382°C, 570°C, 585°C, 593°C, 602°C, and 613°C, respectively [21]. In all of these cases, the concentration of the metal in the sample was <0.2 wt.%. While catalysts are widely used for large-scale production of chemicals and play an important role in biological processes, they are considered as impurities in the case of carbon nanomaterials as they alter their properties and limit the number of potential applications.

Although oxidation has been used to purify carbon materials, oxygen-carbon reactions have been shown to drastically alter physiochemical properties, such as wettability and adsorption characteristics. Moreover, oxidation can easily induce damage to carbon materials or even destroy the sample. This is of particular importance in the case of carbon nanostructures. Thermogravimetric analysis (TGA), which measures changes of mass during oxidation processes, has been widely used to determine the purification conditions [22–24]. However, TGA does not provide information on what type of carbon is removed from the sample or to what extent nanostructures are damaged.

In order to successfully apply oxidation methods to carbon nanomaterials, one has to systematically study their interactions with gases and liquids, monitor changes in structure and composition, and simultaneously follow the reaction kinetics of the different carbon nanostructures. This has been partially achieved for carbon nanotubes [25–27], which have been thoroughly studied, but remains a major challenge for other forms of carbon, including ND or carbon onions.

5.2 Oxidation of Carbon Nanotubes

CNTs are commonly classified into single-walled (SWCNTs) and multi-walled (MWCNTs) nanotubes [28]. SWCNTs consist of a single graphene layer rolled up into a hollow cylinder and are either metallic or semiconducting, whereas MWCNTs are comprised of two, three, or more concentrically arranged cylinders and exhibit only metallic character. Double-wall carbon nanotubes (DWCNTs) are the most basic members of the MWCNT family. The special role of DWCNTs should be emphasized, as they are the link between SWCNTs and the more complex MWCNTs and, therefore, of great interest for a fundamental understanding of these novel nanostructures.

Current synthesis techniques are usually unable to provide large quantities of pure CNTs with well-defined physical and chemical properties [28]. The as-produced material is typically a mixture of different types of CNTs, amorphous carbon, catalyst particles, and defective or damaged tubes, all of which may impair potential applications. Another major challenge for a large number of applications rises from the strong tendency of CNTs to agglomerate and arrange in bundles.

Oxidation provides a great potential for the purification and functionalization of carbon nanomaterials. In situ Raman spectroscopy during oxidation of CNTs enables a time-resolved investigation of the oxidation process and can identify changes in material structure and composition in real time, allowing for an accurate determination of the optimal oxidation conditions.

5.2.1 Thermal Stability and Oxidation Behavior of Carbon Nanotubes

The oxidation behavior of CNT samples depends on various factors. While the thermal stability of an individual CNT is defined by its structure (e.g., diameter, number of walls, defect density), the oxidation behavior of bulk samples containing millions of different nanostructures is mainly determined by the size-distribution of the CNTs, as well as contents of amorphous carbon and other impurities such as metal catalyst and surface functionalities. Due to the complex composition of most CNT samples, an accurate interpretation of thermal analysis data is unfeasible without sufficient information on synthesis conditions and potential post-treatments, such as acid-purification. Even if the composition of the sample is known, determining the reaction kinetics is difficult and remains a major challenge due to the simultaneous contributions of the various factors. Therefore, in addition to thermal analysis, complementary structural characterization methods, such as in situ Raman spectroscopy and HRTEM are required to allow for a correct data interpretation and a better understanding of the reaction kinetics.

A generally accepted oxidation model for CNTs has been proposed by Ayajan et al., wherein the fullerene-like caps are oxidized first, followed by a layer-by-layer gasification of the outer walls [9]. The oxidation of the fullerene-like CNT caps was found to be preferential to oxidation of CNT sidewalls due to the larger curvature and presence of pentagons [8]. However, Shimada et al. revealed that defects within the wall structure (e.g., vacancies, dislocations, cracks) significantly lower the activation energy and may be more reactive than the pentagonal defects of the caps [7]. In case of MWCNTs, activation energies ranging from 225 up to 292 kJ/mol have been measured [29, 30] and were in good agreement with the theoretical predictions [31]. Oxidation studies on SWCNT samples showed activation energies of 119–183 kJ/mol, depending on the material and fitting procedure used during data analysis [24, 29]. It should also be mentioned that the obtained data represents an average value, resulting from many individual CNTs within a broad diameter distribution of most CNT samples. The resistance of SWCNTs toward oxidation is lower than for the larger MWCNTs and bulk graphite. The corresponding activation energies are around 121, 292, and 265 kJ/mole [29]. A higher activation energy of MWCNTs compared to graphite maybe explained by the elimination of reactive edges in nanotubes.

Furthermore, the values of the activation energies depend on the analysis method used. In addition, concentration of defects, homogeneity, and morphology of the sample, presence of catalyst particles, and other synthesis-related factors must be taken into consideration during the analysis of the measured data and interpretation of the results. Therefore, literature data on activation energies of carbon nanomaterials vary widely.

Figure 12.1 summarizes the most important kinetic parameters of the oxidation of different CNTs used in this study. Activation energies of SWCNTs and MWCNTs reported in literature are shown for comparison and represented by the shaded area. Our samples of as-received DWCNTs ($\sim$187 kJ/mol) and MWCNTs ($\sim$195 kJ/mol) were found to exhibit similar activation energies which may not be expected, considering the large differences in size and tube curvature. However, the presence of metal catalysts (Fe) and the large number of structural defects in MWCNTs lower the activation energy and decrease the significance of size and curvature on the thermal stability. The importance of defects and catalyst particles on the oxidation resistance of CNTs is evident when

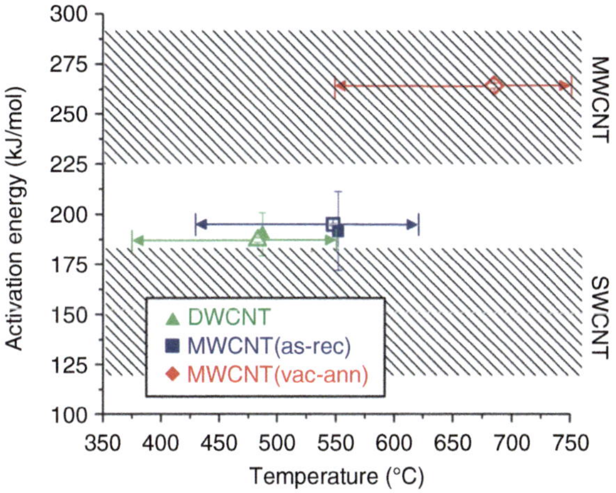

Fig. 12.1 Oxidation kinetics of different types of CNTs. The shaded areas indicate the range of values reported in literature. The *x*-axis represents the temperature at which the maximum weight loss occurs, while the *y*-axis characterizes the activation energy determined using the Achar-Brindley-Sharp-Wendeworth [32] (*large open symbols*) and Arrhenius (*small full symbols*) plots. The *horizontal arrows* indicate the temperature range in which the samples are completely oxidized

comparing as-received and vacuum-annealed MWCNTs. Evaporation of Fe and healing of defects during annealing at high temperatures leads to a 35% increase in the activation energy from 195 to 265 kJ/mol. While one can argue whether the defects are considered as material specific and thus as a characteristic of MWCNTs, these results demonstrate the importance of detailed information on structure and composition of CNT samples for the analysis of reaction kinetics, and may also explain the large variations reported in literature. For the DWCNT sample, catalytic reactions are less likely since the containing metal particles are less active in catalyzing the carbon-oxygen reaction. The measured activation energy is therefore expected to be closer to the actual value of the CNTs. However, in the case of DWCNTs, the reaction heat produced by the exothermic oxidation of larger amounts of amorphous carbon may distort the results. Activation energies of DWCNTs are slightly higher than values reported for SWCNTs, as expected (Fig. 12.1). The activation energy of vacuum-annealed MWCNTs is in sound agreement with literature data, while as-received MWCNTs exhibit smaller activation energies due to the catalytic effect of iron inclusions. Although the activation energies of DWCNTs and as-received MWCNTs were found to be similar, the reaction rate of DWCNTs is significantly higher than that of MWCNTs at any given temperature. As a consequence, the maximum weight loss (highest oxidation rate) of DWCNTs (~485°C) occurs at much lower temperatures compared to as-received MWCNTs (~550°C). Also, the oxidation of the DWCNT sample starts at lower temperatures (~350°C) due to a higher content of amorphous carbon.

5.2.2 Raman Spectra of Carbon Nanotubes

The vibrational modes of an individual SWCNT can be derived from the phonon structure of two-dimensional graphite (graphene) by applying a zone-folding procedure that considers the one-dimensionality of CNTs and a chirality-dependent confinement [33]. Due to CNT's complex structure, their Raman spectra show many size- and chirality-dependent features. In this section, however, only the most prominent first and second order Raman features will be discussed.

The G band is the most intense peak in the Raman spectrum of CNTs (Fig. 12.2) and closely related to the Raman active in-plane E_{2g} mode of graphene where the two atoms of the graphene unit cell are vibrating tangential against each other, giving rise to a single Lorentzian peak at ~1,582 cm^{-1} [34]. Unlike graphite, the G band of a SWCNT consists of two main components centered at ~1,590 cm^{-1} (G^+) and ~1,570 cm^{-1} (G^-), respectively, each consisting of three separate Raman modes (A_{1g}, E_{1g}, and E_{2g}). The G^+ feature is associated with tangential vibrations along the tube axis. The G^- band results from tangential vibrations along the circumferential direction of the CNTs and is softened due to the curvature of the tubes. DWCNTs show G band features comparable to that of SWCNTs [35]. Although less pronounced, the splitting between G^+ and G^- is evident [35]. However, a wider diameter distribution and additional inner tube–outer tube interactions complicate the deconvolution of the G band. Most MWCNT samples show very broad diameter distributions ranging from 5 to >100 nm, and even the smallest inner-tube diameters are usually well above 2 nm. Therefore, curvature effects and confinement-related changes that distinguish the Raman spectra of SWCNTs from that of graphite are less pronounced for MWCNTs, becoming weaker with increasing number of walls [36]. The splitting of the G band is negligible and smeared out due to the overlapping Raman signal of various tube diameters within an individual MWCNT, in addition to ensembles of different MWCNTs commonly present in commercial samples [37]. Due to the different contributions, the G band of MWCNTs predominantly exhibits a weak asymmetric line shape, with a peak maximum close to that of graphite (1,582 cm^{-1}). However, isolated MWCNTs with very small inner-tube diameters were found to show a distinct splitting of the G band, which, in some cases, is even more pronounced than that for SWCNTs [38].

SWCNTs, DWCNTs, or MWCNTs with very small inner-tube diameters show another size-dependent Raman feature in the low-frequency range referred to as radial breathing modes (RBMs) [35, 39, 40]. The RBMs are considered as a clear indicator for the presence of CNTs, since this Raman feature is unique to CNTs and is not observed for other carbon materials. As suggested by the name, the RBM is a bond-stretching, out-of-plane mode, where all carbon atoms vibrate simultaneously in the radial direction. The RBM frequencies are between 100 and 400 cm^{-1} and were found to be inversely proportional to the tube diameter [41–43]. In case of DWCNTs and small-diameter MWCNTs, RBM frequencies higher than 200 cm^{-1} are ascribed to inner tubes while lower frequencies can be associated with both, inner and outer tubes [44, 45]. SWCNTs typically exhibit

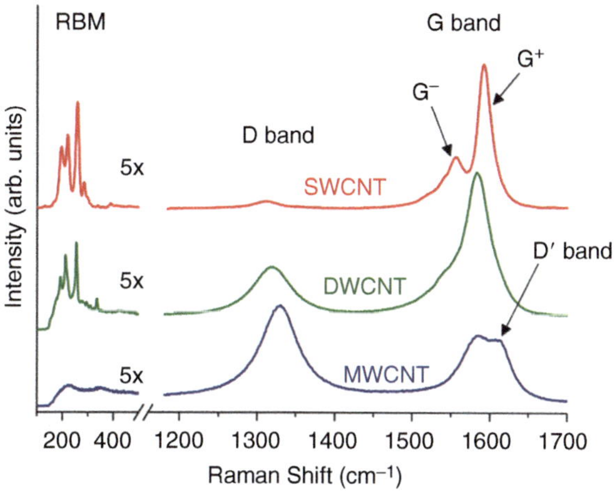

Fig. 12.2 Comparison of high- and low-frequency Raman spectra of SWCNTs, DWCNTs, and small-diameter MWCNTs. Spectra were recorded using 633-nm laser excitation wavelength

a narrower diameter distribution compared to the DWCNTs. Thus, SWCNT samples typically show a lower number of RBM peaks, but with a higher relative intensity compared to the RBM intensity of DWCNTs. Moreover, the RBM spectra of DWCNTs are an overlap of the Raman signals coming from both inner and outer tubes resulting in a wider frequency range, but with a lower intensity. It is important to note that due to the enhancement effect, which selectively amplifies the Raman signal from nanotubes that are in resonance with the incoming laser radiation, RBM intensities recorded using a single excitation wavelength do not reflect the real size distribution of CNTs in the sample.

In addition, CNTs exhibit several Raman features whose frequencies change with changing excitation wavelength. A prominent example for this unusual behavior is the disorder-induced D band which results from a defect-induced double-resonant process [46]. In the "molecular" picture, the D band originates from the breathing vibrations of aromatic rings in the honeycomb lattice. A quantitative description of the D band intensity in graphene was recently derived by Sato et al. [47]. The intensity ratio between D and G band (I_D/I_G) is often used to evaluate the disorder in carbon materials or estimate the amount of defects in the graphitic walls [48]. For example, DWCNTs typically show a larger I_D/I_G ratio compared to SWCNTs. A lower I_D/I_G value for the SWCNT sample suggests a smaller concentration of structural defects and/or amorphous material. Tuinstra et al. proposed to use I_D/I_G as a probe for the size of graphite microcrystals [49]. Their approach has been modified by several authors and was also applied to carbon nanomaterials [50–52]. However, it should be mentioned that the crystallite size estimated by Raman spectroscopy corresponds to homogeneous vibrational domains, not the particle size.

Another disorder-induced Raman peak, known as D' band, can be found around 1,620 cm^{-1} [53]. The D' mode does not exist in pure graphite, but is observed for intercalated graphite compounds and MWCNTs. The peak has been assigned to the in-plane vibrations of the outer parts of graphite domains [54, 55].

5.2.3 Elimination of the D Band from Raman Spectra of Carbon Nanotubes

Since the intensity ratio between the D and G band (I_D/I_G) indicates the degree of disorder in carbon materials, elimination of amorphous carbon and disordered structures upon oxidation can be monitored using in situ Raman spectroscopy. Figure 12.3a shows the in situ Raman spectra of DWCNTs recorded during heating from 25°C to 600°C in air. A linear downshift of all peaks is observed with increasing temperature [45]. The thermal shift of the D, G^+, and G^- bands will be discussed in more detail in the following section. The intensity of the D peak, which results from defects and disordered carbon, starts to decrease at ~430°C until it completely disappears around 510°C. Another important observation is that the intensity changes of D and G band at high temperatures follow different trends (Fig. 12.3a). While the D band intensity decreases, the G band exhibits a continuous increase in intensity within the same temperature range. A quantitative analysis of the Raman data (Fig. 12.3b) shows that the intensity of the G band starts to increase at ~440°C, reaches a maximum around 500°C, and decreases to its original values when reaching 600°C. As shown in Fig. 12.3b, the decrease in D band intensity occurs at temperatures 20–30°C higher than the increase in G band intensity. Therefore, the initial decrease in I_D/I_G appears to be mainly due to the increase of the G band intensity. An explanation for the early intensity increase of G band may be the possible removal of hydrocarbons (the tube synthesis was conducted in a hydrogen-containing atmosphere) and disordered carbon, which were shielding the Raman signal from the CNTs. The fact that the G band's absolute intensity starts to increase around the same temperature as the D band decreases seems to support this hypothesis. Further evidence is provided by the fact that a higher intensity (200–300%) of the G band was observed (same sample spot) after cooling to room temperature in every experiment. When an already oxidized sample was used for the in situ Raman studies, no increase in G band intensity was observed. In this case, there was no disordered carbon on the tube surface which could influence the Raman signal of the tubes. Figure 12.3c compares the high-frequency Raman spectra of the DWCNT sample before and after heating (measured at room temperature). The graph demonstrates that the splitting of the G band is more pronounced after heating. The temperature-induced shift of the peaks is not completely reversible. This can be explained by the removal of amorphous carbon and highly defective CNTs. Since the G^+ peak (~1,590 cm^{-1}) does not depend on the tube diameter [56] its position should not change after oxidation of smaller tubes and their decreasing contribution to the total Raman intensity. The G^+ peak is slightly sharpened, but shows little shift in frequency. However, the G^- band depends on the tube-diameter and is, therefore, affected by the oxidation of the smaller CNTs, leading to irreversible changes in peak positions and larger G band splitting. The RBM range shows no significant changes (Fig. 12.3d), indicating that the overall size distribution of the DWCNT sample has not been changed. New peaks observed between 475 and 690 cm^{-1} were assigned to metal oxides, formed by oxidation of remaining catalyst impurities.

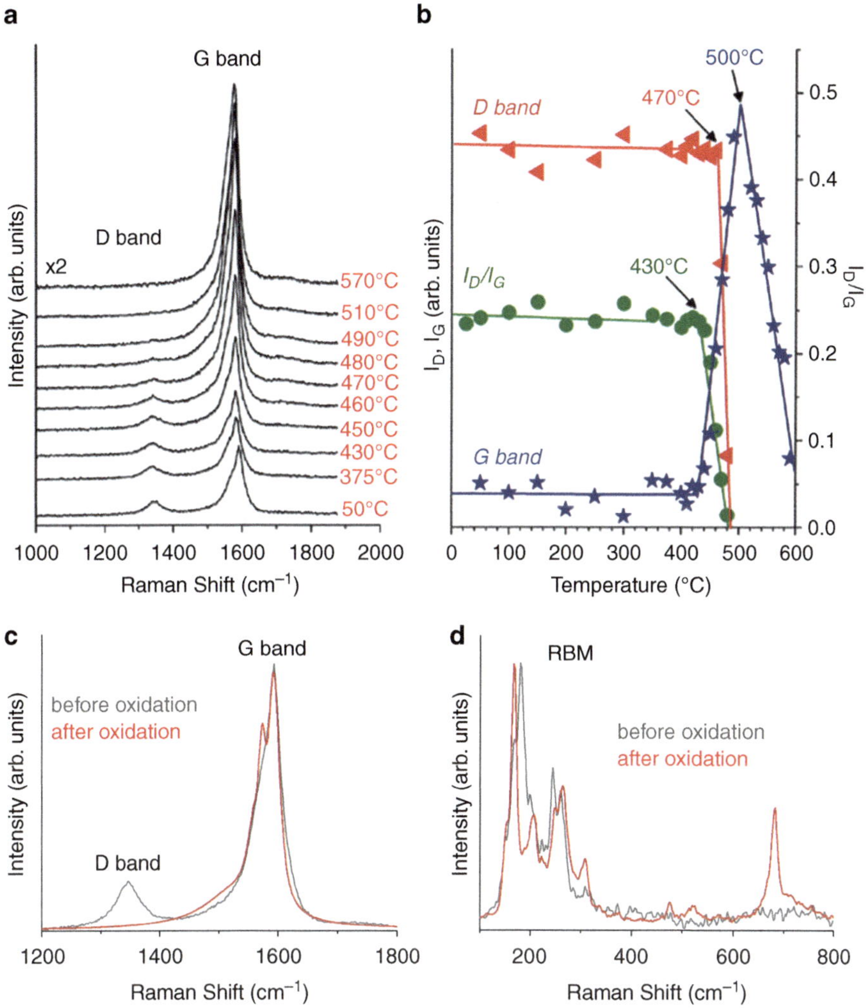

Fig. 12.3 (**a**) In situ Raman spectra of DWCNTs during heating from 25°C to 600°C in air and (**b**) corresponding intensity changes in the D and G bands during oxidation. Comparison of (**c**) high- and (**d**) low-frequency Raman spectra of DWCNT before and after oxidation recorded at room temperature. All spectra were measured using 514-nm laser excitation wavelength

The most noticeable effect during the in situ Raman studies was the near complete disappearance of the disorder-induced D band after oxidation (Fig. 12.3c). These results show that for the DWCNT sample, the D band originates mainly from amorphous carbon present in the sample and not from defects in the wall structure of the nanotubes. While the concentration of defects probably increases during the oxidation, disordered carbon and the associated D band disappear completely. However, it is well known that only metallic CNTs contribute to the D band intensity [57]. Therefore, the absence of any Raman signal between 1,300 and 1,400 cm^{-1} could also indicate a selective removal of metallic CNTs, or the presence of exclusively highly ordered, defect-free DWCNTs.

5.2.4 Nonisothermal Oxidation of Single- and Double-Wall Carbon Nanotubes

Figure 12.4a, b show the in situ Raman spectra of SWCNTs and DWCNTs, respectively, recorded during heating from 25°C to 600°C (in air) using 633-nm laser excitation. Both samples exhibit a linear wavenumber downshift for both D and G bands with increasing temperature. The value of temperature-induced shifts of the G band frequencies in the DWCNTs Raman spectra is between the values of SWCNTs and graphite, as expected [58]. In the case of DWCNTs (Fig. 12.4a), I_D/I_G starts to decrease above 450°C until the D band nearly disappears upon reaching 550°C. In the case of SWCNTs (Fig. 12.4b), no significant changes in I_D/I_G are observed below 400°C. However, between 400°C and 500°C, a decrease in I_D/I_G together with a decrease in both D and G band intensities serve as a clear sign for the oxidation of both disordered carbon and CNTs. Changes in the RBM frequency range after oxidation confirm these assumptions (Fig. 12.4d). The increase in intensity of the peak at 337 cm^{-1} in the Raman spectra of DWCNTs (Fig. 12.4c) after oxidation may be explained by oxidation of outer tubes and the resulting increase in the Raman intensity of the smaller inner tubes. However, it may also originate from remaining metal catalysts and the corresponding oxides formed during heating. Because oxidation of DWCNTs increases the relative amount of catalyst particles in the sample, the contribution of catalyst impurities to the Raman spectrum is expected to increase during heating. We also observed new peaks (475, 520, and 683 cm^{-1}) appearing at temperatures >500°C which can be attributed to cobalt oxide. Similar to DWCNTs, the peak at ~388 cm^{-1} in the Raman spectra of SWCNTs (Fig. 12.4d) is expected to result from catalyst particles or other impurities in the sample. The intensity of this Raman feature does not change after oxidation, while lower frequency peaks, corresponding to larger and more stable SWCNTs, show a noticeable decrease in intensity. The origin of the observed behavior of D and G bands in the case of DWCNTs can be explained by the removal of disordered carbon, which shields the Raman signal of the CNTs. The SWCNT sample, showing no increase in G band intensity, contains less amorphous carbon compared to DWCNTs. Since the SWCNTs were purified after synthesis, we attribute the observed D band intensity to defects in the wall structure of the nanotubes, rather than amorphous carbon. This assumption is supported by the fact that D and G bands behave in a similar manner at temperatures above 400°C. Changes in the RBM range of the SWCNTs after oxidation also affirm this hypothesis (Fig. 12.4d). Small diameter tubes showing higher RBM frequencies start to decrease in intensity at elevated temperatures or even disappear after oxidation.

Figure 12.5 shows the in situ Raman analysis of the low frequency RBM range for DWCNTs, acquired using three different laser excitation wavelengths. The intensity of the peaks between 220 and 280 cm^{-1} decreases constantly during heating from room temperature to 400°C (Fig. 12.5a). Since there are no observable

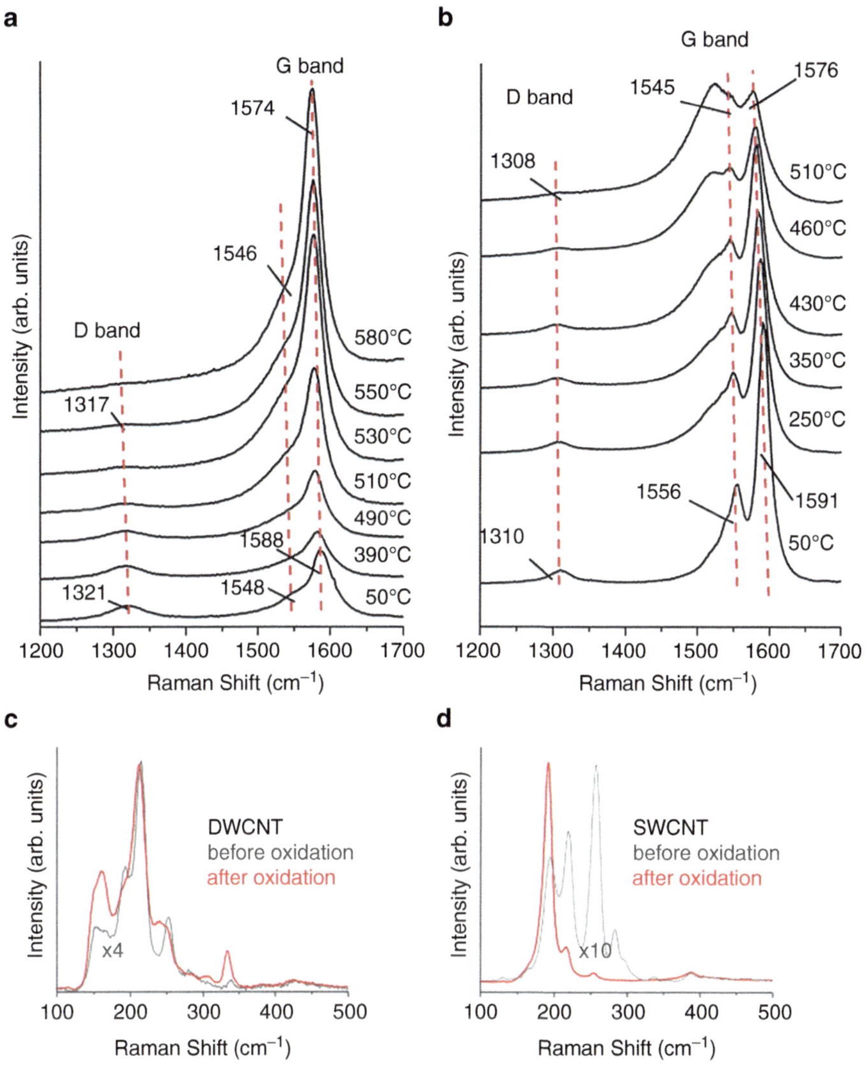

Fig. 12.4 In situ Raman spectra of DWCNTs (**a**) and SWCNTs (**b**) during nonisothermal oxidation in air. RBMs of DWCNTSs (**c**) and SWCNTs (**d**) before and after oxidation. Spectra were recorded using a 633-nm excitation wavelength

changes in the D and G band shape or intensity, the differences cannot be attributed to structural changes or selective oxidation of CNTs. However, RBM frequencies are strongly related to the electronic band structure through the resonance effects. Therefore, temperature-induced changes in electronic density of states may affect the intensity of the RBM signal at elevated temperatures and lead to the observed temperature-dependence. Removal of adsorbed surface species or debundling effects may also contribute to shifts in RBM frequencies.

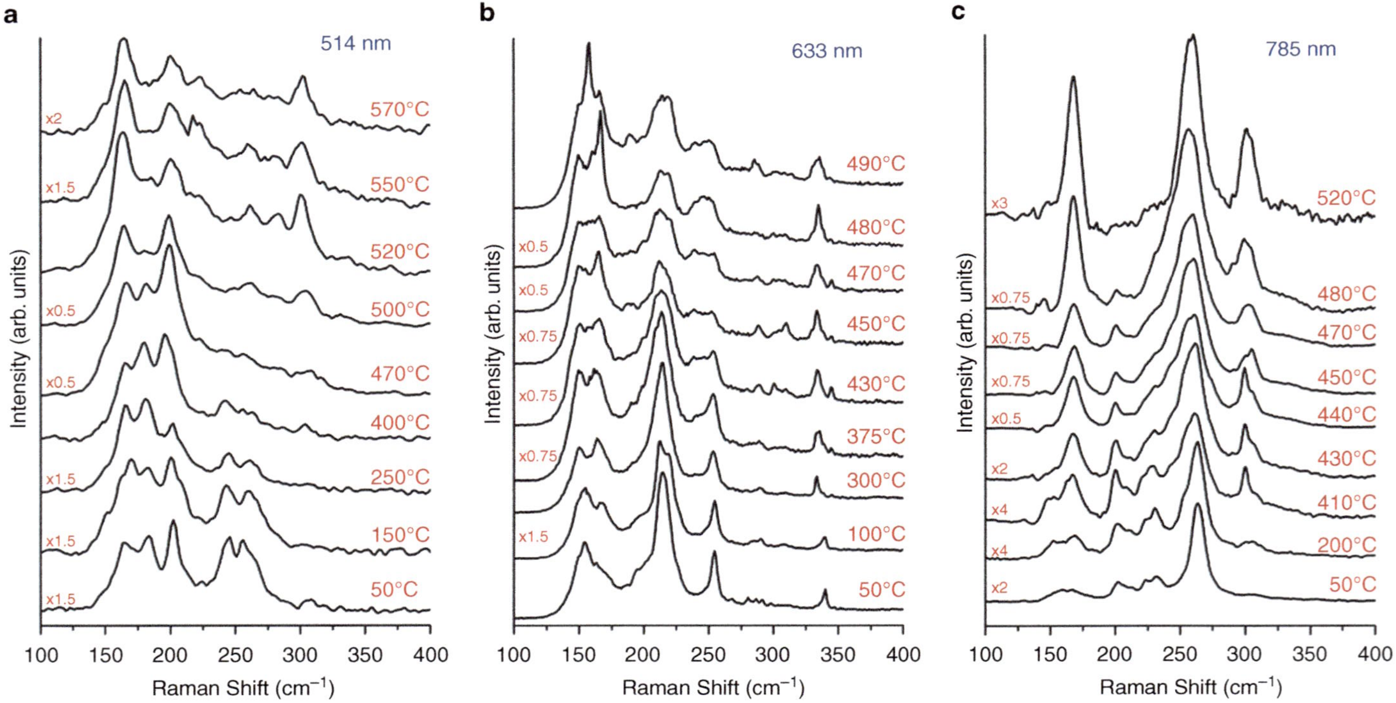

Fig. 12.5 Multiwavelength Raman spectra of DWCNTs recorded in situ using a 514-nm (**a**), 633-nm (**b**), and 785-nm (**c**) laser excitation [44]. Different laser wavelengths lead to different RBM spectra due to resonant enhancement effects

In situ measurements during heating of DWCNTs at 5°C/min up to 400°C, directly followed by cooling to room temperature, show certain RBM peaks decreasing and increasing in intensity while the sample is heated or cooled. The observed changes are completely reversible and somewhat different for frequencies originating from nanotubes with different diameters or when using different excitation wavelengths. Changes in the intensity distribution of the RBM range can thus be caused not only by changing the laser wavelength, but also by changing temperature. This is an important finding, considering the fact that laser-induced sample heating is most likely to occur during Raman analysis of carbon nanomaterials [59]. It should be mentioned that there is also a temperature-induced downshift in the RBM frequencies. However, the shift is only noticeable for some RBM peaks (Fig. 12.5), and it is less pronounced than that observed for D and G bands. The in situ RBM Raman spectra of DWCNTs show the highest intensities in the temperature range between 450°C and 500°C (Fig. 12.5). The small amount of amorphous carbon within the as-received DWCNT sample shields RBM-Raman scattering similar to the G band Raman signal as described in Sect. 5.2.3. After reaching a maximum around 500°C, the RBM intensity decreases due to the oxidation of CNTs. Furthermore, it can be seen that, after reaching temperatures above 400°C, the intensity of several RBM peaks between 300 and 350 cm^{-1} starts to increase (Fig. 12.5a, c). This effect can be explained by the oxidation of outer tubes and the increase in the detection of the Raman scattering of the smaller inner tubes exhibiting higher RBM frequencies. Inner tubes are not oxidized before the outer ones are damaged or destroyed.

5.2.5 Purification of Carbon Nanotubes by Isothermal Oxidation

Figure 12.6a shows changes in I_D/I_G as a function of the oxidation time. Below 365°C no changes are observed and neither amorphous carbon nor other carbon species appear to oxidize under these conditions. This is in agreement with Fig. 12.6b, which shows no changes in the D band area (integrated intensity) below 365°C. The activation energy needed for breaking the carbon bonds is not provided below that temperature. The first changes in I_D/I_G are observed at 370°C (Fig. 12.6a). In these experiments, the ratio decreased from ~0.25 to approximately 0.08. The ratio starts to decrease after the 20–50 min incubation period, which may be required for oxygen chemisorption or the removal of functional groups from the carbon surface, reaches the final value of 0.08 after 220–240 min and does not show further changes with time. In the range of 370–400°C, similar changes in I_D/I_G are observable, suggesting that only the amorphous carbon is oxidized and the nanotubes are not affected by oxidation. The incubation period disappears and I_D/I_G decreases further to zero after a relatively short time as the processing temperature increases from 440°C to 550°C (Fig. 12.6a). Defective and small-diameter CNTs are oxidized in this temperature range leading to the disappearance of the D band (Fig. 12.6b). In the case of 370°C, 375°C, and 400°C, the D band area shows behavior similar to I_D/I_G. The relative value of the D band decreases to about 35% of its original value, while I_D/I_G decreases by ~65%.

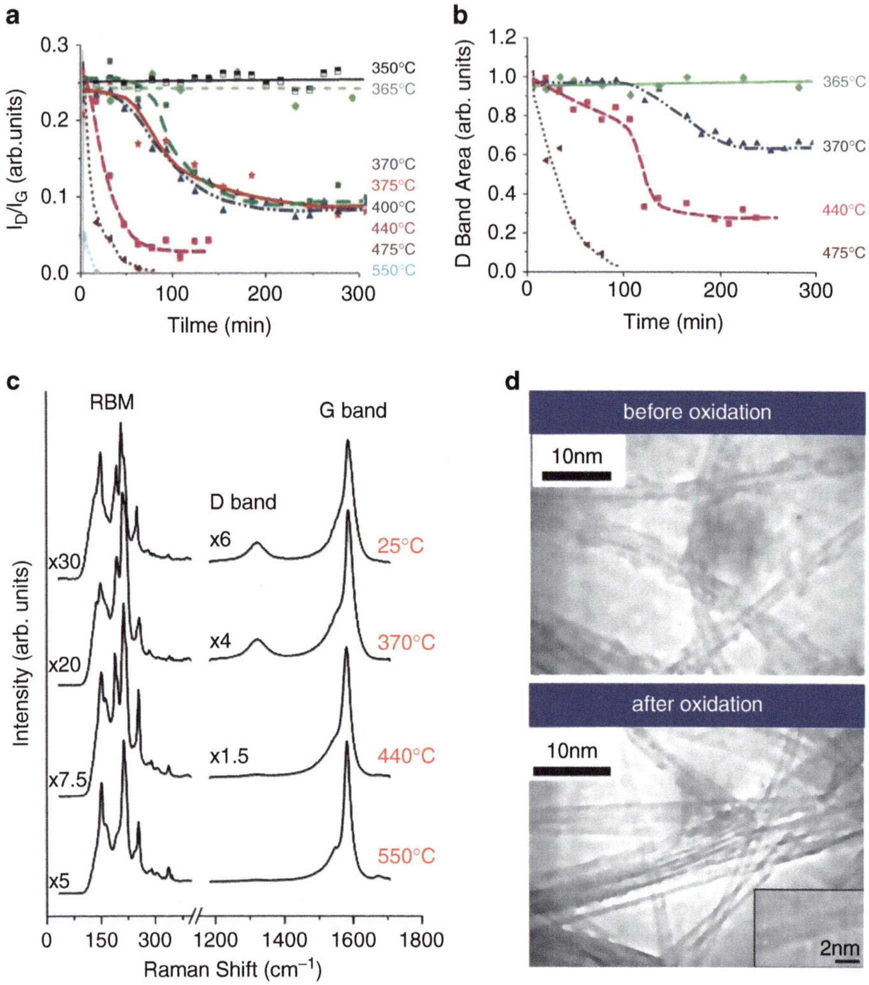

Fig. 12.6 Results of isothermal oxidation of the DWCNTs showing (**a**) the I_D/I_G ratio and (**b**) relative D band intensity. The curves for 350°C and 365°C in (**b**) are superimposed, and idem for 370°C, 375°C, and 400°C. (**c**) Raman spectra of DWCNTs after isothermal oxidation for 5 h at different temperatures, recorded at room temperature. (**d**) HRTEM images of the DWCNT sample before and after oxidation. All Raman spectra were recorded using a 633-nm excitation wavelength

As discussed in Sect. 5.2.3, the decrease of the D band intensity is accompanied by a large increase in G band intensity (not shown). Between the 50th and 150th min, a small decrease in the D band area (10–15%) resulted in a doubling of the G band area, leading to the large decrease of I_D/I_G (Fig. 12.6a). After about 150 min, both the G band area and I_D/I_G reach a maximum and minimum, respectively. The subsequent decrease of the D band changes I_D/I_G significantly since the G band area and intensity are decreasing as well. The isothermal oxidations at

440°C, 470°C, and 550°C exhibit a different behavior (Fig. 12.6a): I_D/I_G starts to decrease during heating since oxidation of the amorphous carbon starts before reaching the final temperature. However, 440°C is not high enough to eliminate the D band in the DWCNTs, but the final I_D/I_G value (0.04) is much lower than the values for 370°C, 375°C, and 400°C. Analysis of Fig. 12.6a, b leads to the conclusion that, along with the amorphous species, smaller and defective tubes are oxidized, resulting in an additional decrease in the D band area at 440°C, compared to isothermal experiments at temperatures between 370°C and 400°C. Please note that the lines for 350°C and 365°C in Fig. 12.6b can be superimposed, and the curves obtained at 370°C, 375°C, and 400°C are almost identical. A further increase of the G band intensity, and thus decrease in I_D/I_G, at temperature above 400°C cannot be assigned to the shielding effect of amorphous carbon. However, the observed behavior may be explained by the fact that the measured G band Raman signal comes primarily from tubes of a particular diameter range, due to resonant effects. Thus, oxidation of smaller tubes results in a higher percentage of resonantly enhanced nanotubes and an increase in the G band Raman intensity. Temperatures above 480°C lead to the complete oxidation of DWCNT samples, showing elimination of D band and time-dependent G band decrease.

Figure 12.6c supports these assumptions. For oxidation temperatures between 350°C and 370°C the RBM frequency ranges show similar shapes and intensity ratios between the peaks within the Raman spectrum recorded after oxidation at room temperature, indicating that no nanotube structures are damaged or removed by oxidation. Temperatures above 440°C lead to changes in the peak shape and intensity ratios of both, the G band and RBM modes, induced by the oxidation of tubes with different diameters. The D band intensity shows no decrease for temperatures below 370°C, while the decrease is similar for temperatures between 370°C and 400°C where it reaches a final value of 65% of its original value (Fig. 12.6b). Above 440°C the final D band intensity, at the completion of the oxidation treatment, depends on the process temperature. Thus, the temperature range for the heating-induced purification of DWCNTs, with respect to the amorphous carbon, is between 370°C and 400°C. These results show that it is possible to develop an effective temperature-dependent oxidation method for nanotubes with different diameters and structural perfection. Oxidation of larger amounts of DWCNTs using a furnace under similar conditions demonstrated the feasibility of scaling up the process. The furnace oxidation at 375°C for 300 min showed a similar decrease in I_D/I_G as after the heating stage experiments, while experiments at 360°C show no changes in the D band intensity, nor in I_D/I_G. Thus, the temperature range between 370°C and 400°C provides the optimal conditions for purification of the investigated DWCNTs. Figure 12.6d shows the HRTEM images of DWCNT samples before and after the purification. Before oxidation, amorphous carbon is present in the sample, typically at the surface of the nanotubes. After the treatment, disordered carbon is completely removed and only catalyst particles are left next to the nanotubes (not shown). While HRTEM does not provide statistically reliable data on the content of defects in the tube walls, it clearly shows that the oxidized tubes are not defect-free and do not look more perfect compared to the nonoxidized DWCNTs. However, these defects did not cause a double-resonance effect and did not produce a D band in Raman spectra, suggesting presence of primarily metallic CNTs. While the weight loss after heating to 600°C was significant, further heating experiments in air have shown that complete removal of disordered carbon leading to disappearance of D band can be achieved by a long-term isothermal treatment at temperatures below 400°C, at a much smaller weight loss. The efficiency of the developed purification procedure was found to be superior to that of acid-based processes and can also be applied to SWCNT and MWCNT samples [44, 60]. Thus, isothermal oxidation of CNTs is a promising method of purification with respect to the removal of non-tubular carbon and defective tubes.

5.2.6 Graphitization, Defect Formation, and Functionalization

Figure 12.7a shows the Raman spectra and HRTEM images of as-received and graphitized (vacuum-annealed at 1,800°C for 3 h) MWCNTs [60]. The as-received tubes are partially surrounded by amorphous carbon and show iron inclusions (catalyst particles), usually at the end of the tubes (inset Fig. 12.7a). They typically exhibit well-ordered graphitic walls with local defects and vary in shape and diameter. Annealing induces a graphitization of amorphous material and healing of structural defects, leading to a well-defined tube surface with a homogenous structure (inset Fig. 12.7a). Furthermore, annealed MWCNTs show no metal inclusions at the tube tips or anywhere else. Catalyst particles are evaporated upon the high-temperature vacuum treatment and removed from the sample [61, 62]. Carbon impurities are graphitized and used to repair the defects in the wall structure of the tube, providing a potential enhancement of electrical and thermal properties. The I_D/I_G ratio of the as-received MWCNTs shows an average value of ~1.8. Due to prior purification steps, the as-received MWCNTs do not contain large amounts of amorphous carbon, and we can assume that the D band intensity mainly results from structural defects in the tubes. The Raman spectrum of the graphitized MWCNTs shows similar, but more narrow features and exhibit a lower I_D/I_G value with an average of ~1.1. Raman spectra were normalized with respect to the D band intensity for comparison purposes. Thus, only statements about relative intensity changes can be made. Since I_D/I_G can be used as a measure of the graphitic nature of carbon materials, we conclude that vacuum annealing (at 1,800°C for 3 h) improves the structural order and the purity of the CNT sample. A lower FWHM of D and G bands in the spectrum of the graphitized MWCNTs confirms this assumption. The vacuum-annealing provides many advantages compared to a chemical purification because it allows the complete removal of metal impurities, including particles trapped inside the tubes. Metal-free nanotubes are desired for a large number of applications, especially biomedical ones in which high purity and defined surfaces are required.

Due to their larger diameters, MWCNTs experience smaller curvature-induced strain on the C=C bonds. Thus, the resistance to oxidation increases, leading to higher activation energies and oxidation temperatures compared to SWCNTs and DWCNTs. In order to determine the temperature range for the oxidation of the

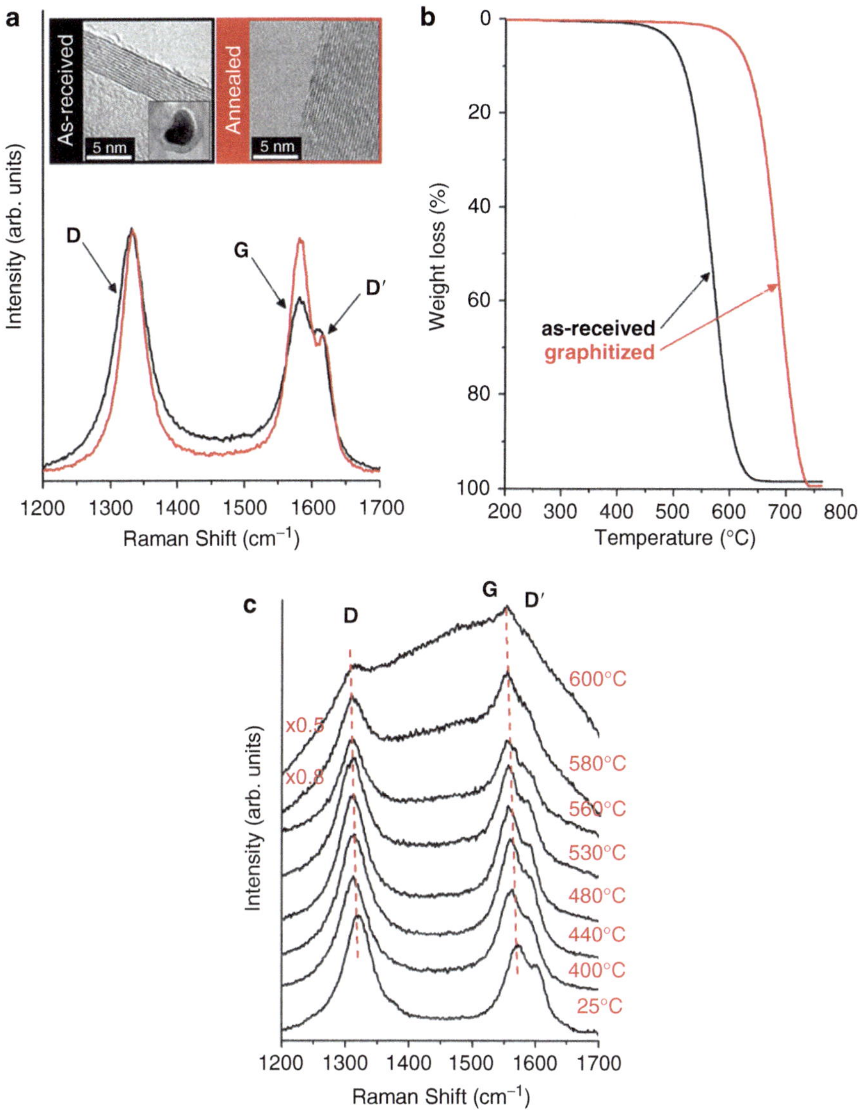

Fig. 12.7 (**a**) Raman spectra and HRTEM images of as-received and vacuum-annealed (graphitized) MWCNTs. As-received nanotubes contain iron particles and amorphous carbon on their surface (inset). (**b**) Weight loss curves (TGA) of as-received, air-oxidized (0.25 h at 550°C), and graphitized MWCNTs. (**c**) In situ Raman spectra of nonisothermal oxidation of as-received MWCNTs. All Raman spectra were recorded using 633-nm laser excitation

as-received and graphitized MWCNTs in air, we performed TGA for both samples (Fig. 12.7b). Oxidation of as-received MWCNTs begins at 440–450°C. The remaining residue of ~2 wt.% results from iron inclusions, which form iron oxide and remain after oxidation of MWCNTs. Graphitized MWCNTs are more resistant against oxidation than as-received MWCNTs, and no oxidation-induced weight loss is observed below ~550°C. The improved thermal stability results from the lower defect density and the removal of iron inclusions.

In situ Raman studies during oxidation of as-received MWCNTs in air showed no significant change in I_D/I_G below ~450°C (Fig. 12.7c), as expected. The intensities of D and G bands started to decrease simultaneously between 450°C and 500°C, while I_D/I_G showed almost no variation indicating a general sample loss without change in ordering. A further temperature increase to ~600°C speeds up the oxidation reaction and increases the sample loss. Similar to SWCNTs and DWCNTs, in situ measurements on MWCNTs revealed a linear wavenumber downshift for all Raman bands with increasing temperature (Fig. 12.7c). The measurements exhibit a downshift of ~0.017 cm^{-1}/°C for D, ~0.025 cm^{-1}/°C for G, and ~0.028 cm^{-1}/°C for D'. As expected, the value of the temperature-induced shift of the G band of MWCNTs is between the values of SWCNTs (0.03 cm^{-1}/°C) and DWCNTs (0.026 cm^{-1}/°C) and high purity graphite (0.024 cm^{-1}/°C). It should be mentioned that the downshift may also be influenced by laser-induced heating. Experiments with graphitized MWCNTs showed similar temperature-induced wavenumber shifts, but no considerable change in the intensity of the Raman bands below 600°C due to the absence of catalyst particles.

To further improve the control over the creation of defects, isothermal oxidation experiments were performed. Figure 12.8a shows the Raman spectra of MWCNTs after isothermal oxidation for 5 h at 400°C, 430°C, 460°C, 500°C, and 530°C in static laboratory air. The spectra were recorded at room temperature after the treatment, normalized with respect to the G band intensity, and displayed with a slight offset for clarity. The Raman spectrum of the nanotubes oxidized at 400°C is similar to that of the as-received tubes, with a slightly lower I_D/I_G of ~1.75. This decrease results from the removal of the amorphous carbon layer at the MWCNTs surface, without oxidizing or damaging the MWCNTs. From 430°C to 530°C, the oxidation reactions occur at the most reactive sites, that is, the tube ends. The graphitic order of the nanotubes is reduced, resulting in an increase in I_D/I_G up to 2.05 at 530°C. The analysis of the spectra showed that I_D/I_G increases almost linearly with temperature between 400°C and 530°C (inset in Fig. 12.8a). No further increase was observed after 530°C and the sample loss after 5 h was very large. Although isothermal measurements allowed a controlled change in I_D/I_G, the overall sample loss after a 5-h oxidation was too high and unacceptable for industrial applications. Thus, in the following section, we report another approach consisting of a "flash oxidation" using a short treatment at high temperature. This approach allows the creation of defects uniformly distributed over the tube surface, without a significant weight loss. At high temperatures, structural defects are created simultaneously everywhere on the nanotubes' surface, but quick cooling inhibits further damage in the nanotubes' structure. Figure 12.8b shows the Raman spectra of the as-received MWCNTs oxidized for 15 min and 30 min at 550°C. I_D/I_G increases from 1.8 to 2.1 (15 min) and 2.3 (30 min). Thus, the values of I_D/I_G after a short-term oxidation at 550°C are

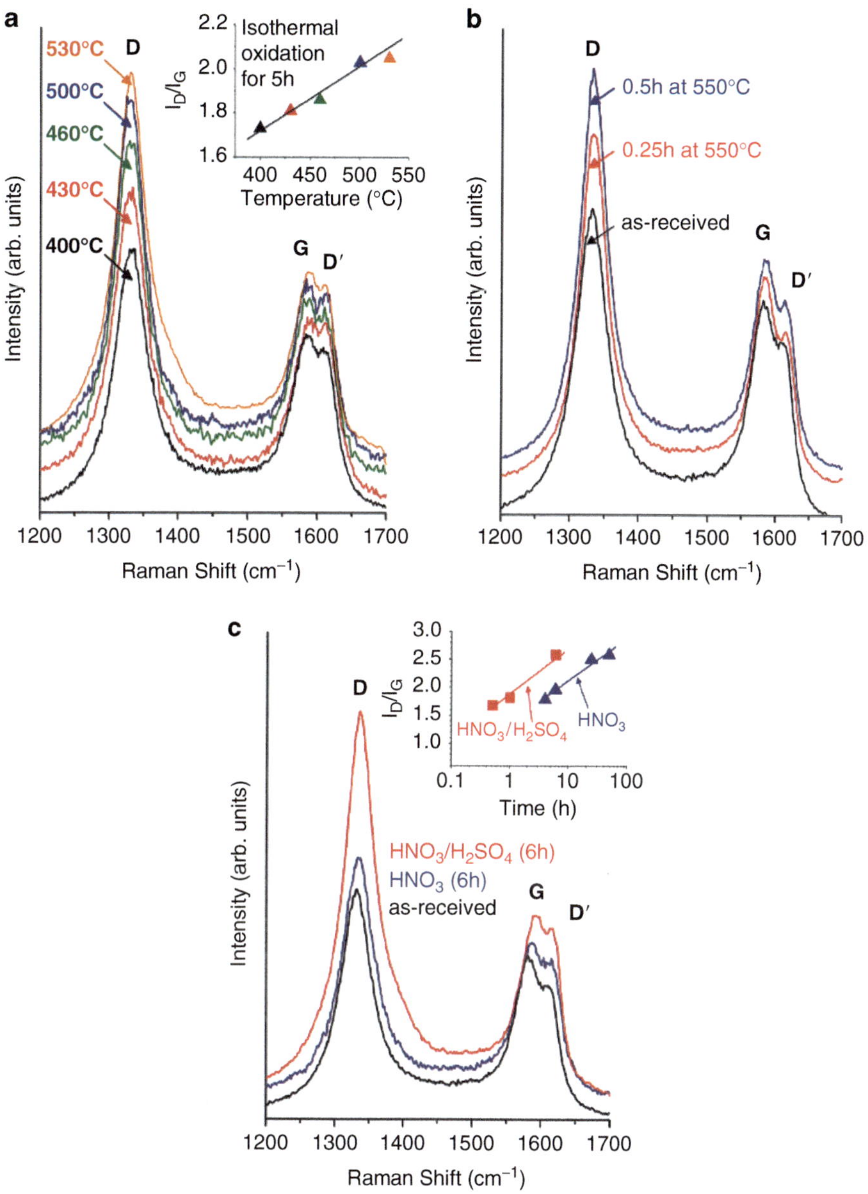

Fig. 12.8 (**a**) Raman spectra of MWCNTs after isothermal oxidation at different temperatures in air, recorded at room temperature. The *inset* shows changes in the ID/IG ratio as a function of oxidation temperature. (**b**) A "flash oxidation" allows a similar increase in ID/IG, but without a significant loss of the sample. (c) Raman spectra and ID/IG ratio (inset) of MWCNTs oxidized in HNO3 and HNO3/H2SO4. Raman spectra were recorded using 633-nm laser excitation

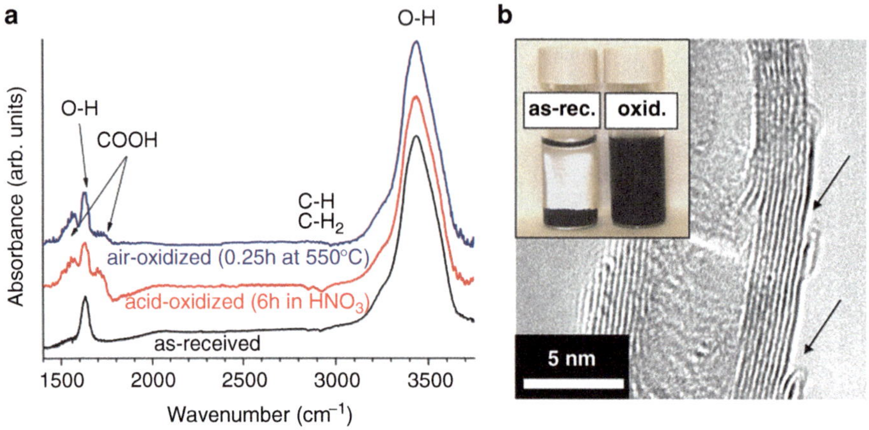

Fig. 12.9 (**a**) FTIR spectra of as-received, air-oxidized, and acid-treated MWCNTs. The assigned functional groups are indicated. (**b**) HRTEM images of oxidized MWCNTs. The oxidation treatment results in the oxidation of the outer walls and introduces defective sites which are subsequently saturated with oxygen-containing functional groups thus improving dispersion in polar solvents such as water (*inset*)

equal or higher than that after 5 h at 530°C ($I_D/I_G \sim 2.05$), indicating a higher defect density at lower sample loss (<10 wt.%) [62]. Therefore, we can conclude that a "flash oxidation" is a promising method to introduce defective sites into MWCNTs.

We also modified the as-received MWCNTs using commonly employed acid-based oxidation procedures. Figure 12.8c compares the Raman spectra of MWCNTs treated for 6 h in HNO_3 and HNO_3/H_2SO_4 at 110°C. While the 6-h HNO_3 treatment results only in a few changes in the Raman spectrum (I_D/I_G remains stable at ~1.8), long-term treatments lead to significant changes. For oxidation periods of 24 and 48 h, I_D/I_G increases to 2.5 and 2.6, respectively. As a comparison, a mixture of concentrated sulphuric and nitric acid was used. This mixture, which is a much stronger oxidizer than HNO_3 alone, gave high I_D/I_G values (~2.6) after only 6 h instead of 48 h required in the case of HNO_3 alone (inset in Fig. 12.8c).

Oxidation-based functionalization of MWCNTs (i.e., COOH grafting) of CNTs can be used to improve the solubility in water and other polar solvents, allowing for homogeneous and stable dispersions. Figure 12.9a shows the FTIR spectra of as-received and air-oxidized MWCNTs, in comparison to acid-treated MWCNTs (8 h at 75°C in HNO_3). The main features in the FTIR spectra of as-received tubes result from O–H vibrations (3,280–3,675 cm^{-1} stretch and 1,640–1,660 cm^{-1} bend) which are assigned to –O–H groups of adsorbed water or covalently bonded functional groups. Since traces of water remaining in the KBr pellets could not completely be removed, even though extensive heating before the measurement (24 h at 100°C) had been performed, we assume that a majority of the O–H vibrations originate from water in the sample rather than functional groups attached to the surface of the MWCNTs.

The FTIR spectra of air-oxidized and acid-treated MWCNTs show some additional features. The broad band at ~1,730 cm^{-1} results from the C=O stretching vibrations of carboxyl and carbonyl groups. The origin of the band at 1,566 cm^{-1} is still controversial [63–65]. Some authors believe that the band results from the in-plane vibrations of the graphitic walls in carbon nanotubes [66, 67]. Others suggest that the spectral features between 1,540 and 1,590 cm^{-1} are also related to carboxyl [68] and carbonyl groups [69]. Our experiments support the latter assumption. Indeed, the IR band at 1,566 cm^{-1} was always accompanied by the 1,730 cm^{-1} band, while it did not appear in the FTIR spectrum of the untreated as-received MWCNTs. The FTIR analysis of carbon samples is very complex and strongly depends on the sample itself, the treatments, and the experimental environment. These circumstances lead to spectral variations. However, FTIR studies showed some distinct differences between the as-received and treated MWCNTs. Both, air-oxidized and acid-treated samples show the presence of C=O bond vibrations, while they were not observed for as-received MWCNTs. Thus, the FTIR analysis further proves the existence of defects in the walls of the nanotubes and the subsequent formation of functional groups. As shown in the inset of Fig. 12.9b, oxidation of MWCNTs leads to significant improvements in solubility and dispersion of CNTs in water.

5.2.7 Comparison of Single-, Double-, and Multi-wall Carbon Nanotubes

The oxidation behavior of SWCNTs and related changes in their Raman spectra are comparable to that of DWCNTs due to similarities in size and structure. Therefore, we will distinguish between large-diameter MWCNTs and small-diameter CNTs. In the following section, DWCNTs were chosen as the representative for small-diameter CNTs. Figure 12.10a compares the oxidation behavior of DWCNTs and MWCNTs during in situ Raman spectroscopy studies. The high oxidation temperatures of vacuum-annealed MWCNTs (defect-free, no catalyst) were beyond the temperature range of the heating stage (25–600°C) and did not allow for any in situ Raman studies.

In general, the oxidation of CNT samples can be divided into three temperature regions (Fig. 12.10a). At temperatures below ~350°C (region I), only water and other adsorbed species are removed from CNT surface, but oxidation of carbon does not occur. Region II is characterized by the removal of amorphous and disordered carbon from the sample. However, the temperatures remain insufficient for CNT oxidation to occur. Above ~400°C (region III), both amorphous carbon and CNTs are oxidized simultaneously. The individual reaction rates depend on both the temperature and the composition of the sample. The transition between region I and II typically occurs around 350–370°C. The transition temperature between region II and III varies from sample to sample, and depends on the relative amount of amorphous carbon and the structural properties of the CNTs. As-received DWCNTs exhibit a lower I_D/I_G ratio than as-received MWCNTs, suggesting a higher structural perfection and/or a lower content of amorphous carbon (region I). In situ Raman studies under isothermal and nonisothermal

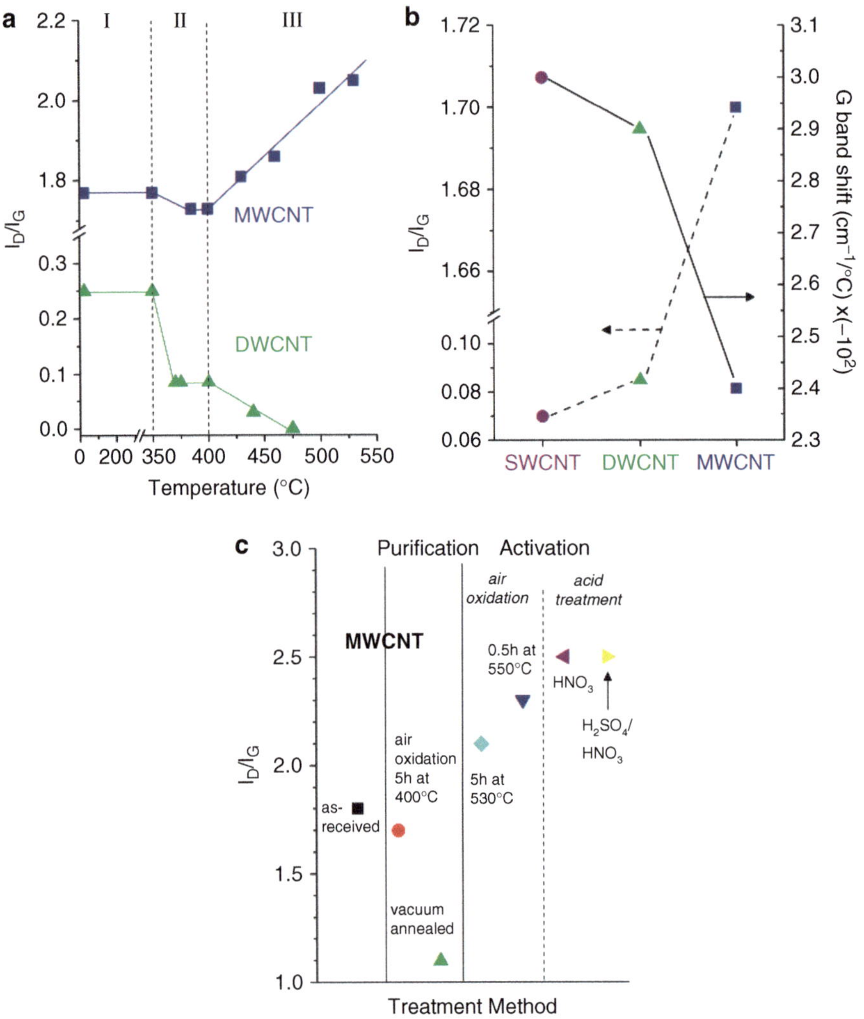

Fig. 12.10 (**a**) In situ Raman studies of I_D/I_G ratio variation during oxidation of DWCNTs and MWCNTs. (**b**) Comparison of G band shift and I_D/I_G ratio of SWCNTs, DWCNTs, and MWCNTs. (**c**) Changes in I_D/I_G of MWCNTs after various oxidation and purification treatments

conditions demonstrated the possibility of selectively removing amorphous carbon from CNT samples between 350°C and 400°C without damaging the tubes. The removal of amorphous carbon from the DWCNT sample upon oxidation revealed a ~65% decrease in I_D/I_G (from 0.25 to 0.09), while in case of MWCNTs the values decreased only by ~2–3% (from 1.77 to 1.73). These results suggest that the majority of the D band intensity in the Raman spectrum of as-received DWCNTs can be ascribed to amorphous carbon, while the D band of MWCNTs originates mainly from structural defects. Thus, only I_D/I_G values measured at ~400°C can be used to evaluate the structural perfection of the CNTs, while the difference in I_D/I_G before and after oxidation at 400°C can be understood as a measure of purity (content of amorphous carbon). Although the contributions of small amounts of amorphous carbon may be insignificant for the weight loss in TGA studies, they strongly affect the Raman spectra of the samples (shielding effect). This is an important finding because it demonstrates that in some cases, the I_D/I_G is not an accurate measure of the structural perfection or the purity of CNT samples, and that additional structural characterization during oxidation is required for a correct interpretation of the I_D/I_G ratio changes and a differentiation between contributions from amorphous carbon and lattice defects. Comparison of Raman spectra of DWCNTs before and after oxidation at different temperatures and in situ studies monitoring changes in Raman intensities in real time showed a selective oxidation of highly defective and/or small-diameter CNTs at temperatures above 400°C, leading to a decrease in I_D/I_G. In contrast, similar studies on MWCNTs revealed a continuous increase in I_D/I_G. The observed behavior can be explained by the large differences in the reaction rate and the selective oxidation of defective and small-diameter DWCNTs. Oxidation of MWCNTs occurs at defective sites, but is less size-selective. Carbon atoms are removed from all MWCNTs simultaneously, leading to higher defect density, while the number of MWCNTs remains almost unchanged.

Figure 12.10b shows the I_D/I_G ratio of SWCNTs, DWCNTs, and MWCNTs after oxidation at 400°C in comparison with the temperature-induced downshift of the G band. It can be seen that the I_D/I_G ratio, which characterizes the defect density of the CNTs, increases with size (number of walls), while the temperature-dependence of the G band decreases. The larger temperature-dependence of the G band of small CNTs results from the curvature-induced strain on the carbon-carbon bonds, which increases the sensitivity to changes in the environment (e.g., temperature, pressure). For the same reason, activation energies are lower for small-diameter CNTs and increase with tube size. Thus, temperature-induced changes in the Raman spectra of CNTs can also be used to evaluate the thermal stability and resistance to oxidation. However, a determination of the exact relationship requires samples with a narrow size-distribution and a high purity (no amorphous carbon, no metal catalyst).

Figure 12.10c shows changes in I_D/I_G of MWCNTs after various treatments, which can be classified into purification (air oxidation at 350–400°C, vacuum-annealing at 1,800°C) and activation (acid treatments, air oxidation at >400°C, flash oxidation at >500°C). Purification techniques lower the content of amorphous carbon without damaging the CNTs, while activation treatments increase the number of defects, leading to lower and higher I_D/I_G values, respectively. If treatment conditions and sample composition are known, I_D/I_G can be used to estimate the defect-concentration, which in turn is a measure of the density of surface functionalities, such as carboxylic groups after H_2SO_4/HNO_3 treatments.

These results demonstrate the great potential of air oxidation for selective purification of CNT samples, and/or a controlled formation of defects accompanied by a surface functionalization with oxygen-containing functional groups.

In addition, vacuum-annealing at 1,800°C was found to induce graphitization of amorphous carbon and healing of structural defects. Catalyst particles were evaporated upon the treatment and completely removed from the sample. Both methods show many advantages compared to a chemical purification (e.g., acid treatments) while providing similar results with respect to purity and control of surface chemistry. Although the I_D/I_G ratio can be used to evaluate purity and defect-concentration of CNT samples, great care must be taken during interpretation of the results, and detailed information on sample composition and synthesis conditions is essential.

5.3 Purification and Size Control in Nanodiamond

Nanocrystalline or ultra-dispersed diamond, often simply referred to as "nanodiamond" (ND), is another very promising carbon nanomaterial with a large number of potential applications [70–73]. ND can be produced either as thin films using chemical vapor deposition (CVD) techniques [71, 72] or as powder via detonation of carbon-containing explosives, such as trinitrotoluene (TNT) and hexogen, in a steel chamber [74, 75]. In this chapter, the term "nanodiamond" refers to ND powders produced by the detonation synthesis. ND powders contain aggregates of primarily ~4–8-nm ND crystals, each consisting of a diamond core partially or completely covered by layers of graphitic and/or amorphous carbon. Due to a large number of surface atoms and a large surface/volume ratio, ND powders exhibit a very high surface reactivity compared to other carbon nanostructures that have mainly basal planes of graphite on their surface. Thus, NDs combine an active surface, featuring a variety of chemically reactive moieties, with the favorable properties of macroscopic diamonds, including the extreme hardness and high Young's modulus, chemical stability, biocompatibility, high thermal conductivity, and electrical resistivity. Moreover, ND is one of few nanomaterials produced in large commercial quantities. Currently, ND is used in composites, as an additive in cooling fluids, lubricants, and electroplating baths [70–73]. However, a large number of other potential applications, including drug delivery, stable catalyst support, transparent coatings for optics, and others still remain under-explored. Most of these applications are hindered by the current inability of the manufacturers to provide ND with well-controlled surface chemistry, as well as the absence of a process that would achieve this control in a research laboratory.

The raw diamond-bearing soot from detonation synthesis consists of ND agglomerates, non-diamond carbon, as well as metals, metal oxides, and other impurities coming from the detonation chamber or the explosives used. Most producers employ liquid phase oxidizers, such as HNO_3, $HClO_4$, H_2SO_4, H_2O_2, or aqueous and acidic solutions of $NaClO_4$, CrO_3, or $K_2Cr_2O_7$, to selectively remove amorphous and graphitic carbon from ND powders [71, 76]. At present, the purification treatment remains the most complicated and expensive stage of the ND production [77]. Moreover, the purity that has been achieved using the

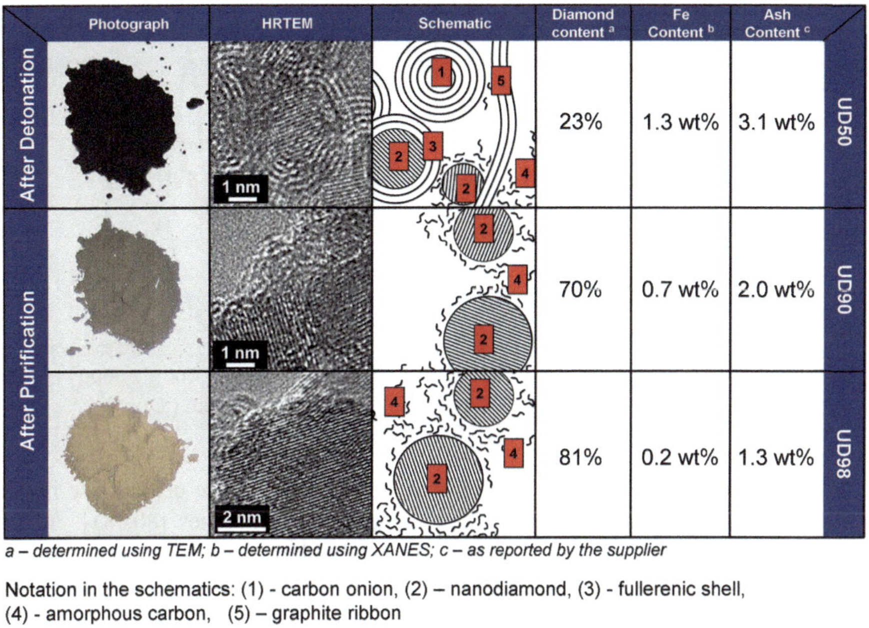

Fig. 12.11 Structure and composition of three different ND powders with diamond contents ranging from <23 wt.% (UD50) to >80 wt.% (UD98) as determined by XANES studies [78]

described methods is still insufficient. Most of the existing methods require the use of either toxic and aggressive substances, or supplementary catalysts, which result in an additional contamination or a significant loss of the diamond phase. Figure 12.11 shows structure and composition of three commercially available ND powders before (UD50) and after purification (UD90, UD98) [78]. Black UD50 is the unpurified raw detonation soot containing mainly non-diamond carbon structures, such as amorphous carbon, carbon onions, and graphite ribbons. The black color of the powder is related to the high content of sp^2 carbon (>70%). UD90 and UD98 are purified by a multi-stage acid treatment using sulfuric (H_2SO_4) and nitric (HNO_3) acid. The detonation soot is immersed in concentrated acid, stirred, and cooked for several hours at elevated temperatures. After cooling, the ND powders are washed in water until neutral (pH ~7) and centrifuged to extract the diamond crystals from the aqueous solution. While these techniques are widely used, they do not provide sufficient purity, leading to the dark-gray color of commercially available ND powders [78].

ND particles with diameters about 4 nm have ~20% of the total number of atoms on the surface. Because the physical properties of nanocrystals are strongly size-dependent, it is crucial to control and accurately measure the crystal size. To some extent, ND crystal size can be controlled by the synthesis conditions, for example, the volume of the detonation chamber [79]. However, it is not something that can be easily changed. Therefore, ND suppliers provide powders of a size that they can

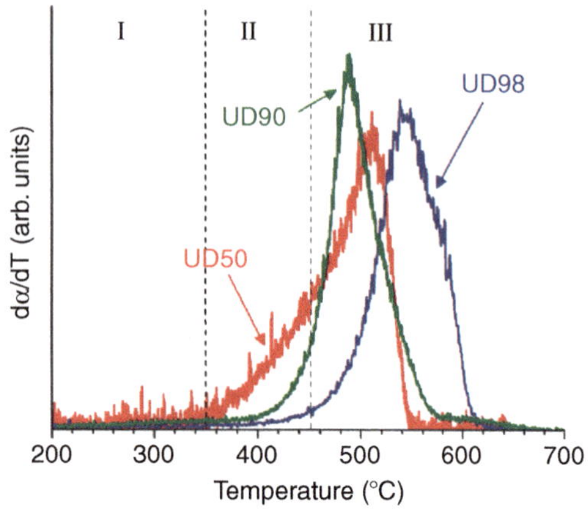

Fig. 12.12 Weight loss rate dα/dT of three different ND powders (UD50, UD90, and UD98). The highest oxidation rate is determined by a maximum in dα/dT

produce. There were only few attempts to control the average crystal size in ND powders, including sintering and other compaction techniques [80, 81]. In this section, we investigate the potential of air oxidation to selectively remove sp^2-bonded carbon from ND powders and introduce a simple, efficient, and environmentally friendly route to purify ND powders. Furthermore we explore the capabilities of air oxidation for adjusting the crystal size in ND powders. Earlier studies on oxidation of ND resulted in an increase of the average crystal size, suggesting the removal of smaller diamond crystals [82, 83]. However, a reduction in average crystal size has yet not been achieved.

5.3.1 Thermal Stability and Oxidation Behavior of Nanodiamond

The oxidation behavior of ND powders strongly depends on the composition and purity of the samples and changes with increasing sp^3 content. Figure 12.12 shows the weight loss rate (dα/dT) of UD50, UD90, and UD98 between 200°C and 700°C. While all three samples are considered ND powders, appearance (see Fig. 12.15) and oxidation behavior are noticeably different. dα/dT was obtained from nonisothermal TGA data and determined using the weight fraction $\alpha=(m_0-m)/m_0$, where m_0 and m are the initial sample weight and sample weight at temperature T, respectively. Similar to CNTs, the oxidation process of ND powders can be divided into three temperature ranges. Below 350°C (range I), oxidation of carbon does not occur. Between 350°C and 425°C (range II), only amorphous and disordered sp^2 carbon are oxidized. At temperatures above ~425°C (range III), both sp^2- and sp^3-bonded carbons are oxidized simultaneously.

The oxidation of UD50 starts around 350°C and reaches the highest weight-loss rate at ~510°C. Oxidation of UD90 does not occur below 400°C; however, the maximum weight loss takes place at lower temperatures (490°C) compared to UD50. This discrepancy can be explained by differences in sample composition. UD50 contains up to 70% of amorphous and graphitic sp^2 carbon surrounding the diamond crystals. In some cases, crystals are encapsulated in fullerenic shells which shield the diamond core against any interaction with their environment. While oxidation of amorphous and disordered sp^2 species in carbon nanomaterials starts around 350°C, the more stable graphitic shells surrounding the diamond cores inhibit diamond oxidation, thus shifting the maximum in the oxidation rate to higher values. Similarly, metal impurities, such as iron, which are known to catalyze the carbon-oxygen reaction, are also encapsulated by graphitic carbons, preventing catalytic reactions at lower temperatures. The temperature window for the oxidation of UD90 is slightly smaller than that of UD50, as expected. Acid-purified UD90 has a remaining sp^2 content of ~30%, but the larger structural variety in UD50 results in broader distribution of oxidation temperatures. UD50 exhibits two different oxidation regimes attributed to removal of amorphous sp^2 carbon at low temperatures and simultaneous oxidation of graphitic carbons and diamond at temperatures above ~425°C, respectively. In case of UD98, both the start (420°C) and the maximum weight loss (540°C) of the oxidation reaction are shifted to higher temperatures. This can be explained by the higher purity of UD98. While the sp^2 carbon content remains rather high (~20%), the extensive acid-purification following the production cycle significantly decreased the amount of iron catalyst in the sample. Oxidation of UD98 is dominated by the diamond phase, but contribution of sp^2 carbons to the total weight loss cannot be distinguished by TGA. The presence of different amounts of metal catalyst and large variations in surface chemistry further complicate the oxidation kinetics.

Isothermal and nonisothermal TGA data can be used to determine the kinetic parameters of the oxidation reactions. Oxidation of UD50 exhibits two different energy regimes, as expected. The temperature range between 370°C and 480°C is dominated by the oxidation of amorphous and graphitic sp^2 carbon, which accounts for roughly 70% of the sample weight. The corresponding activation energy is approximately E_A = 88.5 kJ/mol. Above ~480–490°C, E_A increases to about 190.5 kJ/mol due to oxidation of the remaining diamond crystals. Oxidation of UD90 yields an E_A value of ~223.0 kJ/mol. The activation energy is higher compared to that of UD50 at high temperatures, probably due to the lower Fe content in UD90. The encapsulated Fe particles in UD50 are catalytically inactive in the low-temperature range. However, upon exposure to high temperatures, the surrounding fullerene shells are removed and Fe catalyzes the carbon-oxygen reaction. The activation energy of UD98 (E_A =192.0 kJ/mol) is comparable to that of UD50 at high temperatures, and thus slightly lower than the values measured for UD90. Since purified UD98 contains a smaller amount of Fe and other metal impurities, the lower activation energy cannot be attributed to catalytic effects. A possible explanation might be a larger contribution of diffusion processes since the oxidation occurs at higher temperatures compared to UD90. Another reason for the observed differences could be changes in the surface chemistry of UD98 during extensive acid-purification. Treatments in oxidizing acids such as HNO_3 or H_2SO_4 increase the concentration of oxygen-containing functional groups on the surface of the diamond crystals and lower their activation energy.

The obtained results are in good agreement with data reported in literature. Pichot et al. measured E_A = 142±5 kJ/mol and E_A = 189±26 kJ/mol for the graphitic and the diamond phase, respectively [84]. The ND samples analyzed by Chiganov et al.

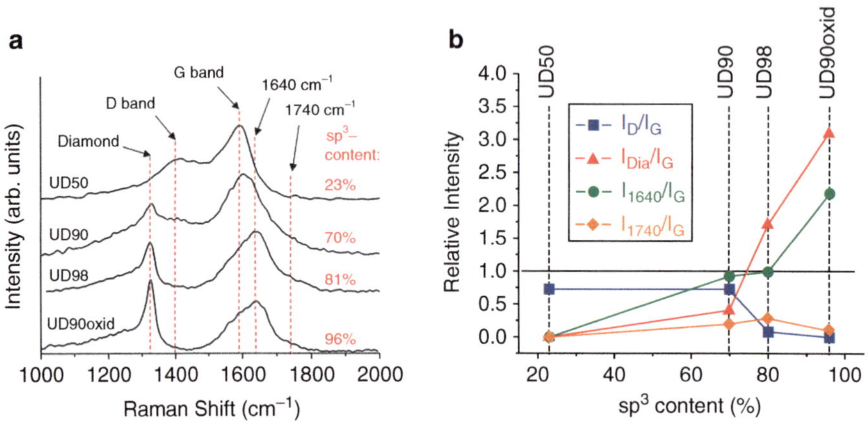

Fig. 12.13 (**a**) Raman spectra of different ND powders with sp^3 contents ranging from 23% to 96%. (**b**) Changes in the intensity of D band, diamond peak, and Raman features at 1,640 and 1,740 cm^{-1}. Intensities were normalized to the G band intensity at 1,590 cm^{-1}

exhibited activation energies of $E_A = 160$ kJ/mol (sp^2 carbon) and $E_A = 180$ kJ/mol (diamond) [2]. The differences in the activation energies of the sp^2 carbon are most likely due to differences in structure and composition of the sp^2 phase. UD50 is the unpurified detonations soot, whereas samples analyzed by Pichot et al. and Chiganov et al. have been partially purified. However, it should be noted that neither of these studies showed a complete removal of Fe and other metal catalysts, making an interpretation of the reported data more difficult. As a consequence, the reported activation energies (140–225 kJ/mol) are distorted by various factors and, therefore, may not accurately represent the diamond phase. However, the activation energies are characteristics of the individual ND samples.

5.3.2 Raman Spectra of Nanodiamond

The most conventional excitation source for Raman spectroscopy, a 514-nm Ar-ion laser, is known to cause a strong fluorescence during the analysis of ND samples. Compared to visible Raman, UV-Raman analysis offers a stronger diamond signal due to the resonance enhancement effect [85]. It is therefore preferred to use UV (244 and 325 nm) excitations for the analysis of ND powders.

The UV Raman spectra of UD50, UD90, and UD98 are shown in Fig. 12.13a [78]. As discussed in the previous section, the composition of ND powders varies largely, depending on the purity of the sample, and diamond (sp^3) contents can range from 20% to 30% (detonation soot) to 80% (acid-purified) and >96% (air-purified). The Raman spectra of ND powders with low sp^3 carbon content, such as UD50, are dominated by the D (1,400 cm^{-1}) and G bands (1,590 cm^{-1}) and exhibit only a weak or no diamond Raman signal due to the high content of graphitic carbon and its larger Raman scattering cross section compared to diamond. In addition, fullerenic shells enclose the diamond crystals and further weaken their Raman signal. The Raman spectra of purified ND powders (UD90, UD98,

UD90oxid) are noticeably different from UD50 (Fig. 12.13a). With increasing sp^3 content, the intensity of the diamond peak (1,326 cm^{-1}) increases, while the D band Raman signal weakens. ND powders with a high sp^3 contents show an asymmetrically broadened diamond peak at ~1,325 cm^{-1} with a shoulder toward lower wavenumbers. Both asymmetric broadening as well as the shoulder at lower wavenumbers are believed to result from the relaxation of the "q=0" selection rule, allowing phonons away from the center of the Brillouin zone to contribute to the Raman spectrum [86]. Spectral changes between 1,500 and 1,800 cm^{-1} are more complex and are often wrongly referred to as upshift of the G band. However, the contribution of sp^2 carbon to the Raman spectra of well-purified ND powders (sp^3 > 80%) is rather small and often overestimated. The broad asymmetric Raman feature is composed of at least three peaks of different origin centered at 1,590, 1,640, and 1,740 cm^{-1} which have been assigned to sp^2 carbon, O–H, and C=O groups, respectively [87]. For detailed analysis, Raman spectra were fitted using six Lorentzian/Gaussian peaks centered at 1,235, 1,300, 1,326 (diamond), 1,400 (D band), 1,590 (G band), 1,640, and 1,740 cm^{-1}.

Figure 12.13b shows the relative intensity changes (peak height) for the different Raman features, with respect to the G band intensity at 1,590 cm^{-1}, as a function of the sp^3 content. The plotted intensities are normalized with respect to the G band intensity at 1,590 cm^{-1}. In general, the intensity ratio between diamond peak and G band (I_{Dia}/I_G) increases with increasing sp^3 content. The sharp increase at ~70% sp^3 may result from the removal of graphitic shells surrounding the diamond core. A similar effect was observed during oxidation of DWCNTs, where amorphous carbon on the outer walls of the nanotubes weakened their Raman signal. The intensity ratio between D and G band (I_D/I_G) decreases with increasing sp^3 content, as expected. Although both Raman bands originate from sp^2 carbon, the D band intensity decreases at a larger rate as it is proportional to the number of rings, while the G band intensity reflects the number of sp^2 pairs. The intensity of the peaks at 1,640 (O–H) and 1,740 cm^{-1} (C=O) is related to the surface chemistry of the ND crystals and may vary for ND powders purified by different oxidation techniques (e.g., acid treatment vs air oxidation).

While in case of CNTs, the intensity ratio between D and G band (I_D/I_G) is often used to evaluate the purity of the sample, great care must be taken when using a similar approach for ND powders. The Raman spectrum of ND is more complex and depends strongly on the composition of the sample. Position, intensity, and number of Raman peaks change with increasing sp^3 or decreasing sp^2 content (Fig. 12.13a). It is important to point out, that in the case of ND powders, the accuracy of peak fitting procedures is limited. The diamond peak is characterized by an asymmetric broadening and a broad shoulder toward lower frequencies. While the diamond peak cannot be fitted using an individual Lorentzian or Gaussian peak, the origin of the shoulder can be understood by considering presence of very small crystals and scattering domains in ND (see Sect. 3.5). Raman peaks between 1,500 and 1,800 cm^{-1} overlap, and in some cases, the individual contributions are not distinguishable. In particular, the position of the G band depends on both ordering and shape of the nanostructure and is different for graphite ribbons, carbon onions, or disordered carbon. Therefore, different levels of purity and varying compositions (e.g., metal catalyst, surface chemistry) further complicate peak fitting and lower the reliability of the results. Moreover, the strong luminescence typically observed for ND powders upon VIS or NIR laser excitation requires extensive background corrections, which can lead to additional errors in data fitting.

5.3.3 In Situ Studies and UV Raman Spectroscopy Characterization

In order to determine the appropriate temperature range for the selective oxidation of sp^2 carbon in ND powders, nonisothermal TGA was performed in air. Figure 12.14a compares the oxidation behavior of UD50, UD90, and UD98, and shows differences in the oxidation rate and the temperature at which the maximum weight loss occurs. At temperatures below 375°C, the oxidation is inhibited or its rate is too low to allow noticeable removal of carbon within a reasonable time frame (range I). At temperatures above 450°C, all kinds of carbon in the sample, including amorphous, graphitic, and diamond phases are quickly oxidized (range III). In the intermediate temperature zone (range II), the oxidation rate shows substantial differences between the samples. While the mass of relatively pure UD98 does not change noticeably, UD50 with a substantial graphitic content shows a significant weight loss.

Figure 12.14b, c shows the results of in situ Raman studies during isothermal oxidation of UD50 for 3 h at 375°C (boundary temperature between range I and II) and 430°C (within range II) in air, respectively. The unpurified ND powder (detonation soot) was chosen because UD50 has a large content of non-diamond carbon and temperature-induced spectral changes during heating are more evident compared to UD90 and UD98. In order to accurately monitor changes in the composition of the sample and to maximize the Raman intensity of amorphous and graphitic carbon, we recorded in situ spectra using 633-nm laser excitation. At 375°C, the boundary temperature between range I and II, no significant changes in the Raman spectra were observed, suggesting the weight loss in range I to result mainly from the removal of adsorbed species, rather than oxidation of carbon. However, at temperatures around ~430°C (range II), a decrease in the G band intensity occurs for longer oxidation times (>1 h), indicating the removal of graphitic carbon during heating. While these results can be used to determine the appropriate temperature range for oxidation of sp^2 carbons in ND powders, they provide no structural information on the diamond phase.

To further optimize the oxidation procedure and avoid loss of ND crystals, we recorded UV Raman spectra of UD50 powders oxidized for 5 h at 375°C, 400°C, 425°C, and 450°C, respectively (Fig. 12.15a). As expected, lower oxidation temperatures (<375°C) are not sufficient to remove amorphous and graphitic carbon. At temperatures above 450°C, oxidized powders become inhomogeneous with respect to the ratio of diamond and non-diamond carbon phases. The Raman intensities of the diamond peak at ~1,325 cm^{-1} and G band at ~1,600 cm^{-1} vary strongly when comparing the UV Raman spectra recorded at different sample spots (data not shown), suggesting that all types of carbon are oxidized simultaneously leading to an inhomogeneous diamond distribution. When using oxidation

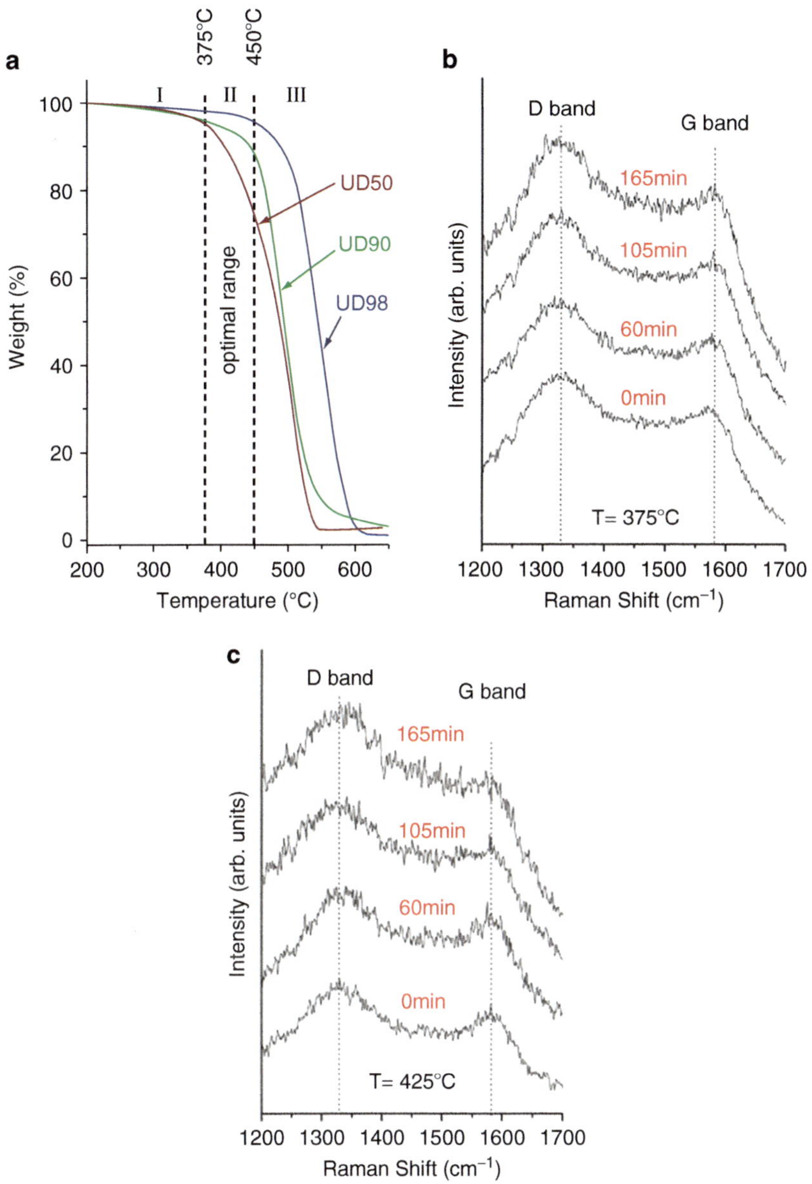

Fig. 12.14 (**a**) Nonisothermal TGA of ND samples in air (Ref. [78]). The graphs are normalized by the sample weight at 200°C. In situ Raman spectra of oxidation of ND (UD50) under isothermal conditions for 3 h at 375 (**b**) and 430°C (**c**), respectively, recorded using 633-nm laser excitation

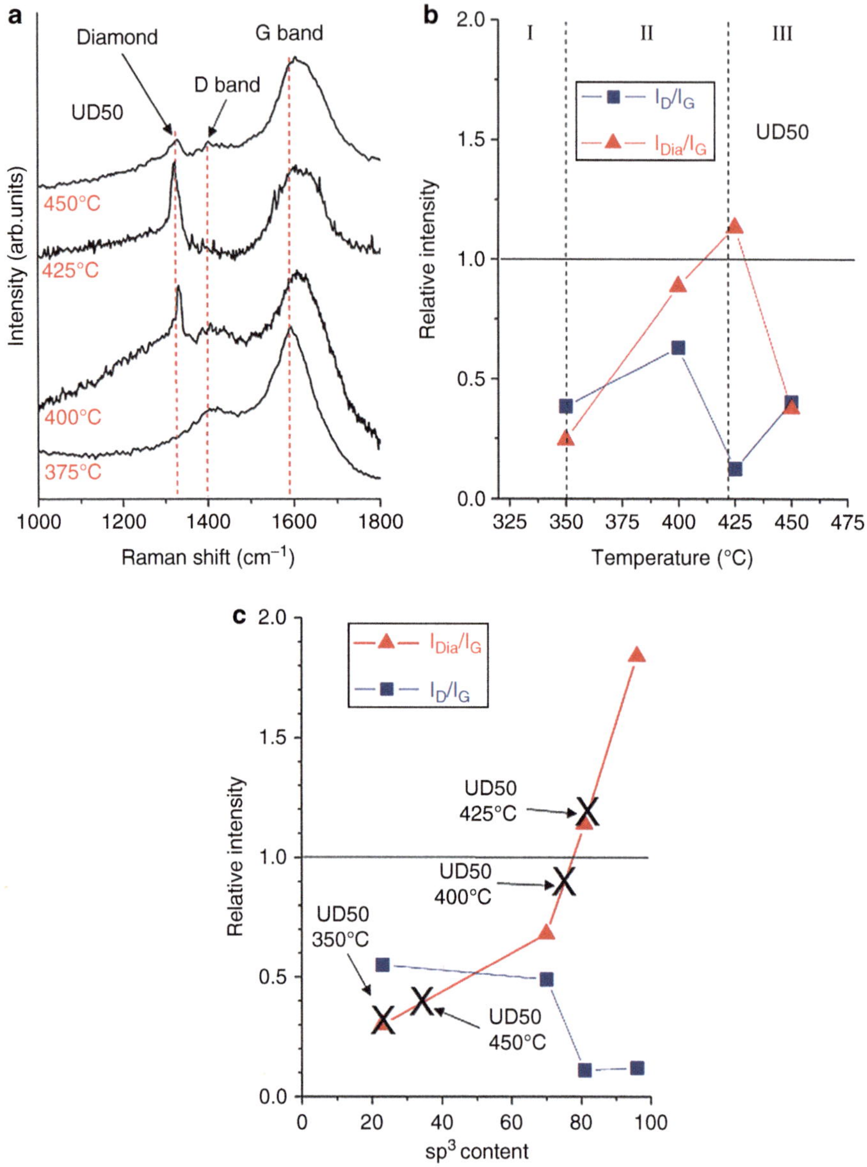

Fig. 12.15 (**a**) UV (325 nm) Raman spectra of UD50 after oxidation for 5 h at 375°C, 400°C, 425°C, and 450°C in air. (**b**) Relative intensity of diamond peak (I_{Dia}/I_G) and D band (I_D/I_G) in the after oxidation of UD50 normalized with respect to the G band intensity at 1,590 cm^{-1}. (**c**) Diamond content of oxidized UD50 (*black cross*) estimated using the relationship between I_{Dia}/I_G and the sp^3 content as determined by XANES for different ND samples (*red triangles*)

temperatures between the extremes (375–450°C), non-diamond carbon can be removed selectively and a significant loss of diamond can be avoided. The Raman spectrum of UD50 oxidized at 400°C demonstrates a substantially enhanced diamond signal. The changes in the corresponding intensity ratios are shown in Fig. 12.15b. While previous intensity ratios were determined by peak fitting procedures, we used a faster and simplified process which directly measures the Raman intensity at the corresponding peak position. This simplification is well justified given the discussed inaccuracy of the fitting procedure. Although the absolute values are different, comparison of intensity ratios determined by peak fitting (Fig. 12.13b) and using the simplified approach (Fig. 12.15c) shows similar trends and reveals the reliability of the obtained data. As discussed above, temperatures below 350°C (range I) are not sufficient for carbon oxidation and the I_{Dia}/I_G ratio is similar to that of the detonation soot (UD50). Between 350°C and 430°C (range II) amorphous and graphitic carbons are oxidized, without significant loss of the diamond phase, leading to a maximum in I_{Dia}/I_G. At oxidation temperatures above 430°C (range III), both sp^2 and sp^3 species oxidize simultaneously, resulting in large inhomogeneities in the sample composition and thus variations in I_{Dia}/I_G.

The diamond content of the oxidized UD50 samples can be estimated by Raman spectroscopy (Fig. 12.15c) using the predetermined relationship between I_{Dia}/I_G and the sp^3 content. The corresponding values are approximately 23%, 74%, 82%, and 34% after oxidation for 5 h at 350°C, 400°C, 425°C, and 450°C, respectively. Thus, the optimal oxidation temperature for purification of UD50 is ~425°C. While the optimal oxidation temperature could be affected by both the sample composition and the experimental conditions, temperatures within the 400–430°C range were found to be most favorable in the present study. Small changes in the oxidation temperature within the given range can be used to find a compromise between the higher purification rate (lower time and cost) and the acceptable weight loss due to minor oxidation of the diamond phase. Oxidation of acid-purified UD90 and UD98 showed similar results and led to further increase in the relative intensity of the diamond peak (see UD90oxid in Fig. 12.13). While air oxidation evidently removes fullerenic shells and other sp^2-bonded carbon impurities from the samples, it also influences the surface chemistry and may affect shape and position of the Raman peaks. The strong upshift of the G peak to ~1,640 cm^{-1} in the oxidized samples (Fig. 12.13a) is likely caused by the formation of hydroxyl groups on sp^2- or sp^3-bonded carbon [87]. A similar upshift of the G band was also observed in disordered diamond-like carbon (also called tetrahedral carbon) with a high content of sp^3-bonded carbon, and explained by resonance phenomena [88]. However, we believe that since ND surface atoms account for over 20% of total atoms in ND particles below 5 nm, surface functionalities are responsible for this peak [87].

5.3.4 Structure and Surface Chemistry of Oxidized Nanodiamond

The determined oxidation conditions were used to purify larger amounts of ND in a chamber furnace to simulate industrial conditions. XANES allowed us to quantify the sp^3 content in ND samples and learn about their bonding structure. As compared to electron energy loss spectroscopy (EELS), XANES is a more quantitative

Table 12.1 Results of the sp^3 content analysis (XANES) of ND samples before and after a 5-h oxidation at 425°C in air

	UD50	UD50 oxidized	UD90	UD90 oxidized	UD98	UD98 oxidized
sp^3 conent,%	22	95	70	90	81	96
sp^2 conent,%	77	5	30	6	19	4
sp^3/sp^2	0.3	19	2.3	16	4.3	24

sp^3 diamond (<5 μm): 97–97.5%
sp^2 graphite (HOPG): 100%

technique that offers a better spectral resolution, minimizes the sample damage, and allows one to obtain an averaged signal from the macroscopic sample. The total weight loss in these experiments was very close to the amount of sp^2 carbon in the samples (see Table 12.1).

Not only did oxidation in air decrease sp^2-bonded carbon in ND samples (UD90 and UD98) pre-purified by acidic treatment by about a factor of 5, but it was also capable of selectively removing graphitic carbon in the soot sample (UD50), thus increasing the sp^3/sp^2 ratio in this sample by nearly two orders of magnitude from 0.3 to 19 (Table 12.1). The purity of the oxidized UD98 is comparable to that of microcrystalline diamond. While some authors claimed a high content (>92%) of diamond in the ND powders, those conclusions were based on X-ray photoelectron spectroscopy (XPS) and X-ray diffraction (XRD) measurements of the diamond/graphite ratio [89] and cannot be considered reliable, because XRD overestimates the diamond content and surface analysis techniques cannot provide good quantitative data for powdered materials.

HRTEM studies fully support results of Raman and XANES analyses, demonstrating elimination of graphitic ribbons, carbon onions, and graphitic shells in UD50 and a substantial decrease in amorphous carbon content in UD90 and UD98 after oxidation. Figure 12.16 shows representative high-resolution micrographs of UD50 and UD90 before and after oxidation for 5 h at 425°C in air. The microstructure of UD98 was very similar to that of UD90 and thus is not presented separately. This is in agreement with XANES measurements showing a similar sp^2/sp^3 carbon ratio for these grades, both before and after oxidation. Thus, oxidation in air is one of the simplest and most efficient ways to purify ND from non-diamond carbon species. However, in addition to purification, air oxidation dramatically changes the surface chemistry of ND.

While Raman spectroscopy is useful to analyze structural features and distinguish between the different carbon species, FTIR spectroscopy was used to determine functional groups and adsorbed molecules on the surface of the carbon. Black and strongly absorbing as-received UD50 shows no detectable FTIR vibrations due to the high content of graphitic and amorphous carbon (Fig. 12.17). The main features in the FTIR spectra of as-received UD90 and UD98 powders are related to C=O (1,740–1,757 cm^{-1}), C–H (2,853–2,962 cm^{-1}), and O–H vibrations (3,280–3675 cm^{-1} stretch and 1,640–1,660 cm^{-1} bend) which can be assigned to –COOH, $-CH_2-$, $-CH_3$, and –OH groups of chemically bonded and adsorbed surface species, mainly resulting from acid-purification [91–93]. The comparison of FTIR spectra of

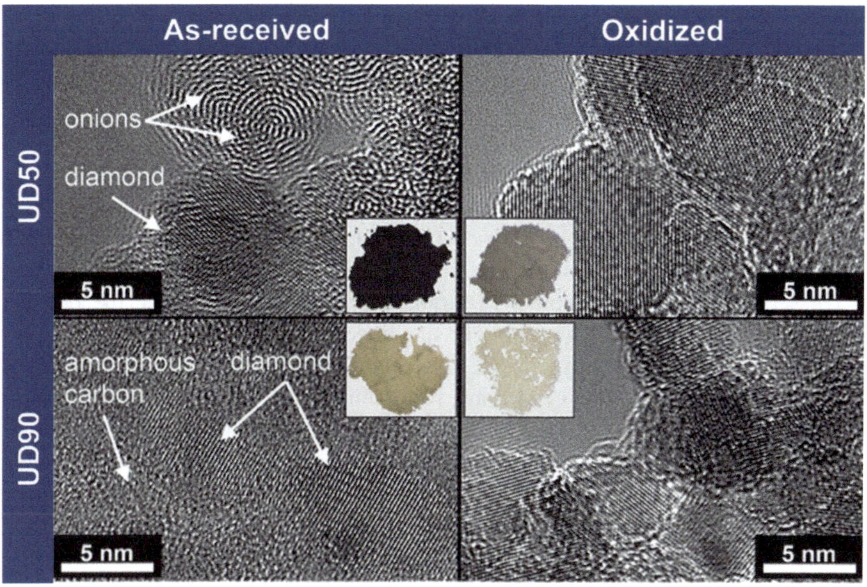

Fig. 12.16 HRTEM images of UD50 and UD90 before and after oxidation for 5 h at 425°C in air. The *insets* show optical photographs of the ND powders

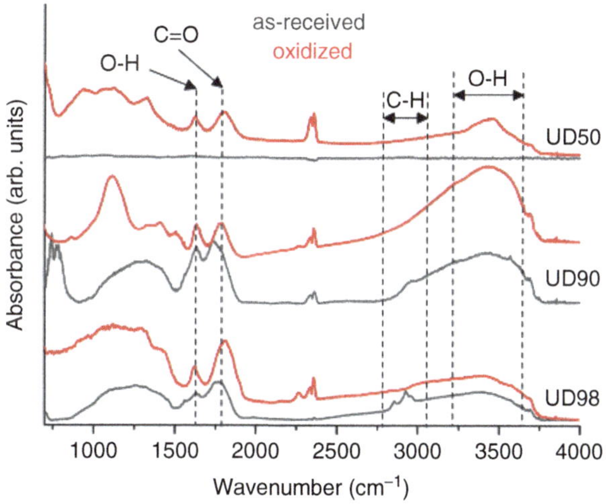

Fig. 12.17 FTIR spectra of as-received and air-oxidized ND samples [90]

purified (red curves) and as-received (gray curves) powders reflects the conversion of a variety of surface functional groups into their oxidized derivatives. After oxidation, $-CH_2-$ and $-CH_3$ groups are completely removed from UD90 and UD98, the amount of –OH groups is increased and C=O vibrations are upshifted by 20–40 cm^{-1} indicating a conversion of ketones, aldehydes, and esters on the surface into carboxylic acids, anhydrides, or cyclic ketones. The most prominent changes in the surface termination after oxidation were found for UD50. Upon the removal of graphitic layers by oxidation, the surface of UD50 became accessible for chemical reactions and is immediately saturated with oxygen or oxygen-containing functional groups.

5.3.5 Control of Nanodiamond Crystal Size

As the dimensions of a crystal are reduced, the increasing surface/volume ratio results in significant changes of properties, especially when entering the lower nanometer range. For example, ND particles with diameters of about 4 nm have ~20% of the total number of atoms on the surface. Because the physical properties of nanocrystals are strongly size-dependent, it is crucial to control and accurately measure the crystal size. To some extent, ND crystal size can be controlled by the synthesis conditions, for example, the volume of the detonation chamber [79]. However, it is not something that can be easily changed. Therefore, ND suppliers provide powders of a size that they can produce. However, there were only few attempts to control the average crystal size in ND powders, including sintering [80, 81] and oxidation [82, 83]. In this section, we describe the potential of air oxidation for adjusting the crystal size of ND powders.

As-received UD90 was oxidized for 2, 6, 17, 26, and 42 h at 430°C, which is the temperature for slow ND oxidation as discussed in Sect. 5.3.3 [94]. Figure 12.18a shows the HRTEM images of two ND powders oxidized at 430°C for 2 and 42 h, respectively. The weight loss due to oxidation was 13% and 74% after 2 and 42 h, respectively. While oxidation for 2 h removes mainly amorphous carbon and other non-diamond species [95], longer oxidation times result in selective oxidation of smaller crystals, thus shifting the size distribution toward larger values.

We determined the size of~200 crystals in each sample in order to estimate the average crystal size (Fig. 12.18b). While 2-h oxidation in air results in an average crystal size of ~7 nm, it is upshifted to ~14 nm after 42-h oxidation. Since crystal growth is not expected under such conditions, large crystals (15–30 nm) observed after oxidation must be present in the pristine powder. However, because they are usually covered by smaller crystals and remain within agglomerates, it is difficult to observe them in HRTEM, indicating the limitations of this characterization technique and the low statistical reliability.

We conducted XRD analysis for a more accurate and statistically significant estimate of changes in the average crystal size. Figure 12.18c compares the XRD patterns of microcrystalline diamond with a ND powder (UD90) oxidized at 430°C for 2 h. While both samples show peaks at ~43.9°, ~75.5°, and ~91.5° corresponding to the [111], [220], and [311] planes of the cubic diamond lattice ($a_C = 3.571$ Å), respectively, the scattering peaks of ND are significantly broadened. By measuring the changes in the linewidth of the diffraction peaks using the Williamson-Hall analysis [96] one can estimate the crystal size and lattice strains, both of which are known to cause broadening of diffraction lines. Using this approach we calculated the average crystal size of the oxidized powders (Fig. 12.18c). The Scherrer equation, often used for size characterization in crystalline materials, neglects potential lattice strains and can lead to values smaller

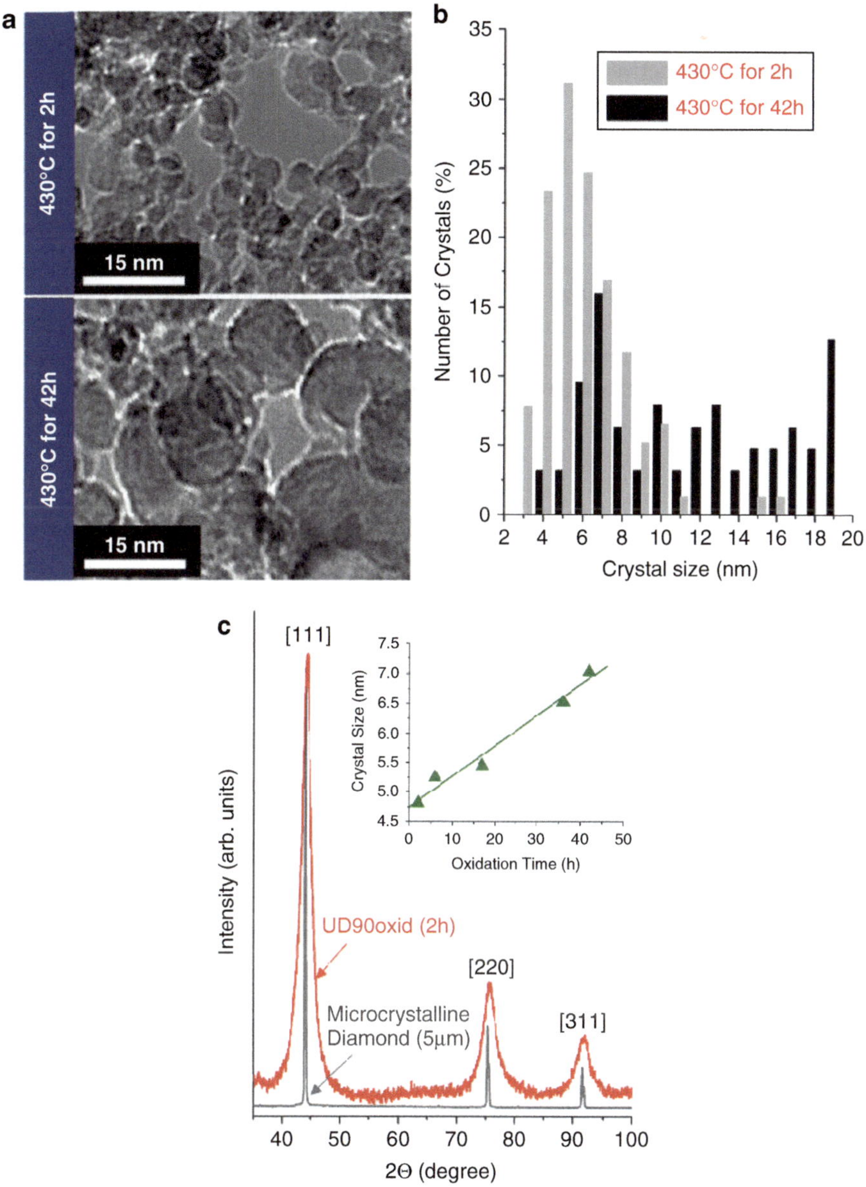

Fig. 12.18 (**a**) HRTEM images of ND powders oxidized at 430°C for 2 h and 42 h. (**b**) Oxidation in air selectively removes small diamond crystals, shifting the crystal site distribution toward larger values. (**c**) XRD pattern of ND powder oxidized at 430°C for 2 h and 42 h, showing the characteristic [111], [220], and [311] diffraction peaks of diamond. The line broadening can be used to estimate the crystal size using the Williamson-Hall analysis. The *inset* shows changes in ND crystal size as a function of oxidation time

than the actual crystal size [97]. However, it is necessary to keep in mind that averaging the contributions from different scattering peaks might further increase the discrepancies between calculated values and the actual crystal size. Palosz et al. demonstrated that diamond nanocrystals typically exhibit lattice strains due to the extensive surface reconstruction in the core-shell nanoparticles, suggesting that size calculations based on the Deby-Scherrer equation such as the Williamson-Hall analysis may oversimplify the diffraction of nanocrystals and should only be considered semi-quantitative [98].

The sample weight of a ND powder is generated by the mass of N individual, diamond crystals. Assuming N to be constant during oxidation, one can express the relative changes in the sample weight through changes in the diameter d of the crystals. If the oxidation rate k is size-independent, the diameter of the ND crystals decreases linearly with oxidation time t and the relative change in mass m can be written as:

$$\frac{m_t}{m_{t=0}} = \left(\frac{d_{t=0} - k \cdot t}{d_{t=0}}\right)^3 \tag{12.1}$$

where the indices "t=0" and "t" represent the initial values and the values at time t, respectively. Figure 12.22a shows the weight loss ($m_t/m_{t=0}$) of UD90 measured using isothermal TGA at 430°C. The weight loss measurement was started after ~300 min of oxidation in order to exclude contributions from amorphous and graphitic sp^2 carbons. The relative sp^3 content increases from approximately 71–96% in this early phase of the oxidation. The weight loss curve was fitted using (12.1), assuming the initial average diameter to be $d_{t=0} = 5$ nm (Fig. 12.19a). The determined oxidation rate ($k = 8.23*10^{-4}$ nm/min) can be used to estimate the time-dependent changes in the diameter of the ND crystals. However, it can be seen that there is only little agreement between the obtained fit and the experimental weight loss, and that the predicted decrease in diameter with increasing oxidation time is in contradiction to data obtained from both HRTEM and XRD analyses.

Considering the broad size distribution (3–50 nm) of detonation-synthesized ND powders, one may expect differences in the oxidation behavior for different crystal sizes. Thus, (12.1) was modified in order to separately account for the weight contributions of small (m') and large (m'') diamond crystals:

$$\frac{m_t}{m_{t=0}} = \frac{\left(d'_{t=0} - k' \cdot t\right)^3 + \left(d''_{t=0} - k'' \cdot t\right)^3}{{d'_{t=0}}^3 + {d''_{t=0}}^3} \tag{12.2}$$

The initial diameter of the small crystals, which represents the majority of the ND crystals, was set to $d'_{t=0} = 5$ nm. The parameters $d''_{t=0}$, k′, and k'' were determined by fitting the weight loss data using (12.2) and found as 9.7 nm, $4.63*10^{-3}$ nm/min, and $1.13*10^{-3}$ nm/min, respectively. The results are shown in Fig. 12.19b and are in good agreement with TGA data. The corresponding changes

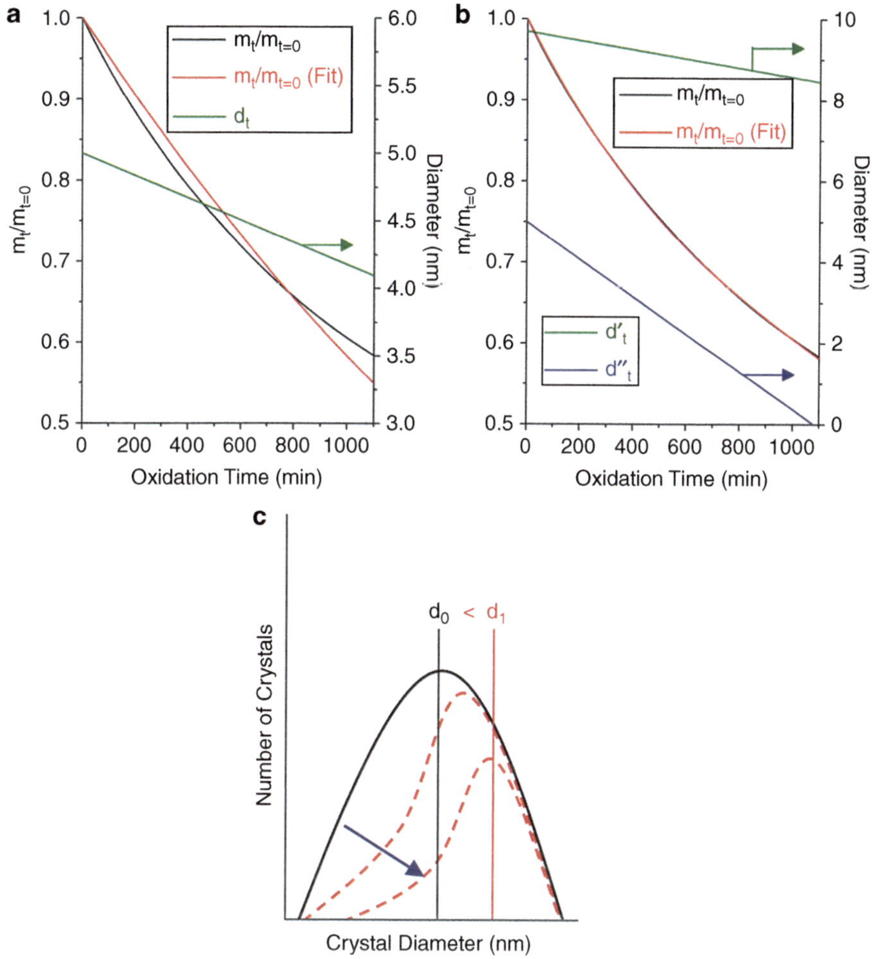

Fig. 12.19 Weight loss of ND powder measured using isothermal TGA at 430°C, normalized to the initial weight. Oxidation rates and corresponding changes in crystal diameter were obtained by fitting the weight loss curve using (12.1) (**a**) and (12.2) (**b**). While the size of the individual ND crystals decrease with time, the higher oxidation rate of smaller crystals shifts the average crystals size (**c**) to larger diameters with increasing oxidation time

in diameter for small (d') and large (d'') ND crystals suggest that while both exhibit a decrease in size, the decrease in diameter occurs much faster for smaller crystals due to their higher oxidation rate ($k' \approx 4{\cdot}k''$). As a consequence, the average crystal size is shifted toward higher values as indicated in Fig. 12.19c. This is in good agreement with experimental data obtained from Raman spectroscopy, XRD, and HRTEM studies.

These results suggest that oxidation in air cannot be used to decrease the average crystal size of ND powders, due to the higher oxidation rate of small ND crystals.

However, the method allows for a controlled increase in the average crystal size with subnanometer accuracy. While the described approach is able to explain the observed increase in the average crystal, it can only be considered as rough approximation and may oversimplify the oxidation behavior of ND samples. In particular, metal catalysts present in the ND samples strongly affect kinetic parameters, such as activation energies and oxidation rate constants [99].

5.3.6 Size Characterization Using Raman Spectroscopy

In addition to structural analysis and purity evaluation, Raman spectroscopy can also be used to estimate the crystal size of nanostructured solids. In most cases size characterization using Raman spectroscopy is based on the phonon confinement model (PCM), which uses changes in Raman frequency and Raman peak shape to estimate the crystal size. Although several attempts have been made to relate confinement-induced changes in the Raman spectrum of ND to the crystal size, the agreement between calculated and experimental data and the accuracy of the fitting procedure are still unsatisfactory. A detailed discussion of the limitations of the PCM and the accuracy of previous studies on ND powders is given in Ref [86].

Raman spectroscopy alone is currently not able to quantitatively measure the average crystal size in ND powders [86]. However, with a better understanding of phonon confinement effects in the Raman spectrum of ND, Raman spectroscopy may be used to accurately measure the average crystal size and determine changes in the size distribution. The following section discusses the effects of crystal size, defect, and size distributions on the Raman spectra of ND and required modifications in the PCM.

Figure 12.20a shows the changes of the diamond Raman peak after oxidation in air at 430°C for 2, 6, 17, 26, and 42 h. The corresponding average crystal sizes were determined by X-ray diffraction (see Fig. 12.18c) and are 4.8, 5.2, 5.5, 6.5, and 7.0 nm, respectively. The confinement-induced asymmetry of the Raman peak decreases with increasing oxidation time, leading to a narrower diamond line (Fig. 12.20a). The intensity of the shoulder around $\sim$1,250 cm^{-1} also decreases with oxidation time, suggesting a possible correlation with the crystal size.

We first determined the ND crystal size of the oxidized powders using the models of Ager et al. [100] and Yoshikawa et al. [101] (Fig. 12.20b). The theoretical Raman spectra were calculated according to:

$$I(w) \cong \int_0^1 \frac{\exp(-\vec{q}^2 L^2/4) \cdot 4\pi \vec{q}^2}{[\omega - \varpi(\vec{q})]^2 - (\Gamma/2)^2} dq \tag{12.3}$$

where q is the phonon wave vector, ω is the Raman frequency, L is the crystal size, and $\omega(q)$ and Γ are the phonon dispersion relation and the natural line width of the zone-center Raman line, respectively.

The approach of Yoshikawa et al. accounts for size-related changes in the phonon lifetime, leading to larger downshift and higher peak asymmetry compared to Ager et al. While the differences are considerably small for larger crystals, they become significant for crystal sizes below 10 nm and are therefore important

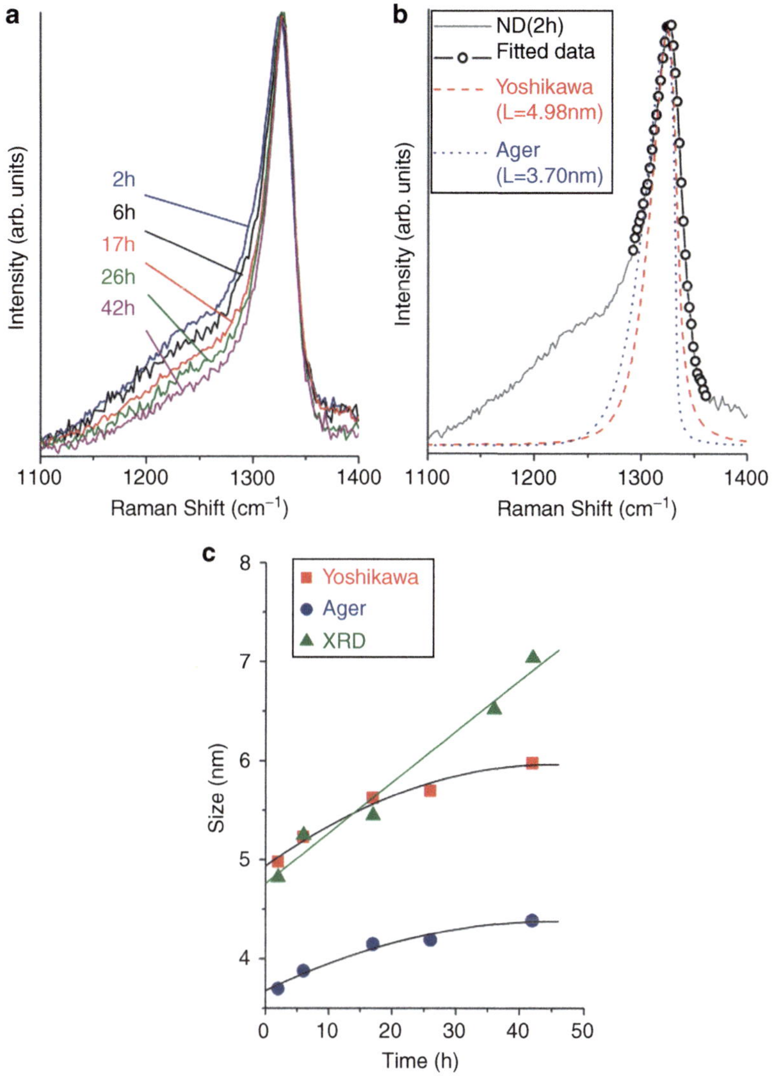

Fig. 12.20 (**a**) Diamond Raman peak recorded after oxidation at 430°C for 2, 6, 17, 26, and 42 h in air. (**b**) Raman spectrum of ND powder oxidized for 2 h at 430°C in air and corresponding peak fit. Data used for peak fitting is indicated by the *circles*. (**c**) Calculated crystal size for ND powders oxidized for 2, 6, 17, 26, and 42 h at 430°C. X-ray diffraction data is shown for comparison

for size-characterization of ND powders. Since both models do not account for the shoulder at ~1,250 cm^{-1}, we can only fit experimental data in the range 1,290–1,360 cm^{-1}. The obtained fits and the corresponding crystal sizes for the ND powder oxidized for 2 h at 425°C, referred to as ND(2 h). Although both models lead to crystal sizes comparable to that obtained from XRD in the range

3–5 nm, the agreement between calculated and recorded Raman spectra is fairly low. The calculated crystal sizes of all oxidized powders are presented in Fig. 12.20c, in comparison to XRD results. While Ager's model leads to slightly lower size values, both approaches show an increase in crystal size with increasing oxidation time, in agreement with XRD data. Crystal sizes calculated using Yoshikawa's model closely match the values obtained from XRD analysis for 2, 6, and 17 h, but are lower for longer oxidation times (>20 nm).

It should be noted that the measured sizes represent average values. HRTEM characterization of the oxidized ND powders revealed a broad size distribution, showing ND crystal sizes of 3–20 nm or larger (see Fig. 12.18a). Faster oxidation of small-diameter ND crystals leads to an increase in the average crystal size. The discrepancies between XRD and Raman-based size calculations observed for longer oxidation times (>20 h) may result from the difference in the effective cross section of both techniques with respect to the increasing contribution of larger ND crystals. The sensitivity toward defects and changes in lattice spacing is also expected to be different for both techniques.

As mentioned earlier, ND powders contain diamond crystals with sizes ranging from 3 to 30 nm. While the majority of the nanocrystals show diameters between 4 and 5 nm, the Raman scattering cross section is proportional to the crystal volume and thus much larger for larger crystals. On the other hand, confinement effects such as the asymmetric peak broadening are more distinct for small crystals (<10 nm). Therefore, experimentally obtained Raman spectra are expected to contain Raman features from both small and large ND crystals. However, a simultaneous contribution to the total Raman intensity leads to a complex line-shape that cannot be fitted using a single peak in the PCM.

Figure 12.21a shows the calculated diamond Raman peaks of 3-, 5-, and 7-nm ND crystals, in comparison to a sample that contains different amounts of 3- and 7-nm crystals, each contributing equally to the overall intensity. The Raman signal of a 3-nm crystal is more than 12 times weaker than that of a 7-nm ND crystal. Thus, although the relative amount of large ND crystals in the sample is small, their contributions to the overall Raman intensity cannot be neglected. However, the calculated Raman intensity of a sample containing only 5-nm ND crystals is very different compared to the spectra resulting from a mixture of 3- and 7-nm crystal, even though both samples exhibit the same average crystal size (5 nm). Considering the volume-dependence of the Raman intensity, one can potentially estimate the contribution of the various crystal sizes to the total Raman intensity and determine the size distribution in the ND powders.

Previous studies on phonon confinement in nanocrystals did not account for possible contributions from lattice defects [100–102]. However, the parameter L in (12.3) represents the coherence length and is, therefore, a measure of the distance between dislocation, vacancies, interstitials, impurities, and other defects within the crystal lattice. The assumption that L represents the crystal size is only valid for defect-free crystals, where the surface is considered to limit the propagation of the phonons. This assumption does not hold for imperfect crystals produced by

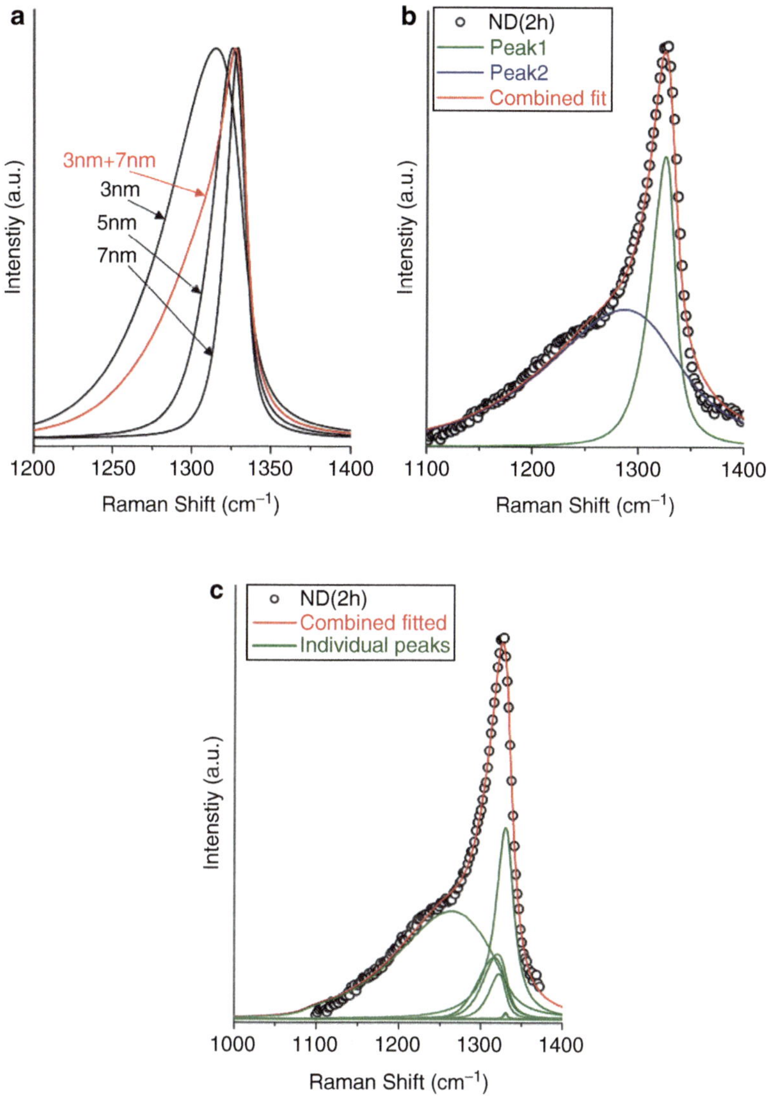

Fig. 12.21 (**a**) Calculated Raman spectra for different ND crystal sizes (*black*) in comparison to that of a sample containing a binary size-distribution (*red*). (**b**) Raman spectrum of oxidized ND (2 h at 430°C) fitted using two peaks to account for contributions from both small and large crystals as well as lattice defects. (**c**) Raman spectrum of oxidized ND (2 h at 430°C) fitted using seven peaks, each representing the contribution of a different dispersion relation

detonation methods which contain cracks and dislocations [86]. Other lattice defects such as impurity interstitials, dumbbell-defects, and vacancies have also been observed before. As a consequence, the coherence length L becomes significantly smaller than the actual ND crystal size.

Therefore, while the results above were produced assuming the ND crystals to be larger than 3 nm ($L > 3$ nm), we now allow vibrational domains (coherence lengths) with $L < 3$ nm to contribute to the Raman spectrum. In addition, the frequency range used for fitting is extended to 1,100–1,400 cm^{-1}.

The fitting results for ND (2 h) are shown in Fig. 12.21b. Two peaks have been used for the analysis to allow for simultaneous contributions of small and larger crystals. It can be seen that the agreement between calculated and measured Raman spectra is significantly improved and the broad shoulder (<1,300 cm^{-1}) can be well fitted using this approach. The corresponding L for Peak 1 and Peak 2 are 6.45 nm and 2.67 nm, respectively. Peak 1 corresponds to L values in the range 6–8 nm and may be assigned to defect-free 6–8-nm ND crystals or to less-defective, larger ND crystals (>8 nm). The L values obtained from Peak 2 are significantly smaller and range between 2.6 and 3.7 nm. Assuming the majority of the ND crystals to be 4–8 nm in size, a coherence length of 2–4 nm may suggest 1–2 defects per ND crystals. While this is in good agreement with HRTEM studies [86, 94], the reported numbers merely represent the average values. Defect-free ND crystals of similar size and crystals with three or more defects have also been observed. Ager et al. and Yoshikawa et al. estimated the crystal size by reducing the three-dimensional integration in (12.3) to a one-dimensional integration over a spherical Brillouin zone (BZ) using an averaged one-dimensional dispersion curve. However, this approximation is only valid for small phonon wave vectors, but oversimplifies the energy dispersion of the phonon modes for larger wave vectors. A coherence length of 2.6 nm would allow phonons far away from BZ center to contribute to the Raman signal. Therefore, while the dispersion relation of Yoshikawa et al. can be used for $L = 4$ nm, coherence lengths below 3 nm require a more accurate description of the energy dispersion of the individual phonon branches.

In order to account for defect contributions and the energy dispersion of the phonon branches we fitted the Raman spectrum of ND (2 h) using seven peaks, each attributed to one of the individual phonon branches. The results are shown in Fig. 12.21c. In general, there is a good agreement between calculated and measured Raman data, suggesting that phonon wave vectors from small vibrational domains can indeed be responsible for the broad shoulder at 1,250 cm^{-1} typically observed in the Raman spectra of ND powders. The corresponding L values are 1.8, 1.9, 2.5, 2.9, 3.9, 11.3, and 17.4 nm. While in these computations each crystal size is only represented by one phonon branch for simplicity reasons, adding the contributions of the remaining six phonon branches for each L value would evidently lead to an improvement in the fitting procedure. Although the described approach is more complex, it provides a possible explanation for the broad shoulder at ~1,250 cm^{-1} and may allow an estimate on the number of defects in the ND crystals.

5.4 Laser-Induced Heating of Carbon Nanomaterials

Carbon nanomaterials show a broad range of optical properties, ranging from transparent in the visible and infrared range for diamond to highly absorbing carbon nanotubes (CNT) and graphite [1]. For example, Yang et al. showed that a low-density CNT array is an almost perfect light absorber, close to an ideal blackbody [103]. The reflectance from these vertically aligned nanotubes has the lowest-ever reported value ($\sim 10^{-7}$) for any known material [103]. Ajayan et al. reported self-ignition of CNTs under a photoflash and observed extensive reconstruction of the carbon material, suggesting temperatures in excess of 1,500°C [104]. Emission of light from carbon nanostructures has also been observed. Nanotube bulb filaments were found to show a higher brightness (at the same voltage) compared to tungsten filaments [105]. Emission from CNTs at higher temperatures is a combination of thermal (blackbody) radiation and photoluminescence that results from electronic transitions [105]. Wang et al. performed experiments on nanostructured carbon materials by using a Crookes radiometer and a 655-nm (35-mW) diode laser, reporting temperatures of ~3,500°C [106, 107].

Understanding interactions of carbon nanostructures with light is particularly important because optical methods are widely used for their characterization, with Raman spectroscopy being the gold standard for analysis of all kinds of carbons [108]. Temperature-induced shifts of Raman bands have been observed in many previous studies [109–112]. They are of particular concern when Raman features are used to determine the diameter distribution of CNTs [113] or measure strain in CNT-polymer composites [114] where an uncontrolled temperature shift can lead to erroneous conclusions. Because of laser-induced heating, Raman spectroscopy can lose its nondestructive character during the analysis of nanostructures that strongly absorb light and have a poor thermal contact between the particles, which is the case for all nanopowders. Raman spectral features, such as the G band of graphitic carbon, can be used to estimate the temperature of the samples during material characterization thus preventing laser-induced oxidation, extensive graphitization, and/or evaporation of carbon nanomaterials.

5.4.1 Effect of Temperature on Raman Spectra

The Raman characterization of different carbon nanomaterials in inert (Ar) atmosphere reveals a strong influence of the laser power (energy density) on the Raman spectra [59]. In general, an increase in the laser power leads to a decrease in the Raman frequency. For example, when using the most common excitation source in Raman spectroscopy – the 514-nm line of an Ar-ion laser – the G Band in the Raman spectrum of carbon onions (Fig. 12.22a) shifts from ~1,594 cm^{-1} (0.1 mW) down to ~1,565 cm^{-1} (0.7 mW). The downshift is related to an increase in the sample temperature and has been measured for other carbon materials including graphite and CNTs.

There exists several ways to measure temperature using Raman spectroscopy. The most common techniques include monitoring the anti-Stokes to Stokes intensity ratio or determining changes in the Raman shift of the peaks. In this study we used the latter method. In order to estimate the laser-induced heating in a carbon onion sample, we determined the relationship between G band position and sample temperature (inset in Fig. 12.22a) by using the lowest laser power possible and Ar atmosphere to prevent sample oxidation. The experiment revealed a downshift of ~0.034 cm^{-1}/°C which is slightly higher than the values reported for graphite

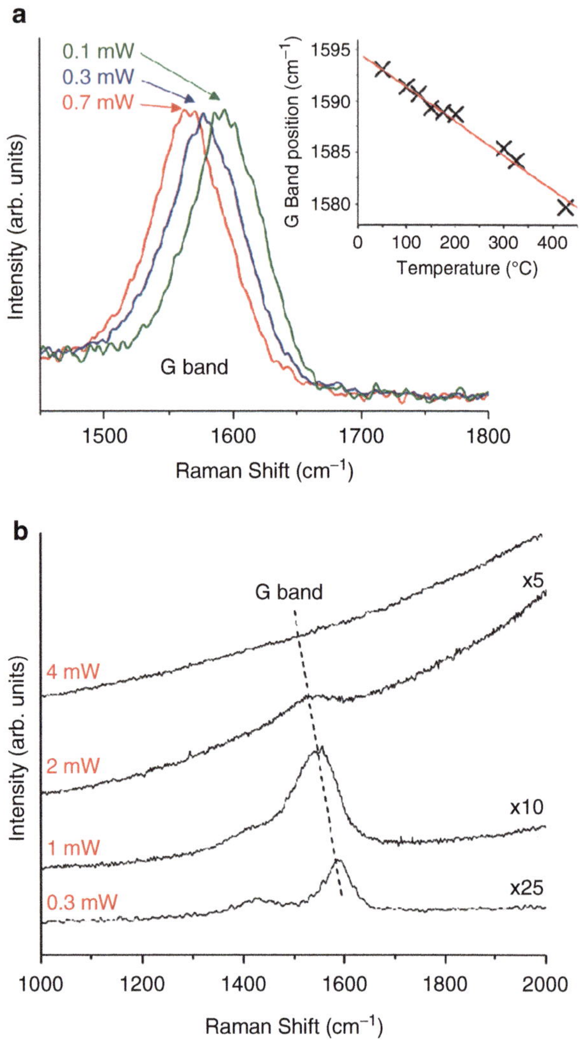

Fig. 12.22 Raman spectra of carbon onions upon laser heating. (**a**) G band recorded at 0.1, 0.3, and 0.7 mW with a 514-nm Argon laser (50x objective). The *inset* shows the temperature dependence of the G band position. (**b**) Raman spectra recorded at 0.3, 1, 2, and 4 mW using a 325-nm UV laser. The intensity of the spectra was adjusted as shown in the figure [59]

(0.024 cm^{-1}/°C) [115] and CNTs (0.025–0.030 cm^{-1}/°C) [44, 45, 60, 87], due to stronger bending of graphitic layers in carbon onions. Based on these results, G band positions of 1,578 and 1,565 cm^{-1} indicate temperatures of approximately 500°C and 890°C, respectively. It should be noted that these estimates assume the laser heating at 0.1 mW to be negligible, which is most likely not the case. Since oxidation of most carbon nanostructures starts well below 500°C, the presence of oxygen would inevitably lead to material burning and change the sample composition within the excited volume. A laser power of 0.3 mW (400 W/cm^2) is lower than the values commonly used for Raman analysis. Raman analysis of the same sample upon 325-nm UV laser excitation in Ar atmosphere showed a similar

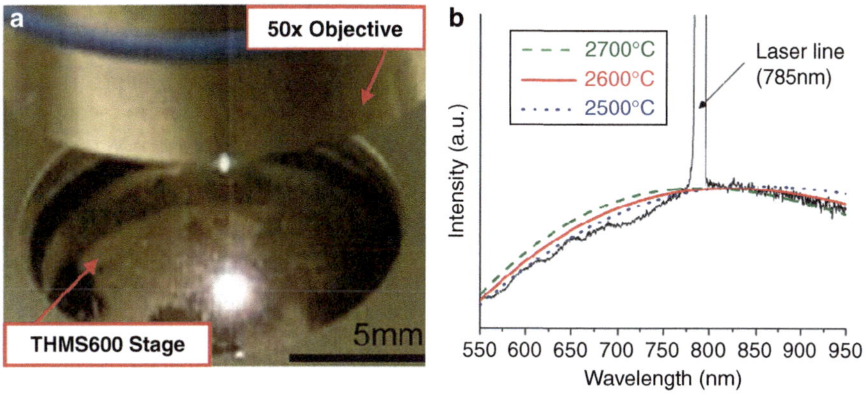

Fig. 12.23 (**a**) Photograph showing light emission of carbon black upon 785-nm laser excitation in an argon atmosphere. The corresponding emission spectrum was recorded using external UV–VIS-NIR spectrometer. The *lines* represent the calculated blackbody emission curves of 50-nm carbon black particles at different temperatures

energy-dependence (Fig. 12.22b). In addition, Raman spectra revealed an increasing background intensity at higher laser power which was accompanied by the emission of a bright white light from the excitation spot. As expected, the G band shifts to lower frequencies with increasing laser power. While one could accurately identify the G band position at lower powers (0.3 and 1 mW), light emission was not observed. At a high laser power, the Raman spectra are increasingly overshadowed by a broad background radiation as indicated by the magnification number and determining peak positions becomes difficult. The temperatures calculated from the G band positions are 242°C (1,586 cm^{-1}), 1,545°C (1,543 cm^{-1}), and 2,000°C (1,528 cm^{-1}), respectively. It should be noted, that while there is a linear relationship between the G band position and temperature between 25 and 500°C (inset in Fig. 12.22a), this assumption might oversimplify the behavior at higher temperatures (> 1,000°C). Moreover, the recorded Raman spectra represent an average over the excited sample volume. Therefore, local temperatures within the sample may be slightly lower or higher compared to the measured average value.

5.4.2 Laser-Induced Heating and Light Emission

Laser heating (thermal emission) increases with increasing laser power, and is less dependent on the excitation wavelength compared to photoluminescence or other light emitting electronic transitions, which require certain minimum photon energies. While the photon energy (in eV) of red light is much lower, the energy density (in W/cm^2) of the diode laser used in this study is greater compared to the UV source used. In order to determine the nature of the emitted light we changed the excitation wavelength from 325 nm (3.81 eV, max. 1,300 W/cm^2) to 785 nm (1.58 eV, max. 7,500 W/cm^2). Figure 12.23a shows the light emission of nanocrystalline graphite (particle size ~50 nm) under 785-nm excitation. Indeed, the intensity of the light was found to be stronger compared to 325 nm, revealing a thermal nature of the emitted light (blackbody radiation). The absence of sharp emission lines also eliminates plasma as a possible emission source. On the other hand, photoluminescence usually shows significantly lower intensities and should not be dependent upon the environment. It should, therefore, occur even when the sample is immersed in water. This was not observed in our experiments.

Assuming the emitted light to be mainly thermal radiation, one can determine the sample temperature using Planck's radiation law. Conventional Raman spectrometers are not suitable for recording emission spectra over a broad wavelength range due to the dispersive operation mode and slow moving optics, which increases the probability of intensity changes during acquisition and decreases the accuracy of temperature data measurements. Moreover, conventional Raman spectrometers do not have second order filtering and the interpretation of results becomes difficult.

The spectral composition of the emitted light was recorded using an external UV-NIR spectrometer. The temperature is determined by comparing the corrected experimental data [59, 116] with the calculated emission spectrum [59, 117, 118]. The blackbody spectrum of the nanocrystalline graphite sample was calculated for three different temperatures. Based on the maxima of the emission curve, the local sample temperature was ~2,600°C (Fig. 12.23b).

Laser heating and light emission can be different for different sample spots and depend on both the structure of the carbon and bulk density of the sample. For example, while carbon onions, carbon black, and ND exhibit strong heating and show light emission at almost each spot excited by the laser beam, its occurrence is much lower for MWCNTs. The main difference between MWCNTs and the other nanostructures is the size. While the particle size (gyration radius) of ND, carbon black, or carbon onions is under 100 nm in any dimension, MWCNTs have a comparable size in cross section, but are at least two orders of magnitude longer. In that case, the energy transferred from the laser beam to the nanoparticles is dissipated much faster and over a larger area, thus reducing local heating.

5.4.3 Structural Changes upon Laser Excitation

While melting of graphitic carbon is not expected at temperatures below 3,500°C, sublimation, graphitization or other heat-induced structural changes have been observed at temperatures well below 2,500°C [119–123]. In order to determine possible changes in structure and composition of carbon nanomaterials upon laser excitation, we recorded the Raman spectra of a nanocrystalline graphite before and after light emission induced by 785-nm excitation (Fig. 12.24a). The main features in the Raman spectrum are two peaks centered at ~1,340 cm^{-1} (D band) and 1,580 cm^{-1} (G band), respectively. After light emission, formation of a black film was observed around the excited sample area, while some carbon material within the excited volume was removed. The Raman spectrum recorded at the initial excitation spot shows almost no changes. Peak shape and intensity ratio between D and G band are similar to the spectrum of nanocrystalline graphite. However, the Raman spectrum of the film is quite different and shows significant changes. While the D band and G band are well separated in the Raman spectrum of nanocrystalline

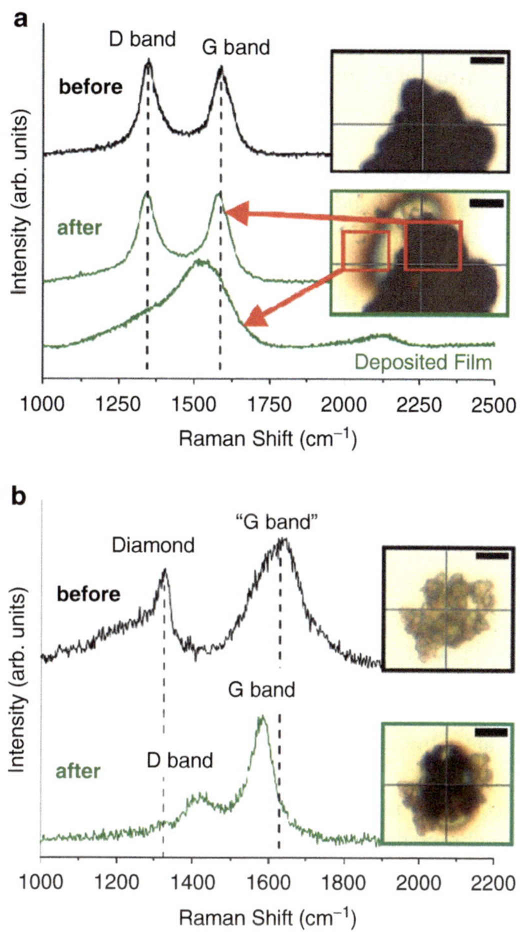

Fig. 12.24 Raman spectra of carbon nanomaterials and corresponding optical images of (**a**) carbon black (514-nm excitation) and (**b**) nanodiamond (325-nm laser excitation) before and after laser-induced light emission, recorded in Ar atmosphere. Carbon black is evaporated by the laser and redeposited as amorphous carbon around the excitation spot. ND is evaporated or transformed into graphitic carbon upon irradiation. The *scale bar* in the photograph is 15 μm

graphite, they overlap in the spectrum recorded from the film and are noticeably broadened. The G band position is shifted to lower wavenumbers indicating the presence of amorphous species and an increase in disorder. Shape and position of the peaks are similar to that of amorphous carbon (>80% sp^2) [52]. These results suggest the evaporation of carbon upon laser excitation and its subsequent redeposition around the excitation spot. While earlier estimates, based on emission spectra, show temperatures lower than the melting temperature of carbon, sublimation of carbon typically occurs at ambient pressure below the melting point.

Evaporation and redeposition of carbon occur at the very first moment upon laser excitation. During light emission, no further changes are observed. The laser heats and evaporates carbon black, reaching temperatures of >3,000°C within the focal point of the beam. Carbon material around the focus spot is heated too, but the energy density of the laser is insufficient to reach the evaporation temperature.

The emitted light is thermal radiation of the crystals that are slightly out of focus and reach temperatures of up to 2,600°C, but are not evaporated.

Structural changes during light emission were also observed for ND powder (Fig. 12.24b). The Raman spectrum (325 nm) of the ND powder before emission shows two characteristic features: a downshifted and broadened, with respect to the Raman mode of single crystal diamond (1,332 cm^{-1}), diamond peak at ~1,325 cm^{-1} and a broad band centered at ~1,620 cm^{-1} resulting from overlapping G band of sp^2 carbon and OH groups [87]. Unlike the G band in graphitic materials, this band consists of at least three separate peaks centered at 1,585, 1,640, and 1,740 cm^{-1}, respectively. After light emission, the Raman spectrum shows the characteristic features of graphitic carbon. The G band shows a lower peak width and is downshifted to ~1,580 cm^{-1}. The intensity of the diamond peak significantly decreases, while the D band appears in the Raman spectrum after light emission, indicating a conversion of ND into graphitic carbon upon laser excitation. Optical images of the sample before and after light emission support this assumption. The color of the ND powder changes from gray-brown to black, suggesting the formation of sp^2-bonded carbon species. It is well known that graphitization of ND and the formation of onion-like carbon and polygonized particles occur upon heating above ~1,100–1,200°C [121, 122]. In addition, similar to the nanocrystalline graphite sample, the formation of a thin film was observed around the excited sample spot, indicating significantly higher temperatures in the focused spot of the beam.

These results demonstrate that nanostructured carbon materials can be evaporated or undergo phase changes/transformations upon laser excitation of significantly lower intensity compared to bulk materials. Even the power of laser pointers commonly used during conference presentations or in the class room (up to 100 mW) is enough to heat carbon samples up to 3,500°C, evaporate and redeposit graphitic nanocarbons or convert nanodiamond into carbon onions, when focused to a spot size of several micrometers.

5.5 Conclusions and Future Perspective

In situ Raman spectroscopy analysis of isothermal and nonisothermal oxidation of DWCNTs in air showed a decrease in the intensity of the D band starting around 370°C, followed by complete D band elimination at 440°C. The oxidation process produced the purest CNTs ever reported, which were free of amorphous carbon and highly defective tubes, while the removal of amorphous material was not accompanied by tube damage. In situ Raman measurements allowed us to determine the different contributions to the D band feature and show the relationship between D band, G band, and RBM Raman modes in the Raman spectra of DWCNTs upon heating. The described approach thus provides an efficient purification method for DWCNTs and SWCNTs, which is also selective to tube diameter and chirality. While oxidation of MWCNTs did not significantly decrease the D band intensity below 450°C, oxidation in air can be an effective route to control the number of defects on outer walls of the MWCNTs. The intensity ratio between the D and G bands can be used to monitor relative changes in the concentration of defective sites, and to create a high density of carboxyl groups with a moderate sample loss.

Raman-assisted oxidation studies on detonation-synthesized ND demonstrated the possibility of selectively removing amorphous and graphitic sp^2-bonded carbon from ND powders by heating in air. The optimal temperature range for oxidation of the ND powders studied is 400–430°C. Oxidation at 425°C increased the content of sp^3-bonded carbon up to 96%, as determined by XANES. The weight loss was roughly equal to the initial content of sp^2 carbon in the sample, suggesting little or no loss of the diamond phase. The purity of ND thus became comparable to that of microcrystalline diamond, and was the highest ever reported for ND powders. Metal impurities, which were initially protected by carbon shells in the as-received samples, become accessible after oxidation and can be completely removed by further treatment in diluted acids. The presented technique is also capable of significantly improving the quality of diamond samples which underwent prior acid-purification treatments without appreciable loss of the diamond phase.

Furthermore, we have shown that oxidation in air can also be used to control the average crystal size in ND powders with subnanometer accuracy. Three different characterization techniques were used for measuring the crystal size because such analysis is very complex for nanocrystals. While HRTEM is able to visualize ND crystals, the calculated size distributions are statistically not reliable and the average size is often overestimated. Agglomeration and difficulties in sample preparation do not allow accurate estimates on average crystal size values. XRD, which directly probes the crystalline structure of a material, is more reliable in terms of statistics and average values, but lattice distortion and strain can interfere with size effects in XRD pattern and lead to an incorrect interpretation of the results.

Raman spectroscopy, which is also used to measure the crystal size of nanostructured solids through the phonon confinement model (PCM), provides only semi-quantitative results for size measurements in ND powders due to insufficient understanding of the Raman spectra of ND and a lack of agreement between theoretical predictions of the model and experimental Raman data. However, taking into account the broad size distribution of ND powders and the contributions of lattice defects, a significant improvement in the predictions of the model was achieved. However, a correct interpretation of Raman data and quantitative size measurements still requires additional information on sample structure and composition. Therefore, a combined use of various characterization techniques such as XRD, HRTEM, and Raman spectroscopy can be recommended for a reliable determination of the average size of ND crystals and their distribution.

In summary, we have shown that oxidation in air provides a powerful route to purify carbon nanostructures, but can also be used as an efficient tool for size control and surface modification. In situ Raman spectroscopy studies under isothermal and nonisothermal conditions allow a detailed and time-resolved investigation of changes in structure and composition during the oxidation process.

Using in situ Raman spectroscopy, we were able to monitor the oxidation of different carbon nanostructures and identify their optimum purification conditions, with little or no sample loss.

For an industrial-scale production of carbon nanomaterials, it is important to use a simple and environmentally friendly purification method to selectively remove sp^2-bonded carbon from nanodiamond and amorphous carbon from nanotubes with minimal or no loss of diamond or nanotubes. In contrast to current purification techniques, which usually use mixtures of oxidizing acids, controlled air oxidation does not require the use of toxic or aggressive chemicals, catalysts, or inhibitors, thus opening avenues for numerous new applications of carbon nanomaterials.

References

1. Gogotsi Y (ed) (2006) Nanomaterials handbook. CRC Press, Baco Raton
2. Chiganov AS (2004) Selective inhibition of the oxidation of nanodiamonds for their cleaning. Phys Solid State 46(4):595–787
3. Evans EL, Griffiths RJM, Thomas JM (1971) Kinetics of single-layer graphite oxidation: evaluation by electron microscopy. Science 171(3967):174–175
4. Lee SM, Lee YH, Hwang YG, Hahn JR et al (1999) Defect-induced oxidation of graphite. Phys Rev Lett 82(1):217–220
5. Yang RT, Wong C (1981) Kinetics and mechanism of oxidation of basal-plane on graphite. J Chem Phys 75(9):4471–4476
6. Yang RT, Wong C (1981) Mechanism of single-layer graphite oxidation: evaluation by electron micriscopy. Science 214(4519):437–438
7. Shimada T, Yanase H, Morishita K, Hayashi J-I et al (2004) Points of onset of gasification in a multi-walled carbon nanotube having an imperfect structure. Carbon 42(8–9): 1635–1639
8. Mazzoni MSC, Chacham H, Ordejón P, Sánchez-Portal D et al (1999) Energetics of the oxidation and opening of a carbon nanotube. Phys Rev B 60(4):R2208
9. Ajayan PM, Ebbesen TW, Ichihashi T, Iijima S et al (1993) Opening carbon nanotubes with oxygen and implications for filling. Nature 362(6420):522–525
10. Navarro MV, Seaton NA, Mastral AM, Murillo R (2006) Analysis of the evolution of the pore size distribution and the pore network connectivity of a porous carbon during activation. Carbon 44(11):2281–2288
11. Pastor-Villegas J, Duran-Valle CJ (2002) Pore structure of activated carbons prepared by carbon dioxide and steam activation at different temperatures from extracted rockrose. Carbon 40(3):397–402
12. Rodriguez-Reinoso F, Molina-Sabio M (1992) Activated carbons from lignocellulosic materials by chemical and/or physical activation: an overview. Carbon 30(7):1111–1118
13. Rosca ID, Watari F, Uo M, Akasaka T (2005) Oxidation of multiwalled carbon nanotubes by nitric acid. Carbon 43(15):3124–3131
14. Urbonaite S, Wachtmeister S, Mirguet C, Coronel E et al (2007) EELS studies of carbide derived carbons. Carbon 45(10):2047–2053
15. Williams PT, Reed AR (2006) Development of activated carbon pore structure via physical and chemical activation of biomass fibre waste. Biomass Bioenergy 30(2):144–152
16. Rodríguez-Reinoso F, Molina-Sabio M, González MT (1995) The use of steam and CO_2 as activating agents in the preparation of activated carbons. Carbon 33(1):15–23
17. Munoz-Guillena MJ, Illan-Gomez MJ, Martin-Martinez JM, Linares-Solano A et al (1992) Activated carbons from Spanish coals. 1. Two-stage carbon dioxide activation. Energy Fuels 6(1):9–15

18. Huang ZH, Kang FY, Yang JB, Liang KM et al (2002) Effect of CO in activating gas on the pore structure of activated carbon fiber with CO_2 activation. J Mat Sci Lett 22(4):293–295
19. McKee DW (1981) Catalyzed gasification of carbon. In: Walker PL, Thrower P (eds) Chemistry. Marcel Dekker, New York
20. Vannice MA (2005) Kinetics of catalytic reaction. Springer, New York
21. McKee DW (1970) Metal oxides as catalyst for the oxidation of graphite. Carbon 8(5):623–626
22. Li J, Zhang Y (2005) A simple purification for single-walled carbon nanotubes. Physica E Low Dimens Syst Nanostruct 28(3):309–312
23. Saxby JD, Chatfield SP, Palmisano AJ, Vassallo AM et al (1992) Thermogravimetric analysis of buckminsterfullerene and related materials in air. J Phys Chem 96(1):17–18
24. Illekova E, Csomorova K (2005) Kinetics of oxidation in various forms of carbon. J Therm Anal Calorim 80(1):103–108
25. Gajewski S, Maneck HE, Knoll U, Neubert D et al (2003) Purification of single walled carbon nanotubes by thermal gas phase oxidation. Diamond Relat Mater 12(3–7):816–820
26. Moon JM, An KH, Lee YH, Park YS et al (2001) High-yield purification process of singlewalled carbon nanotubes. J Phys Chem B 105(24):5677–5681
27. Yang CM, Kaneko K, Yudasaka M, Iijima S (2002) Effect of purification on pore structure of HiPco single-walled carbon nanotube aggregates. Nano Lett 2(4):385–388
28. Gogotsi Y (ed) (2006) Nanotubes and nanofibers. CRC Press, Boca Raton
29. Brukh R, Mitra S (2007) Kinetics of carbon nanotube oxidation. J Mater Chem 17(7): 619–623
30. Ajayan PM, Lijima S (1993) Capillarity-induced filling of carbon nanotubes. Nature 361(6410):333–334
31. Park YS, Choi YC, Kim KS, Chung D-C et al (2001) High yield purification of multiwalled carbon nanotubes by selective oxidation during thermal annealing. Carbon 39(5):655–661
32. Sharp JH, Wentworth SA (1969) Kinetic analysis of thermogravimetric data. Anal Chem 41(14):2060–2062
33. Saito R, Dresselhaus G, Dresselhaus MS (1998) Physical properties of carbon nanotubes. Imperial College Press, London
34. Jorio A, Saito R, Dresselhaus G, Dresselhaus MS (2004) Determination of nanotubes properties by Raman spectroscopy. Philos Trans Roy Society A-Math Phys Engin Sci 362(1824):2311–2336
35. Puech P, Flahaut E, Bassil A, Juffmann T et al (2007) Raman bands of double-wall carbon nanotubes: comparison with single- and triple-wall carbon nanotubes, and influence of annealing and electron irradiation. J Raman Spectrosc 38(6):714–720
36. Zhao X, Ando Y, Qin LC, Kataura H et al (2002) Characteristic Raman spectra of multiwalled carbon nanotubes. Physica B-Cond Mat 323(1–4):265–266
37. Rao AM, Jorio A, Pimenta MA, Dantas MSS et al (2000) Polarized Raman study of aligned multiwalled carbon nanotubes. Phys Rev Lett 84(8):1820–1823
38. Zhao XL, Ando Y, Qin LC, Kataura H et al (2002) Multiple splitting of G-band modes from individual multiwalled carbon nanotubes. Appl Phys Lett 81(14):2550–2552
39. Benoit JM, Buisson JP, Chauvet O, Godon C et al (2002) Low-frequency Raman studies of multiwalled carbon nanotubes: experiments and theory. Phys Rev B 66(7):073417-1–073417-4
40. Zhao XL, Ando Y, Qin LC, Kataura H et al (2002) Radial breathing modes of multiwalled carbon nanotubes. Chem Phys Lett 361(1–2):169–174
41. Jorio A, Saito R, Hafner JH, Lieber CM et al (2001) Structural (n, m) determination of isolated single-wall carbon nanotubes by resonant Raman scattering. Phys Rev Lett 86(6):1118–1121
42. Ci LJ, Zhou ZP, Yan XQ, Liu DF et al (2003) Resonant Raman scattering of double wall carbon nanotubes prepared by chemical vapor deposition method. J Appl Phys 94(9):5715–5719

43. Henrard L, Hernandez E, Bernier P, Rubio A (1999) van der Waals interaction in nanotube bundles: consequences on vibrational modes. Phys Rev B 60(12):R8521–R8524
44. Osswald S, Flahaut E, Gogotsi Y (2006) In situ Raman spectroscopy study of oxidation of double- and single-wall carbon nanotubes. Chem Mater 18(6):1525–1533
45. Osswald S, Flahaut E, Ye H, Gogotsi Y (2005) Elimination of D-band in Raman spectra of double-wall carbon nanotubes by oxidation. Chem Phys Lett 402(4–6):422–427
46. Thomsen C, Reich S (2000) Double resonant Raman scattering in graphite. Phys Rev Lett 85(24):5215–5217
47. Sato K, Saito R, Oyama Y, Jiang J et al (2006) D-band Raman intensity of graphitic materials as a function of laser energy and crystallite size. Chem Phys Lett 427(1–3):117–121
48. Maultzsch J, Reich S, Thomsen C, Webster S et al (2002) Raman characterization of boron-doped multiwalled carbon nanotubes. Appl Phys Lett 81(14):2647–2649
49. Tuinstra F, Koenig JL (1970) Raman spectrum of graphite. J Chem Phys 53(3):1126–1130
50. Matthews MJ, Pimenta MA, Dresselhaus G, Dresselhaus MS et al (1999) Origin of dispersive effects of the Raman D band in carbon materials. Phys Rev B 59(10): R6585–R6588
51. Knight DS, White WB (1989) Characterization of diamond films by Raman-spectroscopy. J Mater Res 4(2):385–393
52. Ferrari AC, Robertson J (2000) Interpretation of Raman spectra of disordered and amorphous carbon. Phys Rev B 61(20):14095–14107
53. Ferrari AC (2007) Raman spectroscopy of graphene and graphite: disorder, electron–phonon coupling, doping and nonadiabatic effects. Solid State Commun 143(1–2):47–57
54. Dresselhaus MS, Dresselhaus G (1982) Light scattering in solids III. Springer, Berlin\ Heidelberg
55. Solin SA, Caswell N (1981) Raman-scattering from alkali graphite-intercalation compounds. J Raman Spectrosc 10(1):129–135
56. Jorio A, Filho AGS, Dresselhaus G, Dresselhaus MS et al (2002) G-band resonant Raman study of 62 isolated single-wall carbon nanotubes. Phys Rev B (Condensed Matter Mater Phys) 65(15):155412–155419
57. Maultzsch J, Reich S, Thomsen C (2001) Chirality-selective Raman scattering of the D mode in carbon nanotubes. Phys Rev B 6412(12):121407-1–121407-4
58. Ci LJ, Zhou ZP, Song L, Yan XQ et al (2003) Temperature dependence of resonant Raman scattering in double-wall carbon nanotubes. Appl Phys Lett 82(18):3098–3100
59. Osswald S, Behler K, Gogotsi Y (2008) Laser-induced light emission from carbon nanoparticles. J Appl Phys 104(7):074308-1–074308-6
60. Osswald S, Havel M, Gogotsi Y (2007) Monitoring oxidation of multiwalled carbon nanotubes by Raman spectroscopy. J Raman Spectrosc 38(6):728–736
61. Banhart F (1999) Irradiation effects in carbon nanostructures. Rep Prog Phys 62(8): 1181–1221
62. Behler K, Osswald S, Ye H, Dimovski S et al (2006) Effect of thermal treatment on the structure of multi-walled carbon nanotubes. J Nanopart Res 8:615–625
63. Fanning PE, Vannice MA (1993) A drifts study of the formation of surface groups on carbon by oxidation. Carbon 31(5):721–730
64. Kuhlmann U, Jantoljak H, Pfander N, Bernier P et al (1998) Infrared active phonons in single-walled carbon nanotubes. Chem Phys Lett 294(1–3):237–240
65. Zawadzki J (1980) Ir spectroscopic investigations of the mechanism of oxidation of carbonaceous films with HNO_3 solution. Carbon 18(4):281–285
66. Kastner J, Pichler T, Kuzmany H, Curran S et al (1994) Resonance Raman and infrared spectroscopy of carbon nanotubes. Chem Phys Lett 221(1–2):53–58
67. Liu M, Yang Y, Zhu T, Liu Z (2005) Chemical modification of single-walled carbon nanotubes with peroxytrifluoroacetic acid. Carbon 43(7):1470–1478
68. Chen X-H, Chen C-S, Xiao H-N, Chen X-H et al (2005) Lipophilic functionalization of multi-walled carbon nanotubes with stearic acid. Carbon 43(8):1800–1803

69. Zawadzki J (1989) Infrared-spectroscopy in surface-chemistry of carbons. Chem Phys Carbon 21:147–380
70. Dolmatov VY (2001) Detonation synthesis ultradispersed diamonds: properties and applications. Uspekhi Khimii 70(7):687–708, in Russian
71. Gruen DM, Shenderova OA, Vul AY (eds) (2005) Synthesis, properties and applications of ultrananocrystalline diamond, vol 192, NATO Science series. Series II: Mathematics, physics and chemistry. Springer, Dordrecht\Berlin\Heidelberg\New York, p 401
72. Shenderova OA, Gruen DM (eds) (2006) Ultrananocrystalline diamond: synthesis, properties, and applications. William-Andrew, Norwich
73. Shenderova OA, McGuire G (2006) Nanocrystalline diamond. In: Gogotsi Y (ed) Nanomaterials handbook. CRC Taylor and Francis Group, Boca Raton\London\New York, pp 203–237
74. Danilenko VV (2003) Synthesizing and sintering of diamond by explosion. Energoatomizdat, Moscow, p 272
75. Danilenko VV (2004) On the history of the discovery of nanodiamond synthesis. Phys Solid State 46(4):595–599
76. Dolmatov VY (2003) Ultradisperse diamonds of detonation synthesis: production, properties and applications. State Polytechnical University, St. Petersburg
77. Dolmatov VY (2001) Detonation synthesis ultradispersed diamonds: properties and applications. Russian Chem Rev 70:607
78. Osswald S, Yushin G, Mochalin V, Kucheyev S et al (2006) Control of sp2/sp3 carbon ratio and surface chemistry of nanodiamond powders by selective oxidation in air. J Am Chem Soc 128:11635–11642
79. Danilenko VV (2005) Specific features of synthesis of detonation nanodiamonds. Combust Explos Shock Waves 41(5):577–588
80. Yushin GN, Osswald S, Padalko VI, Bogatyreva GP et al (2005) Effect of sintering on structure of nanodiamond. Diamond Related Mater 14(10):1721–1729
81. Osswald S, Gurga A, Kellogg F, Cho K et al (2007) Plasma pressure compaction of nanodiamond. Diamond Related Mater 16(11):1967–1973
82. Gordeev S (2004) Do small nanodiamonds exist? J Superhard Mater 6:34–36, in Russian
83. Gordeev SK (2006) Carbon composites based nanodiamonds. In: Nanocarbon and nanodiamond. St. Petersburg, Russia
84. Pichot V, Comet M, Fousson E, Baras C et al (2008) An efficient purification method for detonation nanodiamonds. Diamond Related Mater 17:13–22
85. Ferrari AC, Robertson J (2004) Raman spectroscopy of amorphous, nanostructured, diamond-like carbon, and nanodiamond. Phil Trans Roy Soc Lond A 362:2267–2565
86. Osswald S, Mochalin VN, Havel M, Yushin G et al (2009) Phonon confinement effects in the Raman spectrum of nanodiamond. Phys Rev B 80(7):075419
87. Mochalin V, Osswald S, Gogotsi Y (2009) Contribution of functional groups to the Raman spectrum of nanodiamond. Chem Mater 21(2):273–279
88. Ferrari AC, Robertson J (2004) Raman spectroscopy of amorphous, nanostructured, diamond-like carbon, and nanodiamond. Philos Trans Roy Soc Lond Ser A-Math Phys Eng Sci 362(1824):2477–2512
89. Huang F, Tong Y, Yun S (2004) Synthesis mechanism and technology of ultrafine diamond from detonation. Phys Solid State 46(4):616–619
90. Mochalin VN, Osswald S, Portet C, Yushin G et al (2008) High temperature functionalization and surface modification of nanodiamond powders. Mater Res Soc Symp Proc 1039: 1039-P11-03
91. Ji SF, Jiang TL, Xu K, Li SB (1998) FTIR study of the adsorption of water on ultradispersed diamond powder surface. Appl Surf Sci 133(4):231–238
92. Jiang T, Xu K (1995) FTIR study of ultradispersed diamond powder synthesized by explosive detonation. Carbon 33(12):1663–1671
93. Kuznetsov VL, Aleksandrov MN, Zagoruiko IV, Chuvilin AL et al (1991) Study of ultradispersed diamond powders obtained using explosion energy. Carbon 29(4–5):665–668

94. Osswald S, Havel M, Mochalin V, Yushin G et al (2008) Increase of nanodiamond crystal size by selective oxidation. Diamond Related Mater 17(7–10):1122–1126
95. Osswald S, Yushin G, Mochalin V, Kucheyev SO et al (2006) Control of sp(2)/sp(3) carbon ratio and surface chemistry of nanodiamond powders by selective oxidation in air. J Am Chem Soc 128(35):11635–11642
96. Williamson GK, Hall WH (1953) X-ray line broadening from filed aluminium and wolfram. Acta Metallurgica 1(1):22–31
97. Zhang Z, Zhou F, Lavernia EJ (2003) On the analysis of grain size in bulk nanocrystalline materials via X-ray diffraction. Metall Mater Trans A 34A(6):1349–1355
98. Palosz B (2006) Structure of nano-crystals: the key to understanding the unique properties of nano-materials. In: Mitura NS, Niedzielski P, Walkowiak B (eds) Wydawnictwo Naukowe. Nanodiam, Warszawa, pp 129
99. Alcaniz-Monge J, Linares-Solano A, Rand B (2002) Mechanism of adsorption of water in carbon micropores as revealed by a study of activated carbon fibers. J Phys Chem B 106(12):3209–3216
100. Ager JW, Veirs DK, Rosenblatt GM (1991) Spatially resolved Raman studies of diamond films grown by chemical vapor deposition. Phys Rev B 43(8):6491
101. Yoshikawa M, Mori Y, Obata H, Maegawa M et al (1995) Raman scattering from nanometer-sized diamond. Appl Phys Lett 67(5):694–696
102. Yoshikawa M, Mori Y, Maegawa M, Katagiri G et al (1993) Raman-scattering from diamond particles. Appl Phys Lett 62(24):3114–3116
103. Yang ZP, Ci LJ, Bur JA, Lin SY et al (2008) Experimental observation of an extremely dark material made by a low-density nanotube array. Nano Lett 8(2):446–451
104. Ajayan PM, Terrones M, de la Guardia A, Huc V et al (2002) Nanotubes in a flash – ignition and reconstruction. Science 296(5568):705
105. Wei JQ, Zhu HW, Wu DH, Wei BQ (2004) Carbon nanotube filaments in household light bulbs. Appl Phys Lett 84(24):4869–4871
106. Wang SM, Hu LG, Lu ZD, Zhao DM et al (2003) White and bright radiation from nanostructured carbon. Chinese J Optoelectronics Laser 14(2):215–220
107. Wang S, Hu L, Zhang B, Zhao D et al (2005) Electromagnetic excitation of nano-carbon in vacuum. Opt Express 13(10):3625–3630
108. Ferrari AC, Robertson J (2004) Raman spectrsocopy in carbons: from nanotubes to diamond. Philos Trans Roy Soc A 362:2267–2565
109. Fischbach DB, Couzi M (1986) Temperature-dependence of Raman-scattering by disordered carbon materials. Carbon 24(3):365–369
110. Huang FM, Yue KT, Tan PH, Zhang SL et al (1998) Temperature dependence of the Raman spectra of carbon nanotubes. J Appl Phys 84(7):4022–4024
111. Ni ZH, Fan HM, Fan XF, Wang HM et al (2007) High temperature Raman spectroscopy studies of carbon nanowalls. J Raman Spectrosc 38(11):1449–1453
112. Song L, Ma JW, Ren Y, Zhou WY et al (2008) Temperature dependence of Raman spectra in single-walled carbon nanotube rings. Appl Phys Lett 92(12):121905–121905-3
113. Kukovecz A, Smolik M, Bokova S, Kataura H et al (2005) Determination of the diameter distribution of single-wall carbon nanotubes from the Raman G-band using an artificial neural network. J Nanosci Nanotechnol 5(2):204–208
114. Zhao Q, Wagner HD (2004) Raman spectroscopy of carbon-nanotube-based composites. Phil Trans Roy Soc Lond A 362(1824):2407–2424
115. Rao AM, Richter E, Bandow S, Chase B et al (1997) Diameter-selective Raman scattering from vibrational modes in carbon nanotubes. Science 275(5297):187–191
116. Keyvan S, Rossow R, Romero C (2006) Blackbody-based calibration for temperature calculations in the visible and near-IR spectral ranges using a spectrometer. Fuel 85(5–6):796–802
117. Chang H, Charalampopoulos TT (1990) Determination of the wavelength dependence of refractive indices of flame soot. Phil Trans Roy Soc Lond A 430(1880):577–591

118. Rohlfing EA (1988) Optical-emission studies of atomic, molecular, and particulate carbon produced from laser vaporization cluster source. J Chem Phys 89(10):6103–6112
119. Savvatimskiy AI (2005) Measurements of the melting point of graphite and the properties of liquid carbon (a review for 1963–2003). Carbon 43(6):1115–1142
120. Kim HS, Shioya M, Takaku A (1999) Sublimation and deposition of carbon during internal resistance heating of carbon fibers. J Mater Sci 34(18):4613–4622
121. Portet C, Yushin G, Gogotsi Y (2007) Electrochemical performance of carbon onions, nanodiamonds, carbon black and multiwalled nanotubes in electrical double layer capacitors. Carbon 45(13):2511–2518
122. Qiao ZJ, Li JJ, Zhao NQ, Shi CS et al (2006) Graphitization and microstructure transformation of nanodiamond to onion-like carbon. Scripta Materialia 54(2):225–229
123. Behler K, Osswald S, Ye H, Dimovski S et al (2006) Effect of thermal treatment on the structure of multi-walled carbon nanotubes. J Nanoparticle Res 8(5):615–625

Multiplexed SERS for DNA Detection 13

Karen Faulds

1 Overview

Surface enhanced resonance Raman scattering (SERRS) is an analytical technique with several advantages over competitive techniques in terms of improved sensitivity and selectivity. We have made great progress in the development of SERRS as a quantitative analytical method, in particular for the detection of DNA. However, one of the main advantages over fluorescence and other optical detection techniques is the ability to multiplex.

The enhancing surface which is used is crucial and in this case suspensions of metallic nanoparticles will be the focus since they allow quantitative detection to be achieved and are compatible with solution-based diagnostics assays. When using metallic nanoparticles for SE(R)RS, aggregation is required to massively increase the enhancement achieved and the aggregation conditions used are crucial to the success of the analysis. To obtain successful SERRS of DNA, the use of spermine as an aggregating agent is required. The nature of the label which is used, be it fluorescent, positively or negatively charged also effects the SERRS response and these conditions are again explored here. We have clearly demonstrated the ability to identify the components of a mixture of five analytes in solution by using two different excitation wavelengths and also of a 6-plex using data analysis techniques.

This book chapter will explore the ability of SE(R)RS to detect multiple analytes in a mixture, that is, its ability to multiplex. The detection of DNA will be used as the focus for this discussion however the capabilities of SE(R)RS can be easily extended to any biomolecule, or indeed, any mixture of analytes with distinct spectral fingerprints.

K. Faulds
Centre for Molecular Nanometrology, Department of Pure and Applied Chemistry, WestCHEM, University of Strathclyde, 295 Cathedral Street, Glasgow, UK

Challa S. S. R. Kumar (ed.), *Raman Spectroscopy for Nanomaterials Characterization*,
DOI 10.1007/978-3-642-20620-7_13, © Springer-Verlag Berlin Heidelberg 2012

2 Introduction

Surface enhanced resonance Raman scattering (SERRS) [1–3] is a highly sensitive technique, so sensitive in fact that single molecule detection has previously been reported [4, 5]. It is a very attractive technique for the detection of biomolecules because it produces molecularly specific spectra which make it feasible to easily identify the components of a mixture in a single analysis without extensive separation procedures [6].

SERRS is an advancement over normal Raman scattering which is an inherently weak process with only around one in every million photons being Raman scattered. Used correctly, it can combine many of the advantages of Raman spectroscopy such as molecular specificity and selective identification of a species in situ, with the additional advantages of high sensitivity. SERRS involves adsorbing the analyte of interest onto a roughened metal surface and interrogating the surface using a Raman spectrometer. The most commonly used metals for SERS measurements are gold or silver [3, 7]. The reason for this are that the electronic properties of gold and silver are suitable for SERS as they have surface plasmons which lie in the visible region of the electromagnetic spectrum which coincide with the commonly used Raman excitation wavelengths.

One of the main advantages of SERS, as well as its inherent sensitivity, is in the nature of the spectroscopic data obtained. SERS is a vibrational technique therefore the spectra obtained have sharp, molecularly specific peaks which are unique to the molecule being analyzed. This is in contrast to other highly sensitive optical detection methods such as fluorescence where the spectra produced are not molecularly specific, broad and cannot be used to assign the structure of the molecule being investigated. It is this highly advantageous feature of SERS which allows us to analyze complex samples, for example, to obtain spectra from analytes contained within complex media since if the spectrum of the analyte is known it can be easily distinguished from background signals from the matrix. The other major advantage, which will be focused on within this chapter is the ability to use these unique, fingerprint spectral features to detect, quantify, and identify the components of a mixture when multiple analytes are present, that is, multiplexing. The multiplexing capabilities of SERS will be discussed in the context of biological samples focusing specifically on the example of DNA, however this capability could be easily extended to any analyte or mixture of analytes.

The labeling of biological components to aid in their detection is well established and common practice using a wide range of techniques including micro arrays, ELISA, separation science and real time PCR. The most widely used labels are fluorescent or chemiluminescent in nature. These labels generally offer a high degree of sensitivity, down to single molecule detection levels, therefore they are the most widely used in biological characterization and diagnostics [8]. The same labels can be used to label and detect biomolecules using SERRS as well as using non-fluorescent labels allowing more labeling strategies to be used which are potentially simpler, more extensive and as yet underdeveloped. To achieve maximum surface enhancement the labels must have a chromophore which is coincident with the excitation frequency and the ability to adsorb onto a suitable metal surface [9]. Many compounds of interest do not possess these properties but addition of a SERRS active label can achieve this and this approach has been used successfully for a number of targets including DNA [10–12]. The use of a metal surface quenches any fluorescence emitted meaning that commonly available fluorescent labels can be used for SERRS. Previously, SERRS detection of oligonucleotides has made use of a covalently attached label followed by a biological event then detection.

2.1 Surface Enhanced Raman Scattering

Surface Enhanced Raman Scattering (SERS) was first reported by Fleischman et al. in 1974 [1]. During experiments using Raman spectroscopy to detect pyridine at a silver electrode, it was noted that the Raman scattering was greatly increased when the surface of the electrode was roughened. Work a few years later, in 1977, by Jeanmaire and Van Duyne [2], and Albreicht and Creighton [13] demonstrated that the Raman scattering from pyridine absorbed on a roughened silver surface was a factor of 10^6 greater than the same amount of pyridine in solution. This large enhancement in signal stimulated great interest in the technique and it was discovered that the effect was not unique to pyridine. It has subsequently been shown that it is possible to obtain SERS from a large number of molecules as long as they are absorbed onto a roughened metal surface.

Surface enhanced resonance Raman scattering (SERRS) can be considered to be a combination of two processes, resonance Raman and SERS, and it was first reported by Stacy and Van Duyne [14] in 1983. As a consequence of this, the enhancement in signal that is observed is greater than either of these two processes, with SERRS spectroscopy an enhancement in scattering three to four orders of magnitude greater than SERS is observed. This in turn corresponds to an enhancement of up to 10^{14} in the scattering from some molecules compared to normal Raman scattering.

For an analyte to exhibit SERRS, it must have a chromophore and by tuning the frequency of the laser excitation to the absorption maxima of the analyte, as well as the surface plasmon of the metal substrate, very large enhancements in signal can be observed. This also gives a degree of selectivity as colored compounds with absorption maxima close to the laser excitation will be selectively enhanced over other species that may be present.

2.2 Mechanism of SERS

Since the discovery of the surface enhancement effect, it has been the subject of much debate as to what the origins of the effect are. It is generally understood that there are several mechanisms which are responsible for the observed enhancement. There are two main theories: electromagnetic enhancement and charge-transfer or chemical enhancement which will be mentioned only briefly here.

The collective excitation of the electron cloud of a conductor is known as a plasmon, if the excitation is confined to the surface of the conductor it is known as a surface plasmon. For the excitation of surface plasmons by light, surface roughness or curvature is required. The electromagnetic field of the light at the surface can be greatly increased when the surface plasmon is excited. This results in the amplification of both the incident and scattered and this is the basis of the electromagnetic SERS mechanism [15].

Although electromagnetic enhancement can explain important features of the SERS effect, it does not take into account in any way the chemical structure of the analyte species. Therefore, chemical enhancement models take into account the structural properties of the analyte and it is thought to operate independently from electromagnetic enhancement. Chemical enhancement can result from charge-transfer or bond formation between the metal and the analyte molecule which can result in an increase in the polarizability, α, of the molecule. Therefore, a surface complex between the analyte and the metal must form before chemical enhancement can occur.

It is very difficult to separate the two enhancement effects but this has been attempted by measuring the chemical enhancement from smooth metal surfaces where it was found that the chemical enhancement effect was small [16, 17]. This enhancement process is wholly adsorbate dependent and requires an analyte with a functionality capable of chemi- or physisorption to the metal surface. There is evidence for both these effects and it is widely believed that the enhancement may occur from a combination of these two effects.

2.3 Metal Surface

A wide range of metals have been used to obtain surface enhancement; lithium, palladium, cadmium, and nickel, however, the most commonly used are silver [4, 7, 18], gold [19–22], and copper [23, 24] since they tend to give the largest enhancement in signal and have surface plasmons which lie in the visible region of the electromagnetic spectrum which coincide with the commonly used Raman excitation frequencies.

A wide range of techniques have been used to obtain the roughened metal surface that is required for SERS/SERRS. The original experiments used electrochemical methods to produce roughened metal electrodes [2–4] and this method is still often used. Fiber optics [25], silver coated filter papers [26], nitric acid etching of silver [27], silver coated titanium dioxide [28], and alumina [29], and polymers sol-gels containing silver [30–32] have also been used. Laserna et al. [33] and Rowlen et al. [34] have published papers comparing the effectiveness of some of these SERS substrates.

However, one of the most commonly used substrates for SE(R)RS is colloidal nanoparticle suspensions of the metal, and the most commonly used metals are silver and gold. This is due to the fact that silver tends to give the greatest enhancement in Raman scattering [35, 36], however gold nanoparticles are often used due to their ease of synthesis and they are often favored for work with laser excitations in the near infrared [21]. Colloidal suspensions of nanoparticles are a particularly attractive substrate for SERS/SERRS as they are relatively simple to make, stable and offer a good degree of reproducibility [3, 37]. The colloid provides a fresh surface for each analysis reducing contamination, which can be a problem due to the high sensitivity of the technique. The colloids are relatively inexpensive and can be made in a batch process that can make enough for several analyses.

The most commonly used suspensions of metal nanoparticles tend to be synthesized by reduction of a metal salt by a reducing agent such as citrate, borohydride, etc., resulting in colloidal suspensions of nanoparticles that tend to be fairly monodisperse and stable over time [3, 7]. Much research has gone into the preparation of silver colloids to give the most reproducible and sensitive colloid. Such citrate reduced silver colloids are thought to be stable for around 6 months due to the colloid being stabilized by a layer of citrate molecules which are bound to the metals surface. This would suggest that citrate reduced silver colloid will have a net negative charge which can make it difficult for negatively charged analytes to get good surface attachment and thus give good SERS/SERRS. Citrate reduced gold colloids are also thought to be stabilized by a citrate layer [38, 39].

This chapter will focus almost exclusively on the use of metal nanoparticles to provide the enhancing surface, however the ability to detect multiple analytes simultaneously can equally be applied to any SERS active metal substrate. Also, as previously stated, the focus here is on the multiplexed detection of DNA since it is a good example of an area where there is a strong need for multiplexing capabilities, however the research reported here could equally apply to another biological system, or indeed, any multianalyte system.

3 Experimental and Instrumental Methodology

Silver Nanoparticles Preparation. A colloidal suspension of citrate reduced silver nanoparticles was prepared using a modified Lee and Meisel [8] procedure.

Sample Preparation. All samples were prepared for SERRS analysis using the following amounts of reagents, 10 μl of dye-labeled oligonucleotide, 10 μl of 0.1 mol dm^{-3} spermine, 250 μl of water, and 250 μl of citrate reduced silver nanoparticles.

Concentration Studies. Concentration studies were carried out using the dye-labeled oligonucleotides. The oligonucleotides were diluted to various concentrations using sterile water. Samples were prepared as above. The samples were analyzed within a minute of the addition of the silver colloid, and each oligonucleotide concentration was analyzed five times. The spectra obtained were the result of a 10 s accumulation time with the spectrometer grating centered at 1,400 cm^{-1}. The spectra obtained were baseline corrected using the GRAMS/32 software and the average peak height of the strongest peak in the spectrum was normalized to the silicon standard peak and plotted against the concentration of labeled oligonucleotide.

Multiplexing. Multiplexing was carried out using six (or five in the case of dual wavelength multiplexing) dye-labeled oligonucleotides. The labels used were TAMRA, ROX, HEX, TET, FAM, and Cy3. The multiplex samples were all prepared using initial stock solutions of labeled oligonucleotides that were prepared to be at a concentration of 10^{-7} mol dm^{-3}. The multiplex samples were then prepared by making solutions containing every possible combination of the 6 labeled oligonucleotides resulting in 64 samples. In the multiplex sample mixtures, water was used to replace missing oligonucleotides in the matrix samples, thus allowing the overall concentration of the labeled oligonucleotides in each sample to remain the same. The final concentration of each oligonucleotide in the multiplex sample was 1.92×10^{-9} mol dm^{-3}. All samples were prepared for SERRS analysis using the following amounts of reagents, 60 μl of dye-labeled oligonucleotide, 10 μl of spermine tetrahydrochloride (0.1 mol dm^{-3}, Sigma-Aldrich), 190 μl of distilled water and 250 μl of citrate reduced silver nanoparticles. The samples were analyzed within 1 min of the addition of the colloid and spermine and five replicates of each multiplex concentration were prepared and analyzed in a random fashion. The spectra obtained were the result of a 1 s accumulation time with the spectrometer grating centered at 1,400 cm^{-1}.

Instrumentation. The following Raman instrumentation were used; a Renishaw Model 100 probe system with a 514.5 nm argon ion laser, utilizing a ×20 objective to focus the laser beam into a 1 cm plastic cuvette containing the sample and a Renishaw Microscope System 1000 with a 632.9 nm helium-neon laser utilizing a Ventacon macrosampler to focus the laser beam into a 1 cm plastic cuvette. However, the same dye-labeled oligonucleotide detection limits can be obtained with low cost Raman instrumentation such as the DeltaNu Advantage 532 nm system, with a compromise in the peak resolution which may slightly influence multiplexing capabilities.

4 Key Research Findings

4.1 SERRS of Labeled Oligonucleotides

Detecting target DNA sequences is an extremely important task in various areas of molecular biology, and is a key step to many modern techniques of disease state analysis. Current methodologies require the detection of a specific sequence, or sequences, within a mixture. The most common way to do this is to attach a label to the DNA sequence of interest thus allowing the presence of the label to be indicative of the presence of a specific DNA sequence. The labeled sequence then acts as a probe to the target sequence. In the majority of cases, the polymerase chain reaction (PCR) is used to amplify the DNA sequence of interest during which a hybridization event takes place that allows the labeled probe sequence to be detected and hence infer the presence of the target. The detection is normally then carried out using a spectroscopic method, most commonly fluorescence spectroscopy, in which case the label attached to the sequence will be fluorescent in nature. However, there are several drawbacks to using fluorescence as a detection technique. The main problem is the nature of the fluorescence emission spectrum which is broad and gives limited characteristic information about the target analyte. This makes the detection of multiple analytes in a mixture difficult due to the large spectral emission overlap that occurs from more than one fluorophore. In practice, using a single excitation light source, only four labels are generally detected at once, three if an internal standard is used unless some sort of physical separation method is employed. Thus, to increase the amount of data obtained per experiment and reduce the number of separate measurements required for DNA analysis, it is desirable to increase the number of DNA sequences that can be detected simultaneously, without separation, in a single experiment.

4.2 Practical Detection of DNA by SERRS

As already stated, to achieve the maximum SERRS enhancement, the analyte molecule must contain a chromophore in the visible region which coincides with the Raman excitation wavelength. DNA does not naturally have this property therefore it has to be modified to allow spectroscopic detection. The chemistry involved in labeling DNA is well characterized since these labeling strategies are routinely used in standard fluorescence detection methodologies.

Custom modifications have previously been developed whereby a non-fluorescent chromophore can be attached to the DNA sequence to provide a strong SE(R)RS signature which is indicative of the DNA sequence present. This has been done previously using DABCYL, phthalocyanines and black hole quenchers (BHQs) as well as specifically designed simple azo dyes which contain moieties to aid in their binding ability to metal surfaces such as the benzotriazole motif which has been shown to be very effective at complexing onto silver nanoparticles [12, 13, 40, 41].

However, the simplest method is to use commercially available fluorophores as the label. This approach is much less synthetically challenging as the labels are commercially available and the procedures for attachment of these labels to sequences of DNA or biomolecules is well characterized and understood. The benefit of this approach is also that there are many fluorophores available which have varying absorption maximum therefore they can be chosen to coincide with the Raman excitation wavelength of your choice. The fact that these molecules are fluorescent does not present the same problem as would be the case in conventional Raman spectroscopy since in surface enhancement the fluorescence emitted by the fluorophore is quenched by the metal surface. The other main advantage of using commercially available fluorophores is that there are a large amount of fluorescent labels available, all with unique spectral fingerprints which make them ideal for multiplexing. For example, the structure and the spectral signature of DNA sequences labeled with FAM, TET, and HEX are given in Fig. 13.1 and it can be seen that although the structure of the three labels are very similar, varying only in

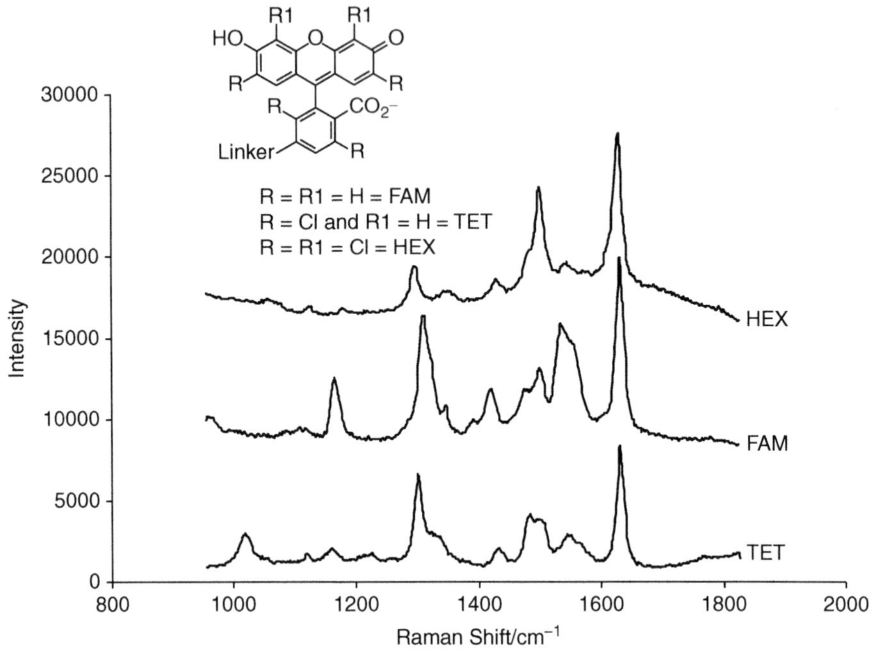

Fig. 13.1 SERRS spectra of FAM, TET, and Hex labeled DNA sequences obtained using silver nanoparticle suspensions and a 514 nm laser excitation

the amount of chlorines on the ring structure, their SERRS spectra are all unique. This opens a vast library of labels which can be used for multiplexing.

The other main requirement to obtain SE(R)RS is that the analyte molecule must come in contact with or be very close to the metal surface used for enhancement. The majority of the work discussed here uses citrate reduced silver nanoparticles which have a net negative charge in aqueous solution due to a layer of citrate that exists on the surface of the silver particles [38]. Since DNA is overall negatively charged, due to the phosphate groups present in the DNA backbone, it is unable to absorb efficiently onto the surface of the silver colloid. However, the surface charge of the nanoparticles can be altered by addition of a positively charged species which will change the effective surface charge of the nanoparticles due to electrostatic layering. A range of species such as ions, poly-L-lysine, and spermine have been investigated for their ability to allow adsorption of negatively charged DNA onto negatively charged nanoparticles, however the polyamine, spermine has proven to be most effective [42]. Spermine hydrochloride has been shown to interact with the DNA backbone and neutralize the negatively charged phosphate backbone of the DNA allowing it to come down onto the metal surface [43]. The spermine also has the advantage that, when in excess, it will also aggregate the metal nanoparticles to give higher enhancement of the Raman scattering due to the increased electromagnetic enhancement experienced through the coupling of the plasmons when the nanoparticles interact. Therefore, spermine serves a dual purpose in these types of

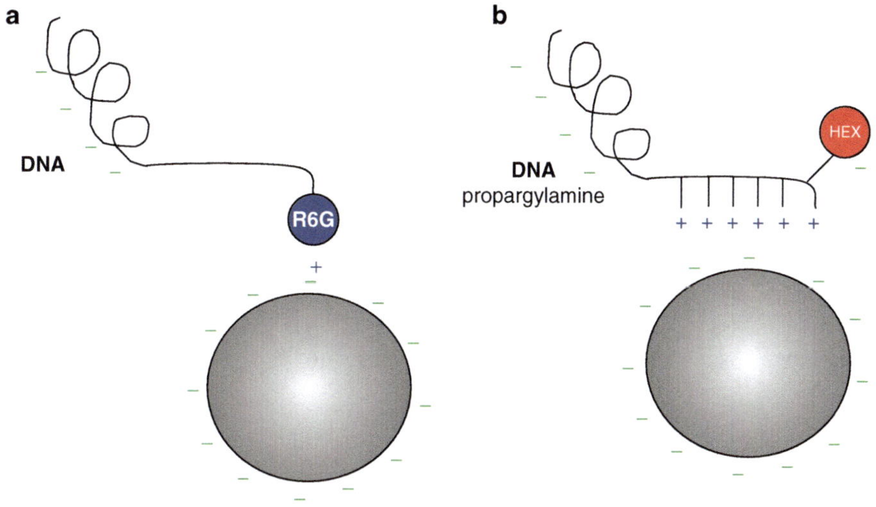

Fig. 13.2 Schematic representation of the process required to achieve detection of labeled DNA by SERRS. Represented is (**a**) an oligonucletide labeled with R6G which is already positively charged, therefore no surface modification is required to allow adsorption onto a silver nanoparticle surface and excited with laser light to give a SERRS spectrum and (**b**) an oligonucleotide modified with a negatively charged dye label, in this case HEX, which requires further modification of the DNA with propargylamine to introduce a region of positive charge to allow adsorption to the metal nanoparticle

studies, namely it promotes surface adsorption of the negatively charged oligonucleotide probe and also aggregates the nanoparticles to provide the higher surface enhancement from the aggregated nanoparticles as opposed to the individual nanoparticles.

The dye labels can also be chosen to aid in the adsorption process to the metal surface, that is, the dye can be specifically designed to have a metal surface complexing group as well as a chromophore such as is the case with azo dyes synthesized to have a benzotriazole group [12, 13] or the charge on the label itself can aid in this adsorption process. Some fluorophores will have a net positive charge in aqueous solution, for example, R6G, ROX, and TAMRA and therefore no further modification is required to allow oligonucleotides modified with these labels to attach to the negatively charged silver surface (Fig. 13.2a). However, some fluorescent labels will have a net negative charge in aqueous solution, for example, Cy3 and FAM therefore further modification of the oligonucleotide is required for effective surface adsorption to occur. Propargylamine modification of DNA sequences utilizing negatively charged dye labels has previously been reported and initially involved the addition of six modified nucleobases at the 5′-terminus close to the dye label (Fig. 13.2b) [44]. When placed in aqueous solution the terminal primary amine groups of the propargylamines will be protonated resulting in a positive charge, allowing the DNA to more effectively adsorb onto the negative silver surface. It has recently been discovered that we do not require six

Table 13.1 SERRS limits of detection for a range of dye-labeled oligonucleotides using citrate reduced silver nanoparticles. No value indicates the experiment was performed but the limit of detection is higher than the base concentration of each dye (~1 × 10^{-8} M). The fluorescence detection limits were not measured (N.M.) for all of the dyes hence the not measured entries

		λ_{ex}: 514.5 nm	λ_{ex}: 632.8 nm		λ_{ex}: 785 nm	Fluorescence LoD
Dye label	λ_{max} (nm)	Silver (mol dm^{-3})	Silver (mol dm^{-3})	Gold (mol dm^{-3})	Silver (mol dm^{-3})	(mol dm^{-3})
FAM	492	2.7×10^{-12}	2.0×10^{-9}	–	–	6.5×10^{-8}
R6G	524	1.2×10^{-12}	1.1×10^{-10}	–	–	3.5×10^{-8}
Cy 3	552	2.6×10^{-10}	1.5×10^{-10}	–	–	4.6×10^{-9}
ROX	585	8.1×10^{-11}	3.3×10^{-11}	1.1×10^{-9}	–	2.3×10^{-8}
BODIPY TR-X	588	1.3×10^{-10}	7.9×10^{12}	4.9×10^{-10}	–	N.M.
PtcCo	625	–	3.2×10^{-11}	N.M.	–	Not fluorescent
PtcAl	640	–	2.8×10^{-11}	N.M.	–	Not fluorescent
PtcZn	680	1.4×10^{-10}	3.2×10^{-11}	N.M.	–	Not fluorescent
Cy5.5	683	–	5.2×10^{-12}	7.3×10^{-11}	–	N.M.

propargylamine-modified nucleobases and that double the SERRS signal can be obtained by adding one modified nucleobase, however at least two are required to get the optimal surface absorption [45]. Therefore, the action of spermine combined with either a positively charged dye or a negatively charged dye and the propargylamine-modified bases allows good absorption of DNA and hence successful SERRS to be obtained.

4.3 Sensitivity of SERRS

When oligonucleotides are designed and analyzed using the conditions described above, it is possible to achieve extremely sensitive detection of labeled oligonucleotides using metal nanoparticles [36, 43]. It is possible to achieve quantitative, linear SERRS responses when the concentration of the labeled oligonucleotide is kept below monolayer coverage of the nanoparticle surface.

A direct comparison between the detection limits using SERRS and fluorescence, when the labels on the DNA are fluorescent, has been carried out, thus allowing a direct comparison of the two techniques. The limits of detection of five different labeled oligonucleotides are shown in Table 13.1 using a range of excitation frequencies and silver nanoparticles [46]. The fluorescence detection limits were calculated using routinely available fluorometers, specifically quantitative PCR instrumentation, thus giving a direct comparison as they are instruments currently found in functioning molecular diagnostic laboratories. The SERRS detection limits were also calculated using a range of different spectrometers, including Raman microscopes and fiber optic-based systems and the detection limits were approximately the same for each of the different Raman systems, regardless of cost or configuration. This indicated that, although there is a compromise in terms of peak resolution when the specification of the spectrometer is lowered, the absolute sensitivity in terms of limits of detection does not appear to be compromised.

The main point to note from Table 13.1 is that under these conditions SERRS is much more sensitive than fluorescence with SERRS detection limits generally at least three orders of magnitude more sensitive than fluorescence [46]. The other feature of the data presented in the table is that there is a clear correlation between the λ_{max} of the dye and the excitation frequency used. This confirms how important the resonance contribution of the label is to the sensitivity of the detection system and that for maximum sensitivity, a resonance contribution with the dye label as well as a resonance contribution with the surface plasmon of the aggregated silver or gold nanoparticles is necessary for maximum enhancement in signal to be achieved. For example, the lowest detection limits obtained for R6G, which has a λ_{max} of 524 nm, are achieved when an excitation of 514.5 nm is used and this detection limit increases by as much as two orders of magnitude when we move away from resonance and use 632.8 nm excitation. It should also be noted that silver nanoparticles tend to give lower detection limits than gold, even when higher excitation wavelengths are used.

These results highlight that, if the SERRS conditions are carefully and correctly chosen, SERRS is an extremely powerful and sensitive technique. It has improved sensitivity over fluorescence, resulting in detection limits that are generally at least three orders of magnitude lower. However, one of the main advantages of SERRS over fluorescence, or any other optical detection technique for that matter, is that coupled with the sensitivity SERRS produces spectra which are molecularly specific and give a pattern of sharp peaks. Thus, SERRS has a huge advantage over fluorescence when analyzing mixtures of analytes since SERRS spectra contain more spectral features to distinguish the analytes whereas fluorescence spectra of mixtures tend to have broad overlapping features which are more difficult to discern. The ability of SERRS to multiplex, or to detect multiple analytes within the same sample using one analysis will be investigated in the remainder of this chapter using DNA as the example.

4.4 Multiplexing

As well as the inherent sensitivity of SERRS and the ability to obtain quantitative detection of dye-labeled oligonucleotides, one of the main advantages of SERRS is the ability to multiplex. In the context of DNA detection, we use the term multiplexing to mean the ability to detect multiple labels attached to different DNA sequences at the same time, but without using any physical separation procedures. Therefore, the resultant SERRS signal from the label will be indicative of the presence or absence of a particular sequence of DNA. The sharp vibrational fingerprint spectra obtained using SERRS are ideally placed to allow us to separate

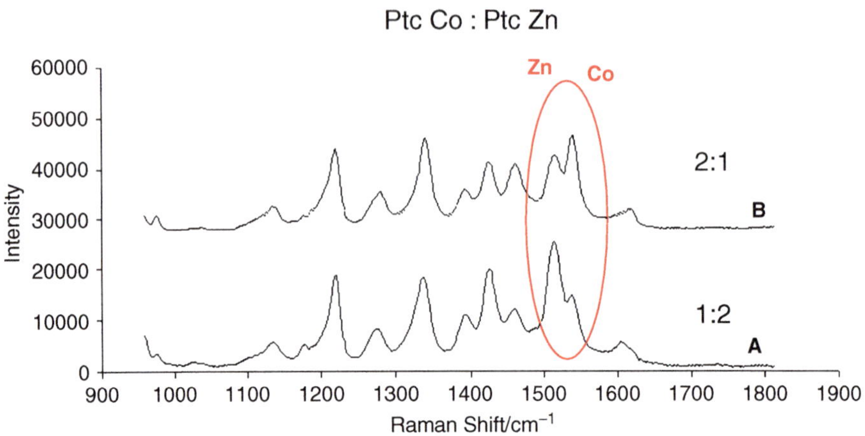

Fig. 13.3 Spectra of mixtures of DNA labeled with phthalocyanine with a cobalt metal center (PtcCo) and zinc metal center (PtcZn) with the following ratios: A –1:2 and B – 2:1. Spectra obtained using 632.8 nm laser excitation with a 10 s accumulation

the components of a mixture using spectroscopy. This can either be simply done visually when there are, for example, three carefully chosen labels in the mixture, however when the mixture contains more components separation becomes much more complex and impossible by eye and it therefore requires mathematical methods to separate and identify the components of the multiplex. Obviously, if an array-based format is used then the multiplexing capability can be increased infinitely by employing spatial separation, however this chapter will focus on the solution-based multiplexing capability of SERRS and the separation of the spectral features of the components.

We have previously shown how phthalocyanines can be used as unique SERRS labels for DNA and demonstrated how they can be used to successfully discriminate between two different DNA sequences. Phthalocyanines have porphyrin structures with metal centers and by varying the metal center the optical properties of the phthalocyanine can be altered to change their absorbance properties or to become fluorescent or non-fluorescent [41]. Cobalt and zinc phthalocyanines were used to provide different labels attached to oligonucleotides which could then be quantified in terms of their SERRS response and when mixed in a multiplex in different ratios each could be easily identified (Fig. 13.3).

Initial work on the duplex detection of fluorescently labeled oligonucleotides was carried out using HEX and R6G labeled oligonucleotides where it was possible to discriminate the ratio of each oligonucleotide sequence present based on the SERRS signal from the labels [47]. A 3-plex of fluorescently labeled oligonucleotide sequences has also been detected by eye, however in this case a lab-on-a-chip format was used. In this example, microfluidics chips were generated from PDMS and DNA sequences labeled with Cy3, FAM, and TET were introduced into the chip, the SERRS signals were measured at a point further down the channel [48].

A microfluidics approach allowed simultaneous detection of three different DNA sequences corresponding to different strains of the Escherichia coli bacterium.

We have also recently carried out a 5-plex of labeled oligonucleotide sequences where we managed to identify five different labeled oligonucleotides in a mixture by careful choice of the label and by using two excitation wavelengths [49]. The sequences used corresponded to a range of different targets. FAM, Cy5.5, and BODIPY TR-X were used to label a universal reverse primer, rhodamine 6G (R6G) was used to label a probe for HPV, and ROX to label a probe to the VT2 gene of *E. coli* 157. The labels were carefully chosen since they have different absorbance maxima (Table 13.1) and because they have unique SERRS spectra.

The spectra obtained from each of the individual labels are shown in Figs. 13.4a and 13.3b, using two different laser excitation frequencies. It can be clearly seen that each label gave a distinctive spectrum, however since the dye labels have different absorbance maxima they will not all be in resonance with the same laser excitation frequency and this property can be exploited to produce a very sensitive and selective method for detecting each of these labels within a mixture of the others using two different laser excitation frequencies. Figure 13.4a shows the spectra of the five labeled oligonucleotides when an excitation wavelength of 514.5 nm was used. Only three of the dye-labeled oligonucleotides (R6G, FAM, and ROX) gave an intense spectrum at this wavelength and this is due to them being in resonance at this excitation wavelength (Table 13.1). Figure 13.4b shows the spectra of the same five dye-labeled oligonucleotides using an excitation wavelength of 632.8 nm, again only three of the labels gave an intense spectrum at this wavelength, however in this case, the labels were Cy5.5, BODIPY TR-X, and again, ROX (ROX could be detected at both excitation wavelengths since it has two absorbance peaks, one at 585 nm and one at ~530 nm which will be in resonance with the 514.5 nm laser).

The multiplex spectra obtained using each excitation frequency are also shown in Fig. 13.3. Using 514.5 nm excitation two identifying bands from FAM and rhodamine 6G can be clearly distinguished and that from the multiplexed spectrum of the same mixture at 632.8 nm excitation identifying bands from ROX, Cy5.5, and BODIPY TR-X can be clearly distinguished. The bands chosen allowed simple and fast identification of the oligonucleotides in the mixture by looking for the presence, or absence, of these key marker bands. The bands had sufficiently different Raman shifts to allow this multiplexed identification to be done by eye.

This approach is very powerful since it is possible to manipulate the resonance contribution of the labels to our advantage; however, this is not always possible, particularly when we wish to increase the multiplexing capability or we only have one excitation frequency available. Clearly, multiplexing could be greatly increased by the use of data analysis techniques and it then becomes necessary to use chemometrics methods to separate the components of the multiplex. Figure 13.5a shows the individual spectra of six dye-labeled oligonucleotides and Fig. 13.5b shows the multiplex spectrum of a 6-plex and the difficulty in separating the components of the multiplex by eye becomes apparent due to the lack of visually distinguishable peaks. Therefore, in this case it becomes necessary to adopt a multivariate analysis

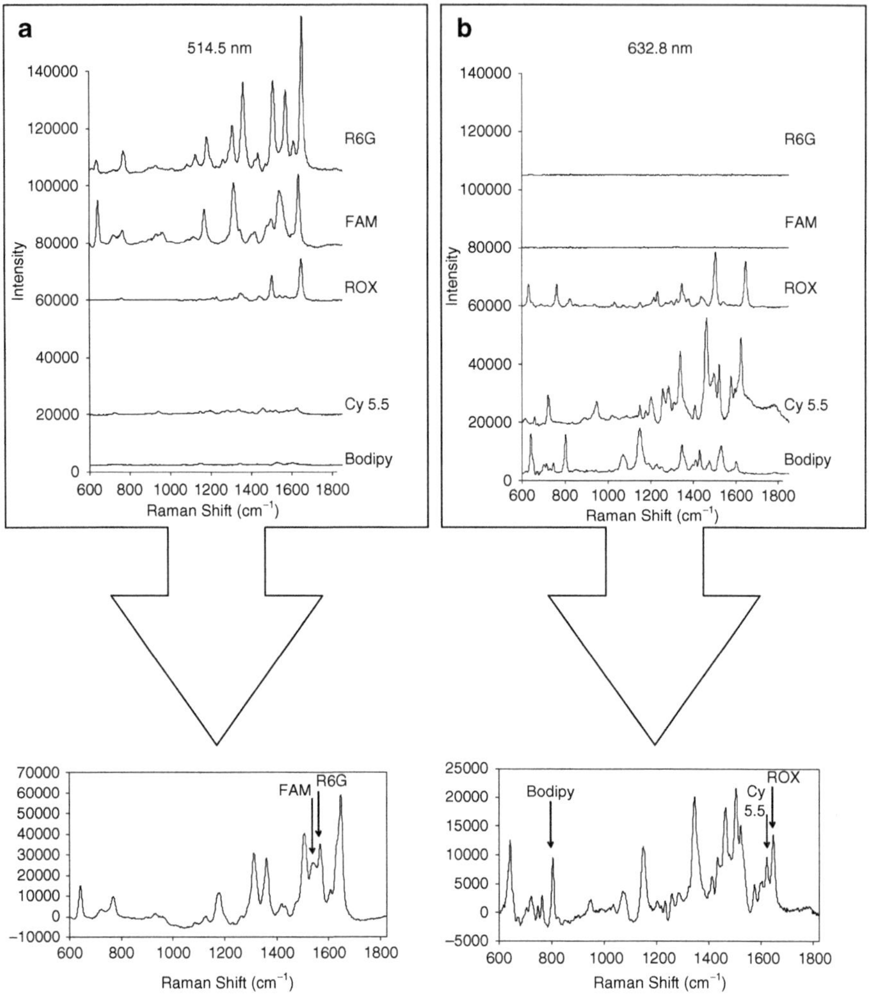

Fig. 13.4 SERRS spectra of each individual dye-labeled oligonucleotide at a concentration of 1.82×10^{-9} mol dm^{-3} obtained using (**a**) 514.5 nm and (**b**) 632.8 nm laser excitation (*top* spectra). The multiplexed SERRS spectra of the 5-plex using (**a**) 514.5 nm and (**b**) 632.8 nm excitation (*bottom* spectra) are also shown

(MVA) approach rather than looking for specific discriminatory Raman bands [50, 51]. In MVA, the whole of the SERRS spectrum is considered.

Using this approach the first multiplexed simultaneous detection of six different DNA sequences, corresponding to different strains of the *Escherichia coli* bacterium, each labeled with a different commercially available dye label (ROX, HEX, FAM, TET, Cy3, or TAMRA) was reported [52]. In this study, both exploratory discriminant analysis and supervised learning, by partial least squares (PLS) regression, were used and the ability to discriminate whether a particular labeled

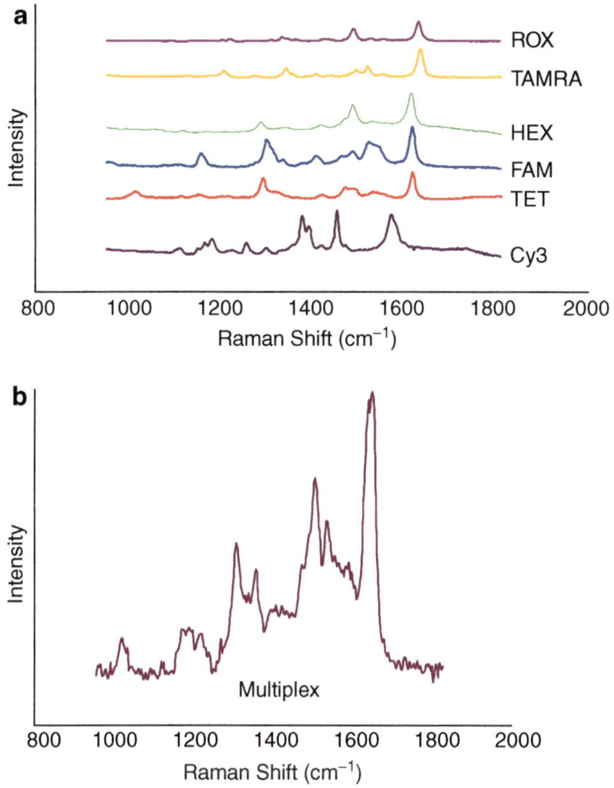

Fig. 13.5 Schematic representation of the use of chemometrics to deconvolute the multiplex spectrum of six dye-labeled oligonucleotides. (**a**) Shows the individual spectra of six dye-labeled oligonucleotides and (**b**) shows the spectrum of the mixture of the six labeled oligonucleotides

oligonucleotide was present or absent in a mixture was achieved using PLS with very high sensitivity (0.98–1), specificity (0.98–1), accuracy (range 0.99–1), and precision (0.98–1).

4.5 Multiplexed Assays

All of the work described above involved the direct detection of labeled DNA sequences, however to generate useful detection methodologies, it is required to carry out some molecular biology to create a meaningful assay which will detect a specific, target DNA sequence of interest in a genuine biological sample. For the detection of DNA targets, the most obvious way to do it is to use a PCR-based approach to amplify and detect the presence of a specific sequence. As stated previously, there are studies in the literature relating to array-based formats for specific DNA detection by SERRS, however, this chapter is focusing on nanoparticle and solution phase-based detection systems and as such the assays reported here relate mainly to this type of format.

An early study for the detection of the mutational status of the cystic fibrosis transmembrane conductance regulator gene used an amplification refractory mutation system (ARMS) approach combined with SERRS [53]. The assay proved to be a good example of the ability to provide a selective 2-plex identification of different gene sequences from genuine patient examples. There are three different possibilities for the genetic status of this particular gene: the wild type, where both alleles are normal, the heterozygote, where one allele is mutated and one is normal, and the fully homozygote mutant where both alleles have the mutated DNA sequence. Specific labeled primers were designed to amplify the region relating to this mutation and used in a multiplexed PCR assay. The PCR amplicons were then identified by SERRS after removal of any unincorporated primers. This allowed the successful determination of the mutational status of these particular samples without the additional separation steps commonly performed in the fluorescence assays [54].

Recent work has developed an assay which has the potential to be used in a closed-tube, homogeneous, and multiplexed assay format [55]. The assay is based on the observation that double stranded DNA has a lower affinity for the surface of silver nanoparticles than single stranded DNA (Fig. 13.6). Therefore, when a single stranded probe sequence labeled with a SERRS active label is added to silver nanoparticles a strong SERRS signal is obtained. However, when the complementary target sequence is present it will hybridize to the labeled probe sequence resulting in a duplex which has a lower affinity for the metal surface resulting in a much reduced SERRS signal as a result of the presence of target DNA. This results in reduced SERRS signal compared to when no target or noncomplementary target is present. The lower affinity of the dsDNA for the metal surface is thought to be due to the increased electrostatic repulsion of the exposed negative phosphate backbone of dsDNA compared to ssDNA.

This assay was used for the detection of three genes which are associated with MRSA and three different SERRS labels namely FAM, HEX, and TAMRA were used to identify these sequences. Using this approach it was possible to detect not only the exact complement DNA but also PCR product. It was also possible to detect every possible combination of the three sequences, present or absent, within a mixture demonstrating the multiplexing potential of SERRS for use in homogeneous molecular diagnostics assay.

Vo-Dinh et al. have previously developed a surface-based assay to detect the BRCA1 gene that codes for breast cancer susceptibility. This approach was used to generate an array of capture sequences immobilized on a 9 nm silver layer on a glass surface using a thiolated DNA capture sequence. The rhodamine labeled BRCA1 target was then captured on the surface by hybridization to the complementary target sequence. This format was only used for the detection of one sequence of DNA however it could be easily extended to give spatial multiplexing of multiple targets with different labels [56, 57]. A more recent approach to DNA detection makes use of *molecular sentinels* which are based upon molecular beacons, Fig. 13.7a [58]. The molecular sentinel is comprised of a region

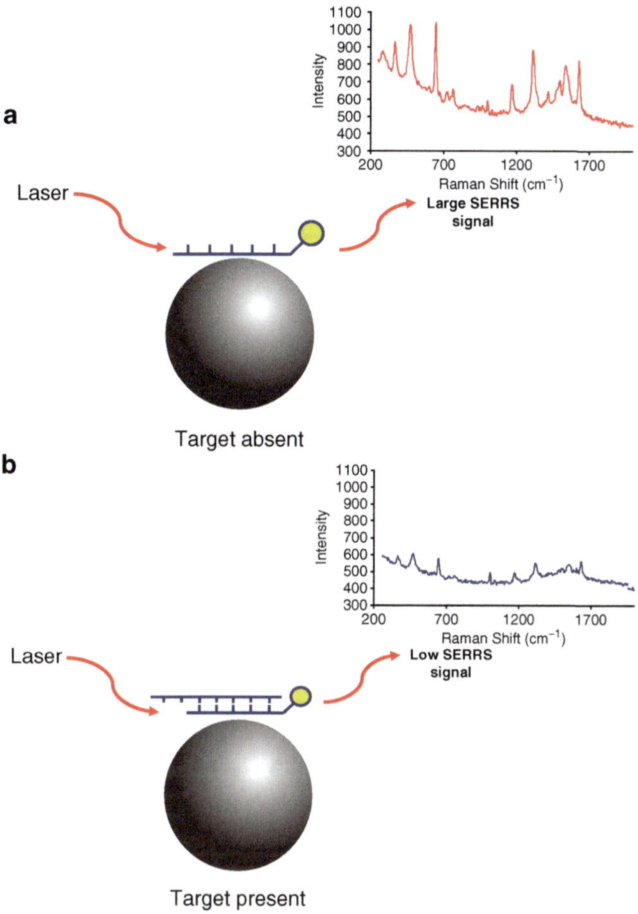

Fig. 13.6 Schematic representation of the SERRS detection assay. (**a**) When target DNA is not present the dye-labeled DNA probe is free to adsorb on the surface of the silver nanoparticles resulting in an intense SERRS signal (**b**) in the presence of target DNA the probe will hybridize to its complement and in this case, the dsDNA is not adsorbed onto the surface of the silver nanoparticles resulting in a reduction in the SERRS signal

complementary to the target DNA sequence and a self-complementary stem that holds the sentinel in a closed, hairpin loop conformation. One of the stems of the hairpin is functionalized with a thiol group to allow it to be attached to the surface of a silver nanoparticle and the other end is functionalized with a Raman reporter. In the closed conformation, an intense SERS signal is observed due to the close proximity of the reporter to the metal surface. However, in the presence of the target DNA sequence the sentinel will open resulting in the Raman reporter becoming distal to the surface of the metal nanoparticle. Therefore, this approach results in a reduction in the SERRS signal obtained upon binding of target DNA. The molecular sentinels have been used to achieve the multiplexed detection of two

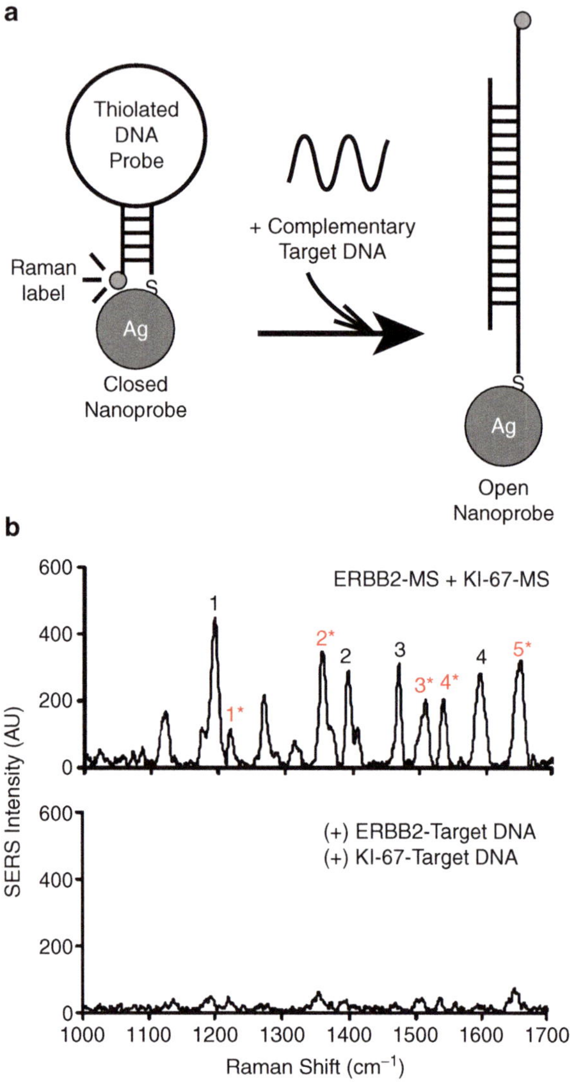

Fig. 13.7 The operating principle of the SERS-based molecular sentinel (MS) nanoprobes. (**a**) The MS nanoprobe is composed of a Raman-labeled DNA hairpin probe and a silver nanoparticle. In the absence of the complementary target DNA, a strong SERS signal is observed due to the hairpin conformation adopted by the MS nanoprobe (*left*: closed state). (**b**) In the presence of the complementary target DNA, the hairpin conformation of the MS nanoprobe is disrupted and the SERS signal is quenched due to the physical separation of the Raman label from the surface of the silver

genes which were biomarkers for breast cancer by creating two sentinels, one labeled with Cy3 and TAMRA (Fig. 13.7b). However, both of these assays have the distinct disadvantage that they are negative assays, that is, the presence of the target DNA sequence results in a decrease rather than an increase in signal.

Recent work by Lowe et al. used the ligase detection reaction (LDR) to detect SNPs in a multiplexed manner using SERS [59]. The so called LDR-SERS approach involved the functionalization of an oligonucleotide primer (which binds downstream of the SNP) with an amine to allow it to attach to the surface of nanoparticles after the ligation. A second oligonucleotide primer sequence incorporating a Raman active fluorophore and a discriminatory base at the 3′ which binds adjacent to the first primer was then used. Since the ligation reaction will only occur when the upstream primer matched perfectly with the template, this means that a Raman signal will only be observed when ligation of the two primers occur since only in this case will the fluorophore be close to the surface of the nanoparticle to give enhancement. Using this approach, a multiplex of two SNP samples were detected, the wild type discriminating primer (labeled with TAMRA) and the mutant (labeled with FAM), which could be detected by eye.

Mirkin et al. have reported the detection of DNA using a microarray-based format. A microarray chip was spotted with capture DNA strands, complementary to half of the target DNA sequence [60]. Gold nanoparticles were functionalized with the other half of the sequence of DNA complementary to the target sequence; however, in this case, a fluorophore, initially Cy3, was attached to the end of the DNA strand in close proximity to the nanoparticles surface. After capture of the target sequence, silver staining was used to further enhance the Raman scattering signal and detection was carried out using a 633 nm laser excitation. Using this approach, it was possible to detect six different target DNA sequences by using different fluorophores for each of the sequences used to functionalize the nanoparticle surface. However, this approach used a spatial multiplex approach to physically separate each of the targets onto different areas of the chip and SERS was used to ensure that there was no nonspecific binding by checking there was only one SERS spectrum per array spot. Using this approach, detection limits of 20 fmol were obtained. During this study, however, the simultaneous detection of SNP targets was carried out. Two RNA targets which bind to the same capture strand of DNA, but have a single base mismatch that lies within the region of the probe sequence, were used to functionalize the nanoparticle; therefore two different sequences labeled with two different fluorophores were used to detect the two target RNA strands. It was possible to use the relative change in the SERS signal of the multiplex of the two reporters to give the relative ratios present in each of the targets.

Irudayaraj et al. have reported the multiplex detection of up to eight different non-fluorescent nanoparticles functionalized with one sequence of DNA [61]. In this approach, a thiolated sequence of DNA was used to functionalize the surface of gold nanoparticles and then non-fluorescent Raman reporters were added to the surface of the nanoparticle to code them with a SERS signal. It was reported that multiplex detection of two, four, and eight differently labeled nanoparticles could be detected in one analysis. However, only one DNA sequence was used in this study to label all the different "flavors" of nanoparticles and the detection of a specific, target DNA sequence by SERS was not reported. However, it was possible to observe the change in surface plasmon by UV–Vis spectroscopy when two batches of nanoparticles functionalized with complementary sequences were hybridized together.

The same group has also reported a multiplexed assay based upon the Mirkin approach but in this case using an array format with non-fluorescent Raman reporters and the reporter molecule was added directly to the surface of the nanoparticle rather than to the end of the sequence of DNA [62]. Thiolated DNA capture probes were immobilized on the surface of a gold coated glass slide whereupon target DNA was added and allowed to hybridize to the capture probes. The SERRS reporter was then added in the form of gold nanoparticles functionalized with DNA and non-fluorescent Raman reporters, prepared in the same way as described previously [62], the attached DNA sequences were complementary to half of the target sequence, completing the sandwich. A silver enhancement solution was then added to enhance the signal. Using this approach, it was possible to detect four DNA sequences specific to the BRCA1 alternative splice variants by labeling the gold nanoparticles with four different reporters indicative of four different sequences. Detection limits of 1 fmol were also achieved. The same array format was also used to monitor gene expression in cancer cell lines [63]. In this case, a multiplex of two splice junction variants of the BRCA1 breast cancer susceptibility were identified simultaneously by SERS after extraction of mRNA from cancer cells and carrying out DNA/RNA hybridization assay followed by S1 nuclease digestion to remove any single stranded nucleic acids leaving only the target DNA/RNA duplex, before carrying out the SERS array assay as before.

Another approach we have recently developed is also based upon the use of DNA functionalized nanoparticles, however in this case the sequence specific hybridization event is used to "turn on" the SERRS signal through controlled assembly of nanoparticles into aggregates [64]. Assembly of gold nanoparticles by a biological interaction has previously been reported by a number of groups, however in this work silver nanoparticles rather than gold were used due to their high extinction coefficient and the increased level of surface enhancement obtained from silver compared to gold. This has the added benefit that the silver staining step in the previously assays, which was required to increase the enhancement, is no longer required. This study utilizes a solution-based methodology similar in concept to work previously reported by Mirkin et al. with gold in the absence of any SERS [60]. Silver nanoparticles were functionalized with a specifically designed SERRS dye, 3, 5-dimethoxy-4-(6′-azobenzotriazolyl)phenol, which contains a chromophore, a benzotriazole group which complexes strongly to silver metal and a negatively charged phenolic group to prevent nonspecific aggregation of the nanoparticles. The nanoparticles were then further functionalized with 5′ thiolated oligonucleotide sequences, complementary to half of the target sequence. Another batch of nanoparticles were also synthesized with the dye and the other half of the probe sequence complementary to the target. Upon addition of target DNA, the hybridization to the two complementary probe sequences occurred which resulted in the nanoparticles being brought close together, that is, the nanoparticles were assembled due to the hybridization event, resulting in a large increase in the SERRS signal due to the increase electromagnetic enhancement experienced by the dye label due to the aggregation event.

This was further extended to allow the multiplex detection of two different target sequences of DNA using two different Raman reporter dyes to code for different probe sequences on the nanoparticles. In this advancement, three batches of

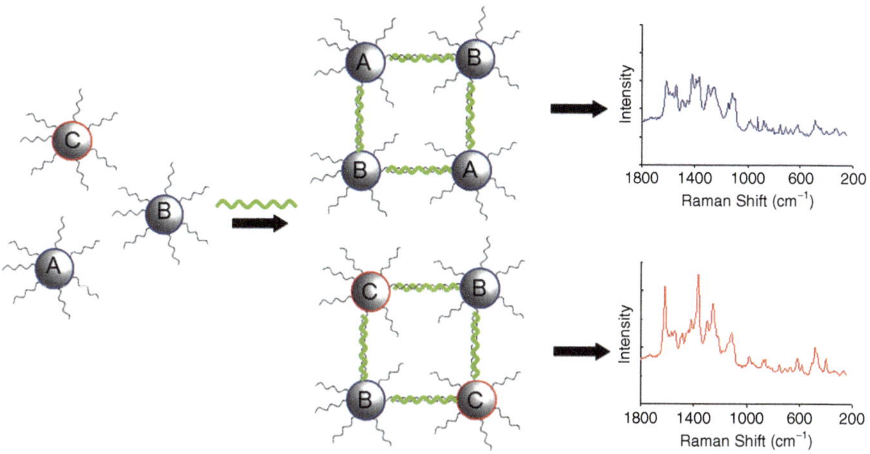

Fig. 13.8 Selective enhancement of specific Raman signals through DNA hybridization. The assay consisted of three different conjugates. Each had a different, noncomplementary oligonucleotide (sequence A, sequence B, and sequence C). Conjugates A and B were functionalized with dye 1, denoted in *blue*, and conjugate C was functionalized with dye 2, denoted by *red*. These three conjugates were mixed together at 30 pM final concentration. When a target complementary to A and B is added, only dye 1 is enhanced (*top right*), and when a target complementary to B and C is added, the spectrum for dye 1 and dye 2 is enhanced (*bottom right*)

nanoparticles were prepared, each functionalized with a different sequence of DNA and coded with two different dye labels, dye 1 and dye 2. Two different target sequences were used to assemble the three sequences such that one hybridization event resulted in SERRS signals being obtained from one dye label only, dye 1, whereas the second hybridization event resulting on SERRS spectra being obtained from both dye 1 and dye 2, see Fig. 13.8. This is the first time that nanoparticles assemblies have been used to turn on the SERRS effect due to a biological interaction as well as demonstrating the potential to use this effect to simultaneously detect multiple target sequences in one analysis.

This chapter has so far solely discussed the multiplexed detection of DNA by SERS as a demonstration of its potential in this area. However, since the multiplexing potential of SERS, as well as its inherent sensitivity, is one of its greatest advantages it has been explored in other areas as well as in DNA analysis.

4.6 Other Multiplexed Formats

Recent work has been published on the use of SERS reporters for in vivo analysis. In 2008, Shuming Nie published work on the in vivo targeting of tumors in live mice [65]. The SERRS particles consisted of 60 nm gold nanoparticles functionalized with a Raman reporter dye molecule and then stabilized with thiolated polyethylene glycols (PEGs). Targeted SERS nanoparticles were prepared by having a mixed monolayer of thiolated PEG and a heterofunctional thiolated PEG with a carboxylic acid terminal group. The carboxylic acid group allowed the covalent addition of the ScFv antibody, which binds to the EGFR receptor of cancer cells, using EDC coupling. Using these functional nanoparticles, it was possible to target the cancer cell in vivo by locating the nanoparticles using SERS. However, this was not carried out to detect multiple targets in a multiplexed manner.

Using a different nanoparticles approach Gambhir et al. used commercially available "nanotags" consisting of gold nanoparticles functionalized with non-resonant Raman reports stabilized with a silica coating [66]. Two differently labeled SERS nanotags were injected into a mouse at three different injection sites and it was possible to detect all three using their SERS signature. Multiplexing was also used to study the fate of two different SERS, one functionalized with PEG and one without, where it was found that both the PEGylated and non-PEGyated nanotags accumulated in the liver to the same extent. In a similar study, the same group were able to carry out multiplex detection of ten different nanotags in vivo, spatially separated at different injection sites [67]. During this study, the simultaneous detection of five different tags was carried out to determine the accumulation of the nanotags within the liver. During this study, the potential application of the nanotags for diagnostics application at depth within tissue, was discussed for example, for the laparoscopic detection tumors, however, this was not demonstrated. However, in a recent publication using surface enhanced spatially offset Raman (SESORS) the detection of functionalized nanoparticles, SERS signals could be detected at a depth of between 15 and 25 mm in tissue [68]. This clearly demonstrates the great potential of SERS for targeted in vivo diagnostics at clinically relevant depths.

Another recent article by Kneipp et al. demonstrates the use of multiplexing within live cells [69]. Two different SERS particles, aggregates of metal nanoparticles functionalized with two different reporter molecules could be imaged within cells using SERS. The data obtained was analyzed using cluster methods and principal components analysis to detect the two SERS reporters within the live cells.

5 Conclusions and Future Outlook

SERS is an extremely sensitive and selective technique; however, one of its main advantages over other detection techniques is the ability to detect multiple analytes simultaneously. Although the potential to provide this type of multiplexed analysis is often discussed with reference to SERS there are actually few publications that successfully demonstrate this in practice. However, in recent years the multiplexing capability has been more widely explored by researchers within the field resulting in data to support this capability of the technique.

This chapter has demonstrated the advantages of using SERS as an analytical technique for the detection of dye-labeled oligonucleotides when compared to other commonly used techniques. The experimental procedures and design of DNA probes which allows extremely sensitive and selective detection to be achievable have been outlined with an emphasis on the multiplexing capability. The range of labels available for SERS analysis, the ability to multiplex and subsequently successfully analyze multiplex data has been examined. These aspects have been discussed and indicate how SERRS can be obtained in a quantitative, multiplexed manner from meaningful targets. This is now leading to the development of quantitative methodologies for a range of specific targets using SERRS detection including proteins, DNA, and small molecules. The key feature of SERS which provides the largest advantage over competitive techniques is the multiplexing capability and this looks set to be exploited for in the future in terms of research and commercial applications. For instance, the use of SERRS detection in genuine molecular diagnostic assays is still in its infancy but shows excellent promise and a number of different avenues are currently under investigation by several groups to exploit SERRS as a highly informative, bioanalytical technique for quantitative, ultrasensitive analysis.

References

1. Fleischmann M, Hendra PJ, McQuillan AJ (1974) Raman spectra of pyridine adsorbed at a silver electrode. Chem Phys Lett 26(2):163–166
2. Jeanmarie DL, Van Duyne RP (1977) Surface Raman spectroelectrochemistry: part I. Heterocyclic, aromatic, and aliphatic amines adsorbed on the anodized silver electrode. J Electroanal Chem 84:1–20
3. Hildebrandt P, Stockburger M (1984) Surface-enhanced resonance Raman spectroscopy of Rhodamine 6G adsorbed on colloidal silver. J Phys Chem 88(24):5935–5944
4. Emory SR, Nie S (1997) Probing single molecules and single nanoparticles by surface-enhanced Raman scattering. Science 275:1102–1106
5. Kneipp K, Wang Y, Kneipp H, Perelman LT, Itzkan I, Dasari RR, Feld M (1997) Single molecule detection using surface-enhanced Raman scattering (SERS). Phys Rev Lett 78:1667–1670
6. Munro CH, Smith WE, White PC (1995) Qualitative and semi-quantitative trace analysis of acidic monoazo dyes by surface enhanced resonance Raman scattering. Analyst 120:993–1003
7. Lee PC, Meisel D (1982) Adsorption and surface-enhanced Raman of silver and gold sols. J Phys Chem 86:3391–3395
8. Li H, Ying L, Green JJ, Balasubramanian S, Klenerman D (2003) Ultrasensitive coincidence fluorescence detection of single DNA molecules. Anal Chem 75:1664–1670
9. Rodger C, Smith WE, Dent D, Edmondson J (1996) Surface-enhanced resonance-Raman scattering: an informative probe of surfaces. J Chem Soc Dalton 5:791–799
10. Isola NR, Stokes DL, Vo-Dinh T (1998) Surface enhanced Raman gene probe for HIV detection. Anal Chem 70:1352–1356
11. Graham D, Brown R, Smith WE (2001) SERRS detection of PNA and DNA labelled with a specifically designed benzotriazole azo dye. Chem Commun 11:1002–1003
12. Brown R, Smith WE, Graham D (2003) Synthesis of a benzotriazole azo dye phosphoramidite for labeling of oligonucleotides. Tetrahedron Lett 44(7):1339–1342
13. Albrecht MG, Creighton JA (1977) Anomalously intense Raman spectra of pyridine at a silver electrode. J Am Chem Soc 99:5215–5217

14. Stacy AM, Van Duyne RP (1983) Surface enhanced Raman and resonance Raman spectroscopy in a non-aqueous electrochemical environment: tris (2,2′-bipyridine)ruthenium(II) adsorbed on silver from acetonitrile. Chem Phys Lett 102:365–370
15. Campion A, Kambhampati P (1998) Surface-enhanced Raman scattering. Chem Soc Rev 4:241–250
16. Schultz G, Janik-Czachor M, Van Duyne RP (1981) Surface enhanced Raman spectroscopy: a re-examination of the role of surface roughness and electrochemical anodization. Surf Sci 104:419–434
17. Jiang X, Campion A (1987) Chemical effects in surface-enhanced Raman scattering: pyridine chemisorbed on silver adatoms on Rh (100). Chem Phys Lett 140(1):95–100
18. Creighton JA, Blatchford CG, Albrecht MG (1979) Plasma resonance enhancement of Raman scattering by pyridine adsorbed on silver or gold sol particles of size comparable to the excitation wavelength. J Chem Soc Faraday Trans II 75:790–798
19. Blatchford CG, Campbell JR, Creighton JA (1982) Plasma resonance – enhanced Raman scattering by absorbates on gold colloids: the effects of aggregation. Surf Sci 120(2):435–455
20. Kneipp K, Dasari RR, Wang Y (1994) Near-infrared surface-enhanced Raman scattering (NIR SERS) on colloidal silver and gold. Appl Spectrosc 48(8):951–955
21. Grabar KC, Freeman RG, Hommer MB, Natan MJ (1995) Preparation and characterization of au colloid monolayers. Anal Chem 67(4):735–743
22. Grabar KC, Brown KR, Keating CD, Stranick SJ, Tang S-L, Natan MJ (1997) Nanoscale characterization of gold colloid monolayers: a comparison of four techniques. Anal Chem 69(3):471–477
23. Curtis CC, Duff DG, Edwards PP, Jefferson DA, Johnson BFG, Kirkland AI, Wallace AS (1988) Preparation and structural characterization of an unprotected copper sol. J Phys Chem 92:2270–2275
24. Huang HH, Yan FQ, Kek YM, Chew CH, Xu GQ, Ji W, Oh PS, Tang SH (1997) Synthesis characterization, and nonlinear optical properties of copper nanoparticles. Langmuir 13:172–175
25. Volkan M, Stokes DL, Vo-Dinh T (2000) Surface-enhanced Raman of dopamine and neurotransmitters using sol–gel substrates and polymer-coated fiber-optic probes. Appl Spectrosc 54(12):1842–1848
26. Laserna JJ, Campiglia AD, Winefordner JD (1988) Surface-enhanced Raman spectrometry on a silver-coated filter paper substrate. Anal Chim Acta 208:21–30
27. Ruperez A, Laserna JJ (1994) Surface-enhanced Raman spectrometry on a silver substrate prepared by the nitric acid etching method. Anal Chim Acta 291:147–153
28. Bello JM, Stokes DL, Vo-Dinh T (1989) Titanium dioxide based substrate for optical monitors in surface-enhanced Raman scattering analysis. Anal Chem 61:1779–1783
29. Walsh RJ, Chumanov G (2001) Silver coated porous alumina as a new substrate for surface-enhanced Raman scattering. Appl Spectrosc 55(12):1695–1700
30. Bell SEJ, Spencer SJ (2001) Disposable, stable media for reproducible surface-enhanced Raman spectroscopy. Analyst 126:1–3
31. Bharathi S, Fishelson N, Lev O (1999) Direct synthesis and characterization of gold and other noble metal nanodispersions in sol–gel-derived organically modified silicates. Langmuir 15:1929–1937
32. Saegmueller B, Brehm G, Schneider S (2000) In situ generated photolytic silver in a gelatin matrix: an approach for high-throughput SERS spectroscopy applying microtiter plates. Appl Spectrosc 54(12):1849–1856
33. Perez R, Ruperez A, Laserna JJ (1998) Evaluation of silver substrates for surface-enhanced Raman detection of drugs banned in sport practices. Anal Chim Acta 376:255–263
34. Norrod KL, Sudnik LM, Rousell D, Rowlen KL (1997) Quantitative comparison of five SERS substrates: sensitivity and limit of detection. Appl Spectrosc 51(7):994–1001
35. McKenzie F, Faulds K, Graham D (2010) Mixed metal nanoparticle assembly and the effect on surface-enhanced Raman scattering. Nanoscale 2:78–80

36. Stokes RJ, Macaskill A, Lundahl PJ, Smith WE, Faulds K, Graham D (2007) Quantitative enhanced Raman scattering of labelled DNA from gold and silver nanoparticles. Small 3(9):1593–1604
37. Munro CH, Smith WE, Garner M, Clarkson J, White PC (1995) Characterization of the surface of a citrate-reduced colloid optimized for use as a substrate for surface-enhanced resonance Raman scattering. Langmuir 11:3712–3720
38. Chow MK, Zukoski CF (1994) Gold sol formation mechanisms: role of colloidal stability. J Colloid Interface Sci 165:97–109
39. Biggs S, Mulvaney P, Zukoski CF, Greiser F (1994) Study of anion adsorption at the gold-aqueous solution interface by atomic force microscopy. J Am Chem Soc 116:9150–9157
40. MacAskill A, Chernonosov A, Koval V, Lukyanets E, Fedorova O, Smith WE, Faulds K, Graham D (2007) Quantitative surface-enhanced resonance Raman scattering of phthalocyanine labelled oligonucleotides. Nucleic Acids Res 35:e42
41. Faulds K, Fruk L, Robson DC, Thompson DG, Enright A, Smith WE, Graham D (2006) A new approach for DNA detection by SERRS. Faraday Discuss 132:261–268
42. Faulds K, Smith WE, Graham D (2004) Evaluation of surface enhanced resonance Raman scattering (SERRS) for highly sensitive and quantitative DNA analysis. Anal Chem 76:412–417
43. Basu HS, Marton LJ (1987) The interaction of spermine and pentamines with DNA. Biochem J 244:243–246
44. Graham D, Smith WE, Linacre AMT, Munro CH, Watson ND, White PC (1997) Selective detection of deoxyribonucleic acid at ultra low concentrations by SERRS. Anal Chem 69(22):4703–4707
45. Faulds K, MacKenzie F, Graham D (2007) Evaluation of the number of modified bases required for quantitative SERRS from labelled DNA. Analyst 132:1100–1102
46. Faulds K, Barbagallo RP, Keer JT, Smith WE, Graham D (2004) SERRS as a more sensitive technique for the detection of labelled oligonucleotides compared to fluorescence. Analyst 129:567–568
47. Graham D, Mallinder BJ, Smith WE (2000) Surface-enhanced resonance Raman scattering as a novel method of DNA discrimination. Angew Chem Int Ed 39(6):1061–1063
48. Frances T, Docherty FT, Clark M, McNay G, Graham D, Smith WE (2004) Multiple labelled nanoparticles for bio detection. Faraday Discuss 126:281–288
49. Faulds K, Mackenzie F, Smith WE, Graham D (2007) Quantitative simultaneous multianalyte detection of DNA by dual- wavelength surface-enhanced resonance Raman scattering. Angew Chem Int Ed 46(11):1829–1831
50. Brereton R (2003) Chemometrics: data analysis for the laboratory and chemical plant. Wiley, Chichester
51. Manly BFJ (1994) Multivariate statistical methods: a primer. Chapman & Hall/CRC Press, New York
52. Faulds K, Jarvis R, Smith WE, Graham D, Goodacre R (2008) Multiplexed detection of six labelled oligonucleotides using surface enhanced resonance Raman scattering (SERRS). Analyst 2008(133):1505–1512
53. Graham D, Mallinder BJ, Whitcombe D, Smith WE (2001) Surface enhanced resonance Raman scattering (SERRS) – a first example of its use in multiplex genotyping. ChemPhysChem 2(12):746–748
54. Graham D, Mallinder DJ, Whitcombe D, Watson ND, Smith WE (2002) Single multiplex genotyping by surface-enhanced resonance Raman scattering. Anal Chem 74:1069–1074
55. MacAskill A, Crawford D, Graham D, Faulds K (2009) DNA sequence detection using surface enhanced resonance Raman spectroscopy in a homogeneous multiplexed assay. Anal Chem 81(19):8134–8140
56. Allain LR, Vo-Dinh T (2002) Surface-enhanced Raman scattering detection of the breast cancer susceptibility gene BRCA1 using a silver-coated microarray platform. Anal Chim Acta 469:149–154

57. Mustafa C, Stokes D, Allain LR, Vo-Dinh T (2003) Surface-enhanced Raman scattering substrate based on a self-assembled monolayer for use in gene diagnostics. Anal Chem 75:6196–6201
58. Wang H-N, Vo-Dinh T (2009) Multiplex detection of breast cancer biomarkers using plasmonic molecular sentinel nanoprobes. Nanotechnology 20:065101 (6 pp)
59. Lowe AJ, Huh YS, Strckland AD, Erickson D, Batt CA (2010) Multiplex single nucleotide polymorphism genotyping utilizing ligase detection reaction coupled surface enhanced Raman spectroscopy. Anal Chem 82(13):5810–5814
60. Cao YC, Jin R, Mirkin CA (2002) Nanoparticles with Raman spectroscopic fingerprints for DNA and RNA detection. Science 297(5586):1536–1540
61. Sun L, Yu C, Irudayaraj J (2007) Surface-enhanced Raman scattering based nonfluorescent probe for multiplex DNA detection. Anal Chem 79:3981–3988
62. Sun L, Yu C, Irudayaraj J (2008) Raman multiplexers for alternative gene splicing. Anal Chem 80:3342–3349
63. Sun L, Irudayaraj J (2009) PCR-free quantification of multiple splice variants in a cancer gene by surface-enhanced Raman spectroscopy. J Phys Chem B 113:14021–14025
64. Graham D, Thompson DG, Smith WE, Faulds K (2008) Control of enhanced Raman scattering using a DNA-based assembly process of dye-coded nanoparticles. Nat Nanotechnol 3(9):548–551
65. Qian X, Peng X-H, Ansari DO, Yin-Goen Q, Chen GZ, Shin DM, Yang L, Young AN, Wang MD, Nie S (2008) In vivo tumor targeting and spectroscopic detection with surface-enhanced Raman nanoparticle tags. Nature Biotechnol 26:83–90
66. Keren S, Zavaleta C, Cheng Z, de la Zerda A, Gheysens O, Ghambhir SS (2008) Noninvasive molecular imaging of small living subjects using Raman spectroscopy. Proc Natl Acad Sci USA 105:5844–5849
67. Zavaleta CL, Smith BR, Walton I, Doering W, Davis G, Shojaei B, Natan MJ, Ghambhir SS (2009) Multiplexed imaging of surface enhanced Raman scattering nanotags in living mice using noninvasive Raman spectroscopy. Proc Natl Acad Sci USA 106(32):13511–13516
68. Stone N, Faulds K, Graham D, Matousek P (2010) Deep-SERRS: demonstration of deep Raman spectroscopy for non-invasive detection of conjugated SERRS nanoparticles buried within 25 mm of mammalian tissue. Anal Chem 82:3969–3973
69. Matschulat A, Drescher D, Kneipp J (2010) Surface-enhanced Raman scattering hybrid nanoprobe multiplexing and imaging in biological systems. ACS Nano 4(6):3259–3269

Raman Spectroscopy of Iron Oxide Nanoparticles

14

Maria A. G. Soler and Fanyao Qu

1 Definition of the Topic

Superparamagnetic iron oxide particles and their Raman spectra.

2 Overview

Superparamagnetic iron oxide (SPIO) particles with tailored surface chemistry have been intensively investigated due to great potential applications, for instance, electromagnetic shielding, spintronics, data storage, catalysis, transformers, biomedicine, and chemical and biological sensors. All these applications require that these nanoparticles have high magnetization values and small size with overall narrow particle size distribution. This chapter summarizes the recent advances in SPIO characterization employing Raman spectroscopy. After a brief introduction about Raman technique and iron oxide materials, the chapter focuses on representative results about the versatility of Raman spectroscopy in phase identification and phase transitions in SPIO systems as well in the characterization and optimization of the synthesis processes.

3 Introduction

The magnetic, chemical, mechanical, optical, and electrical properties of materials change as their size approaches nanoscale and as the percentage of atoms at the

M.A.G. Soler (✉) • F. Qu
Instituto de Física, Universidade de Brasília, Brasília, Brazil

Challa S. S. R. Kumar (ed.), *Raman Spectroscopy for Nanomaterials Characterization*, DOI 10.1007/978-3-642-20620-7_14, © Springer-Verlag Berlin Heidelberg 2012

material surface becomes significant [1, 2]. Nanoparticle size confers unique size-dependent properties such as surface plasmon resonance in some metal particles, quantum confinement in semiconductor quantum dot, and superparamagnetism in magnetic materials. There are many examples of macroscopic manifestation of materials properties due to nanosized effects: Copper nanoparticles smaller than 50 nm are super hard compared to bulk copper, and they do not exhibit the same flexibility and ductility as the bulk copper [3]; optical emission tunability of semiconductor quantum dots [4]; the zero-remanence combined with high saturation magnetization of transition-metal nanosized particles [5]. In addition to these changes in material properties, when the size of materials approaches the nanoscale, they also present novel chemical properties and unusual visualization. For instance, in the macroscale, gold is a yellow-colored inert element. Gold nanoparticles, however, become red-colored and extremely reactive in solution, and they can be used as catalysts to speed up reactions.

Since nanoparticles form a bridge between bulk materials and molecular structures, they are of great scientific interest. For example, superparamagnetic iron oxide (SPIO)-based materials, such as nanosized magnetite (Fe_3O_4), maghemite (γ-Fe_2O_3), and cobalt ferrite (CoF_2O_4), have received increasing attention from areas as different as physics and medicine, driven by their unusual and/or enhanced fundamental properties [6–8]. The superparamagnetic behavior of single-domain magnetic particles is similar to that of paramagnetic materials, losing their net magnetization when the applied magnetic field is removed but presenting remarkably high saturation magnetization below the Curie temperature (T_c). Moreover, SPIO-based nanoparticles also offer many attractive applications in biomedicine. Firstly, since their sizes are controllable in the range of a few nanometers up to tens of nanometers, one can fabricate them with dimensions smaller than or comparable to those of a cell (10–100 μm), a virus (20–450 nm), a protein (5–50 nm), or a gene (2 nm wide and 10–100 nm long). In addition, magnetic nanoparticles can be functionalized with specific molecules to make them interact with or bind to a biological entity. Secondly, the magnetic nanoparticles can be manipulated by uniform external magnetic field. This way they can be prompted to deliver a parcel, such as an anticancer drug, or a cohort of radionuclide atoms, to a targeted region of the body, such as a tumor [9]. Thirdly, the magnetic nanoparticles can be made to respond resonantly to a time-varying magnetic field, with advantageous results related to the transfer of energy from the exciting field to the nanoparticle [10]. However, it is important to keep in mind that the size, charge, and surface chemistry of the magnetic particles, strongly affect both the blood circulation time and bioavailability of the nanoparticles within the body. Thus, all these biomedical and engineering applications require that nanoparticles possess high magnetization values and have size smaller than 100 nm with an overall narrow particle size distribution. Therefore, control and design of nanoscale structures and their magnetic properties are challenging issues in both basic and applied magnetism; where, the complexity of the local magnetic structures such as surfaces, interfaces, and domain walls plays a key role. Since the changes of these local structures lead to variations of phonon propagation which can be probed by Raman spectroscopy, which can be used as an effective nondestructive technique to understand magnetic phenomena of SPIO nanoparticles. It is also very useful in the search for new magnetic materials with desirable properties for biomedical and device applications.

4 Experimental and Instrumentation Methodologies

4.1 Raman Spectroscopy

When photons interact with matter, such as when a sample is illuminated with monochromatic light, it can be reflected, absorbed, or scattered. Raman spectroscopy deals with the interaction between light and matter when the light is scattered [11]. The electrons in an atom may be excited to virtual and excited states. There are two types of scatterings: elastic and nonelastic, as shown in Fig. 14.1. The most common scattering is elastic scattering, called Rayleigh scattering, in which scattered photons (purple arrows) have the same wavelength as the incident light (red arrows). While in nonelastic scattering, also called Raman scattering, the incident photon interacts with matter and its wavelength is either shifted to a lower (redshift) or higher (blueshift) value. Redshifted photons occur in Stokes scattering where the initial vibrational state is the ground state, excitation takes place to a virtual state and decays to a real vibronic state with higher energy than the initial state. Therefore, the scattered photon has a lower energy than the incident photon. In anti-Stokes scattering, the initial state is the excited state and the scattered photon has higher energy than the incident one. At room temperature, the population of vibration excited states is low according to the Boltzmann population of states; hence, although the Stokes and anti-Stokes spectra contain the same frequency information, the Stokes spectrum is always more intense than the anti-Stokes spectrum and commonly used in Raman spectroscopy technique.

Because Raman spectrum stems from the bonds vibrations, it provides an intrinsic nano-probing and offers a "bottom-up" approach of nanostructured materials that comes as a good complement to other techniques such as transmission electron microscopy, X-ray diffraction, and infrared, and Mössbauer spectroscopy. Since almost no sample preparation is needed, Raman technique [12, 13] is commonly used to investigate nanomaterials. This could provide the phase identification and, possibly, size estimation [14].

4.2 μ-Raman Spectroscopy

μ-Raman spectroscopy provides information on the microstructure of analyzed samples and improves the spatial resolution. Raman scattering measurements can be performed at room temperature using several laser sources. The experimental

Fig. 14.1 Illustration of linear Raman scattering processes

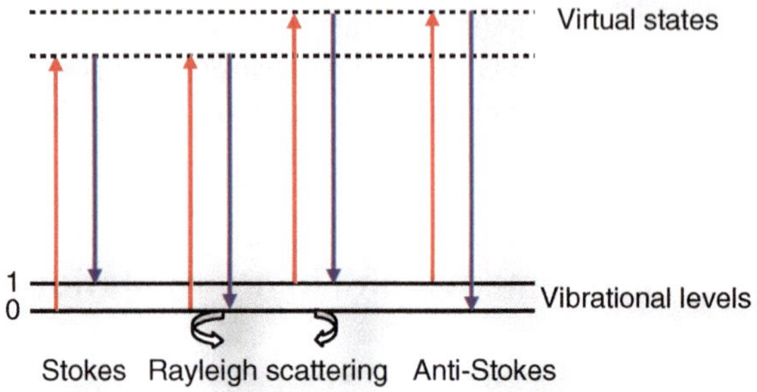

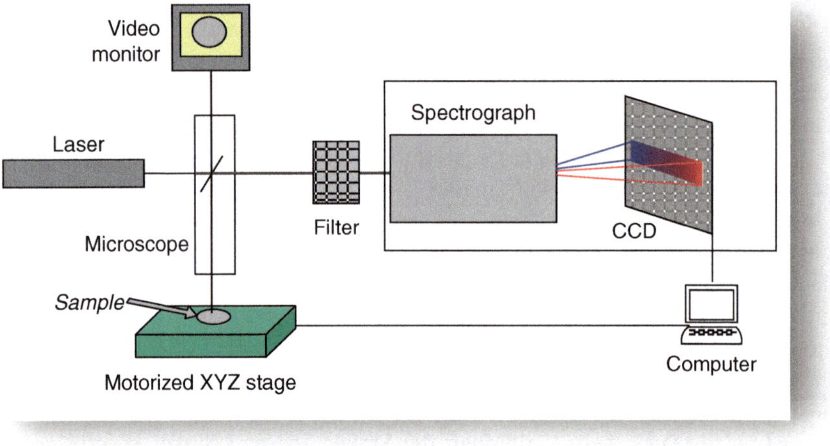

Fig. 14.2 Illustration of a conventional micro-Raman setup

μ-Raman setup is based on an inverted optical microscope with an x–y scan stage. The linearly polarized incident laser beam is reflected by a dichroic beam splitter and focused by a high numerical aperture (NA) on the sample surface through the transparent glass substrate. The scattered light is collected with the same objective and is detected in two distinct ways. In the first, the scattered beam is transmitted through the beam splitter and filtered by a bandpass filter in order to remove all the scattered light with wavelengths outside the range in which the Raman band to be studied is centered on. Then, the filtered scattered beam is detected by a single-photon counting avalanche photodiode (APD). In the second option, the scattered beam is transmitted through the beam splitter and filtered by a notch filter in order to remove the Rayleigh component of the scattered light. The signal is then detected by a combination of a spectrograph and a cooled charge-coupled device (CCD); see Fig. 14.2.

5 Key Research Findings

5.1 Special Features of Iron Oxide–Based Materials

The magnetism of solid is originated from the electrons of atoms. Electrons, which possess both charge and spin, present both orbital motions and spin rotations. Therefore, the magnetic moment, named Bohr magnetron, can be generated from either the orbital or spin motion of a single electron. Although many electrons in a solid are dislocated by orbital motions and spin orientations around the nuclei, only two electrons with up and down spins can occupy one orbit. Therefore, most of the magnetic moments originated from spins cancel each other out and do not contribute to the magnetic moment of a solid. However, *d*-orbit of transition elements and *f*-orbit of rare earth elements are not fully occupied. According to Hund's rule, the orbits are occupied by electrons in such a way that the total spin becomes maximum. Therefore, both transition and rare earth elements exhibit magnetic moments due to the spin orientation of electrons.

In fact, all materials are magnetic to some extent, with their response depending on their atomic structure and temperature. They may be conveniently classified in terms of their volumetric magnetic susceptibility, χ, where $\vec{M} = \chi\vec{H}$ describes the magnetization ($\vec{M}$) induced in a material by an applied field $\vec{H}$. Most materials display little magnetism even in the presence of an applied field. These are classified either as paramagnets, having randomly oriented dipoles that can be aligned with applied field, for which χ falls in the range of 10^{-6} to 10^{-1}, or as diamagnets that do not have magnetic moment in the absence of an external field, presenting χ in the range of -10^{-6} to -10^{-3}. However, some materials exhibit ordered magnetic states and are magnetic even in the absence of an external magnetic field due their atomic orbits not being fully occupied. The magnetism of solids depends on how these atomic magnetic moments are aligned in crystals. They are classified as ferromagnets, ferrimagnets, and antiferromagnets, where the prefix refers to the nature of the coupling interaction between the electrons within the material. In ferromagnetic material, for example, the exchange force causes magnetic moments aligning in one direction, producing spontaneous magnetization in the material. On the other hand, in antiferromagnetic material, the exchange interaction induces antiparallel ordering of the nearest neighboring atomic spins, then the magnetic moments are canceled in the crystal and spontaneous magnetization is lost. Ferrimagnetic materials, however, possess different antiparallel ordered magnetic moments; the material exhibits spontaneous magnetization [15]. Since the magnetic order is disturbed by thermal agitation, ferromagnets, antiferromagnets, and ferrimagnets lose their magnetic order at a temperature being higher than the magnetic transition temperature called "Curie temperature." In addition, magnetic properties can be largely tailored by changing the magnetic material size. For instance, in ferromagnetic and ferrimagnetic bulk materials, one often sees hysteresis, which is an irreversibility in the magnetization process that is related to the pinning of magnetic domain walls at impurities or grain boundaries within the material, as well as intrinsic effects such as the magnetic anisotropy of the crystalline lattice. This gives rise to open M–H curves, called hysteresis loops. The shape of these loops is strongly influenced by the dimensionality (volume) of the sample. In large samples (of the order of micron size or more) there is a multi-domain ground state which leads to a narrow hysteresis loop since it takes relatively little field energy to make the domain walls move; while in smaller samples, there is a single-domain ground state which leads to a broad hysteresis loop. At even smaller sizes (of the order of tens of nanometers or less) of the sample, one can see superparamagnetism, where the magnetic moment of the particle as a whole is free to fluctuate in response to thermal energy, while the individual atomic moments maintain their ordered state relative to each other. At this limit, the orientation of the magnetization no longer splits into smaller domains, but instead maintains the magnetic structure of a single domain. For spherically shaped particles, a rough estimation of the critical diameter (D_c) below which the material could be considered a single domain was first obtained by Kittel [16]. D_c is typically in the range of 10–20 nm which is approximately evaluated by $D_c = 2A^{1/2}/M$, where A is the exchange constant [17]. The typical magnetic moment associated to a single-domain particle is orders of magnitude higher (10^2–10^4 μ_B) than electron magnetic moment ($\mu_e \sim \mu_B$), where μ_B is the Bohr magnetron. In the superparamagnetic regime, the orientation of the nanoparticle magnetic moment thermally fluctuates around the easy axis direction. However, thermal energy promotes reorientation of the magnetic moment against the magnetocrystalline energy barrier, resulting in fluctuation of the magnetic moment in a time scale of nanoseconds. The typical relaxation time ascribed to the average time that the magnetic moment takes for reorientation with respect to the easy axis was calculated for non-interacting single-domain particles and is referred to as the Néel relaxation time (τ_N) [18]: $\tau_N = \tau_o \exp(KV/k_B T)$, where τ_o is typically in the range of 10^{-11} to 10^{-9} s, k_B is the Boltzmann constant, K is the effective anisotropy constant, and V is the particle volume. KV represents the effective magnetocrystalline energy barrier. For a particular observation time (τ_{obs}), typical of a given experimental technique, the blocking temperature (T_B) is defined as [19]: $T_B = \frac{KV}{k_B \ln(\tau_{obs}/\tau_o)}$. At blocking temperature, the magnetization reversal of an assembly of identical (monodisperse) single-domain particles goes from blocked (with hysteresis cycle) to unblocked (non-hysteresis cycle), the latter presenting the superparamagnetic behavior.

SPIO-based materials present ferrimagnetic ordering. The ferrimagnetic cubic ferrites display the spinel structure formed by a nearly close-packed face-centered cubic array of oxygen anions with interstices partly filled by metallic cations. Cubic ferrites present chemical composition MFe_2O_4, where M denotes a divalent transition metal such as Fe, Co, Ni, Mn, Cu, Zn, and Cd. X-ray studies show that the unit cell contains a total of 24 metal atoms: 8 divalent plus 16 trivalent cations. There are two kinds of crystal interstices differing in coordination, given rise to the two magnetic sublattices in cubic ferrites, namely, tetrahedral (A) and octahedral (B) interstices (sites). Two extreme divalent versus trivalent cation occupancy are used to classify cubic ferrites as normal or inverse spinel structures. In normal spinel structures, all N ions occupy the *8-A* sites of O_h^7 space group and have point symmetry T_d (tetrahedral coordination), and each one is surrounded by four oxygen ions forming a tetrahedron (tetrahedral sites). The trivalent metal ions (Fe) of a normal spinel occupy the 16 d-sites of O_h^7 space group and have point symmetry D_{3d} (octahedral coordination), and each one is surrounded by six oxygen ions forming an octahedron (octahedral sites). The oxygen ions occupy the 32 e-sites with site symmetry of C_{3v}. In contrast, in the inverted spinel arrangement, the tetrahedral sites are occupied by half of the trivalent metal ions, while the octahedral sites are filled by divalent and by trivalent metal ions equally. Among cubic ferrites, magnetite (Fe_3O_4) is the most investigated iron oxide. It has a cubic-inverse-spinel structure and presents ferrimagnetic ordering if temperature is above the Verwey transition temperature T_V (120 K) and below 850 K. In this situation, magnetite cation distribution is described by the chemical formula $Fe_A^{3+}[Fe^{2+}Fe^{3+}]_B(O^{2-})_4$, which indicates that in the inverse spinel structure, the *A* sites are occupied by Fe^{3+} ions, whereas the *B* sites are occupied by equal numbers of $Fe^{2+}$and Fe^{3+}. When the Verwey transition occurs, the symmetry of the crystal decreases to orthorhombic. Maghemite (γ-Fe_2O_3) can be thought of as an iron-deficient form of magnetite, with the chemical formula $V_{1/3}Fe_{8/3}O_4$, where V represents a crystal vacant site [20]. In cubic ferrites, only if each sublattice contains just one type of metal ions and all N ions are in tetrahedral sites and all Fe ions are in octahedral sites, composing the direct spinel arrangement, both translational and local symmetries corresponding to the O_h^7 space group strictly apply. Actually, details of the ferrite preparation route determine the final arrangements of metal ions in the spinel structure, and thus, a signature of mixed cubic ferrites materials is very often observed. Mixed ferrites represent the intermediate arrangement between normal and inverted structures, and the degree of inversion is a key parameter used to describe mixed ferrites [21].

Nanosized cubic ferrites exhibit magnetic properties markedly different from their bulk counterparts, such as the already mentioned superparamagnetism behavior. Due to the intrinsic superparamagnetic behavior, high chemical stability, enhanced biological compatibility, and inexpensive and simple chemical synthesis routes of iron oxide nanoparticles, they are widely used in biomedical applications. For instance, SPIO systems are used as stable colloidal dispersions of nanoparticles in a continuous liquid phase named magnetic fluid (MF) or ferrofluid [22]. Under a magnetic field gradients, stably dispersed magnetic nanoparticles drag the carrier liquid, and the system acts macroscopically as a single unique liquid phase. The biocompatible magnetic fluid usually contains iron oxide nanoparticles dispersed in biological medium, providing enough biocompatibility as supported by in vitro and in vivo tests.

In order to produce stable MF samples for biological and medical applications, one needs to start with the preparation of surface-functionalized, biocompatible, and highly magnetized SPIO. In addition, most of the applications require samples containing nanoparticles with average sizes smaller than 20 nm, preferably presenting narrow particle size distribution, the morphological homogeneity assuring uniform physical and chemical properties. In regard to SPIO synthesis, knowing the influence of the synthesis parameters plus the fine control upon them allows one the proper control of the nanosized particle characteristics, such as core phase, size and shape profile, and chemical and structural stability, presents major challenges. These characteristics will determine the ultimate magnetic properties and therefore the applications of nanosized particles. Spinel ferrite nanoparticles can be prepared by several chemical and physical routes. Reviews of different techniques used in the preparation of magnetic nanoparticles have been recently published, for instance, Refs. [23] and [24].

5.2 Fundamentals of Raman Spectroscopy Applied to Nanomaterials

In a perfect crystal, the momentum conservation law leads to a selection rule of the first-order Raman scattering of a photon, governed by the equation $\mathbf{q} = 0$, where $\mathbf{q}$ denotes the phonon wavevector. This implies that Raman spectrum of a perfect crystal is only contributed by phonons at the center of the Brillouin zone. However, when the size of the crystal is reduced, translational symmetry of crystalline materials is broken. Then the previous rule ($\mathbf{q} = 0$) is no longer valid. Consequently, not only the vibrational modes at $\mathbf{q} = 0$ but also the modes at $\mathbf{q} \neq 0$ contribute to the Raman spectrum [25]. In addition, the symmetry breaking at nanoparticle boundaries also results in the appearance of specific surface and interface vibrational contributions. These two factors are often neglected in Raman spectroscopy of bulk materials but they become very significant in nanocrystals, where the concentration of nanoparticle boundaries is very high. Finally, a reduction of the dimension of matter domains down to the nanometer scale also induces quantum confinement of the electronic and vibrational wavefunctions. Then, in most cases, the Raman peak shifts progressively to lower energies, and the lineshape becomes gradually broader and asymmetric (on the lower energy side) as the structure size becomes smaller. Therefore, nanoparticles present different electronic and vibrational properties with respect to bulk systems. For instance, the vibrational properties of nanoparticles strongly depend on their morphology, i.e., structural geometry, symmetry, pore diameter, skeleton size, etc. Figure 14.3 shows Raman shifts of Si nanocrystals embedded in SiO_2 matrices as a function of the inverse particle diameter, reproduced from ref. [26]. Notice that the Raman shifts are inversely proportional to diameter (d) of nanoparticles within the accuracy of the experiment. The similar behavior had also been reported for Ag, Au, and CdS nanocrystals [27, 28].

As far as the theoretical description is concerned, the confined acoustic phonons in an elastic sphere were first theoretically studied by Lamb [29]. He derives two types of confined acoustic modes, spheroidal and torsional modes. The frequencies of these two modes are proportional to the sound velocities in particles and inversely proportional to the particle size. The spheroidal mode is characterized by the quantum number $l \geq 0$, while the torsional modes are characterized by $l \geq 1$. From the symmetry arguments, Raman-active modes are spheroidal modes with $l = 0$ and 2. The $l = 0$ mode is purely radial with spherical symmetry and produces totally polarized spectra, while the $l = 2$ mode is quadrapolar and produces partially

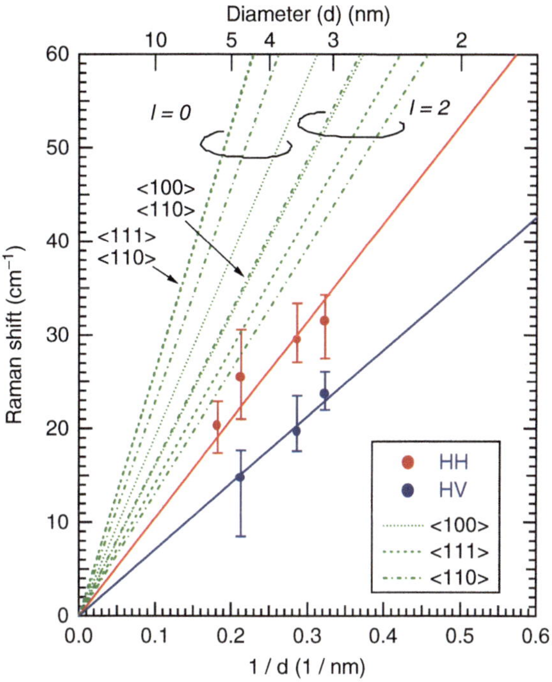

Fig. 14.3 Raman shift of polarized (*filled circles*) and depolarized (*open circles*) Raman spectra of Si nanocrystals embedded in SiO_2 matrices as a function of the inverse particle diameter. *Solid lines* are the results of the least squares fitting. *Dotted*, *broken*, and *dash-dotted lines* are the results of theoretical calculations for the confined acoustic modes (spheroidal modes with $l = 0$ and 2) assuming the sound velocities in <100>, <111>, and <110 > directions, respectively (Reprinted with permission from [26])

depolarized spectra. As it is widely accepted, low wave number Raman scattering of the acoustic vibrational modes of nanoparticles, which allows the simultaneous determination of the Raman frequency, intensity, and linewidth through the energy shift of the Raman peak and the correspondent line broadening, is a simple method to determine nanoparticle size [30, 31]. To help understand the experimental Raman spectra, we summarize the models that have been widely used to determine the size distribution of nanoparticles. In an effort to avoid the complicated mathematics, the frequency shift and damping of modes induced by the effect of matrix on the nanoparticle vibrations are neglected.

In the compounds, if the crystal unit cell contains N atoms, then $3N$ degrees of freedom result in 3 acoustic phonons and $3N-3$ optical phonons. These phonons can propagate in the lattice of a single crystal as a wave and exhibit dispersion depending on their wavelength or equivalently their wavevector in the Brillouin zone. As already mentioned, in ideal single crystals, only zone-center optical phonons can be observed by Raman spectroscopy. However, this $\boldsymbol{q} = 0$ selection rule is relaxed due to a broken translational symmetry of nanoparticles where the propagation is restricted in all three directions. To characterize this behavior of vibrational modes in nanoparticles, a phenomenological phonon confinement is introduced in Raman spectrum simulation. In a Stokes scattering process in an infinite crystal, the wavevector difference $\overrightarrow{k} = \overrightarrow{k}_L - \overrightarrow{k}_S$ is transferred to a phonon with wavevector $\overrightarrow{q_0}$, where $\overrightarrow{k}_L$ and $\overrightarrow{k}_S$ are incident and scattered laser wavevectors, respectively. Then the wavefunction of a phonon in this case can be described by $\Phi(\vec{q_0}, \vec{r}) = u(\vec{q_0}, \vec{r}) \exp(-i\vec{q_0} \cdot \vec{r})$, where $u(\vec{q_0}, \vec{r})$ is the Block function with the periodicity of the lattice. However, a plane-wave-like phonon wavefunction cannot properly describe the propagation of the phonon because it cannot propagate beyond the crystal surface. Under this consideration, one must multiply the phonon wavefunction with a confinement function or envelope function $W(r,D)$, which decays to a very small value close to the boundary. Then the vibrational wavefunction of the nanoparticle is approximated by:

$$\Psi(\vec{q_0}, \vec{r}) = W(r, D) u(\vec{q_0}, \vec{r}) \exp(-i\vec{q_0} \cdot \vec{r}) = \Psi^*(\vec{q_0}, \vec{r}) u(\vec{q_0}, \vec{r}), \quad (14.1)$$

where

$$\Psi^*(\vec{q_0}, \vec{r}) = W(r, D) \exp(-i\vec{q_0} \cdot \vec{r}). \quad (14.2)$$

In order to study the effect of confinement induced by a restricted region in nanocrystal on Raman spectrum, $\Psi^*(\vec{q_0}, \vec{r})$ is expanded in a Fourier series:

$$\Psi^*(\vec{q_0}, \vec{r}) = \int C(\vec{q_0}, \vec{q}) \exp(i\vec{q} \cdot \vec{r}) d^3q \quad (14.3)$$

where the Fourier coefficients $C(\vec{q}_0, \vec{q})$ are given by

$$C(\vec{q_0}, \vec{q}) = \frac{1}{(2\pi)^3} \int \Psi^*(\vec{q_0}, \vec{r}) \exp(-i\vec{q} \cdot \vec{r}) d^3r \quad (14.4)$$

It is noted that the phonon wavefunction is a superposition of plane waves with $\vec{q}$ vectors centered at $\vec{q_0}$. In the literature, several weighting functions such as Gaussian functions, sinc, and exponential functions have been extensively used to describe the confinement functions. The choice of type of weighting function depends upon the material property of nanoparticles. Here, we present a brief review about calculated Raman spectra of spherical nanoparticle of diameter D based on these three confinement functions. In an effort to describe the realistic Raman spectrum more properly, particle size distribution is taken into account. Then the Raman intensity $I(\omega)$ can be calculated by:

$$I(\omega) \propto \sum_{i=1}^{m} \int_0^{\infty} \rho(D) \mathrm{d}D \int \frac{|C(\vec{q_0}, \vec{q})|^2}{[\omega - \omega_1(q)]^2 + (\Gamma/2)^2} d^3q, \quad (14.5)$$

where $\rho(D)$ is the nanoparticle size distribution and Γ is the intrinsic mode line width. The sum is carried over m dispersion curves $\omega_i(q)$, depending on mode degeneration m. As one-phonon Raman scattering probes zone-center phonons, one can establish $\vec{q_0} = 0$. From the expression of $I(\omega)$, it is noted that the Raman scattering intensity is described by a continuous superposition of Lorentzian curves with bandwidth Γ and weighted by the $|C(\vec{q})|^2$. Hence, the key part of the model is the phonon weighting function characterized by parameter α. For Gaussian type weighting function $W(r,D) = \exp(-\alpha r^2/D^2)$, the corresponding Fourier coefficients are given by:

$$|C(\vec{q})|^2 = \frac{D^6}{16(2\pi)^2\alpha^3} \exp(-q^2D^2/2\alpha). \quad (14.6)$$

While for sinc type of confinement function

$$W(r,D) = \begin{cases} \dfrac{\sin(2\pi r/D)}{2\pi r/D}, & r < D/2 \\ 0, & r \geq D/2 \end{cases} \quad W(r,D) = \begin{cases} \dfrac{\sin(2\pi r/D)}{2\pi r/D}, & r < D/2 \\ 0, & r \geq D/2 \end{cases}, \quad (14.7)$$

which gives a vibrational amplitude exactly equal to 0 at the boundary, the corresponding Fourier coefficients are given by:

$$|C(\vec{q})|^2 = \frac{4D^4}{(2\pi)^4} \frac{\sin^2(qD/2)}{q^2(4\pi^2 - q^2D^2)^2}. \quad (14.8)$$

The choice of a sinc function is based on the assumption that the weighting function may be analogous to the wavefunction of an electron in a hard sphere potential. It is interesting to argue that although the combination of these Raman spectrum simulations and Raman measurements has been widely used to determine nanoparticle size and to help identify the vibrational modes of semiconductor nanoparticles [26], it has not yet been applied to investigate the magnetic nanoparticles. It will be very useful to understand the dependence of Raman spectrum on the magnetic nanoparticle size.

5.3 The Vibrational Spectra of Iron Oxide Materials

Raman spectra provide useful information about the internal structure, bonding, and state-of-order of materials ranging from bulk crystals to nanoscale particulates. Contributions from Raman spectroscopy for probing the vibrational properties of iron oxides have been available in the literature for a long time [32]. The coexistence of different phases would prevent the correct interpretation of the Raman spectrum of the iron oxides and sometimes even leads to a misunderstanding for experimental observations. In this section, an overview about the Raman spectra of spinel iron oxide crystals will be presented, aiming to find a regulation for an assignment of vibrational spectra, which serves as a base for nanomaterial investigations.

The group theory analysis of the vibrational spectrum of the spinel structure [32] reveals that the crystal symmetry is cubic, correspondent to the space group O_h^7 $(Fd3m)$ with eight formulas per unit cell. The primitive cell is rhombohedral with two formula units per cell. The vibrational spectrum of the spinel is expected to exhibit five Raman-actives modes and four infrared (IR)-active bands (4 T_{1u}). Thus, the predicted Raman active modes are A_{1g}, E_g, and three T_{2g} modes ($E_g + E_g + 3T_{2g}$). The presence of an inversion center in the centrosymmetrical space group $Fd3m$ implies mutual exclusion of Raman and IR activities for the same vibrational mode. Furthermore, the bands in IR and Raman spectroscopy are sensitive to coordination geometry and oxidation states. For example, the inverse spinel lattice of γ-Fe_2O_3, as already mentioned, is iron deficient, resulting in observed sample preparation–dependent IR and Raman spectra.

Verble investigated Raman spectra of crystal magnetite as a function of temperature, aiming to study the role played by optical phonons in the Verwey transition [33]. Raman spectra were recorded using an Ar laser excitation at 514.5 or 488 nm. Raman trends obtained at 77 K show the presence of five phonon modes at 300, 320, 420, 560, and 680 cm^{-1}. A molecular model taking into account the tetrahedral symmetry of molecular units in the spinel crystal is used to describe normal mode motions of the FeO_4 tetrahedron as following: $T_{2g}^1 \rightarrow$ translatory movement of the whole FeO_4; $T_{2g}^2 \rightarrow$ asymmetric stretch of Fe and O; E_g and $T_{2g}^3 \rightarrow$ symmetric and asymmetric bends of oxygen with respect to Fe, respectively; and $A_{1g} \rightarrow$ symmetric stretch of oxygen atoms along Fe–O bonds. Observed Raman features for Fe_3O_4 are related to T_{2g} symmetry (300, 320, and 560 cm^{-1}), E_g symmetry (420 cm^{-1}), and A_{1g} symmetry (680 cm^{-1}) [33]. Note that these data were taken in a temperature below the Verwey transition in which magnetite structure is changed to an orthorhombic phase. Room-temperature polarized Raman measurements carried out on Fe_3O_4 single crystals, using the 514.5 nm line of an Ar laser as excitation, showed the presence of four Raman-active phonons [34]. The strongest peak at 670 cm^{-1} phonon was assigned to the A_{1g} mode and the features observed at 193, 308, and 540 cm^{-1} to the predicted three T_{2g}.

Highly focused laser used in the micro-Raman experiments could induce a temperature increase. The local temperature of a heated spot can be calculated from Raman spectra using the observed temperature-induced shifts of phonon wavenumbers, namely, Stokes/anti-Stokes method, quasi-harmonic approximation for the shifts of wavenumbers, and calculation of temperature from the thermodynamic Grüneisen parameter [35]. Raman experiments as a function of the laser excitation intensity showed that material degradation frequently occurs under high excitation intensity, and may lead to misinterpretation of spectra. The sensitivity to laser intensity seems to depend on surface morphology, at least in the case of magnetite [36]. In addition, different oxidation pictures for the single crystal and the powder (approximate grain size 0.3–1 μm) magnetite samples were observed. For the powdered Fe_3O_4, the power threshold for the oxidation is significantly lower than for the crystal ones [35].

Tetrahedral symmetry is a type of arrangement also found for maghemite and for inverse spinels such as MFe_2O_4(M = Mg, Ni, Fe), so that A_{1g} band in the 670–710 cm^{-1} region in their Raman spectra corresponds to the stretching modes of tetrahedral units, while Raman modes present in the 450–620 cm^{-1} ferrite spectra are dominated by octahedral groups [37]. However, A_{1g} mode at about 615 cm^{-1} observed in MFe_2O_4 (M = Mn, Zn) might be assigned to the motion of an octahedron [37].

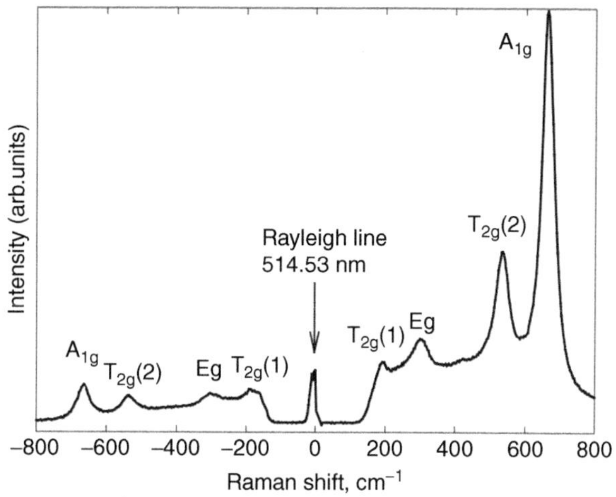

Fig. 14.4 Raman spectrum of magnetite with the symmetry assignment (Reprinted with permission from Ref. [39])

IR- and Raman-active modes of magnetite, maghemite, and hematite were calculated using atomistic simulation [38]. Calculated wavenumbers for hematite (α-Fe_2O_3) that belongs to the D^6_{3d} crystal space group are two A_{1g} (228 and 447 cm^{-1}) and five E_g modes (200, 294, 327, 391, and 466). In the Raman spectra of hematite (α-Fe_2O_3) bulk, recorded under the 632.8 nm He–Ne laser line, was observed the predict two A_{1g} at 227, and 497, and five E_g modes at 246, 293, 299, 411, and 612. The feature at 1,320 cm^{-1} was assigned as a two-magnon scattering arising from the interaction of two magnons created on antiparallel closed spins [36]. In addition Raman spectrum of γ-Fe_2O_3 bulk, recorded under the 632.8 nm He–Ne laser line, displayed three broad bands at around 350, 500, and 700 cm^{-1} [36], which are in agreement with the calculated ones (356, 493, and 695 cm^{-1}) [38].

Raman features of Fe_3O_4 are sample specific, depending on non-stoichiometry, cation distribution, defects, and also on the experimental conditions. In spite of this, the Raman spectra of magnetite are distinct from those obtained from maghemite [36]. Moreover, group theory predicts number and symmetries of Raman and IR modes, their intensities are derived from lattice dynamics calculations. In addition, polarized Raman investigation on oriented Fe_3O_4 single crystal has been conducted by Shebanova and Lazor [39]. They also calculated the relative polarizability intensities aiming to assign the observed Raman bands of Fe_3O_4, reproduced in Fig. 14.4. From Fig. 14.4, it is possible to note that the structure at 193 cm^{-1} is weak, but its presence is confirmed on the anti-Stoke side of the spectrum. Based on polarized experiments on the oriented Fe_3O_4 single crystal, they proposed the following assignment for the vibrational spectrum of the observed modes: T^1_{2g} for 193 cm^{-1}, E_g for 306 cm^{-1}, T^2_{2g} for 450–490 cm^{-1}, and T^3_{2g} for 538 cm^{-1}, and A_{1g} for 668 cm^{-1}, that are listed in Table 14.1. Moreover, they detected four out of five predicted Raman bands and inferred the location of the fifth, which is an unobserved phonon mode, in the range of 450–490 cm^{-1}.

A detailed investigation about the growth of cobalt ferrite single crystals through a borax flux method and its characterization was reported by Wang and Ren [40].

Table 14.1 Comparison of observed Raman modes of Fe_3O_4 crystal and nanoparticle and corresponding assignment

Fe_3O_4 crystal [39]	Fe_3O_4 [44]	Fe_3O_4 calculated [38]	Assignment [38, 44]
		231	T_{1u}
193	179	241	T_{2g}
306	315	296	E_g
	270	326	T_{1u}
		360	T_{1u}
		491	T_{1u}
	374		γ-Fe_2O_3
450–500	445	505	T_{2g}
	501		γ-Fe_2O_3
538	(530)[a]	581	T_{2g}
	565		T_1
	627		γ-Fe_2O_3
668	664	666	A_{1g}
	712		Fe(III)[b]

[a]Value in parentheses is IR data
[b]Oxidation of Fe(II) at octahedral sites

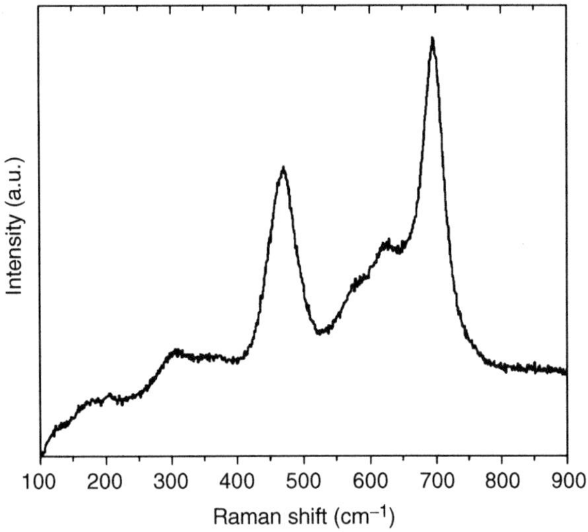

Fig. 14.5 Raman spectra of an as-grown $CoFe_2O_4$ single crystal measured at room temperature (Reprinted with permission from Ref. [40]. Copyright Elsevier B.V. (2006))

The Raman spectrum of cobalt ferrite single crystals excited by 532 nm laser line, attained at room temperature, is reproduced in Fig. 14.5. By performing the fitting of the Raman spectra displayed in Fig. 14.5, it would be possible to observe Raman bands at 183, 304, 469, 573, 627, and 696, listed in Table 14.2. Apart from Fe_3O_4, γ-Fe_2O_3, and $CoFe_2O_4$, others ferrites such as $NiFe_2O_4$ and $Fe_{3-x}Zn_x$ O_4 which

Table 14.2 Raman modes of $CoFe_2O_4$ crystal and nanoparticles

$CoFe_2O_4$ crystal [40]	$CoFe_2O_4$ [50]	$CoFe_2O_4$ [44]	$CoFe_2O_4$ [45]	$CoFe_2O_4$ [47]
532 nm	514 nm	632 nm	632 nm	514 nm
183	220	181	183	186
304	310	306	304	312
		368 439		
469	467	476	474	477
		524		
573		563	560	
627	624	629	615	634
696	690	686	682	687
		714		

exhibit much richer Raman spectra with a number of additional peaks have also been extensively studied by Raman spectroscopy. Changes in Raman spectra as a function of doping have been found.

In short, maghemite Raman features depend on sample preparation, once it is directly related to the degree of crystallinity of the material. Maghemite phase assignment is straightforward, since expected Raman modes as calculated by Chamritski and Burns [38], and attained from experiments by de Faria et al. are in agreement, and characterized by three broad bands around 350, 500, and 700 cm^{-1} [36]. Likewise for the assignment of the five predict of magnetite Raman modes we suggest to follow what is proposed by Shebanova and Lazor [39] in their investigation comprising polarized experiments on the oriented Fe_3O_4 single crystal: A_{1g} for 668 cm^{-1}, E_g for 306 cm^{-1}, and T_{2g}^{1-3} for 193 cm^{-1}, 450–490 cm^{-1}, and 538 cm^{-1}, respectively.

In conclusion, the Raman features observed and calculated for iron oxide crystals have been used as reference to identify Raman modes in their counterpart nanomaterials. Furthermore, it is known that the high power density from a laser excitation source can excessively heat a sample during a Raman experiment, as discussed previously. This effect becomes even more important for micro-Raman experiments of nanomaterials, where laser beams are focused to a spot size with a diameter of only a few micrometers, and nanoparticulates do not dissipate heat well. Moreover, an increase in the local sample temperature may cause a frequency shift in the Raman bands, or it may cause material degradation as the result of oxidation, recrystallization, order-disorder transitions, phase transition, or decomposition.

5.4 Phase Identification and Phase Transition in Iron Oxides

The electronic, structural, and magnetic properties are strongly modified near and at the surface of nanosized cubic ferrite particles. The origin of the surface effects is mainly attributed to the breaking of the translational crystal symmetry at the nanoparticle boundary, which leads to an emergence of specific surface and interface vibrational modes. In addition, the nanoparticulated materials may comprise fine single crystal or polycrystalline particles, they may even aggregate in some special cases. For instance, some nanoparticles are composed of a crystalline core surrounded by either a partially crystalline or amorphous shell. Then, at nanoscale, new effects arise, and several phenomena come into play. When the particle size becomes comparable to the phonon mean free path, the reciprocal of the particle diameter becomes a significant fraction of the Brillouin zone, which induces a relaxation of selection rules, as already mentioned [41]. As the size further decreases, in addition to the disorder at atomic scale, the surface effects play more and more important role in determination of Raman spectrum due to the high surface/volume ratio.

The chemical and structural properties of SPIO particles depend strongly on the details of the sample preparation protocol. The characteristics of the end SPIO particle dispersed in a hosting template (liquid or solid) may also be affected by the particle-template interaction, including aging effects. Therefore, Raman spectroscopy may be used as a complementary technique of IR, XRD, and Mössbauer spectroscopy, or at least one of these characterization techniques, to study the dependence of chemical and structural characteristics of SPIO particles on synthesis routes. In order to control the end magnetic properties of iron oxide nanoparticles and the stability of the resulting magnetic colloids, in this section, we will address the effects of the sample preparation process such as oxidation of magnetite to maghemite and nanoparticle surface passivation on Raman spectra characteristics.

5.4.1 Phase Identification

There are many active efforts to develop new kinds of magnetic nanomaterials. To do so, phase identification such as ensuring the information about whether or not existence of antiferromagnetic hematite (α-Fe_2O_3), and amorphous non-stoichiometric oxyhydroxides, together with magnetite (Fe_3O_4) and maghemite (γ-Fe_2O_3) plays a very important role.

Raman spectroscopy as a reliable technique has been applied in identification of the phase of as-prepared sample and/or in differentiation of the nanomaterials synthesized by distinct chemical routes. For instance, the Raman spectra of nanoparticles reported by Thünemann et al. displaying four broad bands around 350, 500, 700, and 1,400 cm^{-1} [42] were used as fingerprint of γ-Fe_2O_3 in sample identification. Three methods, including synthesis by microwave refluxing [43] of functionalized maghemite nanoparticles in aqueous medium by grafting triethoxysilane monomethylene-PEG (Si-mPEG) onto maghemite nanoparticles, were compared. γ-Fe_2O_3 characteristic features were detected, indicating the presence of maghemite in all of studied samples. Raman trends show that the intensities of grafted maghemite III bands are stronger than those of grafted maghemite II, indicating that microwave heating tends to improve the crystallinity of the particles [43].

The molecular composition of magnetite depends strongly on the preparation protocol as well as on experimental conditions used in their Raman analysis.

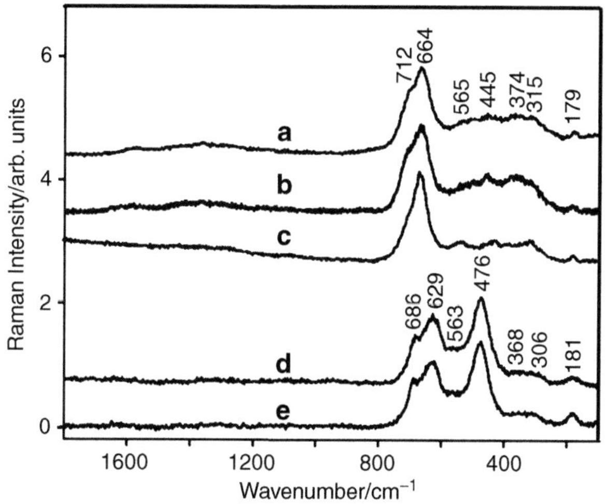

Fig. 14.6 Raman spectra excited at 632.8 nm of (**a**) Fe_3O_4, (**b**) Fe_3O_4/SO, (**c**) Fe_3O_4/CO, (**d**) $CoFe_2O_4$, and (**e**) $CoFe_2O_4$/SO (Reprinted with permission from Ref. [44]. Copyright American Chemical Society (2009))

The structural properties of coated and uncoated Fe_3O_4 and $CoFe_2O_4$ have been studied by means of Raman Spectroscopy, and surface-enhanced Raman scattering (SERS) measurements [44]. These ferrites were surface modified by the adsorption of fatty acids derived from soybean and castor oil, following dispersion in cyclohexane. The FTIR and Raman spectra of bare and surface treated Fe_3O_4 and $CoFe_2O_4$ nanoparticles show that the number of observed vibrational modes is not compatible with the expected O_h^7 symmetry, since IR-only active phonon modes were observed in the Raman spectra and vice versa (see Fig. 14.6). The assigned Raman modes attained from these curve-fitted spectra are presented in Table 14.1, where it is possible to observe that they detected at least ten phonons in the Raman spectrum of magnetite-based particles. The Raman mode observed near 710 cm^{-1} refers to the oxidation of Fe(II) centers at octahedral sites, indicating that the samples were partially oxidized. Moreover, phonons appearing in the Raman spectra, such as the T_{1u} modes near 270 and 560 cm^{-1}, expected to be IR-active, and the A_{1g} phonon observed in the IR spectra are due to breaking of selection rules as expected for nanosized samples. These findings show experimental evidence of the proposed quantum effects in nanomaterials.

Raman spectra recorded for $CoFe_2O_4$ nanoparticles are reproduced in Fig. 14.6, whose fitted bands are presented in Table 14.2. There are at least three Raman features that are also observed in the IR spectra [44], as discussed in the previous assignments for magnetite. In addition, to remove the discrepancy in the literature about the relative intensity of the phonons observed near 474, 620, and 680 cm^{-1}, the authors also recorded the Raman spectra of $CoFe_2O_4$ excited by laser lines of 514.5 and 647.1 nm, respectively, using Na_2SO_4 as an internal standard for intensity normalization. Raman spectra reproduced in Fig. 14.7 show that the intensity of the phonon at 686 cm^{-1} in the spectrum excited at 647.1 cm^{-1} is nearly half that of the phonon at 475 cm^{-1}. Otherwise, at 514.4 nm excitation, both features have almost the same intensity, indicating some resonance processes involving ground and

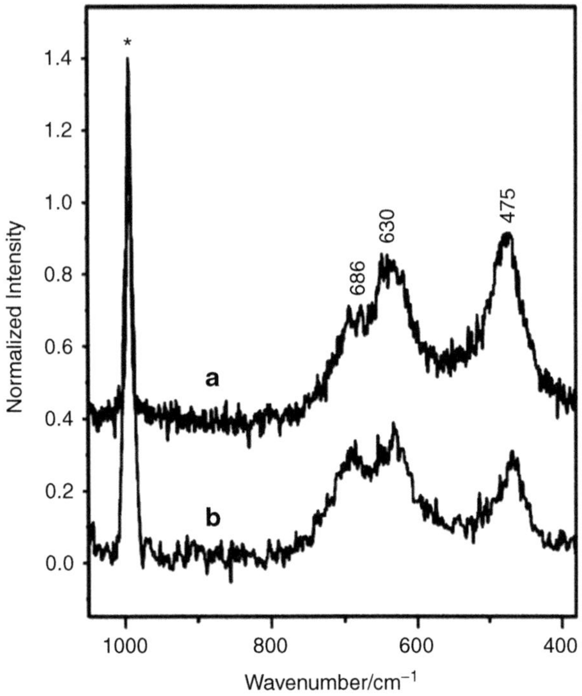

Fig. 14.7 Raman spectra of $CoFe_2O_4$ nanoparticles excited at (**a**) 647.1, and (**b**) 514.5 nm. The *asterisk* is the Raman signal of Na_2SO_4 that was used as an internal standard for intensity normalization (Reprinted with permission from Ref. [44]. Copyright American Chemical Society (2009))

excited electronic states in $CoFe_2O_4$ take place. The SERS measurements of a thin film of $CoFe_2O_4$ deposited on a SERS-active Au electrode at different applied potentials ensure that the Raman features at 630 and 553 cm^{-1} belong to a phonon mode involving Co–O motion, whereas the signals at 686 and 475 cm^{-1} are related to Fe(III)–O motions.

Well-defined features are observed in the Raman spectra of $CoFe_2O_4$ samples (average diameter from 5.0 to 9.1 nm), synthesized by using a high-temperature organometallic decomposition route in the presence of surfactant molecules [45]. Raman spectra reproduced in Fig. 14.8a, b were obtained under an optical excitation at 632.8 nm. Notice that the vibrational modes at 474, 560, 615, and 682 cm^{-1}, listed in Table 14.2, are observed in both Fig. 14.8a and b. Similarly, to the previous discussion, the 682 cm^{-1} and 474 cm^{-1} vibrational modes are assigned to the A_{1g} at tetrahedral sites and a Raman mode involving Co^{2+} at octahedral sites, respectively. It is also noted that there are several differences between Raman spectra displayed for samples IV and V (see Fig. 14.8a, b). Among them, one is the relative intensity ratio between the major peaks at 474 and 682 cm^{-1}. For example, for sample V, the intensity of the feature near 474 cm^{-1} is higher than that of the signal near 680 cm^{-1}, while this behavior is inverted in sample IV, suggesting higher Co^{2+} incorporation in sample V.

The nature of chemical bonding of oleic acid molecules on the active surface sites of the nanocrystals $CoFe_2O_4$ (average diameter ~20 nm), and its influence on

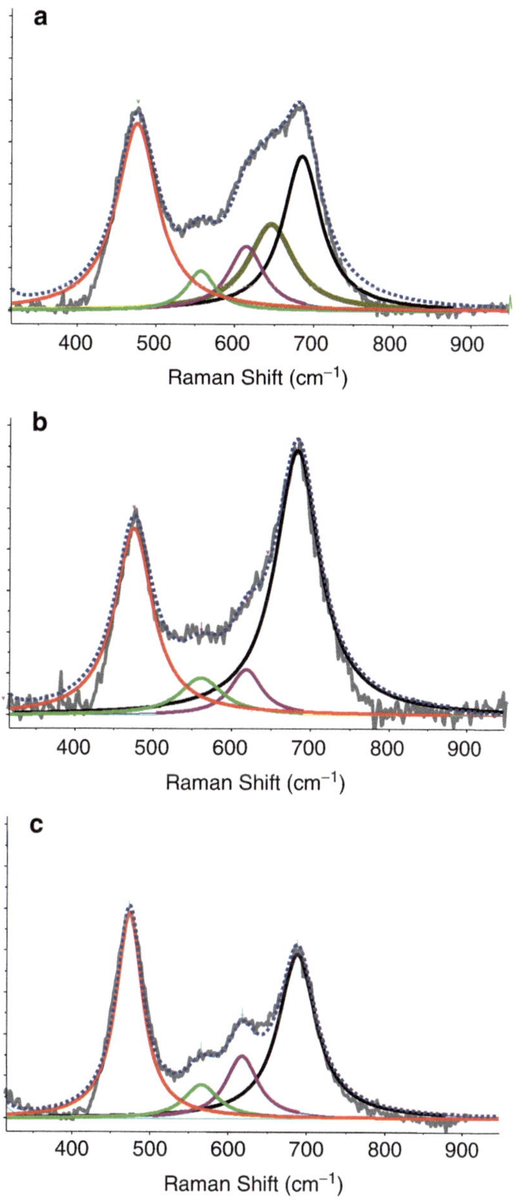

Fig. 14.8 Raman spectra from (**a**) sample V and (**b**) sample IV taken at low-laser power of $\sim 10^2$ W/cm^2. (**c**) A spectrum from the exact spot where spectrum (**a**) was taken but after a fivefold increase in the He–Ne laser power (Reprinted with permission from Ref. [45]. Copyright American Chemical Society (2007))

the high coercivity have been studied [46]. The authors proposed a model in which the OA molecules are strongly bound to the $CoFe_2O_4$ surface atoms, where OA molecules also pose a steric hindrance, creating substantial anisotropy. Raman spectra for oleic acid–grafted $CoFe_2O_4$ samples (pristine), and annealed at 573 K, 973 K, and uncoated sample annealed at 1,473 K, are reproduced in Fig. 14.9, showing six components at 467, 503, 538, 575, 610, and 671 cm^{-1}, for pristine

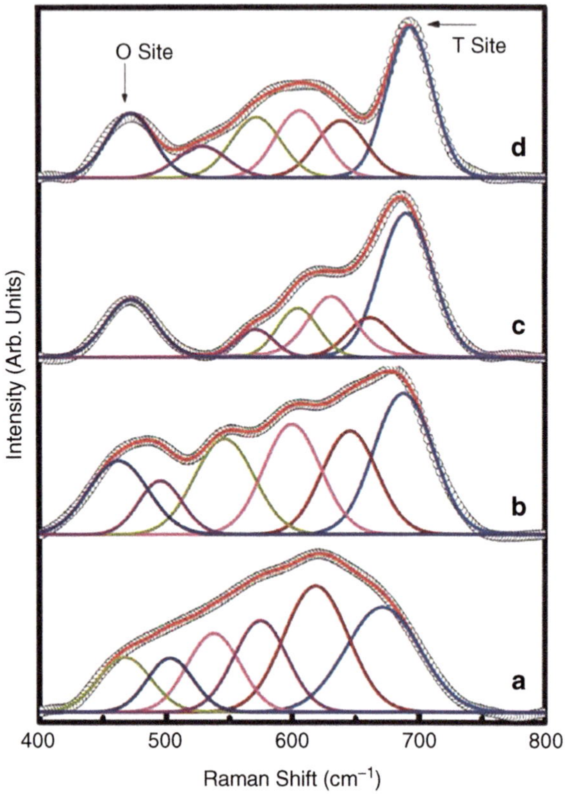

Fig. 14.9 Raman spectra of (0.2 M) oleic acid–capped $CoFe_2O_4$ sample (**a**) pristine, and (**b**) annealed at 573 K, (**c**) at 973 K, and (**d**) uncapped 1473 K annealed (Reprinted with permission from Ref. [46]. Copyright American Chemical Society (2009))

sample, and 473, 529, 572, 607, 639, 662, and 693 cm^{-1}, for uncapped sample that was annealed at 1,473 K. The A_{1g} mode at 693 cm^{-1} in the sample annealed at 1,473 K shifts toward a lower value (671 cm^{-1}) for the pristine sample due to strain. Furthermore, the full width at half-maximum (FWHM) of this T-site related mode is 42.1 for the 1,473 K annealed sample, which was increased to 68.9 cm^{-1} for pristine sample. This broadening of the Raman mode in the pristine sample was attributed to local disorder. Raman analysis suggested that the OA-coated $CoFe_2O_4$ sample is under strain, suggesting the high coercivity was a cumulative effect of disorder of surface spins, large strain, and surface anisotropy at the particle surface due to OA.

Figure 14.10 shows Raman spectra (514-nm Argon laser line) of $CoFe_2O_4$ (average diameter of 14.8 nm) with different Co^{2+} ion concentrations, prepared by the coprecipitation route [47]. Notice that Raman spectrum consists of five broad bands at around 186, 312, 477, 1,330, and 1,580 cm^{-1}, and two strong bands at around 634 and 687 cm^{-1}. Furthermore, the observed phonons are similar to the vibrational modes of bulk maghemite, but with redshifts of peak positions with respect to maghemite, due to the higher mass of Co in comparison with that of Fe. Among the five broad bands, the first three phonon modes at around 186, 312, and 477 cm^{-1} are assigned to T^1_{2g}, E_g and T^3_{2g} modes, respectively, where the T^3_{2g} mode

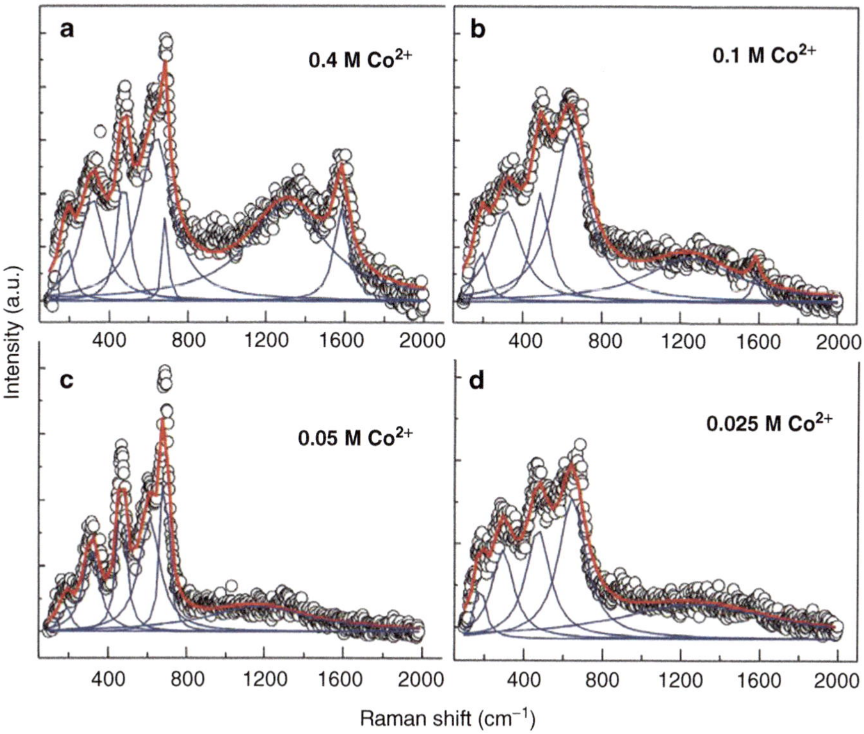

Fig. 14.10 Room-temperature micro-Raman spectra of $CoFe_2O_4$ synthesized at Co^{2+} metal ion concentrations of (**a**) 0.4, (**b**) 0.1, (**c**) 0.05, and (**d**) 0.025 M (Reprinted with permission from Ref. [47]. Copyright American Chemical Society (2010))

is originated from the local symmetry vibrations of metal ions in the octahedral site. Meanwhile, the other two peaks at 1,330 and 1,580 cm^{-1} are assigned to ferrihydrite phase. It is very interesting to note that instead of the expected phonon around 651 cm^{-1} for A_{1g} symmetry vibrations of metal ion in the tetrahedral site and 710 cm^{-1} band of 2-line ferrihydrite, two strong bands appear at around 634 and 687 cm^{-1}. Raman analysis showed the presence of ferrihydrite that was untraceable by XRD, and is responsible for the reduction in the saturation magnetization.

Table 14.2 summarizes typical phonons available in the literature for $CoFe_2O_4$ nanoparticles, compared with that obtained for $CoFe_2O_4$ crystal. As in the case of magnetite, more Raman modes are detected in nanosized samples [44] than in correspondent crystals. Furthermore, some observed phonons possess the characteristics of maghemite vibrational modes, which implies that a possible oxidation process occurs.

Another spinel ferrite is a mixture of two divalent metal ions such as $Zn_xMg_{1-x}Fe_2O_4$ in which Mg^{2+} and Zn^{2+} ratio may be varied. In this case, it is expected that zinc ions replace magnesium ions, in the range of x value from $x = 0$ to 1, and Zn^{2+} ions enter preferentially in tetrahedral positions while the Fe^{3+} ions

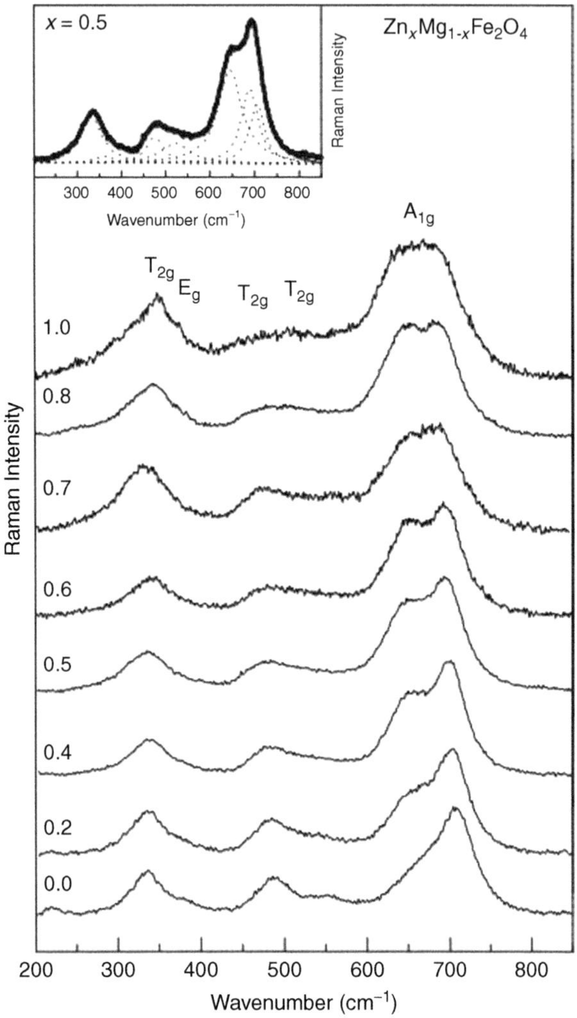

Fig. 14.11 The Raman spectra of $Zn_xMg_{1-x}Fe_2O_4$ ($0 \leq x \leq 1$) samples recorded at room temperature. The *inset* shows the Lorentzian-like assignment to the Raman modes for the sample $x = 0.5$ in accordance with the five active optical modes (Reprinted with permission from Ref. [48])

should be displaced from these sites for the octahedral sites. The effect of Zn^{+2} nonmagnetic ions on structural and magnetic properties of cubic $Zn_xMg_{1-x}Fe_2O_4$ particles (mean diameter of around 40 nm) was studied by means of XRD, Mössbauer spectroscopy, and Raman spectroscopy [48]. The Raman spectra of $Zn_xMg_{1-x}Fe_2O_4$ ($0 \leq x \leq 1$) particles are reproduced in the Fig. 14.11, in which there are five active modes with characteristic of the cubic spinel space group O_h^7 ($Fd3m$), whose fitting is displayed in the inset of Fig. 14.11. According to the fitted spectrum, the band in the 650–710 cm^{-1} region, related to the tetrahedral sublattice having A_{1g} symmetry, splits into three different energy values due to the different mass of the three ions (Zn^{2+}, Mg^{2+}, and Fe^{3+}). Among them, the lightest ion (Mg^{2+}) responds for the phonon peaking at 706 cm^{-1}(for $x = 0.5$) whereas the heaviest one (Zn^{2+}) is related to the 645 cm^{-1}, and the mode at 689 cm^{-1} was associated with Fe^{3+} ion in the tetrahedral sublattice. It was also observed that the intensity of the A_{1g} (Mg) mode decreases as the Mg content decreases, while an opposite behavior was observed for A_{1g} (Zn) mode. The results indicated that Mg, Fe, and Zn ions are present in both sites of the spinel structure.

5.4.2 Phase Transition

The chemical and structural stability of both citrate-coated magnetite and cobalt ferrite NPs, subjected to a high intensity laser field, was investigated by micro-Raman spectroscopy [49]. The room temperature Raman spectra were recorded under a 514-nm Argon ion laser line excitation for the intensity range of 0.7–70 mW. The best Lorentzian-function fitting of the 0.7 mW spectra revealed the presence of five structures at 200, 315, 467, 624, and 688 cm^{-1}; and 195, 360, 505, 661, and 715 cm^{-1} for citrate-coated $CoFe_2O_4$, and citrate-coated Fe_3O_4 samples, respectively. In contrast to the $CoFe_2O_4$-based sample, the Fe_3O_4-based sample presents stronger laser excitation intensity dependence. Besides, its peak positions at 0.7 mW are in agreement with the data of bulk maghemite, indicating that the stoichiometry of nominal Fe_3O_4-based sample is closely related to the maghemite stoichiometry. Also, the nominal $CoFe_2O_4$-based sample presents some similar peaks to that characteristic of maghemite, as reported in ref. [49]. In addition, when the intensity of excitation laser was increased by two orders of magnitude, the citrate-coated $CoFe_2O_4$-based sample presents high chemical and structural stability, while the citrate-coated Fe_3O_4-based sample that possesses a similar Raman spectrum with maghemite changes significantly due to a phase transition to hematite.

Raman spectroscopy has been applied to investigate the structural stability of cobalt ferrite–based nanosized sample (diameter of 8.6 nm) after performing steps of nanoparticle dispersion to prepare MFs. Nanomaterial samples were investigated as coprecipitated, after passivation, and peptized at 0.25 and 0.75 mol/L [50]. Among them, the Raman spectra of the samples peptized at 0.25 and 0.75 mol/L are reproduced in Fig. 14.12a, b, respectively. The Lorentzian-function fitting showed that Raman spectrum of the former sample is composed of five structures at 220, 310, 467, 624, and 690 cm^{-1}, but the latter is built up of seven lines peaking at 190, 300, 340, 475, 516, 610, and 680 cm^{-1}. These findings showed that samples peptized at different conditions might show different responses to laser annealing process. For instance, the cobalt ferrite nanoparticles peptized at 0.25 mol/L to prepare colloidal samples have higher structural stability than the one peptized at 0.75 mol/L.

The physical mechanisms associated with the phase transition of γ-Fe_2O_3 nanoparticles induced by laser irradiation have been studied by El Mendili et al., by performing in situ micro-Raman measurements of both as-prepared γ-Fe_2O_3 nanoparticles (about 4 nm in size) as a function of laser intensity and annealed samples [51]. Raman spectrum for the as-prepared nanoparticles showed the presence of three broad peaks at 350, 500, and 720 cm^{-1} and magnon mode in the neighborhood of 1,300 cm^{-1}. Such a Raman spectrum obeys the symmetry rules established for the inverse spinel structure with a tetragonal sublattice distortion,

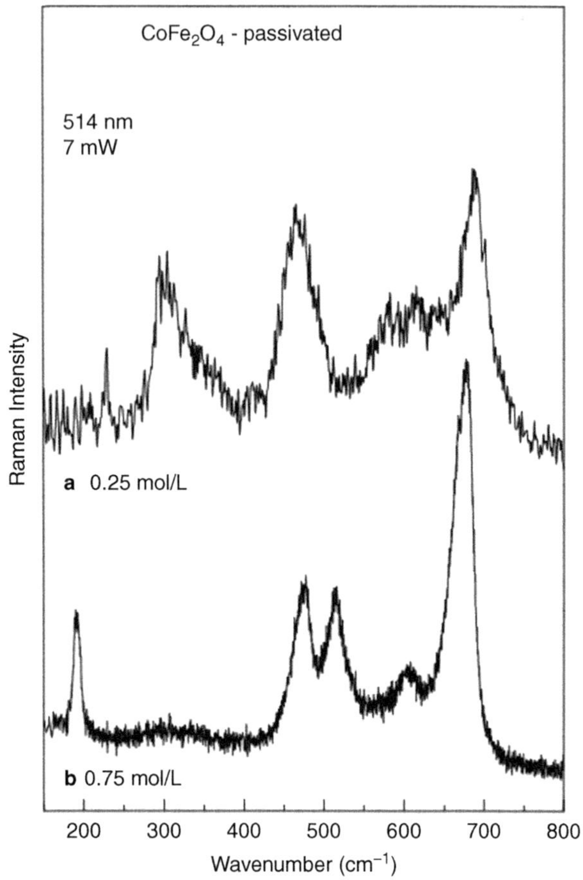

Fig. 14.12 Raman spectra of $CoFe_2O_4$ samples (**a**) peptized at 0.25mol/L perchloric acid) and (**b**) peptized at 0.75mol/L perchloric acid. Reprinted with permission from Ref. [50]. Copyright Elsevier B.V. (2004)

and the lack of resolution of the spectrum suggests a rather poor degree of crystallinity of the small maghemite nanoparticles. Moreover, the phase transition from maghemite into hematite is caused by local heating due to laser irradiation with an increase of grain size of nanoparticles.

5.4.3 Magnetite Oxidation Process

Raman chemical imaging can be employed to access the homogeneity and the structural stability in terms of oxidation rates, onset of hematite, and organic contamination of as-precipitate and oxidized iron oxide samples. Oxidation-related Raman features have been established by comparative study of bulk oxides and nanoparticles attained in two different oxidation states, suggesting that the solid nanophase synthesized had a mixed magnetite-maghemite composition [52].

A detailed investigation about the chemical and structural composition of maghemite (diameters in the range of 10–20 nm) based MF samples, prepared through the oxidation of magnetite nanoparticle, the latter synthesized by the coprecipitation route, was reported [53]. The precipitate containing the

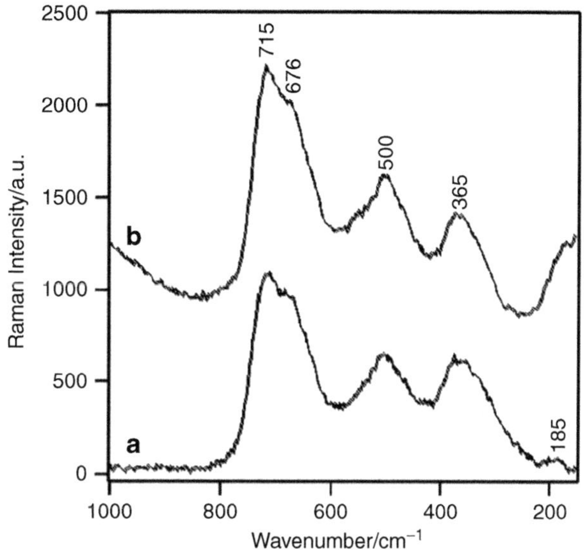

Fig. 14.13 Raman spectra of γ-Fe_2O_3 MF in the solid (**a**) and in solution (**b**) phases (Reprinted with permission from Ref. [53]. Copyright Elsevier B.V. (2007))

nanoparticles was submitted to the treatment as described in the literature [54], and dispersed as MF. Raman spectra of the maghemite-based MF in the solid phase (powder) and dispersed as MF show characteristic Raman features of maghemite (see Fig. 14.13). In addition, surface Raman spectroscopy (SERS) and cyclic voltammetry was employed to sample characterization, indicating that the prepared maghemite nanoparticles dispersed as MF do not contain magnetite in their composition. Moreover, nanoparticle is composed of a nuclei of γ-Fe_2O_3, characterized by a broad SERS features near 700 (composed by a peak at 662 and a shoulder at ~710 cm^{-1}), 500, and 365 cm^{-1}; a surface layer containing a non-stoichiometric oxy-hydroxide characterized by a SERS at 712 cm^{-1}, and also δ-FeOOH, characterized by two broad features peaking at 663, and 410–420 cm^{-1}, and may also contain $Fe(OH)_3$.

Two different routes of oxidation of nanosized magnetite using progressive oxidative stress producing different native iron oxides, which were surface-functionalized using oleic acid, suspended in insulating mineral oil and evaluated in terms of their colloidal stability [55]. Raman spectra attained as a function of the laser excitation intensity (0.5–16.1 mW, using a 514 Argon laser line) for all uncoated and OA-coated samples show that at lower laser excitation intensities, the Raman spectra present the characteristic bands of maghemite. However, the onset of the hematite Raman peaks at the laser excitation intensity of 3.2 mW in all uncoated samples, indicating that the different oxidation treatments performed did not affect the maghemite-uncoated samples' response under laser annealing. However, Raman spectra as a function of laser intensity for nanosized maghemite surface coated with OA present different features. Figure 14.14 displays Raman features of OA-grafted sample, exhibiting the presence of the iron oxide core, OA adsorbed layer, and organic fragments. This spectrum was attained at 0.5 mW laser

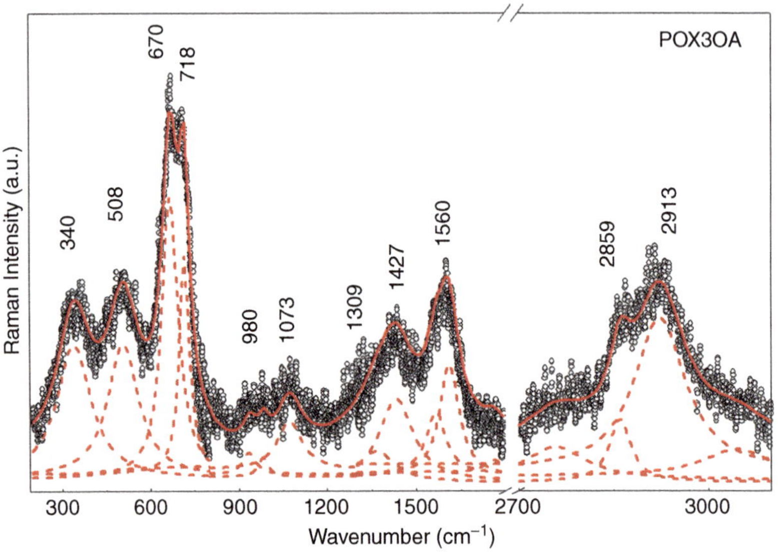

Fig. 14.14 Room-temperature Raman spectrum of POX3OA (Reprinted with permission from Ref. [55]. Copyright American Chemical Society (2010))

excitation intensity, avoiding any possible sample degradation due to laser annealing that revealed the presence of structures at 340, 508, 670, 718, and 1,427 cm^{-1}. Furthermore, the Raman bands observed at 980 and 1,309 cm^{-1} were attributed to $-CH_2-$ modes associated to organic fragments of OA species whereas Raman bands peaking at 1,073 and 1,560 cm^{-1} may suggest disordered carbon and/or a mixture of ligands at the nanoparticle surface. In addition, the Raman modes observed at 2,859 and 2,913 cm^{-1} can be attributed to symmetric and asymmetric $-CH_2-$ stretching modes, as observed in the Raman spectrum of pure OA.

Raman spectra of the OA-coated samples POX1OA and POX3OA with similar total mass loss, but different decomposition profiles, under different laser excitation intensities, are reproduced in Fig. 14.15a and b [55]. It is observed that at 0.5 mW laser excitation intensity the Raman features of these two samples are quite similar, revealing the characteristic bands of maghemite in the range of lower wavenumber (150–1,400 cm^{-1}), and bands related to the OA-coating in the higher (2,800–3,000 cm^{-1}) spectral range. However, the Raman spectrum of sample POX1OA recorded under 1.6 mW laser excitation intensity (see Fig. 14.15a) reveals new features at 222 and 287 cm^{-1}, which are characteristic Raman modes of hematite (α-Fe_2O_3), demonstrating the coexistence of maghemite and hematite phases due to laser annealing. It is also noted that the intensity of the Raman band around 1,600 cm^{-1}, which is attributed to disordered carbon, increases. However, the characteristic symmetric and asymmetric $-CH_2-$ stretching modes peaking at 2,859 and 2,913 cm^{-1} disappear from the spectrum of sample POX1OA.

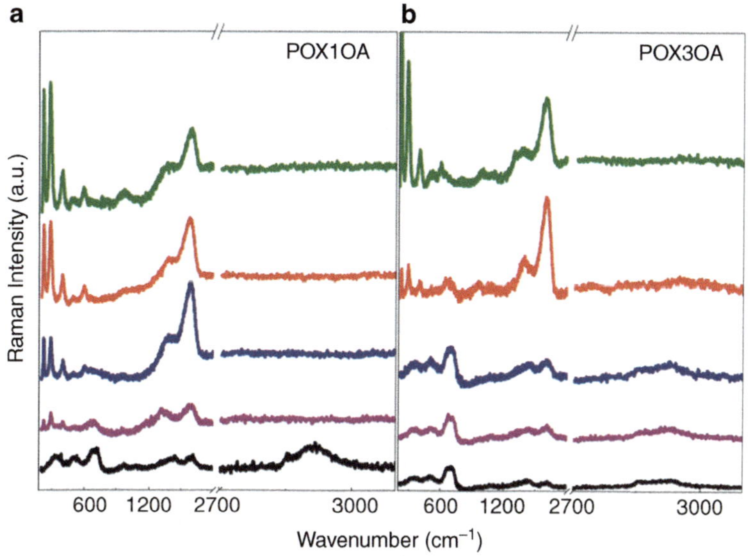

Fig. 14.15 Room-temperature Raman spectra of samples (**a**) POX1OA and (**b**) POX3OA, under the laser excitation intensity of 0.5, 1.6, 3.2, 6.4, and 16.1 mW (from *bottom* to *top*) (Reprinted with permission from Ref. [55]. Copyright American Chemical Society (2010))

Differently, the Raman spectrum of sample POX3OA (Fig. 14.15b) show the onset of the hematite phase only at 6.4 mW laser excitation intensity. In addition, under 6.4 mW laser excitation intensity, the Raman spectrum of sample POX3OA still exhibits the characteristic OA-coating bands ($-CH_2-$ stretching modes) in the higher (2,800–3,000 cm^{-1}) spectral range while the disordered carbon band intensity (around 1,600 cm^{-1}) increases. These findings can be discussed on the basis of the amount of chemisorbed OA species at the nanoparticle surface; once sample POX1OA presents lower OA grafting coefficient (1.2 nm^{-2}) compared to sample POX3OA (2.6 nm^{-2}) the optically induced phase change from maghemite to hematite occurs at a lower laser excitation intensity. As far as the optically induced phase change threshold is concerned, the same trend in other studied samples was observed. These findings suggested that the nanoparticle structural stability against laser irradiation depends on the chemisorbed grafting coefficient: the higher the chemisorbed grafting coefficient the higher the phase stability against laser annealing [55].

5.5 Passivation Process of Cobalt Ferrite Nanoparticles

The effect of the low pH medium on the core chemical stability and aging of SPIO particles suspended as ionic MF samples was investigated in such way that the surface passivation's effect of the as-precipitate sample was addressed following

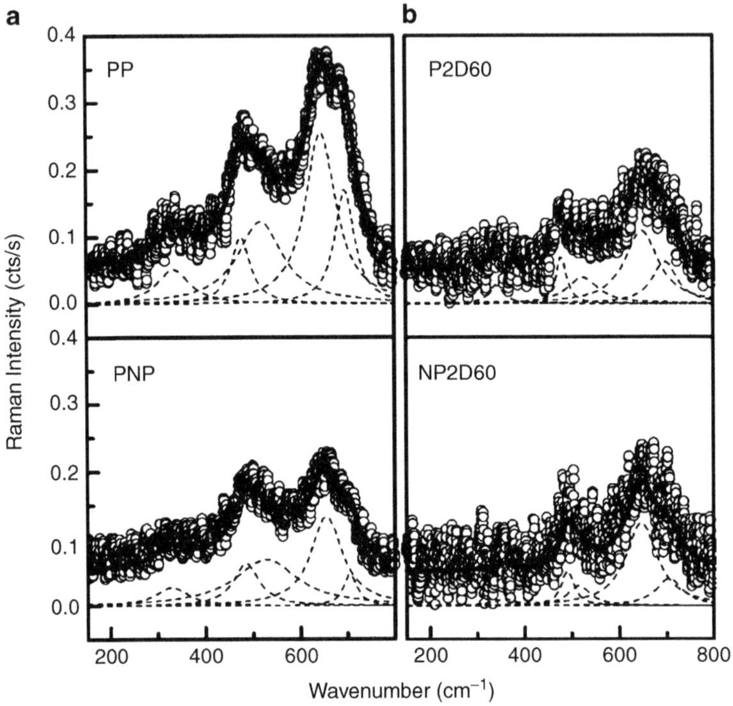

Fig. 14.16 Raman spectra recorded for (**a**) PP and PNP samples and (**b**) P2D60 and NP2D60 samples. The data were recorded at 0.2 mW laser intensity. For comparison, all spectra were in the same scale (Reprinted with permission from Ref. [56]. Copyright American Chemical Society (2007))

the sequence of the preparation route of several low pH ionic MF samples based on $CoFe_2O_4$ nanoparticles (8 nm average diameter) [56]. Figure 14.16 presents the Raman spectra of both passivated (PP) and non-passivated (PNP) samples. Notice that the Raman features show that the crystallinity of sample PP is better than PNP. Furthermore, the Raman intensity of the PP sample is about two times larger than that of the PNP sample. This unexpected result can be understood by studying the Raman spectra as a function of laser intensity. In addition, XRD and Raman measurements of PNP and PP samples annealed at 100°C, 200°C, 350°C, 500°C, and 700°C during 30 min [57] showed that the Raman intensity associated to the PP sample is stronger than that of the PNP sample, up to a laser excitation intensity of 5 mW. Above this excitation intensity value, new Raman peaks emerge for the passivated sample, indicating the onset of the hematite phase induced by the laser annealing. Hence, passivation induces the formation of an iron-rich shell around the passivated cobalt ferrite nanoparticle. Furthermore, the changes observed in the Raman spectra (sharper and intense Raman peaks) and the increase of the lattice parameter of the PP samples suggested that the surface passivation process may favor exchanges of Co for Fe atoms, proving that the surface passivation process accomplished modification of the cobalt ferrite nanoparticle core as well.

Figure 14.16 (right panel) shows the effect of the aging process of the magnetic fluid samples obtained with the passivated and non-passivated nanoparticles on their Raman features. MF Stock dispersions were aged for 60 days, and dialyzed to produce the powder samples P2D60 and NP2D60, respectively. As shown in Fig. 14.16 (right panel), the Raman spectra of samples P2D60 and NP2D60 are quite similar, in spite of the observed reduction of the crystalline quality of sample P2D60 to a greater extent in comparison with the sample NP2D60. This underlying physics can be understood in the following way: In the nanoparticle aging process, Co ion dissolution rate is higher than Fe ion dissolution rate. As a result, during the samples' aging the lattice vacancies were accumulated, which results in poorer crystalline quality of both P2D60 and NP2D60, compared to the as-prepared samples (left panel). Furthermore, the Raman spectra of samples P2D60 and NP2D60 recorded under higher laser intensities (above 3.0 mW) show characteristic peaks of hematite. Hence, the stoichiometry shift ($[Fe]/[Co] > 2$) due to the aging process may favor the material's phase change due to the absence of Co ions in the stable crystal structure of the stoichiometric cobalt ferrite nanoparticles.

5.6 Raman Spectroscopy of Nanocomposites

Raman spectroscopy is also applied to study the interaction of $CoFe_2O_4$ nanoparticles with the host matrix in $CoFe_2O_4$-SiO_2 nanocomposites prepared using sol-gel method [58]. NP concentration and thermal treatment dependent Raman spectra of these samples show that the Si–O–Fe (Co) bonds are observed when high concentration of the $CoFe_2O_4$ NPs is introduced into the sol-gel mixture, or when the samples containing low concentration of $CoFe_2O_4$ NPs are annealed at high temperatures.

Raman spectroscopy has been applied to probe the presence of iron oxide nanomaterials in electrospray-deposited iron oxide nanoparticle films [59]. Raman trends recorded under a 532 nm laser line excitation show the characteristic (350, 500, and 700 cm^{-1}) of maghemite and no significant difference between the spectrum of maghemite powder (particle size <50 nm), used to film preparation, and that of the electrospray-deposited film without annealing. Thus, deposition did not cause any physical transformation or degradation of the particles. The onset of the hematite Raman peaks occur for film annealed at 600°C, and phase transformation from maghemite to hematite takes place at an annealing temperature of 700°C. In addition, the annealing temperature strongly impacts on the iron oxide NP grain size and consequently the Raman spectrum. For instance, both position and width of the 660 cm^{-1} modes are sensitive to the annealed particle size of the hematite. In addition, the spectral separation between the 225 and 660 cm^{-1} Raman modes is a useful parameter for future calibration of the crystallite size in the electrospray-deposited, post-annealed hematite films [59].

5.7 Raman Spectroscopy of Magnetic Colloidal Suspensions

Magnetic colloidal suspensions, known as ferrofluid or magnetic fluid (MF), combine colloidal with magnetic properties, which gives rise to unusual characteristics and wide variability of applications. The stability of the colloid, including the chemical and structural stability of the nanoparticle core, is an extremely important parameter in the preparation, storage, and applications of magnetic nanoparticles in colloidal form. The resulting properties of magnetic colloids depend on the specific characteristics of the SPIO-based material and on the corresponding surface functionalization. Furthermore, the physicochemical properties of the molecular layer adsorbed onto the nanoparticle surface play a key role in the determination of the MF colloidal stability [55]. In order to overcome Van der Waals and magnetic dipole attractions, the steric and electrostatic repulsion mechanisms have been used to improve colloidal stability while preparing magnetic fluids [60]. Evidently, electrostatic plus steric repulsion may work together against nanoparticle clustering in a new class of water-based MF samples. The ligand's ionizable functional groups, which are not involved in the complexation of the surface particle acidic or basic sites, face the outside solvent and add an extra component to the colloidal stability of the magnetic colloid through electrostatic charges. These surface-functionalized nanoparticles can be dispersed in saline media, originating stable water-based samples, appropriate to be used in biological and medical applications. The long-term stability of surfactant-stabilized MF samples depends on the surface grafting coefficient and how efficiently the adsorbed molecules are attached to the nanoparticle surface.

5.7.1 Interface Between Nanoparticle Surface and Carrier Liquid

Monitoring the molecular surface layer chemisorbed at the SPIO particle suspended in magnetic colloids is a key issue in the engineering of magnetic nanoparticles for different applications. The usefulness of the Raman spectroscopy in the study of particles suspended in MF samples was first demonstrated by Morais et al. in the analysis of the microscopic structures of the first molecular layer which dresses the nanoparticle surface, i.e., the interface nanoparticle surface-carrier liquid [61]. The approach they used to investigate the properties of the molecular layer chemisorbed at the surface of uncoated and surfactant-stabilized (with different species) ferrofluid samples [61–63] involved both the OH-bending and the OH-stretching Raman modes associated to the hydroxyl group which is chemisorbed at the nanoparticle's surface. Liquid water shows five OH-stretching modes (v^s,v^a, v_d^s,v_d^a, and v_b), originating from various amounts of hydrogen bonding, in the range of 3,000–4,000 cm^{-1}, which serves as the model picture showed in Fig. 14.17. The two OH-stretching Raman modes in the higher end frequency of the spectra (v^s and v^a) describe non-hydrogen-bonded (OH–) modes, where $v^s$and v^adenote symmetric and antisymmetric OH-stretching modes, respectively. In contrast, the three Raman components (v_d^s,v_d^a, and v_b) at lower frequencies describe hydrogen-bonded modes. v_d^s and v_d^a refer to symmetrically OH-stretching modes, whereas v_b refers to the antisymmetric OH-stretching mode [64, 65]. In order to discuss the quenching of

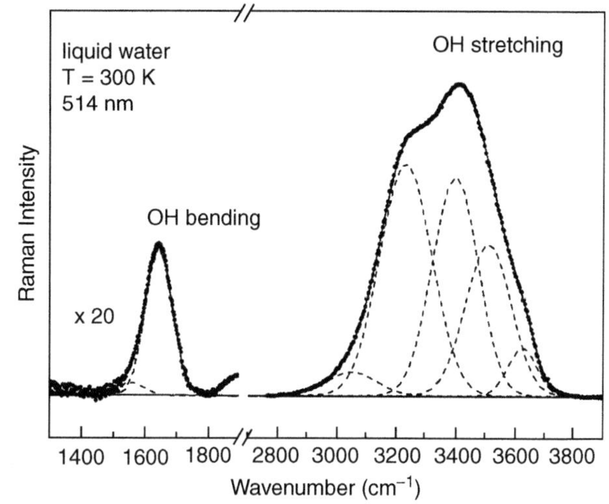

Fig. 14.17 Raman spectrum of liquid water in the OH-bending region (*left-hand side*) and in the OH-stretching region (*right-hand side*). Symbols are experimental data, *solid lines* represent the best numerical fit, and *dashed lines* represent the Gaussian-shaped Raman modes (Reprinted with permission from Ref. [63]. Copyright Elsevier B.V. (2001))

Raman lines and the Raman shift observed in the Raman modes associated to the hydrogen bonding strength involving the chemisorbed OH-group, a comparison of the Raman features obtained from liquid water with that from uncoated (see Fig. 14.18) and functionalized nanoparticles (see Fig. 14.19) dispersed as ferrofluids, in the stretching region, was performed. The suppression of all symmetric OH-stretching phonons from the UFMs (see Fig. 14.18) is an obvious consequence of the vibrational quenching effect due to the replacement of a hydrogen atom from water by the magnetic nanoparticle surface. Based on this model picture, the hydroxyl Raman modes correspond to the OH-group firmly bonded to the NP surface. Furthermore, the hydrogen bonding Raman modes are due to the hydrogen bonding established between the surface OH-group (Me^{n+}) and the first water molecular layer from the medium.

In addition, as observed in the reproduced Raman trends obtained for coating MF samples (see Fig. 14.19), in the WBO-CMF, WBD-CMF, and HBO-CMF spectra, only two modes (ν_b and ν^a) at the higher energy side of the Raman spectra were fitted. The presence of only two vibrational modes in these three samples, instead of five (Fig. 14.17) or three Raman modes observed in the UMF's, can be partly explained provided that vibrational quenching effect is assumed. In fact, the coating layer, which is chemisorbed at the NP surface, inhibits water molecules from the medium to bind to the OH-group at the NP surface and suppresses the ν_d^a Raman mode from the CMF's. Therefore, the two Raman modes left behind three of the CMF's (WBO-CMF, WBD-CMF, and HBO-CMF) are the antisymmetric non-hydrogen bonding (ν^a) plus the antisymmetric hydrogen bonding (ν_b). The quenching of an extra Raman mode (ν_b) from the HBS-CMF sample, however, is mainly attributed to the lack of hydrogen bonding between the residual OH-group chemisorbed at the NP surface and either the long carbon chain from the dodecanoic acid or from the nonpolar solvent. Therefore, the only Raman mode

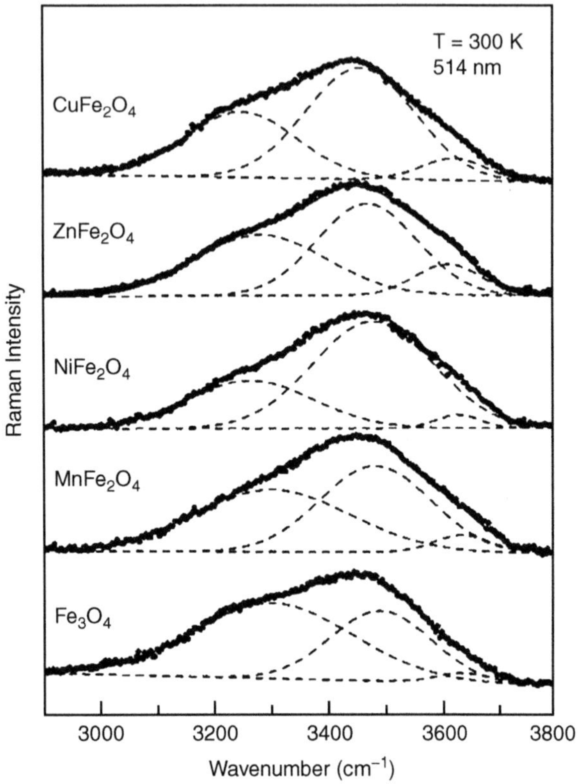

Fig. 14.18 Raman spectra of the UMF's in the OH-stretching region. Symbols are experimental data, *solid lines* represent the best numerical fit, and *dashed lines* represent the Gaussian-shaped Raman modes (Reprinted with permission from Ref. [63]. Copyright Elsevier B.V. (2001))

left behind in HBS-MF sample is the asymmetric non-hydrogen bonding one (ν^a). Note that the possibility of a hydrogen bonding between the residual OH-group chemisorbed at the NP surface, and the OH-group from the ethoxylated polyalcohol (or *N*-oleoylsarcosine), accounts for the presence of the ν_b Raman mode in the WBO-CMF, WBD-CMF, and HBO-CMF. A similar approach has been used to study monolayer and bilayer magnetite-based magnetoliposomes (ML) by investigating the microscopic details of the interaction between the chemically active nanoparticle surface and the polar head group of the inner layer phospholipids [66]. In this case, surface-functionalized SPIO particles can also be hosted into the liposome cavity creating ML, being adequate to be used in biological and medical applications [67]. Similar to surfactant-stabilized MF samples already discussed, the quenching of the symmetric OH-stretching modes from the Raman spectra of the ML samples indicated that chemisorbed OH-groups at the magnetite surface dominated the light scattering process. The reduction of the hydrogen bonding strength observed in the ML samples with respect to the liquid water indicates a weaker interaction between the OH-group at the nanoparticle surface and the polar head groups from the phospholipid layer. Furthermore, such interaction is weaker in the bilayer ML than in the monolayer ML sample.

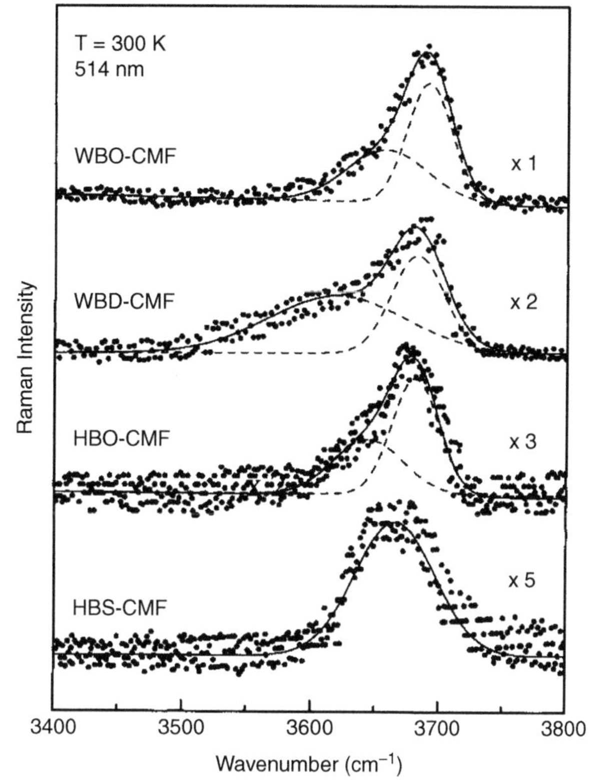

Fig. 14.19 Raman spectra of the CMFs in the OH-stretching region. Symbols are experimental data, *solid lines* represent the best numerical fit, and *dashed lines* represent the Gaussian-shaped Raman modes (Reprinted with permission from Ref. [63]. Copyright Elsevier B.V. (2001))

5.7.2 Raman Spectroscopy of Magnetic Colloids in the Presence of Magnetic Field

Raman spectroscopy has been also applied to study the behavior of water-based MF samples containing surface-functionalized nanosized particles with a bilayer-shell of dodecanoic acid plus marlipal, in the presence of magnetic field [68]. Analysis of Raman trends as a function of temperature, magnetic field intensity, and nanoparticle concentration shows that the overall intensity of the Raman signal decreases with increasing field intensity and with increasing particle concentration. In addition, at higher fields, a broad peak appears in the spectra at about 4,400 cm^{-1}, due to a magnetic excitation. All these effects could not be observed at lower particle concentrations. In the presence of an external field, there is a local increase in the concentration of magnetic particles due to the formation of an ordered structure induced by an enhancement of the magnetic dipole–dipole interaction among particles.

5.7.3 Investigation of the Iron Oxide Core Properties

Figure 14.20 shows the Raman spectra of colloidal dispersions of Fe_3O_4/SO, Fe_3O_4/CO, and $CoFe_2O_4$/SO nanoparticles in cyclohexane whose surface functionalized with carboxylic acids derived from soybean (SO) oil, or carboxylic acids from

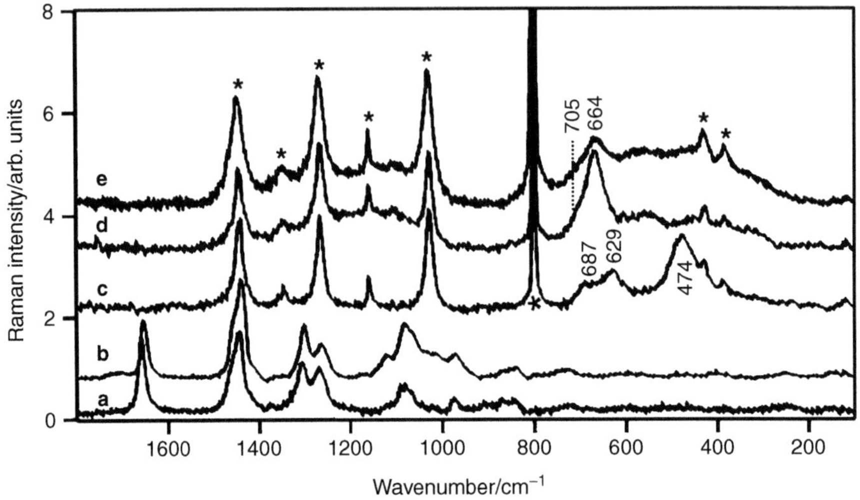

Fig. 14.20 Raman spectra of soybean (a), and castor oil (b) fatty acids and $CoFe_2O_4$/SO (c), Fe_3O_4/CO (d), and Fe_3O_4/SO magnetic fluids in cyclohexane. *cyclohexane peaks (Reprinted with permission from Ref [44]. Copyright American Chemical Society (2009))

castor oil (CO), excited at 632.8 nm [44]. In addition, Raman spectra of surfactants are also displayed in Fig. 14.20. The peaks with the same wavenumbers and relative intensities as displayed in the Raman spectra of solid samples also emerged in Raman spectra of their correspondent MFs.

A study of the effect of the laser intensity (0.9–9.0 mW) on the structural stability and chemical composition of magnetite dispersed as MF, compared with the Raman features of solid phase (powder) of the fluids, was performed using Raman spectroscopy [69]. Magnetite nanoparticles were obtained by the coprecipitation technique and dispersed in two types of beta-cyclodextrin suspensions. Powder samples undergo phase transition from magnetite to hematite at laser power of 1.95 mW, while the same nanoparticles dispersed as ferrofluids undergo transformation at 9 mW. Furthermore, Raman spectra revealed that the main phase of the magnetic core in the fluids is magnetite displaying the characteristic A_{1g} at 670 cm^{-1}. Also, a second phase is present at the nanoparticle's surface with Raman spectroscopy unveiling maghemite-like and small fractions of goethite-like structures.

In brief, the investigation of ferrofluids through Raman spectroscopy permits to access the physical and chemical properties of both solid and liquid phases. Comparison between the Raman spectra obtained from liquid water and the Raman spectra obtained from the uncoated and coated nanoparticles dispersed as ferrofluid provides information about the interface nanoparticle surface-carrier liquid [63]. The onset of the hematite phase in core magnetite dispersed as magnetic fluids was followed by Raman spectroscopy, and the results showed that the laser intensity at which the phase-transition takes place was higher for nanoparticles in the colloid than that for the same core as powder samples [69].

6 Conclusions and Future Perspectives

The Raman spectroscopy has been widely used in the characterization of SPIO-based materials in the last decade. It has been demonstrated that Raman spectroscopy helps to refine the currently existing protocols as well as to develop new synthesis routes. Furthermore, the Raman spectroscopy has also contributed to the development of advanced instrumentations improving the characterization techniques, which are in high demand for the synthesis of materials for nanomedicine and spintronic applications. To engineer magnetic nanoparticles for different applications, we have discussed key physical and chemical properties of SPIO systems revealed by Raman spectroscopy measurement, including to monitor the molecular layer adsorbed at the SPIO surface, examine the chemical and structural properties of core, identify the phase, and verify phase transition in both solid and liquid phase samples. Certainly, Raman spectroscopy is an important tool for preparation of high-quality magnetic nanomaterials with tailored surface chemistry to be applied in the fast-growing field of nanomedicine. Moreover, SPIO-based nanomaterials provide unique platforms for the integration of various imaging modalities together with therapeutic functionalities.

References

1. Shi J, Gilder S, Babcok K, Awschalom DD (1966) Magnetic clusters in molecular beams, metals, and semiconductors. Science 271:937–941
2. Reynolds PJ (1993) On custer and clustering: From atoms to fractals. North-Holland, Amsterdam
3. Buzea C, Pacheco II, Robbie K (2007) Nanomaterials and nanoparticles: sources and toxicity. Biointerphases 2:Mr17–Mr71
4. Alisatos AP (1996) Semiconductor clusters, nanocrystals, and quantum dots. Science 271:933–937
5. Bean CP, Livingston JD (1959) Superparamagnetism. J Appl Phys 30:120S–129S
6. Batlle X, Labarta A (2002) Finite-size effects in fine particles, magnetic and transport properties. J Phys D: Appl Phys 35:R15–R42
7. Bedanta S, Kleemann W (2009) Supermagnetism. J Phys D: Appl Phys 42:013001
8. Guimarães AP (2009) Principles of nanomagnetism. Springer, Berlin
9. Pankhurst QA, Thanh NKT, Jones SK, Dobson J (2009) Progress in applications of magnetic nanoparticles in biomedicine. J Phys D: Appl Phys 42:224001
10. Hergt R, Dutz S, Müller R, Zeisberger M (2006) Magnetic particle hyperthermia, nanoparticle magnetism and materials development for cancer therapy. J Phys Condens Matter 18:S2919–S2934
11. Raman CV, Krishnan KS (1928) A new type of secondary radiation. Nature 121:501–502
12. Smith E, Dent G (2005) Modern Raman spectroscopy – a practical approach. Wiley, Chichester
13. Gardiner DJ, Graves PR (eds) (1989) Practical Raman spectroscopy. Springer, Berlin
14. Gouadec G, Colomban P (2007) Raman spectroscopy of nanomaterials: how spectra relate to disorder, particle size and mechanical properties. Prog Cryst Growth Charact Mater 53:1–56
15. Crangle J (1977) The magnetic properties of solids. Edward Arnold, London
16. Kittel C (1946) Theory of the structure of ferromagnetic domains in films and small particles. Phys Rev 70:965–971

17. Frei EH, Shtrikman S, Treves D (1957) Critical size and nucleation field of ideal ferromagnetic particles. Phys Rev 106:446–455
18. Néel ML (1949) Théorie du trainage magnétique des ferromagnétiques engrains fins avec applications aux terres cuites. Ann Geophys 5:99
19. Nunes WC, Folly WSD, Sinnecker JP, Novak MA (2004) Temperature dependence of the coercive field in single-domain particle systems. Phys Rev B 70:014419
20. Ferguson GA Jr, Hass M (1958) Magnetic structure and vacancy distribution in γ-Fe_2O_3 by neutron diffraction. Phys Rev 112:1130–1131
21. Krupicka S, Novák P (1982) Oxide spinels. In: Wohlfarth EP (ed) Ferromagnetic materials. North-Holland, Amsterdam
22. Blums E, Cebers A, Maiorov MM (1985) Magnetic fluids. Walter de Gruyter, Berlin
23. Jeong BU, Teng X, Wang Y, Yang H, Xia Y (2007) Superparamagnetic colloids: controlled synthesis and niche applications. Adv Mater 19:33–60
24. Lu A-H, Salabas EL, Schüth F (2007) Magnetic nanoparticles, synthesis, protection, functionalization, and application. Angew Chem Int Ed 46:1222–1244
25. Faraci G, Gibilisco S, Russo P, Pennisi AR, La Rosa S (2006) Modified Raman confinement model for Si nanocrystals. Phys Rev B 73:033307
26. Fujii M, Kanzawa Y, Hayashi S, Yamamoto K (1996) Raman scattering from acoustic phonons confined in Si nanocrystals. Phys Rev B 54:R8373–R8376
27. Fujii M, Nagareda T, Hayashi S, Yammanoto K (1991) Low-frequency Raman scattering from small silver particles embedded in SiO_2 thin films. Phys Rev B 44:6243–6248
28. Fujii M, Nagareda T, Hayashi S, Yamamoto K (1992) Raman scattering from acoustic phonons confined in microcrystals: small gold and silver particles embedded in SiO_2 thin films. J Phys Soc Jpn 61:754–755
29. Lamb H (1881) On the vibrations of an elastic sphere. Proc London Math Soc 13:189–212
30. Richter H, Wang ZP, Ley L (1981) The one phonon Raman spectrum in microcrystalline silicon. Solid State Commun 39:625–629
31. Campbell IH, Fauchet PM (1986) The effects of microcrystal size and shape on the one phonon Raman spectra of crystalline semiconductors. Solid State Commun 58:739–741
32. White WB, De Angelis BA (1967) Interpretation of the vibrational spectra of spinels. Spectrochim Acta 23A:985–995
33. Verble JL (1974) Temperature-dependent light-scattering studies of the Verwey transition and electronic disorder in magnetite. Phys Rev B 9:5236–5248
34. Gasparov LV, Tanner DB, Romero DB, Berger H, Margaritondo G, Forró L (2000) Infrared and Raman studies of the Verwey transition in magnetite. Phys Rev B 62:7939–7944
35. Shebanova ON, Lazor P (2003) Raman study PF magnetite (Fe_3O_4), laser-induced thermal effects and oxidation. J Raman Spectrosc 34:845–852
36. de Faria DLA, Silva SV, de Oliveira MT (1997) Raman microspectroscopy of some iron oxides. J Raman Spectrosc 28:873–878
37. Kreisel J, Lucazeau G, Vincent HJ (1998) Raman spectra and vibrational analys of $BaFe_{12}O_{19}$ hexagonal ferrite. J Solid State Chem 137:127–137
38. Chamritski I, Burns G (2005) Infrared- and Raman-active phonons of magnetite, mghemite, and hematite, a computer simulation and spectroscopic study. J Phys Chem B 109:4965–4968
39. Shebanova ON, Lazor P (2003) Raman spectroscopic study of magnetite ($FeFe_2O_4$), a new assignment for the vibrational spectrum. J Solid State Chem 174:424–430
40. Wang WH, Ren X (2006) Flux growth of high-quality $CoFe_2O_4$ single crystals and their characterization. J Crystal Growth 289:605–608
41. White WB (2005) The structure of particles and the structure of crystals, information from vibrational spectroscopy. J Ceramic Process Res 6:1–9
42. Thünemann AF, Schütt D, Kaufner L, Pison U, Möhwald H (2006) Maghemite nanoparticles protectively coated with poly(ethylene imine) and poly(ethylene oxide)-*block*-poly(glutamic acid). Langmuir 22:2351–2357

43. Hu L, Hach D, Chaumont D, Brachais C-H, Couvercelle J-P, Percheron A (2009) Comparison of various methods of grafting of modified-PEG onto maghemite nanoparticles in aqueous medium including synthesis by microwave refluxing. J Sol-Gel Sci Technol 49:277–284
44. Jacintho GVM, Brolo AG, Corio P, Suarez PAZ, Rubim JC (2009) Structural investigation of MFe_2O_4 (M = Fe, Co) magnetic fluids. J Phys Chem C 113:7684–7691
45. Shemer G, Tirosh E, Livneh T, Markovich G (2007) Tuning a colloidal synthesis to control Co^{2+} doping in ferrite nanocrystals. J Phys Chem C 111:14334–14338
46. Limaye MV, Singh SB, Date SK, Kothari D, Raghavendra Reddy V, Gupta A, Sathe V, Choudhary RJ, Kulkarni SK (2009) High coercivity of oleic acid capped $CoFe_2O_4$ nanoparticles at room temperature. J Phys Chem C 113:9070–9076
47. Ayyappan S, Mahadevan S, Chandramohan P, Srinivasan MP, Philip J, Raj B (2010) Influence of Co^{2+} ion concentration on the size, magnetic properties, and purity of $CoFe_2O_4$ spinel ferrite nanoparticles. J Phys Chem C 114:6334–6341
48. da Silva SW, Nakagomi F, Silva MS, Franco A Jr, Garg VK, Oliveira AC, Morais PC (2010) Effect of the Zn content in the structural and magnetic properties of $ZnxMg_1-xFe_2O_4$ mixed ferrites monitored by Raman and Mösbauer spetroscopies. J Appl Phys 107:09B503-1–09B503-3
49. da Silva SW, Melo TFO, Soler MAG, Lima ECD, da Silva MF, Morais PC (2003) Stability of citrate-coated magnetite and cobalt-ferrite nanoparticles under laser irradiation, a Raman spectroscopy investigation. IEEE Trans Magn 39:2645–2647
50. Soler MAG, Melo TFO, da Silva SW, Lima ECD, Pimenta ACM, Garg VK, Oliveira AC, Morais PC (2004) Structural stability study of cobalt ferrite-based nanoparticle using micro Raman spectroscopy. J Magn Magn Mater 272–276:2357–2358
51. El Mendili Y, Bardeau J-F, Randrianantoandro N, Gourbil A, Greneche J-M, Mercierb A-M, Grasset F (2010) New evidences of *in situ* laser irradiation effects on γ-Fe_2O_3 nanoparticles, a Raman spectroscopic study. J Raman Spectrosc 42(2):239–242. doi:10.1002/jrs.2762
52. Chourpa I, Douziech-Eyrolles D, Ngaboni-Okassa L, Fouquenet J-F, Cohen-Jonathan S, Soucé M, Marchais H, Dubois P (2005) Molecular composition of iron oxide nanoparticles, precursors for magnetic drug targeting, as characterized by confocal Raman microspectroscopy. Analyst 130:1395–1403
53. Jacintho GVM, Corio P, Rubim JC (2007) Surface-enhanced Raman spectra of magnetic nanoparticles adsorbed on a silver electrode. J Electroanal Chem 603:27–34
54. Tourinho FA, Franck R, Massart R (1990) Aqueous ferrofluids based on manganese and cobalt ferrites. J Mater Sci 25:3249–3254
55. Viali WR, Alcantara GB, Sartoratto PPC, Soler MAG, Mosiniewicz-Szablewska E, Andrzejewski B, Morais PC (2010) Investigation of the molecular surface coating on the stability of insulating magnetic oils. J Phys Chem C 114:179
56. Soler MAG, Lima ECD, da Silva SW, Melo TFO, Pimenta ACM, Sinnecker JP, Azevedo RB, Garg VK, Oliveira AC, Novak MA, Morais PC (2007) Aging investigation of cobalt ferrite nanoparticles in low pH magnetic fluid. Langmuir 23:9611–9617
57. Melo TFO, da Silva SW, Soler MAG, Lima ECD, Morais PC (2006) Investigation of surface passivation process on magnetic nanoparticles by Raman spectroscopy. Surf Sci 600:3642–3645
58. da Silva SW, Pedroza RC, Sartoratto PPC, Rezende DR, da Silva Neto AV, Soler MAG, Morais PC (2006) Raman spectroscopy of cobalt ferrite nanocomposite in silica matrix prepared by sol-gel method. J Non-Cryst Solids 352:1602–1606
59. Modesto Lopez LB, Pasteris JD, Biswas P (2009) Sensitivity of micro-Raman spectrum to crystallite size of electrospray-deposited and post-annealed films of iron-oxide nanoparticle suspensions. Appl Spectrosc 63:627–634
60. Rosensweig RE (1985) Ferrohydrodynamics. Cambridge University Press, Cambridge
61. Morais PC, da Silva SW, Soler MAG, Sousa MH, Tourinho FA (1999) Raman study of ionic water-based copper and zinc ferrite magnetic fluids. J Magn Magn Mater 201:105–109

62. Morais PC, da Silva SW, Soler MAG, Buske N (2000) Raman investigation of uncoated and coated magnetic fluids. J Phys Chem A 104:2894–2896
63. Morais PC, da Silva SW, Soler MAG, Buske N (2001) Raman spectroscopy in magnetic fluids. Biomol Eng 17:41–49
64. Scherer JR, Go MK, Kint S (1974) Raman spectra and structure of water from −10° to 90°. J Phys Chem 78:1304–1313
65. Carey DM, Korenowski GM (1998) Measurement of the Raman spectrum of liquid water. J Chem Phys 108:2669
66. Soler MAG, da Silva SW, Melo TFO, De Cuyper M, Morais PC (2002) Raman spectroscopy of magnetoliposomes. J Magn Magn Mater 252:415–417
67. Soenen SJH, Hodenius M, De Cuyper M (2009) Magnetoliposomes, versatile innovative nanocolloids for use in biotechnology and biomedicine. Nanomedicine 4:177–191
68. Weber JE, Goñi AR, Pusiol DJ, Thomsen C (2002) Raman spectroscopy on surfacted ferrofluids in a magnetic field. Phys Rev E 66:021407
69. Slavov L, Abrashev MV, Merodiiska T, Gelev Ch, Vandenberghe RE, Markova-Deneva I, Nedkov I (2010) Raman spectroscopy investigation of magnetite nanoparticles in ferrofluids. J Magn Magn Mat 322:1904–1911

Micro-Raman Spectroscopy of Nanostructures

15

Ramesh Kattumenu, Chang H. Lee, Valery N. Bliznyuk, and Srikanth Singamaneni

1 Description of the Topic

Micro-Raman spectroscopy (μRS) involves acquiring spatially resolved Raman spectra by combining the conventional Raman spectrometer with a microscopic tool, typically an optical microscope. This chapter introduces the basic methodology of micro-Raman spectroscopy and presents an overview of its application to organic and inorganic nanostructures using specific examples from literature.

2 Overview

Raman spectroscopy is a unique nondestructive tool for probing the structure and properties of a wide variety of organic and inorganic materials. Raman spectroscopy is based on the inelastic scattering of a monochromatic light that provides molecular fingerprint related to the vibrational, rotational, and other low frequency transitions of molecules [1, 2]. Micro-Raman spectroscopy (μRS), is also called Raman microscopy, involves acquiring spatially resolved Raman spectra by combining the conventional Raman spectrometer with a microscopic tool, typically an optical microscope [3]. With the introduction of commercial micro-Raman

R. Kattumenu • C.H. Lee • S. Singamaneni (✉)
Department of Mechanical Engineering and Materials Science, School of Engineering and Applied Sciences, Washington University, St. Louis, MO, USA

V.N. Bliznyuk
Materials Science and Engineering Department, College of Engineering and Applied Sciences, Western Michigan University, Kalamazoo, MI, USA

Challa S. S. R. Kumar (ed.), *Raman Spectroscopy for Nanomaterials Characterization*,
DOI 10.1007/978-3-642-20620-7_15, © Springer-Verlag Berlin Heidelberg 2012

systems, this technique has emerged as a powerful and handy analytical tool for the characterization of a wide variety of nanostructures over the last decade.

In this chapter, we present recent advances in the application of μRS in probing organic and inorganic nanostructures. We do not intend to provide a comprehensive literature review, instead highlight several key examples that demonstrate the breadth of the applications of μRS in characterizing nanostructured materials. Following a brief introduction to the technique, we present an overview of the instrumentation underlining the similarities and differences compared to conventional Raman spectroscopy. In the following section, we present the application of μRS to characterize a wide variety of nanomaterials which include polymeric, biological, inorganic and hybrid materials. The combination of μRS with surface-enhanced Raman scattering (SERS) that broadens the applications of μRS is also discussed. We conclude with a brief summary and outlook of the μRS applications to characterize a broader range of nanomaterials.

3 Introduction

μRS is a powerful nondestructive and noncontact method for characterizing organic and inorganic nanomaterials. Spectral information obtained using conventional Raman spectroscopy is spatially averaged over a large area (~1 mm^2), whereas μRS enables the collection of spectral information with submicron lateral and vertical resolution. With the use of a high magnification objective, the laser beam can be focused into a spot with an effective diameter (considering a Gaussian footprint of the beam) as small as ~200 nm, providing excellent resolution, limited only by the far-field diffraction limit. The unique combination of high chemical specificity, excellent lateral and vertical resolution, and nondestructive nature of μRS makes it a powerful analytical tool for the characterization of nanostructured materials.

4 Instrumentation and Experimental Methodology

4.1 Overview of the Instrumentation

A highly simplified schematic of micro-Raman spectrometer is shown in Fig. 15.1 [3]. The design is based on a confocal optical microscope equipped with an objective with a large numerical aperture (as high as 1.4) and a pinhole in the intermediate focal plane. An intense light source such as a laser with a narrow line width (~0.2 nm) is used to illuminate the surface under investigation. The light source can be built into the spectrometer or an external source can be used for illumination. Modern Raman spectrometers generally include holographic gratings for improved excitation light rejection, notch filter for reflected and Rayleigh scattered light rejection, and a liquid nitrogen or Peltier-cooled CCD to reduce variations in dark current at the detector. A small local area of the sample is probed,

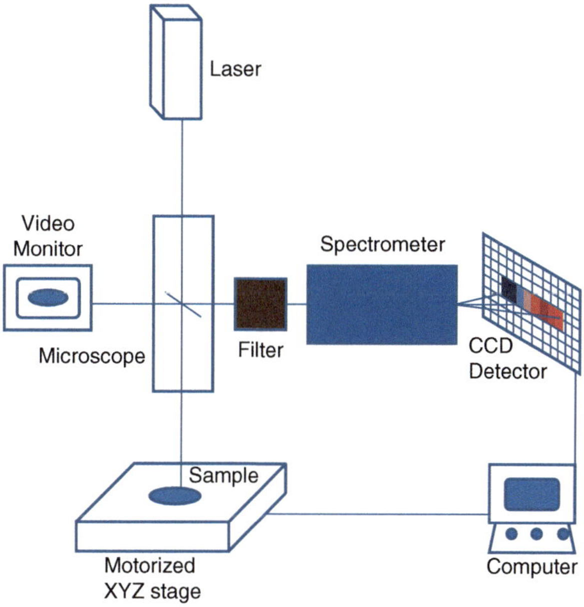

Fig. 15.1 Simplified schematic of a set up of micro-Raman spectrometer (Reprinted from [3])

where the CCD detects only photons coming from a narrow spatial domain of the material [4]. The filtered light is analyzed as a function of the probed position, where the signal is read by the detector and the intensity of each frequency is measured by an individual pixel on the array. The intensity of the Raman signal is a function of four factors: material properties (absorptivity, reflectivity, and the intrinsic strength of the Raman modes), source laser power, the width of the spectrometer admission slit and the width of the resolution slit. Based on the intensity and frequency of the Raman signal received, spatially resolved structure and properties of the material are identified.

Owing to the submicron spatial resolution achieved with μRS, precise areas of interest can be chosen to identify and analyze physical and chemical properties of the materials under investigation. A motorized X, Y and Z stage with a nominal resolution as small as ~10 nm is used for scanning the surface for lateral and depth information. Raman spectral mapping complements the visual image obtained with the microscope by providing information about the variation in chemical (or physical) properties of a heterogeneous surface.

Modern μRS systems are accompanied with powerful spectral acquisition and analysis software, which enables the creation of 1D (cross section), 2D, and 3D maps of various features from the 1D, 2D, or 3D array of spatially resolved Raman spectra. Various features that can be routinely mapped include intensity variations of specific peaks (by plotting the user-defined peak intensity or integrated area under the peak), intensity ratio of two different bands, peak position (by user-defined peak fitting routines such as Gaussian, Lorentzian), and peak widths. The obtained images can be further processed to highlight the spatial variations of the acquired spectra. For example, Boolean maps, which present a binary representation of the desired attribute (such as intensity, peak width range), are obtained by specifying the threshold values of the parameter. The Boolean maps are extremely helpful to visualize the phase distribution of a heterogeneous region as highlighted in some examples in Sect. 5.

4.2 Resolution Criteria

Dictated by the diffraction limit of light, the lateral resolution of a far-field optical microscope is defined as the smallest distance between two features that can still be resolved in the image and is given by Rayleigh criterion as [5]

$$R = 0.61\left(\frac{\lambda}{NA}\right)$$

where λ is the wavelength of incident light and NA (=nsinθ) is the numerical aperture, n is the refractive index of the medium, and θ is the angle subtended by the optics. The depth of the field or vertical resolution of μRS is approximated by Conrady expression as [6]

$$\Delta Z = \left(\frac{\lambda}{n\sin^2\theta}\right)$$

Under normal conditions ($n = 1$ for air, $\lambda = 632.8$ nm, $NA = 0.95$), the typical lateral and depth resolutions would be about 400 and 700 nm, respectively. Using the smallest visible wavelength (~400 nm) and a high numerical aperture ($n = 1.515$, NA ~1.4) one can estimate the highest lateral resolution as 200 nm (based on Abbe criterion) and the smallest field depth as 400 nm. However, if a series of spectra are recorded at very close equidistant locations, a reduced spot size (considering the Gaussian profile of the beam) is obtained through a convolution of the spot profile.

5 Key Research Findings

5.1 Mapping Chemical Composition and Phase of Nanomaterials

Polymer nanocomposites and polymer blends are an extremely important class of materials due to the expected synergistic enhancement of properties and potential multi-functionality. However, the immiscibility of most of the polymers results in poor interfacial interaction between the individual components which severely affects the final properties. A deeper insight into the spatial heterogeneity and morphology of the individual components at a microscopic level and their influence on the macroscopic properties is important for their rational design (such as choice and volume fraction of individual components, surface chemistry, and processing conditions) of the multicomponent systems. While conventional Raman spectroscopy provides information specific to chemical species, it lacks spatial resolution to distinguish different domains, typically on the order of a few microns, found in polymer blends and composites. μRS, on the other hand, provides sub-micron spatial resolution to precisely choose the area of interest to carry out chemical analysis. The chemical mapping of polymer blends or composites based on their characteristic Raman peaks could be used to probe the domains of phase-separated polymer blends with sizes as small as ~200 nm [7]. In this section, we present a few selected examples of the use of μRS in the analysis of polymer blends and composites.

In a study by Huan et al., the spatial distribution of poly(ethylene terephthalate) (PET) and high density poly(ethylene) (HDPE) in a PET/HDPE blend was probed by confocal Raman mapping [8]. The relative distribution of HDPE and PET components in the blends illustrated in Fig. 15.2 were obtained by plotting the Raman images generated by the variation of the scattering bands at 1,062 and 1,725 cm^{-1} belonging to the CH_2 groups in the backbone of HDPE and the $C = O$ groups of PET, respectively. The bright portions in Fig. 15.2a, b correspond to the relatively higher signal intensity of HDPE (1,062 cm^{-1}) and PET (1,725 cm^{-1}) which are almost complementary to each other. The spectra (Fig. 15.2c) collected from the selected points (a, b, c in Fig. 15.2a) show a gradual decrease in peak intensity at 1,062 cm^{-1} and a corresponding increase in 1,725 cm^{-1} peak while traversing from point a to c. The distribution of the components in the blends was further enhanced using "ratio image" approach by plotting the ratio of the peak intensity at 1,062 and 1,725 cm^{-1} as a function of position instead of the absolute intensity as shown in Fig. 15.2d.

Quantitative chemical analysis in different phases of polymer blends can be done from information gathered on structural units, end groups, and additives using μRS. Overbeke et al. showed that the micro-domains formed during curing and phase separation in an epoxy-thermoplastic blend could reveal the thermoplastic–modified epoxy networks by comparing to calibrated Raman data obtained from standard blend samples [9]. Hashida et al. showed the distribution of crystallites within each phase of the blend and mapped the degree of crystallinity at various locations of poly(hexamethylene adipate) (PHMA) and poly(hexamethylene sebacate) (PHMS) blends [10]. PHMA and PHMS have similar chemical structures but exhibit different degrees of miscibility when blended with poly(propylene glycol) (PPG), resulting in phase separation when mixed binary or ternary. The crystallization process is directly related to the variation in composition associated with the phase separation that can be characterized on the basis of the measured intensity for conformation-sensitive Raman peaks.

Li et al. studied the quality and activity of a catalytic system used in carbon nanotube (CNT) growth process using μRS [11]. By localized Raman mapping, the variation of the active metallic species (Fe, Co) over the oxide surface ($CaCO_3$) was quantified and analyzed. In this study, it was observed that the compositional variation in the active species over the oxide surface affected the yield and morphology of the CNT. Mayo et al. employed μRS to study the phase separation of polymer-functionalized single-walled carbon nanotubes (SWCNTs) within

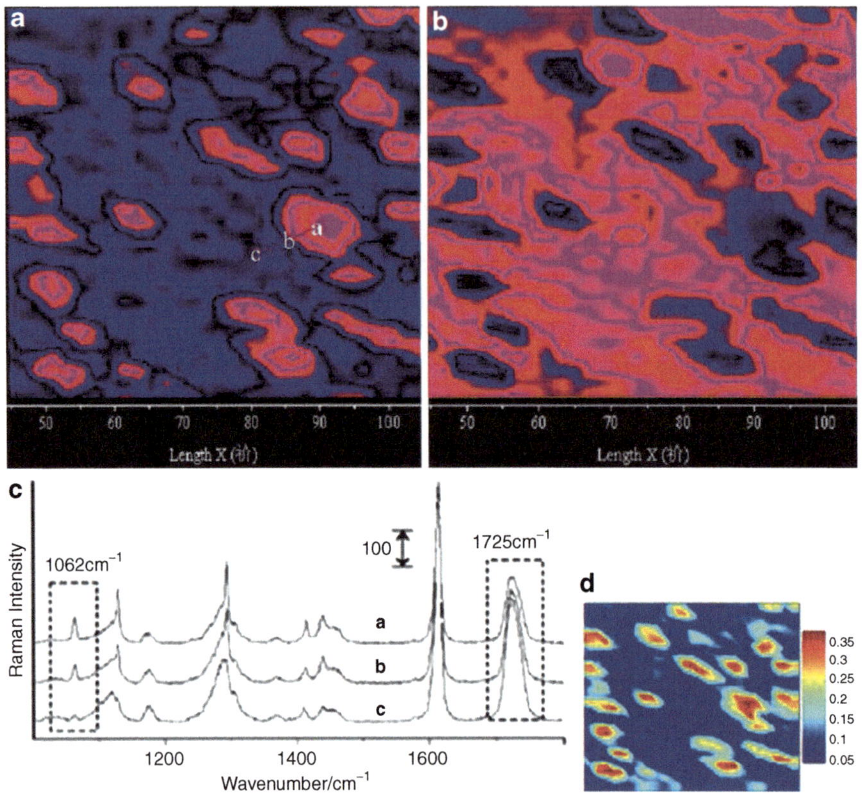

Fig. 15.2 Raman images generated by plotting the variation of the scattering band at 1,062 cm^{-1} and the band at 1,725 cm^{-1}, to illustrate the relative distribution of HDPE (**a**) and PET (**b**) components. (**c**) The corresponding Raman spectra from selected points a, b, c. (**d**) The ratio image for 80% PET/20% HDPE polymer blend prepared without maleic anhydride. The *dark portion* represents PET-abundant regions and the *bright portion* represents HDPE-abundant regions (Reprinted from [8])

a polymer blend [12]. Polystyrene (PS) and poly(methyl methacrylate) (PMMA) blends were used as a means to segregate the polymer-functionalized CNTs to the polymer phase that had high affinity. A Raman map of the phase-separated PS/PMMA polymers containing PS-functionalized SWCNT is depicted in Fig. 15.3. The distribution of the polymer-functionalized SWCNT was found to favor the respective functionalized polymer domain.

Chou et al. employed polarized Raman spectroscopy to rapidly identify the growth orientation and phase of vanadium oxide nanowires [13]. Vanadium oxide nanowires were synthesized using physical vapor deposition in a hot wall horizontal tube without a catalyst similar to ZnO nanobelt vapor-solid growth mechanism [14]. Variations in phases of the nanowires were identified as VO_2 and V_2O_5 along the tube reactor using Raman microscopy. In their geometry, the nanowires were aligned along the laboratory Y-axis direction and the Raman intensity was

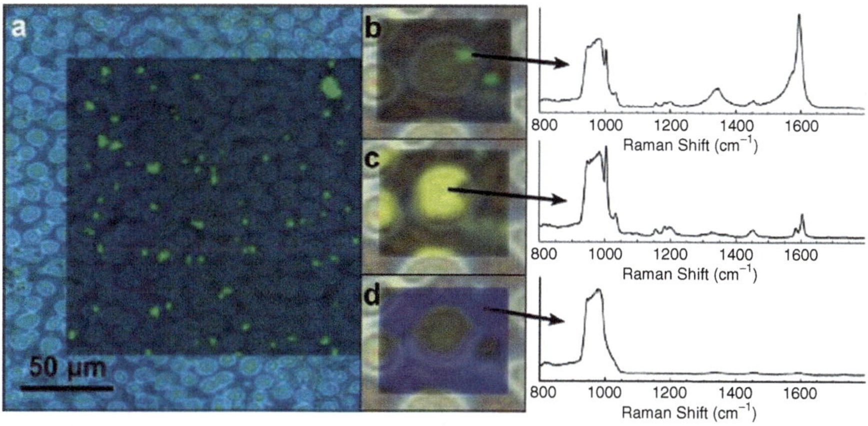

Fig. 15.3 (**a**) Raman map of PS-functionalized SWCNT highlighted in *green*. Insets indicate the signal from the three components: SWCNT (**b**), PS (**c**), PMMA (**d**). The corresponding Raman spectra are shown on the right (Reprinted from [12])

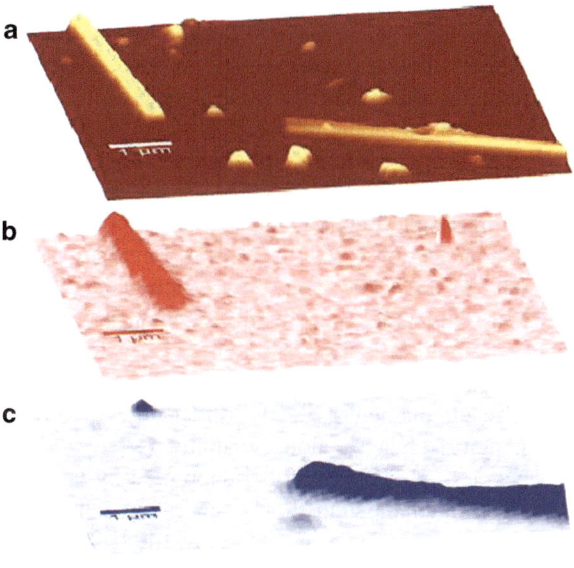

Fig. 15.4 (**a**) AFM topographic images of VO_2 and V_2O_5 nanowires. (**b**) Map of integrated Raman signal (615–625 cm^{-1}) in the same region identifying VO_2 nanowire. (**c**) Map of integrated Raman signal (989–999 cm^{-1}) identifying V_2O_5 nanowire (Reprinted from [13])

measured for light polarized parallel (YY) and perpendicular (XX) to the nanowire axis. The azimuthal orientation of the V_2O_5 nanowires had an influence on the Raman intensity but not the angular dependence on analyzer polarization. The various low-index growth directions were distinguished by comparing the relative intensity of the YY and XX polarizations for each Raman mode. The A_g mode of VO_2 nanowires was distinguished by mapping the mode at 620 cm^{-1} and the V_2O_5 nanowires at 994 cm^{-1} as shown in Fig. 15.4. The Raman intensity of the low-index growth directions of V_2O_5 and VO_2 were higher in the [010] and [100] directions, respectively. The orientation of nanostructures in composites is explained in detail in Sect 5.4 of this chapter.

5.2 Probing Structure and Properties of Low-Dimensional Carbon Systems

Low-dimensional graphitic carbon systems such as CNTs (1D confinement) and graphene (2D confinement) received considerable attention over the last two decades owing to their rather unique electrical, mechanical, and thermal properties [15–18]. μRS proved to be valuable in studying these highly ordered carbon materials due to the characteristic vibrational modes of carbon-carbon networks, which provide information of the structure and properties of these nanostructures. With a relatively high spatial resolution, μRS is a unique tool to study the structure and electronic properties and their variations down to individual nanostructures and in some cases variations within individual nanostructures. The important Raman bands in these structures are: *G* band, which is the tangential graphite-like band at ~1,580 cm^{-1}, known to be sensitive to the electronic nature (metallic vs. semiconducting) and the strain of the C–C bonds; *D* band (known as a defect mode) at ~1,350 cm^{-1}, which is indicative of substitutional heteroatoms, vacancies, and grain boundaries; *G′* band (2,500–2,800 cm^{-1}) is the second-order harmonic of the *D* band and is highly sensitive to the strain on the C–C bonds and structure (number of layers in a graphene flake); and radial breathing mode (RBM), which arises when all the carbon atoms vibrate in phase and in the radial direction, and occurs at frequency range of ~160–300 cm^{-1} [19]. The frequency of RBM mode was found to be inversely proportional to the diameter of a CNT [16]. In this section, examples of micro-Raman investigation of CNTs and graphene are illustrated to provide the analytical aspects of this powerful nondestructive technique.

G, *D*, and RBM bands provide substantial information of CNT structure such as chirality and diameter [20, 21]. Anderson et al. directly demonstrated the chirality (or helicity) changes along spatially isolated SWCNT by mapping the RBM frequencies, which was further supported by *G* band frequency and shape. They have identified the transition from semiconducting-to-metal and metal-to-metal chiralities at the single nanotube level [22]. In addition, they observed increased Raman scattering due to local defects associated with the structural transition, and further, determined the spatial extent of the transition to be ~40–100 nm. Kobayashi et al. observed that micro-Raman spectra of suspended SWCNT were much more intense than those grown on substrates; thus, they were able to isolate and selectively study the structural features of the SWCNT suspended on pillars made of silicon (Si) and silicon oxide (SiO_2) [21]. By studying the intensity ratio of *D* and *G* bands, they were able to evaluate the defect density and noted that the SWCNTs grown on the Si pillars contained more defects.

When combined with an imaging technique such as scanning electron microscopy (SEM), μRS provides enough resolution to probe the electronic characteristic of CNT down to individual nanotubes [23]. Typically, CNTs exhibit a broadened Breit-Wigner-Fano *G*–band shape in the Raman spectra when they are metallic due to the presence of free electrons in the conduction band; on the other hand, semiconducting CNTs show a Lorentzian *G*–band shape [24–26]. By analyzing the *G* band obtained from micro-Raman spectra, the distribution of metallic and semiconducting CNTs can be mapped, which provides valuable information regarding the synthesis and processing conditions. For example, LeMieux et al. used μRS to observe *G* band and RBMs to characterize the electronic nature and chirality of CNTs that were deposited on functionalized surfaces, which were devised as CNT network field-effect transistors [27]. In this study, μRS corroborated device results as a function of surface chemistry to show that CNTs with a specific electronic type were absorbed onto the functionalized surfaces.

Ferrari et al. showed that the G' band exhibits change in shape, width, and position for an increasing number of graphene layers while the *G* band exhibited a small downshift [28]. Similarly, Graf et al. investigated single- and few-layer graphene flakes using μRS, and found that the width of the G' band was highly sensitive to the crossover from single- to double-layer graphene by showing that the single peak for single-layer graphene split into multiple peaks for the double layer as shown in Fig. 15.5(a–f) [29].

By carefully analyzing the double resonance G' band in the form of electronic band of a graphene monolayer, Faugeras et al. observed that graphitized carbon-terminated surfaces of silicon carbide (SiC) were electronically well-decoupled, and therefore could be considered to be graphene multilayers with the appearance of Dirac-like electronic states [30]. From this observation, they suggested that functional graphene-based devices could also be developed from graphite layers epitaxially grown on silicon carbide. In another study, Lee et al. showed that for multilayer-graphene structures (mono-, bi-, tri-, etc.) grown on SiC, the full width at half maximum (FWHM) of the G' band could be used to assign the number of layers fabricated on SiC [31].

Ganganahalli et al. studied the electrochemical reduction of graphene oxide (GO) by in situ monitoring the *G* band of GO using μRS, wherein the *G* ($\sim$1,610 cm^{-1}) band of GO shifted to $\sim$1,585 cm^{-1}, confirming the reduction of GO to graphene [32]. Furthermore, by monitoring the changes in the intensity ratio of *D* to *G* bands, the defect concentration in the reduced graphene oxide could be estimated.

The analysis of the shape, intensity, and shift of these Raman bands allows evaluating the electronic state, structural deformation, and defect density in CNTs and graphene. When complemented by other high-resolution microscopy techniques such as atomic force microscopy (AFM) or X-ray diffraction, μRS offers valuable insight into the structure and properties of these low-dimensional carbon systems.

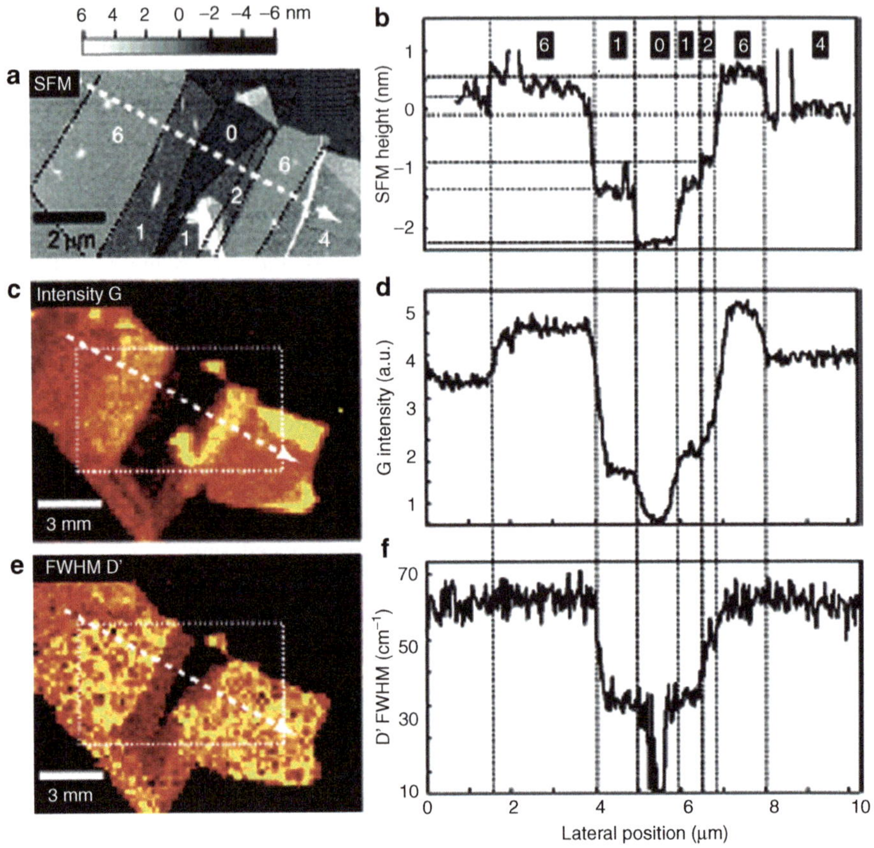

Fig. 15.5 (**a**, **b**) Scanning force microscopy (SFM) micrograph and cross-sectional plot (indicated with the *white dashed arrow*; lateral average over 400 nm) of a few-layer graphene flake with central sections down to a single layer. (**c**) Raman maps showing the integrated intensity of the *G* line (*dashed square* corresponding to the SFM image in (**a**)) and (**e**) the full width at half maximum (FWHM) of the D' line. The related cross sections (**d**, **f**) are aligned (*vertical dashed lines*) with the height trace (Reprinted from [29])

5.3 Probing Crystallinity and Crystal Orientation

5.3.1 Probing Crystallinity of Polymers

The sensitivity of Raman spectroscopy to crystallinity is due to the conformational changes in polymers occurring during the transformation of polymer chains from amorphous domains to that of three-dimensional ordered, crystalline domains [33]. μRS could be used to quantify the crystalline fractions of spherulites by analyzing the characteristic Raman bands in the region of interest.

The morphological changes during the crystallization of PET were studied using μRS [3]. PET, a semicrystalline polymer, has three possible morphological species:

amorphous, ordered amorphous, and true crystalline configuration. Crystallization of PET (glycol and terephthalate segments) involves conformational changes from amorphous domains to transcrystalline structure. PET exhibits gauche glycol segments with disordered terephthalate groups in the amorphous phase, trans glycol and disordered terephthalate groups in the ordered amorphous phase, and an all trans structure with the carbonyl groups coplanar with the benzene ring in the crystalline form [34]. The conformational changes during crystallization can be observed as narrowing of carbonyl band at 1,725 cm^{-1} (corresponding to the rotation of carbonyl group with respect to benzene ring) and the growth of the bands at 1,095 cm^{-1} (combination of stretching vibrations of the ester, glycol, and ring units of the trans configuration of PET) and 1,000 cm^{-1} (trans glycol moiety) [35].

The crystalline content of semicrystalline PET can be determined by the peak intensity ratio of bands at 1,117 and 1,095 cm^{-1}. Highly crystalline samples exhibit a large peak at 1,095 cm^{-1}, whereas amorphous samples show a shoulder on the 1,117 cm^{-1} peak [36]. The increase in the intensity of the 1,095 cm^{-1} peak upon crystallization was assigned to the stretching mode of the C-C trans conformation and the 1,117 cm^{-1} to the crystalline phase gauche conformations. The plot of the intensity ratio of 1,095/1,117 cm^{-1} and the FWHM of the carbonyl stretching mode at 1,725 cm^{-1} gave a linear relationship with a correlation coefficient >0.99 indicating that the two spectroscopic changes measured the same molecular phenomenon [34]. The FWHM of the Raman intensity peak at 1,725 cm^{-1} correlated well with density and gave a good estimate of the degree of crystallinity in the sample [36, 37]. Highly crystalline regions had a narrow peak width whereas amorphous regions had a broader peak width.

The crystalline fractions of polymer within spherulite formed by isotactic polypropylene (iPP) were quantified using μRS. Gatos et al. examined in their study the crystalline profile of isotactic polypropylene (iPP) by analyzing the characteristic Raman band at 809 cm^{-1} that indicates high amorphous content in conjunction with an optical micrograph [38]. Raman spectra were taken in steps of 10 μm on a spherulite with a diameter of 80 μm. This yielded the degree of crystallinity along the spherulite diameter as shown in Fig. 15.6. The interior of the spherulite was further studied using AFM revealing the fibril packing and morphology of the spherulite. In another study, Martin et al. used Raman microscopy to differentiate between α- and β-crystalline polymorphs of iPP. They investigated the β-crystalline morphology of iPP by analyzing the shift in the 842 cm^{-1} peak to generate structural map of one β-spherulite among α-spherulites. They noted that the contour of the β-spherulite had a concave edge and could be well identified among the α-spherulites [39].

5.3.2 Probing Crystal Orientation in Inorganic Nanostructures

It is well known that the electrical and optical properties and the performance of the electronic and optoelectronic devices based on inorganic nanostructured materials critically depend on the composition, crystalline structure, and orientation; hence, these parameters must be known for nanostructures integrated into

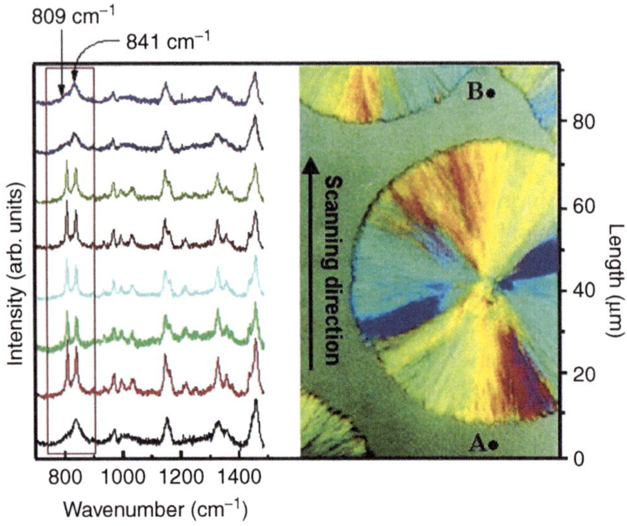

Fig. 15.6 Raman spectra recorded along the diameter of an iPP spherulite during isothermal crystallization from the melt at $T_c = 130°C$. The variation in Raman bands at 809 cm^{-1} and 841 cm^{-1} indicating the amorphous and crystalline content of iPP are matched with their positions on the polymer (Reprinted from [38])

nanodevices. X-ray analysis is commonly used for probing the crystallinity and lattice dimensions of relatively large quantities of nanostructures. However, one of the major limitations is that the technique is not applicable for probing the crystalline orientation of an individual nanostructure. On the other hand, transmission electron microscopy (TEM) and electron diffraction has been the primary tool for identifying the growth direction and crystal facets of individual inorganic nanowires and nanobelts [40]. However, TEM requires special sample preparation and cannot be easily applied to nanostructures integrated into the devices. μRS clearly offers an opportunity to address the crystalline orientation and other physical properties of the nanostructures without any special sample preparation and directly in device configuration. Raman spectra obtained at a given polarization with respect to the orientation of the crystal lattice is known to have distinct features based on the Raman selection rules. While a detailed discussion of the selection rules is beyond the scope of this chapter, we will briefly discuss the Raman spectra of ZnO, which belongs to the wurtzite crystal structure and use it as an example to demonstrate the utility of μRS to identify the crystal orientation of nanostructures.

ZnO exhibits a hexagonal structure belonging to *C*6*mc* space group [41]. The ZnO structure is comprised of alternating planes of tetrahedrally coordinated O^{2-} and Zn^{2+} ions, stacked along the *c*-axis. One important feature of ZnO is the polar surfaces, with the most common polar surface being the basal plane. The oppositely charged Zn^{2+} and O^{2-} ions produce positively charged (0001) and negatively charged (000$\bar{1}$) surfaces, respectively. In the simplest example of a ZnO nanobelt, the fast growth direction can be either [2$\bar{1}\bar{1}$0], [01$\bar{1}$0] or [0001] [42, 43].

Wurtzite ZnO structure with four atoms in the unit cell has a total of 12 phonon modes (one longitudinal acoustic (LA), two transverse acoustic (TA), three longitudinal optical (LO), and six transverse optical (TO) branches). The optical phonons at the Γ point of the Brillouin zone in their irreducible representation belong to A_1 and E_1 branches that are both Raman and infrared active, the two nonpolar E_2 branches are only Raman active, and the B_1 branches are inactive (silent modes). Furthermore, the A_1 and E_1 modes are each split into LO and TO components with different frequencies. For the A_1 and E_1 mode lattice vibrations, the atoms move parallel and perpendicular to the c-axis, respectively. On the other hand, E_2 modes are due to the vibration of only the Zn sublattice (E_2-low) or O sublattice (E_2-high). The expected Raman peaks for bulk ZnO are at 101 cm^{-1} (E_2-low), 380 cm^{-1} (A_1-TO), 407 cm^{-1} (E_1-TO), 437 cm^{-1} (E_2-high), and 583 cm^{-1} (E_1-LO).

Singamaneni et al. have demonstrated high-resolution confocal Raman microscopy as a nondestructive technique for fast and unambiguous identification of the localized crystal orientation of individual ZnO nanostructures [44]. The two important features in the Raman spectra of ZnO which clearly distinguish the c-axis are the intensity of A_1(TO) band and the width of the E_2 band. Figure 15.7a, b show the SEM and the Raman map of E_2 band intensity of two ZnO nanobelts cross each other. Raman mapping clearly revealed that the intensity A_1(TO) band (Fig. 15.7c) and the width of the E_2 band (Boolean map with the thresholds as marked in Fig. 15.7d) can be employed to identify the growth direction of the nanobelt. Apart from these two features, the waveguiding of the Raman scattered signal from the substrate (enhanced parallel to the c-axis and suppression perpendicular to the c-axis) has been revealed (seen in Fig. 15.7e, f).

These features of the Raman bands of the ZnO nanostructures can be extremely powerful for the in situ identification of orientation of ZnO nanostructures employed in a converse piezoelectric actuator directly in an assembled state [45]. While their study focused on ZnO nanostructures, the authors noted that the general features (Raman bands and the waveguiding effect) described are equally applicable to other wurtzite type nanostructures and the approach suggested might serve as a universal tool for the versatile characterization of GaN, ZnS, and CdSe from the wurtzite family, which are utilized for optoelectronics, lasing, and piezoelectricity.

Recently, it has been demonstrated that confocal Raman microscopy can be employed to obtain information on the phase, growth direction and radial crystallographic orientation of GaN nanowires [46, 47]. Fan et al. have demonstrated the orientation-dependent micro-Raman measurements on single wurtzite CdS nanowires with an average diameter of 60 nm [48]. The Raman spectra of nanowires were compared with those of bulk ribbon. As opposed to the Raman spectra from the bulk samples, the intensity of most Raman bands of the nanowires exhibited significant dependence on the polarization of the excitation with respect to the nanowire axis. The Raman bands were found to be polarized along the longitudinal axis because of the shape-induced depolarization effect, similar to

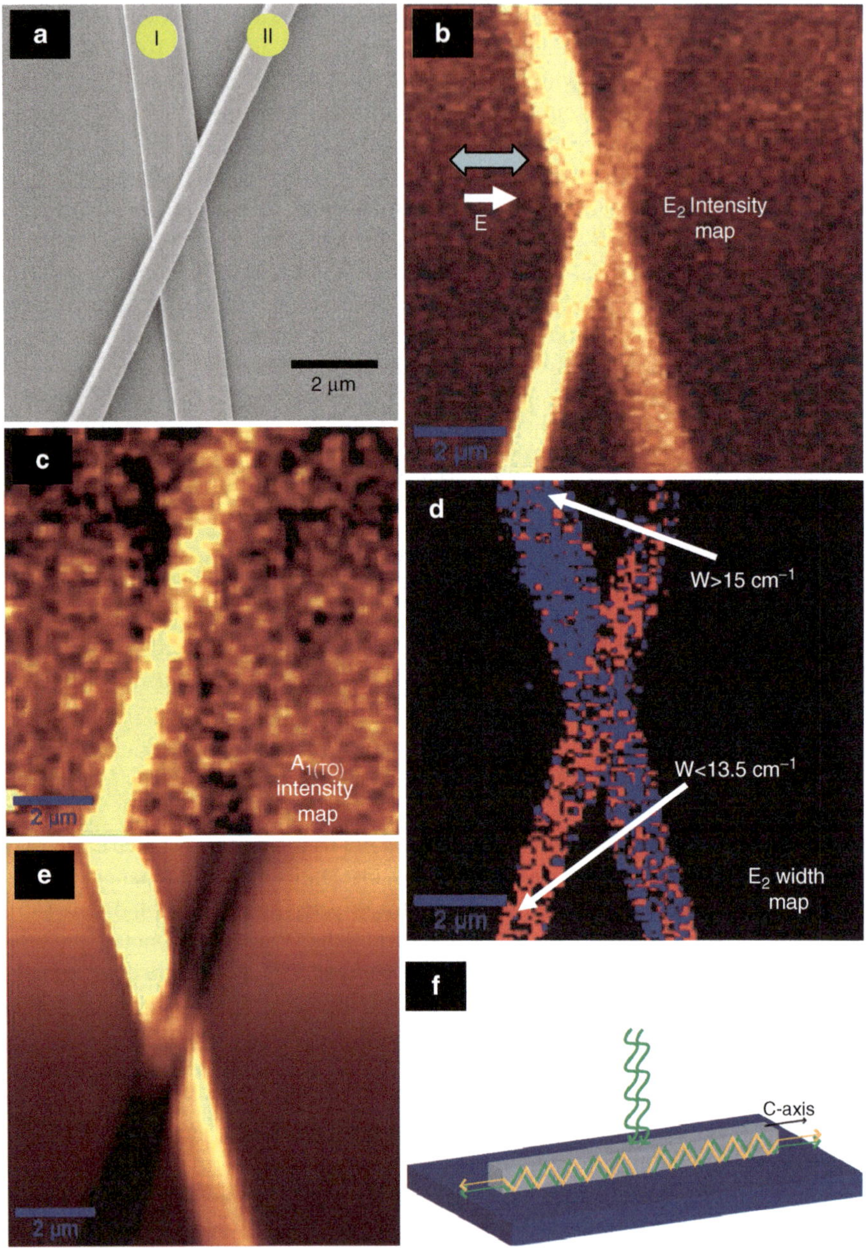

Fig. 15.7 (**a**) Scanning electron microscopy (SEM) image of two intersected ZnO nanobelts. Raman intensity map of two nanobelts shown in (**a**). (**b**) E_2 mode. (**c**) A_1(TO) mode. (**d**) Boolean map of the full width at the half maximum of the E_2 mode. (**e**) intensity map of the silicon signal at 520 cm^{-1} showing the modulation in different belts. (**f**) Schematic shows the waveguiding of the Raman scattered light from the substrate along the c-axis (Reprinted from [44])

polarized photoluminescence. From a detailed angular dependencies of various Raman bands, the authors noted that the LO, 2LO, and multiphoton bands were strongly affected by the anisotropic geometry of the nanowires, while the E_2 band remained insensitive owing to the crystal symmetry of the nanowire.

5.4 Orientation of Nanofillers in Composites

CNTs are excellent candidate as reinforcing phase in many composite materials due to their high stiffness and strength [49]. Though CNTs can be used as reinforcing filler, there are problems like the degree of control of their orientation or alignment in the matrix, matrix-nanotube load transfer efficiency that are yet to be overcome. Fabrication of anisotropic mechanical and electrical properties of polymer/CNT composites requires orientation and well-controlled distribution of CNTs in the polymer matrix [50–52]. μRS has emerged as a sensitive local probe for the orientation of nanostructures in polymer fibers and composites. Polarized Raman spectroscopy is a powerful tool that can be used to obtain molecular orientation distributions. By measuring the angular distribution of the Raman scattering intensity using a monochromatic plane-polarized beam and an analyzer, the molecular orientation distributions in polymers can be studied. Since the vibrational mode possesses an individual differential polarizability ellipsoid that determines the mode of vibration, polarized Raman spectroscopy is also useful for assigning vibrational modes [53]. The G-band intensity ratio, with polarization parallel and perpendicular to the fiber axis, at about 1,592 cm^{-1} is taken as a measure of single wall CNT orientation in composites and fibers [54–56].

In a recent study, Deng et al. examined the dispersion of CNTs in poly (p-phenylene terephthalamide) (PPTA) composite fibers using Raman scattering intensity mapping along the fiber [57]. They examined the distribution of CNTs by taking the relative intensity of nanotube G-band to PPTA 1,610 cm^{-1} band (I_G/I_{1610}) by scanning over the fiber surface. The intensity ratio was found to be uniform at different positions of the fiber indicating that the nanotubes were well dispersed. The orientation of the structural units of the composite had an influence on the mechanical properties. The rigid-rod like polymer chains, coupled with high crystallinity and high orientation impart the PPTA/CNT fibers with high modulus and tensile strength. Molecular dynamic simulations done by Yang et al. have demonstrated that strong interfacial adhesion exists between the nanotubes and polymers containing aromatic ring on their backbone and hence there is stress transfer from the matrix to the nanotube [58].

In another study by Chae et al., orientation and exfoliation of SWCNTs in polyacrylonitrile (PAN) fibers by gel spinning was studied using Raman spectroscopy, and it was based on the G-band intensity ratio of polarized light and the second- and fourth-order orientation parameters of SWCNT [53]. The Raman intensity ratio for beam-polarized parallel and perpendicular to the fiber axis was 42 for gel spun fiber but 38 for conventional solution spinning and that SWCNT orientation was slightly higher (0.915) than conventional spun fiber (0.90).

However, SWCNT orientation in PAN/SWCNT gel spun fiber was less than polymer orientation in fibers like Kevlar, Zylon, and Spectra where typical orientation factor value was ~0.99.

5.5 Internal Stress Monitoring with Nanoscale Resolution

The mechanical properties of the nanocomposites strongly depend on their structure, orientation of the filler, phase separation, and processing conditions. Hence, there is a need for in situ nondestructive characterization technique to probe the internal stress in nanocomposite structures. The shortcomings of many conventional techniques such as low resolution, destructive measurements, complex modeling and applicability to only certain class of materials are overcome by using μRS owing to the sensitivity and nondestructive measurement for monitoring internal stress in various materials [59].

The mechanism of stress transfer between CNTs and polymers in a nanocomposite was studied using Raman microscopy to detect interfacial failure. The Raman response under tensile strain varies for SWCNT in nanocomposites due to debundling and nanotube coupling within the bundles. Mu et al. studied the frequency shift of the *G* band of SWCNT/PMMA composite fibers under axial stress and probed the mechanism of load transfer from amorphous, linear glassy polymer matrix to small SWCNT bundles that have no strong, specific interactions with the polymer [60]. Within a small strain regime, there was a linear stress transfer to SWCNT and the shift in *G* peak and stress–strain curve were reversible. At higher strain there was no linear stress transfer due to the weak interfacial interactions between PMMA and SWCNT, indicating deterioration of the mechanical properties. Load transfer in such composites was limited by the strength of adhesive interactions. Mechanical response to the onset of slipping at the PMMA/SWCNT interface could be examined by using Raman microscopy in situ with mechanical testing.

In another study, Ko et al. studied bent CNT arrays by combined AFM and μRS analysis [61]. They observed systematic lower frequency shift of the tangential *G* mode, and suggested that the shift was attributed to tensile strain of the bending nanotube arrays resulting in the loosening of C–C bonds in the outer shells of the curled nanotubes. Singamaneni et al. have also employed the combination of AFM and confocal Raman microscopy to monitor the internal stress distribution in periodic polymer microstructures [62, 63]. The authors have demonstrated that mechanical instabilities in cross-linked *bisphenol* A Novolak epoxy nanoporous structures fabricated using interference lithography (IL) exhibited dramatic pattern transformation under external or internal stresses. In particular, they observed that polymerization of a rubbery component (polyacrylic acid) in a square array of cylindrical pores of the IL structure resulted in the transformation of the cylindrical pores into mutually orthogonal ellipses along both (10) and (01) of the square array. The AFM images clearly reveal the structural transformation at a microscopic level. The pattern transformation can be related to bending of the struts in alternate directions (along (10) and (01) directions) and the rotation of the nodes in clockwise and anticlockwise directions. On the other hand, confocal Raman microscopy was employed to monitor the stress distribution at the submicron scale in the pristine and transformed structures to complement AFM imaging. The elastoplastic nature of the IL material deformation locked in the mechanical instabilities after the release of the external stress with internal stresses dissipated to a great extent as was confirmed by micro-mapping with Raman microscopy. The authors noted that this is in sharp contrast with the reversible instabilities in elastomeric solids, in which the transformed structures exhibit stress concentration in localized highly deformed elements.

Beecham et al. have demonstrated that both temperature and stress fields can be simultaneously measured for phosphorus-doped polysilicon microheater devices [64]. In MEMS devices and high band gap semiconductors, high temperature and the presence of stress can adversely affect device performance or reliability. Temperature or stress dependence measurements using μRS are based on the variation in the Stokes peak positions. However, under thermomechanical loading conditions the change in the Raman peak position can also be due to the simultaneous effects of temperature and stress. μRS was used to measure both stress and temperature parameters by monitoring the position and the line width of the Stokes-shifted peak due to a relative independence of the Stokes line width on stress. In a later study, the same authors quantified the magnitude of operational thermoelastic stress that evolves in a GaN transistor through simultaneous use of Raman signal's Stokes peak position and line width. They found the experimentally measured shifts to be in a close agreement when compared with the finite element model predictions [65].

In a recent study, Spolanek et al. used μRS to indirectly measure stress of a thin aluminum film on a poly-Si, silicon nitride, and aluminum multilayer structure as a function of temperature by using the silicon layer as a strain gage. Although aluminum is Raman inactive, by measuring the difference between the thermally induced peak shift of a silicon wafer and the measured peak shift of the multilayer structure, the thermal stresses in the different layers (including Raman inactive layers) were calculated. The observed peak shifts were caused by thermal stresses that originated due to the mismatch of the thermal expansion coefficients of the different layers in the multistructure. The authors were able to monitor the average stress and stress distribution simultaneously [66].

The grain orientation and internal stress of a multicrystalline silicon material were also evaluated by μRS [67]. Determination of stresses in silicon wafers with known crystallographic orientations can be done by measuring the corresponding Raman peak shift. In multicrystalline silicon, the determination of stresses becomes complicated for arbitrary grain orientations. The Raman intensity depends on the polarization direction of incident and scattered light and on the crystallographic grain orientations. The intensity dependence on position was used in determining the crystallographic grain orientations and the stress tensors were calculated by the Raman peak shift. Three polarization settings were used so that one of the three phonon modes dominated the other two. The measured frequency shifts were assigned to the dominant modes within the region of interest and correlated with the local stress tensor close to the crack tip which was under tensile stress.

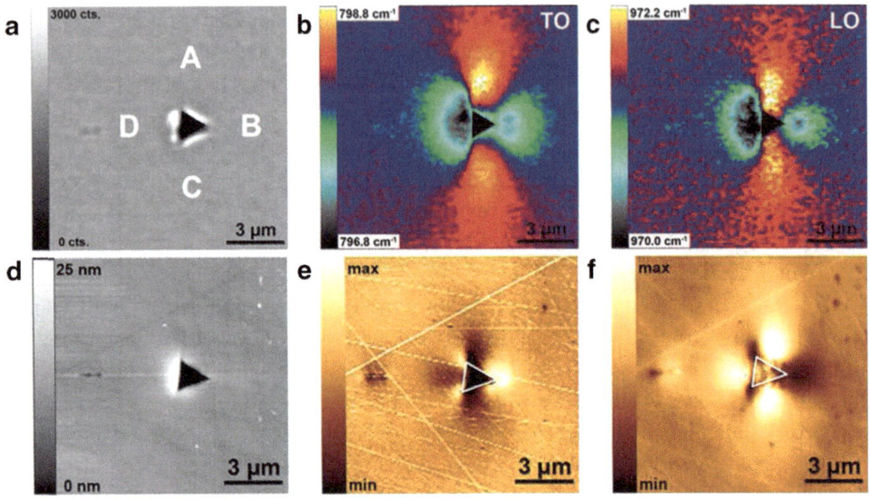

Fig. 15.8 Confocal Raman microscopy and SNOM images of a locally stressed SiC crystal (**a**) Rayleigh intensity map. (**b**), (**c**) Spectral position maps of fitted TO and LO phonon lines obtained by fitting a Lorentzian peak. (**d**) Topography of the indent. (**e**) SNOM amplitude for $\omega_{IR} = 924\ \mathrm{cm}^{-1}$. (**f**) SNOM amplitude for $\omega_{IR} = 944\ \mathrm{cm}^{-1}$ (Reprinted from [68])

Inhomogeneous stress distributions that arise at the grain boundaries in polycrystalline silicon wafers were detected by fan-shaped dislocation clusters close to the grain boundaries. The dislocation clusters are formed during the solidification process and relieve stresses produced in the system.

In another study, Gigler et al. mapped a nanoindent region made in a SiC crystal, by IR-type-scanning near-field optical microscope (SNOM) and μRS in terms of local residual stress fields [68]. They have found that the stress-induced shifts of the LO phonon frequencies as Raman images and the related shift of the phonon-polariton near-field resonance as IR-SNOM images were in contrast to each other around the nanoindentation sites as can be seen in Fig. 15.8. The magnitudes of the shift in phonon frequencies probed by Raman scattering were directly related to local stress similar to that detected by near-field phonon-polariton resonance in SNOM. With the above combination of optical techniques, they were able to resolve the smallest features down to the size of nanocracks. While the SNOM technique revealed stress distribution near the surface, Raman scattering provided valuable information on the subsurface stress distribution. Thus, by combining both analytical techniques, a highly effective assessment of local stress distributions with the full high-resolution spectral information could be obtained.

Chen et al. investigated the effect of strain on the phonon modes of bent InP nanowires [69]. The nanowires were bent by pushing on them with an AFM tip. Line scans were performed along the wire with a step size of 200 nm and spectral mappings were recorded. Raman images were generated from the acquired data by integrating the signal over a frequency band covering the two phonon modes at $304\ \mathrm{cm}^{-1}$ (TO mode) and at $344\ \mathrm{cm}^{-1}$ (LO mode) for the hexagonal nanowires.

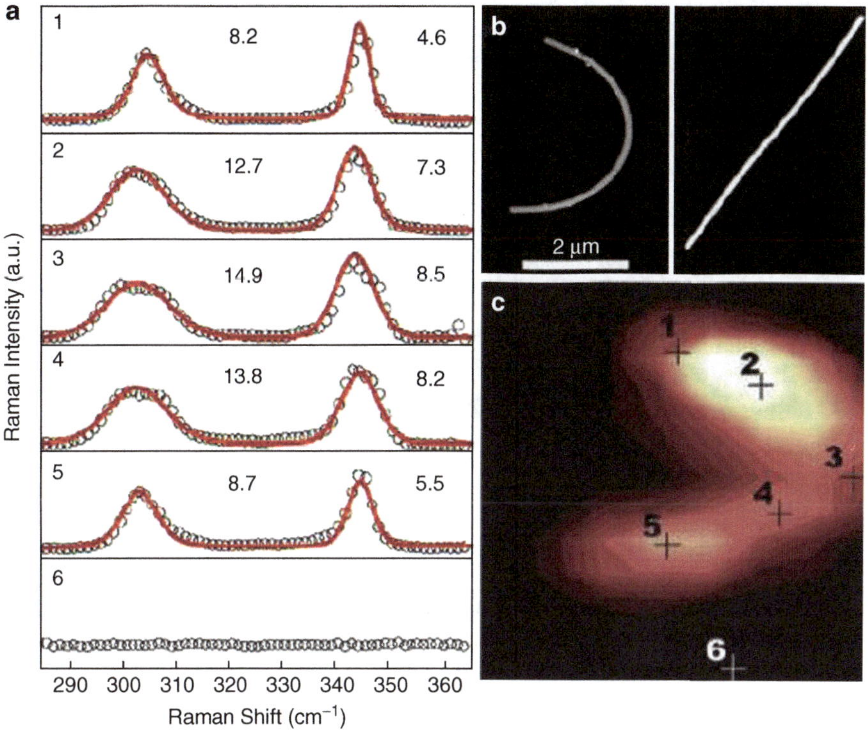

Fig. 15.9 (**a**) Normalized Raman spectra acquired at five different locations along the bent nanowire. (**b**) AFM image of the nanowire after and before bending. (**c**) Raman image generated by integrating the signal over a frequency band covering the TO and LO phonon modes (Reprinted from [69])

Raman spectra were extracted from the resulting optical images for several positions within the scan range as shown in Fig. 15.9. The spectra of the bent wire with the highest curvature (consequently highest strain) were broadened compared to the straight (nonstrained) wire as seen in Fig. 15.9a. The corresponding Raman maps for the strains induced in the nanowire are shown in Fig. 15.9b. The nanowires experienced compressive strain on the inner side of the curve and tensile strain on the outer side leading to a frequency shift and broadening of the peak [70]. For a 1% bending strain of the nanowire, the TO and LO peaks broadened by 1% and 0.2%. The origin of the phonon energy variations was due to the compressive and tensile strains inside the nanowires in accordance with the deformation potential theory.

5.6 Micro-Raman as a Tool to Understand SERS

SERS involves the dramatic enhancement of Raman scattering of the molecules adsorbed on nanostructured metal surface [71]. It is currently widely accepted that

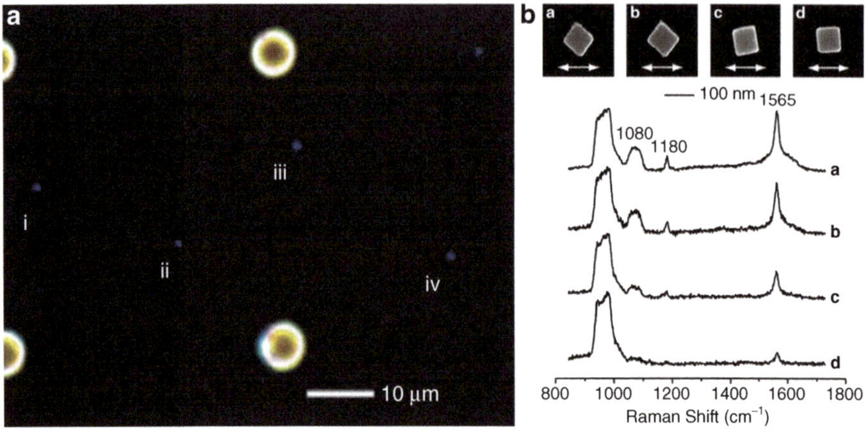

Fig. 15.10 (**a**) Dark field optical image of the Au nanocubes deposited on Si substrate with micropatterened markers. (**b**) SEM images of the Au nanocubes and the Raman spectra corresponding to the relative orientation of the polarization and nanocubes shown in the SEM images (Reprinted from [77])

the enhancement of the Raman signal in SERS sprouts from two distinct contributions, namely, electromagnetic enhancement and chemical enhancement [72]. The dramatic enhancement of the intensity of the Raman spectra from the molecule adsorbed on a nanostructured metal surface, which seems to be simple for experimental realization, has been intensely investigated to understand the underlying physical phenomenon responsible for the effect [72–76].

Micro-Raman is a unique tool to address several fundamental aspects of SERS such as nature of the so-called hot spots (interstices of closely separated metal nanostructures), dependence of the enhancement on the orientation of the anisotropic nanostructures and aggregates with respect to polarization of the incident light. In this section, we will briefly highlight some of the important examples, where micro-Raman was employed to provide unique insight into these aspects.

Xia and coworkers have demonstrated that the polarization of light plays an important role for nanoparticles with anisotropic shapes, especially with truncated corners [77]. Optical dark-field mode imaging was employed to identify individual silver nanocubes deposited on silicon substrate as shown in Fig. 15.10a. Raman spectra were collected from the nanocubes, which were oriented in different directions with respect to the laser polarization. Subsequent SEM imaging of the same cubes enabled the authors to directly correlate the orientation of the nanocubes with respect to the light polarization and the SERS enhancement factor. They observed dramatic variation in SERS intensity when the nanocubes were oriented at different angles relative to the polarization of excitation laser as shown in Fig. 15.10b. SERS spectra of 1,4-benzenedithiol adsorbed on Ag nanocubes oriented in different directions showed different intensities with respect to the light polarization direction. The individual nanocubes with sharp corners were the most

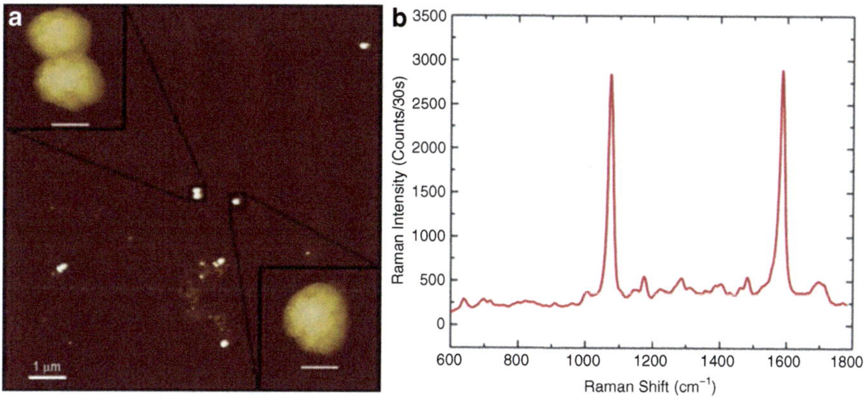

Fig. 15.11 (**a**) AFM image showing the single and dimers of Au nanoparticles and (**b**) surface-enhanced micro-Raman spectrum from the dimers of Au nanoparticles (Reprinted from [78])

active when they were oriented in such a way that their diagonal axes (corner to corner) were parallel to the polarization of the applied light source. On the other hand, the same nanocubes oriented with one of their faces parallel to the laser polarization were much less SERS active (Fig. 15.10b).

Although it has been predicted long back that dimers of nanoparticles result in much higher SERS enhancement compared to the individual particles, an experimental verification of the same using μRS was done recently. Halas and coworkers have obtained Raman spectra from individually and randomly formed dimers of gold nanoparticles [78]. They have employed p-mercaptobenzoic acid as analyte to reveal that the individual Au nanoparticles did not exhibit significant SERS intensity, while nanosphere dimers exhibited a strong enhancement (shown in Fig. 15.11). These experimental observations were confirmed by theoretical finite-difference time-domain studies. Furthermore, they have also noted the dependence of SERS enhancement on the excitation polarization with respect to the dimer axis. Polarization along the interparticle axis resulted in maximum enhancement, confirming previous studies.

In a very recent study, Yan et al. employed μRS to understand the effect of nanoparticle cluster size and intercluster spacing in a periodic array on the SERS enhancement [79]. They fabricated well-defined nanoparticle cluster arrays using template-guided self-assembly, which enabled control of the average number of nanoparticles in the clusters and edge-to-edge separation. SERS enhancement was found to increase with the increase in cluster size up to four particles after which it was found to saturate. Furthermore, they have shown that strong near-field interactions between individual clusters resulted in significant enhancement when the intercluster spacing was less than 200 nm. Their findings clearly highlight the importance of the multi-length-scale aspects for rational design of SERS substrates with high and more importantly reproducible SERS enhancement.

There are numerous examples where micro-Raman technique was employed to probe the enhancement of individual plasmonic nanostructures. Gunawidjaja and coworkers have used confocal Raman microscopy to map the SERS enhancement of a silver nanowire decorated with gold nanoparticles, termed nanocob [80, 81]. They have noted that the SERS enhancement of the nanocob structure was much higher compared to the individual silver nanowires and the enhancement produced by the nanocobs was highly sensitive to the orientation of a nanocob with respect to the direction of the excitation light polarization. μRS and Raman microscopy have emerged as powerful tools to gain a fundamental understanding of SERS and enable the design of highly SERS efficient metal nanostructures.

5.7 Biological Applications

The wealth of quantitative information that can be obtained in a noninvasive manner using μRS have included studies that identify subcellular reactions, organic pigments, imaging and tissue characterization [82, 83]. As the presence of water is not detrimental to Raman spectra, media in biological sample preparation and growth can permit cell sampling to be fixed, dried, analyzed alive, or even be measured in vivo circumventing the need for excisional biopsies [84, 85]. The "biochemical fingerprint" of Raman bands represents the vibrational modes of molecules in a cell within the region of interrogation [86]. The use of μRS is not limited to tissues and cells, but also in determining structure and conformation of proteins and nucleic acids in viruses [87].

Using spectral comparison and dilution studies, heterogeneity in a specific region can be determined by the use of data analysis methods like hierarchical clustering analysis (HCA), principle component analysis (PCA), etc. For each pixel, spectral images are created based on the intensity of individual peaks by assigning colors to identify different regions in a heterogeneous sample specimen. Such approaches can be used to evaluate the amount of cancerous tissue relative to noncancerous tissue [88, 89]. The transition from a normal cell to a diseased state is accompanied by changes in the variety of biomolecules that can be indiscriminately probed using Raman microscopy. This yields spectral signatures enabling differentiation between normal and cancerous cells [90, 91].

In another study by Uzunbajakava et al., nonresonant Raman imaging of protein distribution in two different cell types, peripheral blood lymphocytes (PBLs) and eye lens epithelial cells (LECs) was analyzed using confocal Raman microspectroscopy [86]. The difference in protein distribution in the cell nuclei of both PBL and LEC could be observed in the Raman images. Other techniques like TEM were used to analyze the chromatin distribution in LECs and fluorescence method was used for chromatin compactness. However, these techniques could not provide information on the single-cell level without using specific stains or labels. With the application of μRS, the boundaries of higher and lower protein intensities were clearly distinguishable with a size of 1 μm for the territories in the nucleus of mature PBLs. The nucleus of the LECs appeared to be more homogenous. The detectable number of counts for the high frequency protein band ($\sim$2,900 cm^{-1})

was four to five times higher than the corresponding band (~1,451 cm^{-1}) in the fingerprint region. In another study by the same authors, the distribution of DNA, RNA, and lipids in human cells were visualized using Raman imaging [86].

Raman maps of nuclei generated by their spatial distribution in cells were used to investigate the chemical changes caused by a disease. The white blood cells in the cerebrospinal fluid were investigated for the diagnosis of diseases in central nervous system (for the identification of bacterial meningitis) [92, 93]. Raman maps of yeast cells were recorded to differentiate various strains where the mapping approach proved valuable to eukaryotic cells owing to their molecular compartmentalization [94, 95]. The advantages of μRS have been realized in a number of studies dealing with different types of cancer, e.g., colon cancer, skin tumors, oral cancer, and breast cancer [96–100]. These studies demonstrated the prospects for Raman microscopy as a useful tool for early cancer diagnosis. To highlight the differences between cancerous and healthy cells rather than individual cells, Raman microscopy can be utilized in combination with chemometric analysis [84].

6 Summary and Future Perspective

The unique combination of rich molecular information (physical and chemical), high spatial resolution, nondestructive nature, and simplicity makes μRS an extremely valuable tool for characterization of nanostructures. As illustrated by numerous examples in this chapter, μRS has been applied to a wide variety of nanostructures. For instance, μRS has emerged to be an indispensible tool in the characterization of low-dimensional carbon nanostructures such as carbon nanotubes and graphene. There are only few recent reports in the literature where μRS has been applied to individual inorganic nanostructures such as ZnO, GaN nanowires to probe the crystalline orientation of the nanostructures in a nondestructive and in-device state. We believe that the application of μRS technique is still in its infancy especially in the context of characterizing individual nanostructures.

One of the major impediments for the application of μRS to broader set of nanomaterials (apart from the strong Raman scatterers such as CNTs, graphene) is the small cross section for normal Raman scattering process. In particular, the case of organic nanostructures, low scattering cross section combined with the need to limit the laser source power density (to avoid photodegradation of organic materials) poses a significant challenge. One way to overcome this obstacle is in combining μRS with SERS to enhance the Raman scattering from the nanostructures. In fact, there are already few reports which exploit this method [79–82]. A major challenge in this approach is in the fabrication of reliable, low-cost SERS substrates with uniform enhancement over large areas. We expect that such substrates, which are being extensively pursued, would dramatically broaden the application of μRS to polymeric and biological nanostructures.

References

1. Raman CV, Krishnan KS (1928) A new type of secondary radiation. Nature 121:501–502
2. Ferraro JR, Nakamoto K (1994) Introductory Raman spectroscopy. Academic, Boston
3. Gouadec G, Colomban P (2007) Raman spectroscopy of nanomaterials: how spectra relate to disorder, particle size and mechanical properties. Prog Cryst Growth Ch 53:1–56
4. Sourisseau C (2004) Polarization measurements in macro- and micro- Raman spectroscopies: molecular orientations in thin films and azo-dye containing polymer systems. Chem Rev 104:3851–3891
5. Rayleigh L (1879) Phil Mag 8:261
6. Conrady AE (1960) Applied optics and optical design, part 2. Dover, New York
7. Schmidt U, Hild S, Ibach W, Hollricher O (2005) Characterization of thin polymer films on the nanometer scale with confocal Raman AFM. Macromol Symp 230:133–143
8. Huan S, Lin W, Sato H, Yang H, Jiang J, Ozaki Y, Wu H, Shen G, Yu R (2007) Direct characterization of phase behavior and compatibility in PET/HDPE polymer blends by confocal Raman mapping. J Raman Spectrosc 38:260–270
9. van Overbeke E, Devaux J, Legras R, Carter JT, McGrail PT, Carlier V (2003) Phase separation in epoxy-copolyethersulphone blends: morphologies and local characterization by micro-Raman spectroscopy. Polymer 44:4899–4908
10. Hashida T, Jeong YG, Hua Y, Hsu SL (2005) Spectroscopic study on morphology evolution in polymer blends. Macromolecules 38:2876–2882
11. Li Z, Little R, Dervishi E, Saini V, Xu Y, Biris AR, Lupu D, Trigwell S, Saini D, Biris AS (2008) Micro-Raman spectroscopy analysis of catalyst morphology for carbon nanotubes synthesis. Chem Phys 353:25–51
12. Mayo JD, Behal S, Adronov A (2009) Phase separation of polymer-functionalized SWNTs within a PMMA/Polystyrene blends. J Polym Sci Polym Chem 47:450–458
13. Cou JY, Lensch-Falk JL, Hemesath ER, Lauhon LJ (2009) Vanadium oxide nanowire Phase and orientation analyzed by Raman spectroscopy. J Appl Phys 105:034310-1–034310-6
14. Wang ZL (2004) Functional oxides nanobelts-materials, properties and potential applications in nanosystems and biotechnology. Annu Rev Phys Chem 55:159–196
15. Bliznyuk VN, Singamaneni S (2009) Polymer carbon nanotube composites. Encyclopedia of nanoscience and nanotechnology, (2012), ASP, Ed. H. S. Nalwa
16. Tasis D, Tagmatarchis N, Bianco A, Prato M (2006) Chemistry of carbon nanotubes. Chem Rev 106:1105–1136
17. Compton OC, Nguyen SB (2010) Graphene oxide, highly reduced graphene oxide, and graphene: versatile building blocks for carbon-based materials. Small 6:711–723
18. Allen MJ, Tung VC, Kaner RB (2010) Honeycomb carbon: a review of graphene. Chem Rev 110:132–145
19. Jorio A, Saito R, Dresselhauss G, Dresselhaus MS (2004) Determination of nanotube properties by Raman spectroscopy. Phil Trans R Soc Lond A 362:2311–2336
20. Kobayashi K, Takagi D, Ueno Y, Homma Y (2004) Characterization of carbon nanotubes suspended between nanostructures using micro-Raman spectroscopy. Physica E 24:26–31
21. Dresselhaus MS, Eklund P (2000) Phonons in carbon naotubes. Adv Phys 49:705–814
22. Anderson N, Hartschuh A, Novotny L (2007) Chirality changes in carbon nanotubes studied with near-field Raman spectroscopy. Nano Lett 7:577–582
23. Gomez-De Arco L, Lei B, Cronin BS, Zhou C (2008) Resonant micro-Raman spectroscopy of aligned single-walled carbon nanotubes on a-plane sapphire. Appl Phys Lett 93:123112
24. Jorio A, Pimenta MA, Souza Filho AG, Saito R, Dresselhaus G, Dresselhaus MS (2003) Characterizing carbon nanotube samples with resonance Raman scattering. New J Phys 5:139.1–139.17
25. Dresselhaus MS, Dresselhaus G, Jorio A, Souza Filho AG, Saito R (2002) Raman spectroscopy on isolated single wall carbon nanotubes. Carbon 40:2043–2061

26. Fantini C, Pimenta MA, Dantas MSS, Ugarte D, Rao AM, Jorio A, Dresselhaus G, Dresselhaus MS (2001) Micro-Raman investigation of aligned single-wall carbon nanotubes. Phys Rev 63:161405-1–161405-4
27. LeMieux MC, Roberts M, Barman S, Jin YW, Kim JM, Bao Z (2008) Self-sorted, aligned nanotube networks for thin film transistors. Science 321:101–104
28. Ferrari AC, Meyer JC, Scardaci V, Casiraghi C, Lazzeri M, Mauri F, Piscanec S, Jiang D, Novoselov KS, Roth S, Geim AK (2006) Raman spectrum of graphene and grahene layers. Phys Rev Lett 97:187401-1–187401-4
29. Graf D, Molitor F, Ensslin K, Stampfer C, Jungen A, Hierold C, Wirtz L (2007) Spatially resolved Raman spectroscopy of single- and few-layer graphene. Nano Lett 7:238–242
30. Faugeras C, Nerriere A, Potemski M, Mahmood A, Dujardin E, Berger C, de Heer WA (2008) Few-layer graphene on SiC, pyrolitic graphene: a Raman scattering study. Appl Phys Lett 92:011914-1–011914-3
31. Lee DS, Riedl C, Krauss B, von Klitzing K, Starke U, Smet JH (2008) Raman spectra of epitaxial graphene on SiC and of epitaxial graphene transferred to SiO_2. Nano Lett 8:4320–4325
32. Ramesha GK, Sampath SE (2009) Electrochemical reduction of oriented graphene oxide films: an in situ Raman spectroelectrochemical study. J Phys Chem C 113:7985–7989
33. Bulkin BJ, Lewin M, DeBlase FJ (1985) Conformational change, chain orientation, and crystallinity in poly (ethylene terapthalate) yarns: Raman spectroscopic study. Macromolecules 18:2587–2594
34. Fleming OS, Kazarian SG (2004) Confocal Raman microscopy of morphological changes in poly(ethylene terepthalate) film induced by supercritical CO_2. Appl Spectrosc 58:390–394
35. Melveger AJ (1972) Laser-Raman study of crystallinity changes in poly(ethylene Terepthalate). J Polym Sci Part A-2 10:317–322
36. Grasselli JG, Bulkin BJ (1991) Analytical Raman spectroscopy. Wiley, New York
37. Rossier JS, Bercier P, Schwartz A, Loridant S, Girault HH (1999) Topography, crystallinity and wettability of photoablated PET surfaces. Langmuir 15:5173–5178
38. Gatos KG, Minogianni C, Galiotis C (2007) Quantifying crystalline fraction within polymer spherulites. Macromolecules 40:786–789
39. Martin J, Bourson P, Dahoun A, Hiver JM (2009) The β- spherulite morphology of isotactic polypropylene investigated by Raman spectroscopy. Appl Spectrosc 63:1377–1381
40. Ding Y, Wang ZL (2004) Structure analysis of nanowires and nanobelts by transmission electron microscopy. J Phys Chem B 108:12280–12291
41. Ozgur U, Alivov Y, Liu C, Teke A, Reshchikov MA, Dogan S, Avrutin V, Cho SJ, Morkoc H (2005) A comprehensive review of ZnO materials and devices. J Appl Phys 98:041301-1–041301-103
42. Wang ZL (2008) Oxide nanobelts and nanowires – growth, properties and applications. J Nanosci Nanotechol 8:27–55
43. Pan ZW, Dai ZR, Wang ZL (2001) Nanobelts of semiconducting oxides. Science 291:1947–1949
44. Singamaneni S, Gupta M, Yang R, Tomczak M, Naik RR, Wang ZL, Tsukruk VV (2009) Nondestructive in situ identification of crystal orientation of anisotropic ZnO nanostructures. ACS Nano 3:2593–2600
45. Hu Y, Gao Y, Singamaneni S, Tsukruk VV, Wang ZL (2009) Converse piezoelectric effect induced transverse deflection of a free-standing ZnO microbelt. Nano Lett 9:2661–2665
46. Pauzauskie PJ, Talaga D, Seo K, Yang P, Lagugné-Labarthet F (2005) Polarized Raman confocal microscopy of single gallium nitride nanowires. J Am Chem Soc 127:17146–17147
47. Livneh T, Zhang J, Cheng G, Moskovits M (2006) Polarized Raman scattering from single GaN nanowires. Phys Rev B 74:035320-1–035320-10
48. Fan HM, Fan XF, Ni ZH, Shen ZX, Feng YP, Zou BS (2008) Orientation-dependent Raman –spectroscopy of single wurtzite CdS Nanowires. J Phys Chem C 112:1865–1870

49. Coleman JN, Khan U, Blau WJ, Gun'ko YK (2006) Small but strong: a review of the mechanical properties of carbon nanotube-polymer composites. Carbon 44:1624–1652
50. Zhu J, Peng HQ, Rodriguez-Macias F, Margrave JL, Khabashesku VN, Imam AM, Lozano K, Barrera EV (2004) Reinforcing epoxy polymer composites through covalent integration of the functionalized nanotubes. Adv Funct Mater 14:643–648
51. Blond D, Barron V, Ruether M, Ryan KP, Nicolosi V, Blau WJ, Coleman JN (2006) Enhancement of modulus, strength, and toughness in poly(methyl methacrylate)-based composites by the incorporation of poly(methyl methacrylate)- functionalized nanotubes. Adv Funct Mater 16:1608–1614
52. Chae HG, Minus ML, Kumar S (2006) Oriented and exfoliated single wall carbon nanotubes in polyacrylonitrile. Polymer 47:3494–3504
53. Tanaka M, Young RJ (2006) Polarised Raman spectroscopy for the study of molecular orientation distribution in polymers. J Mater Sci 41:963–991
54. Haggenmueller R, Zhou W, Fischer JE, Winey KI (2003) Production and characterization of polymer nanocomposites with highly aligned single-walled carbon nanotubes. J Nanosci Nanotechnol 2(1–2):105–110
55. Potschke P, Brunig H, Janke A, Fischer D, Jennichen D (2005) Orientation of multiwalled carbon naotubes in composites with polycarbonate by melt spinning. Polymer 46: 10355–10363
56. Zhao Q, Wagner HD (2004) Raman spectroscopy of carbon-nanotube-based composites. Phil Trans R Soc Lond A 362:2407–2424
57. Deng L, Young RJ, Zwaag SVN, Picken S (2010) Characterization of the adhesion of single-walled carbon nanotubes in poly(P-Phenylene Terepthalamide) composite fibers. Polymer 51:2033–2039
58. Yang M, Koutsos V, Zaiser M (2005) Interactions between polymers and carbon nanotubes: a molecular dynamics study. J Phys Chem B 109:10009–10014
59. Dietrich B, Dombrowski KF (1999) Experimental challenges of stress measurements with resonant micro-Raman spectroscopy. J Raman Spectrosc 30:893–897
60. Mu M, Osswald S, Gogotsi Y, Winey KI (2009) An in situ Raman spectroscopy study of stress transfer between carbon nanotubes and polymer. Nanotechnology 20:335703-1-7
61. Ko H, Pikus Y, Jiang C, Jauss A, Hollricher O, Tsukruk VV (2004) High-resolution Raman microscopy of curled carbon nanotubes. Appl Phys Lett 85:2598–2600
62. Singamaneni S, Bertoldi K, Chang S, Jang J-H, Young S, Thomas EL, Boyce M, Tsukruk VV (2009) Bifurcated mechanical behavior of deformed periodic porous solids. Adv Funct Mater 19:1426–1436
63. Singamaneni S, Bertoldi K, Chang S, Jang J-H, Thomas EL, Boyce M, Tsukruk VV (2009) Instabilities and pattern transformation in periodic, microporous, elasto-plastic solids. ACS Appl Mater Interfaces 1:42–47
64. Beecham T, Graham S, Kearney SP, Phinney LM, Serrano JR (2007) Simultaneous mapping of temperature and stress in microdevices using micro-Raman spectroscopy. Rev Sci Instrum 78:061301-1–061301-9
65. Beecham T, Christensen A, Green DS, Graham S (2009) Assessment of stress contributions in GaN high electron mobility transistors of differing substrates using Raman spectroscopy. J Appl Phys 106:114509-1–114509-9
66. Wermelinger T, Charpentier C, Yuzsek MD, Spolanek R (2009) Measuring stresses in thin metal films by means of Raman microscopy using silicon as a strain gage material. J Raman Spectrosc 40:1849–1857
67. Becker M, Scheel H, Christiansen C, Strunk HP (2007) Grain orientation, texture, and internal stress optically evaluated by micro-Raman spectroscopy. J Appl Phys 101:063531-1–06351-10
68. Gigler A, Huber AJ, Bauer M, Ziegler A, Hillenbrand R, Stark R (2009) Nanoscale residual stress-field mapping around nanoindents in SiC by IR s-SNOM and confocal Raman microscopy. Opt Express 17:22351–22357

69. Chen J, Conache G, Pistol ME, Gray SM, Borgstrom MT, Xu HX, Xu HQ, Samuelson L, Hakanson U (2010) Probing strain in bent semiconductor nanowires with Raman spectroscopy. Nano Lett 10:1280–1286
70. Zheng K, Han XD, Wang LH, Zhang YF, Yue YH, Qin Y, Zhang XN, Zhang Z (2009) Atomic mechanisms governing the elastic limit and the incipient plasticity of bending Si nanowires. Nano Lett 9:2471–2476
71. Ko H, Singamaneni S, Tsukruk VV (2008) Nanostructiured surfaces and assemblies as SERS media. Small 4:1576–1579
72. Otto A, Mrozek I, Grabhorn H, Akemann W (1992) Surface enhanced Raman-scattering. J Phys Condens Matter 4:1143–1212
73. Otto A (2002) What is observed in single molecule SERS, and why? J Raman Spectrosc 33:593–598
74. Campion A (1985) Raman spectroscopy of molecules adsorbed on solid surfaces. Annu Rev Phys Chem 36:549–572
75. Moskovits M (1985) Surface-enhanced spectroscopy. Rev Mod Phys 57:783–826
76. Kambhampati P, Child CM, Foster MC, Campion A (1998) On the chemical mechanism of surface enhanced Raman scattering: experiment and theory. J Chem Phys 108:5013–5026
77. McLellan JM, Li Z-Y, Siekkinen AR, Xia Y (2007) The SERS activity of a supported Ag nanocube strongly depends on its orientation relative to laser polarization. Nano Lett 7:1013–1017
78. Talley CE, Jackson JB, Oubre C, Grady NK, Hollars CW, Lane SM, Huser TR, Nordlander P, Halas NJ (2005) Surface-enhanced Raman scattering from individual Au nanoparticles and nanoparticle dimer substrates. Nano Lett 5:1569–1574
79. Yan B, Thubagere A, Premasiri WR, Ziegler LD, Dal Negro L, Reinhard BM (2009) Engineered SERS substrates with multiscale signal enhancement: nanoparticle cluster arrays. ACS Nano 3:1190–1202
80. Gunawidjaja R, Peleshanko S, Ko H, Tsukruk VV (2008) Bimetallic nanocobs: decorating silver nanowires with gold nanoparticles. Adv Mater 20:1544–1549
81. Gunanwidjaja R, Kharlampieva E, Choi I, Tsukruk VV (2009) Bimetallic nanostructures as active Raman markers: gold-nanoparticle assembly on 1D and 2D silver nanostructure surfaces. Small 5:2460–2466
82. Zhang G, Moore DJ, Flach CR, Mendelsohn R (2007) Vibrational microscopy of skin: from single cells to intact tissue. Anal Bioanal Chem 387:1591–1599
83. Bocklitz T, Putsche M, Stuber C, Kas J, Niendorf A, Rosch P, Popp J (2009) A comprehensive study of classification methods for medical diagnosis. J Raman Spectrosc 40:1759–1965
84. Mariani MM, Lampen P, Popp J, Wood B, Deckert V (2009) Impact of fixation on in vitro cell culture lines monitored with Raman spectroscopy. Analyst 134:1154–1161
85. Uzunbajakava N, Lenferink A, Kraan Y, Willekens B, Vrensen G, Greve J, Otto C (2003) Nonresonant Raman imaging of protein distribution in single human cells. Biopolymers 72:1–9
86. Krafft C (2004) Bioanalytical applications of Raman spectroscopy. Anal Bioanal Chem 378:60–62
87. Lambert PJ, Ahitman AG, Dyson OF, Akula SM (2006) Raman spectroscopy: the gateway into tomorrow's virology. Virol J 3(51):1–8
88. Keller MD, Kanter EM, Mahadevan-Jansen A (2006) Raman spectroscopy for cancer diagnosis. Spectroscopy 21(11):33–41
89. Lasch P, Diem M, Naumann D (2004) FT-IR microspectroscopic imaging of prostate tissue sections. Proc SPIE 5321:1–9
90. Jackson M, Sowa MG, Mantsch HH (1997) Infrared spectroscopy: a new frontier in medicine. Biophys Chem 68(1–3):109–125
91. Gazi E, Dwyer J, Lockyer N, Gardner P, Vickerman JC, Miyan J, Hart CA, Brown M, Shanks JH, Clarke N (2004) The combined application of FTIR microspectroscopy and ToF-SIMS imaging in the study of prostate cancer. Faraday Discuss 126:41–59

92. Harz M, Kiehntopf M, Stockel S, Rosch P, Deufel T, Popp J (2008) Analysis of single blood cells for CSF diagnostics via a combination of fluorescence staining and micro-Raman spectroscopy. Analyst 133:1416–1423
93. Harz M, Kiehntopf M, Stockel S, Rosch P, Straube E, Deufel T, Popp J (2009) Direct analysis of clinical relevant single bacterial cells from cerebrospinal fluid during bacterial meningitis by means of micro-Raman spectroscopy. J Biophoton 2:70–80
94. Rosch P, Harz M, Schitt M, Popp J (2005) Raman spectroscopic identification of single yeast cells. J Raman Spectrosc 36:377–379
95. Rosch P, Harz M, Peschke KD, Ronneberger O, Burkhardt H, Popp J (2006) Identification of single eukaryotic cells with micro-Raman spectroscopy. Biopolymers 82:312–316
96. Krafft C, Ramoji AA, Bielecki C, Volger N, Meyer T, Akimov D, Rosch P, Schmitt S, Dietzek B, Petersen I, Stallmach A, Popp J (2009) A comparative Raman and CARS imaging study of colon tissue. J Biophoton 2:303–312
97. Krafft C, Steiner G, Beleites C, Salzer R (2009) Disease recognition by infrared and Raman spectroscopy. J Biophoton 2:13–28
98. Malini R, Venkatakrishna K, Kurien J, Pai KM, Rao L, Kartha VB, Krishna CM (2006) Discrimination of normal, inflammatory, premalignant and malignant oral tissue: a Raman spectroscopy study. Biopolymers 81:179–193
99. Redd DC, Feng ZC, Yue KT, Gansler TS (1993) Raman spectroscopic characterization of human breast tissues: implications for breast cancer diagnosis. Appl Spectrosc 47:787–791
100. Shafer-Peltier KE, Haka AS, Fitzmaurice M, Crowe J, Myles J, Dasari RR, Feld MS (2002) Raman microspectroscopic model of human breast tissue: implications for breast cancer diagnosis in vivo. J Raman Spectrosc 3:552–563

Tip-Enhanced Raman Spectroscopy 16

Norihiko Hayazawa, Alvarado Tarun, Atsushi Taguchi, and
Kentaro Furusawa

1 Definition of the Topic

This chapter describes Raman spectroscopy with a high spatial resolution beyond the diffraction limit of light.

2 Overview

Near-field scanning optical microscopy (NSOM) has been developed as a combination of scanning probe microscope and optical microscope in which the spatial resolution is determined by scanning probe microscope resolution while the signals detected are coming from several optical interactions. As a result, NSOM has achieved a higher spatial resolution than that of the classical optical microscopy that uses a conventional lens, which is strictly limited by the diffraction

N. Hayazawa (✉)
Nanophotonics Laboratory, RIKEN, Wako, Saitama, Japan
and
Near-Field NanoPhotonics Research Team, RIKEN, Wako, Saitama, Japan
and
CREST, Japan Science and Technology Agency, Kawaguchi, Saitama, Japan

A. Tarun • K. Furusawa
Nanophotonics Laboratory, RIKEN, Wako, Saitama, Japan

A. Taguchi
Nanophotonics Laboratory, RIKEN, Wako, Saitama, Japan
and
CREST, Japan Science and Technology Agency, Kawaguchi, Saitama, Japan

Challa S. S. R. Kumar (ed.), *Raman Spectroscopy for Nanomaterials Characterization*,
DOI 10.1007/978-3-642-20620-7_16, © Springer-Verlag Berlin Heidelberg 2012

limit of light. In this chapter, recent advances in near-field optical microscopy and spectroscopy are reviewed, particularly metallic tip–based probes. One of the most successful capabilities of the metallic tip is the application to Raman spectroscopy with respect to well-known surface-enhanced Raman scattering (SERS) since the metallic tip can work as an electric field enhancer in this case. Because of the analogy with SERS, near-field detection of Raman scattering using a metallic tip has been recognized as tip-enhanced Raman spectroscopy (TERS). The discussion covers tip-enhancement effects using the metallic tip, several system geometries, and applications of Raman and nonlinear Raman spectroscopy.

3 Introduction

Recently, there have been tremendous developments in the growth technologies of nanoscale materials, which show new characteristic features, distinctly different from those of bulk materials. However, in contrast to the rapid growth of fabrication techniques, the progress in analysis techniques for such nanoscale materials has been slow. In this chapter, we introduce tip-enhanced Raman spectroscopy (TERS) for the characterization of materials at nanoscale. This newly emerging technique, which has an extremely high spatial resolution far beyond the diffraction limits of the probing light, is suitable for analyzing nanomaterials.

The principle of how to achieve a high spatial resolution beyond the diffraction limit of light is based on near-field optics [1]. Near-field scanning optical microscopy (NSOM) is characterized by its super-resolution capability. It can exceed the classical spatial resolution limit, the so-called diffraction limit, due to the wave nature of light [2]. The imaging mechanism of the NSOM is different from the classical optical microscopy, which uses an objective lens. In NSOM, the light intensity is detected as a result of strong electromagnetic interactions between the probe and the sample structure in the near-field via evanescent photons. In this chapter, we will show a simple aspect on how to confine the photons in the near-field and utilize as a nanolight source for spectroscopic studies in Raman spectroscopy. In Sect. 3.1, we will categorize NSOM into aperture and apertureless probes and then the discussion will be focused on the latter, apertureless probes. The advantages of the apertureless probes will be presented in Sect. 3.2. They are discussed in terms of localized surface plasmon polariton (SPP) excitations, which is promising for nanoscale sensing since the signal level becomes smaller as the sample volume is reduced. Section 4 introduces the typical system configurations, including optical geometries, metallic tips for tip enhancement, and feedback schemes for precise tip-sample distance control in the near-field. The most widely developed tip enhancement in Raman spectroscopy (TERS) is discussed in Sect. 5.1, and the extended works in UV and DUV regions are introduced in Sect. 5.2. For higher sensitivity and resolution, tip enhancement applied to nonlinear Raman spectroscopy is introduced in Sect. 5.3. Originated from the successful demonstrations of TERS, other emerging applications of TERS are presented in Sect. 5.4.

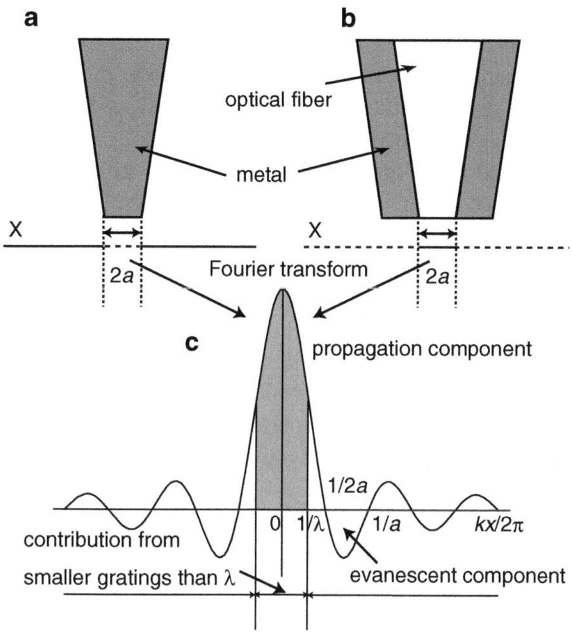

Fig. 16.1 Concept of (**a**) apertureless and (**b**) aperture probes. (**c**) Fourier transform of the tip end

3.1 Beyond the Diffraction Limit: Aperture and Apertureless NSOM Probes

According to Fourier optics [3], a small aperture or a small object with a finite size can be regarded as a superposition of grating sets, which have broad distributions of lattice constants from much smaller to larger than the wavelength of a light. These concepts are illustrated in Fig. 16.1. Small aperture (Fig. 16.1b) and small object (Fig. 16.1a) with the same size (=$2a$, which is much smaller than the wavelength of light) become the same *Sinc* function by Fourier transformation. Among those gratings, the grating with a lattice constant larger than the wavelength can diffract the light to the proper angle that satisfies the diffraction condition. Thus the diffracted light will propagate at a specific direction. On the other hand, the grating with lattice constant smaller than the light wavelength will not satisfy the condition for propagation of a light. Hence, the diffracted light becomes an evanescent field, which cannot propagate from the grating structure and remains localized at the object [4]. With the concept of these evanescent field components, light field is confined and localized either at the aperture or the object, which results in the generation of a nanolight source having the size that corresponds to the diameter (=$2a$) of the aperture or object. This is because the evanescent light components have wave number, k, which is much larger than the propagation light component as illustrated in Fig. 16.1c. The concept of confining light as presented in Fig. 16.1a, b has been realized as so-called apertureless-type near-field microscopes [5–7] and aperture-type near-field microscopes [8], respectively. The typical configuration of the near-field probes is illustrated in Fig. 16.1a, b. Aperture-type probes are generally made of sharpened optical fiber coated with metal in order to block the leakage of light from the taper except at the aperture. Apertureless probes, on the other hand, are generally sharp silicon cantilever tips of atomic force microscope (AFM) coated with metal or sharpened pure metallic wires (see Sect. 4.2 for details). Theoretically, the same spatial resolution can be expected in both the aperture and apertureless near-field microscopes. Higher spatial resolution requires either smaller aperture of the optical fiber depicted in Fig. 16.1b or smaller diameter of the sharpened tip shown in Fig. 16.1a. According to Fig. 16.1c, the wave number becomes higher when a becomes smaller. Also, smaller aperture or tip diameter means more evanescent components. However, for practical reasons, it would be more difficult to detect the weak signals in the nanoscale since smaller apertures make the near-field signal smaller. This is crucial particularly for Raman applications since Raman scattering cross section is extremely small, which results in the use of large aperture (poor resolution) or unrealistic long acquisition time [9, 10]. On the other hand, additional signal enhancement can be expected in the case of apertureless-type probes especially when the probes are made of a metal. This additional signal enhancement is specifically expected when the tip diameter becomes smaller, which corresponds to higher spatial resolution. In the next section, we discuss the mechanism of the electric field enhancement effect by the metallic probe tip.

3.2 Tip Enhancement

When a metallic probe having a nanometric tip diameter is illuminated with an optical field, conductive free electrons collectively oscillate at the surface of the metal (Fig. 16.2). The quantum of the induced oscillation is referred as surface plasmon polariton (SPP) [11]. The concentrated electrons (and the positive charge) at the apex can generate a strong external electric field. Photon energy is confined in the local vicinity of the tip apex. Therefore, the metallic tip works as a photon reservoir [12]. The local electric field is used as a nanolight source to excite photon-matter interaction for various spectroscopic techniques, in particular, Raman spectroscopy in this chapter. The enhancement of Raman scattering by metallic structures has been widely investigated as surface-enhanced Raman scattering (SERS) since the 1970s [13–15]. For the SERS effect, metallic nanoparticles and their aggregates are extensively used in various models to understand the physical behavior of the local electric field and the magnitude of the enhancement [16]. Similarly, in the case of the tip enhancement, the local electric field has been calculated using a variety of numerical methods that are based on electromagnetic mechanism. So far various analytical methods have been utilized for the calculation of electromagnetic field such as finite difference

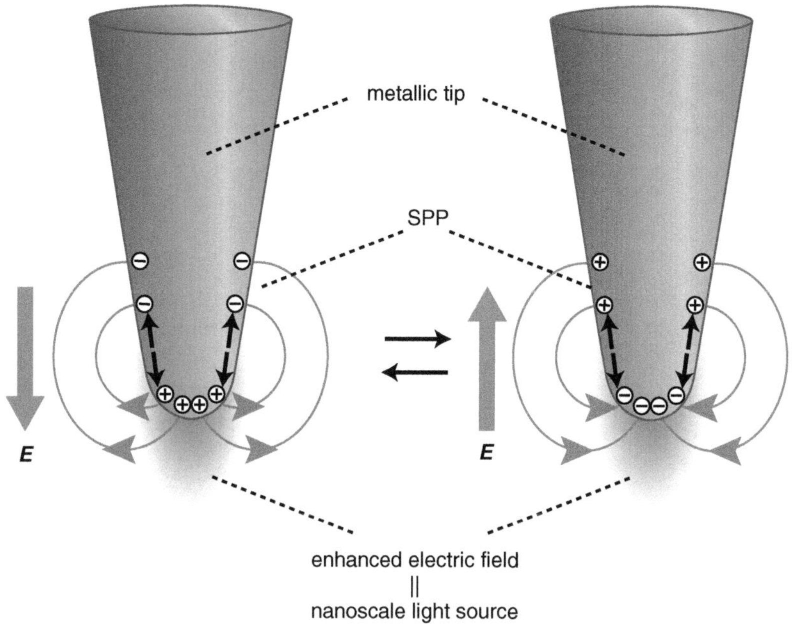

Fig. 16.2 Localized surface plasmon polariton excited at the metallic tip apex

time domain method (FDTD) [17], boundary element method (BEM) [18], finite element method (FEM) [19], and multiple multipole method (MMP) [20]. All of these methods at least qualitatively resulted in the same conclusion for the tip enhancement. Figure 16.3a shows the local electric field distribution near a silver probe tip calculated via FEM derived from the numerical solution of Maxwell's equations. It can be seen that the optical field is highly confined into a tiny volume near the tip end. In addition to the spatial confinement, the optical field is strongly amplified by 15 times and corresponds to 225 times in electric field intensity. Calculation methods or models, however, could vary the factor of the enhancement; nevertheless the size of the enhanced field is comparable to that of the probe tip in any cases [12]. Figure 16.3b shows the same tip but with s-polarized light excitation. The field enhancement is very small, down to only $\sim$3 times, compared with p-polarized light excitation in Fig. 16.3a. The role of SPP is clearly evident in the case of a dielectric tip (Fig. 16.3c) or semiconducting tip (Fig. 16.3d). The maximum field in these cases is 1.7 and 2.6 times, respectively. It has been shown that a metallic probe tip highly localizes and strongly amplifies optical field through the resonance effect of the SPP at the probe tip. This concept has made it possible to optically observe a variety of materials with a nanometric spatial resolution and, thus, is promising for Raman spectroscopy.

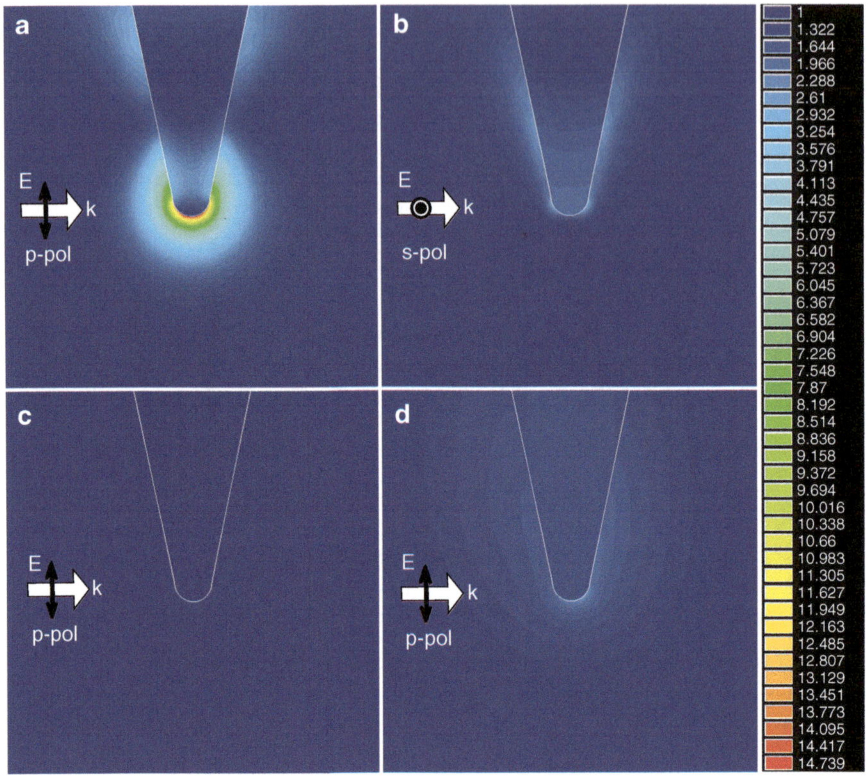

Fig. 16.3 Electric field distribution of a silver tip apex (30 nm ϕ) excited by (**a**) p-polarized and (**b**) s-polarized light (λ: 532 nm). (**c**) and (**d**) are the field distribution of SiO_2 and silicon tip with p-polarized light

4 Basic Methodology

4.1 Optical Geometry: Transmission Mode and Reflection Mode

The basic system configurations so far reported by several groups are based on a combination of scanning probe microscope (SPM) and an inverted optical microscope as illustrated in Fig. 16.4a [21]. An expanded and collimated light field from a visible laser enters into the epi-illumination optics of an inverted optical microscope. As the metallic tip is moved closer to the focused spot generated by a high numerical aperture (NA = 1.4) objective lens, a localized and enhanced electric field is generated at the tip apex as described in Sect. 3. This is because the high NA objective can realize strong longitudinal field along the tip axis, which corresponds to p-polarized light [22–24]. The enhanced and localized electric field at the tip is utilized as a nanolight source for Raman spectroscopic studies as discussed in the following sections. The distance between the sample and the

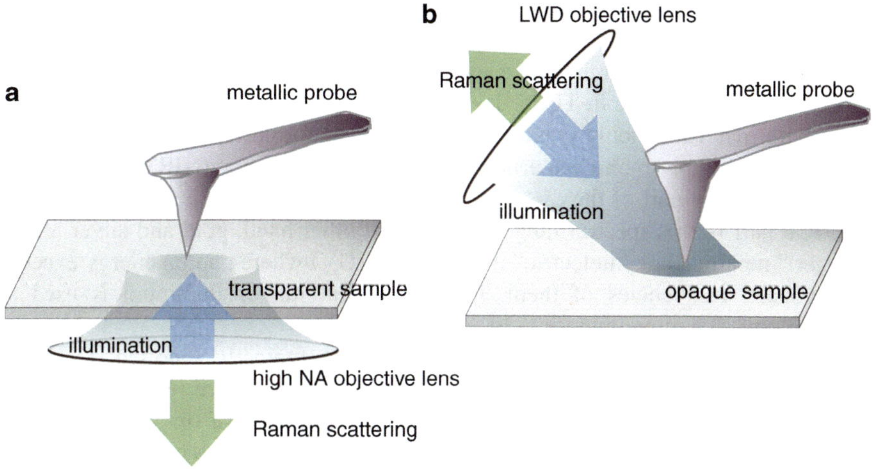

Fig. 16.4 Schematic of (**a**) transmission and (**b**) reflection mode NSOM

metallic tip is regulated by SPM feedback scheme such as atomic force microscope (AFM) and scanning tunneling microscope (STM) (see Sect. 4.3). The sample is scanned with piezoelectric transducers (PZT) in the X–Y plane. Scanning the XY-PZT sample stage while simultaneously detecting the optical signal can perform near-field optical imaging in the nanoscale. The spatial resolution is determined by the size of the enhanced electric field and in most cases corresponds to the tip diameter. It should be pointed out that the system configuration described above has been recently recognized as "transmission mode" [25] because the light has to transmit through the sample. So, the applicability of the system is limited to very thin or transparent samples. In order to apply NSOM for thick or opaque samples, several groups developed the so-called reflection mode [25, 26]. This configuration is illustrated in Fig. 16.4b. While the reflection mode is advantageous for opaque samples, the background signals are relatively higher than transmission mode because a lower NA objective lens with a long working distance is used to illuminate the tip due to the limited spatial clearance around the SPM scanner head and tip. The details of both configurations are described in Ref. [27].

4.2 Metallic Tips for Tip Enhancement

The metallic tip is the most important component consisting TERS system. The tip radius, material, structure, and its SPP resonance have to be properly controlled to perform TERS measurements successfully with high resolution and enhancement in a reproducible manner.

The resolution of the TERS imaging is basically determined by the sharpness of the tip apex. The diameter of the tip is usually required to be the order of several tens of nanometers or even less. The TERS enhancement, on the other hand, comes from the SPP resonance at the tip. Thus, it is also important to control the spectral response of the metallic tip to match the excitation light frequency. The control of the plasmon resonance at the probe apex having the size of the order of nanometer is still a challenging issue in TERS developments.

As tip materials, noble metals such as silver and gold are mostly used. These metals are good plasmonic materials in visible to near-infrared (NIR) because of the small imaginary part of the dielectric function while realizing a negative value of the real part in this spectral range [28]. On the other hand, gold and silver are no longer "metallic" but "dielectric" in ultraviolet (UV) where photon energy exceeds the plasma frequencies of them. In this spectral range, aluminum is used as a plasmonically active material [29, 30].

So far, two types of metallic tips have been typically reported. One is a metal-coated cantilever and the other is etched metallic wire, each of which is used in a suitable combination with the feedback scheme of tip-sample distance control (see Sect. 4.3). The metal-coated cantilever is fabricated conveniently from a commercially available silicon (Si) or silicon nitride (Si_3N_4) cantilever of AFM by depositing thin metallic film onto the probe surface (Fig. 16.5a) [31–33]. Metal with >99.999% purity is thermally evaporated under vacuum condition and deposited onto the probe surface slowly with a rate of less than 1 Å/s in order not to damage the tip apex. The deposition thickness is typically several tens of nanometers. A small diameter of several tens of nanometers at the tip is easily obtained.

The second type of the metallic probe, etched wire (Fig. 16.5b), is used either in STM-based TERS systems [34, 35] or in AFM-based systems operated by a shear-force feedback mechanism in which an etched wire is glued to one of the prongs of a tuning fork [36–38]. Silver wires are etched using perchloric acid, ethanol, and water solution [35]. Gold wires are etched using hydrochloric acid and ethanol solution [39].

The reported Raman enhancement factors experimentally obtained using metallic probe is typically 10^3–10^4 [40]; however, the enhancement factor also depends on the number of molecules contributed to the net enhancement in the size of tip-enhanced field. This fact suggests that even a much higher enhancement factor could be achievable (e.g., $\sim 10^{14}$) similar to single molecule detection by SERS [41]. In any case, the enhancement is related to the SPP of the metallic tip, thus, the spectral investigation of the tip plasmon is quite important. One of these works is the spectral measurement of the light scattered by the metallic tip. Typically the measured spectra shows a single resonance peak with the peak width of approximately 100 nm, and the peak position appears in the visible ~ NIR between 500 and 800 nm when a gold or silver tip is used [42, 43]. A gold tip shows the resonance peak at a relatively longer wavelength compared with silver because of the difference in the dielectric functions [44].

The maximum TERS enhancement factor is expected when the tip plasmon and the excitation laser are spectrally matched. In practice, the operation wavelength is fixed to the available laser facility used for TERS experiments, and the tip plasmon is controlled. One of the techniques to control the SPP resonance wavelength

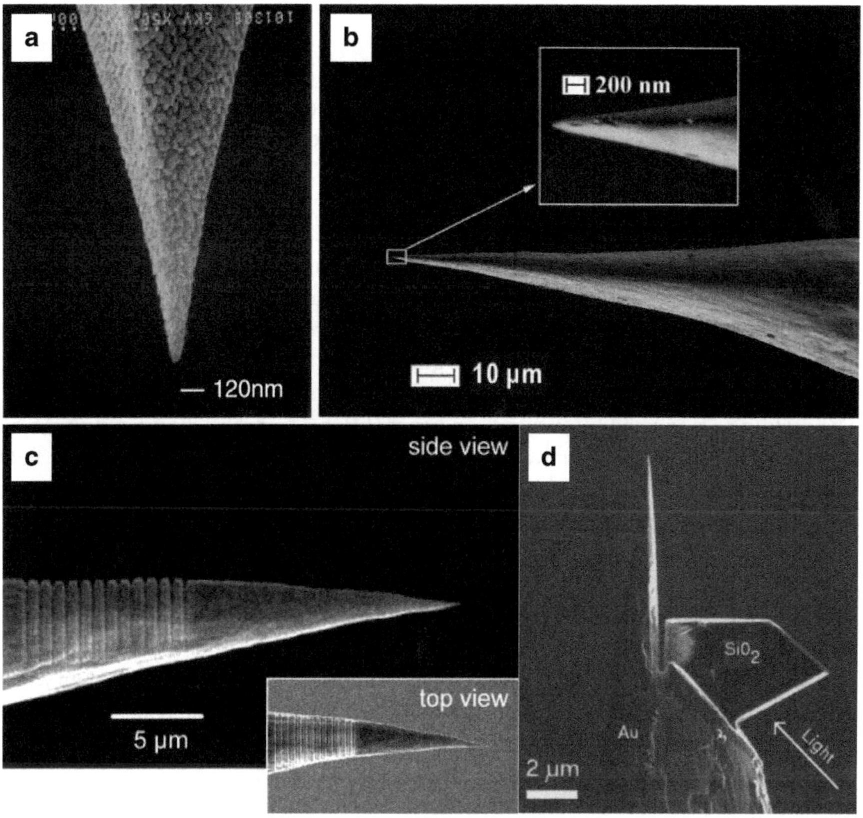

Fig. 16.5 SEM images of various types of metallic probes. (**a**) Ag-coated Si cantilever; (**b**) Ag tip made by wire etching (Reprinted from [37], Copyright 2005, with permission from American Chemical Society); (**c**) metallic probe with a grating fabricated on the shaft (Reprinted from [48], Copyright 2005, with permission from American Chemical Society); (**d**) metallic probe with Otto coupler (Reprinted from [49], Copyright 2005, with permission from American Institute of Physics)

of the metal-coated probe is to modify the refractive index of the probe base materials, e.g., Si [45, 46]. The plasmon resonance wavelength of the metallic nanostructure is known to be affected by its surrounding medium: increasing the refractive index of the surrounding medium shifts the resonance wavelength to the longer side, and vice versa [47]. Substituting the refractive index of Si (n = 4.4) to lower indices such as n = 1.5 (SiO_2) and n = 1.4 (AlF_2) provides the blue-shifted plasmon resonance. Such a probe is used in combination with the blue to green lasers and improved enhancement is observed [45]. The substitution of the refractive index is realized either by covering the surface of the Si probe with the low index material [45], or thermally oxidizing the surface of the Si probe [46]. The advantages of the latter method are that the tip does not become dull after the index modification, and the effective index can be continuously adjusted by the thickness of the oxidized layer, which provides flexibility in optimizing the plasmon resonance wavelength.

Another important aspect in considering the TERS probe is the coupling efficiency between the tip plasmon and the far-field light, because the tip plasmon is excited initially by the light coming from the far-field, and the Raman signal is consequently measured by a detector placed also in the far-field. Two typical configurations are introduced in Sect. 4.1. Increasing the coupling efficiency both in illumination and detection is a straightforward strategy to achieve a higher signal-to-noise ratio in TERS. So far only a few methods have been reported for efficient coupling of illumination. The probe shown in Fig. 16.5c has a grating on the tip body. Incoming far-field light is efficiently converted to the SPP propagating toward the tip apex through the grating [48]. Without the grating, the coupling between the surface plasmon and the far-field light becomes limited because of the wavevector mismatch. The grating is fabricated by focused ion beam (FIB). The probe shown in Fig. 16.5d also aims to fill the gap in wavevector mismatch, but uses an evanescent field generated on a micro-prism fabricated near the probe surface (Otto configuration) [49]. These techniques are especially effective for the probes made by wire etching.

4.3 Feedback Scheme for Tip-Sample Distance

Tip-sample distance control constitutes an essential part of TERS experiments as it determines how sensitively surface properties can be interrogated in conjunction with the properties of the tip used. In TERS experiments, both STM [50] and AFM [51] have so far been used including many variants of AFM such as a shear-force microscope [52]. In STM, electric conductivity of the sample is required for sensing tunneling current and extreme flatness of the sample surface is important to achieve high resolution. In contrast, these conditions are not necessary to AFM since it senses forces acting between the tip and the sample, and hence a wide variety of samples can be studied by using it. All the pioneering works on TERS are based on the contact mode AFM in the year 2000 [31, 33, 36]. Followed by several related TERS works using AFM, Pettinger, et al. reported the TERS based on STM in the year 2002 [35]. However, since the response to the tip-sample distance depends on the sensing mechanism, resulting enhancement effects also vary depending on the approaches used for the tip-sample distance control.

TERS based on an AFM have been the most popular approach for the reasons described above. Initially, contact mode AFMs were used [31–33], where the feedback signal contains primarily a DC component and can typically be detected through the deflection of a laser beam induced by the motion of a cantilever, and which is proportional to the repulsive force applied to the tip [53]. More recent development includes so-called dynamic AFM [54], where the motion of the tip is temporally modulated, taking advantage of the mechanical resonance of the tip. The feedback signal is then obtained through demodulating the sensor signal. Si cantilevers and quartz tuning forks (TFs) are the most popular transducers used. In particular, a TF exhibits a high spring constant, a high quality factor, and good thermal stability. It's also extremely versatile that almost any sorts of tips are compatible with TFs [55–57]. Furthermore, all electrical excitation of a TF is possible without using any additional dithering PZT elements [58]. The possibility of excluding a light source from the transducer subsystem is also an attractive option for TERS since the tip-sample distance control can be made perfectly independent from the laser source used for any spectroscopic experiments.

The demodulated signal may represent either an amplitude or phase of the oscillatory motion of the tip. However, it turns out that the amplitude feedback is not useful with the high quality factor of the mechanical resonance, resulting in slow scan speeds. Although the quality factor control [59] can be used for reducing the excess Q value, the fundamental sensitivity of the transducer has to be compromised. This shortcoming can be prevented by using the phase information as a feedback signal [60, 61]. In order to detect the frequency of the oscillatory motion of the tip, a phase-lock-loop (PLL) is often used [62], with which the magnitude as well as the direction of the force applied to the tip can readily be detected (i.e., the amount of frequency shifts is proportional to the force while the sign of the shifts corresponds to the direction of the force: a positive frequency shift corresponds to a repulsive force while an attractive force leads to a negative frequency shift). By observing the automatic gain control signal within a PLL circuit, the dissipative component of the force can also be detected [63]. (The frequency shift is induced by both conservative and dissipative components of the force.) By using frequency as a feedback signal, it is now possible to make high sensitivity (1pN [52]) compatible with good stability and fast response (>1 kHz [64]). Typical interface circuitry can be found in Ref. [65].

The motion of the tip can either be vertical or lateral to the sample surface. A popular approach taken with vertical motion is the tapping mode, where the tip undergoes a cyclic contact with the sample (with an amplitude of typically >50 nm). The virtue of the tapping mode AFM is that the oscillatory motion of the tip can suitably be used for studying the tip-sample distance dependence of the TERS signals [66, 67] (see Sect. 5.4.2).

The noncontact mode with a small oscillation amplitude (<5 nm) in vacuum, established for atomic resolution AFM [68], still remains to be applied to the TERS experiments, and this may provide an interesting opportunity for TERS since the tip-sample distance achieved in this case would be comparable to that of STM. Note that the force is rather a long-range interaction in ambient conditions, and a short tip-sample distance may only be achieved by detecting the force originating from the short-range interactions observable in vacuum.

The lateral motion of the tip leads to a shear-force generated between the tip and the sample. The oscillation amplitude of the tip is decreased under a constant excitation due to the damping as the tip approaches to the surface. This can be detected at a distance as far as 30 nm away from a sample surface, depending on the oscillation amplitude of the tip [69]. Detailed characteristics of a TF shear-force sensor were also reported and it was shown that the quality factor is decreased as well as that the resonance frequency shifts to higher frequencies as the tip-sample distance is decreased [52]. Due to the jump-in-contact behavior caused by

Table 16.1 Variants of the scanning microscopy methods used for TERS and their tip-sample distance ranges

	STM	AFM			
Origin of the sensor signal	Tunneling current	Vertical force			
		Contact mode	Tapping mode (cyclic contact)	Noncontact mode (attractive force)	Shear-force
Typical tip-sample distance (reported)	<1 nm [71, 72]	Contact	<1 nm (w/demodulation) [66]	N/A	>5 nm [64]

the meniscus forces formed by a surface layer in ambient conditions [64, 69], the oscillation amplitude has to be of the order of a few nanometers in general, and it would be difficult to set the tip-sample distance below 5 nm (although it strongly depends on the sample) in order to avoid any tip damages due to tapping. However, the tip-sample distance dependence of the TERS signals can be studied in a static fashion [70]. This is an important feature when a weak spectroscopic signal has to be detected in TERS experiments.

Although stringent requirements on the flatness of the surface and the conductivity of the sample limit widespread use of STM in TERS experiments, it appears an attractive option for fundamental studies as it allows us to control tip-sample distance at a very short distance, typically less than 1 nm, thanks to the increasing characteristics of the tunneling current as the tip approaches to the surface. For instance, with a bias voltage of 1 V and the tunneling current of 1 nA, the tip-sample distance can be controlled within a sub-nanometer range by retaining a constant tunneling current as a feedback signal [71]. The tunneling current is essentially a DC quantity, which also simplifies the sensing scheme. The bandwidth of the current detection circuit can significantly be decreased, and hence the detection noise can also be reduced. Combined with the metallic substrate, the gap mode plasmons excited between the tip apex and metallic substrate results in strong field enhancement [72]. Furthermore, since an atomically flat substrate does not generate any parasitic surface-enhanced signals, the assessment of the tip-enhancement effect is less ambiguous. A comparative study [73] between STM and AFM has shown that more reproducible experiments can be performed by using STM.

The reported tip-sample distance control methods and the typical values of the resultant controlled distance in TERS experiments are summarized in Table 16.1.

5 Key Research Findings

5.1 Tip-Enhanced Raman Spectroscopy (TERS)

The tip-enhancement effect has been successfully utilized mostly from visible to NIR wavelength, related to the plasma frequency of the noble metals [44] used for efficient tip enhancement. Tip enhancement was initially developed as an elastic optical scattering microscopy wherein the topographic contrast obtained by SPM is converted into an optical contrast [5–7]. The metallic tip or even nonmetallic silicon tip works as a simple nanosized scatterer of the evanescent optical field localized on the nanostructures. A photo detector detects the elastically scattered optical field without a spectrometer (no spectroscopic information) in a point-by-point manner while scanning the sample stage. In this case, the obtained optical contrast is rather close to the topographic contrast and no further information from the materials was obtained. Consequently, this type of near-field microscopy was mainly devoted for detailed analysis of artifacts in optical images. Artifacts which were attributed to the tip movement depend on the topography [74, 75]. The best spatial resolution achieved is 1 nm [76]. Even with this resolution, the technique did not gain attentions probably because of lack of additional information, which is not observable from the topographic contrast. In order to gain rich information from the optical interactions, inelastic scattering such as Raman scattering was employed. This is one of the promising candidates since nanosized metallic structures can enhance the weak Raman scattering cross section by a factor of 10^3 up to 10^{14}, which is now well known as SERS [16, 41, 77]. In 1985, Wessel reported the concept of surface-enhanced optical microscopy in which the laser-irradiated metallic nanoparticles with piezoelectric positioning elements enhance the incident and scattered photons. The frequency-converted photons (Raman scattering) are detected through a spectrometer [78]. Encouraged by Wessel's report and the development of apertureless elastic scattering NSOM, we reported our preliminary experimental results in the conference proceedings on tip enhancement for the application of Raman spectroscopy in 1999 [79]. Immediately after that, three groups independently reported tip-enhanced near-field Raman spectroscopy in 2000 [31, 33, 36]. In these pioneering three cases, the system consists of an inverted optical microscope with AFM head as illustrated in Fig. 16.4a. The enhanced electric field is utilized as a nanoscale light source for Raman spectroscopy. This technique has been recently recognized as tip-enhanced Raman scattering (TERS) because of the analogy to SERS. Note that tip enhancement is not only on the incident light but also on the scattered light. In the case of Raman scattering, the enhancement factor can be approximated as $|E_i|^2|E_s|^2 \sim |E|^4$ assuming that the frequency difference (Raman shift) between incident, E_i, and scattered, E_s, light is negligible [16]. Figure 16.6 shows one example of a TERS image, which is visualized by detecting the intensity of a specific vibrational mode of crystal violet molecule in a mixture of other organic molecules. The observed mode corresponds to C-H out-of-plane bending mode having an energy of 908 cm^{-1}. Clear distribution with a 30 nm spatial resolution is obtained and is not resolved by a topographic contrast [21].

One of the successful applications of TERS is for the characterization of nanomaterials and even for the single molecule detections [37, 80–83]. Among the nanomaterials, carbon nanotubes have been extensively studied by TERS [38, 84] and the physical properties such as chirality, diameter, and defects have been investigated. In terms of a spatial resolution, Novotny's group recently reported

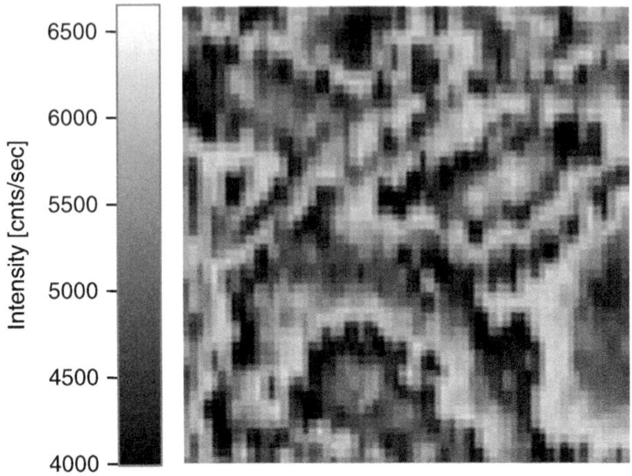

Fig. 16.6 Transmission TERS image (1 × 1 μm^2) of crystal violet molecules observed by C-H out-of-plane bending mode at 908 cm^{-1}. Excitation wavelength is 532 nm. The tip is a silver-coated Si cantilever

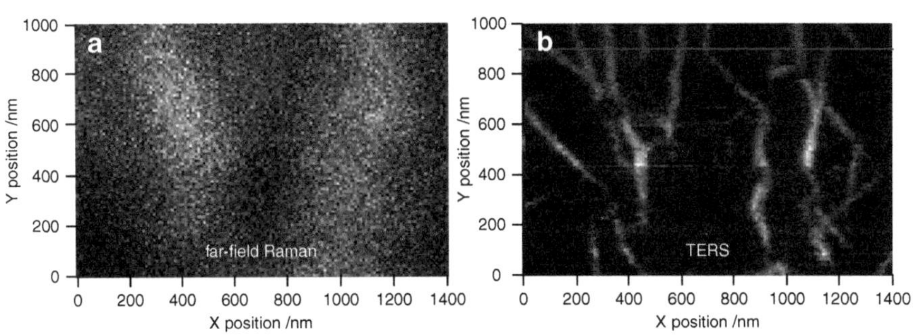

Fig. 16.7 (**a**) Confocal Raman and (**b**) TERS image of a single-walled carbon nanotube. The contrast in the images reflects the local intensity of the Raman G band (1595 cm^{-1}). Excitation wavelength is 532 nm. The tip is a silver coated cantilever tip

14 nm spatial resolution by observing the carbon nanotubes [85] using well-isolated carbon nanotubes. We visualized densely-packed carbon nanotubes with 20 nm spatial resolution as shown in Fig. 16.7. Figure 16.7 clearly shows the resolving power of TERS as compared to confocal Raman detection.

As previously discussed, transmission mode is applicable only to thin/ nanomaterials or transparent samples even though the detection efficiency is very high owing to the capability of high NA objective lens. In order to apply tip enhancement to opaque or bulk samples such as silicon-based materials, reflection mode TERS has been developed [26, 86–90]. The idea of the reflection mode is very simple in a way that both illumination and detection are conducted at the same side of the tip and the samples surface as illustrated in Fig. 16.4b. The illumination can be either normal or oblique to the sample surface. Illumination depends on the availability of the space between objective lens and the SPM head. This critical issue is often solved using a relatively low NA objective lens with a long working

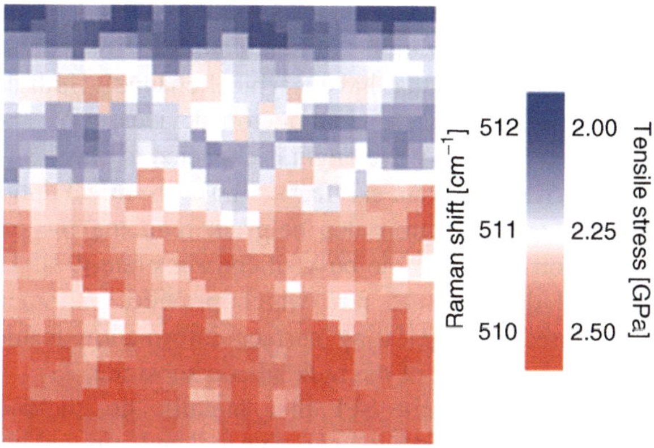

Fig. 16.8 Reflection TERS image (800 × 800 nm^2) of strained silicon. Image is plotted as a Raman shift (not Raman intensity) at each pixel. Excitation wavelength is 442 nm. The tip is a silver-coated Si_3N_4 cantilever

distance. Figure 16.8 shows a TERS image depicting the strain distribution in the nanoscale from strained silicon substrate. Each pixel in the image is plotted not as Raman intensities but as Raman shifts of the observed silicon phonon mode, which give the information of strained lattice constant in the nanoscale [91].

While reflection mode has a big advantage over transmission mode because of the wide variation of the applicable samples, the detection efficiency is quite low because of the fact that low NA objectives are practically mountable in the system. However, this issue could be solved by parabolic mirror, which has a NA approximately 1.0 [92, 93].

5.2 Tip Enhancement in UV and DUV Regions

Since its invention, the development of TERS has been focused on visible to NIR and not much attention has been paid to UV. The primary reason is that both silver and gold that are commonly used in the visible TERS experiments do not show a plasmonic field enhancement in the UV region (see Sect. 4.2). However, achieving tip enhancement in DUV Raman scattering will make a significant contribution especially in the nano-bioimaging technology because proteins and nucleic acids exhibit an electric resonance in the DUV region [94]. The resonantly excited Raman scattering process provides quite a large Raman scattering cross section compared with the nonresonant Raman process. The resonantly enhanced Raman signal can be further increased by the tip-enhancement effect at the very localized region. Moreover, in DUV excitation, the fluorescence and Raman scattering hardly interfere with each other because they are spectrally well separated. Thus, high signal-to-noise ratio is expected. Combining the advantages of DUV resonance Raman scattering with tip-enhancement technique to obtain information in the nanoscale will make DUV resonance Raman analysis a valuable technique in biomolecular research.

The importance of UV-TERS also comes from the increasingly large demands of Raman characterizations of semiconductor nanodevices such as ever shrinking CMOS transistors that have a typical channel width down to 45 nm. In the TERS characterization of semiconductor devices, UV excitation particularly below 365 nm (=3.4 eV) wavelength is expected to play a significant role because silicon shows a resonant Raman effect owing to the direct optical E_1 gap transition [95, 96]. The E_1 gap transition also helps in reducing the penetration depth of the UV excitation light into the bulk silicon substrate [97]. The penetration depth can be reduced down to ~5 nm, which provides extremely high surface sensitivity; otherwise, the interesting surface information is hidden in the strong Raman background signal of the bulk silicon substrate.

In order to excite SPP in a metallic tip, the metal is required to have an optical property that has a negative real and minimal imaginary part in the dielectric function at the excitation wavelength [98] as discussed in Sect. 4.2. As a plasmonic material in the UV region, aluminum can be used instead of silver and gold. The dielectric function of aluminum shows a reasonably small imaginary part while the real part keeps negative in the UV.

Researches that explore the use of aluminum in UV tip enhancement have just started, although the surface plasmon was first discovered in aluminum [99, 100]. The potential of aluminum as a UV Raman enhancer was firstly observed in SERS experiments using 244 nm DUV excitation, in which a crystal violet aqueous solution placed on a 50 nm thick aluminum surface showed stronger Raman signal [30]. Tip enhancement of DUV resonant Raman scattering of adenine molecules has been demonstrated recently using an aluminum-coated silicon probe with a 266 nm Raman excitation wavelength [29]. Design and further optimization of the plasmon resonance in the aluminum tip are ongoing to realize UV/DUV TERS.

5.3 Tip-Enhanced Nonlinear Raman Spectroscopy

The discussion above was based on spontaneous Raman scattering; however, when tip-enhanced microscopy and spectroscopy are combined with high-order nonlinear optical effects, the spatial resolution can essentially be improved. Photons generated by the nonlinear optical effects are spatially confined in a smaller volume. It is smaller than the volume of the tip-enhanced field coming from the fundamental frequency because the intensity from nonlinear effects is proportional to the high-order powers (square, cube, etc.) of the excitation light intensity. The spatial distribution of the signal emission (harmonic signal, two photon excited fluorescence, etc.) becomes narrower than the intensity distribution of the excitation field. This leads to the reduction of the effective volume where light-matter interaction occurs [101]. Moreover, higher-order optical effects give a much finer spatial response (Fig. 16.9a). In addition to the spatial confinement, a small tip enhancement of the excitation field could lead to a huge enhancement of the emitted signal because of the nonlinear responses. This effect allows reduction of

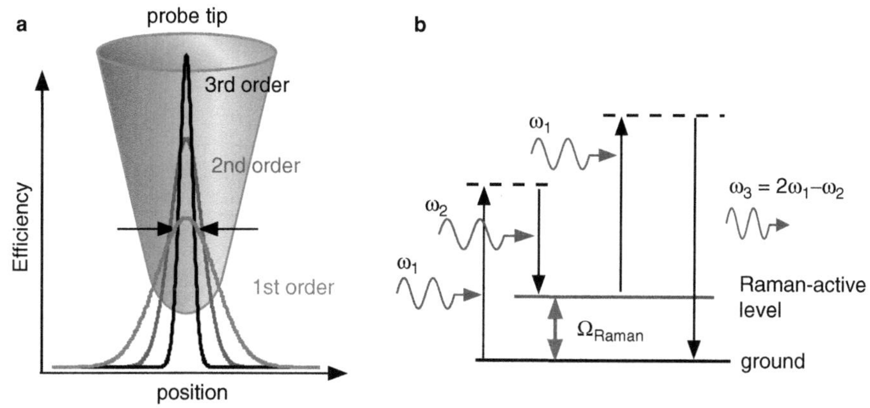

Fig. 16.9 (**a**) Concept of higher photon confinement at the tip apex by a nonlinear optical process. Efficiency of a first-order process such as spontaneous Raman or fluorescence directly reflects the field distribution at the tip apex while the efficiency of a second-order process such as SHG or SFG and a third-order process such as CARS shows further confinement due to the nonlinear response of the material. (**b**) Energy diagram of CARS process

the far-field background. In order to realize molecular-vibration spectroscopy, coherent anti-Stokes Raman scattering (CARS) spectroscopy is employed [102], which is one of the most widely used nonlinear Raman spectroscopes [103]. The CARS spectroscopy uses three incident fields consisting of a pump field (ω_1), a Stokes field (ω_2; $\omega_2 < \omega_1$), and a probe field ($\omega_1{}' = \omega_1$), and induces a nonlinear polarization at the frequency of $\omega_3 = 2\omega_1 - \omega_2$, which is given in a scalar form by

$$P^{(3)}_{CARS}(\omega_3 = 2\omega_1 - \omega_2) = \chi^{(3)} E(\omega_1) E^*(\omega_2) E(\omega_1), \tag{16.1}$$

where $\chi^{(3)}$ represents the third-order nonlinear susceptibility and $E(\omega_1)$ and $E(\omega_2)$ are the electric fields for excitation light. $E^*(\omega_2)$ denotes the complex conjugate of $E(\omega_2)$. The nonlinear susceptibility is expressed by the vibration-resonant term ($\chi^{(3)}_R$) and nonresonant term ($\chi^{(3)}_{NR}$):

$$\chi^{(3)} = \chi^{(3)}_R + \chi^{(3)}_{NR} = \frac{A}{\Omega_{Raman} - (\omega_1 - \omega_2) - i\Gamma} + \chi^{(3)}_{NR}. \tag{16.2}$$

The coefficient of the fraction, A, denotes a constant related to the strength of the vibration, Ω_{Raman} denotes one of the specific molecular vibrational frequencies from a given sample, and Γ corresponds to the spectral bandwidth of the vibration mode. When the frequency difference of ω_1 and ω_2, $\omega_1 - \omega_2$, coincides with Ω_{Raman}, the anti-Stokes Raman signal is resonantly generated. Figure 16.9b shows an energy diagram for the CARS process. One can obtain a CARS spectrum by plotting the CARS signal intensity with sweeping ω_2. The CARS spectrum gives essentially

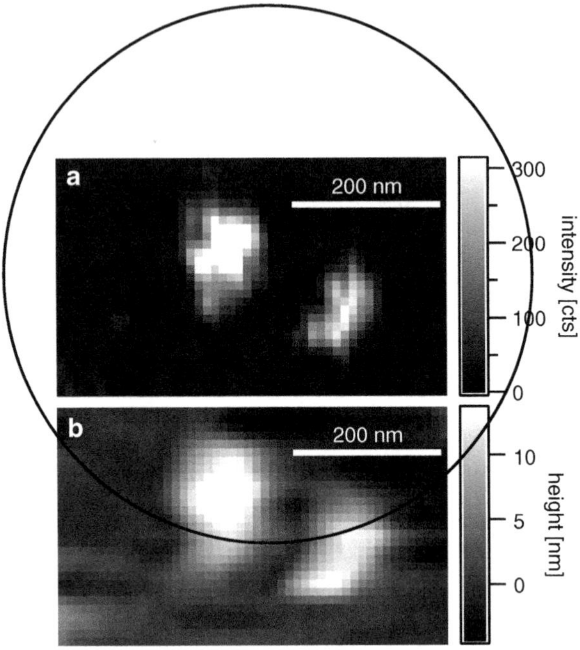

Fig. 16.10 (**a**) Tip-enhanced CARS image of adenine nanoclusters obtained by ring breathing mode of a diazole ring at 1,337 cm^{-1} and (**b**) simultaneously obtained topographic image. *Circle* on the figure denotes a diffraction-limited focused spot. Excitation wavelengths for ω_1 and ω_2 are adjusted to 787 nm and 870 nm, respectively. The tip is a silver-coated Si tip

identical information with spontaneous Raman spectra [104]. Figure 16.10a shows a tip-enhanced CARS image [105] on adenine nanoclusters that is obtained by tuning the frequency difference, $\omega_1-\omega_2$, to a specific vibrational mode (1,337 cm^{-1}) of adenine molecule. Clear distribution of adenine molecules beyond the diffraction limit (indicated by the circle on the figure) is obtained and is in good agreement with the simultaneously obtained topographic image in Fig. 16.10b.

Figure 16.11 shows tip-enhanced CARS images [106] obtained on a DNA network [107] consisting of adenine and thymine hetero structures. The tip-enhanced CARS images at the on- and off-resonant frequencies are shown in Fig. 16.11a, b. The DNA bundles are observed at the resonant frequency in Fig. 16.11a, while they cannot be visualized at the off-resonant frequency in Fig. 16.11b. This indicates that the observed contrast is dominated by the vibrationally resonant CARS signals. Figure 16.11c shows one-dimensional line profiles by tuning the frequency difference, $\omega_1-\omega_2$, to a specific vibrational mode (1,337 cm^{-1}) of adenine molecules. The line profile shows FWHM of 15 nm. Moreover, when the tip is far away from the sample (micro CARS experiment), the signal is totally vanished (see dashed line in Fig. 16.11c). This fact suggests that tip-enhanced CARS is advantageous not only in its high spatial resolution but also in high sensitivity.

Aside from CARS, a second-order nonlinear technique, such as second harmonic generation or sum frequency generation, can also be a suitable spectroscopic technique because of its higher surface selective detection [103], which is arising from $\chi^{(2)}$.

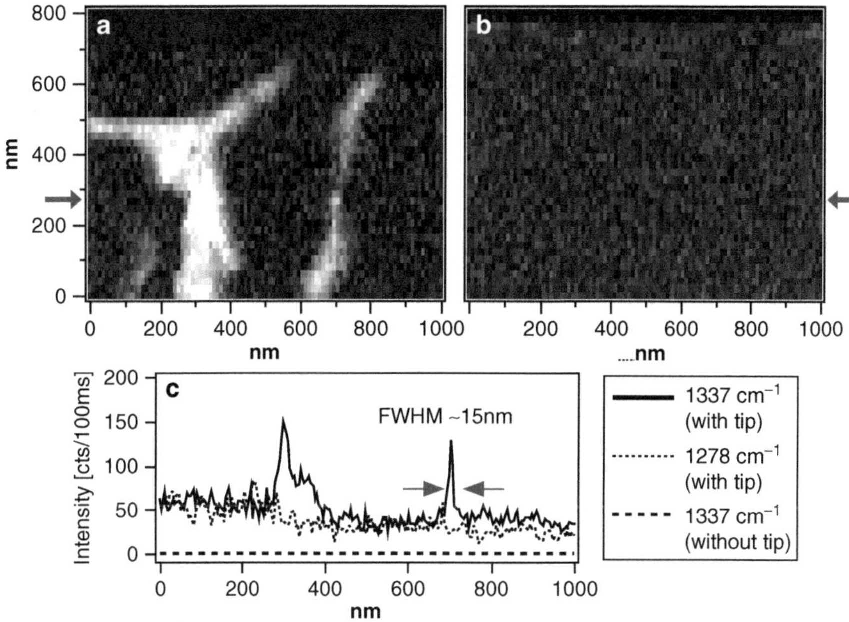

Fig. 16.11 Tip-enhanced CARS images of the DNA network obtained at (**a**) on-resonant frequency (1,337 cm^{-1}) and (**b**) off-resonant frequency (1,278 cm^{-1}). (**c**) Cross-sectional line profiles at the position indicated by the solid arrows. The scanned area is 1,000 nm × 800 nm (Reprinted from [106], Copyright 2004, with permission from American Physical Society)

5.4 Applications of TERS

5.4.1 Tip-Pressure and Tip-Heating Effects

In TERS, the apex of the metal tip interacts with sample molecules in several ways: electromagnetically, chemically, mechanically [82, 108], thermally, and, when tip is biased like STM, electrically. Though these interactions operate independently from each other, it can also be manipulated to occur in pairs or in groups. In this section, we will focus on the separate mechanical and thermal interactions of the tip with the sample molecules. Mechanical interactions occur when the tip is pushed against the sample molecules or vice versa (Fig. 16.12a). Thermal interactions are when heat generated at the tip is transferred by conduction and radiation (Fig. 16.13a). The Joule heat generated at the tip is primarily due to the absorption of light, which is proportional to the imaginary part of the dielectric function of the tip. These two interactions can occur intentionally or unintentionally during TERS measurement. The effects of tip-pressure and tip-heating in the spatial resolution and sensitivity of TERS is investigated along with the application of these two interactions to characterize the mechanical and thermal properties of nanoscale materials. These two unique techniques will assist in accelerating the slow progress in the development of tools to characterize nanoscale devices.

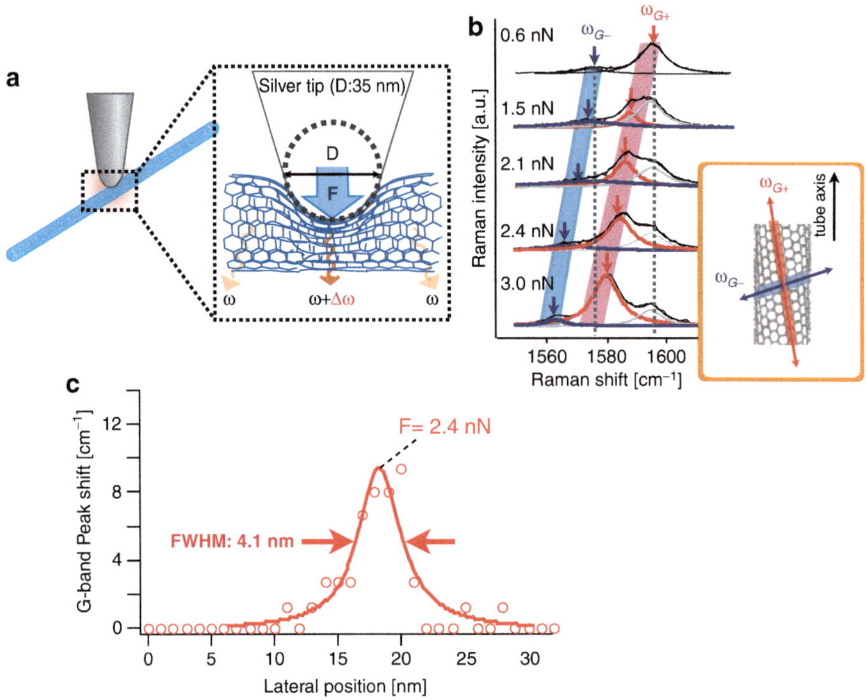

Fig. 16.12 (**a**) Schematic of tip-pressure-assisted TERS; (**b**) G-band Raman spectra of SWNT at different applied force (inset: stretching associated to G-band peaks); (**c**) peak shift of the G-band as a function of tip position. The FWHM, which is proportional to the spatial resolution, is ~4.1 nm

Tip-Pressure-Assisted TERS

Tip-pressure-assisted TERS can be achieved using AFM, in which the applied tip-force onto the sample can be precisely controlled [109]. This force is used as an external pressure that can be applied selectively to a small volume with precision onto the sample through the apex of a sharp metallic tip. The local distortions that arise from the externally applied pressure modify the molecular vibrations of only the pressurized molecules directly in contact with tip. Yano et al. reported an extremely high spatial resolution down to 4 nm in tip-pressure-assisted TERS using isolated single-walled carbon nanotubes and adenine nanocrystals. In the method used, conventional TERS techniques were combined with a controlled amount of tip-pressure to mechanically perturb a few molecules in the sample. This was accomplished by precisely controlling the contact area between the sample and the tip (Fig. 16.12a). The optical response of the perturbed molecules was measured using the capability of TERS to enhance weak optical responses coming from molecular-level interactions. It relies on pressure-dependent shifts in the vibrational frequencies of Raman modes resulting from mechanical distortions of the molecules in the sample. Therefore, the high spatial resolution was achieved

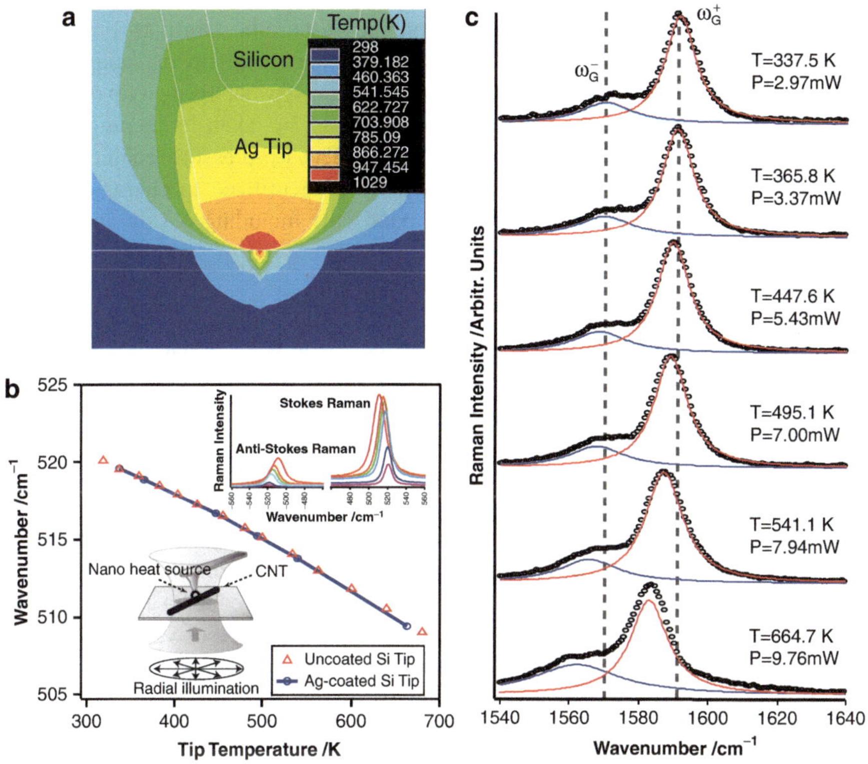

Fig. 16.13 (**a**) Thermal distribution of tip and sample calculated using finite element method [111, 112]. (**b**) Silicon Raman peak of the tip as a function of temperature calculated from the ratio of stokes and anti-stokes Raman as shown in upper inset. Lower inset shows the simplified schematic of tip-heating-assisted TERS. (**c**) G-band Raman spectra at different incident laser power for varying the temperature

by sensing the localized pressure applied by the sharp metal tip by means of Raman spectra. Figure 16.12b shows the G-band TERS spectra of SWNT at different tip-applied forces ($0.6 \leq F \leq 3.0\,\text{nN}$). The spectral peak of the G-band downshifts with increasing applied pressure. The large amount of observed Raman mode shift (more than 10 cm^{-1}) confirms that the optical response of the pressurized SWNT can be experimentally distinguished from unpressurized SWNT. In order to experimentally determine the resolution, one-dimensional imaging was performed by raster scanning the tip in a 1 nm step under a constant tip-applied force of 2.4 nN along a line crossing the SWNT. Figure 16.12c shows the change in the G-band peak relative to initial (unpressurized) peak position as a function tip position. The FWHM from the best fit (red line) is approximately 4 nm, which is even smaller than the tip diameter of ~35 nm. The flexibility of the technique was also tested using adenine nanocrystals that have much weaker elasticity compared to SWNTs. Similar to SWNT, a resolution of ~4 nm was obtained. Although the measured resolution is 4 nm, it is clear that the sensitivity of the optical response in tip-pressure is as high as 1 nm based on the discernible shift at 1 nm step. The risk of damaging or irreversibly deforming the metal tip or sample when applying pressure results in some limitation, particularly when using a tip or sample inclined to such deformation. In any case, care must be taken in keeping the tip-force within the elastic limit of the sample and tip. This tip-pressure technique is not only useful for its high resolution, but is also being utilized to characterize the mechanical properties of materials in the nanoscale including elastic and plastic deformations [109, 110].

Tip-Heating-Assisted TERS

Tip-heating-assisted TERS makes use of the Joule heat generated at the tip apex of a metal-coated tip by transferring controlled amounts of thermal energy to a localized region of the sample in order to obtain a temperature-sensitive process or thermal properties at that location alone [111–113]. Similar to the tip-pressure technique, vibrational frequencies of heated molecules can be different from unheated molecules. The enhanced Raman signal from the area in contact with the tip is easily resolved from the far-field Raman of the surrounding areas as the contact area's signal is shifted as a result of the thermal energy. Utilizing this difference, tip-heating TERS can improve not only the sensitivity of TERS by reducing the amount of background signal [112], but also the thermal characterization of various materials requiring site-selective and controllable nanoscale heat sources.

Tip-heating-assisted TERS requires the measurement of the temperature at the tip apex and validation of heat transfer from tip to sample for it to be useful. To attain this, we determined the temperature at the tip by monitoring the ratio of intensities of the Stokes and anti-Stokes Raman of silicon from a bare and metal-coated silicon tip at different powers of excitation light (inset Fig. 16.13b). The heat transfer from the tip to the sample was demonstrated by cutting SWNT using a bare silicon tip [114]. After confirmation from the results above, we performed a thermal characterization of SWNT at different temperatures using a metal-coated silicon tip. Figure 16.13c shows the Raman spectra at different temperatures (obtained from varying laser power, P, $2.9 \leq P \leq 9.7\,\text{mW}$). The G-band peak clearly shows downshift with increasing temperature. This observed shift in the TERS of SWNT is attributed primarily to the direct conduction of heat from the tip and not attributed to the absorption of light by the CNT itself because of the very small far-field downshift even at 9.7 mW, which responds to ~350 K [115]. The key advantage of tip-heating-assisted TERS detection is that local near-field detection ensures a nearly uniform thermal gradient by selectively enhancing only a smaller volume below the tip. The conductivity of the sample does not affect much the ability of the technique to produce a nearly uniform temperature gradient in the area of importance for the TERS reading. This is because the spatial resolution in TERS is proportional to the fourth power of the electric field whereas the heat generated (temperature rise) is to the square of the electric field. The disadvantage of this technique is that the enhancement factor varies with temperature due to changes in the optical properties of metals at elevated temperatures [116, 117]. It is also evident that the effect of temperature increase on tip enhancement is dependent on the excitation frequency relative to the position of the local plasmon frequency at room temperature [118]. Also, it is possible to structurally damage the tip or sample during heating. If the application allows, it is easy to avoid these drawbacks by operating at temperatures below the melting point of the sample and tip and also waiting for the tip and sample to be stabilized before measurement [114, 119]. This tip-heating TERS technique, with its ability to characterize both optically and thermally with high precision, may open up opportunities in fields of heat-assisted biochemical and physiochemical reactions and can assist in the development of new photothermal devices.

It is clear that utilizing additional active imaging techniques such as tip-pressure and tip-heating can create breakthroughs in the nano-imaging and nano-analysis of materials. This type of characterization is of increasing importance in the emerging field of nanotechnology as new materials are engineered.

5.4.2 Distance-Dependent Detection of TERS

In TERS microscopy and spectroscopy, the tip enhancement due to the SPP resonance plays the most essential role both for signal sensitivity and spatial resolution. However, the tip-enhancement effect is not the only one affecting Raman spectra. There coexist other interaction mechanisms between a metal tip and sample molecules, chemical interactions similar to SERS [120–122], and, in addition, mechanical interactions (see Sect. 5.4.1). The latter two interactions show up only when sample molecules are in a close vicinity of a tip. In the TERS system using a contact mode AFM, an experimentally observed TERS spectrum is a complex combination of the contributions of these three interactions, which makes it difficult to interpret experimental TERS spectra. Therefore, elucidation and discrimination of the tip-sample interactions are of scientific and practical importance. This can be realized by measuring a tip-sample distance dependence of TERS, since those three interaction mechanisms have different dependencies on the tip-sample distance. The active control of the distance between the tip and sample is a unique feature only possible in TERS not in SERS. Two system configurations, time-gated detection and time-gated illumination, are described below.

TERS System Using Tapping Mode AFM with Time-Gated Detection

In the first scheme using the time-gated detection, as shown in Fig. 16.14a, TERS signals from molecules are detected by an avalanche photodiode through a monochromator. A dual-gated photon counter captures the Raman scattering signal of a certain vibrational mode while the gates are triggered synchronously with the tapping oscillation of the AFM cantilever [66]. The concept was first demonstrated in an apertureless near-field fluorescence microscope [123], in which spectral information is not a big deal. The modification to the TERS system in Fig. 16.14a is a substitution of a tapping mode AFM and a photon-counting APD for the contact mode AFM and the CCD detector, respectively. A metal-coated tip is longitudinally vibrated at a certain frequency over sample molecules. This technique is specialized for probing distance-dependent change of TERS

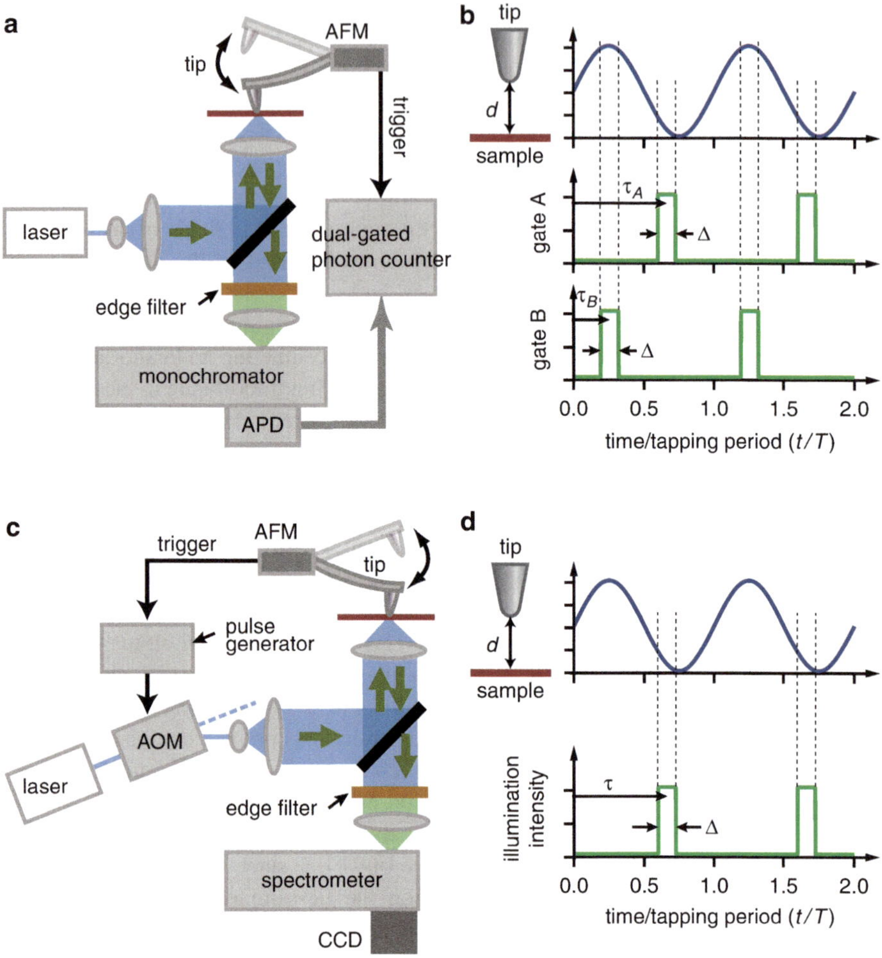

Fig. 16.14 (**a**) Setup for TERS with time-gated detection and its (**b**) schematic of the dual-gating photon-counting scheme (Reprinted from [66], Copyright 2007, with permission from American Institute of Physics). (**c**) Setup for TERS with time-gated illumination and its (**d**) schematic of the gating relative to the tip oscillation (Reprinted from [67], Copyright 2009, with permission from American Physical Society)

intensity, as a single channel detector is used. Figure 16.14b shows a schematic of a dual-gated photon-counting process utilized for the measurement of the dependence of TERS on the longitudinal distance (d) between a tip and a sample. The sinusoidal electric signal (upper curve in the figure) from the AFM controller corresponds to the longitudinal vibration of the tip at a certain frequency. The two gates (gate A and gate B) are independently opened within the tapping period. The time delay of gate A (τ_A) is swept relative to the other (τ_B) fixed at the time when the tip is farthest from the sample surface. The Raman intensity measured

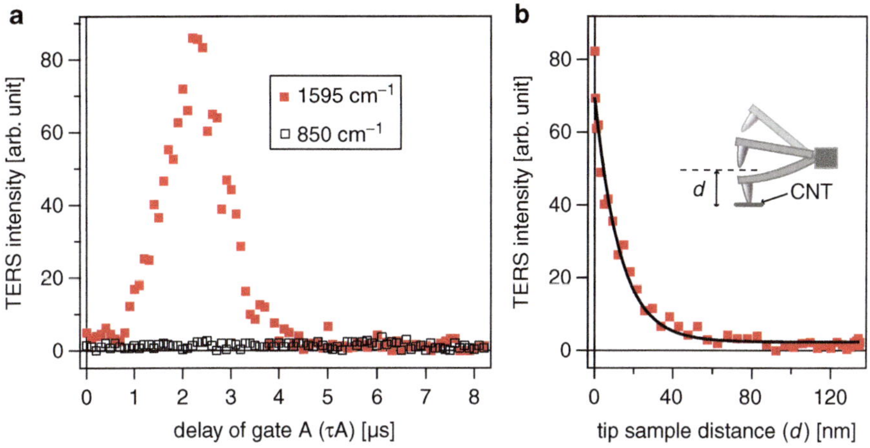

Fig. 16.15 (**a**) TERS intensity of carbon nanotubes measured while scanning one of two gates through the entire tapping oscillation (filled square: a vibrational frequency of the G-band at 1,595 cm^{-1}; open square: no vibrational frequency at 850 cm^{-1}). (**b**) Distance dependence of TERS intensity of the G-band derived from (**a**). The delay time is converted to tip-sample distance by taking into account the harmonic oscillation of the tip (Reprinted from [66], Copyright 2007, with permission from American Institute of Physics)

through gate A (I_A) includes both near-field and far-field Raman components from the sample, while the intensity measured through gate B (I_B) includes only the far-field Raman scattering component. Therefore, we can extract pure near-field contribution by subtracting I_B from I_A. Sweeping τ_A over the tapping period, one could obtain the tip-sample distance dependence of the TERS intensity. Figure 16.15 shows the typical tip-sample distance dependence of TERS intensity of SWNTs.

TERS System Using Tapping Mode AFM with Time-Gated Illumination

The first scheme sufficiently meets one's demand when intensity change is of major interest. However, this scheme uses a single channel detector detecting only a Raman signal at a specific wave number, enabling us to only probe intensity changes. Therefore, it is not easy to investigate changes in spectral shape, such as peak frequency shift, emergence of new peaks, and relative intensity change, which are induced by chemical and mechanical interactions between a metal tip and sample molecules. For this particular purpose, another scheme using a time-gated illumination has been developed [67]. The excitation laser intensity is time-gated in this scheme so that one can record an entire Raman spectrum by a multichannel detector (CCD camera). Figure 16.14c schematically shows the configuration of this scheme. An acousto-optic modulator (AOM) is utilized to switch the excitation laser intensity such that the tip-sample arrangement is selectively illuminated only for a particular tip-sample distance. Figure 16.14d illustrates the time course of the sinusoidal oscillation of the tip and the synchronized opening of the time-gate. By selecting a particular value of the time delay (τ) of the time-gate, one can

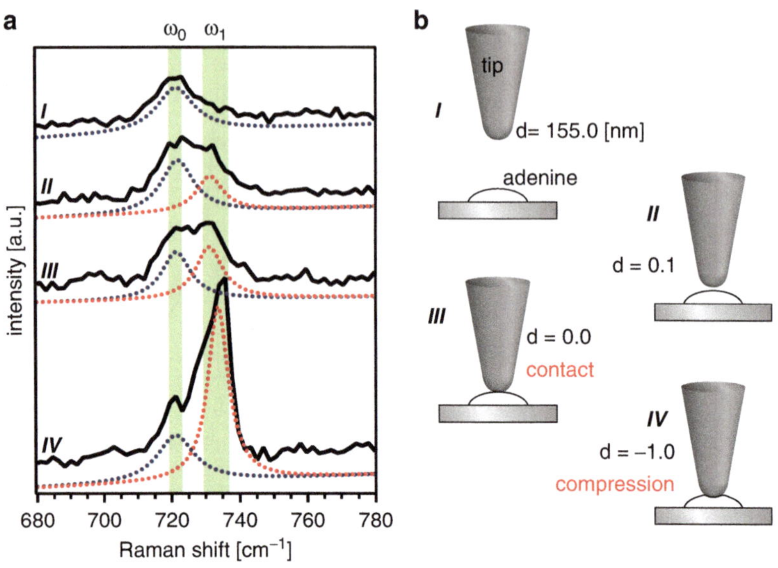

Fig. 16.16 (**a**) Raman spectra of an adenine nanocrystal at several tip-sample distances indicated in the schematic of (**b**). The peaks marked by ω_0 and ω_1 represent the unperturbed ring breathing mode (~720 cm^{-1}) [107] and the frequency shifted one, respectively. Deconvoluted Lorentzian peaks for each spectrum are also shown by the dotted curves

preselect a desired tip-sample distance. TERS spectra corresponding to a desired tip-sample distance can be recorded with high accuracy. By sweeping the value of τ, one can acquire a distance-dependent dataset of TERS spectra. Figure 16.16 shows the typical tip-sample distance dependence of TERS spectra of adenine nanocrystals. Spectral shift is clearly observed as the distance changes [82, 108].

6 Summary and Future Perspective

Owing to the enhancement virtue of a metallic probe tip as a surface enhancer for the SERS effect, TERS allows us to obtain localized molecular vibrational information without averaging the signal inside the diffraction-limited focused spot, and has not only high spatial resolution but also high signal sensitivity compared with far-field spectroscopy, which requires many more molecules to compensate for the small Raman scattering cross section. The achieved spatial resolution so far reported by several groups is around 10 nm. However, the number of materials observed by a tip-enhanced technique is still limited. In this sense, TERS has not yet been "truly" a microscope with 10 nm spatial resolution. In the next decade, TERS is expected to be a true "nano"-scope, which can extensively apply the best spectroscopic technique to each physical/chemical and biological study. For this purpose, both optical/spectroscopic development and SPM development (mostly in ambient condition) are strongly required. The combination of the aforementioned techniques, including tip enhancement, polarization control, tip-pressurized/heating effect, and nonlinear processes, is quite promising for higher spatial resolution and sensitivity. In addition, fabrications and optimizations of specific tips for higher tip enhancement or findings of new near-field effects are also very important factors for the future development of tip-enhanced spectroscopy.

References

1. Novotny L, Hecht B (2006) Principles of nano-optics. Cambridge University Press, Cambridge
2. Born M, Wolf E (1999) Principles of optics, 7th edn. Cambridge University Press, Cambridge
3. Goodman JW (1996) Introduction to Fourier optics. McGraw-Hill, New York
4. de Fornel F (2001) Evanescent waves from Newtonian optics to atomic optics. Springer, Berlin
5. Inouye Y, Kawata S (1994) Near-field scanning optical microscope with a metallic probe tip. Opt Lett 19:159
6. Bachelot R, Gleyzes P, Boccara AC (1995) Near-field optical microscope based on local perturbation of a diffraction spot. Opt Lett 20:1924
7. Zenhausern F, O'Boyle MP, Wickramasinghe HK (1994) Apertureless near-field optical microscope. Appl Phys Lett 65:1623
8. Pohl DW, Denk W, Lanz M (1984) Optical stethoscopy: Image recording with resolution $\lambda/20$. Appl Phys Lett 44:651
9. Tsai DP, Othonos A, Moskovits M, Uttamchandani D (1994) Raman spectroscopy using a fiber optic probe with subwavelength aperture. Appl Phys Lett 64:1768
10. Webster S, Bachelder DN, Smith DA (1998) Submicron resolution measurement of stress in silicon by near-field Raman spectroscopy. Appl Phys Lett 72:1478
11. Raether H (1988) Surface plasmon polaritons on smooth and rough surfaces and on gratings. Springer, Berlin
12. Kawata S (ed) (2001) Near-field optics and surface plasmon polaritons. Springer, New York
13. Fleischmann M, Hendra PJ, McQuillan AJ (1974) Raman spectra of pyridine adsorbed at a silver electrode. Chem Phys Lett 26:163
14. Jeanmaire DL, Van Duyne RP (1977) Surface Raman spectroelectrochemistry. Part I: heterocyclic, aromatic, and aliphatic amines adsorbed on the anodized silver electrode. J Electroanal Chem 84:1
15. Albrecht MG, Creighton JA (1977) Anomalously intense Raman spectra of pyridine at a silver electrode. J Am Chem Soc 99:5215
16. Chang RK, Furtak TE (eds) (1981) Surface enhanced Raman scattering. Plenum, New York
17. Furukawa H, Kawata S (1998) Local field enhancement with an apertureless near-field-microscope probe. Opt Commun 148:221
18. Demming F, Jersch J, Dickmann K, Geshev PI (1998) Calculation of the field enhancement on laser-illuminated scanning probe tips by the boundary element method. Appl Phys B 66:593
19. Kottmann JP, Martin OJF, Smith DR, Schultz S (2001) Non-regularly shaped plasmon resonant nanoparticle as localized light source for near-field microscopy. J Microscopy 202:60
20. Novotny L, Bian RX, Xie XS (1997) Theory of nanometric optical tweezers. Phys Rev Lett 79:645
21. Hayazawa N, Inouye Y, Sekkat Z, Kawata S (2002) Near-field Raman imaging of organic molecules by an apertureless metallic probe scanning optical microscope. J Chem Phys 117:1296

22. Hayazawa N, Inouye Y, Kawata S (1999) Evanescent field excitation and measurement of dye Fluorescence in a metallic probe near-field scanning optical microscope. J Microsc 194:472
23. Bouhelier A, Beversluis M, Hartschuh A, Novotny L (2003) Near-Field Second-Harmonic Generation Induced by Local Field Enhancement. Phys Rev Lett 90:013903
24. Hayazawa N, Saito Y, Kawata S (2004) Detection and characterization of longitudinal field for tip-enhanced Raman spectroscopy. Appl Phys Lett 85:6239
25. Hayazawa N, Saito Y (2007) Tip-enhanced spectroscopy for nano investigation of molecular vibrations. In: Bhushan B, Kawata S (eds) Applied scanning probe methods VI. Springer, Berlin, p 257
26. Sun WX, Shen ZX (2003) Apertureless near-field scanning Raman microscopy using reflection scattering geometry. Ultramicroscopy 94:237
27. Tarun A, Hayazawa N, Motohashi M, Kawata S (2008) Highly efficient tip-enhanced Raman spectroscopy and microscopy of strained silicon. Rev Sci Instrum 79:013706
28. Palik ED (ed) (1991) Handbook of optical constants of solids. Academic, Orlando
29. Taguchi A, Hayazawa N, Furusawa K, Ishitobi H, Kawata S (2009) Deep-UV tip-enhanced Raman scattering. J Raman Spectrosc 40:1324
30. Dörfer T, Schmitt M, Popp J (2007) Deep-UV surface-enhanced Raman scattering. J Raman Spectrosc 38:1379
31. Hayazawa N, Inouye Y, Sekkat Z, Kawata S (2000) Metallized tip amplification of near-field Raman scattering. Opt Commun 183:333
32. Hayazawa N, Inouye Y, Sekkat Z, Kawata S (2001) Near-field Raman scattering enhanced by a metallized tip. Chem Phys Lett 335:369
33. Anderson MS (2000) Locally enhanced Raman spectroscopy with an atomic force microscope. Appl Phys Lett 76:3130
34. Pettinger B, Picardi G, Schuster R, Ertl G (2000) Surface enhanced Raman spectroscopy: Towards single molecule spectroscopy. Electrochemistry (Jpn) 68:942
35. Pettinger B, Picardi G, Schuster R, Ertl G (2002) Surface-enhanced and STM-tip-enhanced Raman spectroscopy at metal surfaces. Single Mol 5:285
36. Stöcke RM, Suh YD, Deckert V, Zenobi R (2000) Nanoscale chemical analysis by tip-enhanced Raman spectroscopy. Chem Phys Lett 318:131
37. Zhang W, Yeo BS, Schmid T, Zenobi R (2007) Single molecule tip-enhanced Raman spectroscopy with silver tips. J Phys Chem C 111:1733
38. Hartschuh A, Sánchez EJ, Xie XS, Novotny L (2003) High-resolution near-field Raman microscopy of single-walled carbon nanotubes. Phys Rev Lett 90:095503
39. Ren B, Picardi G, Pettinger B (2004) Preparation of gold tips suitable for tip-enhanced Raman spectroscopy and light emission by electrochemical etching. Rev Sci Instrum 75:837
40. Yeo BS, Zhang W, Vannier C, Zenobi R (2006) Enhancement of Raman signals with silver-coated tips. Appl Spectrosc 60:1142
41. Nie S, Emory SR (1997) Probing single molecules and single nanoparticles by surface-enhanced Raman scattering. Science 275:1102
42. Neacsu CC, Steudle GA, Raschke MB (2005) Plasmonic light scattering from nanoscopic metal tips. Appl Phys B 80:295
43. Mehtani D, Lee N, Hartschuh RD, Kisliuk A, Foster MD, Sokolov AP, Cajko F, Tsukerman I (2006) Optical properties and enhancement factors of the tips for apertureless near-field optics. J Opt A Pure Appl Opt 8:S183
44. Johnson PB, Christy RW (1972) Optical constant of the noble metals. Phys Rev B 6:4370
45. Yeo BS, Schmid T, Zhang W, Zenobi R (2007) Towards rapid nanoscale chemical analysis using tip-enhanced Raman spectroscopy with Ag-coated dielectric tips. Anal Bioanal Chem 387:2655
46. Taguchi A, Hayazawa N, Saito Y, Ishitobi H, Tarun A, Kawata S (2009) Controlling the plasmon resonance wavelength in metal-coated probe using refractive index modification. Opt Exp 17:6509

47. Malinsky MD, Kelly KL, Schatz GC, Van Duyne RP (2001) Nanosphere lithography: effect of substrate on the localized surface plasmon resonance spectrum of silver nanoparticles. J Phys Chem B 105:2343
48. Ropers C, Neacsu CC, Elsaesser T, Albrecht M, Raschke MB, Lienau C (2007) Grating-coupling of surface plasmons onto metallic tips: a nanoconfined light source. Nano Lett 7:2784
49. Sánchez EJ, Krug JT II, Xie XS (2002) Ion and electron beam assisted growth of nanometric Si_mO_n structures for near-field microscopy. Rev Sci Instrum 73:3901
50. Binnig G, Rohrer H, Gerber C, Weibel E (1982) Surface studies by scanning tunneling microscopy. Phys Rev Lett 49:57
51. Binnig G, Quate CF, Gerber C (1986) Atomic force microscope. Phys Rev Lett 56:930
52. Karrai K, Tiemann I (2000) Interfacial shear force microscopy. Phys Rev B 62:13174
53. Meyer G, Amer NM (1988) Novel optical approach to atomic force microscopy. Appl Phys Lett 53:1045
54. Morita S, Wisendanger R, Meyer E (eds) (2002) Noncontact atomic force microscopy. Springer, Berlin
55. Rensen WHJ, van Hulst NF, Ruiter AGT, West PE (1999) Atomic steps with tuning-fork-based noncontact atomic force microscopy. Appl Phys Lett 75:1640
56. Hoppener C, Molenda D, Fuchs H, Naber A (2002) Scanning near-field microscopy of a cell membrane in liquid. J Microsc 210:288
57. Gottlich H, Stark RW, Pedarnig JD, Hecki WM (2000) Noncontact scanning force microscopy based on a modified tuning fork sensor. Rev Sci Instrum 71:3104
58. Grober RD, Acimovic J, Schuck J, Hessman D, Kindlemann PJ, Karrai K, Tiemann I, Manus S (2000) Fundamental limits to force detection using quartz tuning forks. Rev Sci Instrum 71:2776
59. Antognozzi M, Szczelkun MD, Humphris ADL, Miles MJ (2003) Increasing shear force microscopy scanning rate using active quality-factor control. Appl Phys Lett 82:2761
60. Albrecht TR, Grutter P, Horne D, Rugar D (1991) Frequency modulation detection using high-Q cantilevers for enhanced force microscope sensitivity. J Appl Phys 69:668
61. Ruiter AG, Veerman JA, van der Werf KO, van Hulst NF (1997) Dynamic behavior of tuning fork shear-force feedback. Appl Phys Lett 71:28
62. Giessibl FJ (1998) High-speed force sensor for force microscopy and profilometry utlizing a quartz tuning fork. Appl Phys Lett 73:3956
63. Ueyama H, Sugawara Y, Morita S (1998) Stable operation mode for dynamic nonconctact atomic force microscopy. Appl Phys A 66:S295
64. Giessibl FJ, Tortonese M (1997) Self-oscillating mode for frequency modulation noncontact atomic force microscopy. Appl Phys Lett 70:2529
65. Jersch J, Maletzky T, Fuchs H (2006) Interface circuits for quartz crystal sensors in scanning probe microscopy applications. Rev Sci Instrum 77:083701
66. Yano T, Ichimura T, Taguchi A, Hayazawa N, Verma P, Inouye Y, Kawata S (2007) Confinement of enhanced field investigated by tip-sample gap regulation in tapping-mode tip-enhanced Raman microscopy. Appl Phys Lett 91:121101
67. Ichimura T, Fujii S, Verma P, Yano T, Inouye Y, Kawata S (2009) Subnanometric near-field Raman investigation in the vicinity of a metallic nanostructure. Phys Rev Lett 102:186101
68. Giessibl FJ (1995) Atomic resolution of the silicon (111)-(7×7) surface by atomic force microscopy. Science 267:68
69. Gregor MJ, Biome PG, Schofer J, Ulbrich RG (1996) Probe-surface interaction in near-field optical microscopy: the nonlinear bending force mechanism. Appl Phys Lett 68:307
70. Roth RM, Panoiu NC, Adams MM, Osgood RM, Neacsu CC, Raschke MB (2006) Resonant-plasmon field enhancement from asymmetrically illuminated conical metallic-probe tips. Opt Exp 14:2921
71. Julian Chen C (1993) Introduction to scanning tunneling microscopy. Oxford University Press, New York

72. Pettinger B, Ren B, Picardi G, Schuster R, Ertl G (2004) Nanoscale probing of adsorbed species by tip-enhanced Raman spectroscopy. Phys Rev Lett 92:096101
73. Picardi G, Nguyen Q, Schreiber J, Ossikiovski R (2007) Comparative study of atomic force mode and tunneling mode tip-enhanced Raman spectroscopy. Eur Phys J Appl Phys 40:197
74. Hecht B, Bielefeldt H, Inouye Y, Pohl DW, Novotny L (1997) Facts and artifacts in near-field optical microscopy. J Appl Phys 81:2492
75. Hatano H, Inouye Y, Kawata S (1997) Near-field optical microscope with a multiheight scanning–imaging mode. Opt Lett 22:1532
76. Zenhausern F, Martin Y, Wickramasinghe HK (1995) Scanning interferometric apertureless microscopy–Optical imaging at 10 angstrom resolution. Science 269:1083
77. Kneipp K, Wang Y, Kneipp H, Perelman LT, Itzkan I, Dasari RR, Feld MS (1997) Single molecule detection using surface-enhanced Raman scattering (SERS). Phys Rev Lett 78:1667
78. Wessel J (1985) Surface-enhanced optical microscopy. J Opt Soc Am B 2:1538
79. Inouye Y, Hayazawa N, Hayashi K, Sekkat Z, Kawata S (1999) Near-field scanning optical microscope using a metallized cantilever tip for nanospectroscopy. Proc SPIE 3791:40
80. Domke KF, Zhang D, Pettinger B (2006) Toward Raman fingerprints of single dye molecules at atomically smooth Au(111). J Am Chem Soc 128:14721
81. Neacsu CC, Dreyer J, Behr N, Raschke MB (2006) Scanning-probe Raman spectroscopy with single-molecule sensitivity. Phys Rev B 73:193406
82. Hayazawa N, Watanabe H, Saito Y, Kawata S (2006) Towards atomic site-selective sensitivity in tip-enhanced Raman spectroscopy. J Chem Phys 125:244706
83. Steidtner J, Pettinger B (2008) Tip-Enhanced Raman Spectroscopy and Microscopy on Single Dye Molecules with 15 nm Resolution. Phys Rev Lett 100:236101
84. Hayazawa N, Yano T, Watanabe H, Inouye Y, Kawata S (2003) Detection of an individual single-wall carbon nanotube by tip-enhanced near-field Raman spectroscopy. Chem Phys Lett 376:174
85. Anderson N, Hartschuh A, Cronin S, Novotny L (2005) Nanoscale vibrational analysis of single-walled carbon nanotubes. J Am Chem Soc 127:2533
86. Saito Y, Motohashi M, Hayazawa N, Iyoki M, Kawata S (2006) Nanoscale characterization of strained silicon by tip-enhanced Raman spectroscope in reflection mode. Appl Phys Lett 88:143109
87. Lee N, Hartschuh RD, Mehtani D, Kislius A, Maguire JF, Green M, Foster MD, Sokolov AP (2007) High contrast scanning nano-Raman spectroscopy of silicon. J Raman Spectrosc 38:789
88. Nguyen Q, Ossikovski R, Schreiber J (2007) Contrast enhancement on crystalline silicon in polarized reflection mode tip-enhanced Raman spectroscopy. Opt Commun 274:231
89. Picardi G, Nguyen Q, Ossikovski R, Schreiber J (2007) Polarization properties of oblique incidence scanning tunneling microscopy–tip-enhanced Raman spectroscopy. Appl Spectrosc 61:1301
90. Poborchii V, Tada T, Kanayama T, Geshev P (2009) Optimization of tipmaterial and shape for near-UV TERS in Si structures. J Raman Spectrosc 40:1377
91. Hayazawa N, Motohashi M, Saito Y, Ishitobi H, Ono A, Ichimura T, Verma P, Kawata S (2007) Visualization of localized strain of a crystalline thin layer at the nanoscale by tip-enhanced Raman spectroscopy and microscopy. J Raman Spectrosc 38:684
92. Steidtner J, Pettinger B (2007) High-resolution microscope for tip-enhanced optical processes in ultrahigh vacuum. Rev Sci Instrum 78:103104
93. Zhang D, Wang X, Braun K, Egelhaaf H, Fleischer M, Hennemann L, Hintz H, Stanciu C, Brabec CJ, Kern DP, Meixner AJ (2009) Parabolic mirror-assisted tip-enhanced spectroscopic imaging for non-transparent materials. J Raman Spectrosc 40:1371
94. Efremov EV, Ariese F, Gooijer C (2008) Achievements in resonance Raman spectroscopy review of a technique with a distinct analytical chemistry potential. Anal Chim Acta 606:119

95. Holtz M, Duncan WM, Zollner S, Liu R (2000) Visible and ultraviolet Raman scattering studies of $Si_{1-x}Ge_x$ alloys. J Appl Phys 88:2523
96. Ogura A, Yamasaki K, Kosemura D, Tanaka S, Chiba I, Shimidzu R (2006) UV-Raman spectroscopy system for local and global strain measurements in Si. Jpn J Appl Phys 45:3007
97. Moutanabbir O, Reiche M, Hähnel A, Erfurth W, Motohashi M, Tarun A, Hayazawa N, Kawata S (2010) UV-Raman imaging of the in-plane strain in single ultrathin strained silicon-on-insulator patterned structure. Appl Phys Lett 96:233105
98. Bohren CF, Huffman DR (1983) Absorption and scattering of light by small particles. Wiley, New York, Chap 5
99. Powell CJ, Swan JB (1959) Origin of the characteristic electron energy losses in aluminum. Phys Rev 115:869
100. Ritchie RH (1957) Plasma losses by fast electrons in thin films. Phys Rev 106:874
101. Kawata S, Sun HB, Tanaka T, Takada K (2001) Finer features for functional microdevices. Nature 412:697
102. Hayazawa N, Ichimura T, Hashimoto M, Inouye Y, Kawata S (2004) Amplification of coherent anti-stokes Raman scattering by a metallic nanostructure for a high resolution vibration microscopy. J Appl Phys 95:2676
103. Shen YR (1984) The principles of nonlinear optics. Wiley, New York
104. Levenson MD, Kano SS (1988) Introduction to nonlinear optical spectroscopy. Academic, New York
105. Ichimura T, Hayazawa N, Hashimoto M, Inouye Y, Kawata S (2004) Application of tip-enhanced microscopy for nonlinear Raman spectroscopy. Appl Phys Lett 84:1768
106. Ichimura T, Hayazawa N, Hashimoto M, Inouye Y, Kawata S (2004) Tip-enhanced coherent anti-stokes raman scattering for vibrational nanoimaging. Phys Rev Lett 92:220801
107. Tanaka S, Maeda Y, Cai L, Tabata H, Kawai T (2001) Application of tip-enhanced microscopy for nonlinear Raman spectroscopy. Jpn J Appl Phys 40:4217
108. Watanabe H, Ishida Y, Hayazawa N, Inouye Y, Kawata S (2004) Tip-enhanced near-field Raman analysis of tip-pressurized adenine molecule. Phys Rev B 69:155418
109. Yano T, Verma P, Saito Y, Ichimura T, Kawata S (2009) Pressure-assisted tip-enhanced Raman imaging at a resolution of a few nanometres. Nature Photon 3:473
110. Yano T, Inouye Y, Kawata S (2006) Nanoscale uniaxial pressure effect of a carbon nanotube bundle on tip-enhanced near-field Raman spectra. Nano Lett 6:1269
111. Downes A, Salter D, Elfick A (2006) Heating effects in tip-enhanced optical microscopy. Opt Exp 14:5216
112. Tarun A, Motohashi M, Hayazawa N, Kawata S (2008) Highly efficient tip-enhanced Raman spectroscopy in depolarized configuration. Proc SPIE 7033:70330C
113. Malkovskiy AV, Kisliuk AM, Barrios CA, Foster MD, Sokolov AP (2009) Tip-induced heating in apertureless near-field optics. J Raman Spectrosc 40:1349
114. Tarun A, Hayazawa N, Kawata S (2010) Site-selective cutting of carbon nanotubes by laser heated silicon tip. Jpn J Appl Phys 49:025003
115. Chiashi S, Murakami Y, Miyauchi Y, Maruyama S (2008) Temperature dependence of Raman scattering from single-walled carbon nanotubes: undefined radial breathing mode peaks at high temperatures. Jpn J Appl Phys 47:2010
116. Pang YS, Hwang HJ, Kim MS (1998) Reversible temperature dependence in surface-enhanced Raman scattering of 1-propanethiol adsorbed on a silver island film. J Phys Chem B 102:7203
117. Leung PT, Hider MH, Sánchez EJ (1996) Surface-enhanced Raman scattering at elevated temperatures. Phys Rev B 53:12659
118. Chiang H–P, Leung PT, Tse WS (2000) Remarks on the substrate-temperature dependence of surface-enhanced Raman scattering. J Phys Chem B 104:2348
119. Zhang W, Schmid T, Yeo B, Zanobi R (2008) Near-field heating, annealing, and signal loss in tip-enhanced Raman spectroscopy. J Phys Chem C 112:2104

120. Otto A, Mrozek I, Grabhorn H, Akemann W (1992) Surface-enhanced Raman-scattering. J Phys 4:1143
121. Kambhampati P, Child CM, Foster C, Campion A (1998) On the chemical mechanism of surface enhanced Raman scattering: Experiment and theory. J Chem Phys 108:5013
122. Lombardi JR, Birke RL (2008) A unified approach to surface-enhanced Raman spectroscopy. J Phys Chem C 112:5605
123. Yang TJ, Lessard GA, Quake SR (2000) An apertureless near-field microscope for fluorescence imaging. Appl Phys Lett 76:378

Raman Spectroscopy for Characterization of Semiconducting Nanowires 17

Gregory S. Doerk, Carlo Carraro, and Roya Maboudian

1 Definition of the Topic

Raman scattering behavior germane to semiconductor nanowires (NWs) and the application of Raman spectroscopy to characterize the structure, composition, strain, and temperature of individual semiconductor nanowires with submicron resolution are discussed.

2 Overview

This chapter covers the electromagnetic radiation scattering behavior of semiconductor nanowires as it impacts their Raman scattering properties, as well as how the basic theory of Raman scattering in semiconductor crystals may be applied to the corresponding nanowire forms, with specific examples from the recent literature. The discussion then turns to more specific ways in which Raman spectroscopy can be used to precisely determine crystal structure, crystal size (or quality), composition, impurity concentration, strain, and temperature with submicron resolution, as demonstrated in recent literature examples.

G.S. Doerk
IBM Almaden Research Center, 650 Harry Road, San Jose, CA, USA

C. Carraro • R. Maboudian (✉)
Department of Chemical and Biomolecular Engineering, University of California, Berkeley, 201 Gilman Hall, Berkeley, CA, USA

Challa S. S. R. Kumar (ed.), *Raman Spectroscopy for Nanomaterials Characterization*,
DOI 10.1007/978-3-642-20620-7_17, © Springer-Verlag Berlin Heidelberg 2012

3 Introduction

Nanowires are high-aspect-ratio solid structures with sizes confined to several hundred nanometers or less in two dimensions (i.e., an effective diameter less than several hundred nanometers), but much longer in the third. While the methods commonly used to synthesize nanowires were initially studied and developed in the 1960s and 1970s [1, 2], investigation of their properties and prototypical device demonstrations did not begin in earnest until after the emergence of carbon nanotubes as an academic research interest in the 1990s [3, 4]. Since then there has been an explosion of research both in synthesizing nanowires (and related nanostructures) and in characterizing their properties. Nanowires (NWs) are usually larger in diameter than carbon nanotubes, and their three-dimensional crystal structures—rather than a structure based upon rolled-up graphene as found in carbon nanotubes—render physical behavior that is at times somewhat less striking; it also makes their possible integration into new technology more straightforward.

As a probe of lattice vibrations, Raman spectroscopy is very sensitive to intrinsic crystal properties and extrinsic stimuli, especially in semiconductors. It may be employed to study crystal structure and quality, crystal orientation, optical interactions, chemical composition, phases, dopant concentration, surface and interface chemistry, and local temperature or strain. As an optical technique, important sample information may be obtained rapidly and nondestructively with minimal sample preparation. Submicron lateral resolution is possible with the use of confocal lenses. These features have made it a vital tool for research labs studying semiconductor-based technologies. They also are increasingly important for the study of semiconductor NWs fabricated by both top-down and bottom-up approaches since many of the common characterization methods used with bulk crystals or thin films cannot be applied to NWs in a direct manner.

In this chapter, we focus on the application of Raman spectroscopy to the study of semiconductor NWs. We begin with a discussion of the details of Raman spectroscopy germane to NW samples and then move on to a survey of how Raman spectroscopy can be used to characterize semiconductor NWs both qualitatively and quantitatively. Each section will be supplemented with examples from the scientific literature. Given the wide range of applications for Raman spectroscopy, this chapter is necessarily brief. However, at an atomic level, most NWs are not fundamentally different from their bulk crystal counterparts and the prior gains in understanding the Raman spectra of bulk crystals remain extremely helpful, since much of the underlying physics is the same. Therefore, we also refer the reader to the chapter "Characterization of semiconductors by Raman spectroscopy" in *Analytical Raman Spectroscopy* [5], to the book *Scattering of Light by Crystals* by W. Hayes and R. Loudon [6], and various volumes of *Light Scattering in Solids* edited by M. Cardona and G. Guntherodt [7].

As a matter of terminology, while the term "nanowire" is on occasion reserved for cases in which the diameter is strictly less than 100 nm, this definition is arbitrary. Various properties may be substantially different in NWs with diameters over 100 nm compared to those measured in bulk crystals of the same material depending on the relevant length scales for the physical processes involved. For instance, optical interactions with dielectric NWs typically do not vary with diameter in a monotonic fashion when their diameters are on the order of the wavelength of the radiation. This extends the area of interest for these studies to a diameter range well above 100 nm. Therefore, in this chapter, we do not maintain a strict naming convention, and apply the term "nanowire" to those structures with a diameter less than one micron and which may exhibit notably different properties from their bulk crystal counterparts.

4 Basic Methodology

4.1 Raman Scattering from Semiconductor Nanowires

The profoundly different optical interactions in subwavelength dielectric wires noted in the introduction lead to considerable differences in Raman scattering intensities from bulk crystal samples as well as strong polarization dependencies. Such phenomena arise from the effects of geometric confinement on light scattering due to the boundary mismatch of the dielectric function at optical frequencies.

Under the semiclassical theory of Raman scattering, the inelastically scattered light is produced by a modulation of the dielectric polarization through an excitation of the crystal with a finite momentum and energy. Raman scattering is witnessed only with the existence of a nonzero derivative of the dielectric susceptibility tensor with respect to the excitation amplitude, which for normal (vibrational) Raman scattering in crystals is provided by the creation (Stokes process) or annihilation (anti-Stokes process) of phonons. In Raman scattering from crystals, both energy and momentum must be conserved. For first-order Raman scattering, this means

$$\omega_s = \omega_i \pm \omega_j \tag{17.1a}$$

$$\boldsymbol{k}_s = \boldsymbol{k}_i \pm \boldsymbol{q}_j \tag{17.1b}$$

where ω refers to angular frequency, $\boldsymbol{k}$ refers to a wave vector $(2\pi/\lambda)$ of light, $\boldsymbol{q}$ is the phonon wave vector, and λ is the wavelength of light. The subscripts i, s, and j refer to the incident light, scattered light, and the phonon mode, and the (+) or (−) sign refers to an anti-Stokes or Stokes process, respectively. Since the maximum phonon wave vector is achieved in the limit where the scattered wave vector is equal in magnitude and opposite in direction (perfect backscatter) to the incident wave vector, q must be less than $4\pi/\lambda_i$. The Brillouin zone on the other hand is at its maximum value for π/d, where d is the appropriate lattice constant. For visible light ($\lambda >$ ~300 nm), the maximum phonon wave vector is limited to values of approximately 10^{-3} relative to the extent of the first Brillouin zone, or essentially zero for the sake of most calculations, giving rise to what is often referred to as a "selection rule" (not to be confused with polarization selection rules) for first-order Raman scattering:

$$q_j \approx 0 \tag{17.2}$$

Given their position at the center of the Brillouin zone and their nonzero frequency at zero (or near zero) momentum, the mode for these phonons is commonly referred to as the zone-center optical mode.

For Stokes processes involving phonons, the intensity I of scattered light is given by [6]:

$$I_s \propto \lambda_s^{-4} |\hat{e}_i \cdot \chi_j \cdot \hat{e}_s|^2 (n_j + 1) I_i \tag{17.3}$$

The $\hat{e}$ term refers to a unit polarization vector, χ is the first-order derivative of the susceptibility tensor with respect to the phonon amplitude, also known as the Raman tensor, and n_j is the phonon occupation number for the jth mode, given by:

$$n_j = [exp(\hbar\omega_j/kT) - 1]^{-1} \tag{17.4}$$

where $\hbar$ is Planck's constant, k is Boltzmann's constant, and T is absolute temperature. The proportionality is essentially the same for anti-Stokes processes except that $n_j + 1$ is replaced by n_j. The terms except I_i are part of the Raman scattering cross section $\sigma_R(\omega, \omega_j)$.

At this point, we may take note of the salient features of this relation and their implications for Raman scattering experiments. First, since the frequency of the scattered light is shifted from the incident light frequency by a much smaller phonon frequency, the incident and scattered light frequencies are approximately the same value. Their wavelengths are then also approximately the same, and the Raman scattering intensity may be treated as possessing a similar fourth power dependency on the incident light wavelength, which is a characteristic feature for scattering cross sections at frequencies below the main absorption band of the scattering material [6]. The second term involving the Raman tensor reflects the symmetry of the scattering medium. For crystals, this is of particular importance as the cross section may be required to disappear for certain polarization unit vector combinations. Careful analysis of intensities for various phonon lines in the Raman spectra may provide valuable information about the crystal symmetry (and hence structure). The phonon occupation number terms introduce an explicit dependence of the scattered intensity on sample temperature. Though the presence of other components to the scattered intensity in (17.3) as well as factors involving the experimental apparatus will make a firm quantitative link between temperature and intensity precarious, the ratio between Stokes and anti-Stokes intensities is far less ambiguous. We will return to the possibility of measuring local temperature from Raman spectra later.

As shown, (17.3) implies that the Raman scattered intensity is proportional to the incident radiation intensity, and in turn on the square of the incident electric field amplitude. However, since Raman scattering depends on the dielectric polarization within the crystal, it is the internal field which must actually be considered, and the internal field is altered from the incident field by boundary conditions at the surface that require continuity of the tangential electric field and normal electric

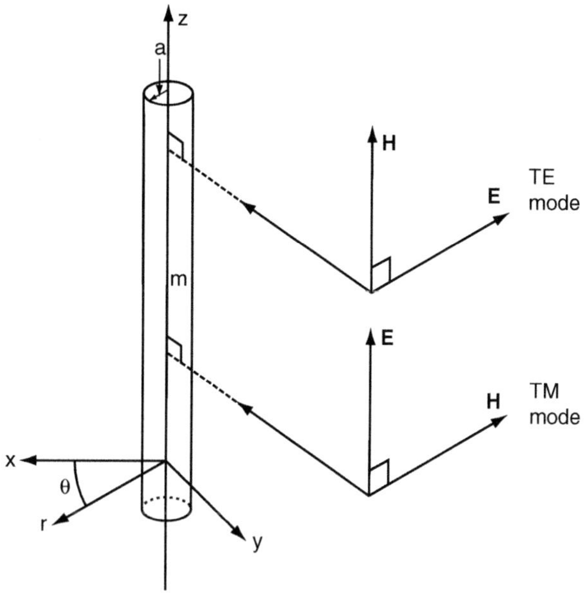

Fig. 17.1 Coordinates for scattering of electromagnetic radiation at normal incidence by a cylinder with radius a and complex refractive index m, with the cylinder axis parallel to the z-direction. Cylindrical coordinates are z, r and θ

displacement vectors. Due to the unique geometries and sizes of nanowires, a large dielectric mismatch between the wire and its surrounding medium requires special calculations of the internal electric field. These calculations may be performed using computational methods such as finite difference approaches [8] or the discrete dipole approximation [9]; they may also be performed analytically by solving Maxwell's equations using separation of variables. This approach is completely analogous to Mie scattering theory for spherical scatterers. Analytical calculations explain a number of experimental phenomena exhibited in the Raman scattering from individual nanowires.

We consider the cylindrical nanowire geometry shown in Fig. 17.1, with an incident plane wave normal to the cylinder axis and with an amplitude E_o. This is the simplest case to solve analytically and the one most often treated in experimental spectroscopic investigations of single nanowires. Possible orientations of linearly polarized incident light with respect to the wire axis are bounded by two cases. The first is the transverse magnetic (TM) polarization where the electric field is polarized parallel to the wire axis, and the second is the transverse electric (TE) polarization where the electric field is polarized perpendicularly to the wire axis. In TM polarization, the condition of continuity of the tangential electric field is expected to maximize the internal field, while in TE polarization, the dielectric mismatch should suppress the internal field. The incident plane wave may be expanded in cylindrical functions as:

$$E_{inc} = E_o\hat{e} \sum_{n=-\infty}^{\infty} (-i)^n e^{in\theta} J_n(k_o r), \tag{17.5}$$

where $\hat{e}$ is a unit vector in the direction of the field polarization, $k_o = 2\pi/\lambda_i$ is the propagation constant in free space, and J_n is the Bessel function of integral order n.

In the normal incidence case presented here, the internal and scattered waves are of the same mode and polarization as the incident wave (mixing of the polarizations occurs when the incident wave is at an oblique angle). This yields two limiting cases for the internal electric field [10]:

$$E_{\text{int}}^{TM} = E_o\hat{z}\sum_{n=-\infty}^{\infty}(-i)^n e^{in\theta} d_n J_n(mk_o r) \tag{17.6a}$$

$$E_{\text{int}}^{TE} = E_o\sum_{n=-\infty}^{\infty}(-i)^n c_n e^{in\theta}\left[\frac{in}{r}J_n(mk_o r)\hat{r} - mk_o J_n'(mk_o r)\hat{\theta}\right] \tag{17.6b}$$

where $m = \sqrt{\varepsilon}$ is the complex index of refraction (ε is the dielectric constant). By applying the surface boundary conditions that require continuity of tangential components of $\boldsymbol{E}$ and the magnetic field intensity $\boldsymbol{H}$, the coefficients c_n and d_n may be related to the far-field scattering coefficients a_n and b_n:

$$c_n = \frac{J_n'(k_o a) - a_n H_n'(k_o a)}{J_n'(mk_o a)} \tag{17.7a}$$

$$d_n = \frac{J_n(k_o a) - b_n H_n(k_o a)}{J_n(mk_o a)} \tag{17.7b}$$

Here, H_n refers to the Hankel function of the second kind and prime denotes differentiation with respect to the argument. The values of a_n and b_n are given by

$$a_n = \frac{J_n(k_o a)J_n'(mk_o a) - mJ_n'(k_o a)J_n(mk_o a)}{H_n(k_o a)J_n'(mk_o a) - mH_n'(k_o a)J_n(mk_o a)} \tag{17.8a}$$

$$b_n = \frac{mJ_n(k_o a)J_n'(mk_o a) - J_n(mk_o a)J_n'(k_o a)}{mH_n(k_o a)J_n'(mk_o a) - H_n'(k_o a)J_n(mk_o a)} \tag{17.8b}$$

The volume-averaged intensity of the internal field per unit length is obtained by integrating the squared field amplitudes over the NW cross section, and the presence of Bessel functions in the result that are functions of the reduced optical radius, $mk_o a = 2\pi ma/\lambda$, reveals that the internal field is modulated in a way that depends explicitly on the NW diameter and the wavelength of incident light. When the incident light is TM polarized with respect to the NW axis, the modulation can be considered an enhancement. Cao et al. [11] used the Mie scattering model to explain an observed enhancement in the first-order Raman signal intensity per unit volume for Si NWs and nanocones when compared to bulk Si. This enhancement can approach magnitudes of 10^3 as shown in Fig. 17.2. If one defines

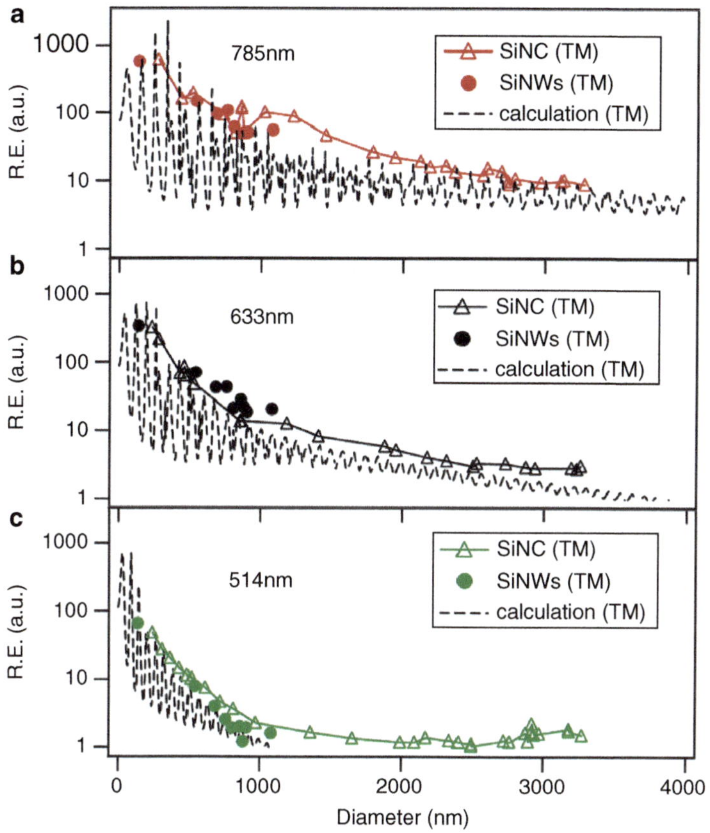

Fig. 17.2 Experimental and calculated Raman Enhancement (R.E.) in reference to bulk Si for silicon NWs and nanocones at three different excitation wavelengths (With permission from reference [11]. Copyright (2006) by the American Physical Society)

a parameter $Q(\omega_i) = \bar{I}_{int}/\bar{I}_{bulk}$ to represent the enhancement in the internal field intensity of the NW compared with the bulk crystal at the incident light frequency, then the inelastic scattering is expected to be enhanced by this factor in a directly proportional manner through the dielectric polarization. However, the Raman scattered radiation will also be enhanced by the same structural resonance, giving rise to a second enhancement factor for the Raman scattered light $Q_{Raman}(\omega_s)$. Because the frequencies of the scattered light and the incident light are nearly the same, it is reasonable to assume that $Q_{Raman}(\omega_s) \approx Q(\omega_i)$, so that the total Raman enhancement factor goes as the second power of the internal TM field intensity (or as the fourth power in E_{int}^{TM}).

This calculated enhancement models the experimentally observed enhancements reasonably well. However, the experimental values consistently occur at the upper end of the calculated enhancement curves and reflect none of the oscillatory behavior expected. For the Si nanocones, the diameter changes by more than the

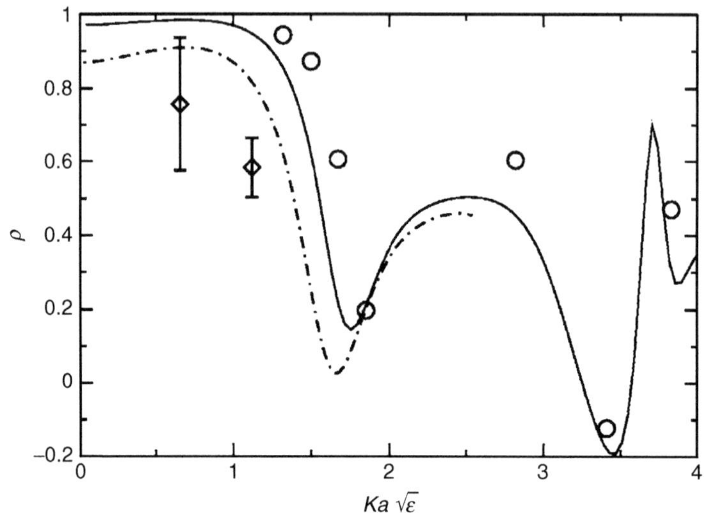

Fig. 17.3 Polarization anisotropy ratio as a function of reduced optical radius $ka\sqrt{\varepsilon}$, where ε is the dielectric constant of the material. *Solid* (*broken*) *lines* are the theoretical curves for Si (SiC) NWs. *Circles* (*diamonds*) are the experimental values for Si (SiC) NW optical phonon (TO) *lines* (With permission from reference [12]. Copyright (2006) by the American Physical Society)

period of the enhancement oscillations over the width of the laser spot and so the fluctuations in Raman enhancement may be smeared out to reflect only the upper edge. This does not account for the lack of similarly high Raman enhancement values for the Si NWs though. Another key feature is that the Raman scattering is enhanced less in comparison with bulk Si as the wavelength of the incident light is decreased. With decreasing wavelength, the "bulk-like" nonresonant Raman scattering exhibiting a λ^{-4} dependence contributes more, in turn diminishing the enhancement from the optical structural resonance.

The Mie scattering model also helps explain the strong polarization anisotropy or "antenna effects" [9, 12] in the Raman scattering of semiconductor NWs, so named because of the selective Raman scattering of TM-polarized light by way of this structural resonance modulation. This behavior is quantified in the polarization anisotropy ratio, calculated by an averaging of the squared field amplitudes of the product of incident and scattered fields [12], determined from a Mie-type solution:

$$\rho = \frac{\left\langle |E^{TM}|^4 \right\rangle - \left\langle |E^{TM}|^2 |E^{TE}|^2 \right\rangle}{\left\langle |E^{TM}|^4 \right\rangle + \left\langle |E^{TM}|^2 |E^{TE}|^2 \right\rangle} \tag{17.9}$$

As shown in Fig. 17.3, the calculated behavior is in excellent agreement with values for Si and SiC NWs (TO phonon line) measured as the ratio of the difference between polarized and depolarized Raman signal intensities over their sum. For NWs in the backscattering configuration with the excitation radiation at normal incidence, the term "polarized" refers to the case where both the incident and scattered radiation are polarized along the NW axis; in the "depolarized" case, the incident and scattered radiation are polarized parallel and perpendicular to the NW axis, respectively (or vice versa). The polarized configuration gives the maximum signal intensity as both the internal field from the incident light and the Raman scattering field are maximized in the TM polarization. In the depolarized configuration, the internal field is still maximized, but this is countered by the fact that the light collected by the spectrometer is from the TE polarization where the Raman scattering field is suppressed.

The general case of linearly polarized light at normal incidence to the NW axis can be treated analytically by introducing the polarization angle θ between the incident electric field polarization vector and the NW axis ($\theta = 0°$ for TM and $\theta = 90°$ for TE). Accounting for the internal field and Raman scattering field enhancements, the collected Raman scattering intensity will be in part proportional to $\cos^4 \theta$ [9] as θ is rotated from the TM polarization. If the polarization of the scattered light is not analyzed, only the modulation of the internal field is considered, and the Raman signal proportionality is with $\cos^2 \theta$ [12]. The clear and practical implication of these findings for Raman investigations of single NWs is that the signal is maximized in the TM polarization. This becomes increasingly important as the diameter is decreased and the corresponding Raman scattering cross section becomes exceedingly small (even despite the relative enhancements over bulk crystals).

In conjunction with the large surface-to-volume ratio in semiconductor NWs, the polarization response in their Raman scattering affords an exquisite sensitivity to the structure and composition at the wire surface. A clear example has been demonstrated for the case Raman scattering from SiC wires that exhibit quantities of carbon above the stoichiometric balance in some cases. This carbon may exist in a tetrahedrally coordinated diamond cubic configuration in the volume of the NW, as nanocrystalline graphite at the wire surface, or as a combination of the two. The former is characterized by the degenerate zone-center optical phonon with a frequency at ~1,330 cm^{-1}, while the latter is marked by a peak at ~1,580 cm^{-1} (G-peak) and a broader dispersive peak in the range of 1,330–1,400 cm^{-1} (D-peak). Distinction of disordered carbon within the wire from that on the surface is difficult based on the peak positions alone. However, since the polarization behavior in semiconductor NWs arises due to optical structural modulation of the internal fields, the Raman signal suppression in the TE polarization primarily affects volume scattering modes and not surface scattering modes. In this way, the TE polarization grants a much higher sensitivity to surface modes relative to volume modes. As a volume mode, the diamond cubic signal will be suppressed along with the bulk SiC signal for depolarized configuration, whereas the signal from nanocrystalline graphite at the surface will remain largely unaffected, showing that polarization analysis reveals the degree of carbonization of the SiC NW [13].

5 Key Findings

5.1 Chemical and Structural Characterization

One of the most valuable features of the Raman scattering from semiconductor crystals is that it provides immediate evidence about the crystal composition, structure, and quality. Identification of the crystal is quickly assessed by the most prominent peaks, which belong to the zone-center optical phonons as a result of the momentum "selection rule" (17.2) for first-order Raman scattering. In Si, Ge, and diamond, these lines are in the vicinity of 520, 300, and 1,330 cm^{-1}, respectively. In GaAs, the zone-center optical phonon line is split into a LO line at 292 and a TO line at 269 cm^{-1}, respectively. This splitting arises from the fact that the LO mode can scatter light via its macroscopic longitudinal electric field [6] and is a common feature of polar zinc blende crystals. For certain polarization configurations however, only one of the two lines may be observed as a consequence of the polarization selection rules imposed by the Raman tensor, through (17.3).

For consistency with notation found in many other sources, we express (17.3) in a simpler manner:

$$I_{s,j} = C_j |\hat{e}_i \cdot \boldsymbol{R}_j \cdot \hat{e}_s|^2 \tag{17.10}$$

where $\boldsymbol{R}_j$ is the Raman tensor and C_j is a proportionality constant. The symmetry of the Raman tensors for particular crystal classes (i.e., point groups) can be derived via group theory. We will not address this topic here but we note that the Raman tensors for all the different three-dimensional crystal symmetries may be found in Hayes and Loudon [6].

For the case of cubic zone-center optical phonons, the Raman tensor is of rank three, with a second-rank tensor for phonon components polarized along x, y, or z Cartesian coordinate directions:

$$\begin{bmatrix} 0 & 0 & 0 \\ 0 & 0 & d \\ 0 & d & 0 \end{bmatrix} : (x), \tag{17.11a}$$

$$\begin{bmatrix} 0 & 0 & d \\ 0 & 0 & 0 \\ d & 0 & 0 \end{bmatrix} : (y), \tag{17.11b}$$

and

$$\begin{bmatrix} 0 & d & 0 \\ d & 0 & 0 \\ 0 & 0 & 0 \end{bmatrix} : (z), \tag{17.11c}$$

The LO phonon is polarized along q, where the direction of q as defined by (17.1b). The contribution of longitudinal phonons may be obtained by the projection of the Raman tensor element along the phonon unit vector $\hat{q}_j$:

$$I_{s,j,L} = C_j\{(\hat{e}_i \cdot R_j \cdot \hat{e}_s) \cdot \hat{q}_j\}^2 \tag{17.12}$$

where the subscript L represents the longitudinal phonon. The transverse contribution to the overall intensity is given by the difference between (17.10) and (17.12).

The importance of these selection rules in studying semiconductor nanowires through Raman spectroscopy may be made evident by example. Many zinc blende NWs are thermodynamically constrained to grow in a $\langle 111 \rangle$ crystallographic direction. At normal incidence and in the backscattering configuration, the polarized signal from the zone-center optical phonons for these NWs is proportional to $(4/3)d_{TO}^2$, while the depolarized signal is proportional to $(1/3)d_{TO}^2$. The LO line should be entirely absent, and a factor of four suppression of the TO signal beyond what is predicted by Mie scattering is to be expected due to crystal symmetry. The situation changes significantly when we instead examine NWs with a $\langle 110 \rangle$ growth direction, which become increasingly common as the NW diameters are reduced below a certain threshold diameter [14]. The polarized signal is proportional to d^2, with the fractional contribution from LO phonons ranging from one third to one depending on the orientation of the incident light wave vector. The depolarized intensity depends on the direction of the incident wave vector, but is entirely TO in character. The polarization dependencies of the TO and LO mode intensities give a clear way to distinguish between low index zinc blende NW crystal orientations. Unfortunately, Raman spectroscopy cannot provide the same information about diamond cubic crystals due to the degeneracy of the zone-center optical phonon.

Some semiconductor materials that have been synthesized in NW geometry such as GaN possess a wurtzite crystal structure [15] (point group C_{6v}). Moreover, many compound semiconductor NWs that exhibit only the zinc blende structure in bulk crystals may possess zinc blende or wurtzite crystal structures, and in some regions both in an alternating fashion along the wire axis [16]. Since the wurtzite structures only exist for some materials in the NW form, full phonon dispersion relations are not always available. However, the wurtzite unit cell length along the [0001] axis (the wurtzite NW growth axis common with the [111] zinc blende NW growth axis) is twice the [111] zinc blende unit cell length, so an approximate phonon dispersion may be obtained by folding the [111] direction zinc blende phonon dispersion from half of the Brillouin zone back to the zone center (Γ point). The end result for wurtzite Raman spectra is that the existing optical phonon lines from zinc blende spectra are retained while two additional lines are added that are native to the wurtzite structure [17].

Based on this, Spirkoska et al. [18] were able to use Raman spectroscopy to distinguish between individual GaAs NWs with majority zinc blende or majority wurtzite structures. Figure 17.4 shows the Raman spectra from three different GaAs NWs with different wurtzite contents. Spectrum α corresponds to a pure zinc blende wire and as expected for a [111] growth direction, only the TO peak is observed.

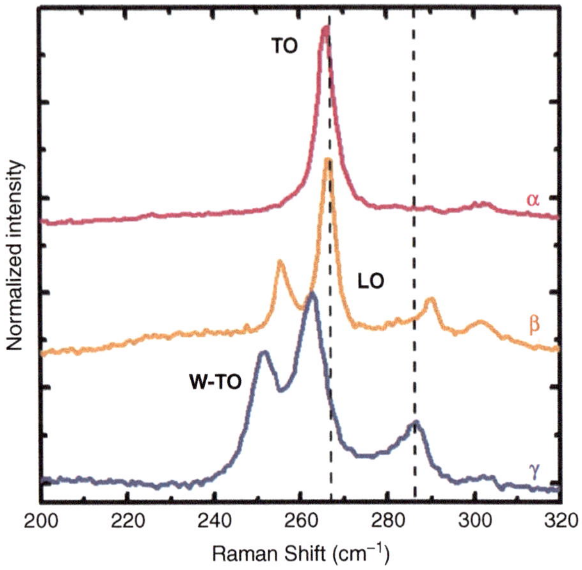

Fig. 17.4 Raman spectra from three individual GaAs NWs with fractional zinc blende/wurtzite structures. From *top* to *bottom*, the spectra correspond to a full zinc blende structure (α), an approximately 30% wurtzite structure (β), and an approximately 70% wurtzite structure (With permission from reference [18]. Copyright (2009) by the American Physical Society)

Spectrum β, which is for a NW with a majority zinc blende content begins to show a wurtzite line at ~255 cm^{-1} corresponding to the E_2 mode which is the wurtzite equivalent TO. The zinc blende LO mode also appears, which is attributed to a possible relaxation of the Raman selection rules due to the presence of numerous rotational twins and saw tooth faceted surfaces. Spectrum γ reflects a majority wurtzite structure, and is similar to spectrum β except that the wurtzite TO intensity is increased. All the peaks are also red shifted by ~4–5 cm^{-1} from spectrum β. This actually brings the wurtzite TO line closer to the value predicted by the zone folding approximation, indicating that this line was likely blue shifted in spectrum β. Due to the dependence of peak position on stress (to be discussed later in this chapter), this implies that the minority wurtzite regions in NW β are under compression due to lattice mismatch with the zinc blende regions, while the minority zinc blende regions in NW γ are under tension for the same reason. This spectral information can also be used with piezo-stage positioning control to create two-dimensional spatial Raman maps of single NWs, in turn determining the extent of wurtzite- or zinc blende–rich regions with submicron resolution [17].

Analyses of Raman spectra can be used to understand similar phenomena witnessed in Si NWs involving the emergence of ordered (or disordered) {111} stacking fault arrays parallel to the growth axis. By performing correlated transmission electron microscopy (TEM) and Raman spectroscopy (Fig. 17.5), Lopez et al. [19] were able to show that the ordering of these stacking fault arrays was equivalent to the inclusion of non-3C diamond cubic polytypes, either of the diamond hexagonal (2H) polytype or the rhombohedral Si (9R) polytype. TEM and correlated Raman mapping showed that disordered fault regions, polytype regions, and defect-free regions may all exist in the same NW. Finally, using the folded-zone approximation to index lines found in the Raman spectra, their carefully correlated TEM and Raman polarization analyses suggest that the raised regions at ~506 cm^{-1} between polytype Si modes at ~495 and ~517 cm^{-1} may be due to disorder in the stacking fault regions rather than the E_{1g} mode as previously suggested [20]. The combined Raman and TEM characterization work described in the preceding paragraphs illustrates that Raman spectroscopy and TEM are highly complementary techniques. It also attests to the fact that characterization of single NWs gives us a much deeper understanding of NW structures and growth processes, such as the emergence of different crystal structures within single NWs, that may be obfuscated if the NWs are examined only as ensembles.

Even in NWs that do not deviate from the crystal structure expected from their bulk crystalline counterparts, Raman spectroscopy provides a simple and rapid way to estimate crystal quality. In crystalline Si for example, the only allowed first-order Raman scattering is from the zone-center phonons ($q \approx 0$) leading to a sharp Lorentzian line shape at ~521 cm^{-1}. This $q \approx 0$ selection rule (17.2) is only valid if there is long-range order in the crystal so that momentum conservation is preserved. When the average crystal domain size is severely reduced, phonons are confined to domain boundaries (by which they are elastically scattered) which invalidates (17.1b) and results in a breakdown of the $q \approx 0$ rule. The momentum range for phonons participating in first-order Raman scattering is then extended further from the Brillouin zone center as the average domain size is decreased. For amorphous Si with no long-range order, optical phonons across the entire Brillouin zone may participate and the sharp Lorentzian at ~521 cm^{-1} becomes a very broad one centered at ~480 cm^{-1}. In Raman spectra for the intermediate cases of micro- and nanocrystalline Si, the zone-center line is red shifted and asymmetrically broadened to lower energy, the extent of which is determined by the average crystal domain size (Fig. 17.6).

A very commonly used model for the effect of phonon confinement was developed initially by Richter et al. [21] and Campbell and Fauchet [22] (RCF model), and adapted by various other researchers for their particular sample analyses. Under this model, the Raman intensity for the optical mode(s) confined to crystal domains of average diameter D may be expressed as [23]:

$$I(\omega) \propto \int_{BZ} \frac{|C(0,q)|^2 d^3q}{[\omega - \omega(q)]^2 + [\Gamma_o/2]^2} \qquad (17.13)$$

where Γ_o is the intrinsic crystalline material line width, $\omega(q)$ is the phonon dispersion, and $|C(0, q)|^2$ is the squared amplitude of the Fourier-transformed confinement function. Typically a Gaussian confinement function is assumed so that

$$|C(0,q)|^2 = exp\left[-\left(\frac{qD}{4\pi}\right)^2\right]. \qquad (17.14)$$

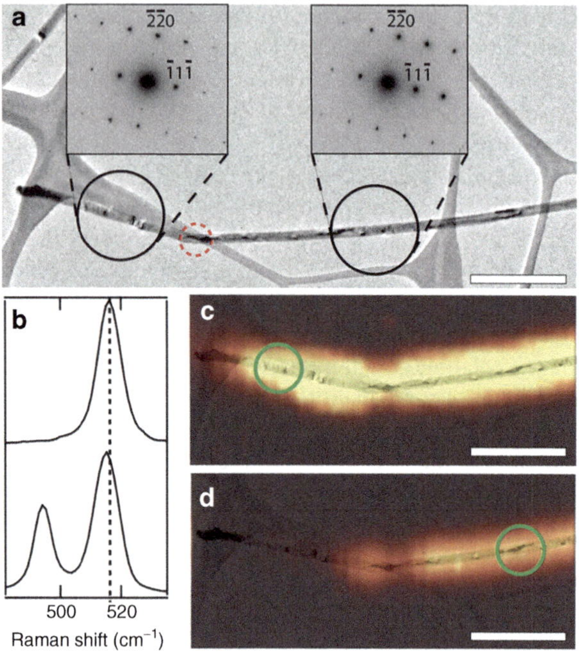

Fig. 17.5 (**a**) Transmission electron microscope image of a Si NW with corresponding selective area diffraction patterns indicating a change in growth direction from {111} to {112} after a kink (*left* to *right*). (**b**) Raman spectra from regions *marked circles* in (**c**) and (**d**) (*top* and *bottom*, respectively). (**c**) Raman intensity map of the diamond cubic region characterized by the ~519 cm^{-1} band. (**d**) Raman intensity map of the 2H polytype region formed by {111} stacking faults, characterized by the ~495 cm^{-1} band (With permission from reference [19]. Copyright (2009) by the American Chemical Society)

Richter et al. initially derived the model for microcrystalline Si, though its application should be general to many semiconducting materials. The model also assumes a spherical domain shape, though fairly simple adjustments may be applied to the confinement function (17.14) for domains with other geometries [22].

In the last half decade of the twentieth century, there was a great deal of interest in observing confined optical phonons in Si NWs, for which it was natural to apply the RCF model. Early results found that the model fits the experimental data reasonably well [24, 25] but there were conspicuous discrepancies. The most notable one was that the actual diameters of the Si NWs examined were far larger than those diameters for which the RCF function is expected to apply. Piscanec et al. [23] resolved many of the inconsistencies through a sustained investigation of the effect of laser heating on the Raman spectra of Si NWs, showing that if the laser power is not low enough, the spectra will reflect a spurious phonon confinement. However, other factors such as the sample diameter distribution, individual NW crystalline quality, inhomogenous laser heating [26], stress (for example from bent NWs) [27], and the morphology and termination of the NW surfaces render a quantitative understanding of the confinement effect by diameter reduction in Si NWs incomplete.

5.2 Quantitative Property Measurements with Raman Spectroscopy

In the previous section, we demonstrated how Raman spectra give rapid chemical and structural information about NWs in a facile manner. While the information that may be obtained through the ways shown above is extremely valuable, the emphasis has been on qualitative characterization. As a noncontact, nondestructive technique with submicron resolution, Raman spectroscopy could be even more valuable if used toward quantitative measurements on single NWs, and indeed, this is an active area of research. Here we will illuminate how Raman spectroscopy may be used to perform quantitative measurements of local composition, temperature, and strain in single NWs.

Due primarily to the predominance of Si in mainstream electronics, the properties of Si_xGe_{1-x} alloys have been particularly well studied. It is also an exemplary system for Raman studies because both elements are fully soluble in the other for the entire composition range, and it exhibits three distinguishable modes in Raman spectra corresponding to Si–Si, Ge–Ge, and Si–Ge atomic vibrations. For bulk crystals and thin films, the fractional composition may be determined by the relative integrated intensity $I_{Ge\text{–}Ge}/I_{Si\text{–}Si}$ [28] though this will exhibit diminishing accuracy for compositions (either Ge or Si) below 20% or if it becomes difficult to resolve one of the peaks below the noise floor. More often, composition is measured by the position of the peaks. Each mode exhibits a composition dependence, though the strongest peak, which also exhibits the strongest composition dependence, belongs to the Si–Si vibrational mode. For unstrained Si_xGe_{1-x} the Si–Si mode frequency $\omega_{Si\text{–}Si}$ depends linearly on the composition x and may be expressed as [28, 29]:

$$\omega_{Si-Si} = \omega_{Si} - 68x \tag{17.15}$$

where ω_{Si} is the optical phonon frequency of pure crystalline Si (~520–521 cm^{-1}).

This expression has been applied to Si_xGe_{1-x} NWs, where the local nature of confocal Raman measurements (confined to the laser spot diameter after focusing, on the order of ~1 μm or less) has revealed axial composition gradients whether intentional [30] or unintentional [31]. Care must be taken, however, when using Raman spectroscopy to quantitatively measure local composition in Si_xGe_{1-x} NWs because while the results for bulk crystals and thin films have been found to hold for Si_xGe_{1-x} NWs in general, they exhibit a slight redshift in $\omega_{Si\text{–}Si}$, especially at high Ge concentrations [32] (Fig. 17.7). In part this could be attributed to tensile strain, with a phonon frequency response that may be accounted for by inserting an additional linear term proportional to strain into the right side of (17.15) (for small strains) [28]. Raman spectroscopy can also be used to measure the composition of semiconductor NWs of ternary alloys such as $Zn_xCd_{1-x}Se$ [33] and $CdS_{1-x}Se_x$ [34]. Based on previous film measurements, NWs composed of other ternary alloy semiconductors like $In_xGa_{1-x}N$ [35] or $Al_xGa_{1-x}As$ [36] may also be compositionally accessible to Raman spectroscopy.

Besides measuring binary or ternary alloy composition, Raman spectroscopy evinces an acute sensitivity to impurity concentrations sometimes well below ~1% levels. This makes Raman spectroscopy especially well suited to measuring electrically active dopant concentrations in NWs since techniques such as scanning spreading resistance, capacitance-voltage, or Hall measurements that have been

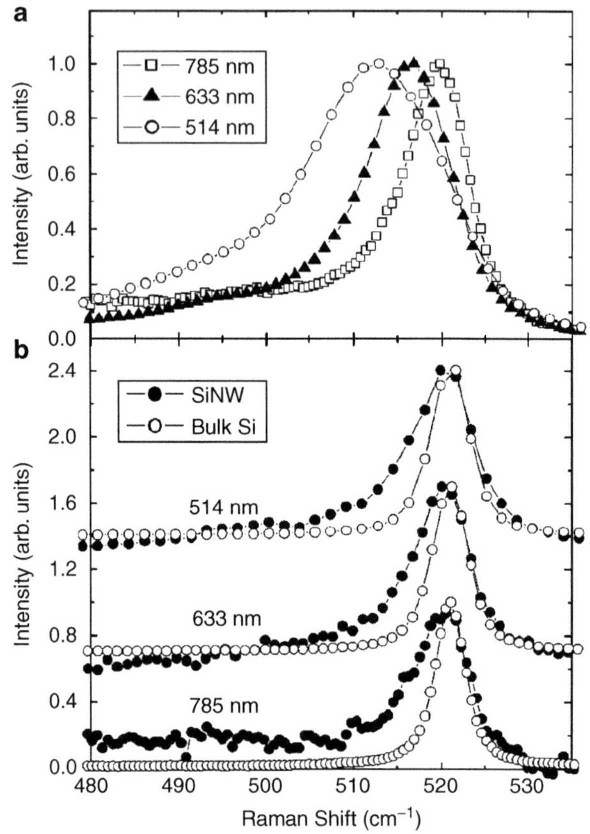

Fig. 17.6 (**a**) Raman spectra of small diameter Si NWs (<~15 nm) with three excitation wavelengths at ~3 mW power, which should induce different degrees of local heating according to silicon's absorptivity. (**b**) Comparison of Raman spectra from bulk Si and Si NWs at three excitation wavelengths and a laser power (0.02 mW) low enough to not induce local heating in the Si NWs. The asymmetry of optical phonon line for Si NWs may be attributed to phonon confinement (With permission from reference [23]. Copyright (2003) by the American Physical Society)

applied to thin film or bulk samples are extremely difficult if not impossible to apply to nanowires. Since the electrical measurements readily available to nanowires like transconductance or 4-point current–voltage measurements are complicated by factors such as contact resistances, surface interface states [37], or dopant deactivation [38], independent spectroscopic techniques possess added value. The impact of dilute impurities on Raman spectra may originate from different physical effects and thus may emerge as different functional dependencies; it may therefore be necessary to consider different material-impurity combinations independently. Nonetheless, this may be seen as an advantage of Raman spectroscopic electrical characterization as the presence of different electrically active impurities may be distinguishable in some cases. We will discuss a few of the more prominent spectral phenomena that have been successfully used to study electrically active impurities in semiconductors.

In the Raman spectra of Si, the line shape of the optical phonon peak is quite sensitive to degenerate p-type doping as a result of interference of Raman scattering from a continuum of electronic intravalence band transitions (electronic Raman scattering) with that from discrete phonon excitations (vibrational or phonon

Raman scattering) [39]. This continuum-discrete interference is known as the Fano effect and is a general phenomenon seen in a number of different systems. In p-type silicon, it is made possible by the fact that the intravalence energy gaps overlap in energy with the optical phonon energy, so when the Fermi level cuts across the heavy-hole and light-hole energy bands, there will be a multitude of electronic transitions whose energy range surrounds that of the single optical phonon energy. Extracting doping information from this effect may be achieved by fitting the altered Lorentzian peak to a Fano line shape (Fig. 17.8):

$$I(\omega) = A\frac{(q+\varepsilon)^2}{1+\varepsilon^2} \tag{17.16a}$$

where

$$\varepsilon \equiv \frac{\omega - \omega_o}{\Gamma} \tag{17.16b}$$

where ω_o is the phonon frequency and Γ is the line width, both of which are renormalized due to electronic Raman scattering. The q term is commonly referred to as an asymmetry parameter and its reciprocal is proportional to the magnitude of the ratio of the Raman tensor for electronic scattering ($\boldsymbol{R}_e$) to the Raman tensor for phonon scattering ($\boldsymbol{R}_p$) [39]:

$$\left|\frac{1}{q}\right| = \left|\frac{\boldsymbol{R}_e}{\boldsymbol{R}_p}\right| \tag{17.17}$$

In virtue of this, the reciprocal of the asymmetry parameter, $1/q$, is approximately proportional to the free carrier concentration in the range of 10^{19} cm^{-3} and higher. The q parameter is also a function of the laser wavelength. This may be understood by the fact that the discrete phonon Raman scattering occurs below the resonant state, and thus, its intensity follows the λ^{-4} dependence given in (17.3), while the electronic Raman scattering is an above-resonant Raman process (with regard to the intravalence band transitions) for visible excitation wavelengths and, thus, expresses no strong dependence. Therefore, as the excitation wavelength is reduced, the intensity of electronic Raman scattering is muted relative to phonon Raman scattering. As a result, it is often appropriate to use longer wavelength excitation lines when trying to obtain clear quantitative doping information from the Fano effect, as long as calibration measurements may be performed. Both ω_o and Γ are dependent on the doping level as well, and independent of the excitation wavelength [39], but are also affected by strain [40] and temperature [41]. Similar doping dependent Fano effects have been observed in the Raman spectra of highly boron-doped diamond [42] and gallium-doped germanium [43].

In the case of p-type doping in Si, the Fano effect is evident in the asymmetric broadening of the optical phonon line to higher frequency. There is indeed

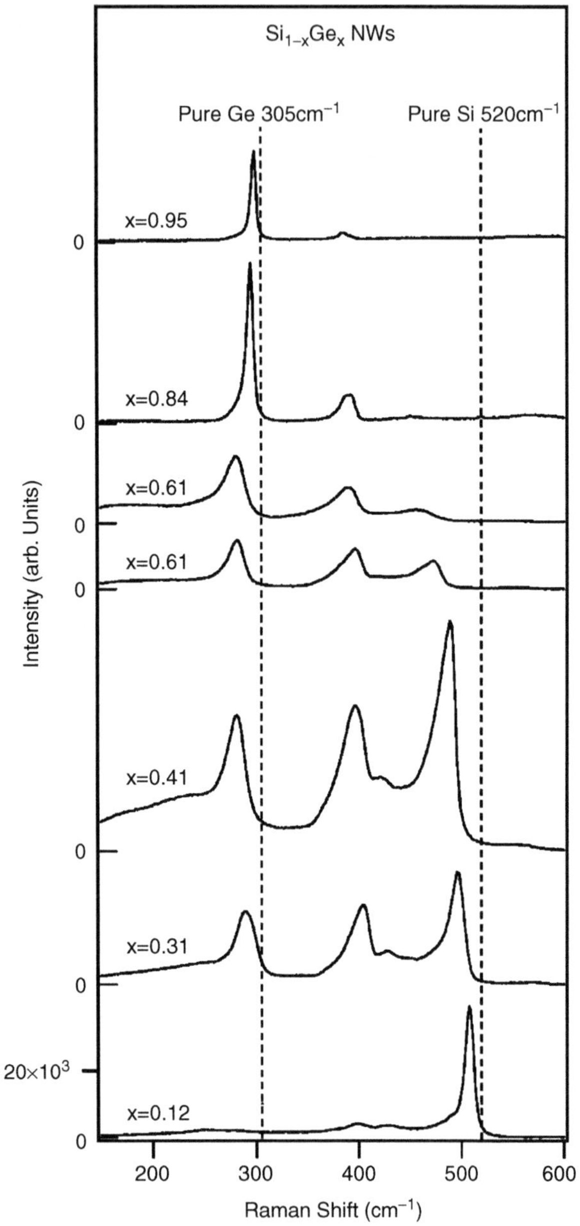

Fig. 17.7 (**a**) Raman spectra of ensembles of $Si_{1-x}Ge_x$ (approximately ~100 in each) with varying Ge atomic fraction (x) (With permission from reference [32]. Copyright (2008) by the American Chemical Society)

a matching Fano effect for n-type doping in Si, but its use suffers from two difficulties. First, it is characterized by an asymmetric broadening of the optical phonon line to lower (not higher) frequency, mathematically represented in a negative q value. While this conveniently distinguishes it from p-type doping, it also means that n-type doping is more likely to be convoluted with temperature,

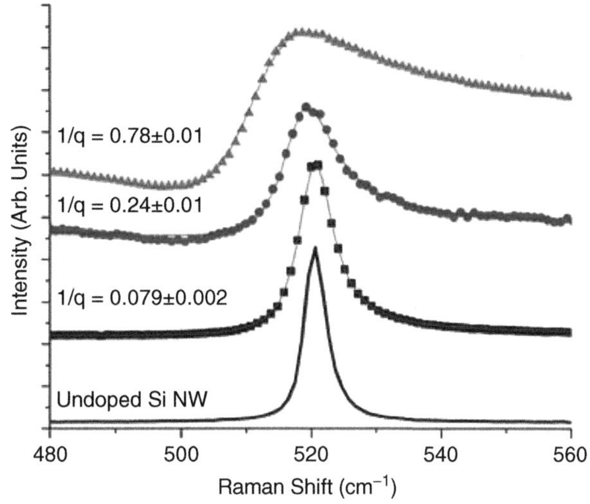

Fig. 17.8 Raman spectra from individual Si NWs (excitation wavelength of 633 nm) with different active boron concentrations. Fano fitting renders $1/q$ values of (bottom to top) ~0.079, ~0.24, and ~0.78, corresponding respectively to approximate active boron concentrations of ~1–2 × 10^{19}, ~5 × 10^{19}, and greater than 1 × 10^{20} cm^{-3} (Adapted with permission from reference [47]. Reproduced by permission of the Royal Society of Chemistry (http://dx.doi.org/10.1039/10.1039/C0NR00127A))

strain, and confinement (or crystal quality) effects. Secondly, the n-type Fano effect is much weaker, usually not emerging in measured Raman spectra until doping levels exceed 10^{20} cm^{-3} [44].

The Fano effect has been put to good use in studying both in situ and ex situ boron doping in Si NWs. It is a convenient and rapid way to characterize active boron concentrations in individual Si NWs without the use of extensive lithography that may take hours or days and is frequently unsuccessful in enabling accurate single NW electrical measurements. The ability to create Raman maps with submicron resolution also makes it possible to perform axial dopant profiling. In this way, Raman spectroscopy has been used to reveal axial doping nonuniformity in as-grown Si NWs due to uncatalyzed conformal deposition of highly doped Si on the NW sidewalls [45, 46] (Fig. 17.9). Inversely, Raman mapping has also been employed to demonstrate the high axial doping uniformity achieved via ex situ boron doping [47].

Another way to assess the quantity of impurities in a semiconductor is through local vibrational modes (LVMs). When impurities occupy substitutional sites, they give rise to local vibrations at a different frequency from the host lattice. The intensity ratio of the LVM modes to normal vibrational modes is proportional to the mean concentration of that impurity in the volume probed. This conveniently permits distinction between concentrations of different impurities, but careful calibration measurements are necessary if absolute impurity concentration values are desired. In Si (and its NWs), this has seen limited use since the LVMs of common dopants like boron are at frequencies in close proximity to peaks in the second-order Si Raman spectra [48, 49]. For other semiconductor NWs on the other hand, LVMs are beginning to see more use. For example, impurity Si can occupy either Ga or As sites in GaAs, acting as either an n-type or p-type impurity respectively. The LVM for Si in a Ga substitutional site has been assigned to 384 cm^{-1}, while a peak at 393 cm^{-1}

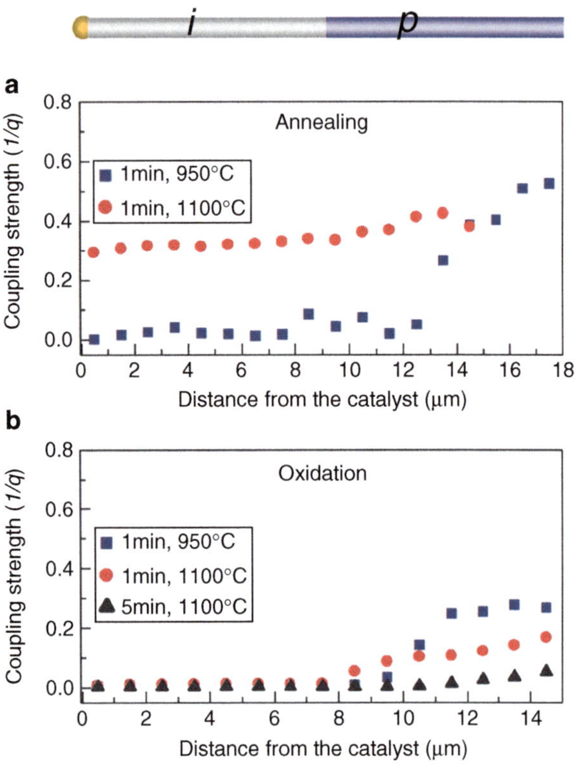

Fig. 17.9 Fano asymmetry parameter ($1/q$) as a function of distance from the catalyst for (**a**) annealed and (**b**) oxidized Si NWs that were in situ boron doped in the lower half and intrinsic in the upper half (With permission from reference [46]. Copyright (2009) by the American Chemical Society)

corresponds to the Si LVM at an As substitutional site. Correlating the ratio of the Si LVM peak intensity to the TO peak intensity with 4-point probe electrical measurements along the same wires, Dufouleur et al. were able to show that during in situ Si doping of GaAs NWs grown by molecular beam epitaxy (MBE), the Si is predominantly incorporated as part of a highly doped epitaxial GaAs shell (similar to results seen for in situ boron doping of Si NWs described above) and that Si more often substituted for As sites [50] (Fig. 17.10). However, Hilse et al. found that Si was incorporated into both Ga and As sites based on the LVMs in the Raman spectra for in situ doping of GaAs NWs also grown by MBE, suggesting that the incorporation route may be very process dependent [51].

For polar semiconductors, the ionic character of the primitive unit cell results in a coupling of the LO phonons with the longitudinal plasma oscillations of the free electrons or plasmons, which can lead in turn to the appearance of coupled phonon-plasmons modes in their Raman spectra on either side of the LO peak (designated by a (+) or a (−) sign). While the frequency of these modes provides a simple way to estimate n-type dopant concentration as well as carrier depletion effects [52], their use in studying NWs has been rather limited, most likely due to the fact that longitudinal modes are generally forbidden when NWs are examined in the back-scattering configuration at normal incidence.

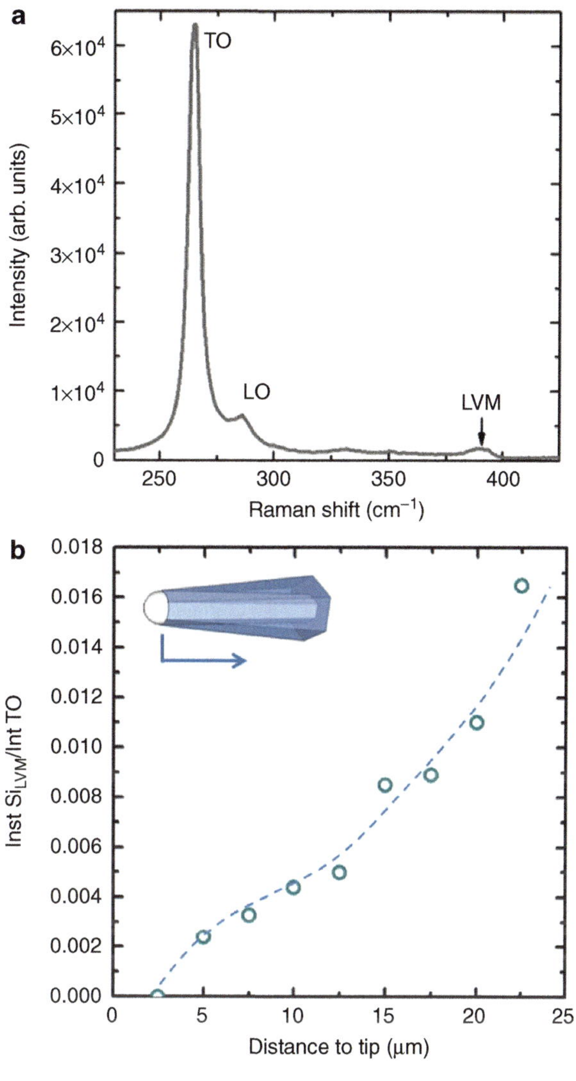

Fig. 17.10 (**a**) Raman spectrum from a single GaAs NW doped with Si exhibiting a LVM at 393 cm^{-1} corresponding to Si incorporation at As sites. (**b**) Intensity ratio of the LVM to TO modes as a function of distance to the NW tip (With permission from reference [50]. Copyright (2010) by the American Chemical Society)

One of the most valuable features of Raman spectroscopy is the well-known effect of local strain on the optical phonons (at $q \approx 0$). The most basic approach to the theory of lattice vibrations assumes that interatomic forces in the crystal are linear functions of the interatomic displacement so that they obey a form of Hooke's Law. Under this harmonic approximation, the frequency ω for mode j is given by:

$$K_{jj}^{0} = \bar{m}\omega_j^2 \qquad (17.18)$$

where $\bar{m}$ is the reduced mass of the two atoms and K_{jj}^{o} is the effective spring constant (or force constant) for mode j in the absence of strain. In analogy with a simple beam whose resonant frequency changes with the length of the beam, the phonon frequency may be expected to change with strain. This is introduced as an added contribution to the spring constant from terms representing its first-order derivatives with respect to strain (ε_{lm}) [40]:

$$\frac{\partial K_{jk}}{\varepsilon_{lm}}\varepsilon_{lm} = K^{(1)}_{jklm}\varepsilon_{lm} = K^{(1)}_{jkml}\varepsilon_{ml} \tag{17.19}$$

The derivative terms are known as the phonon deformation potentials, and each is a component of a full phonon deformation potential matrix $\boldsymbol{K}^{(1)}$. For cubic crystals, symmetry considerations reduce the number of independent phonon deformation potentials to three:

$$K^{(1)}_{1111} = K^{(1)}_{2222} = K^{(1)}_{3333} = \bar{m}p, \tag{17.20a}$$

$$K^{(1)}_{1122} = K^{(1)}_{2233} = K^{(1)}_{1133} = \bar{m}q, \tag{17.20b}$$

$$K^{(1)}_{1212} = K^{(1)}_{2323} = K^{(1)}_{3333} = \bar{m}r \tag{17.20c}$$

The relationship between strain and frequency is found in the eigenvalues (λ_j) of the matrix representing the changes in force constants under strain from their unstrained values, $\Delta\boldsymbol{K}$. The explicit connection between the eigenvalues and frequency is:

$$\lambda_j \equiv \omega_j^2 - \omega_{jo}^2 \approx 2\omega_{jo}\Delta\omega \tag{17.21}$$

where ω_j is the strained frequency, ω_{jo} is the unstrained frequency, and $\Delta\omega = \omega_j - \omega_{jo}$. Applying $\boldsymbol{K}^{(1)}$ to the strain tensor referred to the cubic crystallographic axes, $\hat{x} = [100]$, $\hat{y} = [010]$, and $\hat{z} = [001]$, the following characteristic equation may be used to obtain values for λ_j.

$$\begin{vmatrix} p\varepsilon_{xx} + q(\varepsilon_{yy} + \varepsilon_{zz}) - \lambda_j & 2r\varepsilon_{xy} & 2r\varepsilon_{xz} \\ 2r\varepsilon_{xy} & p\varepsilon_{yy} + q(\varepsilon_{zz} + \varepsilon_{xx}) - \lambda_j & 2r\varepsilon_{yz} \\ 2r\varepsilon_{xz} & 2r\varepsilon_{yz} & p\varepsilon_{zz} + q(\varepsilon_{xx} + \varepsilon_{yy}) - \lambda_j \end{vmatrix} = 0. \tag{17.22}$$

Note, that up to three independent values of λ_j may be obtained, reflecting the splitting of the degeneracy for phonon mode j brought upon by breaking the normal crystal symmetry. For a given stress tensor (determined by a stress model or a well-designed experimental setup), the strains that exist in the cubic reference system may be determined by the corresponding compliance tensor, $\boldsymbol{S}$. Cerdeira et al. have derived the linear relationships between the zone-center optical phonon frequency and stress for uniaxial stress along the $\langle 100\rangle$, $\langle 110\rangle$, and $\langle 111\rangle$ crystallographic directions in cubic crystals [40]. For $\langle 100\rangle$ and $\langle 111\rangle$ stress (or strain), the zone-center phonon splits into a nondegenerate phonon (the "singlet") and a doubly degenerate one (the "doublet"). For the $\langle 110 \rangle$ direction, the zone-center optical phonon degeneracy is completely lifted. Finally, a similar approach to that outlined above may be applied to wurtzite crystals referred to crystal axes where $\hat{x} \| [11\bar{2}0]$, $\hat{y} \| [1\bar{1}00]$, and $\hat{z} \| [0001]$ [53, 54] or a close variant [55].

Straining a crystal will begin to alter its crystal symmetry, thus affecting relative intensities from different phonon modes according to the Raman tensor. However, the effect of strain is usually small and the strained Raman tensor may be treated as the unstrained Raman tensor changed by linear proportionality to the appropriate components of the $\Delta \boldsymbol{K}$ matrix [56]. This means that in most cases, the unstrained Raman tensor gives a very clear idea of what phonon lines will be observable in Raman spectra of the strained crystal.

In polar zinc blende semiconductors, strain effects may usually be assumed to be independent of the normal LO-TO splitting though the phonon deformation potentials for TO and LO phonons are not necessarily identical [40]. For uniaxial stress in a $\langle 111 \rangle$ direction (or a $\langle 100 \rangle$ direction) with Raman spectra obtained in the backscattering configuration, one doublet component is associated with the LO phonon, while the other doublet and the singlet components are associated with TO phonons. A splitting of the TO phonon ($\Delta\omega_{TO}$) may be expected due *only to shear strain*, though it may not always be resolved. The behavior of TO and LO modes under strain was used by Montazeri et al. [57] to show how Raman spectroscopy may be used to estimate hydrostatic and shear strains in GaAs-Gap core-shell semiconductor NWs, where the GaAs core is expected to be under compressive strain (and the GaP shell under tensile strain) due to lattice mismatch (Fig. 17.11). Specifically they fitted two Lorentzians curves of equal line widths to the GaAs TO peak, and one Lorentzian curve to the LO peak in the Raman spectra of these individual NWs. The strain was treated as being radially symmetric about the $\langle 111 \rangle$ NW axis giving results similar to $\langle 111 \rangle$ uniaxial strain, and was assumed to be uniform across the entire core radius, which they supported with theoretical modeling. The obtained stain values were then used to estimate the expected change in the core band gap according to electron deformation potential theory [58, 59] and the results agreed well with both experimental photoluminescence spectra measured from the same NWs as well as theoretical calculations using strain-dependent $\boldsymbol{k} \cdot \boldsymbol{p}$ theory [60]. In general, this example demonstrates well the potency and utility of using Raman spectroscopy to measure strain in semiconductor NWs quantitatively.

Many other examples of stress or strain measurements through Raman spectroscopy are still primarily qualitative [18, 27]. Much of this stems from the fact that Raman spectroscopy provides only limited additional information (generally only in the form of frequency shifts) from potentially complicated strain distributions. Furthermore, care must be taken when extracting stresses from measured Raman shifts as key mechanical properties such as Young's modulus (which is related to the compliance or stiffness matrix elements) may be diameter dependent in NWs [61]. Still, Raman mapping with submicron spatial resolution and careful polarization analyses may help clarify the piezospectroscopic properties of semiconductor NWs in ongoing research.

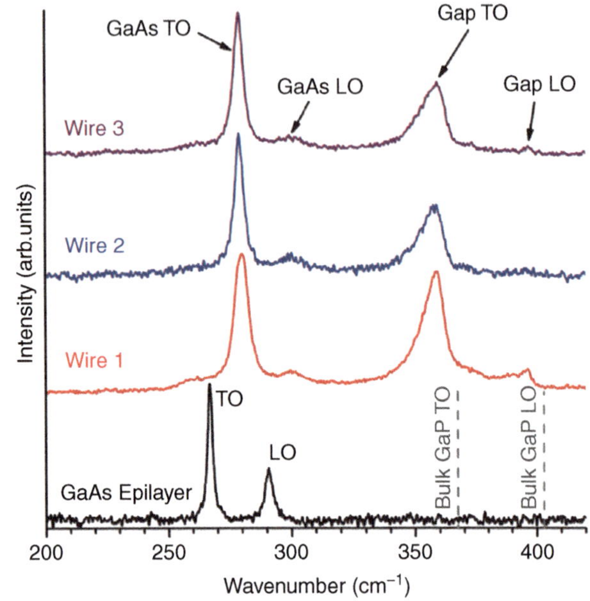

Fig. 17.11 Raman spectra of three individual GaAs/GaP core-shell NWs and an unstrained GaAs epilayer at room temperature. LO and TO frequencies of GaP at expected room temperature are also provided for convenience (With permission from reference [57]. Copyright (2010) by the American Chemical Society)

Another intriguing quality of Raman spectroscopy is its capability to measure local temperature quantitatively and precisely. This is possible in two distinct ways, arising due to two different characteristics of the Raman spectra in crystalline solids. The first characteristic is the presence of the phonon occupation number in the Raman scattering cross section in accordance with (17.3). While the relation to temperature of the strict intensity of a particular phonon peak is obfuscated by the numerous other components of the Raman scattering cross section, taking the ratio of integrated intensities of the Stokes (I_S) and anti-Stokes (I_{AS}) peaks provides the following relationship by which to measure temperature:

$$\frac{I_S}{I_{AS}} = A\left(\frac{\omega_S}{\omega_{AS}}\right)^3 \exp\left(\frac{hc\omega}{kT}\right) \tag{17.23}$$

where ω_S and ω_{AS} are the Stokes and anti-Stokes frequencies, ω is the phonon frequency (we have dropped the subscript j), and A is a constant determined by the light absorption of the sample and the scattering cross sections at the Stokes and anti-Stokes frequencies that should be calibrated for the sample. For bulk samples, the value of A is close to $\sim$1 due to the fact that the Stokes and anti-Stokes frequencies differ from the frequency of the excitation line by a relatively small amount, making this a convenient measure of local temperature that is not dependent on extraneous factors such as stress or crystal quality. However, the application to NWs is not as straightforward. First, the resonant light interactions of NWs means that the value of A may change for any individual NW as a function of its diameter or the laser wavelength. Secondly, the accuracy of the method is

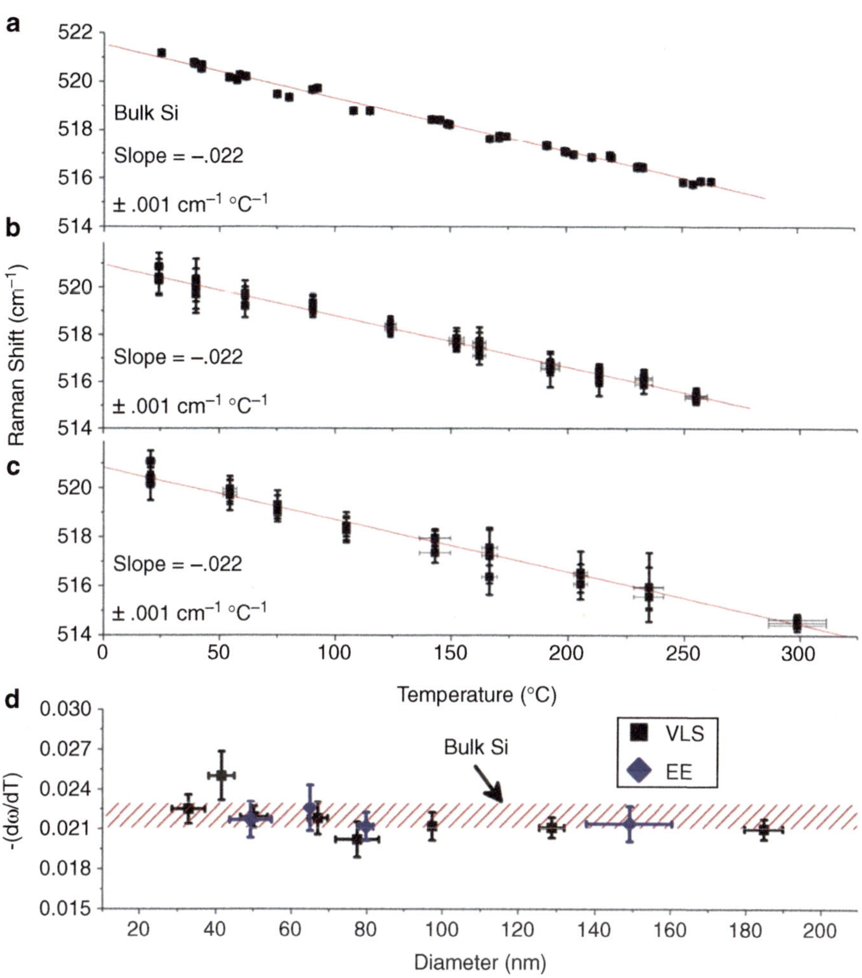

Fig. 17.12 Temperature dependence of the optical phonon frequency for (**a**) bulk Si, (**b**) a 50-nm-diameter Si NW grown by the vapor-liquid–solid (VLS) method [1] and (**c**) a 50-nm-diameter Si NW formed through Ag-catalyzed electrochemical etching (EE) [63]. (**d**) Slopes of the optical phonon frequency versus temperature for individual Si NWs of both types with diameters greater than ~30 nm. They match the same value for bulk Si within experimental error (With permission from reference [62]. Copyright (2009) by the American Physical Society)

significantly reduced when it becomes difficult to resolve the anti-Stokes peak, which is often a problem for small diameter NWs (<~50 nm) or when low laser powers are used to minimize local heating.

A more precise measurement of local sample temperature is made possible by the anharmonicity of the crystal vibrational potential energy. Phonon-phonon interactions that reduce individual phonon lifetimes, phonon softening, and thermal expansion give rise to increasing peak width and a change in peak frequency (typically negative) with increasing temperature. The variability of peak intensities under real experimental conditions and the factors such as doping or confinement that can affect peak shape and width make the dependence on peak width an imprecise measurement for temperature when applied to NWs. The peak position on the other hand is much more easily and accurately determined. Above room temperature and for a range of a few hundred degrees K, the relationship between phonon frequency and temperature is typically linear, though calibration measurements are generally necessary for different types of samples. For NWs, local heating by the laser can confound the accurate determination of this relationship; the easiest and most direct way to overcome this is to measure the position of the peak of interest as a function of laser power and to set the laser power below which the position does not change as the maximum for further thermal measurements. Using this approach, Doerk et al. performed measurements of the frequency of the zone-center optical phonon as a function of temperature for individual Si NWs down to 30 nm in diameter and found that this relationship is the same as for bulk Si [62] (Fig. 17.12). Importantly, this is true for Si NWs synthesized by two different routes, with very different thermal conductivities [64].

Raman thermometry, the measurement of local sample temperature through collecting Raman spectra from that sample, is becoming increasingly important for determining key thermal properties of nanomaterials as evidenced by the measurement of thermal conductivity in carbon nanotube bundles [65] and graphene [66, 67] via Raman spectroscopy. A recent example of the use of Raman thermometry takes advantage of the local nature of the temperature measurement to determine the thermal conductivities of single cantilevered Si NWs. By intentionally heating the NW with the laser and measuring the temperature at the same spot, an axial temperature profile is extracted from a Raman map around the NW. From an estimate of the absorption cross section (using for example a Mie-type solution to Maxwell's equations as described previously in this chapter) and the slope of the temperature profile, the thermal conductivity of the NW may be measured without influence of thermal contact resistances [68]. This highlights a way in which the submicron temperature resolution of Raman thermometry may be a key characterization technique in studying the thermal properties of NWs and related nanostructures in the future.

6 Conclusions and Future Directions

In this chapter, we have discussed the unique interactions of electromagnetic radiation with semiconductor NWs that lead to resonant absorption and scattering, the importance of Raman selection rules and phonon confinement in determining the crystal structure of NWs, and the ways in which Raman spectroscopy can be used to measure composition, strain, and temperature quantitatively with submicron resolution. These qualities of Raman spectroscopy are already commonly employed in the characterization of semiconductor NWs, and one may anticipate Raman spectroscopy to be used even more widely as the applications to NW characterization are better outlined. The nondestructive nature of Raman measurements will likely lead to even more applications when combined concurrently with other in situ experiments that chemically modify the NW or apply electrical currents or mechanical strain.

Another exciting development in Raman spectroscopy is the emergence of apertureless near-field Raman spectroscopy, also known as tip-enhanced Raman spectroscopy (TERS). By guiding a sharp tip (less than 10s of nanometers in radius) of a noble metal material illuminated by a laser along the sample surface, the local electric field is substantially enhanced in the near-field region of the tip [69]. This in turn enhances the Raman scattering intensity only in the area of the tip. For semiconductor NWs, the most direct benefit will be the dramatically improved resolution to less than $\sim$100 nm, well beyond diffraction-limited resolution. With this improved resolution, it may be possible to measure impurity concentrations across nanoscale junctions along the NW axis, to profile strain or composition along the radial direction of the NW, and, in some cases, to determine the fraction of different crystal structures with nanoscale precision in ambient conditions, though new Raman selection rules for near-field Raman scattering must be employed arising from the superposition of the far-field selection rules and the polarization-dependent tip enhancement [70]. All together, the value of Raman spectroscopy will be amplified, potentially opening new avenues for research.

References

1. Wagner RS, Ellis WC (1964) Vapor-liquid-solid mechanism of single crystal growth. Appl Phys Lett 4:89–90
2. Givargizov EI (1975) Fundamental aspects of VLS growth. J Cryst Growth 31:20–30
3. Iijima S (1991) Helical microtubules of graphitic carbon. Nature 354:56–58
4. Iijima S, Ichihashi T (1993) Single-shell carbon nanotubes of 1-nm diameter. Nature 363:603–605
5. Pollak FH (1991) Characterization of semiconductors by Raman spectroscopy. In: Grasselli JG, Bulkin BJ (eds) Analytical Raman spectroscopy, vol 114, Chemical analysis series. Wiley, New York, pp 137–214
6. Hayes W, Loudon R (1978) Scattering of light by crystals. Wiley, New York
7. Cardona M, Guntherodt G (eds) (1975–1984) Light scattering in solids, vol I–IV, Topics in applied physics. Springer, Berlin/Heidelberg/New York
8. Cao LY, White JS, Park JS, Schuller JA, Clemens BM, Brongersma ML (2009) Engineering light absorption in semiconductor nanowire devices. Nat Mater 8:643–647
9. Chen G, Wu J, Lu QJ, Guitierrez HRH, Xiong Q, Pellen ME, Petko JS, Werner DH, Eklund PC (2008) Optical antenna effects in semiconducting nanowires. Nano Lett 8:1341–1346
10. Kerker M (1969) The scattering of light and other electromagnetic radiation. Academic, New York
11. Cao LY, Nabet B, Spanier JE (2006) Enhanced Raman scattering from individual semiconductor nanocones and nanowires. Phys Rev Lett 96:157402
12. Frechette J, Carraro C (2006) Diameter-dependent modulation and polarization anisotropy in Raman scattering from individual nanowires. Phys Rev B 74:161404
13. Frechette J, Carraro C (2006) Resolving radial composition gradients in polarized confocal Raman spectra of individual 3 C-SiC nanowires. J Am Chem Soc 128:14774–14775
14. Schmidt V, Senz S, Gosele U (2005) Diameter-dependent growth direction of epitaxial silicon nanowires. Nano Lett 5:931–935

15. Pauzauskie PJ, Talaga D, Seo K, Yang PD, Lagugne-Labarthet F (2005) Polarized Raman confocal microscopy of single gallium nitride nanowires. J Am Chem Soc 127:17146–17147
16. Caroff P, Dick KA, Johansson J, Messing ME, Deppert D, Samuelson L (2009) Controlled polytypic and twin-plane superlattices in III-V nanowires. Nat Nanotechnol 4:50–55
17. Zardo I, Conesa-Boj S, Peiro F, Morante JR, Arbiol J, Uccelli E, Abstreiter G, Fontcuberta i Morral A (2009) Raman spectroscopy of wurtzite and zinc-blende GaAs nanowires: polarization dependence, selection rules, and strain effects. Phys Rev B 80:245324
18. Spirkoska D, Arbiol J, Gustafsson A, Conesa-Boj S, Glas F, Zardo I, Heigoldt M, Gass MH, Bleloch AL, Estrade S, Kaniber M, Rossler J, Peiro F, Morante JR, Abstreiter G, Samuelson L, Fontcuberta i Morral A (2009) Structural and optical properties of high quality zinc-blende/wurtzite GaAs nanowire heterostructures. Phys Rev B 80:245325
19. Lopez FJ, Hemesath ER, Lauhon LJ (2009) Ordered stacking fault arrays in silicon nanowires. Nano Lett 9:2774–2779
20. Cao LY, Laim L, Nabet B, Spanier JE (2005) Diamond-hexagonal semiconductor nanocones with controllable apex angle. J Am Chem Soc 127:13782–13783
21. Richter H, Wang ZP, Ley L (1981) The one-phonon Raman spectrum in microcrystalline silicon. Solid State Commun 39:625–629
22. Campbell IH, Fauchet PM (1986) The effects of microcrystal size and shape on the one phonon Raman spectra of crystalline semiconductors. Solid State Commun 58:739–741
23. Piscanec S, Cantoro M, Ferrari AC, Zapien JA, Liftzhitz Y, Lee ST, Hofmann S, Robertson J (2003) Raman spectroscopy of silicon nanowires. Phys Rev B 68:241312
24. Li B, Yu D, Zhang SL (1999) Raman spectral study of silicon nanowires. Phys Rev B 59:1645–1648
25. Wang RP, Zhou GW, Liu YL, Pan SH, Zhang HZ, Yu DP, Zhang Z (2000) Raman spectral study of silicon nanowires: high-order scattering and phonon confinement effects. Phys Rev B 61:16827–16832
26. Adu KW, Xiong Q, Gutierrez HR, Chen G, Eklund PC (2006) Raman scattering as a probe of phonon confinement and surface optical modes in semiconducting nanowires. Appl Phys A 85:287–297
27. Chen JN, Conache G, Pistol ME, Gray SM, Borgstrom MT, Xu HX, Xu HQ, Samuelson L, Hakanson U (2010) Probing strain in bent semiconductor nanowires with Raman spectroscopy. Nano Lett 10:1280–1286
28. Tsang JC, Mooney PM, Dacol F, Chu JO (1994) Measurement of alloy composition and strain in thin Ge_xSi_{1-x} layers. J Appl Phys 75:8098–8108
29. Alonso MI, Winer K (1989) Raman spectra of c-$Si_{1-x}Ge_x$ alloys. Phys Rev B 39:10056–10062
30. Yang JE, Park WH, Kim CJ, Kim ZH, Jo MH (2008) Axially graded heteroepitaxy and Raman spectroscopic characterizations of $Si_{1-x}Ge_x$ nanowires. Appl Phys Lett 92:263111
31. Nishimura C, Imamura G, Fujii M, Kawashima T, Saitoh T, Hayashi S (2008) Raman characterization of Ge distribution in individual $Si_{1-x}Ge_x$ alloy nanowires. Appl Phys Lett 93:203101
32. Lu QJ, Adu KW, Gutierrez HR, Chen GG, Lew KK, Nimmatoori P, Zhang X, Dickey EC, Redwing JM, Eklund PC (2008) Raman scattering from $Si_{1-x}Ge_x$ alloy nanowires. J Phys Chem C 112:3209–3215
33. Venugopal R, Lin PI, Chen YT (2006) Photoluminescence and Raman scattering from catalytically grown Zn_xCd_{1-x} Se alloy nanowires. J Phys Chem B 110:11691–11696
34. Liang YQ, Zhai L, Zhao XS, Xu DS (2005) Band gap engineering of semiconductor nanowires through composition modulation. J Phys Chem B 109:7120–7123
35. Kuykendall T, Ulrich P, Aloni S, Yang P (2007) Complete compositional tunability of InGaN nanowires using a combinatorial approach. Nat Mater 6:951–956
36. Abstreiter G, Bauser E, Fischer A, Ploog K (1978) Raman spectroscopy – a versatile tool for characterization of thin films and heterostructures of GaAs and $Al_xGa_{1-x}As$. Appl Phys 16:345–352

37. Schmidt V, Senz S, Gosele U (2007) Influence of the Si/SiO_2 interface on the charge carrier density of Si nanowires. Appl Phys A 86:187–191
38. Bjork MT, Schmid H, Knoch J, Riel H, Riess W (2009) Donor deactivation in silicon nanostructures. Nat Nanotechnol 4:103–107
39. Cerdeira F, Fjeldly TA, Cardona M (1973) Effects of free carriers on zone-center vibrational modes in heavily doped p-type Si. II. Optical modes. Phys Rev B 8:4734–4745
40. Cerdeira F, Buchenauer CJ, Pollak FH, Cardona M (1972) Stress-induced shifts of first-order Raman frequencies of diamond- and zinc-blende-type semiconductors. Phys Rev B 5:580–593
41. Balkanski M, Wallis RF, Haro E (1983) Anharmonic effects in light scattering due to optical phonons in silicon. Phys Rev B 28:1928–1934
42. Locher R, Wagner J, Fuchs F, Maier M, Gonon P, Koidl P (1995) Optical and electrical characterization of boron-doped diamond films. Diam Relat Mater 4:678–683
43. Olego D, Cardona M (1981) Self-energy effects on the optical phonons of heavily doped p-GaAs and p-Ge. Phys Rev B 23:6592–6602
44. Chandrasekhar M, Renucci JB, Cardona M (1978) Effects of interband excitations on Raman phonons in heavily doped n-Si. Phys Rev B 17:1623–1633
45. Imamura G, Kawashima T, Fujii M, Nishimura C, Saitoh T, Hayashi S (2008) Distribution of active impurities in single silicon nanowires. Nano Lett 8:2620–2624
46. Imamura G, Kawashima T, Fujii M, Nishimura C, Saitoh T, Hayashi S (2009) Raman characterization of active B-concentration profiles in individual p-type/intrinsic and intrinsic/p-type Si nanowires. J Phys Chem C 113:10901–10906
47. Doerk GS, Lestari G, Liu F, Carraro C, Maboudian R (2010) Ex situ vapor phase boron doping of silicon nanowires using BBr_3. Nanoscale 2:1165–1170
48. Fukata N (2009) Impurity doping in silicon nanowires. Adv Mater 21:2829–2832
49. Temple PA, Hathaway CE (1973) Multiphonon Raman spectrum of silicon. Phys Rev B 7:3685–3697
50. Dufouleur J, Colombo C, Gamma T, Ketterer B, Uccelli E, Nicotra M, Morral AFI (2010) P-doping mechanism in catalyst-free gallium arsenide nanowires. Nano Lett 10:1734–1740
51. Hilse M, Ramsteiner M, Breuer S, Geelhaar L, Riechert H (2010) Incorporation of the dopants Si and Be into GaAs nanowires. Appl Phys Lett 96:193104
52. Schwartz GP, Gualtieri GJ, Dubois LH, Bonner WA, Ballman AA (1984) Free carrier reduction in vacuum-annealed S-, Sn-, and Ge-doped (100) InP. J Electrochem Soc 131:1716–1720
53. Darakchieva V, Paskova T, Schubert M, Arwin H, Paskov PP, Monemar B, Hommel D, Heuken M, Off J, Scholz F, Haskell BA, Fini PT, Speck JS, Nakamura S (2007) Anisotropic strain and phonon deformation potentials in GaN. Phys Rev B 75:195217
54. Irmer G, Brumme T, Herms M, Wernicke T, Kneissl M, Weyers M (2008) Anisotropic strain on pnonons in a-plane GaN layers studied by Raman scattering. J Mater Sci Mater Electron 19:S51–S57
55. Briggs RJ, Ramdas AK (1976) Piezospectroscopic study of the Raman spectrum of cadmium sulfide. Phys Rev B 13:5518–5529
56. Narayan S, Kalidindi SR, Schadler LS (1997) Determination of unknown stress states in silicon wafers using microlaser Raman spectroscopy. J Appl Phys 82:2595–2602
57. Montazeri M, Fichenscher M, Smith LM, Jackson HE, Yarrison-Rice J, Kang JH, Gao Q, Tan HH, Jagadish C, Guo YN, Zou J, Pistol ME, Pryor CE (2010) Direct measure of strain and electronic structure in GaAs/GaP core-shell nanowires. Nano Lett 10:880–886
58. Nolte DD, Walukiewicz W, Haller EE (1987) Band-edge hydrostatic deformation potentials in III-V semiconductors. Phys Rev Lett 59:501–504
59. Qiang H, Pollak FH (1990) Piezo-photorereflectance of the direct gaps of GaAs and $Ga_{0.78}Al_{0.22}As$. Solid State Commun 76:1087–1091
60. Pistol ME, Pryor CE (2008) Band structure of core-shell semiconductor nanowires. Phys Rev B 78:115319

61. Zhu Y, Xu F, Qin QQ, Fung WY, Lu W (2009) Mechanical properties of vapor-liquid-solid synthesized silicon nanowires. Nano Lett 9:3934–3939
62. Doerk GS, Carraro C, Maboudian R (2009) Temperature dependence of Raman spectra for individual silicon nanowires. Phys Rev B 80:073306
63. Peng KQ, Yan YJ, Gao SP, Zhu J (2002) Synthesis of large-area silicon nanowire arrays via self-assembling nanoelectrochemistry. Adv Mater 14:1164–1167
64. Hochbaum AI, Chen RK, Delgado RJ, Liang WJ, Garnett EC, Najarian M, Majumdar A, Yang PD (2008) Enhanced thermoelectric performance of rough silicon nanowires. Nature 451:163–167
65. Hsu IK, Pows MT, Bushmaker A, Aykol M, Shi L, Cronin SB (2009) Optical absorption and thermal transport of individual suspended carbon nanotube bundles. Nano Lett 9:590–594
66. Balandin AA, Ghosh S, Bao WZ, Calizo I, Teweldebrhan D, Miao F, Lau CN (2008) Superior thermal conductivity of single-layer graphene. Nano Lett 8:902–907
67. Ghosh S, Bao WZ, Nika DL, Subrina S, Pokatilov EP, Lau CN, Balandin AA (2010) Dimensional crossover of thermal transport in few-layer graphene. Nat Mater 9:555–558
68. Doerk GS, Carraro C, Maboudian R (2010) Single nanowire thermal conductivity measurements by Raman thermography. ACS Nano 4:4908–4914
69. Novotny L, Stranick SJ (2006) Near-field optical microscopy and spectroscopy with pointed probes. Annu Rev Phys Chem 57:303–331
70. Berweger S, Raschke MB (2009) Polar phonon mode selection rules in tip-enhanced Raman scattering. J Raman Spectrosc 40:1413–1419

Optical Tweezers for Raman Spectroscopy 18

Lianming Tong, Kerstin Ramser, and Mikael Käll

1 Definition of the Topic

The integration of laser tweezers with Raman spectroscopy for optical manipulation and spectroscopic analysis of individual micro- and nanoscopic objects in physics, chemistry, and the life sciences.

2 Overview

Like all optical spectroscopies, Raman scattering is about utilizing photons for *studying* a sample. But the light—matter interaction can also be used to *manipulate* the sample under study. Here we review the powerful combination of optical tweezers and Raman spectroscopy as applied to small particles in solution. Apart from the special case of surface-enhanced Raman scattering (SERS), which we cover at the end of this chapter, "Raman tweezers" has not yet been utilized extensively within nanoscience. For sake of completeness and inspiration, we therefore first review the significant amount of work that has been published on particles in the micron range, in particular on biological cells. A brief outlook concludes the chapter.

L. Tong • M. Käll (✉)
Department of Applied Physics, Chalmers University of Technology, Göteborg, Sweden

K. Ramser
Department of Computer Science and Electrical Engineering, Luleå University of Technology, Luleå, Sweden

Challa S. S. R. Kumar (ed.), *Raman Spectroscopy for Nanomaterials Characterization*,
DOI 10.1007/978-3-642-20620-7_18, © Springer-Verlag Berlin Heidelberg 2012

3 Introduction

Optical tweezers, or laser tweezers, is a spin off from research on linear and angular momentum transfer between light and small particles. Already in 1970, Ashkin showed that the radiation pressure from a focused laser beam significantly affected the dynamics of micrometer-sized transparent and neutral particles [1]. This finding was soon developed into what is now known as optical tweezers, meaning a single-beam gradient-force trap for manipulation of small particles in solution. The subsequent four decades has seen a continuous and rapid development of optical trapping techniques and applications. Important examples include the development of holographic optical tweezers, a variety of biophysics measurements on single biomacromolecules, cell-sorting applications, the development of techniques for optical rotation of small particles, optical binding and aggregation of particles, and trapping in subwavelength fields created by plasmonic nanostructures. The reader is referred to a number of excellent reviews for more information on some of these developments [2–6]. Thorough theoretical descriptions of basic optical trapping effects can be found in the resource letter by Lang and Block [7] and in the review by Jonas and Zemanek [8].

An exact calculation of the optical forces affecting an arbitrary particle in an arbitrary laser field can in principle be done using the so-called Maxwell stress tensor technique, but it requires an exact solution of Maxwell's equation for the particular particle-field configuration under study. Fortunately, a number of approximate theories exist that can be used to explain the optical tweezing effect and to obtain reasonably accurate estimates of optical forces and trapping potentials. If the trapped object is dielectric and much larger than the wavelength of the trapping laser, then trapping can be understood from a ray-optics approach. Refraction of light corresponds to a change in the momentum carried by the light. According to Newton's third law, an equal and opposite momentum change, being proportional to the light intensity, is transferred to the object. In particular, simple geometrical arguments show that if a focused light beam refracts in a transparent object, then transfer of linear momentum from the refracted photons leads to two force components on the object, one pointing toward the focal point, the so-called gradient force, and one pointing along the optical axis, known as the "scattering force." Both forces are proportional to the refractive index contrast between the object and the surrounding medium, and to the intensity of the laser light. Importantly, if a Gaussian laser beam profile is considered, it is mainly the peripheral rays that contribute to the gradient force while photons traveling along the central part of the beam acts as a photonic "fire hose," pushing the particle in the direction of light propagation. Stable trapping close to the focal point can only be achieved if the gradient force can overcome the scattering force, which thus requires a large fraction of peripheral rays. In practice, this means that the laser beam needs to be tightly focused using a microscope objective with high numerical aperture (NA). Naturally, the absolute gradient force also has to be large enough to overcome thermal forces in the form of Brownian motion.

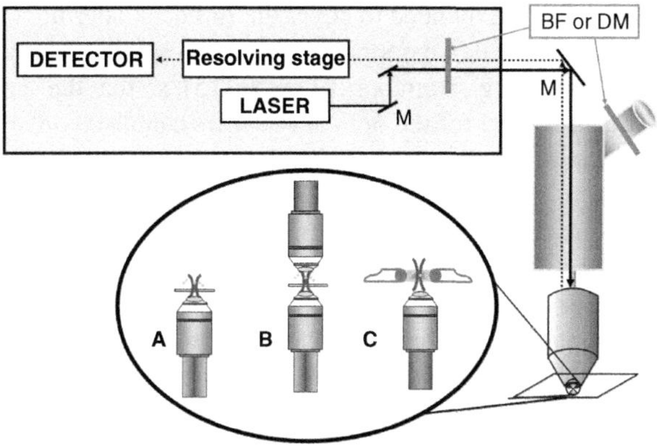

Fig. 18.1 Schematic of typical Raman tweezers setups. The laser light for Raman excitation is guided into the microscope objective and Raman scattering is collected in backscattering geometry in all cases. The excitation and Raman light is separated using a blocking filter (BF), typically a holographic notch filter, or a dichroic mirror (DM), typically a so-called edge filter. The Raman scattered light is spectrally resolved by an optical spectrometer comprising of a resolving stage, typically based on tunable reflection gratings, and a sensitive detector, typically a cooled CCD chip. Optical trapping can be performed using the Raman excitation laser or a separate laser beam using a single (**a**) or two separate objectives (**b**). The optical trap can also be created through two counter propagating laser beams, for example from optical fibers (**c**). It is also necessary to use BFs or DMs to shield the eyes of the operator—a particularly important point in "home built" Raman tweezers using invisible near-infrared laser radiation

In contrast to dielectric beads or cells, nanoparticles are in the Rayleigh limit, meaning that they are much smaller than the optical wavelength. This implies that they essentially behave as induced dipoles in the applied optical field. An analysis of the optical force in the Rayleigh limit starts from the Lorentz force and leads to an expression for the gradient force that is proportional to the particle polarizability, and hence to the volume of the particle [9–11]. If the particle is absorbing, i.e., the polarizability contains an imaginary part, there is also a radiation pressure component that destabilizes the particle, while the actual scattering force appears as the first finite-size correction to the Rayleigh approximation. Unfortunately, many objects that are interesting for optical tweezing experiments have sizes in between the regions of validity for the Rayleigh and ray-approximations, i.e., sizes between $\sim$0.1 and $\sim$10 λ, which call for more complete electromagnetic descriptions [12, 13]. Nieminen et al. have developed a computational toolbox that can be used for quantitative estimates of optical trapping effects in this region [14].

4 Experimental and Instrumental Methodology

Figure 18.1 shows a schematic of the most common types of Raman tweezers setups. Most optical tweezers solutions are based on a single laser beam with a Gaussian intensity profile focused by a high NA microscope objective. It is important that the beam is expanded to cover the full back aperture of the objective, so that the full NA is utilized. An optical trap can also be created using two counter propagating laser beams, e.g., from optical fibers [15], so that the scattering force is canceled out. The laser used for the optical tweezers can also conveniently be used to excite Raman scattering. However, many Raman studies require resonant excitation using visible or even ultraviolet wavelengths, while optical tweezing is best performed using near-infrared light to minimize laser-induced heating of the aqueous medium or the sample itself. Two lasers can then be used, one for Raman excitation and one for optical trapping, which means that the setup has to be equipped with suitable filters or dichroic mirrors in order to separate the tweezers and Raman light paths. Some experiments may even require the use of two separate microscope objectives facing each other, one for Raman microscopy and one for laser tweezing [16]. In any case, it is important to remember that a working Raman tweezers based on separate laser beams requires perfect overlap between the respective laser foci, which for widely separate excitation and tweezing wavelengths can be complicated due to, for example, chromatic aberrations. It is also possible to generate multiple optical traps using various optical interference patterns [17, 18], but this has not yet been used in combination with Raman spectroscopy.

In ordinary Raman tweezers setups, the detected Raman spectra contain signals from both the trapped object and the surrounding solution. Naturally, this is particularly troublesome if important "fingerprint" Raman bands overlap with strong Raman or fluorescent features from the solution. A recent solution to this problem takes full advantage of the possibility to move the trapped object relative to the Raman probe beam [19, 20]. Two separate laser beams were used, one for optical trapping and one for Raman excitation. The spatial position of the trapping beam was modulated slightly at a certain frequency f while the Raman excitation beam was kept at a fixed position. Since the Raman signal is maximized, and the background minimized, when the trapped object is in the center of the excitation beam, the Raman signal will be modulated at a frequency of $2f$. Moreover, the Raman signal and the solution background will be phase-shifted by 90°. This means that essentially background-free Raman spectra (or pure background spectra) can be recorded using a phase sensitive lock-in amplifier detecting at $2f$.

5 Key Research Findings

5.1 Raman Spectroscopy and Laser Trapping Performed on Dielectric Particles in the Micrometer Range

In 1986, Ashkin and coworkers reported on the first successful single-beam gradient-force trap, or laser tweezers, for dielectric particles [21]. They were able to trap particles of glass, silica, and polystyrene (PS) in the range from 25 nm to 10 μm in water. Optical trapping techniques have since been integrated to a range of different optical read-out techniques to enable various types of single particle investigations [8]. The first combination of micro-Raman spectroscopy and optical tweezers was presented in 1994, when Urlaub et al. investigated a polymerization reaction in optically trapped emulsion particles [22], while Raman tweezers entered the field of biophysics in 2002 with studies of single cells and organelles [23–25]. The first articles on Raman tweezers dealt mainly with engineering aspects, i.e., how to integrate optical tweezers to Raman instrumentation, but today, a decade later, several comprehensive reviews on Raman studies of optically trapped dielectrics have already appeared. These include reviews on aerosol particle studies, which are important for atmospheric chemistry and physics, combustions science, drug delivery, and epidemiology [26–29] as well as overviews on eukaryotic cell studies, such as cell sorting, biomolecular reactions of cells to environmental changes, and cancer detection [29–32]. This section gives a brief overview on these subjects with an emphasis on biological particles and light-induced effects on trapped biological materials. For clarity, the studies are summarized in tables after a brief introduction of important aspects of the subject.

5.1.1 Optical Trapping and Raman Spectroscopy of Aerosols, Gas Bubbles, and Polymorphs

Chemical and physical properties of aerosols are conventionally investigated by performing ensemble measurements under controlled environmental conditions. Important parameters include droplet size, refractive index, evaporation dynamics, and chemical composition [26]. Many of these parameters can be evaluated at the *single* particle level using Raman microscopy. Hence, it may not come as a surprise that the first Raman tweezers studies dealt with micro droplets and aerosols and focused on changes in morphology and chemical compositions due to environmental changes. In many studies, both elastic (Mie) and Raman scattering were measured to get more conclusive results. For theory and introduction, we refer to the reviews [26, 28, 33]. A brief overview of studies is given in Table 18.1.

5.1.2 Raman Tweezers Applied to Biological Cells

Table 18.2 gives an overview over the most important Raman tweezers studies on biological cells regarding experimental design, object, the relevant Raman peaks, and the results. The vast variety of topics demonstrates that Raman tweezers provides an excellent tool for research on the single cell/particle level. For example, fast screening for comparisons and detection of healthy and infected cells, as well as sorting of different cell types, can be performed. Because of the rich but complicated Raman spectra of complex biological components, such as DNA, RNA, lipids, and proteins [68, 69], such studies can be greatly facilitated by statistical tools. In particular, multivariate methods applying principal component analysis and linear discriminate analysis have been shown to be valuable for discrimination between different organelles or cell types with high accuracy. These tools are most helpful since the spectral differences between healthy or infected cells often lie in the intensity of certain Raman peaks, i.e., upon infection some cellular components

Table 18.1 Raman tweezers studies on dielectric particles

Reference	Setup	Object	Relevant Raman lines	Result
[34]	Optical trap at 1,064 nm, Raman 514.5 nm	Gas bubbles	N_2 at 2,340 cm^{-1} O_2 at 1,560 cm^{-1} Palmitic acid at C-C 1,060 and 1,124 cm^{-1}, CH_2 twist at 1,291 cm^{-1}	N_2 and O_2 composition of air bubble mixture was determined. Liquid/gas interface of the bubbles revealed a thin layer of palmitic acid
[35]	Same laser for Raman and one optical tweezers 514.5 nm	Microdroplets	CH-stretching region 2,800–3,000 cm^{-1} and C = C at 1,628 cm^{-1}	Mie and Raman scattering of microdroplet polymerization. Size, refractive index, and morphology are determined. Droplet evaporation and chemical reactions are studied
[36]	Same laser for Raman and one optical tweezers tunes from 730 nm to 780 nm	Toluene microdroplet containing *p*-cresol (MD)	*p*-cresol at 785, 823, 843, 1003, and 1,030 cm^{-1}	Single MD could be observed and a difference in solubility for the *p*-cresol in the bulk solvent and in the MD solvent was seen
[37]	Same laser for Raman and one optical tweezers 647 nm	Polymorphic forms of carbamazepine (CBZ) in microliter fluid volumes	CBZ dehydrate 259, 445, 470 cm^{-1} CBZ form I 264 and 459 cm^{-1} CBZ form II 254, 272, 445, 470, 484 cm^{-1}	Recrystallization of CBZ in methanol under temperature control
[38]	Same laser for Raman and one optical tweezers 647.1 nm	Polystyrene (PS) particles 200 nm–10 μm	PS Raman peak at 1,000 cm^{-1}	Sensitivity and spatial selectivity as a function of particle size were investigated
[39]	Same laser for Raman and one optical tweezers 514,5 nm	Aerosol droplets	OH stretching at 2,900–3,000 cm^{-1}	Limiting conditions of optical forces and capillary forces dominating the aerosol coagulation event are explored by varying the optical trap force and by characterizing the coagulation of different droplets. Dual trap is used

increase, while others decrease. Spectral shifts, on the other hand, are rare. However, studies have shown that intensity variations among individual cells of the same type were small in comparison to their alterations caused by infections or cancer development. Indeed, high sensitivities and reproducibility for detection of

Table 18.2 Raman tweezers studies on biological materials

Reference	Experimental design	Object	Relevant Raman peaks	Results
Studies related to cancer				
[40]	Same laser for Raman and one optical tweezers 780 nm	Healthy and tumor infected cells causing Kaposi's sarcoma–associated herpesvirus (KSHV)	Intensities of the bands at 1,004, 1,093, and 1,663 cm^{-1} were altered	Detection of cancer infected cells. Large study with 85% sensitivity and 92.5% specificity in blind tests to diagnose cancer. Biological relevance of KSHV infection upon cell signaling was investigated
[41]	Same laser for Raman and one optical tweezers 633 nm	Neoplastic and normal hematopoietic cells	DNA (backbone below 800 cm^{-1} nucleic acids), protein (phenylalanine, 1,002 cm^{-1}, amide I, 1,650 cm^{-1}, and amide II 1,570 cm^{-1}), lipids, CH_2 wagging, 1,200–1,400 cm^{-1}	Detection of cancer infected cells, $p < <0.05$ sensitivity 98.3% for cancer detection and 97.2% of the cells are correctly analyzed
[42]	Same laser for Raman and one optical tweezers 782.5 nm	Epithelial cells colorectal cancer	Increases at 940.988, 1004, 1096.433, 1209.015, 1265.006, 1304.081, 1341.605, 1437.853, 1520.245, 1655.304, and 1672.795 cm^{-1}, decreases below 720 cm^{-1}	Detection of cancer infected cells, sensitivity 86.3% specificity 86.3%
[43]	Same laser for Raman and one optical tweezers 633 nm	Leukemia cells, T and B lymphocytes	DNA, RNA, protein, and lipid vibrations at 785, 1,230, 1,305, and 1,660 cm^{-1}	Detection of cancer infected cells, 95% of normal cells and 90% of cancerous cells were correctly identified
[44, 45]	Either a double microscope configuration with 1,064 nm for optical trapping and 514.5 nm or same laser for Raman and one optical tweezers 514.5 nm	Urological cells, prostate cancer, and bladder cells lines	Nucleic acids, lipids, carbohydrates, proteins	Fixation protocol and classification of urological cells are investigated. For SurePath fixed cells, sensitivity >72% and specificity 90%
				For formaline fixed cells sensitivity >93%, specificity 98%

(*continued*)

Table 18.2 (continued)

Reference	Experimental design	Object	Relevant Raman peaks	Results
[46]	Same laser for Raman and one optical tweezers 633 nm	Fixation of leukemia cells	Mainly DNA at 785, 1,093, 1,373, 1,575 cm^{-1} (ring breathing mode)	Fixation is important for correct analysis, methanol surpasses paraformaldehyde
			Changes also seen in 678 (nucleic acid), 1003 (phenylalanine), 1093(PO_2), 1126 (C-N), 1337, 1447, 1605, and 1615 (C = C) and 1,650 cm^{-1} (amide I)	
Lipids, vesicles, and membranes				
[27]	Same laser for Raman and one optical tweezers 647.1 nm	Four different types of phosphatidylcholine lipids	C-C (1,050-1,150 cm^{-1}) and C-H (2,800–3,200 cm^{-1}) stretching modes	Four types of phospholipids could be distinguished and investigated at different temperatures (gel to liquid-crystal phases). Ion transfers through the membrane were studied
[47]	Same laser for Raman and one optical tweezers 633 nm	Very low-density lipoprotein, a subgroup of the human triglyceride-rich lipoprotein (TGRL)	Highly ordered saturated fatty acids 1,000–1,150 cm^{-1}	Feasibility of monitoring TGRL during interaction with lipoprotein lipase was demonstrated. Composition and distribution were studied
			Unsaturated fatty acids 1,655 cm^{-1}	
[48]	Same laser for Raman and one optical tweezers 780 nm	Pancreatic β-cell exposed to high glucose	Intensity of phenylalanine at 1,002 cm^{-1}, and lipids at 1,085, 1,445, and 1,655 cm^{-1} were increased	The cells under basal conditions 2.8 mmol/l glucose culture medium were measured when exposed to high glucose 20 mmol/l. Proteins and lipids were affected due to β-cell protein and lipid synthesis

Yeast cell and bacterial spores				
[49]	Same laser for Raman and one optical tweezers 810 nm	Yeast, *E. coli* and *Enterobacter aerogenes*	Changes of phenylalanine at 1,004 cm^{-1}	Raman spectra of native and heat denaturated cells show that conformational studies of proteins within cells can be studied by Raman tweezers
[50]	Same laser for Raman and one optical tweezers 787 nm	Two different kinds of *Bacillus stearothermophilus* (bacterial spores)	–	SERS investigation of bacterial spores regarding spore coat composition in two. Discrimination between the two was possible
	Same laser for Raman and one optical tweezers 488 nm	Bacterial spores, polystyrene, and silica beads [51]	Amide III at 1,245 cm^{-1} amide I at 1,655 cm^{-1}, phenylalanine at 1,001 cm^{-1}, Ca-DPA at 658, 822, 1013 1,395, and 1,572 cm^{-1}	Rapid identification of bacterial spores that could be used for cancer detection or concentration analysis
[52]	Same laser for Raman and one optical tweezers 780 nm	*Schizosaccharomyces pombe* yeast cell	Decrease of intensity in amide I (1,656 cm^{-1}) and amide III (1,270 cm^{-1}) bands and alterations in tyrosine band at 850 cm^{-1}	Alterations in hydration state of protein could be observed over time
[53]	Same laser for Raman and one optical tweezers 785 nm	*Saccharomyces cerevisia*, *Escherichia coli (E. coli)*	Shift of phenylalanine at 1,002 cm^{-1} to 965 cm^{-1}, indicates uptake of C-13	Feasibility of studying unculturable microorganisms is demonstrated
[54]	Same laser for Raman and one optical tweezers 785 nm	2,6-pyridine dicarboxylic acid (Ca-DPA) in bacteria spores	Ca-DPA at 824, 1,017, 1,395, and 1,572 cm^{-1}	Raman and elastic light scattering intensity (ESLI) investigation on spore germination dynamics gives new insight into the process
[55]	Same laser for Raman and one optical tweezers 785 nm	2,6-pyridine dicarboxylic acid (Ca-DPA) in bacteria spores	Ca-DPA at 1,017 cm^{-1}	Raman spectra of the Ca-DPA peak can be used as quantitative standard measure to reveal concentration and characteristic spore information

(*continued*)

Table 18.2 (continued)

Reference	Experimental design	Object	Relevant Raman peaks	Results
Blood-related diseases, red blood cells, and hemoproteins				
[16, 56]	Double microscope configuration. One optical trap 830 nm, Raman, 488.0, 514.5, and 568.2 nm. One optical trap 1,064 nm, Raman 514.5 nm	Hemoglobin (Hb) within red blood cells (RBCs)	Oxidation state markers between 1,300 and 1,640 cm^{-1}	Different wavelengths for resonance Raman on Hb are investigated Oxygenation cycle of a red blood cell in a microfluidic system is demonstrated
[57]	One optical trap 830 nm Raman 413.1 nm	Neuroglobin (Ngb) overexpressed in *E. coli*	Oxidation state marker ν_4 at 1,361 and 1,374 cm^{-1}	Ngb over-expressed in *E. coli* is transferred back to deoxygenated state, even under oxygen-rich conditions. Ngb seems to respond to a reducing enzymatic system
[58]	One optical trap 1,064 nm Raman 532 nm	Normal and thalassemia-infected RBCs [59, 60]. A review of the field is given in	ν_{34} at 1,580 and 1,588 cm^{-1} is dramatically modified in infected cells (1,588 cm^{-1} to 1,583 cm^{-1} or 1,581 cm^{-1} depending on infection type). Reduction or disappearance of ν_{10} band for infected cells	Raman tweezers can separate healthy and infected RBCs. Mechanical properties are investigated as well
[61]	Dual trap 1,064 nm Raman 785 nm	Hb in RBCs	Polarization order of bands not affected by oxygenation 670 cm^{-1}, 750 cm^{-1}, 970 cm^{-1}, 1,000 cm^{-1}, 1,120 cm^{-1}, 1,400 cm^{-1}, 1,440 cm^{-1}, 1,545 cm^{-1}, and 1,580 cm^{-1}	RBC deformation and polarization state of Hb in an optical dual trap, stretched RBCs show higher packing
[62]	Same laser for Raman and one optical tweezers 785 nm	Hb in RBCs	S = S and C-S at 500, 519, 550, 565, 640, and 665 cm^{-1}	Enhanced glutathione level as a protective response against oxidative stress is observed. By means of PCA, normal and stressed RBCs can be distinguished

Organelles and subcellular structures				
[25]	Same laser for Raman and one optical tweezers 730 nm	Synaptosomes	CH_2 deformation at 1,445 cm^{-1}, Amide I at 1,657 cm^{-1}	Synaptosomes isolated from rat brain neurons dispersed in buffer solution. Appearing beads reveal that synaptosomes include liposomes and proteins
[15]	Optical trapping by two counter propagating diode laser beams 1,070 nm, Raman 785 nm through microscope objective	HK60 human promyelocytic leukemia cells	Cytoplasm 643, 852, 939, 1,660 cm^{-1} nucleus, A, T, G, and C at 785, 830, 895, 1,048, 1,093 cm^{-1} and thymine at 751 cm^{-1}	Nucleus, cytoplasm, and membrane are distinguished in one cell
[63, 64]	Same laser for Raman and one optical tweezers NIR	Chromosomes	725 785, 1,004, 1,094, 1,340, 1,450, and 1,660 cm^{-1}	Generalized discriminate analysis (GDA) was applied on human chromosomes 1, 2, and 3 after G-banding. The analyzing method was disputed by Bak et al. [65] but Ojeda et al. replied
[66]	Same laser for Raman and one optical tweezers 785 nm	Single trapped mitochondria from rat liver	1,654, 1,446, 1,301, and 1,226 and decrease of 1,602 cm^{-1}	Heart and liver mitochondria can be distinguished. Swelling due to Ca_{2+} indicates bioactivity of single mitochondria
[67]	Dual tram 1,064 nm Raman 785 nm	DNA molecules	–	SERS study of DNA bound in between two trapped beads. The solutions contained Ag particles

infected cells have been achieved. When investigating optically trapped cells by Raman, one can focus on one organelle of the cell, like the nucleus, which enhances the reproducibility of correct cell analysis. Differences in size, refractive index, or chemical compositions of different cells are highly valuable for cell-sorting purposes. Sometimes, changes or shifts in the Raman spectra may even give specific information on the reaction of cells to different chemical environments.

5.1.3 Experimental Considerations Influencing the Measurements; Photo-Induced and Thermal Effects on Biological Material

In most cases, NIR wavelengths are chosen for Raman excitation of biological samples in order to minimize fluorescence background, photo-induced chemical changes, and heating. However, biological samples, and especially single cells, can still be very sensitive toward laser illumination. It is therefore important to carefully study the impact of the Raman tweezers experiment for each biological system under investigation. Importantly, even closely related species can have quite different optical behaviors or react differently to preparation protocols. One method to check for laser-induced damage is to investigate spectral changes over time by illuminating the same spot on the sample continuously while taking series of spectra with short integration time. Table 18.3 summarizes results from experiments exploring heating and photo-induced damage caused by laser light from the UV to the NIR wavelength region.

The response of biological material to the interaction with laser light strongly depends on the wavelengths of the trapping laser, as well as on the applied power. The powerful microscope objectives create strong foci with extremely high irradiances. There are basically two events that have to be considered: First, the laser light might cause either direct or indirect heating of the sample, leading to a temperature rise of the sample proportional to the laser intensity and the absorption cross section of the sample. Second, the laser wavelength might trigger chemical reactions within the sample. This effect is extremely important to consider in resonance Raman spectroscopy, which can be very useful for particular biomolecular studies. One such case concerns hemoproteins and studies of blood-related diseases. Since the important heme resonances are in the visible and UV, and intense radiation at these wavelengths generates effective light absorption and, consequently, photochemistry, it is usually necessary to utilize a separate NIR laser beam for trapping while resonance Raman excitation is performed with as little laser intensity as possible. Still, the photo-induced chemistry caused by ultraviolet or visible Raman excitation can be more dramatic and obvious than that generated by the NIR trapping laser, despite that the laser power usually is around 100 times less. Importantly, some molecules might not react directly with the laser light, but damage can still be induced if a sensitizer or oxygen is present. In this case, the laser illumination excites a long-lived triplet state of the sensitizer, which in turn can react with oxygen to produce reactive oxygen species that attack molecules like DNA, proteins, and lipids [76].

Table 18.3 Effects of different laser wavelengths and powers upon cells and organelles

Wavelength (nm)	Irradiance or power	Sample	Observation
<260	0.1–1.1 Wcm^{-2}	Bacterial cells	Good reproducibility, no observed fluorescence [70–72]
257	10 Wcm^{-2}	T47 D mammary tumor cells	No physical damage, good reproducibility [73]
413.1	90 k Wcm^{-2}	Nerve myoglobin within the nerve cord of *A. aculeata*	Photo dissociation of oxygen from the porphyrin ring, [57]. No damage at 8 kWcm^{-2}
48	80 k Wcm^{-2}	Hb within RBCs	Strong fluorescence after 30s, significant decrease of SN ratio [74]
514.5	90 k Wcm^{-2}	Hb within RBCs	Strong fluorescence and oxidation of Hb [75]
514.5	4 M Wcm^{-2}	Isolated chromosomes composed of nucleic acids and proteins	Photochemically induced changes in spectra [76]
514.5	4 M Wcm^{-2}	Pure DNA and protein solutions	No damage [76]
514.5	4 M Wcm^{-2}	Cell nucleus	Reduced cell viability [76]
568.2	80 k Wcm^{-2}	Hb within RBCs	Stable SN ratio, weak alterations in the oxygen-sensitive region of the spectrum [56]
568.2	0.8 M Wcm^{-2}	Nerve myoglobin within the nerve cord of *A. aculeata*	Photo-induced breaking of porphyrin ring to biliverdin [77]
632.8	160 k Wcm^{-2}	Hb within RBCs	No photo-degradation [78]
632.8, 660	8 M Wcm^{-2}	Cell nucleus	Good cell viability [76]
660	6 M Wcm^{-2}	Chromosomes, lymphocytes, granulocytes	No observed damage < 30 s < slight changes in spectra [76]
690	0.8 M Wcm^{-2}	Yeast cells	Heat denaturation of Phe [49]
		E. coli Enterobacter aerogenes bacterial cells	
840–930	Energy	*E. coli*	Weak dependence on λ, linear dependence on power. Lethal energy dose for *E. coli* is 5 J [79]
cw 790–970	88–100 mW	*E. coli* under anaerobic conditions, Chinese hamster ovary cells	Minima of photo-damage at 830 nm and 970 nm [80]
			Maxima at 870 nm and 930 nm [81]
cw 700–1,064	100 mW	*C. elegans* with heat chock promoter	Maxima below 760 nm, at higher λ less photo chemistry but an increase of heating [82]
cw 760	105 mW	Human spermatozoa	Autofluorescence at 760 nm, paralysis after 35 ± 20 s, cell death after 65 ± 20 s
			Linear and two-photon processes, single photon absorption process [83]

(*continued*)

Table 18.3 (continued)

Wavelength (nm)	Irradiance or power	Sample	Observation
cw 830 nm,	120 mW	RBCs	No observable influence after 20 min of trapping [56]
cw 1,064 nm	80–120 mW	RBCs	Linear power dependence of photo-damage due to heating, reducible by flow of cold buffer [16]

5.2 Optical Tweezers for Surface-Enhanced Raman Scattering (SERS)

Metal nanoparticles, typically gold and silver nanoparticles, play an important role in optical tweezers for Raman analysis in the context of surface-enhanced Raman scattering (SERS). In order to obtain high Raman enhancement, metal nanoparticles are usually aggregated either during the trapping or before being subjected to the trapping laser. Probe molecules are pre-functionalized on the surface of the particles, typically by mixing the metal colloid and the analyte solution of certain concentration.

If a metal nanoparticle is irradiated by a laser beam, a strong electromagnetic near-field is induced due to the excitation of the collective oscillations of the conduction electrons known as localized surface plasmon resonances (LSPRs) [84]. For a single spherical metal nanoparticle, the intensity of the induced electric field can be approximated as $E \approx (\varepsilon(\omega) - \varepsilon_0)/(\varepsilon(\omega) + 2\varepsilon_0) \cdot E_0$, where $\varepsilon(\omega)$ is the frequency-dependent dielectric constant of the metal, ε_0 is the dielectric constant of the surrounding medium, and E_0 is the incident electric field. It can be seen that E is much stronger than E_0 when the denominator $(\varepsilon(\omega) + 2\varepsilon_0) \rightarrow 0$. This is the condition for LSPR excitation. For silver and gold, the most common "SERS metals," the corresponding resonance frequencies fall in the blue and green, respectively. Note that near-field enhancement in aggregates, such as particle dimers, or sharp particles can be much stronger than that for single spherical particles due to, for example, near-field coupling effects. Nevertheless, under plasmon resonance excitation conditions, the Raman probe molecules thus experience an enhanced incident-driving field, but the Raman scattering rate is enhanced by approximately the same factor as the induced field. In total, the EM enhancement of Raman scattering is proportional to the product of the excitation and emission enhancement factors, i.e., $\left|\frac{E(\omega_0)}{E_0(\omega_0)}\right|^2 \cdot \left|\frac{E(\omega_{Raman})}{E_0(\omega_{Raman})}\right|^2$, where ω_0 and ω_{Raman} are the incident and Raman scattered frequencies, respectively.

Although metal nanoparticles experience strong radiation pressure and scattering forces due to their high absorption and polarizability, stable trapping of such particles has been demonstrated in both two and three dimensions [74, 85, 86]. As for cells and dielectric particles, a NIR laser is usually used to trap metal nanoparticles while a separate laser, or the trapping laser itself, excites the

Raman probes. SERS can then be performed on optically induced aggregates of the trapped particles. Alternatively, metal nanoparticles can also be attached on micron-sized dielectric beads, which are much easier to trap. Raman probes can be adsorbed on the surface of the metal nanoparticles. In addition, combined with other techniques, such as microfluidics, the applicability of optical tweezers for SERS can be even more expanded.

5.2.1 Optical Aggregation of Metal Nanoparticles for SERS

A particularly important variant of the optical force, interparticle forces, turns out to be crucial for SERS. This effect is similar to the attractive van der Waals force between small particles, which is due to interactions between spontaneously fluctuating dipoles, but the optical interaction is due to coupling between the actual particle dipoles induced by the trapping laser. Due to the interparticle optical forces, metal nanoparticles aggregate in an optical tweezers and produce "hotspots," i.e., particle junctions with intense local fields for SERS. Raman probes can be excited either by the trapping laser or, preferably, by a separate low power beam that does not disturb the trapping.

The simplest form of an aggregate, a dimer of two metal nanoparticles, has been shown to produce high enough field enhancement for efficient SERS detection [87]. The example in Fig. 18.2a shows the dimerization of two silver nanoparticles using optical tweezers to create "hotspots" for SERS. In these experiments, a NIR laser of 830 nm is used for optical trapping, and a separate laser 514.5 nm is used for Raman excitation. Thiophenol molecules are chemically bound to the surface of the silver nanoparticles as Raman probe. A single silver nanoparticle immobilized on a glass slide or trapped by the focused laser can be identified by dark-field imaging as a bluish spot, from which no SERS signal is detected. However, if a dimer is created by moving a trapped particle to an immobilized one, significant SERS signal arises due to the strong near-field coupling effect.

The dimerization is easily understood considering the optical potential created by the trapping laser. Figure 18.2b shows the calculated optical potential experienced by a silver nanoparticle that is free to move in a Gaussian laser focus at a wavelength of 830 nm. The particle is also affected by the optical interparticle force from an immobilized silver particle located at different separations from the laser focus. It is clear that a deep potential minimum is induced when the trapped particle approaches the immobilized one, giving rise to spontaneous optical dimerization and a SERS "hot spot" in the optical trap. Note that the two particles are expected to align parallel to the laser polarization in this case, as has been demonstrated experimentally recently [88].

For larger aggregates, stronger enhancement is expected. The work by Tanaka et al. [89] showed that silver nanoparticles simultaneously trapped by a NIR laser beam also tend to align parallel to the linear polarization, and it exhibited pronounced SERS signal at a very low concentration of probe molecules, 10^{-14} M pseudoisocyanine (PIC) excited by a separate laser beam, whereas no SERS appeared if the trapping laser was not focused in the solution.

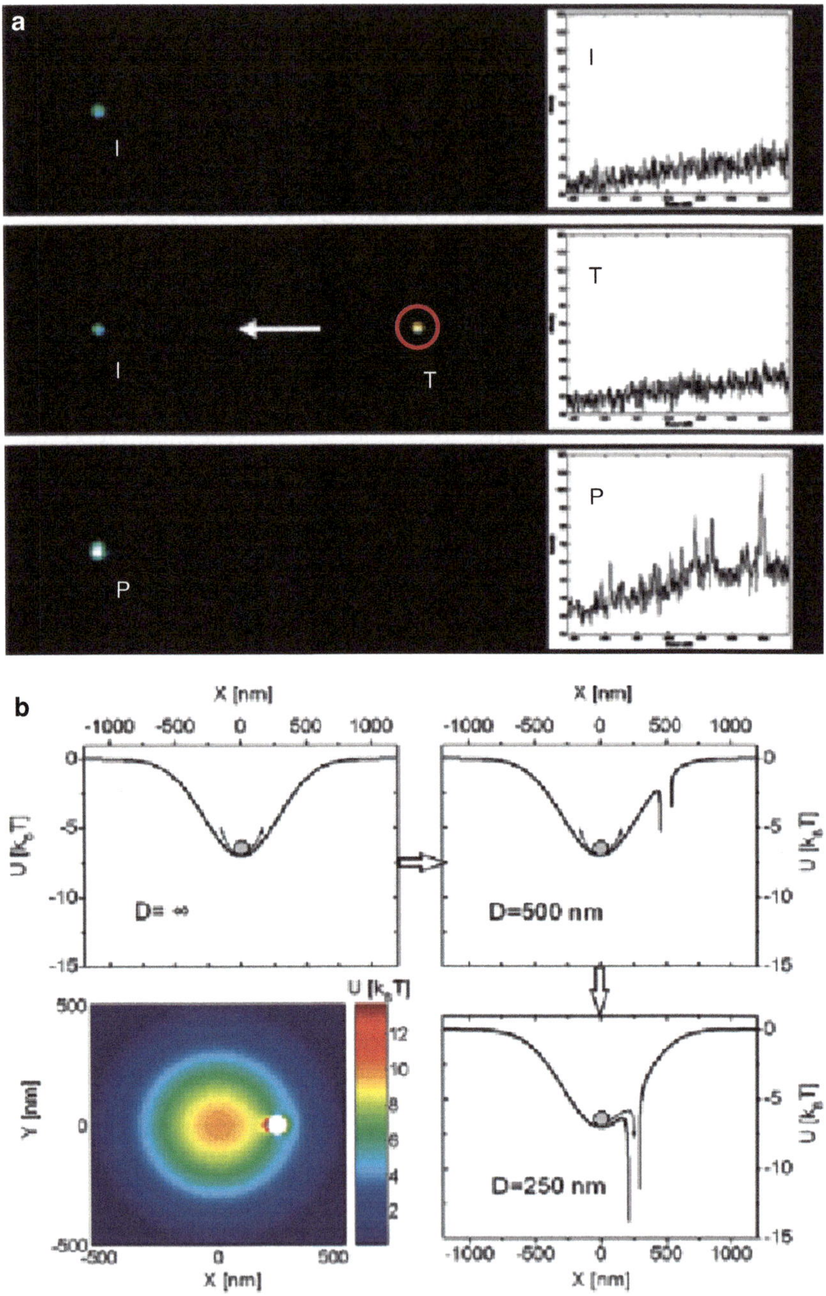

Fig. 18.2 (**a**) The dimerization of silver nanoparticles, dark-field images, and corresponding Raman spectra. (**b**) Simulated optical potential under 830-nm laser irradiation experienced by a trapped AgNP with different separations to an immobilized one

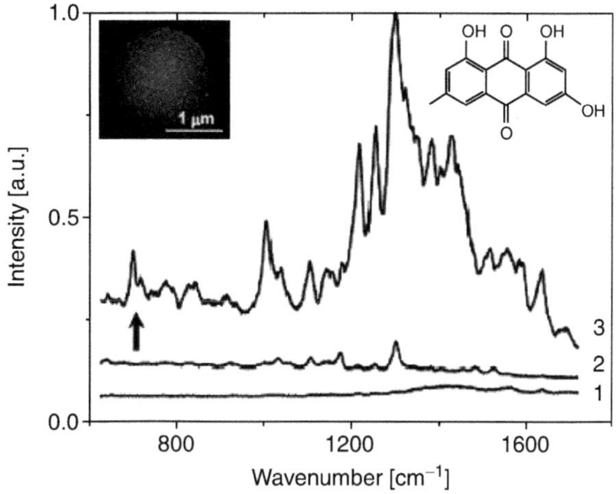

Fig. 18.3 SERS spectra from silver nanoparticles decorated silica beads. Curve 1: silica beads only. Curve 2: silica beads decorated by silver nanoparticles. Curve 3: silver nanoparticles decorated silica beads with probe molecules. *Left inset*: SEM image of silver nanoparticles decorated silica bead. *Right inset*: molecular structure of emodin

5.2.2 SERS from Metal Nanoparticles Immobilized on Optically Trapped Beads

Due to the high dissipative force, stable trapping of metal nanoparticles requires high NA objectives and high laser power. In order to facilitate the optical trapping of metal nanoparticles without compromising their SERS capability, an alternative approach is to decorate metal nanoparticles on dielectric beads that are much easier to trap [90, 91]. Petrov et al. [90] reported the optical trapping of silica beads, a few microns in diameter, decorated with 20–40 nm silver nanoparticles bound on their surfaces. The silica beads were firstly functionalized using aminopropyltrimethoxysilane (APTMS) and then mixed with a silver colloid. The surface density of nanoparticles could be adjusted by changing the reaction conditions, such as the rotation rate of the vial during the binding process. Strong SERS spectra were detected if probe molecules (emodin in this example) were added (curve 3 in Fig. 18.3), while no signal appeared for non-decorated silica beads (curve 1) and only weak peaks from silica show up for silver nanoparticles decorated beads without probe molecules (curve 2). The laser beam used for both optical trapping and Raman excitation was 785 nm. In a slightly different experimental setup where silver nanoparticles decorated silica beads were trapped for SERS. Jordan et al. [91] used a frequency doubling KTiOPO4 (KTP) crystal to produce 532-nm laser light for Raman excitation while the original laser line 1,064 nm line was used for trapping.

5.2.3 Optical Trapping of Raman Active Objects Approached to a SERS-Active Substrate

Instead of creating SERS-active sites in an optical trap, Raman enhancement can also be obtained by moving Raman active probes to the vicinity of a SERS-active substrate using an optical tweezers [50]. As schemed in Fig. 18.4a, gold nanoparticles were pre-immobilized on the surface of a glass slide facing the solution where biological spores were trapped. When the spore is far from the

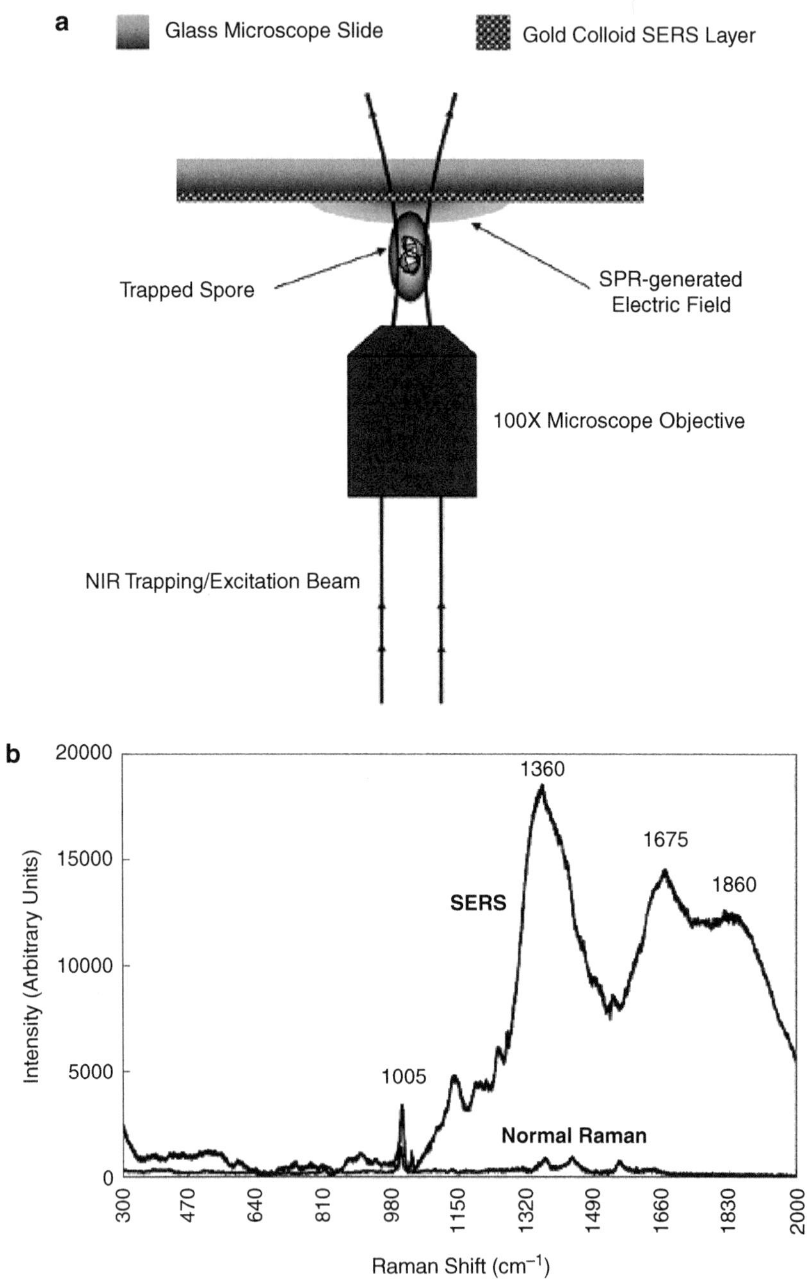

Fig. 18.4 (**a**) Scheme of the experiment. (**b**) SERS and normal Raman spectra of the trapped spore. The SERS spectrum was measured when the spore was approached to the substrate

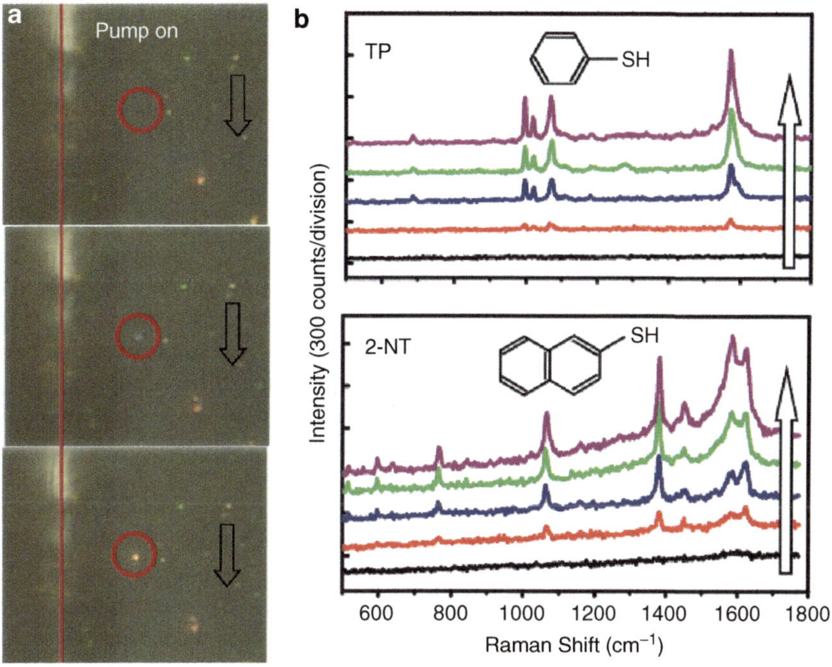

Fig. 18.5 (**a**) Dark-field microscopy images of silver nanoparticles trapped in a microfluidic channel. The trapping wavelength was 830 nm. (**b**) Corresponding SERS spectra from the optically induced nanoparticle aggregates. A separate laser line at 514.5 nm was used to excite the Raman probes

substrate, weak normal Raman spectra were recorded. However, if the spore is moved close to the SERS-active substrate, the Raman intensity increases dramatically, as shown in Fig. 18.4b. Although the features in the SERS spectrum are difficult to assign due to the complex bacterial composition, the surface-enhancement in the Stokes-shifted range is clear.

5.2.4 Optical Tweezers for SERS Integrated with Microfluidics

Raman tweezers exhibits a prominent potential for lab-on-a-chip based chemo/bio sensing [92]. Using microfluidics, two flows, the metal colloid and the analyte solution, can be injected separately and allowed to mix in a common flow pipe before entering the microfluidic channel. The metal nanoparticles incubated by the molecules to be detected can then be trapped and allowed to form aggregates in the channel, thus producing a SERS signal for sensing. Figure 18.5a shows dark-field images of the growing aggregate created by optical tweezers using a NIR laser trapping at 830 nm. The corresponding SERS spectra, excited by the 514.5 nm Ar^+ line, are shown in Fig. 18.5b. Using microfluidics, the analyte flow is interchangeable, enabling consecutive detection of a series of species simply by changing the analyte solutions and keeping the Ag colloid flow. With a Y-shaped microfluidic channel, where metal colloid and analyte solution can flow in parallel, a SERS signal can also be detected at the interface where nanoparticles and analyte molecules are mixed through diffusion [92].

6 Conclusions and Future Perspective

A large number of studies have demonstrated that optical tweezers integrated with Raman spectroscopy have a very high potential for real-time chemical characterization and diagnosis of micron-sized dielectric beads and biological objects. The most important benefit of optical tweezing to Raman spectroscopy is that small particles can be easily isolated from solution, brought to the Raman excitation beam and kept there for sufficiently long time for spectroscopic identification with high signal-to-noise ratio. This enables studies that are impossible using ensemble averaging macro-Raman techniques or Raman microscopy on freely diffusing particles. However, optical tweezing requires high-power laser beams, usually much higher than what is needed for Raman excitation, and great care has to be taken to avoid laser-induced thermal or photochemical destruction of the sample.

Raman tweezers offers unique but as yet largely unexplored possibilities for nanoscience. However, optical trapping is more difficult in the case of nanoparticles because the optical gradient force decreases rapidly with particle volume. Nevertheless, a number of studies on metal nanoparticles, which have sufficiently high polarizability for trapping, have demonstrated a significant potential. In particular, by utilizing the power of surface-enhanced Raman scattering, the sensitivity increases by orders of magnitude and a vast range of small organic molecules attached to metal nanoparticles can be studied in solution. One interesting possibility is to use trapped silver or gold colloids as biosensors within biological cells. Proofs of principle have been demonstrated [74] and gold nanoparticles have been introduced into cells to serve as nanosensors that measure Raman signals from biomolecules in their native environment [93]. There is also a large potential in the combination of Raman tweezers with microfluidic systems [16, 57], so called lab-on-a-chips. In particular, SERS studies based on optical aggregation completely avoids the considerable reproducibility problems associated with chemical aggregation of metal colloids.

References

1. Ashkin A (1970) Acceleration and trapping of particles by radiation pressure. Phys Rev Lett 24:156–159
2. Quidant R, Girard C (2008) Surface-plasmon-based optical manipulation. Laser Photonics Rev 2(1–2):47–57
3. Mehta AD et al (1999) Single-molecule biomechanics with optical methods. Science 283(5408):1689–1695
4. Grier DG (2003) A revolution in optical manipulation. Nature 424(6950):810–816
5. Moffitt JR et al (2008) Recent advances in optical tweezers. Annu Rev Biochem 77:205–228

6. Dienerowitz M, Mazilu M, Dholakia K (2008) Optical manipulation of nanoparticles: a review. J Nanophotonics 2:1–32
7. Lang MJ, Block SM (2003) Resource letter: LBOT-1: laser-based optical tweezers. Am J Phys 71(3):201–215
8. Jonas A, Zemanek P (2008) Light at work: the use of optical forces for particle manipulation, sorting, and analysis. Electrophoresis 29(24):4813–4851
9. Neuman KC, Block SM (2004) Optical trapping. Rev Sci Instrum 75(9):2787–2809
10. Ashkin A (1992) Forces of a single-beam gradient laser trap on a dielectric sphere in the ray optics regime. Biophys J 61:569–582
11. Zhang H, Liu KK (2008) Optical tweezers for single cells. J R Soc Interface 5(24):671–690
12. Barton JP, Alexander DR, Schaub SA (1988) Internal and near-surface electromagnetic-fields for a spherical-particle irradiated by a focused laser-beam. J Appl Phys 64(4):1632–1639
13. Barton JP, Alexander DR, Schaub SA (1989) Internal fields of a spherical-particle illuminated by a tightly focused laser-beam—focal point positioning effects at resonance. J Appl Phys 65(8):2900–2906
14. Nieminen TA et al (2007) Optical tweezers computational toolbox. J Opt A-Pure Appl Opt 9(8):S196–S203
15. Jess PRT et al (2006) Dual beam fibre trap for Raman microspectroscopy of single cells. Opt Express 14(12):5779–5791
16. Ramser K et al (2004) A microfluidic system enabling Raman measurements of the oxygenation cycle in single optically trapped red blod cells. Lab Chip 5:431–436
17. Fallman E, Axner O (1997) Design for fully steerable dual-trap optical tweezers. Appl Opt 36(10):2107–2113
18. Schonbrun E et al (2005) 3D interferometric optical tweezers using a single spatial light modulator. Opt Express 13(10):3777–3786
19. Rusciano G et al (2007) Enhancing Raman tweezers by phase-sensitive detection. Anal Chem 79(10):3708–3715
20. Rusciano G et al (2006) Phase-sensitive detection in Raman tweezers. Appl Phys Lett 89(26):261116
21. Ashkin A et al (1986) Observation of a single-beam gradient force optical trap for dielectric particles. Opt Lett 11(5):288–290
22. Urlaub E et al (1994) Raman investigation of styrene polymerization in single optically trapped emulsion particles. Chem Phys Lett 231:511–514
23. Ajito K, Torimitsu K (2001) Near-infrared Raman spectroscopy of single particles. Trac-Trends Anal Chem 20(5):255–262
24. Xie C, Dinno MA, Li Y (2002) Near-infrared Raman spectroscopy of single optically trapped biological cells. Opt Lett 27(4):249–251
25. Ajito K, Torimitsu K (2002) Laser trapping and Raman spectroscopy of single cellular organelles in the nanometer range. Lab Chip 2:11–14
26. Mitchem L, Reid JP (2008) Optical manipulation and characterisation of aerosol particles using a single-beam gradient force optical trap. Chem Soc Rev 37(4):756–769
27. Cherney DP, Conboy JC, Harris JM (2003) Optical-trapping Raman microscopy detection of single unilamellar vesicles. Anal Chem 75:6621–6628
28. Reid JP (2009) Particle levitation and laboratory scattering. J Quant Spectrosc Radiat Transf 110(14–16):1293–1306
29. Petrov DV (2007) Raman spectroscopy of optically trapped particles. J Opt A-Pure Appl Opt 9(8):S139–S156
30. Snook RD et al (2009) Raman tweezers and their application to the study of singly trapped eukaryotic cells. Integr Biol 1(1):43–52
31. Stevenson DJ, Gunn-Moore F, Dholakia K Light forces the pace: optical manipulation for biophotonics. J Biomed Opt 15(4)
32. Ramser K, Hanstorp D (2010) Optical manipulation for single-cell studies. J Biophotonics 3(4):187–206

33. Cherney DP, Harris JM (2010) Confocal Raman microscopy of optical-rrapped particles in liquids. In: Annual review of analytical chemistry, vol 3, Annual Reviews, Palo Alto, pp 277–297
34. Lankers M et al (1997) Raman investigations on laser-trapped gas bubbles. Chem Phys Lett 277:331–334
35. Kiefer W et al (1997) Raman-Mie scattering from single laser trapped microdroplets. J Mol Struct 408/409:113–120
36. Ajito K, Morita M (1999) Imaging and spectroscopic analysis of single microdroplets containing p-cresol using the near-infrared laser tweezers Raman microprobe system. Sur Sci 428:141–146
37. Anquetil PA et al (2003) Laser Raman spectroscopic analysis of polymorphic forms in microliter fluid volumes. J Pharm Sci 92(1):149–160
38. Bridges TE, Houlne MP, Harris JM (2004) Spatially resolved analysis of small particles by confocal Raman microscopy: depth profiling and optical trapping. Anal Chem 76(3):576–584
39. Buajarern J et al (2006) Controlling and characterizing the coagulation of liquid aerosol droplets. J Chem Phys 125(11):114506
40. Hamden KE et al (2005) Spectroscopic analysis of Kaposi's sarcoma-associated herpesvirus infected cells by Raman tweezers. J Virol Methods 129(2):145–151
41. Chan JW et al (2006) Micro-Raman spectroscopy detects individual neoplastic and normal hematopoietic cells. Biophys J 90(2):648–656
42. Zheng F, Qin YJ, Chen K (2007) Sensitivity map of laser tweezers Raman spectroscopy for single-cell analysis of colorectal cancer. J Biomed Opt 12(3):034002
43. Chan JW et al (2008) Nondestructive identification of individual leukemia cells by laser trapping Raman spectroscopy. Anal Chem 80(6):2180–2187
44. Harvey TJ et al (2008) Spectral discrimination of live prostate and bladder cancer cell lines using Raman optical tweezers. J Biomed Opt 13(6):064004
45. Harvey TJ et al (2009) Classification of fixed urological cells using Raman tweezers. J Biophotonics 2(1–2):47–69
46. Chan JW, Taylor DS, Thompson DL (2009) The effect of cell fixation on the discrimination of normal and leukemia cells with laser tweezers Raman spectroscopy. Biopolymers 91(2):132–139
47. Chan JW et al (2005) Raman spectroscopic analysis of biochemical changes in individual triglyceride-rich lipoproteins in the pre- and postprandial state. Anal Chem 77(18):5870–5876
48. Rong X et al (2010) Real-time detection of single-living pancreatic beta-cell by laser tweezers Raman spectroscopy: high glucose stimulation. Biopolymers 93(7):587–594
49. Xie C et al (2003) Study of dynamical process of heat denaturation in optically trapped single microorgansims by near-infrared Raman spectroscopy. J appl phys 94(9):6138–6142
50. Alexander TA, Pellegrino PM, Gillespie JB (2003) Near-infrared surface-enhanced-Raman-scattering-mediated detection of single optically trapped bacterial spores. Appl Spectrosc 57(11):1340–1345
51. Chan JW et al (2004) Reagentless identification of single bacterial spores in aqueaous solution by confocal laser tweezers Raman spectroscopy. Anal Chem 76:599–603
52. Basar G, Kin S (2008) Monitoring of spectroscopic changes of a single trapped fission yeast cell by using a Raman tweezers set-up. Opt Commun 281(19):4998–5003
53. Huang WE, Ward AD, Whiteley AS (2009) Raman tweezers sorting of single microbial cells. Environ Microbiol Rep 1(1):44–49
54. Peng LX et al (2009) Elastic and inelastic light scattering from single bacterial spores in an optical trap allows the monitoring of spore germination dynamics. Anal Chem 81(10):4035–4042
55. Huang X et al (2010) Probing the machanism and Ca-DPA concentration of individual Bacillus spores using trapping and raman spectroscopy. Spectrosc Spectr Anal 30(8):2151–2156

56. Ramser K et al (2004) Resonance Raman spectroscopy of optically trapped functional erythrocytes. J Biomed Opt 9(3):593–600
57. Ramser K et al (2007) Micro-resonance Raman study of optically trapped *Escherichia coli* cells overexpressing human neuroglobin. J Biomed Opt 12(4):044009
58. Rusciano G et al (2008) Raman tweezers as a diagnostic tool of Hemoglobin-related blood disorders. Sensors 8(12):7818–7832
59. De Luca AC, Rusciano G (2007) Monitoring cellular diseases by Raman tweezers. Nuovo Cimento Della Societa Italiana Di Fisica B-General Physics Relativity Astronomy and Mathematical Physics and Methods. Il Nuovo Cimento B 122(6–7):731–738
60. De Luca AC et al (2008) Spectroscopical and mechanical characterization of normal and thalassemic red blood cells by Raman tweezers. Opt Express 16(11):7943–7957
61. Rao S et al (2009) Raman study of mechanically induced xxygenation state transition of red blood cells using optical tweezers. Biophys J 96(1):209–216
62. Zachariah E et al (2010) Probing oxidative stress in single erythrocytes with Raman tweezers. J Photochem Photobiol B-Biol 100(3):113–116
63. Ojeda JF et al (2007) Chromosomal analysis and identification based on optical tweezers and Raman spectroscopy: reply. Opt Express 15(10):6000–6002
64. Ojeda JF et al (2006) Chromosomal analysis and identification based on optical tweezers and Raman spectroscopy. Opt Express 14(12):5385–5393
65. Bak J, Jorgensen TM (2007) Chromosomal analysis and identification based on optical tweezers and Raman spectroscopy: comment. Opt Express 15(10):5997–5999
66. Tang H et al (2007) NIR Raman spectroscopic investigation of single mitochondria trapped by optical tweezers. Opt Express 15(20):12708–12716
67. Rao S et al (2010) Single DNA molecule detection in an optical trap using surface-enhanced Raman scattering. Appl Phys Lett 96(21):213701
68. De Gelder J et al (2007) Reference database of Raman spectra of biological molecules. J Raman Spectrosc 38(9):1133–1147
69. Lord RC, Yu N-T (1970) Laser-excited Raman spectroscopy of biomolecules *1: I. Native lysozyme and its constituent amino acids. J Mol Biol 50(2):509–524
70. Baek M et al (1988) UV-excited resonance raman-spectra of heat denatured lysozyme and *Staphylococcus epidermidis*. Appl Spectrosc 42(7):1312–1314
71. Baek M, Nelson WH, Hargraves PE (1989) Ultra-violet resonance Raman-spectra of live cyanobacteria with 222.5–251.0-nm pulsed laser excitation. Appl Spectrosc 43(1):159–162
72. Nelson WH, Manoharan R, Sperry JF (1992) UV resonance Raman studies of bacteria. Appl Spectrosc Rev 27(1):67–124
73. Sureau F et al (1990) An ultraviolet micro-Raman spectrometer—resonance Raman-spectroscopy within single living cells. Appl Spectrosc 44(6):1047–1051
74. Prikulis J et al (2004) Optical spectroscopy of single trapped metal nanoparticles in solution. Nano Lett 4(1):115–118
75. Ramser K et al (2003) Importance of substrate and photo-induced effects in Raman spectroscopy of single functional erythrocytes. J Biomed Opt 8(2):173–178
76. Puppels GJ et al (1991) Laser irradiation and Raman spectroscopy of single living cells and chromosomes: sample degradation occurs with 514.5 nm but with 660 nm laser light. Exp Cell Res 195:361–367
77. Ramser K et al (2007) A combined micro-resonance Raman and absorption set-up enabling in vivo studies under varying physiological conditions: the nerve globin in the nerve cord of *Aphrodite aculeata*. J Biochem Biophys Methods 70:627–633
78. Wood BR, McNaughton D (2002) Raman excitation wavelenghts investigations of single red blood cells in vivo. J Raman Spectrosc 33:517–523
79. Mirsaidov U et al (2008) Optimal optical trap for bacterial viability. Phys Rev E 78(2)
80. Neuman KC et al (1999) Characterization of photodamage to *Escherichia coli* in optical traps. Biophys J 77:2856–2863

81. Liang H et al (1996) Wavelength dependence of cell cloning efficiency after optical trapping. Biophys J 70(3):1529–1533
82. Leitz G et al (2002) Stress response in *Caenorhabditis elegans* caused by optical tweezers: wavelength, power and time dependence. Biophys J 82(4):2224–2231
83. Konig K et al (1996) Effects of ultraviolet exposure and near infrared laser tweezers on human spermatozoa. Hum Reprod 11(10):2162–2164
84. Haes AJ et al (2005) Plasmonic materials for surface-enhanced sensing and spectroscopy. Mrs Bull 30(5):368–375
85. Svoboda K, Block SM (1994) Optical trapping of metallic Rayleigh particles. Opt Lett 19(13):930–932
86. Hansen PM et al (2005) Expanding the optical trapping range of gold nanoparticles. Nano Lett 5(10):1937–1942
87. Svedberg F et al (2006) Creating hot nanoparticle pairs for surface-enhanced Raman spectroscopy through optical manipulation. Nano Lett 6(12):2639–2641
88. Tong LM, Miljkovic VD, Kall M (2010) Alignment, rotation, and spinning of single plasmonic nanoparticles and nanowires using polarization dependent optical forces. Nano Lett 10(1):268–273
89. Tanaka Y et al (2009) Surface enhanced Raman scattering from pseudoisocyanine on Ag nanoaggregates produced by optical trapping with a linearly polarized laser beam. J Phys Chem C 113(27):11856–11860
90. Balint S et al (2009) Simple route for preparing optically trappable probes for surface-enhanced Raman scattering. J Phys Chem C 113(41):17724–17729
91. Jordan P et al (2005) Surface-enhanced resonance Raman scattering in optical tweezers using co-axial second harmonic generation. Opt Express 13(11):4148–4153
92. Tong LM et al (2009) Optical aggregation of metal nanoparticles in a microfluidic channel for surface-enhanced Raman scattering analysis. Lab Chip 9(2):193–195
93. Kneipp J, Kneipp H, Kneipp K (2007) Surface-enhanced Raman spectroscopy-based optical labels deliver chemical information from live cells. In: Kneipp K et al (eds) New approaches in biomedical spectroscopy. Amer Chem Soc, Washington, pp 186–199

Portable SERS Sensor for Sensitive Detection of Food-Borne Pathogens

19

Hongxia Xu, Michael Y. Sha, Remy Cromer, Sharron G. Penn, Ed Holland, Gabriela Chakarova, and Michael J. Natan

1 Definition of the Topic

A prototype portable Raman spectrometer has been developed for detection of microbial DNA. The pathogen genomic DNA was captured by probe conjugated magnetic bead and detected by probe conjugated SERS tag under Raman spectrometer.

2 Overview

Nanoplex^{TM} biotags are silica-coated gold nanoparticles that, by virtue of surface-enhanced Raman scattering (SERS) active molecules adsorbed to the metal surface, can be used in numerous assay formats for analyte quantification. Multiple, unique-signal tags can be generated by varying the adsorbed Raman active molecules, thus allowing extensive multiplexed detection to be performed. Importantly, because the tags are excited at near-IR wavelengths, the optical output generated is unaffected by biological matrices offering robust measurement in crude lysates and impure/semi-pure DNA preparations. A novel assay comprising Nanoplex™ biotags, magnetic beads, and a prototype portable Raman spectrometer has been developed for rapid and simultaneous detection of enteric food-borne bacterial pathogens such as *Escherichia coli* O157:H7 and *Salmonella* species with good analytical sensitivity, accuracy, and specificity. Sensitivity in the low picomolar range was obtained with a single capture probe for detection of the Stx1 gene of *E.coli* O157:H7 in a model system. The assay sensitivity was greatly improved to low femtomolar range in real genomic DNA samples by using a multiple probes

H. Xu (✉) • M.Y. Sha (✉) • R. Cromer • S.G. Penn • E. Holland • G. Chakarova • M.J. Natan
Oxonica Inc, 330 Middlefield Rd, Mountain View, CA, USA

Challa S. S. R. Kumar (ed.), *Raman Spectroscopy for Nanomaterials Characterization*,
DOI 10.1007/978-3-642-20620-7_19, © Springer-Verlag Berlin Heidelberg 2012

approach without the aid of amplification. The simplicity of the Nanoplex™ biotag DNA direct assay and the minimal level of operator intervention and training required to achieve exquisitely sensitive DNA detection offers enormous benefit to the nucleic acid testing in the demanding area, such as point-of-care, on-site food quality diagnostics, or field GMO testing.

3 Introduction

Raman scattering is a laser-based optical spectroscopy that, for molecules, generates a fingerprint-like vibrational spectrum with features that are much narrower than fluorescence. Raman scattering can be excited using monochromatic far-red or near-IR light that has photon energies too low to excite the inherent background fluorescence in biological samples. However, Raman scattering is an extremely inefficient process with low scattering cross section. After surface-enhanced Raman scattering (SERS) was recognized 30 year ago, [1–3] which the SERS detection sensitivity is enhanced up to 10^6–10^{12} order of magnitude over the conventional spectroscopy by coupling Raman active molecules to noble metal particles, SERS has been demonstrated to be a powerful analytical tool. Several characteristics of SERS make it an attractive method for the detection of biological samples. Because of the uniqueness of molecular Raman spectra, it is possible to detect several molecular species simultaneously and enable viable multiplexed detection [4, 5]. As Raman scattering, SERS is resistant to photobleaching [6]. It has also been demonstrated that SERS has lower limits of detection even down to the single-molecule level [7–9], by three orders of magnitude in comparison to fluorescence with same commercially available fluorophore [10], and is 200-fold brighter with 20–30 times narrower emission peak than another commonly used nanotag, quantum dot [11]. Particular advantages over traditional assay detection technologies with fluorescence and chemiluminescence are that SERS are excited at near-IR wavelengths, so the optical output generated is unaffected by biological matrices such as blood, tissue, crude lysates, and impure DNA preparations, offering robust measurement in raw biological samples, largely reducing the cost and labor associated with sample preparation and simplifying the assay procedures. With these exquisite sensitivities, SERS transforms Raman into a useful optical label for applications in life science and diagnostics.

Healthcare and disease management rely heavily on diagnostics. To meet the overwhelming demand for diagnostics, SERS has been extensively applied in many areas, such as cancer diagnostics [12], pathogen and biothreat detection [13], and with various techniques, such as in vitro glucose biosensor, [14] real-time in vivo SERS glucose sensing [15], DNA and RNA detection [4, 16, 17], protein immunoassay [18–20], enzyme activity [21, 22], pH measure in cells [23], cell imaging [24–27], tissue immunohistrochemistry [28], lateral flow immunoassay [29], and in vivo tumor detection [11]. Moreover, the unique biological SERS spectra have also been applied for identification of pathogens [30–36] or detection of biological substrates, such as creatinine in human serum or urine samples [37, 38]. However, the robustness, reproducibility, and throughput still remain challenges associated with the label-free and direct detections.

Molecular diagnostics mainly focuses on DNA detection using technology such as DNA hybridization, microarray, PCR, or DNA sequencing. The detection targets include food-borne pathogens, human, plant and animal pathogens, biothreats, and even genetically modified organism (GMO), drugs, and substances of interest in forensics. Since the completion of the sequencing of the human genome was announced [39], DNA microarrays have been increasingly used for many application in molecular diagnostics. The common types of microarrays are either solid (planar)-based or liquid (Bead)–based. Planar microarrays offer the potential for simultaneous detection of many pathogens that are of interest to public health, medicine, homeland security, and veterinary diagnostics. However, its high cost and limited sample throughput render it less attractive in comparison to bead-based arrays [40]. Bead array technology has been widely used for gene expression, DNA methylation, immunoassay, and SNP genotyping [41–47]. A few publications reported the bead-based assay applied in pathogen detection [48–51].

Food-borne and diarrheal diseases are the most common infectious diseases. An estimated 76 million cases of food borne disease occur each year in the United States. The food borne diseases cause suffering and death and impose a financial burden on society. The food borne pathogen *Escherichia coli* O157:H7 (*E. coli* 0157:H7) and *Salmonella* spp were chosen as the targets in our assay system, because both of them are the major causative agents of food borne infectious diseases, and are actively monitored by the Centers for Disease Control & Prevention (CDC) [52–54]. The traditional methods for pathogen detection are labor intensive and time consuming. Here, we report a novel assay comprising Nanoplex™ biotags, magnetic beads, and a portable Raman spectrometer for rapid, sensitive identification of the food-borne pathogens. The assay sensitivity was further improved to low femtomolar range in real genomic DNA samples by using a multiple probes approach without the aid of amplification. Thus, it increased assay sensitivity without introducing amplification bias.

The demands for medical diagnostics with high quality, more information, and reduced cost have been a significant driver for multiplexing assays over single-analyte assays [41, 55]. In recent years, eight multiplexed diagnostic products have been cleared by the US FDA, of which half were molecular diagnostic-based [41]. Several groups reported a SERS-based multiplexed assay for DNA detection using fluorescent dyes as Raman labels, [4, 56] but the far broader range of non-fluorescent Raman tags has been left fairly unexplored. Sun et al. attached the DNA probe and non-fluorescent Raman tags to the surface of gold nanoparticles (DNA-AuP-Rtag) and indicated the possibility for multiplexed detection of complementary DNA targets in solution [57]. However, it remains a challenge to control the loading of non-fluorescent tags and DNA probes on the nanoparticle media. Unaddressed, this issue may cause high CVs and low reproducibility of the assay. To demonstrate the multiplexing capability of Nanoplex biotags, we developed a multiplexed assay for detection of *E. coli* and *S. enteriencic* simultaneously. The assay is rapid with high sensitivity and selectivity, and capable of integration into a portable instrument. The simplicity of the Nanoplex™ biotag technology (called DNA direct assay in this application) and the minimal level of operator intervention and training required to achieve exquisitely sensitive DNA detection offers enormous benefit to the nucleic acid testing in demanding areas, such as point-of-care, on-site food quality diagnostics, or field GMO testing.

4 Experimental and Instrumental Methodology

4.1 Genomic DNA Fragmentation

Escherichia coli O157:H7 genomic DNA (ATCC# 700927D) was purchased from American Type Culture Collection (Manassas, VA). 1 microliters 10X FastDigest buffer and 1 microliters FastDigest enzyme RsaI (One FastDigest® Unit/microliters, Fermentas, MD) were added to the 8 microliters (200 ng/microliters) of genomic DNA sample, mixed gently and incubated at 37°C for 5 min to digest DNA. Digested DNA fragments were denatured by boiling for 3 min and chilling quickly on ice.

4.2 Probes Design and Synthesis

The coding regions of *E. coli* O157:H7 Stx1 gene and *Salmonella enterica* invA gene sequences have been chosen as the target sequences. The pair of capture and detection probes were designed according to the target sequences using the Oligo 6 software (Molecular Biology Insight Inc., CO). Each capture probe was synthesized with an amine group at its 5′ end and a biotin group was attached to the 3′ end of each detection probe. Table 19.1 lists the sequence details of each probe.

4.3 Nanoplex™ Biotag Synthesis and Functionalization

The synthesis of the Nanoplex™ biotags has been described by Mulvaney et al. [5] and Freeman et al. [58]. Briefly, a spherical Au colloid, generally about 50 nm in diameter, is coated first with a layer containing both a glass precursor silane and the Raman label molecule. A thin (few nanometers) silica coating is grown then by slow polymerization of Na silicate providing encapsulation of the Raman reporter molecule on the gold surface. The thin glass shell is subsequently thickened up to 20–30 nm by using a mixture of Ethanol, water, tetraethylorthosilicate (TEOS), and Ammonia. To facilitate the use of these nanoparticles as biological tags, the glass surface is further functionalized with bioreactive groups such as thiols or amines. The Nanoplex™ biotags used in this work were thiolated by replacing 5% of the TEOS with mercaptopropyltrimethoxysilane (MPTMS). A variety of label molecules have been used for the synthesis of different flavors of Nanoplex™ biotags, but for demonstrating their multiplex capability in the DNA direct assay only two flavors were used: SERS-420 (4, 4′-dipyridyl) and SERS-421 (d8-4,4′dipyridyl).

Table 19.1 Probe design from *E. coli* O157:H7 Stx1 and *S. enterica* invA Gene

Probe Name	*Sequence*	*Tm (C)*
Stx1-B1	amine-C12-TTTTTTTTTTTTTTTTTT CTTTTTTCTTTGTTAT CTTTTCAGTTA	68
Stx1-S1_biotin	TACCTTAGACTTCTCGACTGCAAAGACGTA TTTTTTTTTTTTTTTTTT-C12-biotin	79
Stx1-T1	CATACGTCTTTGCAGTCGAGAAGTCTAAGGTAAATTCC TTCGCAACCACATTAACTGAAAAGATAACAAAGAAAAAAGTT	85
Stx1-B2	amine-C12-TTTTTTTTTTTTTTTTTT CTGATGATTGATAGTGG CACAGGGGA	79
Stx1-S2_biotin	TTGTTTGCAG TTGATGTCAG AGGGATAGA TTTTTTTTTTTTTTTTTT-C12-biotin	77
Stx1-T2	GGATCTATCCCTCTGACATCAACTGCAAACAAATTAT CCCCTGTGCCACTATCAATCATCAGTA	87
Stx1-B3	amine-C12-TTTTTTTTTTTTTTTTTT AGCGGTTACATTGTCTGG TGACAGTAG	79
Stx1-S3_biotin	GTTACAGCGTGTTGCAGGGATCAG TTTTTTTTTTTTTTTTTT-C12-biotin	79
Stx1-T3	CGACTGATCCCTGCAACACGCTGTAACGTGGTATAGCT ACTGTCACCAGACAATGTAACCGCTGTTG	91
Stx1-B4	amine-C12-TTTTTTTTTTTTTTTTTT ATCGCCATTCGTTGACT ACTTCTTATCT	75
Stx1-S4_biotin	TGTCGCATAGTGGAACCTCACTG TTTTTTTTTTTTTTTTTT-C12-biotin	78
Stex1-T4	GTCAGTGAGGTTCCACTATGCGACATTAAATCCAGATAAGAA GTAGTCAACGAATGGCGATT	87
Stx1-B5	amine-C12-TTTTTTTTTTTTTTTTTT TGGATGATCTCAGT GGGCGTTCTT	77
Stx1-S5_biotin	ATCTTACATTGAACTGGGGAAGGTTG TTTTTTTTTTTTTTTTTT-C12-biotin	76
Stx1-T5	CTCAACCTTCCCCAGTTCAATGTAAGATCAACATC TTCAGCAGTCATTACATAAGAACGCCCACTGAGATCATCCAGT	89
invA-B1	amine-C12-TTTTTTTTTTTTTTTTTT CTGTCTACTTATACCA TGCTGACCATTGGT	79
invA-S1_Dig	TTGTCGCCCAGATCCCCGCATTG TTTTTTTTTTTTTTTTTT-C12-Dig	80
invA-T1	AACAATGCGGGGATCTGGGCGACAAGACCATCACCAATG GTCAGCATGGTATAAGTAGACAGAG	90

[a]Stx1 sequence comes from *E. coli* 0157:H7 stx1 gene. GenBank accession number: AB083044; invA sequence came from *S. enterica* invasion protein gene. GenBank accession number: U43238
[b]Dig: digoxigenin

4.4 Conjugation of NeutrAvidin to SERS Nanoplex™ Biotags

1.5 mL of thiol-modified Nanoplex™ biotags (OD 24) was mixed with 250 µl of 50 mM phosphate buffer (pH 7.15) in a 2 mL Eppendorf tube. 120 µl of NeutrAvidin (5 mg/mL) and 120 µl of freshly prepared sulfo-SMCC (5 mg/mL in H_2O) were then added to the tube, and reacted with gentle mixing for 2.5 h at room temperature. After incubation, 100 μl of a solution consisting of 10% BSA and 30 mg/mL MESA (in conjugation buffer) was added to the reaction tube and incubated at room temperature for 45 min to quench the excess maleimide groups. To purify the conjugates, the reaction mixture was centrifuged at 1,000 RCF for 15 min. After removing the supernatant, the pellet was resuspended in 1.5 mL of resuspension buffer (20 mM phosphate, pH 7.5/0.1% BSA/0.05% sodium azide). The centrifugation/resuspension steps were repeated four times and the particles were then resuspend to desired concentration (OD 24).

4.5 Magnetic Microsphere Conjugation

A 40 microliters aliquot (8×10^7beads) of 2.8 micrometers magnetic Dynabeads® M-270 carboxylic acid (Invitrogen Inc. CA) was rinsed with 50 mM MES (pH 4.5) twice by magnetic pull in a magnetic bead concentrator. The beads were resuspend within 163 microliters of 50 mM MES (pH 4.5) and 8 microliters 100 miocromole oligo (800 pmole) was then added to the tube. The mixture was kept on ice. A 15 microliters of 20% EDC (1-Ethyl-3-[3-dimethylaminopropyl]carbodiimide Hydrochloride, Pierce Biotechnology, Inc. IL) in 50 mM MES (pH 7.0) was added to the beads/oligo solution. The solution was briefly vortexed and incubated in a shaker for 1 h at 4°C. The oligo conjugated beads were rinsed with 500 microliters of 10 mM PBS four times and resuspended in 100 microliters of 10 mM PBS.

4.6 DNA Hybridization

Five microliters of target oligo or denatured fragmented genomic DNA at various concentrations, 2 microliters (1.6×10^6) magnetic Dynabead capture probe, and 5 microliters 1 miocromole biotinylated detection probe were added to 38 microliters HS 114 buffer (Molecular Research Center, Inc. Cincinnati, OH 45212) and hybridized at 55°C for 20 min. The magnetic beads carrying the hybrids were pulled down by a magnetic pole piece and the hybridization buffer was removed. A 40 microliters (OD 0.6) NeutrAvidin conjugated Nanoplex™ biotag S420 was then added to the pellet, briefly mixed and incubated at room temperature for 10 min. After rising with PBS once, the Raman signal was acquired with a in house built Raman reader (called Nanoplex™ Reader) and analyzed with SenserSee ™ software.

4.7 Raman Instrument Nanoplex™ Reader Design

A Raman spectrometer instrument compatible with the assay format, consisting of an optical system and sample tube holder was developed in house. The sample holder includes a magnetic assembly that produces a focused magnetic field at a precise location on the wall of the sample tube. This location is coincident with the focal point of the Raman optical sensor. In this way, magnetic particles are preconcentrated at this location on the tube such that quantitative measurement of Nanoplex™ biotags associated with the magnetic beads is facilitated. The optical measurement system comprises a Raman spectrometer with an excitation wavelength of 785 nm (Ahura Corp. ECL 785-325 laser module) configured to deliver approximately 20 mW power. An Inphotonics filter head and Ocean optics QE65000 spectrometer complete the optical train.

5 Key Research Findings

A biosensor is an analytical device which can be classified by their bioreceptor such as DNA or antibody or their transducer such as optical, mass-based, or electrochemical [59]. To date, biosensor system cannot detect bacteria at concentration below 10 CFU/mL without bacteria preenrichment of the sample [60]. However, the integration of nanotechnology such as nanoparticles (NPs) into biosensor has held a great promise. Optical transduction can achieve assay LOD from 1 to 1×10^4 CFU/mL, magnetic transduction from 10 to 1×10^5 CFU/mL and electrochemical transduction can get 10 to 4.2×10^2 CFU/mL. However, applying NPs to achieve multiple bacteria identification using biosensor has not yet been demonstrated [60]. In order to meet this need (multiplex) for clinical diagnostics, recently we synthesized multiple SERS tags and had successful applied to in vivo tumor detection [61] and circulation tumor cell detection from the whole blood [62]. In this chapter, we tried to apply these SERS tags to detect multiple pathogens from same sample simultaneously. By applying multiple probes strategy, we demonstrated 1,000-fold increase its sensitivity, its LOD down to 0.001 fM (6×10^2 CFU/mL) by using five DNA probes. More importantly, we also demonstrated that detection *E. coli* and *S. enterica* simultaneously from single sample.

5.1 SERS Tags and Its Detection Instrumentation

Oxonica's Nanoplex™ biotags used in this chapter are comprised of Au nanoparticles, a submonolayer of a reporter molecule adsorbed to the metal surface, and a silica encapsulant [5]. Figure 19.1a shows a TEM image of Nanoplex™ biotags; the gold core about 50 nm in diameter and the silica shell about 20 nm thick. Figure 19.1b is an illustration of the Nanoplex™ biotags. The SERS biotags carry numerous advantages in bioassay applications. Unlike quantum dots, where different sizes of particles have different emission spectra, SERS spectra are independent of particle size although optimized particle size gave higher signal park. Their uniformity (5–10% CV) insures that all tags react similarly including having the same diffusion coefficient. Since SERS spectra are narrow, multiple, unique-signal tags can be generated by varying the adsorbed reporter molecules, thus allowing extensive multiplexed detection to be performed. Figure 19.1c presents Raman spectra from seven different *Nanoplex ™ biotags* illustrating their

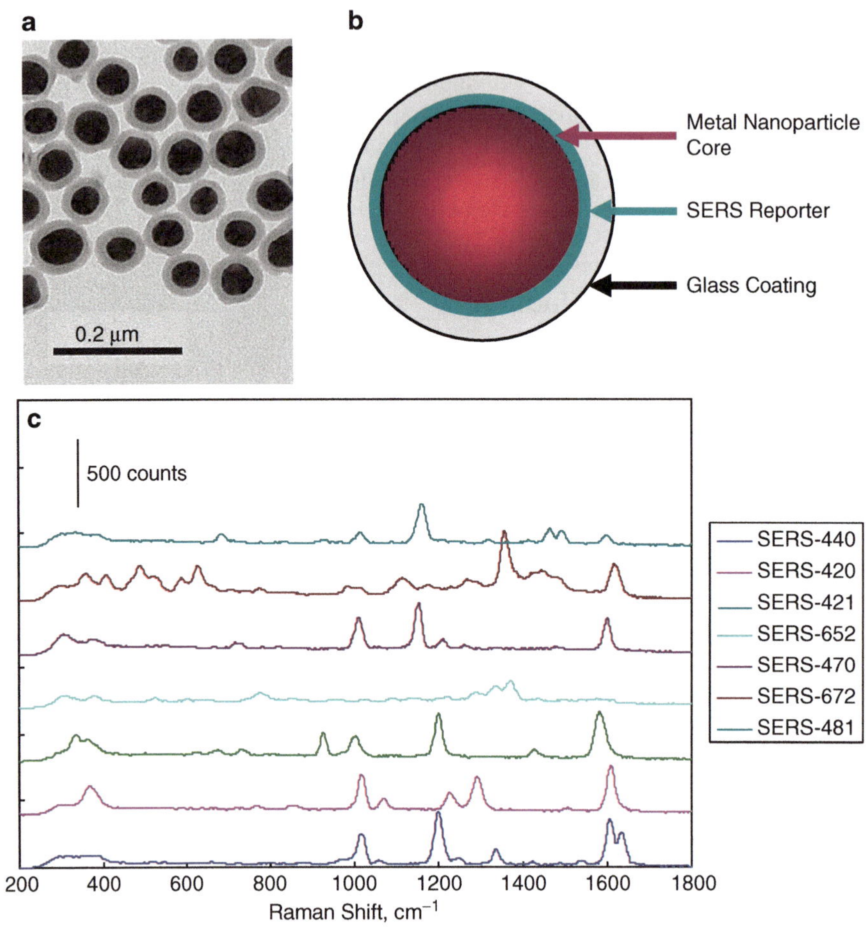

Fig. 19.1 Nanoplex^{TM} biotags. (**a**) TEM image of Nanoplex biotags showing metal core and glass coating. (**b**) Cartoon of a Nanoplex biotag. Metal core responsible for Raman enhancement (~50 nm size) *Thin layer* of label molecules (SERS report) responsible for unique spectrum. Anchored silica shell (~20 nm size) provides stability and sites for biomolecular attachment. (**c**) Raman spectra of multiple Nanoplex biotags

unique Raman spectra. The silica encapsulation provides the physical and optical robustness of SERS tags, such as stability with respect to changes in temperature, pH, ionic strength, with no apparent photobleaching. Such properties are enormous benefits to bioassay development. In addition, the silica shell is an excellent surface for biomolecular attachment. Importantly, because the tags are excited at near-IR wavelengths in which assay reagents and media exhibit low inherent Raman scattering/fluorescence background, the optical output generated by SERS tags is unaffected by biological matrices [62]. Thus as a sensing technique, Raman spectroscopy as a measurement of SERS tags coupled with magnetic bead provides

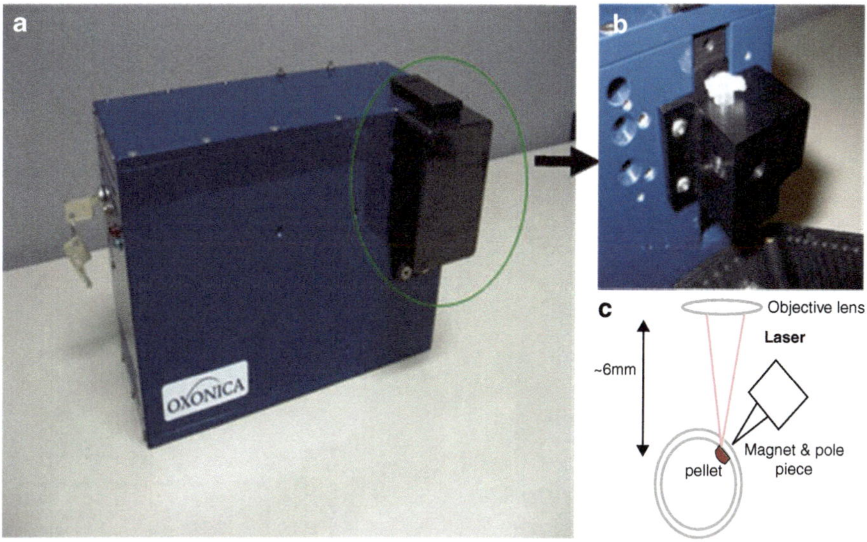

Fig. 19.2 Raman instrument Nanoplex™ Reader. (**a**) the portable Raman instrument. (**b**) the magnet sample holder. (**c**) Line drawing illustrating sample holder function

a robust and sensitive technique for a no-wash homogenous assay in a diverse range of matrices, for example, whole blood, tissue, milk, urine, crude lysates, and impure/semi-pure DNA preparations.

In order to meet the need for point-of-care or point-of-field, obviously a portable Raman sensor is needed for data collection. A Raman instrument (Nanoplex™ Reader) was designed to provide an inexpensive and user-friendly means to acquire Raman spectra. It includes a Raman spectrometer and a magnetic sample holder (Fig. 19.2). Figure 19.2a is a picture of the portable instrument developed in house. It is 8″×5″×3″ in size and less than 4 lbs. The excitation spot size of the spectrometer can be set to 150 or 300 μ depending on the magnetic bead pellet size by substitution of alternative optical fibers, and excitation power is continually adjustable from 5 to 200 mW. Figure 19.2b shows that a reaction tube is inserted into the sample holder in the process of Raman signal acquisition. Figure 19.2c further illustrates the process. The magnetic beads binding with SERS tags are pulled down and pelleted by a magnetic pole piece. By specific arrangement of the sample holder design, the pellet is formed at the focal plane of the spectrometer's objective lens. Thus Raman signals from SERS tags may be recorded from a pellet of concentrated reagents in a consistent manner.

5.2 SERS-Based Microsphere Array in an Oligo Model System

Food-borne diseases are the most common infectious diseases. *E. coli* 0157:H7 is a major food-borne pathogen that causes diarrhea, hemorrhagic colitis, and hemolytic uremic syndrome and is actively monitored by the Centers for Disease Control

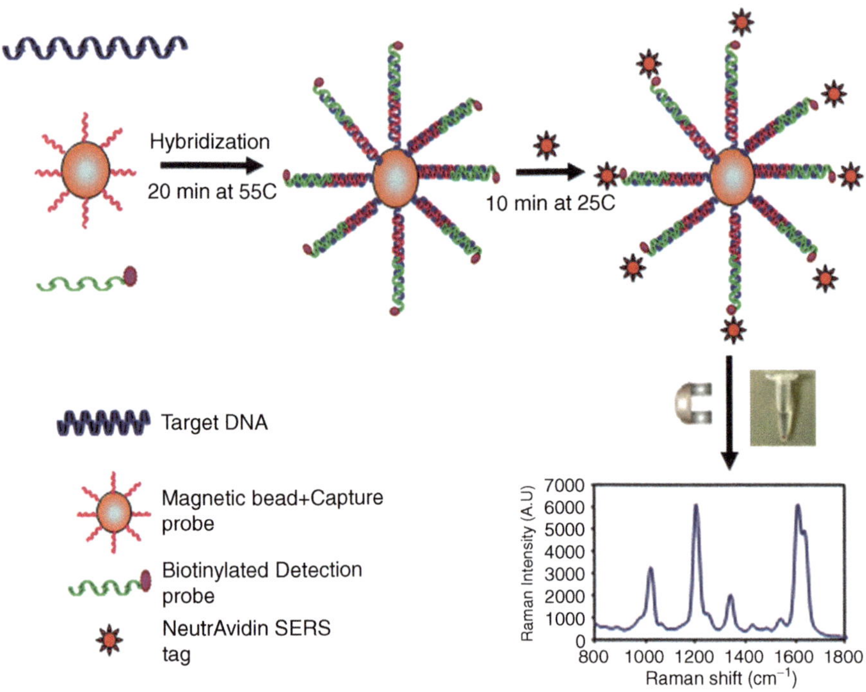

Fig. 19.3 Schematic of an approach for rapid detection of pathogen DNA sequence by SERS-based microsphere array

& Prevention (CDC). The complete genome of *E. coli* O157:H7 is about 5.5 Mb [63]. The most clinical signs of the disease arise as consequence of the production of Shiga-like toxin 1 (Stx1), Stx2, or the combination of these toxins. In this chapter, *E. coli* O157:H7 Stx1 gene sequence thus was chosen as a target sequence. In order to obtain a rapid, sensitive, and specific detection of this bacterial pathogen, we developed an assay platform comprising Nanoplex™ biotags-based microsphere array and a portable Raman spectrometer. Figure 19.3 illustrates the procedure of this assay. Target sequences were first hybridized to magnetic beads-capture probes and biotinylated detection probes in one microtube. Pulling down magnetic beads carrying the hybrids allowed the separation of target sequence from noncomplementary sequences in the sample. The hybrids formed by the target sequence, capture probe and biotinylated detection probe were then bound to NeutrAvidin coated Nanoplex™ biotags. The resultant Raman signal from Nanoplex biotags attached to magnetic beads carrying the hybrids was aquired with a protable Raman Spectrometer as Fig. 19.2 describes. In the initial format, single oligo capture and detection probe was applied to the assay. A capture probe Stx1-B1 and detection probe Stx1-S1-biotin were designed for detecting Stx1 gene of *E. coli* O157:H7 (Table 19.1). To investigate the sensitivity of the assay, the *E. coli* oligo target Stx1-T1 was serially diluted to 1,000, 100, 10, 1 pM, and 0 pM as

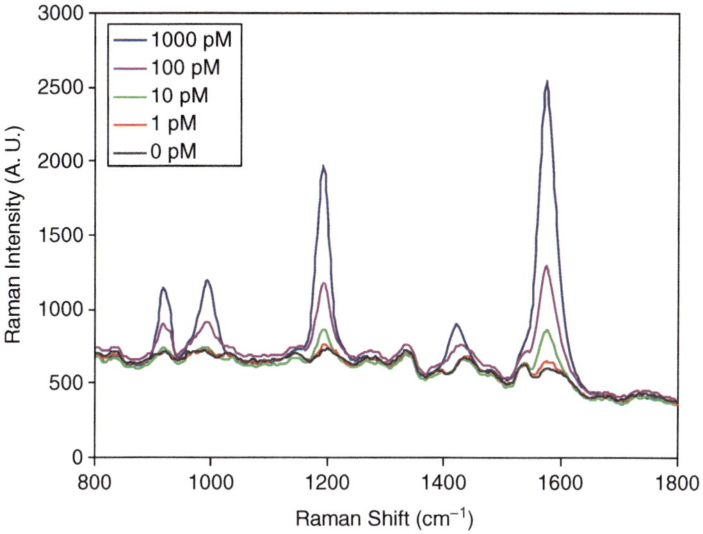

Fig. 19.4 Detection of *E. coli* O157:H7 Stx1 using Nanoplex™ biotag S421. A tenfold dilution series from 1,000 to 1 pM were applied to *E. coli* O157:H7 Stx1 target oligo. The Raman spectra of Nanoplex biotag S421 were collected from the target Stx1 at various concentrations and indicated by a particular color

negative control and hybridized to capture probe conjugated magnetic bead and biotinylated detection probe, then incubated with NeurtrAvidin-Nanoplex™ biotag S421. The Raman signal was acquired from the pulled-down, pelleted magnetic beads. Figure 19.4 shows the Raman spectra collected from the oligo target Stx1-T1 at various concentrations and indicates a full quantitative titration demonstrating expected dose response. Spanning over four orders of magnitude, the assay has a good dynamic range and remains linear at the low concentration range (also see Fig. 19.6a). The negative control has very little background in 1,500–1,600 Raman Shift although same account Biotag was used in the assay. The limit of detection for Stx1-T1 synthetic oligonucleotide was found to be 2 pM with 99.7% confidence (cut off at background + 3 standard deviations) based on experiments run in triplicate.

5.3 Multiple Probes Approach for Increasing the Assay Sensitivity

Unlike the conventional flow cytometry which scans the signal of individual bead, the Nanoplex™ Reader Raman Instrument acquires a total signal from the whole population of the beads within the laser spot (150 μm). Thus, increasing the portion of the beads that carry the specific signal in the laser spot area should improve the limit of detection of the assay. A multiple probe scheme is created to facilitate this characteristic (Fig. 19.5). In the assay, target DNA was first chopped into small

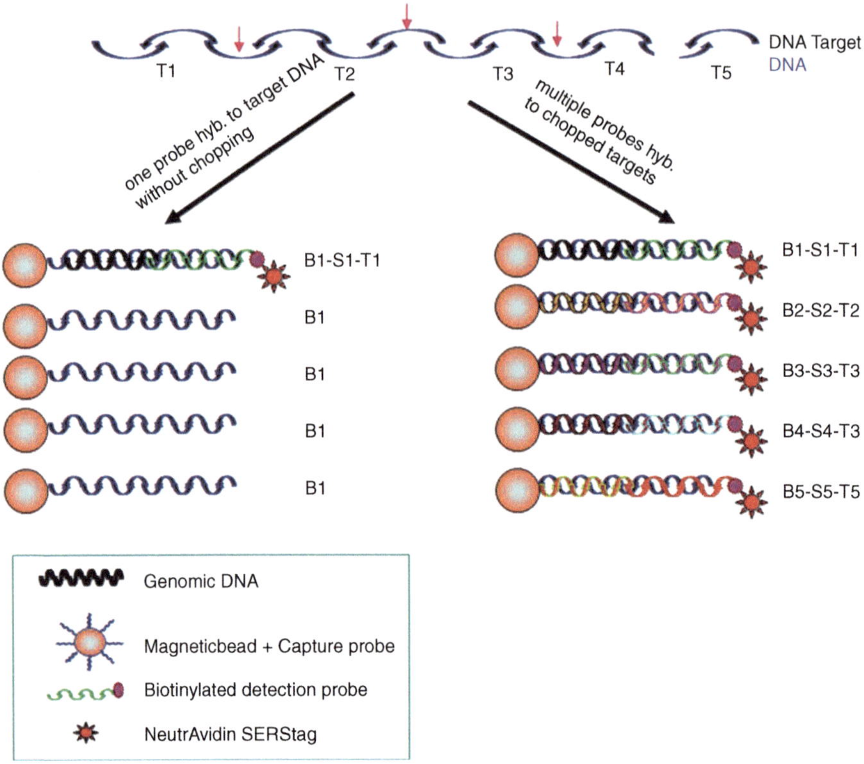

Fig. 19.5 Multiple probes design for path to increased assay sensitivity. Leverage ability to use multiple sequences from a single gene, for increased sensitivity. The *left panel* illustrates the single probe approach. The whole gene sequence is hybridized to a single capture and detection probe. The *right panel* illustrates a multiple probes approach. The target DNA is chopped into five fragments by restriction enzyme. Five different capture probes and detection probes hybridized to its complementary target. *B1* capture probe conjugated to magnetic bead, *S1* detection probe bond to SERS tag, *T1* target fragment. B1-S1-T1 is a pair of capture and detection probes which hybridized to the T1 DNA fragment

pieces by a fast frequent restriction enzyme. Thus, the target DNA concentration (fragment number) was increased accordingly. The multiple probes were then designed according to the fragmented target gene sequences to take advantage of the increased DNA fragments generated by enzyme cutting and each probe will hybrid to its complementary fragment of the target gene. For example, there will be five beads show Biotag signal by this strategy when using five probes (S1–S5, Fig. 19.5, right panel). However, there is only one bead show Biotag signal when using only one probe for detection whole gene sequence. In addition, smaller pareses of DNA sequences are favorable for hybridization efficiency. Therefore, this strategy approach can increase the assay sensitivity. In the multiple probes experiment, each probe was conjugated to the magnetic beads separately, pooled, and hybridized to the digested DNA fragments. In this way, the number of beads carrying positive signal in the laser spot area is greatly increased, particularly in the lower target concentration, thus causing the absolute signal to be further enhanced. Consequently, the limit of detection of the assay will benefit from the modality. Figure 19.6 shows results from comparative experiments using single verses multiple probes in the oligo model system. Figure 19.6a represents the single probe approach. The small window indicates the linear range of the titration and its LOD, about 1.4 pM. Figure 19.6b shows the multiple probes approach. As expected, the hybridization signal was tremendously increased by combining five different oligonucleotide probes and corresponding targets compared to the single probe at the same condition. The small window (Fig. 19.6b) indicates linear signal correlation and its LOD is less than 1 fM with dynamic range greater than 5 log. With the multiple probe approach, the sensitivity of the assay appears to succeed the current technologies, such as SERS PDMS microfluidic chip (LOD 10 pM), [16] bead-based lab-on-a chip (LOD 20.6 nM), [51] gold nanoparticles probe for human genomic DNA (LOD 50 fM), [64] Ag enhancing gold nanoparticles probe for oligonucleotide target (LOD 10–20 fM), [4, 65] and Qbead technology with RNA amplification (LOD 1–100 fM) [42].

5.4 Detection of Stx1 Gene from *E. coli* O157:H7 Genomic DNA

A factor as important as sensitivity is the ability for an assay to work with real world samples rather than synthetic oligonucleotide targets. Therefore, the assay was performed by using *E. coli* O157:H7 genomic DNA from ATCC. Genomic DNA was serially diluted, followed by fragmentation using a fast restriction enzyme and hybridized with a mixture of multiple probes (five probes). The result showed that *E. coli* O157:H7 genomic DNA at 1 fM, 0.1 fM, 0.01 fM or even 0.001 fM (6×10^2 CFU/mL) was detectable when the multiple probes technique was applied (Fig. 19.7). Compared to the oligo model system, sensitivity for genomic DNA seem to be 10–100 fold improved. One of the explanations for this is that the higher copy number of the Stx1 gene in this strain (*E. coli* O157:H7) [66] may be the result of more available target number for hybridization, which resulted in detectable signal even in 0.001 fM. This phenomenon also was observed for genetics modify organism (GMO) detection (unpublished data). This assay shows that the limit of detection of the assay is 6×10^2–6×10^3 CFU/mL, similar to or more sensitive than the current amplification related methods, such as microarray with target amplification (10^4–10^6 CFU/mL), [67, 68] PCR/gel electrophoresis (10^5–10^6 CFU/mL) [69], fiber-optical DNA microarray (10^3–10^4 CFU/mL) [49], or real-time PCR (10^3–10^4 CFU/mL) [9, 70]. This SERS biotag assay requires the minimal level of operator intervention, allows the untrained and inexperienced personnel to perform, and achieves exquisitely sensitive DNA detection. This offers enormous benefit to nucleic acid testing in demanding areas, such as point-of-care diagnostics, on-site food safety testing or field GMO testing. The rapid nature of the test (less than 1 h) offers advantages over today's technologies including being amenable for use in an intra operative setting.

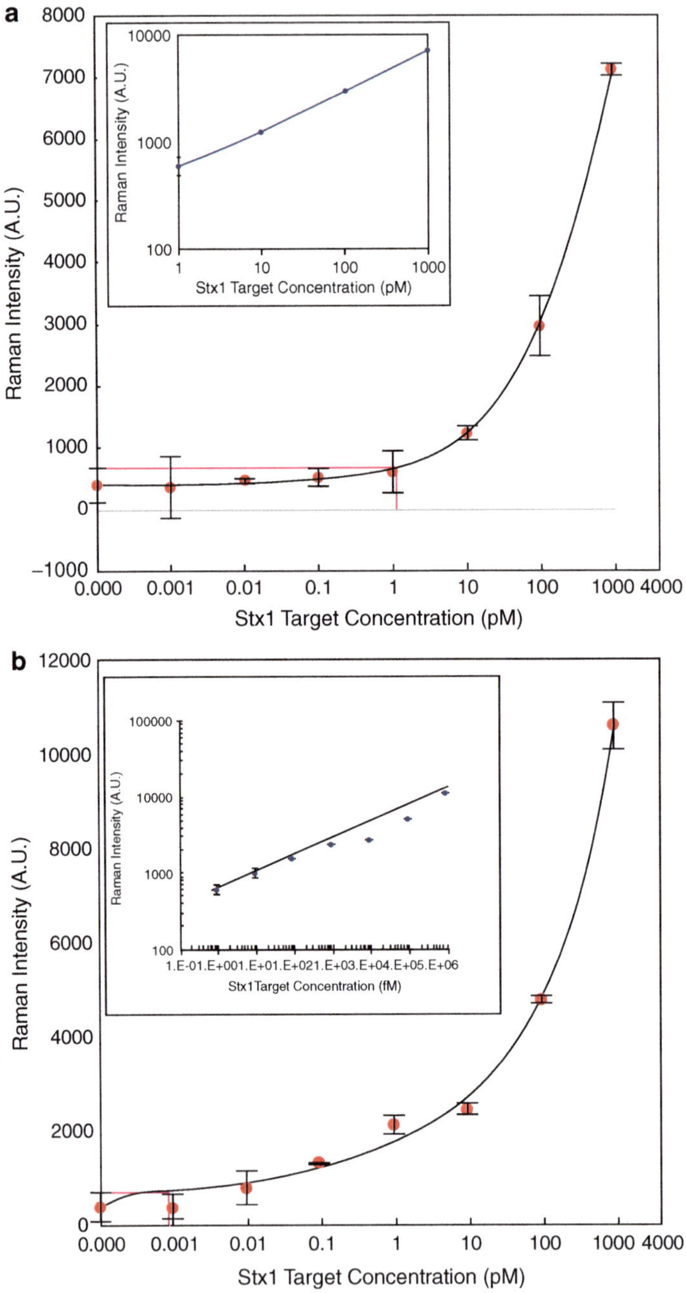

Fig. 19.6 Comparison of assay sensitivity by single and multiple probes. (**a**) Single probe predicted LOD at 1.4 pM (*pine color line*) with 99.7% confidence. (**b**) Multiple probes predicted LOD at 0.88 fM with 99.7% confidence. *Error bars* represent ±1SD, $n = 3$

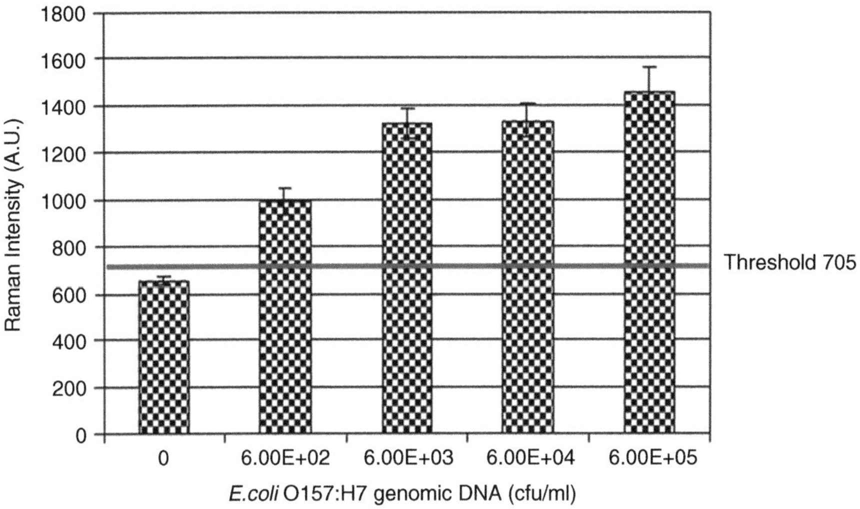

Fig. 19.7 Detection of Stx1 nucleotide sequence from *E. coli* O157:H7 genomic DNA sample. Threshold = Average background signal + 3 standard deviations (99.7% confidence). *Error bars* represent ±1SD, $n = 3$

5.5 Possibility of Multiplexed Assay for Detection of *E. coli* O157:H7 and *S. enterica* DNA

To further illustrate the detection capability of Nanoplex™ biotags, the assay was extended to detect *E. coli* O157:H7 and *S. enterica* simultaneously. In order to prove of the concept, we used single probe for each pathogen in following test. In the multiplexed assay, the capture probe (Stx1-B1), biotinlyated detection probe (Stx1-S1_biotin), and NeutrAvidin-S421 tag were applied to identify *E. coli* O157:H7 *Stx1*. The capture probe (invA-B1), Digoxigenin (Dig) labeled detection probe (invA S1-Dig), anti-dig-S420 tag were used to identify *S. enterica* invA (Table 19.1). After hybridization, the magnetic beads carrying the hybrids were concentrated and the Raman signal was acquired, deconvoluted, and analyzed with in-house software SENSERSee™. Figure 19.8a illustrates the quantitative results from the reactions where either no target was added or 1,000 pM amount of target was added to each tube. The Raman signal for *S. enterica* invA (S420) and for *E. coli* Stx1 (S421) was 2,500 counts and 240 counts, respectively when 1,000 pM invA and 0 pM Stx1 targets were present. The Stx1 signal was equivalent to the level of negative control. Similarly, when 0 pM invA and 1,000 pM Stx1 targets were present, the panels were reversed (300 counts and 3,600 counts). Furthermore, *S. enterica* invA (S420) and *E. coli* Stx1 (S421) were detected at 2,600 and 3,500 counts simultaneously when both targets were present. The data indicate that the positive signal is gene sequence-specific and demonstrates the possibility of multiplexed pathogen detection. With current availability of many different SERS tags in house, we hypothesize that other high risk food-borne pathogens such as

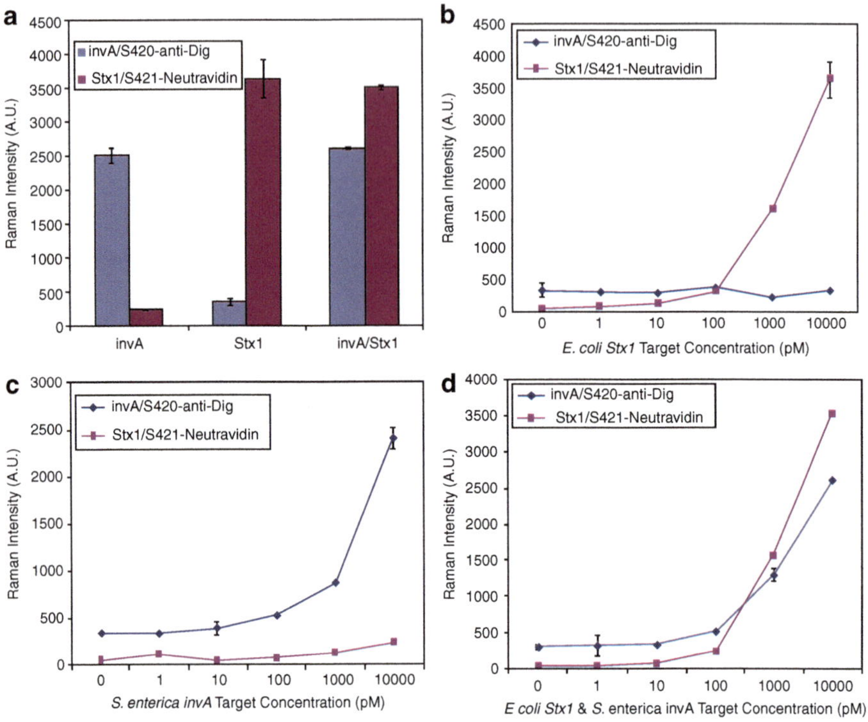

Fig. 19.8 Multiplexed assay for detection of Stx1 from *E. coli* O157:H7 and invA from *S. enterica* in an oligo model system. In the experiments, Stx1 detection probe was labeled with Nanoplex biotag S421 and invA detection probe was labeled with Nanoplex biotag S420. (**a**) Quantitative analysis for multiplexed detection of *E. coli* Stx1 target and *S. enterica* invA target. *Error bars* represent ±1SD, n = 3; B-D: Titration of multiplexed assay for detection of Stx1 from *E. coli* O157:H7 and invA from *S. enterica* using single probe approach in an oligo model system. Stx1 detection probe was labeled with Nanoplex biotag S421 and invA detection probe was labeled with Nanoplex biotag S420. (**b**) S420 (*blue line*) and S421 (*pink line*) Raman signal was acquired from the tubes containing 0, 1, 10, 100, and 1,000 pM *E. coli* Stx1 target alone. (**c**) S420 (*blue line*) and S421 (*pink line*) Raman signal was acquired from the tubes containing 0, 1, 10, 100, and 1,000 pM *S. enterica* invA target alone. (**d**) S420 (*blue line*) and S421 (*pink line*) Raman signal was acquired from the tubes containing both *E. coli* Stx1 and *S. enterica* invA targets at 0, 1, 10, 100, and 1,000 pM. *Error bars* represent ±1SD, $n = 3$

Campylobacter jejuni, *Listeria monocytogenes*, *Shigella flexneri*, *Yersinia enterocolitica*, *E. coli*, and *S.* spp can be detected simultaneously in the multiplexed assay.

To further illustrate the feasibility of a multiplexed assay, a titration of the target analyte was conducted with tenfold serial dilutions (Fig. 19.8). In Fig. 19.8b, *E.coli* Stx1 graph was dose-responsive, but not *S. enterica* when only *E. coli* Stx1 was present. Similarly, *S. enterica* invA showed a dose response, but not *E. coli* Stx1 when only *S. enterica* invA was present (Fig. 19.8c). Furthermore, both *E. coli* Stx1 and *S. enterica* invA showed dose response in the presence of both targets

(Fig. 19.8d). The data indicate that the multiplexed assay is not only DNA sequence-specific, but also can quantitatively and simultaneously identify the different food-borne pathogens. Obviously, the multiplex related labor and sample processing reduction, simplicity, and breadth of testing options on one device, and combinational power of multiple biomarkers, [41] will make the assay even more attractive for the nucleic acids-based diagnostics.

6 Conclusions and Future Perspective

In this study, an assay using Nanoplex™ biotag-based microsphere array for rapid, ultra sensitive, and multiplexed detection of food-borne pathogens has been successfully demonstrated. The optical properties of the SERS biotag allows the detection of pathogen genomic DNA sequences without the need for wash steps or removal of sample mixtures, requires only limited handling steps, and provides operator simplicity. Particularly, the technique shows great promise for the non-amplified detection of genomic DNA. The assay is very rapid and can be completed in less than 1 h. Furthermore, Nanoplex™ biotags provide great advantages for multiplexing. We envision that this platform has the potential to leapfrog current efforts in robotics and microfluidics required for sample preparation and assay miniaturization, by integrating and translating advances from the nanotechnology fields into unique applications in molecular diagnostic, genomics, and drug discovery.

Acknowledgments Thanks to Dr. Davis G. for SENSER See™ programming and Dr. Golightly R. for kindly proofreading this chapter.

References

1. Fleischmann M, Hendra PJ, McQuillan AJ (1974) Raman spectra of pyridine adsorbed at a silver electrode. Chem Phys Lett 26:163–166
2. Jeanmaire DL, Van Duyne RP (1977) Surface Raman spectroelectro chemistry Part I. Heterocyclic, aromatic, and aliphatic amines adsorbed on the anodized silver electrode. J Electroanal Chem 84:1–20
3. Albrecht MG, Creighton JA (1977) Anomalously intense Raman-spectra of pyridine at a silver electrode. J Am Chem Soc 99:5215–5217
4. Cao YC, Jin R, Mirkin CA (2002) Nanoparticles with Raman spectroscopic fingerprints for DNA and RNA detection. Science 297:1536–1540
5. Mulvaney SP, Musick MD, Keating CD, Natan MJ (2003) Glass-coated, analyte-tagged nanoparticles: a new tagging system based on detection with surface-enhanced Raman scattering. Langmuir 19:4784–4790
6. Doering WE, Nie S (2003) Spectroscopic tags using dye-embedded nanoparticles and surface-enhanced Raman scattering. Anal Chem 75:6171–6176
7. Nie S, Emory SR (1997) Probing single molecules and single nanoparticles by surface-enhanced Raman scattering. Science 275:1102–1106

8. Kneipp K, Wang Y, Kneipp H, Perelman LT, Itzkan I, Dasari R, Feld MS (1997) Single molecule detection using surface-enhanced Raman scattering (SERS). Phys Rev Lett 78:1667–1670
9. Le Ru EC, Meyer M, Etchegoin PG (2006) Proof of single-molecule sensitivity in surface enhanced Raman scattering (SERS) by means of a two-analyte technique. J Phys Chem B 110:1944–1948
10. Graham D, Faulds K, Smith WE (2006) Surface-enhanced Raman scattering: physics and applications. Chem Commun (Camb) 42:4363–4371
11. Qian X, Peng XH, Ansari DO, Yin-Goen Q, Chen GZ, Shin DM, Yang L, Young AN, Wang MD, Nie S (2008) In vivo tumor targeting and spectroscopic detection with surface-enhanced Raman nanoparticle tags. Nat Biotechnol 26:83–90
12. Culha M, Stokes D, Allain LR, Vo-Dinh T (2003) Surface-enhanced Raman scattering substrate based on a self-assembled monolayer for use in gene diagnostics. Anal Chem 75:6196–6201
13. Zhang XY, Zhao J, Whitney AV, Elam JW, Van Duyne RP (2006) Ultrastable substrates for surface-enhanced Raman spectroscopy: Al_2O_3 overlayers fabricated by atomic layer deposition yield improved anthrax biomarker detection. J Am Chem Soc 128:10304–10309
14. Shafer-Peltier KE, Haynes CL, Glucksberg MR, Van Duyne RP (2003) Toward a glucose biosensor based on surface-enhanced Raman scattering. J Am Chem Soc 408(125):588–593
15. Dieringer JA, McFarland AD, Shah NC, Stuart DA, Whitney AV, Yonzon CR, Young MA, Zhang XY, Van Duyne RP (2006) Surface-enhanced Raman spectroscopy: new materials, concepts, characterization tools, and applications. Faraday Discuss 132:9–26
16. Park T, Lee S, Seong GH, Choo J, Lee EK, Kim YS, Ji WH, Hwang SY, Gweon DG, Lee S (2005) Highly sensitive signal detection of duplex dye-labelled DNA oligonucleotides in a PDMS microfluidic chip: confocal surface-enhanced Raman spectroscopic study. Lab Chip 5:437–442
17. Sha MY, Penn SG, Freeman RG, Doering WE (2007) Detection of human viral RNA via a combined fluorescence and SERS molecular beacon assay. Nanobiotechnology 3:23–30
18. Cao YC, Jin R, Nam JM, Thaxton CS, Mirkin CA (2003) Raman-dye-labeled Nanoparticle probes for proteins. J Am Chem Soc 125:14676–14677
19. Driskell JD, Uhlenkamp JM, Lipert RJ, Porter MD (2007) Surface-enhanced Raman scattering immunoassays using a rotated capture substrate. Anal Chem 79:4141–4148
20. Gong JL, Liang Y, Huang Y, Chen JW, Jiang JH, Shen GL, Yu RQ (2007) Ag/SiO_2 core-shell nanoparticle-based surface-enhanced Raman probes for immunoassay of cancer marker using silica-coated magnetic nanoparticles as separation tools. Biosens Bioelectron 22:1501–1507
21. Moore BD, Stevenson L, Watt A, Flitsch S, Turner NJ, Cassidy C, Graham D (2004) Rapid and ultra-sensitive determination of enzyme activities using surface-enhanced resonance Raman scattering. Nat Biotechnol 22:1133–1138
22. Ruan C, Wang W, Gu B (2006) Rapid and ultra-sensitive detection of alkaline phosphatase using surface enhanced Raman spectroscopy. Anal Chem 78:3379–3384
23. Talley CE, Jusinski L, Hollars CW, Lane SM, Huser T (2004) Intracellular pH sensors based on surface-enhanced Raman scattering. Anal Chem 76:7064–7068
24. Kneipp J (2006) Surface-enhanced Raman scattering: physics and applications. Top Appl Phys 103:335–349
25. Kim JH, Kim JS, Choi H, Lee SM, Jun BH, Yu KN, Kuk E, Kim YK, Jeong DH, Cho MH, Lee YS (2006) Nanoparticle probes with surface enhanced Raman spectroscopic tags for cellular cancer targeting. Anal Chem 78:6967–6973
26. Hu QY, Tay LL, Noestheden M, Pezacki JP (2007) Mammalian cell surface imaging with nitrile-functionalized nanoprobes: biophysical characterization of aggregation and polarization anisotropy in SERS imaging. J Am Chem Soc 129:14–15
27. Lee S, Kim S, Choo J, Shin SY, Lee YH, Choi HY, Ha SH, Kang KH, Oh CH (2007) Biological imaging of HEK293 cells expressing PLCgamma1 using surface-enhanced Raman microscopy. Anal Chem 79:916–922

28. Sun L, Sung KB, Dentinger C, Lutz B, Nguyen L, Zhang JW, Qin HY, Yamakawa M, Cao MQ, Lu Y, Chmura AJ, Zhu J, Su X, Berlin AA, Chan S, Knudsen B (2007) Composite organic–inorganic nanoparticles as Raman labels for tissue analysis. Nano Lett 7:351–356
29. Doering WE, Piotti ME, Natan MJ, Freeman RG (2007) SERS as a foundation for nanoscale optically detected biological labels. Adv Mater 19:3100–3108
30. Jarvis R, Clarke S, Goodacre R (2006) Surface-enhanced Raman scattering: physics and applications. Top Appl Phys 103:397–408
31. Jarvis RM, Brooker A, Goodacre R (2006) Surface-enhanced Raman scattering for the rapid discrimination of bacteria. Faraday Discuss 132:281–292
32. Jarvis RM, Goodacre R (2004) Rapid discrimination of bacteria using surface enhanced Raman spectroscopy. Anal Chem 76:40–47
33. Zhang XY, Shah NC, Van Duyne RP (2006) Sensitive and selective chem/biosensing based on surface-enhanced Raman spectroscopy (SERS). Vib Spectrosc 42:2–8
34. Daniels JK, Caldwell TP, Christensen KA, Chumanov G (2006) Monitoring the kinetics of *Bacillus subtilis* endospore germination via surface-enhanced Raman scattering spectroscopy. Anal Chem 78:1724–1729
35. Premasiri WR, Moir DT, Klempner MS, Krieger N, Jones G, Ziegler LD (2005) Characterization of the surface enhanced Raman scattering (SERS) of bacteria. J Phys Chem B 109:312–320
36. Shanmukh S, Jones L, Driskell J, Zhao YP, Dluhy R, Tripp RA (2006) Rapid and sensitive detection of respiratory virus molecular signatures using a silver nanorod array SERS substrate. Nano Lett 6:2630–2636
37. Stosch R, Henrion A, Schiel D, Guttler B (2005) Surface-enhanced Raman scattering based approach for quantitative determination of creatinine in human serum. Anal Chem 77:7386–7392
38. Wang TL, Chiang HK, Lu HH, Peng FY (2005) Semi-quantitative surface enhanced Raman scattering spectroscopic creatinine measurement in human urine samples. Opt Quant Electron 37:1415–1422
39. Venter JC et al (2001) The sequence of the human genome. Science 291:1304–1351
40. Call DR (2005) Challenges and opportunities for pathogen detection using DNA microarrays. Crit Rev Microbiol 31:91–99
41. Ling MM, Ricks C, Lea P (2007) Multiplexing molecular diagnostics and immunoassays using emerging microarray technologies. Expert Rev Mol Diagn 7:87–98
42. Eastman PS, Ruan W, Doctolero M, Nuttall R, de Feo G, Park JS, Chu JS, Cooke P, Gray JW, Li S, Chen FF (2006) Qdot nanobarcodes for multiplexed gene expression analysis. Nano Lett 6:1059–1064
43. Bibikova M, Lin Z, Zhou L, Chudin E, Garcia EW, Wu B, Doucet D, Thomas NJ, Wang Y, Vollmer E, Goldmann T, Seifart C, Jiang W, Barker DL, Chee MS, Floros J, Fan JB (2006) High-throughput DNA methylation profiling using universal bead arrays. Genome Res 16:383–393
44. Fan JB, Gunderson KL, Bibikova M, Yeakley JM, Chen J, Wickham GE, Lebruska LL, Laurent M, Shen R, Barker D (2006) Illumina universal bead arrays. Methods Enzymol 410:57–73
45. Fan JB, Hu SX, Craumer WC, Barker DL (2005) BeadArray-based solutions for enabling the promise of pharmacogenomics. Biotechniques 39:583–588
46. Xu H, Sha MY, Wong EY, Uphoff J, Xu Y, Treadway JA, Truong A, O'Brien E, Asquith S, Stubbins M, Spurr NK, Lai EH, Mahoney W (2003) Multiplexed SNP genotyping using the Qbead system: a quantum dot-encoded microsphere-based assay. Nucleic Acids Res 31: e43–e52
47. Steemers FJ, Gunderson KL (2007) Whole genome genotyping technologies on the BeadArray platform. Biotechnol J 2:41–49
48. Song L, Ahn S, Walt DR (2006) Fiber-optic microsphere-based arrays for multiplexed biological warfare agent detection. Anal Chem 78:1023–1033

49. Ahn S, Walt DR (2005) Detection of *Salmonella* spp. using microsphere-based, fiber-optic DNA microarrays. Anal Chem 77:5041–5047
50. Liang Y, Gong JL, Huang Y, Zheng Y, Jiang JH, Shen GL, Yu RQ (2007) Biocompatible core-shell nanoparticle-based surface-enhanced Raman scattering probes for detection of DNA related to HIV gene using silica-coated magnetic nanoparticles as separation tools. Talanta 72(2):443–449
51. Monaghan PB, McCarney KM, Ricketts A, Littleford RE, Docherty F, Smith WE, Graham D, Cooper JM (2007) Bead-based DNA diagnostic assay for chlamydia using nanoparticle-mediated surface-enhanced resonance Raman scattering detection within a lab-on-a-chip format. Anal Chem 79:2844–2849
52. Callaway TR, Elder RO, Keen JE, Anderson RC, Nisbet DJ (2003) Forage feeding to reduce preharvest *Escherichia coli* populations in cattle, a review. J Dairy Sci 86:852–860
53. Voetsch AC, Van Gilder TJ, Angulo FJ, Farley MM, Shallow S, Marcus R, Cieslak PR, Deneen VC, Tauxe RV (2004) FoodNet estimate of the burden of illness caused by nontyphoidal Salmonella infections in the United States. Clin Infect Dis 38(Suppl 3): S127–S134
54. Braden CR (2006) *Salmonella enterica* serotype enteritidis and eggs: a national epidemic in the United States. Clin Infect Dis 43:512–517
55. Marusina K (2007) Evaluating data flow in systems biology. Genet Eng Biotechnol News 27:23–25
56. Faulds K, Fruk L, Robson DC, Thompson DG, Enright A, Smith WE, Graham D (2006) A new approach for DNA detection by SERRS. Faraday Discuss 132:261–268
57. Sun L, Yu C, Irudayaraj J (2007) Surface-enhanced Raman scattering based nonfluorescent probe for multiplex DNA detection. Anal Chem 79:3981–3988
58. Freeman RG, Doering WE, Walton ID, Penn SG, Glenn D, Wong F, Natan MJ (2005) Detection of biomolecules using nanoparticle surface enhanced Raman scattering tags. Proc SPIE 5705:114–121
59. Velusamy V, Arshak K, Korostynska O, Oliwa K, Adley C (2010) An overview of foodborne pathogen detection: in the perspective of biosensors. Biotechnol Adv 474(28):232–254
60. Sanvicens N, Pastells C, Pascual N, Marco M (2009) Nanoparticle-based biosensors for detection of pathogenic bacteria. Trends Anal Chem 28:1243–1252
61. Zavaletaa C, Smitha B, Waltonb I, Doering W, Davisb G, Shojaeib B, Natan M, Gambhira S (2009) Multiplexed imaging of surface enhanced Raman scattering nanotags in living mice using noninvasive Raman spectroscopy. Proc Natl Acad Sci USA 106:13511–13516
62. Sha M, Xu H, Natan M, Crmy R (2008) Surface-enhanced Raman scattering tags for rapid and homogeneous detection of circulating tumor cells in the presence of human whole blood. J Am Chem Soc 130:17214–17215
63. Hayashi T, Makino K, Ohnishi M, Kurokawa K, Ishii K, Yokoyama K, Han CG, Ohtsubo E, Nakayama K, Murata T, Tanaka M, Tobe T, Iida T, Takami H, Honda T, Sasakawa C, Ogasawara N, Yasunaga T, Kuhara S, Shiba T, Hattori M, Shinagawa H (2001) Complete genome sequence of enterohemorrhagic *Escherichia coli* O157:H7 and genomic comparison with a laboratory strain K-12. DNA Res 8:11–22
64. Bao YP, Huber M, Wei TF, Marla SS, Storhoff JJ, Muller UR (2005) SNP identification in unamplified human genomic DNA with gold nanoparticle probes. Nucleic Acids Res 33:e15
65. Xu X, Georganopoulou DG, Hill HD, Mirkin CA (2007) Homogeneous detection of nucleic acids based upon the light scattering properties of silver-coated nanoparticle probes. Anal Chem 79:6650–6654
66. Hopkins KL, Hilton AC (2000) Simultaneous molecular subtyping and shiga toxin gene detection in *Escherichia coli* using multiplex polymerase chain reaction. Lett Appl Microbiol 30:122–125
67. Maynard C, Berthiaume F, Lemarchand K, Harel J, Payment P, Bayardelle P, Masson L, Brousseau R (2005) Waterborne pathogen detection by use of oligonucleotide-based microarrays. Appl Environ Microbiol 71:8548–8557

68. Li Y, Liu D, Cao B, Han W, Liu Y, Liu F, Guo X, Bastin DA, Feng L, Wang L (2006) Development of a serotype-specific DNA microarray for identification of some Shigella and pathogenic *Escherichia coli* strains. J Clin Microbiol 44:4376–4383
69. Call DR, Brockman FJ, Chandler DP (2001) Detecting and genotyping *Escherichia coli* O157: H7 using multiplexed PCR and nucleic acid microarrays. Int J Food Microbiol 67:71–80
70. Jothikumar N, Griffiths MW (2002) Rapid detection of *Escherichia coli* O157:H7 with multiplex real-time PCR assays. Appl Environ Microbiol 68:3169–3171

SERS Spectroscopy and Microscopy 20

Maurizio Muniz-Miranda, Cristina Gellini, and Massimo Innocenti

1 Definition of the Topic

Surface-enhanced Raman spectroscopy coupled with microscopic investigation. Combining SERS and microscopy measurements.

2 Overview

Coupling surface-enhanced Raman scattering (SERS) experiments and microscopic measurements allows obtaining important information on the mechanisms of the Raman enhancement, as well as on the properties of both adsorbates and metal surfaces, where ligand molecules adhere. In particular, confocal SERS microspectroscopy represents an effective development in the performances of the conventional SERS technique, because more intense Raman signals could be obtained from the microscopic analysis of the sample. Moreover, this investigation allows identifying different molecular components in polymeric layers, in rocks, in cells, and tissues, and studying photochemical processes and catalytic reactions. Applications are possible in many research fields, from nanosensors to nanomedicine, from metal corrosion to art restoration.

M. Muniz-Miranda (✉) • C. Gellini • M. Innocenti
Dipartimento di Chimica "Ugo Schiff", Università di Firenze, Via della Lastruccia 3, Sesto Fiorentino, Italy

Challa S. S. R. Kumar (ed.), *Raman Spectroscopy for Nanomaterials Characterization*, DOI 10.1007/978-3-642-20620-7_20, © Springer-Verlag Berlin Heidelberg 2012

3 Introduction

Microscopic investigation allowing molecular recognition assumes fundamental importance in several research fields, as the analysis of biomolecules like nucleic acids and proteins, in applications like clinical medicine and forensic analysis, as well as in preservation of artistic materials or in geochemical and astronomical investigations. On the other hand, the material science concerning sensors and heterogeneous catalysis finds large help from the use of nanostructures that can be analyzed by microscopic measurements. It is of interest in this context to profit from spectroscopic methods able to couple high sensitivity with structural information.

The vibrational spectroscopy allows a valid recognition of different molecules, because from the observed bands representing the vibrational modes, it is possible to achieve a chemical fingerprint to be used for the above-cited applications and many others. In particular, Raman scattering is a rapid, noninvasive, and nondestructive technique applied to the investigation of both liquid and solid samples. Coupling a Raman spectrometer with an optical microscope allows collecting spectra from samples with volumes less than 1 μm^3. Hence, Raman microspectroscopy is able to provide information at molecular level about the composition and the structure of cells or tissues [1]. This technique offers marked advantages with respect to the analogous FTIR (Fourier transform infrared) absorption spectroscopy, which impairs the analysis of hydrated biological samples due to the strong absorption of water in the infrared region. Raman scattering, instead, exhibits a small interference from the aqueous environment, along with a spatial resolution larger than that of the infrared spectroscopy (about 1 μm for Raman, ~ 10 μm for FTIR), because of the minor wavelength used for the laser excitation.

A wide employment of the Raman microspectroscopy, however, is severely limited by the low sensitivity of this technique. The probability of a molecule to produce scattering is much less than the probability of absorption in the IR region. The Rayleigh scattering intensity is typically 10^{-3} to 10^{-5} of the incident intensity, while the Raman scattering only 10^{-7} to 10^{-10}. As a consequence, a Raman spectrum may be measured only when the incident radiation, normally in the visible spectral range, interacts with a large number of molecules (10^{-2} to 10^{-3} M) in solution or with many molecular layers in a solid. The intensities of the Raman bands can be increased if the energy of the exciting radiation is close to that necessary to promote an electronic excitation of the sample molecules, as in the resonance Raman (RR) measurements. Only the vibrational modes associated with the resonating electronic transitions show sizeable enhancements with respect to the normal Raman (NR) bands, allowing identifying the bands of the chromophore groups. RR technique is mainly used to study biological molecules, which present electronic excitations in the visible spectral region, where the more common laser lines are found. On the other hand, exciting in an electronic transition may also cause fluorescence emission and/or photodegradation of the sample, both inconveniences occurring in RR experiments.

The Raman scattering efficiency can be greatly increased by the use of the SERS (surface-enhanced Raman scattering) effect [2–4]. The latter provides huge enhancements for the Raman signals of molecules adhering to surfaces of metals like Ag, Au, and Cu with nanoscale roughness. The observed enhancement factors are in the order of 10^7 with respect to the NR scattering of non-adsorbed molecules. By coupling the SERS spectroscopy with the microscopic investigation, performances and applications of this technique can be largely amplified [5]. Moreover, by adopting a suitable instrumental arrangement, enhancements up to 10^{14}–10^{15} were measured in the single-molecule SERS experiments [6, 7], which allows a microscopic investigation beyond the diffraction limit.

In this chapter, we want to present and discuss the advantages deriving from the microscopic investigation for obtaining SERS spectra. For this aim, it is of basic importance to give a brief illustration of the mechanisms of the SERS effect in relation with the surface roughness, which can be investigated by microscopic techniques.

Two mechanisms usually contribute to the SERS enhancement, both related to the increase in the dipole moment μ induced by the incident radiation, $\mu = \alpha E$, that is, molecular polarizability timed with the radiation electric field.

The electromagnetic effect enhances by some orders of magnitude the local electric field nearby the metallic surface. The chemical effect, instead, derives from polarizability changes when molecules are bound to active sites of the metal substrates. Both mechanisms are related to metal surface irregularities or defects, on nanometric or atomic scale, as shown in Scheme 20.1.

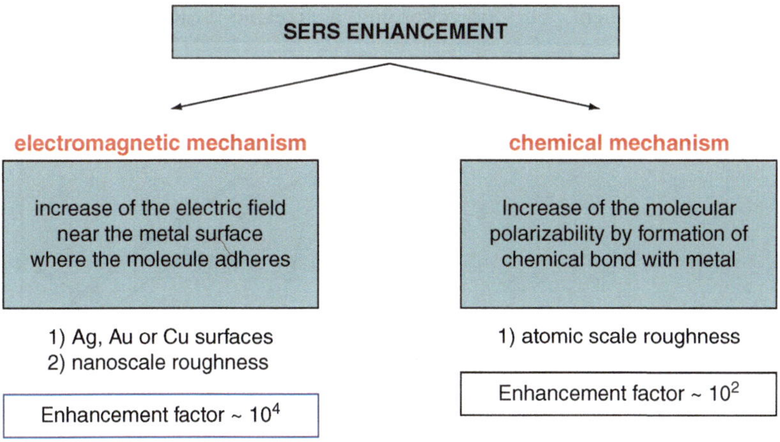

Scheme 20.1 SERS enhancement mechanisms related to the surface roughness

The SERS electromagnetic enhancement originates from the resonance between incident radiation and electronic excitation wave on the metal surface, called "surface plasmon band," as explained below. The resonance condition depends on the dielectric constant of the metal $\varepsilon(\omega) = \varepsilon_1 + i\,\varepsilon_2$, which is a complex function of the frequency ω. The enhancement factor can be expressed as:

$$\mathrm{EF} = \{[\varepsilon(\omega_o) - 1]/[\varepsilon(\omega_o) + 2]\}^2\{[\varepsilon(\omega) - 1]/[\varepsilon(\omega) + 2]\}^2$$

where ω_o and ω are the frequencies of the incident and scattered radiations, respectively. By approximating ω to ω_o, we obtain

$$\mathrm{EF} = [\varepsilon(\omega_o) - 1]^4/[\varepsilon(\omega_o) + 2]^4$$

Finally, if the real part of the dielectric constant, ε_1, is near to -2, as it is for high-reflectivity metals, the enhancement factor becomes:

$$\mathrm{EF} \propto \frac{1}{(\varepsilon_2)^4}$$

In other words, huge enhancements of the scattering response are expected when the imaginary part of the dielectric constant, ε_2, is very small. This condition, which is related to the high reflectivity of the metal, occurs for silver in all visible regions and for gold and copper in the red-light region. Moreover, below 500 nm, the real part of the dielectric constant is near to -2 for Ag, Au, and Cu. In Fig. 20.1, the profiles of the real and imaginary parts of the dielectric constants for silver, gold, and copper are reported versus the wavelength of the electromagnetic radiation.

The condition concerning the dielectric constant is necessary to have electromagnetic enhancement, but for observing a sizeable SERS effect, it is also

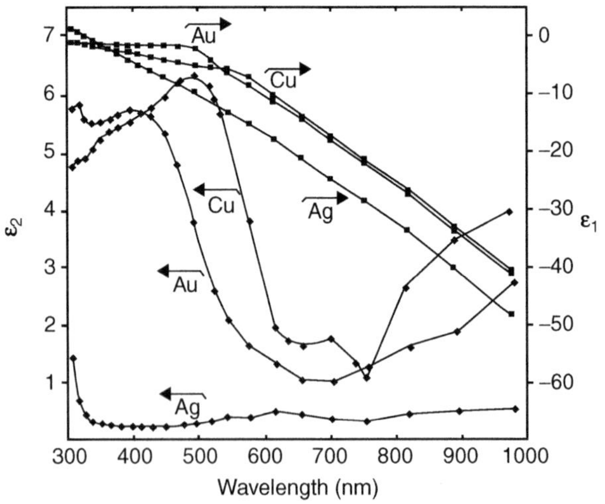

Fig. 20.1 Dependence of the dielectric constant of noble metals on the radiation wavelength

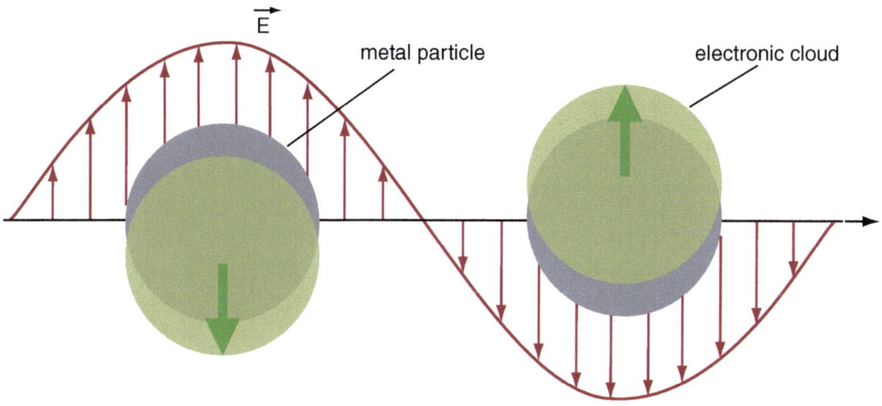

Fig. 20.2 Electronic excitation in metal nanoparticles by radiation electric field

necessary that the metal exhibits a nanostructured surface. Actually, the maximum enhancement is obtained when the frequency of the incident radiation is near to that of the electronic excitation band due to the plasmons localized at the metal surface. The localized plasmons are collective oscillations of the conduction electrons of a metal under electromagnetic irradiation, when they are trapped in structures with dimensions smaller than 100 nm, as shown in the Fig. 20.2. Larger particles generate plasmons whose bands may occur in the infrared region, while too small particles cannot produce collective oscillations.

The surface plasmon resonance (SPR) bands of metals like silver, gold, and copper are observed in the near-UV and visible extinction spectra, depending on the type of metal and the dimension of the nanoparticles. For Ag colloids, single particles with 15-nm average size exhibit a very strong SPR band around 390 nm. The latter moves to longer wavelengths, showing a secondary SPR band, when adsorbed molecules induce aggregation of the metal nanoparticles, as shown in Fig. 20.3 for the adsorption of different benzodiazines [8]. For these ligands, the maximum enhancement occurs when the laser excitation radiations match the SPR bands of nanoparticle aggregates observed at longer wavelengths, as verified by the SERS excitation profiles shown in Fig. 20.3. Actually, for every benzodiazine adsorbed on Ag colloidal particles, the maximum SERS enhancement is obtained with exciting laser lines corresponding to the maximum of the secondary plasmon resonance band, which is located at different wavelengths depending on different particle aggregations.

The electromagnetic mechanism is necessary to observe the SERS effect from adsorbed molecules, providing enhancement factors of more than 10^4. The chemical enhancement mechanism usually gives rise to factors only up to 10^2, but however it is important because both position and relative intensity of the SERS bands closely depend on this effect [9]. The latter is due to the formation of complexes between ligand and metal active sites, which represent atomic-scale defects of the surface. Hence, the chemical SERS effect can be considered mainly due to the resonance between the incident radiation and the electronic excitation of the ligand-metal adduct.

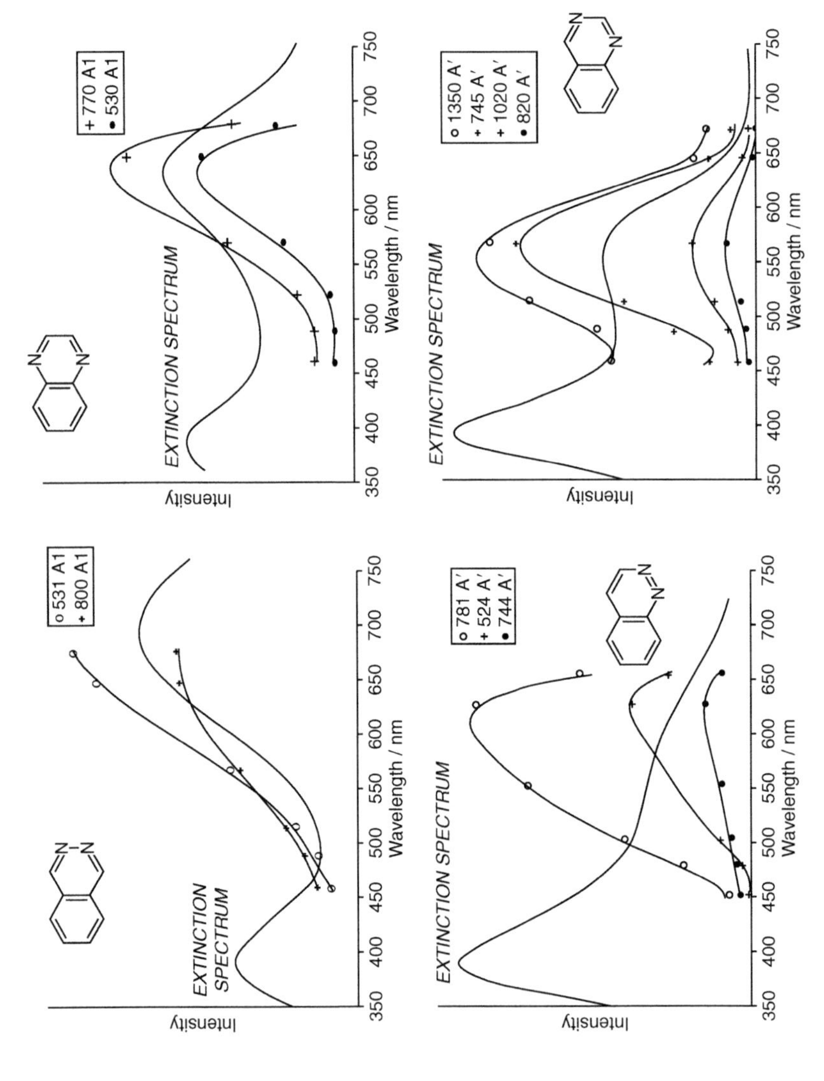

Fig. 20.3 SERS excitation profiles of benzodiazines, compared with the extinction spectra of ligand/Ag colloids

4 Basic Methodology

4.1 Spectroscopic Measurements

The Raman spectra on solid metal substrates were measured using a Renishaw RM2000 micro-Raman apparatus, coupled with a diode laser source emitting at 785 nm. Sample irradiation was accomplished using the 50× microscope objective of a Leica microscope DMLM. The beam power was ~3 mW, the laser spot diameter was adjusted between 1 and 3 μm. Raman scattering was filtered by a double holographic notch filter system and collected by an air-cooled CCD detector. The acquisition time for each measurement was 10 s. All spectra were calibrated with respect to a silicon wafer at 520 cm^{-1}.

SERS spectra in Ag hydrosols were recorded using a Jobin-Yvon HG2S monochromator equipped with a cooled RCA-C31034A photomultiplier and a data acquisition facility. To reduce the thermal effects due to the laser light, a defocused beam with low power (20 mW) was used. Raman data were obtained with exciting lines supplied by Ar^+ and Kr^+ lasers (406.7, 413.1, 457.9, 488.0, 514.5, 520.8, 568.2, 647.1, 676.4 nm) or by He-Ne laser (632.8 nm). All spectra were corrected to account for monochromator and photomultiplier efficiency. Power density measurements were performed with a power meter instrument (model 362; Scientech, Boulder, CO, USA) giving 5% accuracy in the 300–1,000 nm spectral range.

The UV-vis absorption spectra of silver colloids were observed by means of a CARY 5 spectrophotometer.

4.2 Preparation of the Ag Colloids

Stable silver sols were prepared by reduction of $AgNO_3$ (Aldrich, purity 99.998%) with excess $NaBH_4$ (Aldrich, purity 99%), aged a week to prevent the formation of reduction products. The usual pH value of the aqueous suspension was about nine. NaCl (Aldrich purity 99.999%) was added in a small amount (10^{-3} M) to the Ag colloids to improve the SERS effect, without altering the sol stability.

4.3 Filtration of Ag Nanoparticles

Silver hydrosols (10 ml) with chloride anions were filtered by using ANODISC alumina filters (0.02 mm diameter) and a membrane filter holder supplied by Whatman Intern. Ltd. (England). After two successive filtrations, the liquid was free of metal particles, as detected by the total absence of plasmon absorption bands in the UV-vis region. The filter surface resulted coated by a layer of colloidal silver.

4.4 Preparation of Metal Substrates

Smooth plates of silver (1.0-mm thick, 99.9% purity), as supplied by Aldrich, were first treated with a plasma cleaner (Mod. PDC-32G, Harrick Sc. Co., NY, USA) to eliminate traces of preexisting substances, and then, they were roughened by immersion for 30 s in a solution composed of ammonium hydroxide (Aldrich, 28% in water) and hydrogen peroxide (Aldrich, 50% in water) in a 10:1 volume ratio. Alternatively, Ag plates were polished with successively finer grades of alumina powder down to 0.3 mm, then kept in a stirred solution of 30 mM thiourea (Fluka) and 20 mM $Fe(NO_3)_3{\cdot}9(H_2O)$ for 30 s, producing etching of silver with very homogeneous roughness.

4.5 TEM Measurements

Transmission electron microscopy (TEM) measurements on Ag hydrosols were obtained by using a Philips EM 201 instrument with an electron beam emitted at 80 kV, after placing a drop of colloidal sample on a carbon–Cu grid. Large magnifications (up to 46,000) were adopted in order to investigate the metal aggregates at the level of nanoparticles.

4.6 SEM Measurements

Microscopic measurements were performed by using a Quanta 200 ESEM (environmental scanning electron microscopy) instrument (FEY Company), operating in low-vacuum mode, with an electron beam emitted at 25 or 30 kV under 1 Torr (133 Pa) pressure. Solid-state backscatter detector (SSB) allowed collecting backscattered electrons emitted from the samples.

4.7 AFM Measurements

Topography was measured ex situ in a dry nitrogen atmosphere, using a Molecular Imaging AFM (PicoSPM, Molecular Imaging) operating in contact mode, with a commercial Si_3N_4 cantilever (Nanosensors, Wetzlar-Blankenfeld). The samples were characterized by nonfiltered 512×512 pixel images of areas ranging from $1 \times 1\ \mu m^2$ to $10 \times 10\ \mu m^2$.

5 Key Research Findings

5.1 Contribution of the Microscopic Techniques to the SERS Investigation

Owing to the strong link between nanostructured surfaces and SERS effect, it is easily understandable that the various microscopic techniques may offer a valid help in obtaining and interpreting the SERS spectral data. In this section, the importance of TEM (transmission electron microscopy), SEM (scanning electron

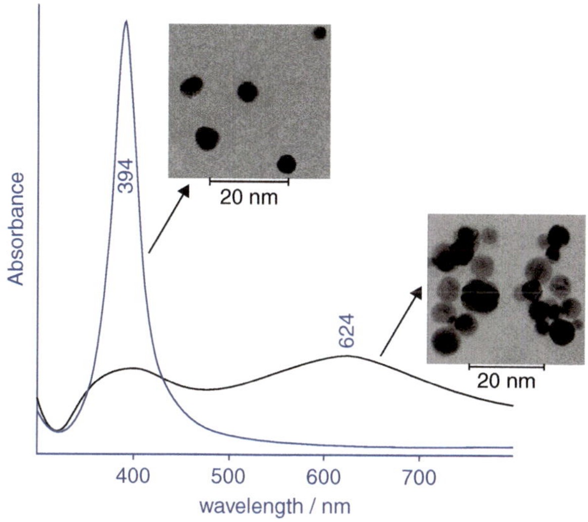

Fig. 20.4 Extinction spectra of pure Ag and adenine/Ag colloids with the corresponding TEM images

microscopy), and AFM (atomic force microscopy) analyses will be briefly presented in relation to some SERS studies on different substrates.

5.1.1 TEM Microscopy

To gain the maximum SERS enhancement in colloidal systems, the exciting radiation must have a wavelength corresponding to the plasmon band of the metal nanoparticles coated by adsorbed ligand. As shown in Fig. 20.4, when adenine is added to the colloidal dispersion, the ligand adsorption induces a fast aggregation of the metal particles, evidenced by the appearing of a broad secondary band at longer wavelength with respect to that of the non-aggregated Ag (about 390 nm).

The aggregation may be explained considering that neutral ligand molecules adsorbed on the metal surface remove charged species from the surface itself, mainly hydroxide anions, which prevents the approach of the colloidal particles [10]. This effect may be monitored by means of TEM measurements that shows isolated spheroidal nanoparticles when ligand is absent; when, instead, the ligand is added, the nanoparticles get closer forming larger aggregates. Moreover, the TEM experiments can explain why a very strong SERS enhancement is observed for adenine: The molecules may be chemisorbed between metal particles, in very effective hot spots producing high Raman signal intensification. The SERS spectrum of adenine adsorbed on silver colloid is shown in Fig. 20.5.

The spectrum has been measured in the presence of 10^{-5} M adenine by using the 632.8-nm exciting line of a He-Ne laser. Although the laser power is low (around 5 mW), the resonance with the plasmonic band at 624 nm (see Fig. 20.4) ensures

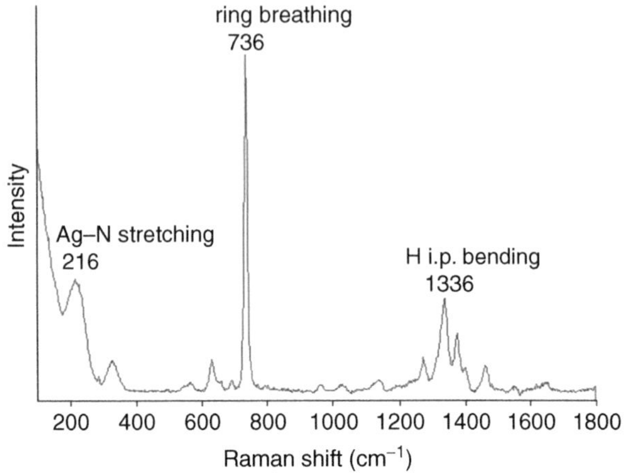

Fig. 20.5 SERS of adenine adsorbed on Ag colloid. Exciting line: 632.8 nm

a very intense SERS spectrum. The strongest enhancement occurs for the ring breathing mode (736 cm^{-1}), indicating that the molecules interact with the silver particles by means of the nitrogen atoms of the aromatic system, as confirmed by the presence of a band at 216 cm^{-1}, due to the stretching mode of the Ag-N bond.

Recently, novel nanomaterials have become a new frontier for SERS experiments, where different metals are collected together to form, for example, bimetallic particles. Thus, the same nanoparticle could be responsible for both SERS effect and catalytic activity. This is the case of the Ag/Pd colloids synthesized by chemical reduction with sodium borohydride ($NaBH_4$) of silver nitrate ($AgNO_3$) and palladium nitrate ($Pd(NO_3)_2$), with a 96:4 Ag/Pd molar ratio [11]. The silver nanoparticles provide the SERS enhancement for the ligand molecules, while palladium may induce catalytic reactions. Also, in this case, TEM microscopy provides an important help to characterize these composite materials. In Fig. 20.6 TEM images at different magnifications are reported for bimetallic Ag/Pd particles, in comparison with those constituted by pure silver. While these latter present spheroidal shapes, bimetallic particles show more irregularities, due to palladium clusters in contact with the silver core surface.

The catalytic activity toward hydrogenation reactions has been indeed observed in such bimetallic colloidal systems. The SERS spectrum of *p*-nitrobenzoate (PNBA) adsorbed on colloidal silver and the SERS spectra observed on Ag/Pd nanoparticles, immediately after the addition of PNBA and after 1 week, are shown in Fig. 20.7 A, B, and C, respectively. In the bimetallic colloid, instead of the SERS spectrum of PNBA (spectrum A), a different spectrum is obtained (spectrum B) that slowly evolves toward a different spectral feature, which becomes predominant after a week (spectrum C). This modification may be related to the initial formation of *p*-aminobenzoate as a result of the catalytic reduction of the nitro group, followed by slow oxidation to azodibenzoate by atmospheric oxygen (see Fig. 20.8).

TEM investigation becomes relevant in forecasting an effective SERS response in other composite materials, made of SERS metal particles joined by nonmetallic

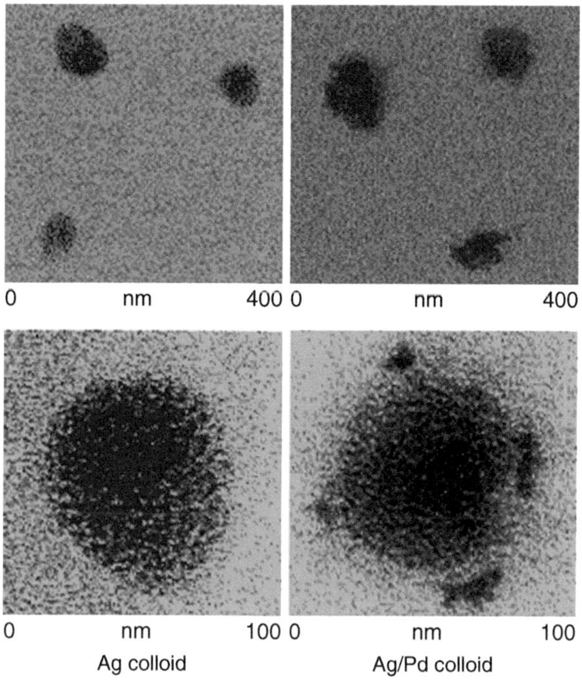

Fig. 20.6 TEM images of Ag and Ag/Pd colloidal nanoparticles at different magnifications

materials. Figure 20.9 shows the TEM images of colloidal silica particles partially coated by silver clusters obtained by photoreduction of Ag^+ ions with laser irradiation in the visible region [12]. The average particle size is less than 100 nm; thus, adsorbed molecules are expected to undergo SERS effect due to the resonance with the plasmons localized at the metallic surface. Actually, this composite material is an efficient SERS platform, as shown in Fig. 20.9 for adsorbed phenazine.

5.1.2 SEM Microscopy

SEM microscopy may be equally used in providing information about the SERS efficiency of mesoporous silica doped with silver nanoparticles, as produced by photoreduction with laser light at 514.5 nm [13]. The SEM micrographs of Fig. 20.10 are related to mesoporous silica deposited on a quartz plate as dry layer. The shining areas refer to nanosized silver aggregates, whose SERS activity is proved by using 1,10-phenanthroline as organic ligand (Fig. 20.11).

The SERS enhancement of adsorbed molecules increases when colloidal particles start to aggregate. The aggregation may be induced by adding chloride anions, which markedly intensify the SERS signal of the adsorbate, generally by 2 or 3 orders of magnitude [14–20]. This activation is important because only for activated Ag colloids, single-molecule SERS spectra were detected. Different explanations were proposed on the basis of the surface complex formation and of the electromagnetic field increase by aggregation of metal colloidal particles [21].

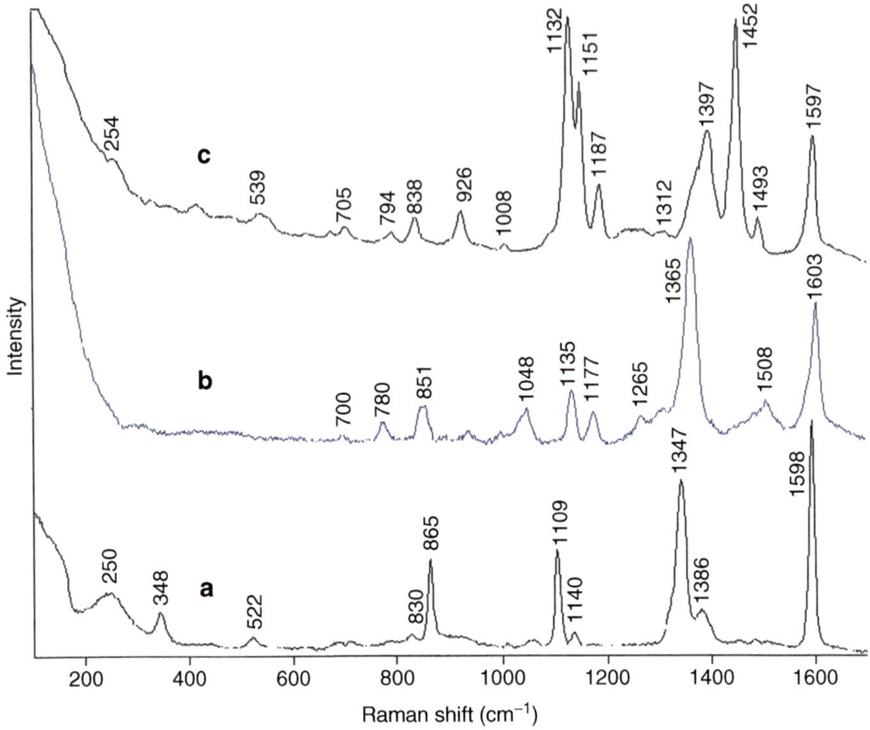

Fig. 20.7 SERS of PNBA: in Ag colloid (A); in Ag/Pd colloid (B); as (B), after a week. Exciting line: 514.5 nm

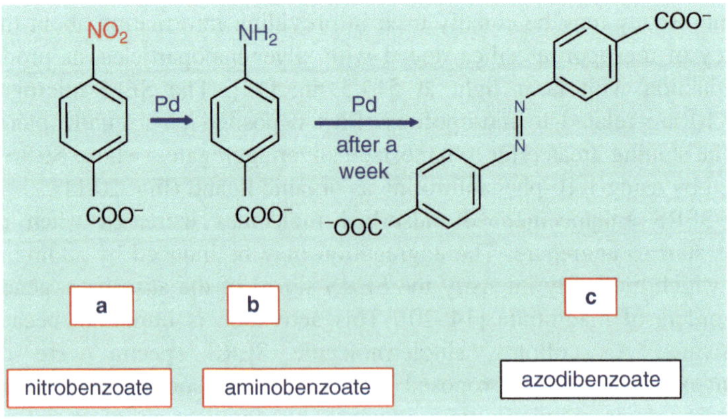

Fig. 20.8 Reduction process of PNBA adsorbed on Ag/Pd colloid

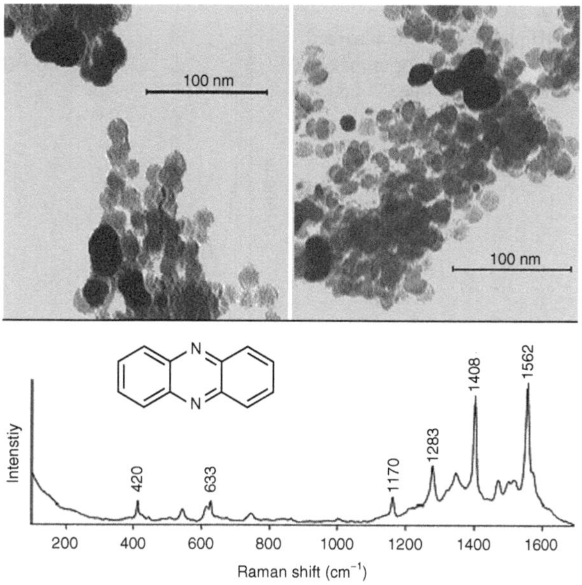

Fig. 20.9 TEM images of Ag/SiO_2 colloidal particles, along with the SERS of phenazine (exciting line: 514.5 nm)

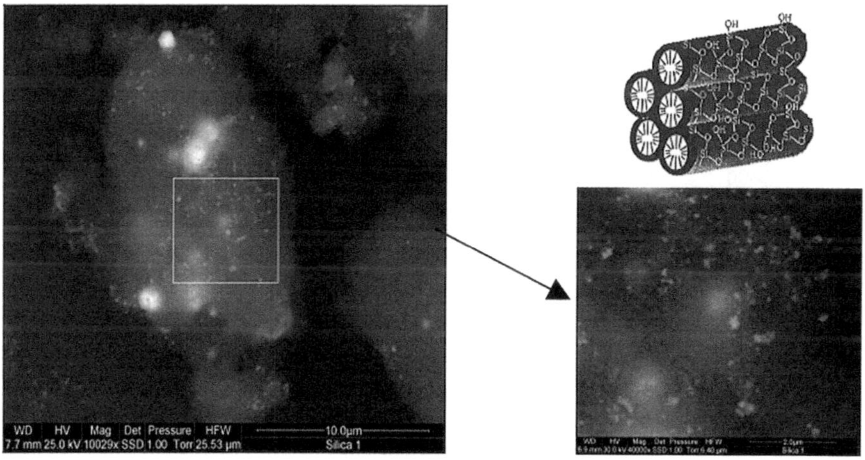

Fig. 20.10 SEM images of Ag-doped mesoporous silica

This effect is depicted combining the SERS analysis with the microscopic investigation, as done in Fig. 20.12, where SEM images of Ag colloidal particles with adsorbed 1H-1,2,3-triazole, deposited as dry layers onto cover glasses [22], are reported together with the SERS spectrum and the SPR (surface plasmon resonance) bands occurring in the UV-vis region.

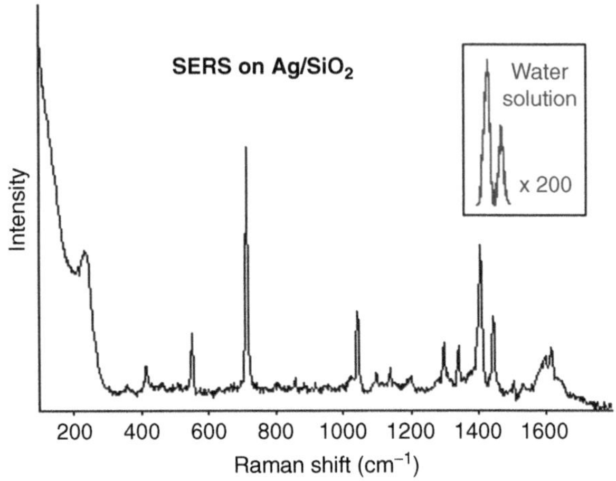

Fig. 20.11 SERS of 1,10-phenanthroline adsorbed on Ag-doped mesoporous silica. Exciting line: 514.5 nm

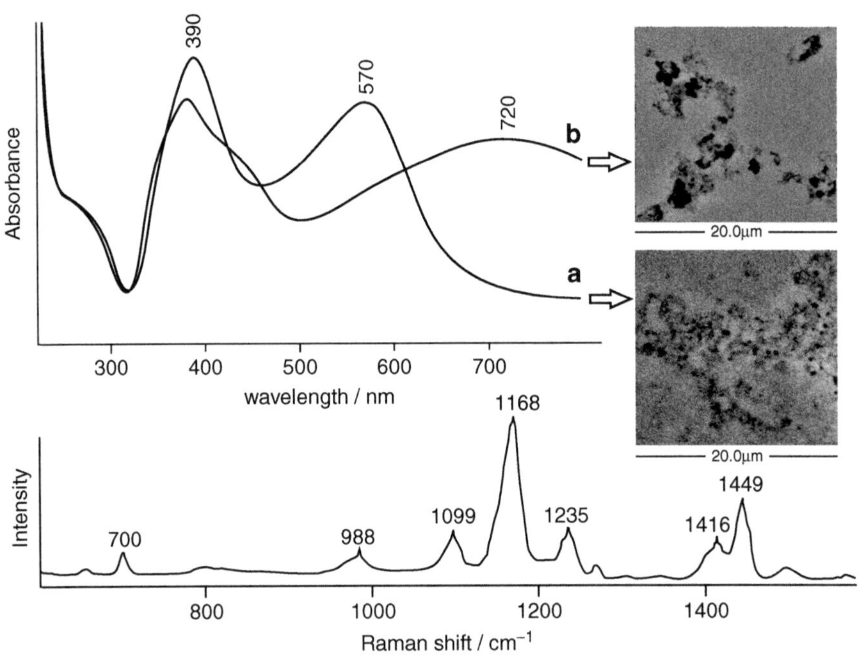

Fig. 20.12 Extinction spectra of Ag colloids with (b) and without (a) chloride anions, with the corresponding SEM images. In the *lower panel*, the SERS of 1H-1,2,3-triazole is shown. Exciting line: 514.5 nm

5.1.3 AFM Microscopy

AFM microscopy is more suited for the morphological study of solid surfaces, such as metallic plates with variable roughness; therefore, it is extremely helpful when used in combination with the SERS investigation. In this way, the SERS

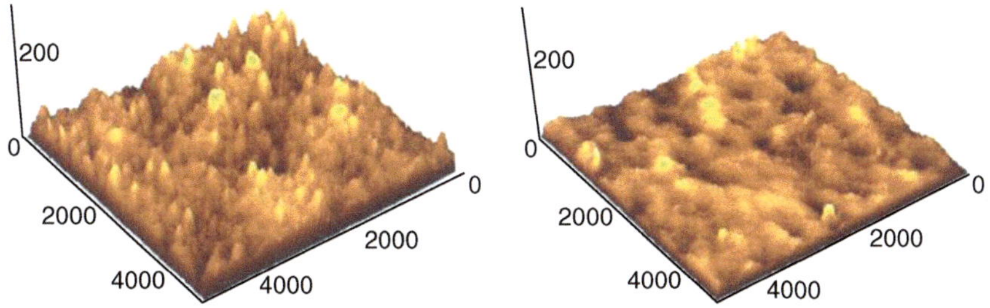

Fig. 20.13 AFM images of alumina filters coated with Ag particles with (*left*) and without (*right*) chloride anions

enhancement measured by spectroscopic experiments can be related to the surface roughness evaluated by means of AFM measurements [23].

Figure 20.13 shows the $5 \times 5\ \mu m^2$ AFM micrographs of silver nanoparticle aggregates with (left) and without (right) chloride anions, as obtained by filtration of colloidal suspensions on alumina. The second sample appears clearly less roughened than the first one. The roughness degree can be estimated directly from the AFM images by the root mean square (RMS) factor, defined as

$$RMS^2 = (1/N)\Sigma_{ij}\left[h_{ij} - \hat{h}\right]^2$$

Here, N is the number of pixels in the image, h_{ij} is the local height at pixel ij, and $\hat{h}$ is the mean height. In the presence of chloride anions, RMS is equal to 28.1 nm, while in the absence of anions, it is only 16.6 nm. By considering that RMS square is proportional to the effective surface area, the ratio between the square of these two values gives a factor of about 3, which roughly corresponds to the increase of the SERS signal, as experimentally observed [23].

When used as dried film spread on surfaces, silver colloidal particles may be employed to activate solid substrates that do not show SERS efficiency. It may be the case of smooth noble metal surfaces, such as Ag, Au, Cu, that do not have the required nanometric roughness, or nonreflecting transition metals, such as Ni, Fe, Pd. For both these two cases, AFM microscopy provides the indispensable information to predict the SERS response before performing the spectroscopic measurements.

This activation method consists in a first stage where a metallic plate is plunged into a water solution of ligand for obtaining the chemisorption of molecules on the metal surface. Then, the plate is accurately washed to remove non-adsorbed molecules and dried. Finally, a drop of silver colloid is deposited onto the surface and left to dry as a thin layer. Thus, the ligand is adsorbed on the plate surface, while the colloidal particles play the role of promoting the electromagnetic enhancement necessary to observe the adsorbate SERS spectrum. An example of this procedure is given in Fig. 20.14: On the left, the AFM image shows a smooth Au surface coated with a diacetylenic polymer monolayer [24], for which no Raman signal of the adsorbate could be observed; on the right, the AFM figure

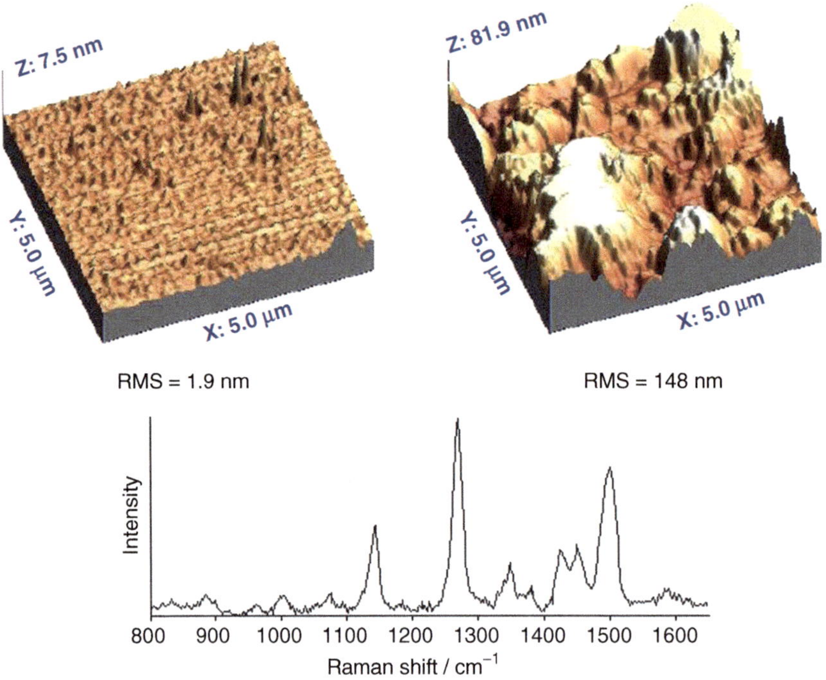

Fig. 20.14 AFM images of a flat Au surface (*left*) with adsorbed diacetylenic polymer, then coated with deposited Ag particles (*right*). In the *lower panel*, the SERS of the polymer is shown. Exciting line: 647.1 nm

shows the gold substrate covered by a dry layer of deposited Ag particles, which provides a quite strong SERS spectrum, as shown in the lower panel of the same figure.

For transition metals, unable to provide SERS enhancement, activation of smooth surfaces by deposition of Ag colloidal particles produces strong SERS signals of adsorbed molecules that could not be observed even after surface roughening treatment.

Figure 20.15 shows the AFM micrographs of a nickel smooth surface, without (upper) and with (lower) Ag nanoparticles [25]. Below, the SERS spectra of 1,10-phenanthroline adsorbed on Ni, Fe, and Pd are shown, after SERS activation by means of the Ag layer.

The same molecule has been used as probe to evaluate the SERS efficiency of substrates where ligand molecules are located in a "sandwich" configuration between two metal surfaces both SERS active [10]. The first one belongs to a silver plate, whose surface roughening was produced via chemical etching with ammonia and hydrogen peroxide; the second is a layer of silver colloidal particles deposited after chemisorption of the ligand molecules, as shown in Fig. 20.16.

The enhancement factor of the Raman signal by depositing Ag particles is estimated to be more than ten (see Fig. 20.17). The strong SERS spectrum observed

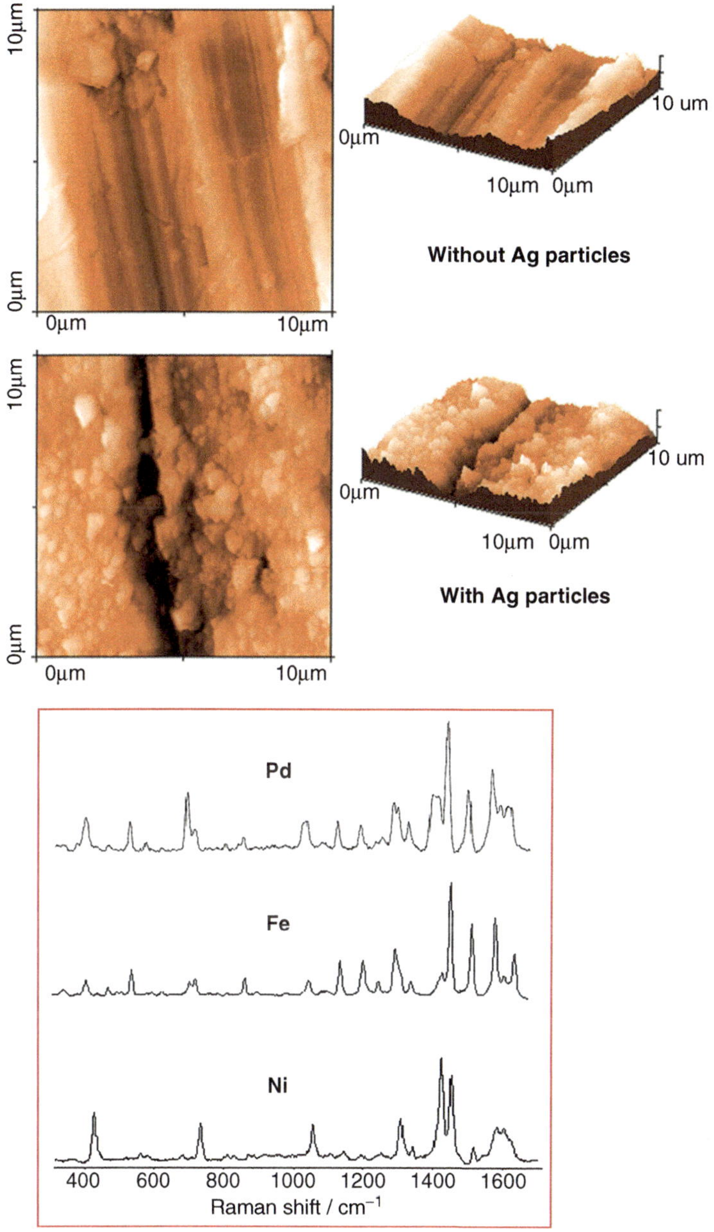

Fig. 20.15 AFM images of a smooth nickel surface (*upper panel*), then coated with deposited Ag particles (*lower panel*). On the *right side*, the SERS spectra of 1,10-phenanthroline adsorbed on Ni, Fe, and Pd smooth surfaces are shown. Exciting line: 514.5 nm

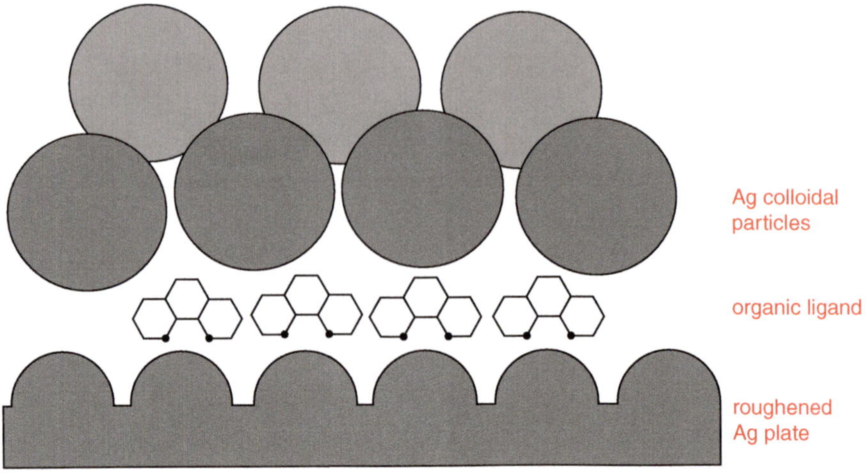

Fig. 20.16 Simplified depiction of a sandwich-like SERS-active Ag substrate. The dimension of the ligand molecules is not realistic

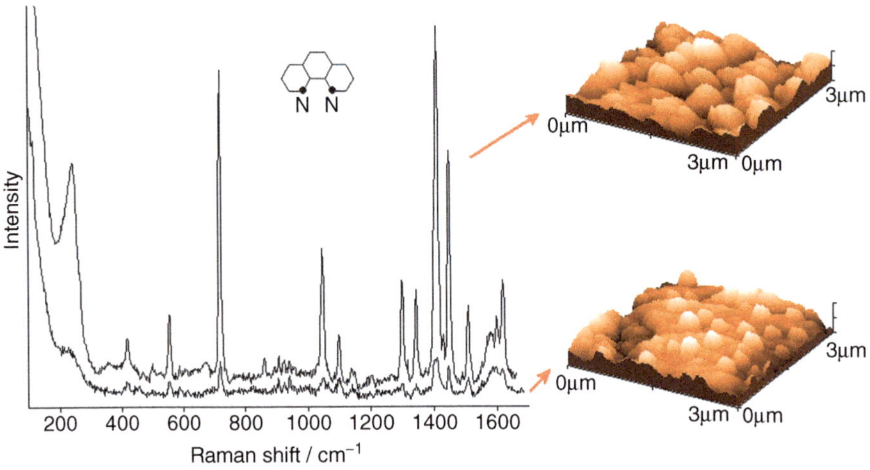

Fig. 20.17 SERS spectra of 1,10-phenanthroline adsorbed on a rough Ag plate and on the sandwich-like Ag substrate, with the corresponding AFM micrographs. Exciting line: 514.5 nm

here for 1,10-phenanthroline can be ascribed to the fact that the molecule is located in *hot spots* that are particularly effective in promoting SERS effect.

Finally, in order to demonstrate the close relationship between microscopic analysis and SERS spectroscopic investigation, we present here a study on the functioning of OLED (organic light-emitting diode) based on poly(3-octylthiophene) (P3OT) deposited on Al electrode [26]. A simultaneous reduction of the emitted power and an increase of the SERS signal of the polymer were observed with current flow, as shown in Fig. 20.18 for the C = C stretching band of the chain.

Fig. 20.18 SERS of P3OT on Al cathode recorded at different bias voltages. Exciting line: 406.7 nm

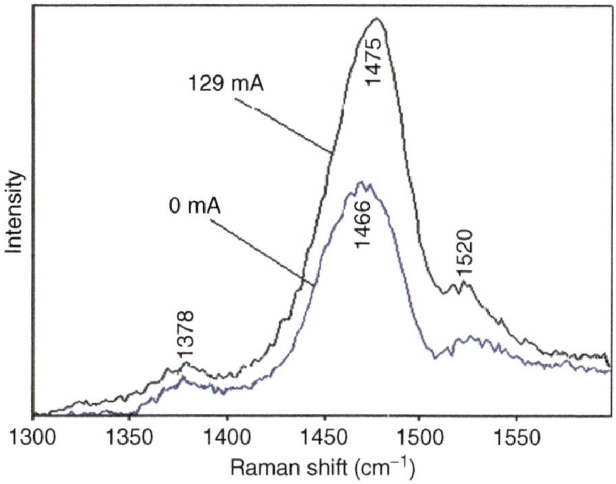

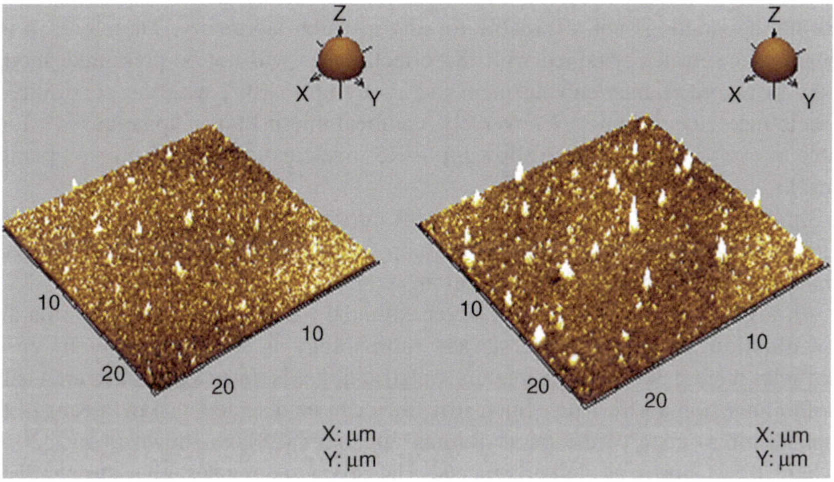

Fig. 20.19 AFM picture of Al cathode before (*left*) and after (*right*) current flow

Figure 20.19 (left side) shows the surface morphology of the Al cathode before current flow. In this case, the RMS roughness evaluated on a 10 × 10 μm^2 area was 7.4 nm with peak-valley fluctuations of 42 nm and an average cluster radius of 138 nm. Figure 20.19 (right side) shows, instead, an image of the same electrode after current flow: an increase of surface roughness corresponding to a RMS value of 9.1 nm, with peak-valley fluctuations of 72 nm, and an average cluster radius of 178 nm is observed. The growth of the SERS signal with current flow can be attributed to the electromagnetic enhancement due to roughening of the Al layer.

5.2 SERS Spectroscopy with Confocal Optical Microscopy

Spectroscopic analysis has become richer in information, thanks to the use of confocal micro-Raman instrumentation, where the same portion of sample observed by means of the microscope is irradiated with the laser excitation to gain the scattering signal. This instrumental apparatus is extremely useful for studying the SERS effect deriving from surface roughness in metals. The previous section has reported on the multiplicity of information that may be acquired by combining spectroscopic measurements with microscopic techniques that explore the surface morphology at nanoscale. However, in that case, the microscopic analysis was performed ex situ, that is, apart from the spectroscopic investigation. By using SERS microspectroscopy, instead, the same analysis is performed in situ, with the same instrument that explores the surface and detects the scattering radiation, even if it is focused in sample areas down to 1 μm^2 spatial resolution.

The Raman techniques combined with AFM microscopic imaging, as for instance TERS (tip-enhanced Raman scattering) spectroscopy [27], allow to analyze surface nanostructures beyond the diffraction limit, but the cost of the instrumental apparatus is not affordable for any research laboratory. Therefore, in this chapter, the results obtained with those techniques will not be presented, though they increased Raman enhancement factors by up to 10^{15}, with the possibility of single-molecule detection. Conversely, confocal micro-Raman apparatus is affordable to every research group allowing SERS investigations with more comparable results.

In this section, it will be shown as most information at nanometer level coming from electronic or atomic force microscopy, previously described, may be easily transferred to micro-SERS investigation.

In Fig. 20.20, AFM images of silver colloidal particles placed on alumina disk by filtration are shown. The surface morphology is formed mainly by metal particles having 30–50 nm diameters, but also larger isolated aggregates are visible with dimension above 1 μm. Such structures can be detected also by means of the optical microscope of the micro-Raman apparatus [28], as shown in Fig. 20.21, where they appear as darker objects. The silver aggregates give strong SERS signals, as shown in the case of adsorbed *p*-nitrobenzoate (PNBA) reported in Fig. 20.22; other regions of the investigated sample, instead, do not give definite SERS spectra. These metal aggregates can also promote catalytic reactions when irradiated by focused 514.5-nm laser line. In this case, the nitro group bands at 865 cm^{-1} (bending) and at 1,348 cm^{-1} (symmetric stretching) disappear, along with the growing of the strong 1,452-cm^{-1} band, assigned to $N = N$ stretching. This band signs the photoreduction of PNBA to azodibenzoate, as hypothesized in the previous section. Irradiating the sample with the 785-nm laser line shows no photoreaction.

In the case of mixed Ag/Pd nanoparticles, shown in the previous section, the TEM measurements pointed out the importance of palladium clusters lying on the Ag nanoparticle surface in promoting the catalytic hydrogenation of adsorbed nitroderivatives. When bimetallic particles are immobilized on quartz plates by

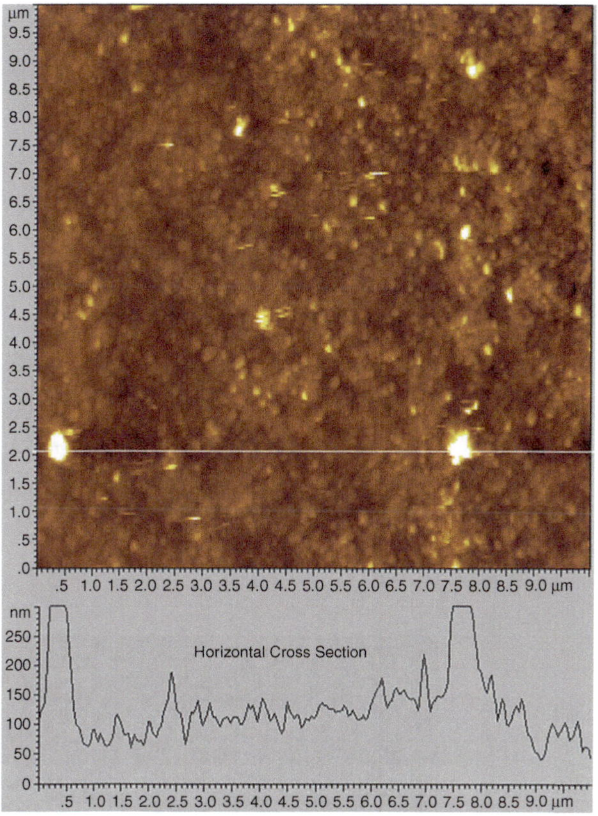

Fig. 20.20 $10 \times 10\ \mu m^2$ AFM image of Ag-coated filter with underlying topographic profile

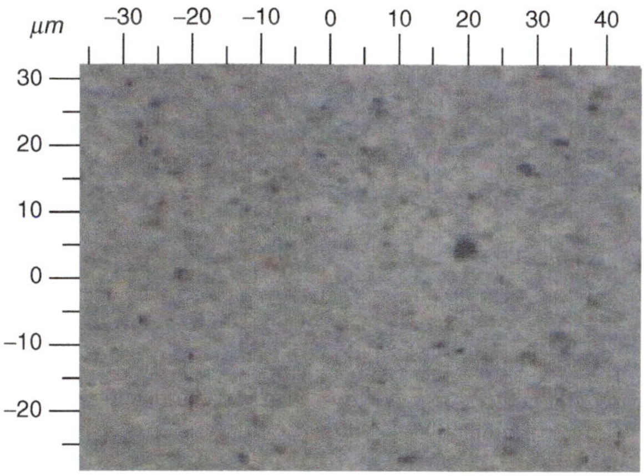

Fig. 20.21 Microscopic image of Ag-coated filter, as obtained by the optical microscope of the Raman instrument

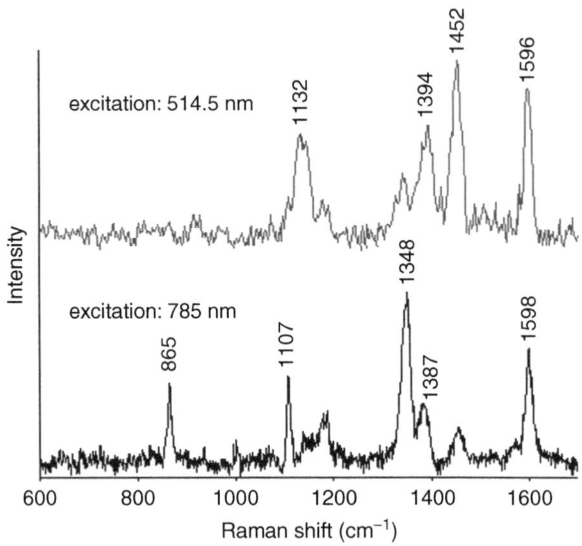

Fig. 20.22 Micro-SERS spectra of PNBA adsorbed on Ag-coated filter

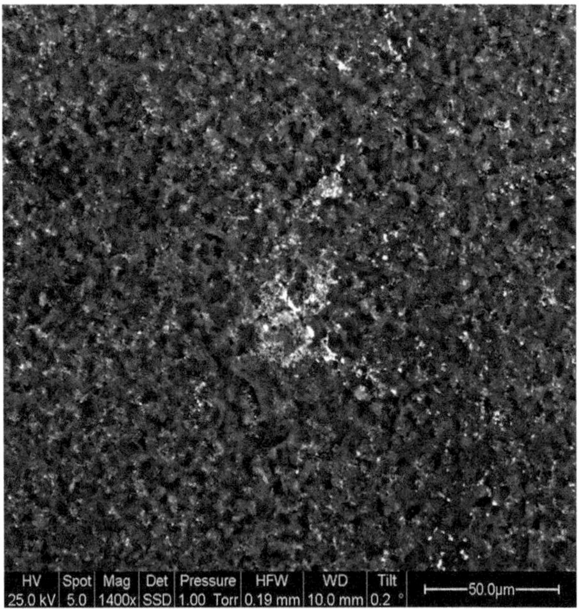

Fig. 20.23 SEM image of a quartz plate coated with immobilized Ag/Pd particles

drying the colloidal suspension, they may be analyzed through the micro-Raman instrumentation. The SEM image of Fig. 20.23 highlights that the bimetallic nanoparticles tend to aggregate in larger structures that become visible with the Raman microscope. The SERS results for *p*-nitrobenzonitrile (NBN) adsorbed on

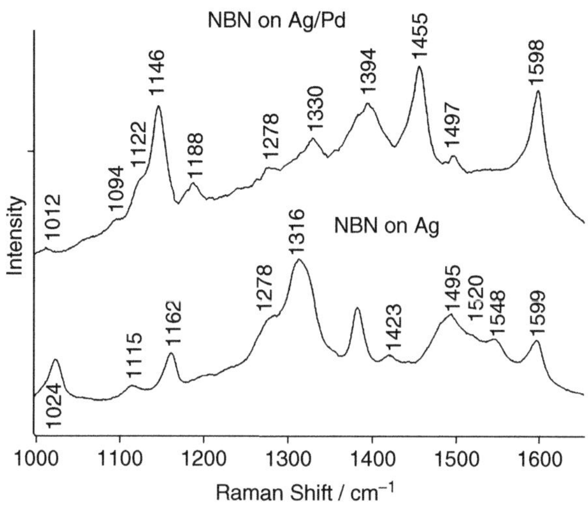

Fig. 20.24 Micro-SERS spectra of NBN adsorbed on Ag and Ag/Pd particles immobilized on quartz. Exciting line: 514.5 nm

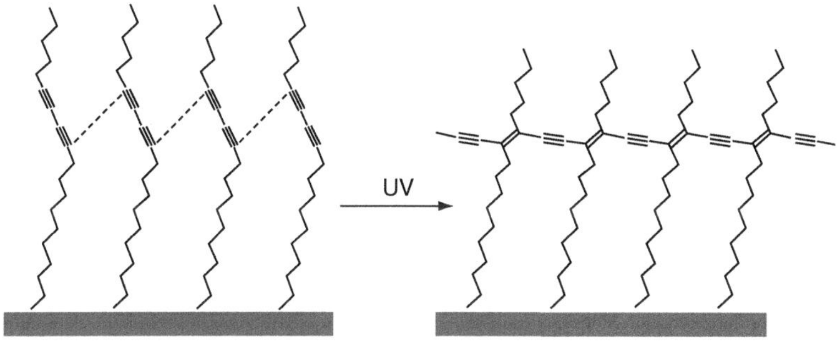

Fig. 20.25 Polymerization of diacetylenic chains adsorbed on metal by UV irradiation

Ag/Pd particles [29] are shown in Fig. 20.24, in comparison with those obtained from pure Ag particles. A catalytic reaction due to the presence of palladium occurs for the adsorbed NBN, which transforms to azodibenzonitrile. Actually, the 1,316-cm^{-1} nitro group band observed on Ag particles is replaced by that at 1,455 cm^{-1}, due to the N = N stretching mode of the azoderivative.

Diacetylenic polymers can be obtained by irradiating the monomer units adsorbed on gold surfaces with UV light [24]. The irradiation induces polymerization between adjacent diacetylenic chains resulting in formation of CC single, double, and triple bonds, as shown in the scheme in Fig. 20.25.

Actually, different polymer forms are obtained under UV light, mainly the blue form, with a more ordered arrangement of the polymer chain, and the red form, which becomes dominant after prolonged irradiation. The presence of these polymers can be revealed by SERS measurements by observing the occurrence of the C = C and C ≡ C stretching modes. The same polymer forms have been observed

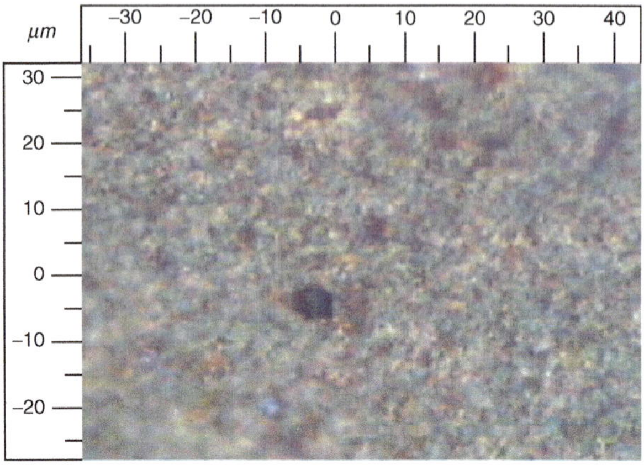

Fig. 20.26 Microscopic image of alumina filter coated with Ag/polymer particles, as obtained by the optical microscope of the Raman instrument

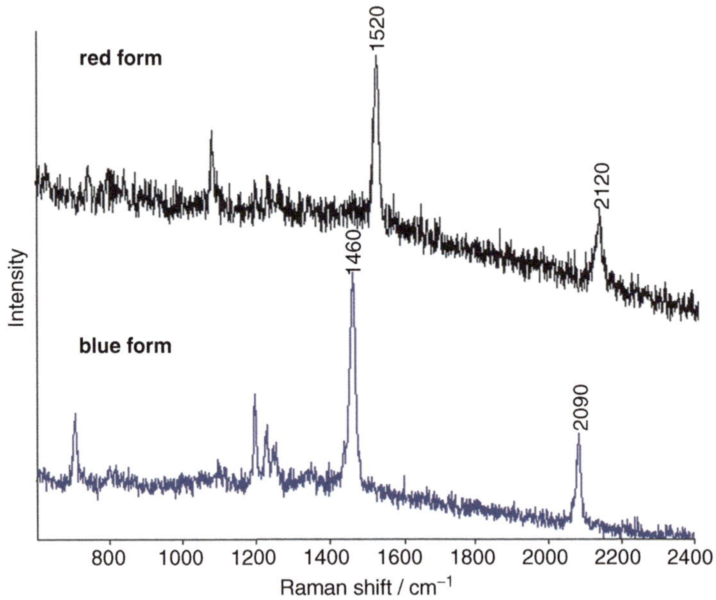

Fig. 20.27 Micro-SERS spectra of the different forms of a diacetylenic polymer adsorbed on Ag particles. Exciting line: 785 nm

also on silver nanoparticles deposited on alumina by filtration and then studied by means of SERS microspectroscopy [30].

The image shown in Fig. 20.26 is obtained with the Raman microscope, which allows to identify the presence of the blue and red forms by observing sample areas with different colors and to obtain separately the SERS spectra of the polymers, as reported in Fig. 20.27.

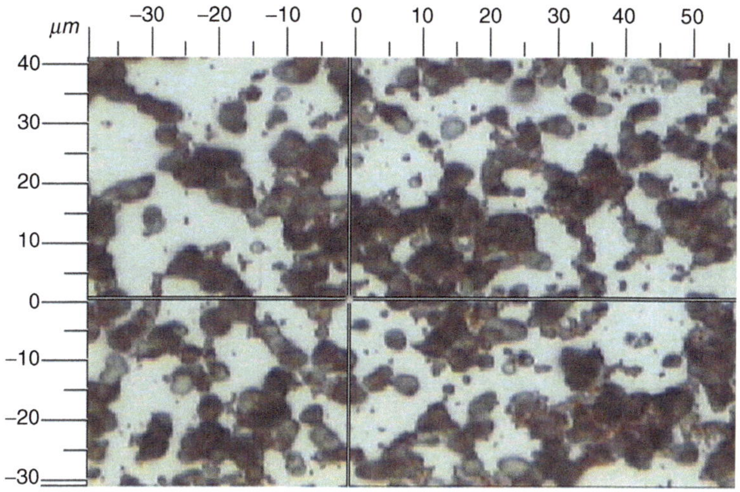

Fig. 20.28 Microscopic image of Ag aggregates after deposition of colloidal suspension, as obtained by the optical microscope of the Raman instrument

The SERS spectrum of the blue form is dominated by the stretching bands of double and triple CC bonds at 1,460 and 2,090 cm^{-1}, respectively. In the red form, the same vibrational modes occur at 1,520 and 2,120 cm^{-1}, due to less electronic conjugation along the polymer chain.

Silver colloidal nanoparticles can be used to activate non-SERS substrates by just depositing them over the substrate surface. This technique is of general use both in the case of metallic as well as nonmetallic surfaces, but when adopted in SERS microspectroscopy, it is very effective. In fact, by means of the Raman microscope, the sample regions, where the deposited Ag nanoparticles are present as large aggregates, can be selected, as shown in Fig. 20.28. There, a stronger SERS response is expected, as verified with different adsorbed ligands [31].

Fabrication of nanostructured surfaces has grown as a field of active research in recent years due to the possibility of controlled deposition of regular patterns on these surfaces. The so-called micro-contact printing (μCP) lets the production of SERS-active solid surfaces characterized by ordered nanostructures with different roughness as a function of the stamp [32]. When nanopatterning is performed on a metal substrate having high reflectivity, such as Ag, Au, or Cu, the optical properties of the surface can be exploited to produce giant Raman enhancements from adsorbed molecules by means of SERS effect. The nanopatterned regions can be identified by Raman microscope. In Fig. 20.29 the micro-SERS spectrum of 1,10-phenanthroline on Ag nanopatterned substrate obtained by μCP technique is shown, together with the AFM images relative to the ordered nanostructures. Since the plasmon bands depend on the shape and dimension of the surface nanostructures, by employing printing supports with variable lateral dimension of the stripes, it is possible to tune the SERS response with respect to the exciting radiation. Conversely, tuning the excitation laser line in the plasmonic band can optimize the SERS response.

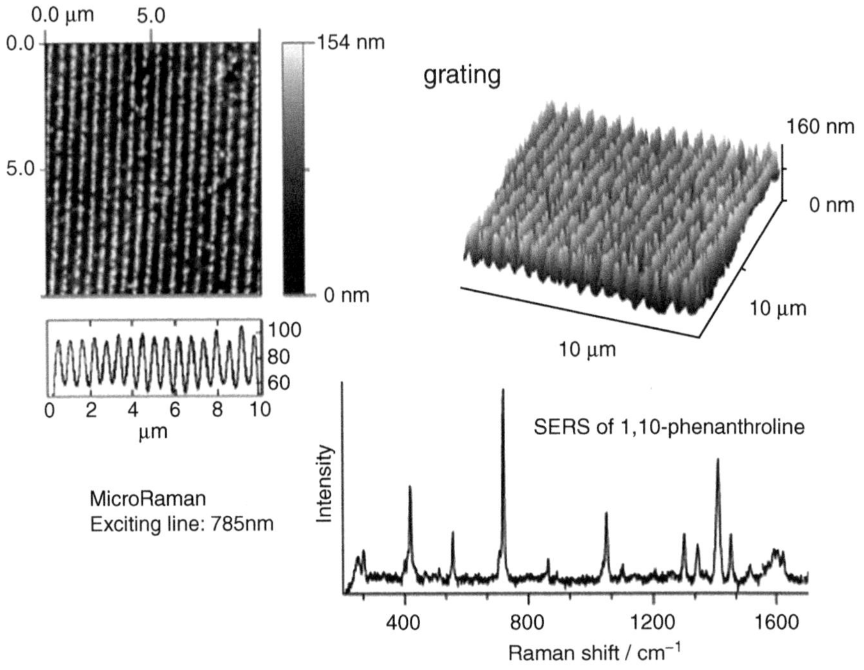

Fig. 20.29 10 × 10 μm^2 AFM images of a nanopatterned Ag surface obtained by micro-contact printing by using a grating stamp. The SERS of 1,10-phenanthroline observed from this substrate is shown *below*

SERS spectroscopy may also provide fundamental information in the mineralogical analysis on rocks and soils, allowing the identification of the different minerals constituting the rock matrix and the chemical and structural composition of these minerals. The spectroscopic exploration of composite rocks may be performed by means of a micro-Raman apparatus; in this way, it should be possible to analyze the response of different micro-specimens distributed all over the surface. The SERS scattering could be greatly enhanced over the fluorescence emission of the rock matrix by depositing Ag colloidal nanoparticles onto the surface of the rock matrix, thus revealing its composition [33]. The effect is ensured by the close vicinity of the Ag particles with the microcrystalline grains embedded in the rock. The SERS spectrum (upper) of clinopyroxene obtained by a microspectroscopic analysis on a rock sample is shown in Fig. 20.30, in comparison with the Raman spectrum of the corresponding single crystal (lower).

6 Applications and Future Perspectives

In this section, some applications of SERS microspectroscopy will be reviewed, by considering that this technique is being liable of new perspectives and evolution in the future of micro-analysis.

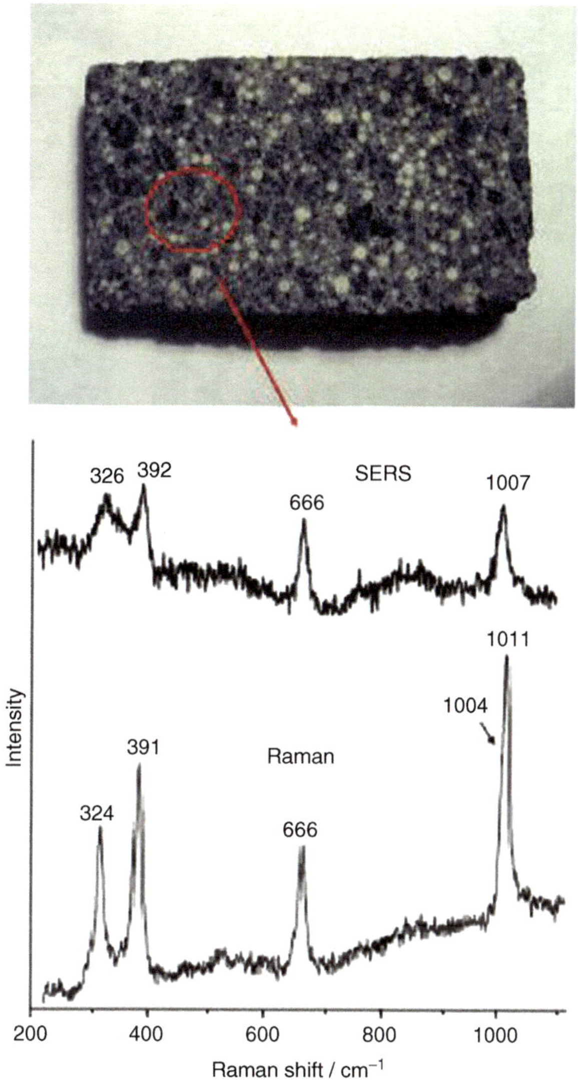

Fig. 20.30 Rock sample image recorded with a digital camera. In the *red circle*, the region wet with Ag colloid is highlighted. The micro-SERS spectrum is compared with the normal Raman spectrum of a pyroxene single crystal. Exciting line: 785 nm

6.1 Nanosensors

Detection systems that are able to detect molecules with a very high efficiency and sensitivity are now in great demand in diverse research fields, ranging from the analytical determination of pollutants to the discovery of biological traces in forensic activities.

The high sensitivity ensured by the SERS microspectroscopy has thus stimulated many efforts in fabricating nanosensors especially suitable for biomedical applications [34]. For these, the challenge is to produce substrates with high SERS

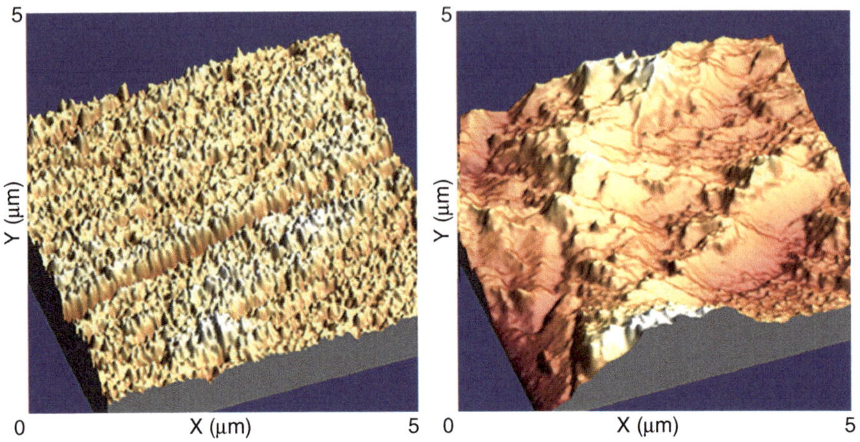

Fig. 20.31 $5 \times 5\ \mu m^2$ AFM images of Ag plates roughened with thiourea and ferric nitrate (*left*) and with ammonium hydroxide and hydrogen peroxide (*right*)

efficiency but at the same time easy to synthesize, be reusable, and be stable with a long half-life. This ensures maximum reproducibility of the spectroscopic data obtained in different laboratories and research groups. Following these requirements, silver SERS-active substrates have been prepared simply by chemical etching of metal plates to get the most effective and uniform surface roughness.

Figure 20.31 shows two silver surfaces after chemical etching, respectively, with thiourea and ferric nitrate on the left, with ammonium hydroxide and hydrogen peroxide on the right. Although the degree of roughness expressed by means of RMS value is high in both the cases, the first one presents a much larger uniformity of nanostructures all over the surface. Hence, the micro-SERS experiments performed on these metal substrates can provide quite reproducible results, because the spectral response is almost constant over the different areas of the examined sample. The SERS efficiency has been verified by adsorption of micro-RNAs with different nucleotide contents, with the trace detection approaching the attomole limit. As a conclusion, these Ag platforms have demonstrated to be suitable biosensors, by recognizing the SERS bands of nucleobases.

6.2 Metal Corrosion

Copper was the first metal used and shaped by mankind, and nowadays, it is employed in different fields for its resistance, ductility, malleability, and electric and thermal conductivities. It is indispensable for the metallurgic industry as well as in water and gas transport, in applications of advanced technology, and in the aerospace industry.

Copper does not usually undergo severe corrosion problems, due to its high electrochemical potential and due to the formation of protective salts on its surface.

However, chemical agents like ammonia or nitric acid, and stray electric currents may give rise to destructive corrosion processes. Since then, protective coatings have been largely used to prevent the copper oxidation, by forming stable, inert, and compact films on the metal surface. Normally, these films are made of organic molecules forming strong Cu-S bonds, such as cysteine [35] and thiolic derivatives of benzoazoles [36] or strong Cu–N bonds, such as triazoles and benzotriazoles [37, 38]. Because of the huge application of copper in many industrial and domestic manufactures, it becomes fundamental to have a very high sensitive technique for testing the integrity level of these protective coatings at the microscopic level. SERS microspectroscopy may be applied in this important field of survey, thanks to the Raman enhancement provided by roughened copper surfaces. Anyway, in the case of smooth copper surfaces, the addition of silver colloidal particle directly onto the copper surface may induce an efficient SERS activation [39].

6.3 Biomedical Applications

SERS microspectroscopy can also be used for the analysis of cells (average dimension 10 μm) to recognize the different components and gain information about their metabolism. SERS enhancement may ensure a giant intensification of the Raman scattering of the proteins contained in living cells, along with a drastic quenching of the fluorescence emission that makes their identification possible and provides structural and biological information on their structure and activity. Early studies were performed in 2006 [40] to localize the PSA antigen in biopsy specimen from patients undergoing prostatectomy. The most promising biomedical applications, however, concern the in vivo micro-SERS detection of important diseases, such as cancer. The first study was reported by Nie and coworkers in 2008 [41] and was realized by means of gold colloidal particles suitably functionalized for selectively binding the epidermal growth factor receptors, overexpressed in many malign tumors.

Other than for diagnostic purposes, SERS-active nanoparticles may be used also for therapeutic aims. Nanohybrids consisting of gold (or silver) colloidal nanoparticles functionalized with drugs may be inserted in living cells and detected by SERS microspectroscopy. Thus, two goals may be reached: first, the drug is localized inside the cellular structures and, second, the action mechanism may be discovered.

6.4 Astrobiology

The inquiry about life traces (past or present) outside Earth represents one of the most fascinating goals of the space exploration toward Mars, Titan, Europa, and other bodies of the solar system.

In both NASA-MSL and ESA-ExoMars projects, an analysis by means of micro-Raman instrumentation is contemplated for the in situ detection of past traces of flowing water and of geologically recent volcanism, indicating the possibility of life in an environment warmer and wetter than previously hypothesized.

The scarce sensitivity generally observed in Raman measurements and the occurrence of fluorescence, which could strongly interfere with the observation of the vibrational bands, are inconveniences that can be overcome, thanks to the SERS effect induced by deposition of silver colloidal nanoparticles. By this method, we have demonstrated [42] that on pyroxene rocks, similar to that present on Mars surface, it is possible by means of SERS microspectrometry to detect the presence of nucleotides in traces less than 1 pg. As substrates we have chosen terrestrial magmatic minerals containing mainly pyroxenes, which present close analogy with those existing in the Martian rocks. On the other hand, pyroxenes represent the most important mineral group in planets and are the major rock-forming silicates on the surface of Mars. Adenine and guanine, nucleobases present in both DNA and RNA strands, and micro-RNA containing adenine (A), guanine (G), and cytosine (C) have been analyzed in this study by SERS microspectroscopy as testing ligands for identifying traces of nucleic acids in Martian rocks and sediments. Adenine, in particular, is a fundamental prebiotic factor [43], which gives rise to strong SERS spectra [44–46]. The micro-Raman apparatus probes different areas of the surfaces without sample manipulation, in contrast with previous Raman experiments [47, 48]. Hence, this study represents a basis for reliable tests of in situ search for life traces in extraterrestrial environments.

Figure 20.32 shows the micro-Raman image of a pyroxene rock wet with a nucleotide solution and successively with silver colloid. The dark regions contain large silver aggregates and give more intense SERS signals. In the lower part of the figure, the SERS spectrum of a micro-RNA sequenced as AGAGACAGCAGA-GACAGACA is shown. The predominant band observed at 735 cm^{-1}, marker of the adenine nucleotide, may be used to detect the presence of biological material. This kind of investigation may be reasonably extended to the extraterrestrial detection of amino acids and oligopeptides, other than being employed for samples transferred onto Earth from planets like Mars or from asteroids. As a conclusion, SERS microspectroscopy may become a frontier technique in the astrobiological research field.

The applications that have been presented above are among the most innovative for SERS microspectroscopy, but others are still developing in very different fields, such as cultural heritage preservation and restoration. To this purpose, the SERS study is very important to identify organic dyes in fibers, paintings, and artistic enamels [49].

Finally, it has to be highlighted that the micro-SERS performances can be greatly implemented by combining the spectroscopic measurements with the "mapping" and "imaging" techniques, already present in the modern instrumentations. In the first case, X-Y motorization of the microscopic analysis allows a spectroscopic scanning of all the points included in a defined sample region. In the second case, the spectra acquired in a complete scan of a region allows to rebuild an image of the region itself as an intensity function of single spectral components.

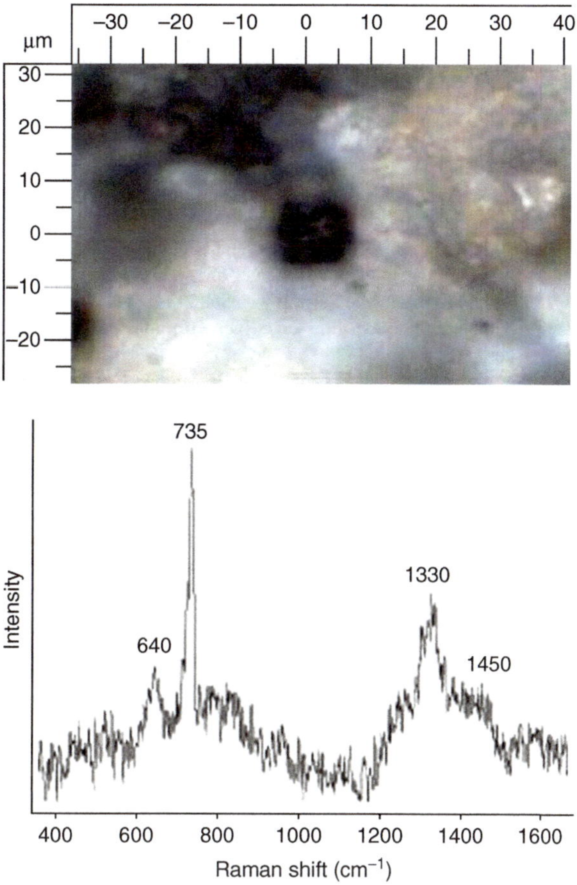

Fig. 20.32 Microscopic image of a pyroxene rock wet with a micro-RNA solution and then with Ag colloid. The micro-SERS spectrum of micro-RNA is shown *below*. Exciting line: 785 nm

However, it should be considered that a focused laser radiation, as that employed in the micro-Raman experiments, may alter the examined sample due to thermal or photochemical effect, leading to severe problems in obtaining reproducible spectra. This is mainly true in the micro-SERS measurements, where monolayers or submonolayers of molecules adsorbed on metal surfaces are investigated. Anyway, an appropriate choice of laser powers, focusing conditions, and wavelengths of the exciting radiation is crucial for overcoming these difficulties.

References

1. Swain RJ, Stevens MM (2007) Raman microspectroscopy for non-invasive biochemical analysis of single cells. Biochem Soc Trans 35:544–549
2. Aroca R (2006) Surface-enhanced vibrational spectroscopy. Wiley, Chichester, UK
3. Kneipp K, Moskovits M, Kneipp H (2006) Surface-enhanced Raman scattering: physics and applications. Springer, Heidelgerg

4. Le Ru EC, Etchegoin PG (2009) Principles of surface-enhanced Raman spectroscopy and related plasmonic effects. Elsevier, Oxford, UK
5. Schlücker S (2009) SERS microscopy: nanoparticle probes and biomedical applications. Chemphyschem 10:1344–1354
6. Kneipp K, Wang Y, Kneipp H, Perelman LT, Itzkan I, Dasari RR, Feld MS (1997) Single molecule detection using surface-enhanced raman scattering (SERS). Phys Rev Lett 78: 1667–1670
7. Nie S, Emory SR (1997) Probing single molecules and single nanoparticles by surface-enhanced Raman scattering. Science 275:1102–1106
8. Muniz-Miranda M, Neto N, Sbrana G (2003) SERS intensities of benzodiazines adsorbed on silver nanoparticles. J Mol Struct 651–653:85–90
9. Muniz-Miranda M, Cardini G, Schettino V (2004) Surface-enhanced Raman spectra of pyridine and pyrazolide on silver colloids: chemical and electromagnetic effects. Theor Chem Acc 111:264–269
10. Muniz-Miranda M, Pergolese B, Bigotto A, Giusti A (2007) Stable and efficient silver substrates for SERS spectroscopy. J Colloid Interface Sci 314:540–544
11. Pergolese B, Bigotto A, Muniz-Miranda M, Sbrana G (2005) Gold/palladium and silver/palladium metallic substrates for surface-enhanced Raman scattering. Appl Spectrosc 59: 194–199
12. Muniz-Miranda M (2003) Silver clusters onto nanosized colloidal silica as novel surface-enhanced Raman scattering active substrates. Appl Spectrosc 57:655–660
13. Muniz-Miranda M, Ottaviani MF (2004) Silver nanoclusters in mesoporous silica, as obtained by visible-laser irradiation. Laser Phys 14:1533–1538
14. Hildebrandt P, Stockburger M (1984) Surface-enhanced resonance Raman spectroscopy of Rhodamine 6G adsorbed on colloidal silver. J Phys Chem 88:5935–5944
15. Hildebrandt P, Keller S, Hoffmann A, Vanhecke F, Schrader B (1993) Enhancement factor of surface-enhanced Raman scattering on silver and gold surfaces upon near-infrared excitation. Indication of an unusual strong contribution of the chemical effect. J Raman Spectrosc 24:791–796
16. Schneider S, Grau H, Halbig P, Freunscht P, Nickel U (1996) Stabilization of silver colloids by various types of anions and their effect on the surface-enhanced Raman spectra of organic dyes. J Raman Spectrosc 27:57–68
17. Bosnick KA, Jiang J, Brus LE (2002) Fluctuations and local symmetry in single-molecule Rhodamine 6 G Raman scattering on silver nanocrystal aggregates. J Phys Chem B 106: 8096–8099
18. Futamata M, Maruyama Y, Ishikawa M (2002) Microscopic morphology and SERS activity of Ag colloidal particles. Vib Spectrosc 30:17–23
19. Doering WE, Nie S (2002) Single-molecule and single-nanoparticle SERS: examining the roles of surface active sites and chemical enhancement. J Phys Chem B 106:311–317
20. Otto A, Bruckbauer A, Chen YX (2003) On the chloride activation in SERS and single molecule SERS. J Mol Struct 661–662:501–514
21. Grochala W, Kudelski A, Bukowska J (1998) Anion-induced charge-transfer enhancement in SERS and SERRS spectra of Rhodamine 6 G on a silver electrode: how important is it? J Raman Spectrosc 29:681–685
22. Pergolese B, Muniz-Miranda M, Bigotto A (2005) Surface enhanced Raman scattering investigation of the halide anion effect on the adsorption of 1,2,3-triazole on silver and gold colloidal nanoparticles. J Phys Chem B 109:9665–9671
23. Muniz-Miranda M, Innocenti M, Foresti ML (2006) Relation between the photoreaction of p-nitrobenzoic acid onto silver-coated filters and the surface roughness, as detected by SERS and AFM. Surf Sci 600:2096–2102
24. Muniz-Miranda M, Giorgetti E, Margheri G, Del Rosso T, Sottini S, Giusti A, Alloisio M (2005) SERS investigation on the polymerization of carbazolyl-diacetylene monolayers on gold surfaces. Macromol Symp 230:67–70

25. Muniz-Miranda M, Pergolese B, Bigotto A, Giusti A, Innocenti M (2007) A Raman and AFM investigation on metal surfaces SERS-activated by silver colloidal nanoparticles. Mat Sci Eng C 27:1295–1299
26. Giorgetti E, Margheri G, Del Rosso T, Sottini S, Muniz-Miranda M, Innocenti M (2004) A study of the degradation of poly(3-octylthiophene)-based light emitting diodes by surface enhanced Raman scattering. Appl Phys B 79:603–609
27. Domke KF, Pettinger B (2010) Studying surface chemistry beyond the diffraction limit: 10 years of TERS. Chemphyschem 10:1365–1373
28. Muniz-Miranda M, Innocenti M (2004) AFM and micro-Raman investigation on filters coated with silver colloidal nanoparticles. Appl Surf Sci 226:125–130
29. Pergolese B, Muniz-Miranda M, Bigotto A (2007) Catalytic activity of Ag/Pd bimetallic nanoparticles immobilized on quartz surfaces. Chem Phys Lett 438:290–293
30. Raimondo C, Alloisio M, Demartini A, Cuniberti C, Dellepiane G, Jadhav SA, Petrillo G, Giorgetti E, Gellini C, Muniz-Miranda M (2009) Photopolymerization in water of diacetylenes chemisorbed onto noble metal nanoparticles: a spectroscopic study. J Raman Spectrosc 40:1831–1837
31. Muniz-Miranda M, Gellini C, Salvi PR, Schettino V, Pagliai M, Innocenti M (2011) Nanostructured Ag platforms as biosensors of nucleobase chains. J Nanosci Nanotechnol 11:1
32. Gellini C, Muniz-Miranda M, Innocenti M, Carlà F, Loglio F, Foresti ML, Salvi PR (2008) Nanopatterned Ag substrates for SERS spectroscopy. Phys Chem Chem Phys 10: 4555–4558
33. Muniz-Miranda M, Gellini C, Bindi L (2009) Surface-enhanced Raman spectroscopy for identifying rock composition. Spectrochim Acta A 73:456–459
34. Kneipp J, Kneipp H, Wittig B, Kneipp K (2010) Novel optical nanosensors for probing and imaging live cells. Nanomedicine 6:214–226
35. Zhang D-Q, Gao L-X, Cai Q-R, Lee KY (2010) Inhibition of copper corrosion by modifying cysteine self-assembled film with alkylamine/alkylacid compounds. Mat Corros 61:16–21
36. Yan CW, Lin HC, Cao CN (2000) Investigation of inhibition of 2-mercaptobenzoxazole for copper corrosion. Electrochim Acta 45:2815–2821
37. Fleischmann M, Mengoli G, Musiani MM, Pagura C (1985) An electrochemical and Raman spectroscopic investigation of synergetic effects in the corrosion inhibition of copper. Electrochim Acta 30:1591–1602
38. Maciel JM, Villamil Jaimes RFV, Corio P, Rubim JC, Volpe PL, Neto AA, Agostinho SML (2008) The characterisation of the protective film formed by benzotriazole on the 90/10 copper–nickel alloy surface in H_2SO_4 media. Corros Sci 50:879–886
39. Pergolese B, Muniz-Miranda M, Bigotto A (2006) Surface-enhanced Raman scattering investigation of the adsorption of 2-mercaptobenzoxazole on smooth copper surfaces doped with silver colloidal nanoparticles. J Phys Chem B 110:9241–9245
40. Schlücker S, Küstner B, Punge A, Bonfig R, Marx A, Ströbel P (2006) Immuno-Raman microspectroscopy: In situ detection of antigens in tissue specimens by surface-enhanced Raman scattering. J Raman Spectrosc 37:719–721
41. Qian X, Peng X-H, Ansari DO, Yin-Goen Q, Chen GZ, Shin DM, Yang L, Young AN, Wang MD, Nie S (2008) In vivo tumor targeting and spectroscopic detection with surface-enhanced Raman nanoparticle tags. Nat Biotechnol 26:83–90
42. Muniz-Miranda M, Gellini C, Salvi PR, Pagliai M (2010) Surface-enhanced Raman micro-spectroscopy of DNA/RNA bases adsorbed on pyroxene rocks as a test of in situ search for life traces on Mars. J Raman Spectrosc 41:12–15
43. Oro J (1961) Mechanisms of synthesis of adenine from hydrogen cyanide under possible. Primitive earth conditions. Nature 191:1193–1194
44. Kneipp K, Kneipp H, Kartha VB, Manoharan R, Deinum G, Itzkan I, Dasari RR, Feld MS (1998) Detection and identification of a single DNA base molecule using surface-enhanced Raman scattering (SERS). Phys Rev E 57:R6281–R6284

45. El Amri C, Baron M-H, Maurel M-C (2003) Adenine and RNA in mineral samples: surface-enhanced Raman spectroscopy (SERS) for picomole detections. Spectrochim Acta A 59:2645–2654
46. Bell SEJ, Sirimuthu NMS (2006) Surface-enhanced Raman spectroscopy (SERS) for sub-micromolar detection of DNA/RNA mononucleotides. J Am Chem Soc 128:15580–15581
47. El Amri C, Baron M-H, Maurel M-C (2004) Adenine adsorption on and release from meteorite specimens assessed by surface-enhanced Raman spectroscopy. J Raman Spectrosc 35:170–177
48. Dunn DS, Sridhar N, Miller MA, Price KT, Pabalan R, Abrajano TA Jr (2007) Development of a surface-enhanced Raman technique for biomarker studies on Mars. Appl Spectrosc 61:25–31
49. Casadio F, Leona M, Lombardi JR, Van Duyne R Identification of organic colorants in fibers, paints and glazes by surface enhanced Raman spectroscopy. Acc Chem Res. doi:10.1021/ar100019q, ACS ASAP

Ultraviolet Raman Spectroscopy of Nanoscale Ferroelectric Thin Films and Superlattices

21

Dmitri A. Tenne

1 Definition of the Topic

Raman spectroscopy with ultraviolet excitation: advantages and recent applications for nanoscale ferroelectric materials.

2 Overview

Ultraviolet Raman spectroscopy has emerged as a powerful technique for characterization of nanoscale materials, in particular, wide-bandgap semiconductors and dielectrics. The advantages of ultraviolet excitation for Raman measurements of ferroelectric thin films and heterostructures, such as reduced penetration depth and enhanced scattering intensity, are discussed. Recent results of application of ultraviolet Raman spectroscopy for studies of the lattice dynamics and phase transitions in nanoscale ferroelectric structures, such as superlattices based on $BaTiO_3$, $SrTiO_3$, and $CaTiO_3$, as well as ultrathin films of $BaTiO_3$ and $SrTiO_3$ are reviewed.

3 Introduction

3.1 Ferroelectricity at the Nanoscale

Ferroelectrics are a class of materials possessing a spontaneous electric polarization, which can be switched between crystallographically defined directions by

D.A. Tenne
Department of Physics, Boise State University, 1910 University Drive, Boise, ID, USA

Challa S. S. R. Kumar (ed.), *Raman Spectroscopy for Nanomaterials Characterization*,
DOI 10.1007/978-3-642-20620-7_21, © Springer-Verlag Berlin Heidelberg 2012

the application of an electric field [1]. Ferroelectric materials have been investigated for various device applications such as nonvolatile memory devices, piezoelectric micro- and nano-electromechanical systems, or tunable microwave devices [2–7]. Recent progress in epitaxial technology of oxide materials [8–10] has opened a new stage in the field, making possible an experimental realization of complex artificial oxide nanoscale heterostructures with an atomic-level thickness control comparable to that developed for semiconductor heterostructures. The advances of epitaxial oxide thin film deposition and the continuous demand for device miniaturization have rapidly moved the science and technology of ferroelectrics toward thin films and multilayer structures at the nanometer scale. High-quality ultrathin films of ferroelectric materials can be grown and combined with other ferroelectrics and such materials as superconducting and magnetic oxides, semiconductors, allowing fabrication of multifunctional structures and opening exciting opportunities for potential device applications. On the other hand, nanoscale ferroelectrics are also fascinating objects from the fundamental physics point of view, since the reduction of the structural dimensions gives rise to new phenomena and properties dramatically different from those of homogeneous bulk ferroelectrics [6, 11–14].

A number of factors can influence the behavior of ferroelectric thin films and multilayer structures with layer thickness at nanometer scale. One of the major factors is strain in epitaxial structures [15]. Recent demonstrations of huge strain effect on ferroelectric properties include changes in the phase diagram [16–22], dramatic enhancement of ferroelectric polarization, and increase of the ferroelectric phase transition temperature T_c [23–27], induced ferroelectricity in non-ferroelectric materials like $SrTiO_3$ or $KTaO_3$ [28–33], or even simple rocksalt binary oxides like BaO ([34], theoretically predicted).

Size effect is another factor that strongly influences the properties of ferroelectric nanostructures, and the issue of a critical size for ferroelectricity has been actively discussed [4, 6, 11–14, 35]. For a long time it was believed that ferroelectricity was suppressed in small particles and thin films [1], and there was a critical size in order of few tens of nanometers below which a spontaneous polarization cannot be sustained in a material. Recent experimental and theoretical studies [36–48] demonstrated that ferroelectricity exists down to vanishingly small sizes, much smaller than previously thought. These studies revealed that the issue of critical size is very complex, and electrical and mechanical boundary conditions play an essential role in nanoscale ferroelectricity.

Periodic structures containing alternating layers of different ferroelectric and non-ferroelectric materials – superlattices (SLs) – have received increased attention. High-quality ferroelectric SLs with nearly atomically sharp interfaces in various perovskite systems have been synthesized and investigated [14, 24–27, 49–56], and also studied theoretically [44, 57–64]. Properties of such structures are not just a simple combination of the properties known for constituent bulk materials, as they are affected by both mechanical (lattice-mismatch-induced strain) and electrostatic boundary conditions at multiple interfaces located within close proximity to each other. Strain engineering is one of the most appealing ways to enhanced properties in ferroelectric materials [15–17], and SLs are especially attractive for that purpose. This is because it is usually much easier to grow a large thickness multilayer film, while maintaining the structure under a coherent strain state (commensurate to the substrate), which is hard or impossible to achieve for a single layer of the same thickness. Varying the thicknesses of individual SL constituents provides an additional opportunity to manipulate its properties. Just as the development of semiconductor epitaxy in the 1970s–1980s enabled to synthesize semiconductor superlattices with monoatomic layer precision and lead to great advances in semiconductor physics and technology, so the advances in fabrication of artificial ferroelectric and dielectric superlattices are expected to promote the progress in fundamental understanding of ferroelectric effects and utilization of these effects in novel electronic devices.

3.2 Lattice Dynamics and Ferroelectricity

Of central importance for understanding the fundamental properties of ferroelectrics is dynamics of the crystal lattice, which is closely related to the phenomenon of ferroelectricity [1]. The soft-mode theory of displacive ferroelectrics [65] has established the relationship between the polar optical vibrational modes and the spontaneous polarization. The lowest-frequency transverse optical phonon, called the soft mode, involves the same atomic displacements as those responsible for the appearance of spontaneous polarization, and the soft mode instability at Curie temperature T_c causes the ferroelectric phase transition. The soft-mode behavior is also related to such properties of ferroelectric materials as high dielectric constant, large piezoelectric coefficients, and dielectric nonlinearity, which are extremely important for technological applications. The Lyddane-Sachs-Teller (LST) relation connects the macroscopic dielectric constants of a material with its microscopic properties – optical phonon frequencies:

$$\varepsilon_0 = \varepsilon_\infty \prod_j \frac{\omega_{LO_j}^2}{\omega_{TO_j}^2} \tag{21.1}$$

where ε_0 and ε_∞ are the static and high-frequency dielectric constants and ω_{LO} and ω_{TO} are the frequencies of longitudinal and transverse optical phonons, respectively. Usually the frequencies of higher optical phonons vary slightly with temperature, and the temperature dependence of ε_0 is mostly determined by the behavior of the lowest-frequency TO mode, the soft mode. The decrease of the soft mode frequency as the temperature approaches to T_c causes a dramatic increase of the dielectric constant. The electric field–induced hardening, that is, increase of the soft-mode frequency [66, 67] is responsible for the dielectric nonlinearity, since, according to the LST relation, the increase of ω_{TO} leads to the decrease of ε_0. Another important materials parameter for applications, the dielectric loss, is also related to the soft-mode behavior. In an ideal ferroelectric crystal, it is determined by the damping of the soft mode through multiple-phonon processes in the paraelectric phase, and dominated by the quasi-Debye contribution in the ferroelectric phase [68–70].

Optical phonons other than the soft mode, the so-called hard modes, can also be used to study ferroelectric and structural phase transitions, since their frequencies and spectroscopic activity are affected by transformations of crystal structure.

Lattice dynamics in bulk perovskite oxide ferroelectrics has been investigated for several decades using neutron scattering [71–77], far infrared spectroscopy [78–83], and Raman scattering. Raman spectroscopy is one of the most powerful analytical techniques for studying the lattice vibrations and other elementary excitations in solids providing important information about the structure, composition, strain, defects, and phase transitions. This technique was successfully applied to many ferroelectric materials, such as bulk perovskite oxides: barium titanate ($BaTiO_3$), strontium titanate ($SrTiO_3$), lead titanate ($PbTiO_3$) [84–88], and others.

More recently, important results were obtained on lattice dynamics of thin films of these materials ([89–106]; see [107] for a review). These results contributed significantly to a fundamental understanding of lattice dynamical properties in thin films of ferroelectric materials. However, the above mentioned results were obtained on films thicker than 150 nm. Study of lattice dynamics in ferroelectric ultrathin films and heterostructures thinner than $\sim$100 nm has been a challenging task. A difficulty in applying conventional Raman spectroscopy (using visible or near-infrared excitation) for thin films of ferroelectrics and other wide-bandgap materials lies in the fact that the visible photon energy is much smaller than the bandgap. Consequently, the absorption of light is extremely weak and the penetration depth is large, allowing light to travel through thin film into the much thicker substrate. Much larger scattering volume of the substrate thus generates overwhelming signals in the Raman spectra.

Several approaches were used to overcome this difficulty. In earlier Raman studies of ferroelectric thin films, the films were grown either on reflective substrates, such as Pt/Si [89] or on substrates with low Raman activity at the frequency of interest, such as Al_2O_3 [90, 91], $KTaO_3$ [92], fused silica [93], and MgO [93–95]. For many applications, however, these are not the substrates of choice in terms of growth of high-quality defect-free epitaxial films and structures because of lattice and thermal expansion mismatch problems, and the crystalline quality of the ferroelectric thin films is often compromised. In the studies of relatively thick (over 100 nm) $SrTiO_3$, $BaTiO_3$, $Ba_xSr_{1-x}TiO_3$ films, the metal-oxide bilayer technique using an opaque layer of a conductive oxide, such as $SrRuO_3$, has been employed to eliminate the substrate contribution [96–102]. The conductive layer reflects a substantial part of the laser beam back into the dielectric film under study, and the rest of the laser light attenuates quickly within the conducting layer. Therefore, Raman scattering from the substrate can be avoided and the signal from the ferroelectric film can be detected. In contrast to thin film growth on substrates like Pt/Si or Al_2O_3, the metal-oxide layer allows the growth of the epitaxial ferroelectric thin films with significantly higher structural quality, which is important for probing intrinsic thin film properties. This technique was also successfully applied for study of the soft modes in $SrTiO_3$ films by far infrared ellipsometry [103].

However, for ultrathin films (<100 nm) and nanostructures, another approach is required. In such thin films, the intensity of Raman scattering using visible excitation sources is extremely low, and even a weak signal from the conductive oxide layer used to block the substrate can prevent the observation of the phonons of the nanometer-thick film. Raman spectroscopy using an ultraviolet (UV) excitation has been recently demonstrated to be a suitable tool for this task [43]. In this chapter, we summarize the advantages of UV Raman spectroscopy for probing nanoscale ferroelectric heterostructures and review recent results obtained using this technique on ultrathin $BaTiO_3$ and $SrTiO_3$ films and $BaTiO_3/SrTiO_3$ superlattices.

4 Basic Methodology

4.1 Basic Principles of Raman Scattering

Raman spectroscopy in the visible range is a well-established technique, widely used for characterization of various materials [108–117]. Raman scattering is the process of inelastic light scattering in which the energy of scattered light $\hbar\omega_s$ is different from that of the incident light $\hbar\omega_i$. The difference is the energy $\hbar\Omega$ of an elementary excitation (molecular or crystal lattice vibration or electronic excitation) participating in the process. Energy conservation yields:

$$\hbar\omega_s = \hbar\omega_i \pm \hbar\Omega \qquad (21.2)$$

Here the "–" and "+" signs correspond to those Raman processes in which an elementary excitation is created or annihilated (Stokes and anti-Stokes processes, respectively). In crystalline solids, the quasi-momentum conservation gives the following relation between the wave vectors $\mathbf{k}_i$ of the incident light, $\mathbf{k}_s$ of the scattered light, and $\mathbf{q}$ of the elementary excitation:

$$\mathbf{k}_s = \mathbf{k}_i \pm \mathbf{q} \qquad (21.3)$$

The largest magnitude of scattering wave vector q occurs in the backscattering geometry:

$$q = \frac{1}{c}(n(\omega_i)\omega_i + n(\omega_s)\omega_s) \qquad (21.4)$$

Here c is the light velocity in vacuum and n is the refractive index. For conventional Raman experiments, when ω_i is in the optical range, Equation (4) yields a q value in the range of 10^6 cm^{-1}, which is much smaller than the wave vector of the Brillouin zone edge ($\sim10^8$ cm^{-1}). Therefore, only near-zone-center excitations can participate in the first-order Raman process.

In a Raman scattering experiment, the frequency, wave vector, and polarization of incident light (usually from a laser source) are known. Analysis of scattered light frequency, polarization, and relative intensity provides information on properties of elementary excitations in the material under study. General description of the Raman process as well as details of Raman instrumentation can be found elsewhere [108–110, 118].

In Raman scattering, there are certain selection rules that imply the criteria which allow determining the activity of a certain elementary excitation (in our case a phonon) in the process and in specified polarization geometry. Energy and quasi-momentum conservation laws (21.2) and (21.3) can also be considered as selection rules. Selection rules of another type are determined by the symmetry of a phonon and the symmetry of the scattering system (crystal lattice). The symmetry of lattice vibrations and Raman selection rules were derived by group theory methods for all zone-center phonons in all 32 crystallographic classes and can be found in Ref. [118]. Here, we mention one characteristic example of a selection rule found in centrosymmetric crystals and important for ferroelectric materials discussed in the following sections. Phonons in such crystals can be of either even or odd parity with respect to the inversion. If the phonon displacement is of odd parity, the Raman scattering vanishes. Odd phonons are active in infrared absorption, since the electric field vector of infrared radiation is also antisymmetric under inversion. Hence, infrared light can interact with odd, but not with even phonons. Therefore in centrosymmetric crystals, the Raman-active phonons (even parity) are not infrared-active, and vice versa. In some crystals there are phonon modes called "silent," which are neither Raman nor infrared-active.

4.2 Advantages of Ultraviolet Excitation for Raman Studies of Wide-Bandgap Thin Film Materials

One of the major directions in recent development of Raman spectroscopy has been the expansion of the technique to the ultraviolet region [119]. Until the 1980s, the most commonly used excitation sources in Raman spectroscopy were lasers emitting in the visible range (argon ion laser with several lines between 457 and 514.5 nm being the most frequently used). Later, ultraviolet Raman spectrometers were developed [120–123]. Various continuous wave laser sources are currently being used, including fundamental lines of Ar^+ (364 and 351 nm), Kr^+ (350.7), and He–Cd (325 nm) lasers, as well as intra-cavity frequency-doubled lines of Ar^+ (257, 244, 238, and 228.9 nm) and Kr^+ (206 nm) lasers. Currently, UV Raman spectroscopy finds applications in various fields of physics, chemistry, biology, and materials science [123].

Despite the advantages of UV excitation for a number of applications including studies of thin films of wide bandgap materials, there have been several technical difficulties, such as lower throughput efficiency, insufficient dispersion, and higher stray light level of UV Raman spectrometers compared to those operating in visible range. These difficulties impeded the wide application of UV Raman spectroscopy, in particular, deep UV (200–300 nm) for the studies of thin films of wide-bandgap materials, such as ferroelectric oxides.

For effective Rayleigh scattering separation and stray light rejection required for near-excitation measurements of low-frequency vibrations such as soft phonon modes, triple monochromators are normally used. However, triple monochromators have low throughput efficiency, and in the UV range only 2–3% (5–7% at best) of scattered light is typically through a triple spectrometer to the detector. Therefore, for the experiments not requiring near-excitation measurements (below 200 cm^{-1} from the excitation laser line), single monochromators are commonly used in combination with holographic notch filters for the Rayleigh light rejection. However, efficient notch filters are not currently available in the UV region. Efficient Raman spectrometers operating in UV, including deep UV range with excitation at 244 nm and even at 224 nm have been reported [120–122]. Asher's group [120, 121] constructed a highly efficient (over 10% throughput at 250 nm) UV micro-Raman spectrometer based on single monochromator combined with specially designed dielectric filters for Rayleigh scattering rejection. A Cassegrain reflective microscope objective was used to increase the efficiency of collecting the scattered radiation. Sands et al. [122] described a confocal and scanning near-field Raman microscope operating at deep UV wavelengths. However, Raman measurements on those systems were limited to high-frequency range (at least 300 cm^{-1}, typically above 500 cm^{-1}). Such a limitation of frequency range is insufficient for studies of many dielectric and semiconductor materials, because many optical phonon lines in these materials lie below 500 cm^{-1}, not to mention the soft phonon modes, which are usually observed below 100 cm^{-1} (as low as 10 cm^{-1}). Therefore, more effective Rayleigh scattering filtering and stray light reduction is required.

Reports of UV Raman spectrometers for characterization of dielectric (ZrO_2, $SrZrO_3$) [124, 125] and wide bandgap semiconductor (SiC and AlGaN) [126] materials demonstrated that these problems can be overcome. These spectrometers were, in fact, triple monochromators using subtractive double monochromator as a filter for stray light reduction. The use of double monochromator as a filter allowed Yashima et al. to measure Raman spectra of ZrO_2 as close as 20 cm^{-1} using 363.8 nm excitation line [124]. Using the same system and excitation wavelength, Fujimori et al. studied the high-temperature phase transition in $SrZrO_3$ and observed the soft phonon modes in the frequency range 110–140 cm^{-1}. Nakashima et al. [126] reported deep UV (using 244 nm excitation) Raman spectra of several wide bandgap materials (SiC, AlGaN) in the frequency range starting as low as 170 cm^{-1}. The reduction of the throughput efficiency caused by using double monochromator as a filter can be compensated in part by the improved throughput of the collecting optics and incident optical path by using low optical loss high-reflectivity dielectric mirrors and objectives. Nakashima et al. [126] reported the throughput efficiency of 34% with deep UV (244 nm) excitation. Currently, double and triple monochromators and efficient multichannel CCD detectors optimized for UV range are available commercially, allowing measurement of UV Raman spectra very close to the excitation laser line (as close as $\sim$10 cm^{-1} with 325-nm excitation). Recent examples of application of UV Raman spectroscopy for characterization of thin films of various wide bandgap materials include GaN [127], InGaN [128], nanocrystalline diamond [129–132], and silicon

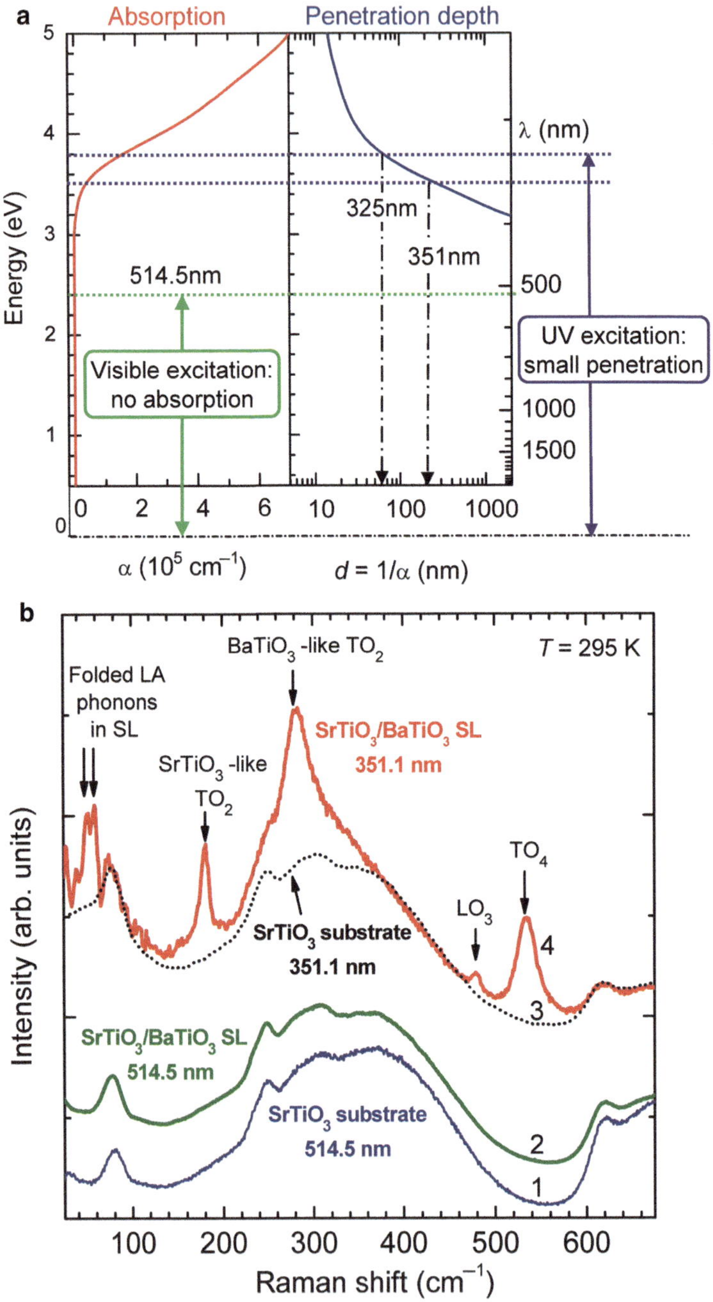

Fig. 21.1 (continued)

nitride [134]. Tisinger et al. reported UV Raman study of epitaxial $SrTiO_3$ films on Si substrates.

There are several advantages of using UV Raman spectroscopy for thin films and nanostructures of wide bandgap materials compared to excitation in the visible region. First, using UV excitation allows probing nanometer-thin films and heterostructures because of greatly reduced penetration depth. As mentioned in Sect. 1.2, a major reason why conventional visible Raman spectroscopy works poorly for ferroelectric thin films is transparency of the materials to visible light leading to the dominance of the substrate signal in spectra. For UV excitation, the photon energy is above the bandgap of ferroelectric oxides such as $SrTiO_3$ or $BaTiO_3$, which leads to a much stronger absorption. This is illustrated in Fig. 21.1a showing absorption and penetration depth of visible and UV light in $BaTiO_3$ (absorption data from [136]). This advantage of UV over visible excitation is clearly demonstrated by Fig. 21.1b showing Raman spectra of a ferroelectric thin film heterostructures: $(BaTiO_3)_5(SrTiO_3)_4 \times 25$ superlattice (SL, one period consists of 5 unit cell-thick $BaTiO_3$ layer followed by 4 unit cell-thick $SrTiO_3$; repeated 25 times; total thickness ~90 nm) measured with visible (514.5 nm) and UV (351.1 nm) excitations. The substrate features completely dominate the 514.5-nm spectrum while they are greatly reduced in the UV (351.1 nm) spectrum, in which peaks of superlattice phonons are clearly observed. Using even shorter, 325-nm excitation wavelength allows further reducing the penetration depth. Raman spectra from $BaTiO_3$ films as thin as 1.6 nm have been detected using this wavelength [48].

Many ferroelectric oxides (such $BaTiO_3$, $PbTiO_3$, $SrTiO_3$) have bandgaps in the range 3.0–3.5 eV and UV excitation near the bandgap can also lead to a resonance enhancement of Raman signals. In quantum-mechanical treatment, the process of inelastic light scattering consists of three transitions: (1) the electronic transition from the ground state $|0\rangle$ to an excited state $|\alpha\rangle$ – creation of an electron–hole pair induced by the absorption of a photon $\hbar\omega_i$; (2) the electronic transition from $|\alpha\rangle$ to another intermediate state $|\beta\rangle$ caused by the interaction with a phonon or another elementary excitation; (3) the transition from $|\beta\rangle$ to the ground state: recombination of the electron–hole pair accompanied by the emission of a scattered photon $\hbar\omega_s$. For the probability of the whole process the third-order perturbation theory yields the following expression [109, 110]:

$$I_s \propto I_i \frac{\omega_s^4 V}{c^4} \left| \sum_{\alpha,\beta} \frac{\langle 0|H_{ER}|\beta\rangle\langle\beta|H_{EP}|\alpha\rangle\langle\alpha|H_{ER}|0\rangle}{(E_\beta - \hbar\omega_s + i\Gamma_\beta)(E_\alpha - \hbar\omega_i + i\Gamma_\alpha)} \right|^2 \times \delta(\hbar\omega_i - \hbar\omega_s - \hbar\Omega) \tag{21.5}$$

Fig. 21.1 (**a**) Schematic of the band structure, absorption coefficient α, and penetration depth d of light in $BaTiO_3$ as compared to the energies of the visible (514.5 nm) and UV (351 and 325 nm) photons. (**b**) Spectra of (21.1) a bare $SrTiO_3$ substrate (*blue curve*), (21.2) a $(BaTiO_3)_5(SrTiO_3)_4 \times 25$ superlattice (*green curve*) measured with visible excitation (514.5 nm), and (21.3) the same superlattice measured with 351.1 nm UV excitation (*red curve*). The *dashed line* (21.4) shows the bare $SrTiO_3$ substrate spectrum measured with 351.1-nm UV excitation. *Arrows* and *labels* indicate the superlattice phonon peaks (After Ref. [43])

Here I_i, I_s are the intensities of incident and scattered light, respectively, and V is the scattering volume. E_α, E_β, Γ_α, Γ_β are the energies and widths of the electronic levels $|\alpha\rangle$ and $|\beta\rangle$, H_{ER}, and H_{EP} are the Hamiltonians of electron-radiation and electron–phonon interaction, respectively. The summation in (21.5) is performed over all possible intermediate states. The formula (21.5) gives an expression for the dominant term of scattering probability. If one takes into account all six possible permutations of the three processes, the expression for total probability will contain five more similar terms [110]. The energy conservation must be obeyed for the whole process, but it is not required for each stage separately. However, if the energy of the incident or scattered light is close to the energy of an electronic transition (i.e., $h\omega_I \approx E_\alpha$ or $h\omega_s \approx E_\beta$) then the scattering intensity is strongly enhanced. This effect is called resonant Raman scattering. The resonant enhancement, combined with the intensity increase resulting from the ω^4 dependence of the normal Raman scattering cross-section, gives the UV another advantage over visible excitation. For bulk dielectric crystals and ceramics, the scattering volume V can be orders of magnitude larger for non-absorbing visible excitation, thus yielding larger overall scattering intensity (unless laser energy is precisely tuned into the electronic or excitonic resonance). For nanoscale films, however, V is often determined by film thickness rather than the penetration depth, so the resonant denominator and ω^4 factor lead to a significantly larger UV Raman signal from thin films of ferroelectric oxides and other wide bandgap materials compared to visible excitation. It should be noted that using deep UV excitation (e.g., 244 nm) for materials like $BaTiO_3$ would lead to extremely short penetration depth, hence very small scattering volume, while moving away from the resonant enhancement. Therefore, Raman signal can become too weak to detect. So, for ferroelectric titanates and materials with similar band gaps, the wavelength range 300–350 nm seems to be optimal for UV Raman excitation.

UV excitation presents another advantage important for characterization of high-temperature phase transitions in ferroelectric materials. A difficulty in measuring Raman spectra of materials at high temperatures using visible excitation is caused by continuous background due to thermal emission, which can be rather intense compared to weak and broadened Raman features. UV excitation shifts Raman spectra to much shorter wavelength range, far away from the peak intensity of thermal radiation. Yashima et al. [124] demonstrated that in the case of conventional visible excitation using the 488-nm line of Ar^+ laser, thermal radiation background became significant even at 900°C, and rapidly increased between 900 and 1100°C, while the UV excitation (363.8 nm) allowed measuring Raman spectra of ZrO_2 with practically no backgrounds up to 1500°C. UV Raman spectroscopy was successfully applied for high-temperature studies of calcium silicates [137, 138], and phase transitions in hafnia at ~2085 K [139]. High-temperature UV Raman spectra of $SrZrO_3$, which, like many ferroelectrics, belongs to the perovskite family, revealed the soft phonon mode behavior causing the cubic-tetragonal phase transition around 1,200°C [125].

Micro-Raman spectroscopy is widely utilized for characterization of semiconductor materials and devices, providing information about local strain, composition, and crystallinity in semiconductor structures at sub-micrometer scale. Raman scattering by phonons and electronic excitations is very sensitive to the local environment and can provide information unavailable by other commonly used materials characterization techniques, such as conventional X-ray diffraction (lacks lateral and depth resolution) and transmission electron microscopy (destructive technique). In this area, UV excitation can also present an advantage over commonly used visible Raman spectroscopy, because of improved lateral and in-depth spatial resolution [126, 134, 140]. Shorter wavelength of the UV light allows significant reducing of the focused laser beam size (theoretical Rayleigh limit for 325-nm laser line and 100$\times$ objective with a numerical aperture 0.9 is $\sim$220 nm, which can be reduced further by using a high numerical aperture immersion lens [140]), while much shorter optical penetration depth (less than 5 nm in Si at 244 nm) allows probing very thin films. However, much stronger absorption of UV radiation can cause significant local heating and potential damage to a material, and care must be taken to avoid these effects, especially when using a microscope, since the laser power density can be much higher than in macro-configuration. For ferroelectric oxide materials, potential laser damage is not an issue for the power levels used in spectroscopy and microscopy. However, local heating effect should be avoided as it can introduce an error when determining phase transition temperatures from spectral data. Using low or moderate power densities (below 10 W/cm^2) is usually sufficient to avoid the laser-induced heating effects even in silicon, which has about ten times stronger absorption in the UV range compared to ferroelectric oxides such as $SrTiO_3$ or $BaTiO_3$ [141].

Recent results of Raman studies utilizing the advantages of ultraviolet excitation for probing nanoscale ferroelectric heterostructures and ultrathin films are reviewed in the next section.

5 Key Research Findings

5.1 Applications of Ultraviolet Raman Spectroscopy to Characterization of Ferroelectric Nanostructures: $BaTiO_3/SrTiO_3$ Superlattices

Until recently, UV Raman spectroscopy has not been widely used for ferroelectric films because of the above mentioned technical difficulties such as lower throughput efficiency, insufficient dispersion, and higher stray light level of UV Raman spectrometers compared with those operating in the visible range. Recent progress in UV Raman instrumentation has made measurement of ferroelectric ultrathin films possible. First measurement of $SrTiO_3$ films grown on Si using 325-nm excitation has been reported [135], however, only room temperature data were presented, and ferroelectric properties were not discussed. Periodic layered heterostructures – $BaTiO_3/SrTiO_3$ superlattices – were the first nanoscale ferroelectric materials studied by UV Raman spectroscopy [43] using UV-optimized triple spectrometers with a multichannel-coupled charge-device detector. The spectrometer (Horiba Jobin Yvon T64000) employed a triple monochromator to provide high-resolution and effective stray light reduction. The 351.1-nm line of an Ar^+ laser and the 325-nm line of a He–Cd laser were used for excitation, and optical paths were optimized to improve the throughput. With these setups, Tenne et al. measured Raman scattering in $BaTiO_3/SrTiO_3$ superlattices as thin as 24 nm, and in single-layer $BaTiO_3$ films as thin as 1.6 nm, that is, four unit cells (Sect. 5.3).

The $[(BaTiO_3)_n/(SrTiO_3)_m] \times N$ superlattice samples were grown by reactive molecular beam epitaxy (MBE) on TiO_2-terminated (001) $SrTiO_3$ substrates. (Here n and m refer to the thickness, in unit cells, of the $BaTiO_3$ and $SrTiO_3$ layers, respectively; N is the total number of periods.) The n and m values were controlled in situ using reflection high-energy electron diffraction oscillations during growth, and confirmed by X-ray diffraction (XRD) and in some samples by high-resolution transmission electron microscopy (HRTEM) [141]. XRD showed excellent epitaxy and crystallinity in the superlattice samples, with all superlattice peaks present in the θ–2θ scan. HRTEM images showed atomically sharp $BaTiO_3/SrTiO_3$ interfaces and accurate periodicity. Growth and structural characterization is described in detail by Soukiassian et al. [56]. According to XRD results, all samples studied were commensurate to the $SrTiO_3$ substrate ($a = 3.905$ Å), meaning that the SrTiO3 layers are strain-free and the $BaTiO_3$ layers are under 2.2% compressive biaxial strain ($a = 3.992$ Å in bulk $BaTiO_3$ at room temperature), except for a 200-nm-thick superlattice $[(BaTiO_3)_8/(SrTiO_3)_4] \times 40$, in which the strain is partially relaxed ($a = 0{:}3946$ Å in-plane). In this case, the $SrTiO_3$ layers are under biaxial tensile strain and the $BaTiO_3$ layers are under biaxial compressive strain.

Figure 21.2 shows room temperature spectra of two SLs, $[(BaTiO_3)_5(SrTiO_3)_4] \times 25$ and $[(BaTiO_3)_8/(SrTiO_3)_4] \times 10$ compared to the spectra of $BaTiO_3$ single crystals in the three different ferroelectric phases and the $SrTiO_3$ substrate spectra at 295 and 5 K. The phonon mode assignment is made by comparison with the spectra of $SrTiO_3$ and $BaTiO_3$ single crystals. Bulk $BaTiO_3$ and $SrTiO_3$ have the same cubic perovskite structure in the high-temperature phase, however, they exhibit different behaviors at lower temperatures. $BaTiO_3$ undergoes 3 consecutive phase transitions from a cubic *3mm* to a tetragonal *4mm* phase at 403 K, then to an orthorhombic *2mm* phase at 278 K, and to a rhombohedral *3m* phase at 183 K [136]. The three low-temperature phases are ferroelectric, and can be distinguished by their characteristic Raman spectra [20]. Bulk $SrTiO_3$ is an incipient ferroelectric (quantum paraelectric), in which the ferroelectric phase transition is suppressed by quantum fluctuations [142]. An antiferrodistortive cubic-tetragonal phase transition occurs in $SrTiO_3$ at 105 K, which involves the rotation of the Ti–O octahedra. The resulting tetragonal structure *4/mmm* is still centrosymmetric and, hence, not ferroelectric.

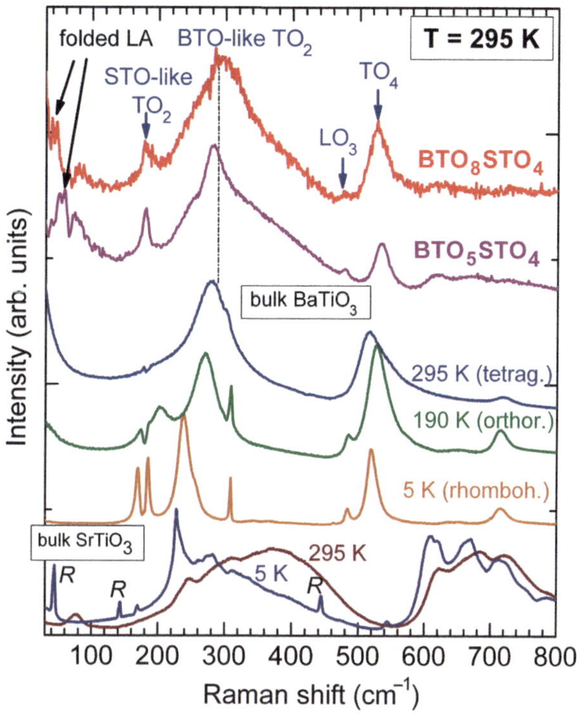

Fig. 21.2 Raman spectra of the $[(BaTiO_3)_5(SrTiO_3)_4] \times 25$ and $[(BaTiO_3)8/(SrTiO3)4] \times 10$ superlattices at 295 K, measured with 351.1-nm laser line. The spectra of $BaTiO_3$ single crystals at 295, 190, and 100 K, corresponding to tetragonal, orthorhombic, and rhombohedral phases, respectively, and the spectra of $SrTiO_3$ single crystals at 295 and 5 K are shown for comparison. The bulk crystal spectra were measured with the 514.5-nm laser line (After Tenne et al. [141])

The perovskite structure has five atoms per unit cell, therefore, there are 12 optical vibrational modes. In the cubic $3mm$ phase, the zone center optical phonons belong to $3F_{1u} + 1F_{2u}$ irreducible representations. Each of the F_u modes is triply degenerate, and all of them are of odd parity with respect to the inversion, therefore, forbidden in the first order Raman scattering. The F_{1u} modes are infrared active, while the F_{2u} modes are silent. Upon transition from the cubic to the tetragonal phase the F_{1u} modes split into A_1 and E modes, and the F_{2u} phonons give rise to B_1 and E modes. The E modes are doubly degenerate. In the orthorhombic phase, the optical phonons belong to A_1, A_2, B_1, and B_2 symmetry. In the lowest-temperature, rhombohedral phase, the modes originating from the cubic F_{1u} phonons are split into A_1 and E modes, while the F_{2u} vibrations produce A_2 and E modes. The symmetric A_1 and E modes are Raman-active. Also, in polar crystals, which perovskite oxides belong to, long-range electrostatic interaction results in additional splitting of each optical phonon branch into transverse (TO) and longitudinal (LO) modes. Since $SrTiO_3$ remains paraelectric and centrosymmetric down to 5 K, its phonons remain Raman inactive, and the spectra of $SrTiO_3$ bare substrate are dominated by the second-order (two-phonon) Raman features [86]. However, the antiferrodistortive phase transition at 105 K leads to the appearance of sharp peaks of R modes in the low-temperature spectra (labeled R in Fig. 21.2). Those are the zone-edge (R point) phonons, which become Raman active via double folding of the Brillouin zone.

The top two spectra in Fig. 21.2 are typical of the UV Raman spectra of $BaTiO_3$/$SrTiO_3$ SLs below T_c, exhibiting strong first order (single-phonon) peaks as labeled in the figure. (Weak second-order (two-phonon) features from the $SrTiO_3$ substrate can be seen between 600 and 700 cm^{-1} and as a background in the range 200–500 cm^{-1}). The lines at about 290 cm^{-1} have similar positions and shapes to the TO_2 modes of A_1 symmetry of $BaTiO_3$ in the tetragonal phase, thus are assigned to the $BaTiO_3$ phonons, indicating that the $BaTiO_3$ layers in the SLs are tetragonal. This is supported by the absence of the sharp peak at 310 cm^{-1} characteristic of the orthorhombic and rhombohedral phases of $BaTiO_3$, but not pronounced in the tetragonal phase. It should be noted that the spectra of superlattices, and thin films described in this and following chapters were measured in backscattering geometry in both parallel $z(x,x)\bar{z}$ and perpendicular $z(x,y)\bar{z}$ polarization configurations, where $z \parallel (001)$ (normal to the film surface), x and y are in-plane directions, for example, (100) and (010). (In Porto notation used here, the letters outside the brackets indicate the propagation directions of the incident and scattered light, and the inside letters refer to the polarization directions of the incident and scattered light, respectively.) However, it was found that the polarized signal dominates the spectra, while almost no signal was observed in $z(x,y)\bar{z}$ geometry. Therefore, although the spectra shown here and in the following sections are unpolarized, the dominant contribution comes from $z(x,x)\bar{z}$ configuration, in which A_1 (LO) modes are active. The presence of the A_1(TO) modes is likely due to deviations from true backscattering along (001) direction.

The frequencies of several phonon branches in $SrTiO_3$ and $BaTiO_3$ are close to each other, and the phonons are not expected to be strongly localized within the thin $SrTiO_3$ and $BaTiO_3$ layers. Such vibrations extend through the whole superlattice. This is the case for the LO_3 and the TO_4 modes, observed at about 478 and 530 cm^{-1}, respectively.

The line at about 180 cm^{-1} is attributed to the $SrTiO_3$-like TO_2 phonons. Its position corresponds closely to the TO_2 line in the electric-field-induced Raman spectrum of $SrTiO_3$ crystals [87] and the weak feature in the 5 K spectrum of the $SrTiO_3$ single crystal. (The first-order Raman lines are visible in these cases because the electric field and defects break the inversion symmetry in the $SrTiO_3$ crystals.) The 180 cm^{-1} line is not from the $SrTiO_3$ substrate because the first-order Raman lines are symmetry-forbidden in bulk $SrTiO_3$. In fact, even the much stronger second-order features of the substrate at 230 and 610 cm^{-1} are barely seen in the UV Raman spectra. Although $BaTiO_3$ also has a feature due to the TO_1 mode of A_1 symmetry at about the same position (177 cm^{-1}), at room temperature (in the tetragonal phase) that feature is about 15–20 times weaker compared with the TO_2 and TO_4 lines, and it has a characteristic interference dip due to the coupling of the TO_1 and TO_2 modes of A_1 symmetry [85]. The 180 cm^{-1} line observed in the spectra of the superlattices is of the same order of magnitude in intensity as the TO_2 and TO_4 lines of $BaTiO_3$ and does not have the dip characteristic of $BaTiO_3$. Therefore, although the contribution of the $BaTiO_3$ layers cannot be absolutely ruled out, we believe this line should be attributed to the phonons of the $SrTiO_3$ layers in the superlattices. The observation of the first-order Raman

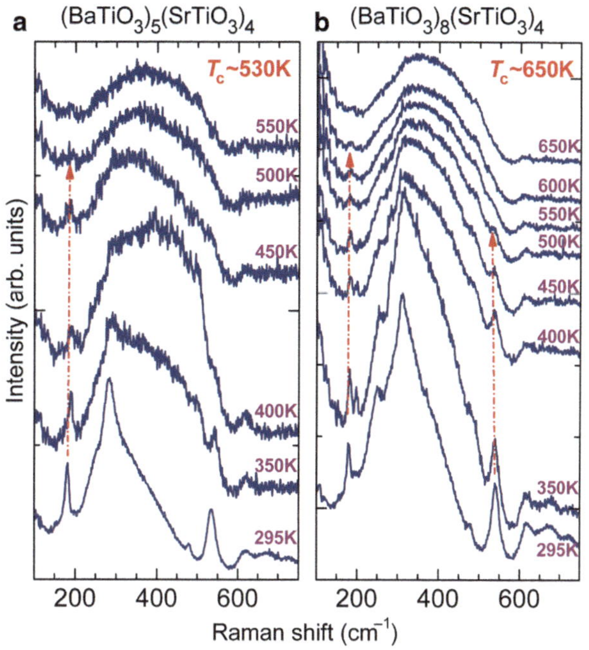

Fig. 21.3 Temperature evolution of UV Raman spectra of SLs $[(BaTiO_3)_5(SrTiO_3)_4] \times 25$, and $[(BaTiO_3)_8(SrTiO_3)_4] \times 10$ ((**a**) and (**b**), respectively). The *arrows* mark the superlattice phonon peaks, whose intensities decrease as the temperature increases and disappear at T_c. After Tenne and Xi [107]

scattering by $SrTiO_3$ phonons indicates that the inversion symmetry is broken, and the $SrTiO_3$ layers in the superlattices are polar. In this case, the polarization in the unstrained $SrTiO_3$ layers is induced by neighboring ferroelectric $BaTiO_3$ layers.

Doublets of folded longitudinal acoustic (LA) phonons due to the superlattice periodicity [143] can also be seen in the Raman spectra of the SLs (indicated by arrows in Fig. 21.2). The positions of the doubled peaks agree well with the first doublet frequencies calculated within the elastic continuum model [144]. The observation of the LA phonon folding suggests that these superlattices possess the requisite structural quality for acoustic Bragg mirrors and cavities to be used for potential coherent phonon generation applications [145–147].

Studying the temperature evolution of UV Raman spectra was demonstrated to be an effective approach to determine the ferroelectric phase transition temperature T_c in ferroelectric ultrathin films and superlattices, which is a critical but challenging step for understanding ferroelectricity in nanoscale systems. The T_c determination from Raman data is based on the above mentioned fact that perovskite-type crystals have no first order Raman active modes in paraelectric phase. Therefore, Raman intensities of the ferroelectric superlattice or thin film phonons decrease as the temperature approaches T_c from below and disappear upon transition into paraelectric phase. Above T_c, the spectra contain only the second-order features, as expected from the symmetry selection rules. This method was applied to study phase transitions in $BaTiO_3/SrTiO_3$ superlattices. Figure 21.3 shows the temperature evolution of Raman spectra for two $BaTiO_3/SrTiO_3$ superlattices. From the shapes and positions of the $BaTiO_3$ lines it follows that the $BaTiO_3$ layers remain in ferroelectric tetragonal phase at all temperatures below the T_c, that is, the low-temperature phases (orthorhombic and rhombohedral) characteristic for bulk $BaTiO_3$ are suppressed by compressive strain imposed on the $BaTiO_3$ layers by the $SrTiO_3$ substrate. The intensities of the first-order superlattice phonons decrease as the temperature increases and disappear at T_c. Above T_c, the spectra contain only the second-order features as expected from the symmetry selection rules. When the $BaTiO_3$ layers are paraelectric, the induced polarization in the $SrTiO_3$ layers also disappears.

By plotting the first-order Raman intensity as a function of temperature, one can determine T_c accurately as the temperature where the intensity of the first-order Raman peaks becomes zero, as shown in Fig. 21.4a and b. For the Tc determination from the SLs spectra shown here, the TO_2 and TO_4 phonon lines (shown by arrows in Fig. 21.3) are the most suitable because they do not overlap with the second-order substrate features. (However, other optical phonon modes can be used in the same manner, provided that they are clearly observed in the spectra of the ferroelectric phase.) The results, with the phonon intensities normalized by the Bose factor $n + 1 = (1 - \exp(-\hbar\omega/kT))^{-1}$ and divided by the intensities of corresponding peaks at 5 K, are shown for four SLs: $[(BaTiO_3)_2(SrTiO_3)_4] \times 40$, $[(BaTiO_3)_5(SrTiO_3)_4] \times 25$, $[(BaTiO_3)_8(SrTiO_3)_4] \times 40$, and $[(BaTiO_3)_8(SrTiO_3)_4] \times 10$. Sample $[(BaTiO_3)_8(SrTiO_3)_4] \times 40$ is partially relaxed (remaining compressive strain in $BaTiO_3$ layers is $\sim$1%), while the other samples are commensurate with the $SrTiO_3$ substrate. Both TO_2 and TO_4 phonons show similar behaviors and the dashed-dotted lines are linear fits to the average of the two modes. The linear fit corresponds to a parabolic decrease of polarization with temperature approaching T_c from below, since Raman intensity is proportional to the square of atomic displacement. The intersection of a dash-dotted line with the horizontal axis is taken as the T_c of the sample.

In comparison with the values of T_c determined from Raman data, the temperature dependence of ferroelectric polarization calculated using a thermodynamic phase-field model [148] is plotted in Fig. 21.4c and d for the same samples as in Fig. 21.4a and b. The model assumed the $BaTiO_3$ and $SrTiO_3$ layers in the SLs to possess their respective bulk elastic and thermodynamic properties. The in-plane lattice constant was commensurately constrained to the $SrTiO_3$ substrate except for the partially relaxed case, and the top surface is stress-free, and a short-circuit electrostatic boundary condition was employed. The corresponding three-dimensional time-dependent Ginzburg-Landau equations were then numerically solved using the perturbation method with semi-implicit Fourier-spectral algorithms [149]. More detail on the phase field modeling can be found elsewhere [148, 150, 151]. The result reveals a spontaneous polarization along the growth direction with multiple 180° domains in the $BaTiO_3$ layers, which induces polarization in the adjacent $SrTiO_3$ layers whose magnitude and distribution vary with the thickness and domain size of the $BaTiO_3$ layers. The spontaneous polarization in the $BaTiO_3$ layers becomes zero at T_c, and the predicted T_c values agree well with those from the Raman data. This is remarkable considering that no fitting parameters from the Raman experiments are used in the calculations.

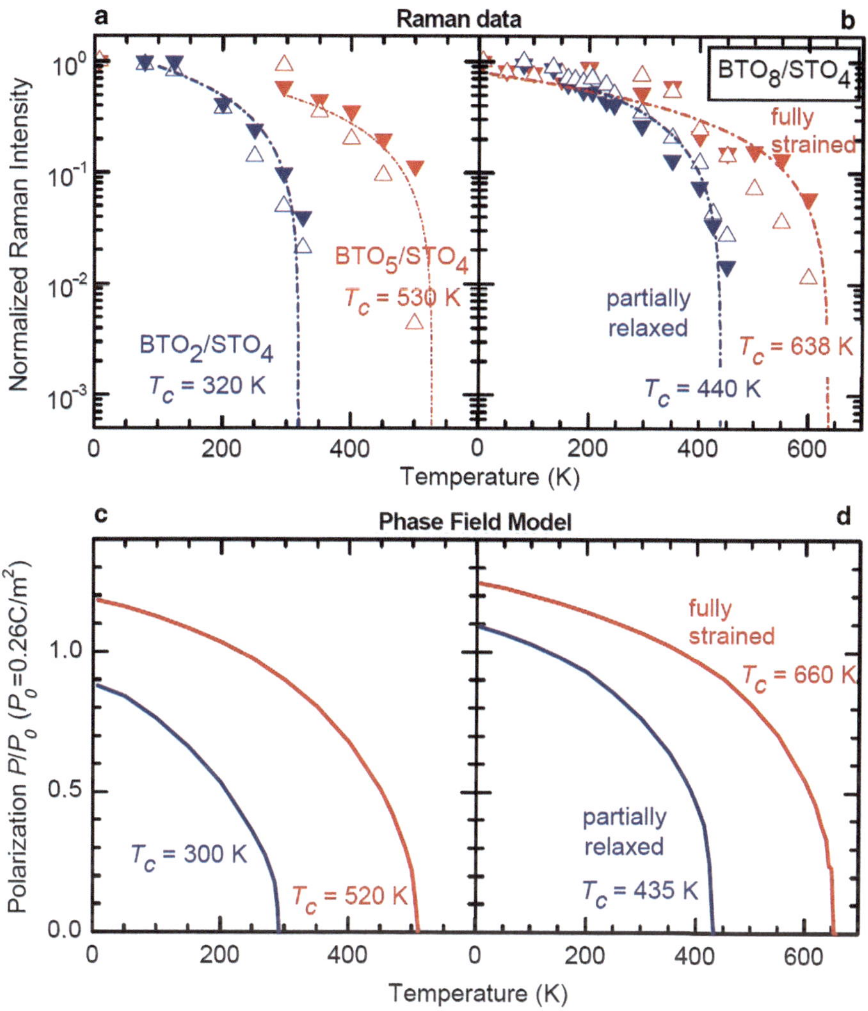

Fig. 21.4 Temperature dependencies of normalized Raman intensities of TO_2 (*solid triangles*) and TO_4 (*open triangles*) phonons for (**a**) SLs $[(BaTiO_3)_2(SrTiO_3)_4] \times 40$ and $[(BaTiO_3)_5(SrTiO_3)_4] \times 25$; (**b**) SLs $[(BaTiO_3)_8(SrTiO_3)_4] \times 40$ and $[(BaTiO_3)_8(SrTiO_3)_4] \times 10$. The *dash-dotted lines* are fits to linear temperature dependence. (**c**) and (**d**) – three-dimensional phase-field model calculations of polarization as a function of temperature in the same superlattice samples. Polarization is given as a fraction of the polarization of bulk $BaTiO_3$, $P_0 = 0.26$ C/m^2 (After Tenne et al. [43])

Figure 21.5 shows the T_c in superlattices determined by the Raman data, variable-temperature XRD, and the phase-field model, as a function of the $BaTiO_3$ and $SrTiO_3$ layer thicknesses. The XRD measurements provide data on the temperature dependence of the lattice parameters, and the observed noticeable change in the slope of the out-of-plane *c*-axis lattice parameter due to the

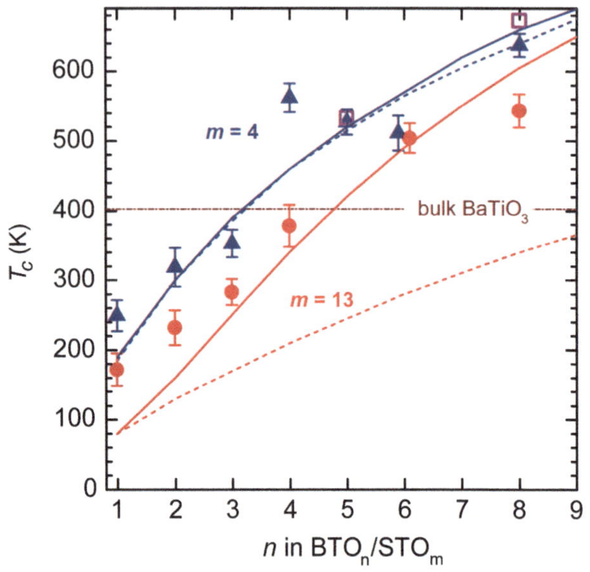

Fig. 21.5 T_c dependence on layer thicknesses n and m in superlattices $(BaTiO_3)_n/(SrTiO_3)_m$. *Blue triangles* and *red circles* are for $m = 4$ and $m = 13$, respectively. *Open squares* show the values obtained from variable temperature X-ray diffraction measurements. *Solid lines* are from the three-dimensional phase-field model calculations, *dashed lines* – simulations assuming a single domain in the $BaTiO_3$ layers. The *dash-dotted line* shows T_c in bulk $BaTiO_3$ (After Li et al. [150])

development of spontaneous polarization in the superlattice is an indication of T_c [141, 151]. The T_c values obtained from XRD agree well with Raman results. As Fig. 21.5 shows, UV Raman spectroscopy results demonstrated that $BaTiO_3$ layers in the superlattices are ferroelectric even when their thickness is only one unit cell, with T_c as high as 250 K. These ultrathin $BaTiO_3$ layers are not only ferroelectric, but are able to polarize the much thicker adjacent non-ferroelectric $SrTiO_3$ layers. T_c increases with increasing $BaTiO_3$ layer thickness n as the dipole-dipole interaction in $BaTiO_3$ layers becomes stronger, while large $SrTiO_3$ layer of thickness m suppresses T_c by reducing the coupling between the $BaTiO_3$ layers. By changing the values of n and m, T_c was tuned from $\sim$150 to 640 K, that is, from 250 K below to 240 K above the bulk value of $BaTiO_3$ (403 K). The higher-than-bulk T_c is due to the strain in the $BaTiO_3$ layers, just as strain enhances T_c in single-layer ferroelectric films [23, 24, 30]. When the strain is partially relaxed in sample $[(BaTiO_3)_8(SrTiO_3)_4] \times 40$, T_c drops to 440 K, almost to the bulk $BaTiO_3$ value (see Fig. 21.4b, d). The partial relaxation leads to the inhomogeneous strain distribution in the superlattice, so XRD does not provide the information about T_c [151], since the temperature dependence of the c lattice parameter does not exhibit a noticeable change in the slope. Raman spectroscopy has a certain advantage here, yielding an average value of T_c (Fig. 21.4b). However, the effect of strain on the phase diagram of superlattices could be better investigated if the superlattices were subject to coherent strain of variable magnitude. With recently developed novel rare earth scandate substrates $ReScO_3$ (Re = Gd, Dy, Sc, and other lanthanide elements) [152, 153], such a systematic variation of strain state in superlattices becomes possible. These scandates have pseudocubic lattice constants that can be tailored in approximately 0.01 Å steps from 3.93 Å (for $HoScO_3$) to 4.05 Å (for $LaScO_3$) by varying the Re element [152]. This is the range of lattice constant of many perovskite ferroelectrics including $BaTiO_3$, (tetragonal at room temperature with $a = 3.9920$ Å, $c = 4.0361$ Å), and $SrTiO_3$, (cubic, $a = 3.905$ Å) [136]. Utilizing these substrates allows varying the in-plane lattice constant from 3.905 Å ($SrTiO_3$ substrate; $BaTiO_3$ under 2.2% compressive biaxial strain, $SrTiO_3$ unstrained) to 3.99 Å ($SmScO_3$ substrate; $BaTiO_3$ nearly unstrained, $SrTiO_3$ under 2.2% tensile strain) and further to 4.016 Å ($NdScO_3$ substrate; both $BaTiO_3$ and $SrTiO_3$ under tensile strain). Growth of $BaTiO_3/SrTiO_3$ superlattices of very high structural quality on some of these substrates was recently demonstrated [56], and preliminary data show potential of UV Raman spectroscopy for characterization of phase diagram in these structures.

Comparison of the experimental results with the phase-field model calculations shows that the three-dimensional phase-field model allowing domain formation (180° domains for the case of fully coherently strained superlattices on $SrTiO_3$ substrates) provides a good description of the Raman data, while simulations assuming a single domain in the $BaTiO_3$ layers yield significantly lower T_c, especially for thicker $SrTiO_3$ layers ($m = 13$). This demonstrates the importance of taking into consideration multiple domains for more accurate description of ferroelectric properties, in particular, prediction of the phase transition temperature in better agreement with the experimental data. UV Raman spectroscopy results supported by the phase fields modeling allow to conclude that ferroelectricity can be very strong in one-unit-cell thick $BaTiO_3$ layers ($T_c \sim 250$ K for $n/m = 1/4$, see Fig. 21.5). The electrical boundary conditions play a critical role. With the highly polarizable $SrTiO_3$ in contact with the $BaTiO_3$ layers, the critical thickness is reduced to a single unit cell. Meanwhile, the mechanical boundary condition imposed by the $SrTiO_3$ substrate leads to strain in the $BaTiO_3$ layers and thus enhanced ferroelectricity. The interplay between the electrical and mechanical boundary conditions enables the tuning of T_c by nearly 500 K. This interplay was further explored in single-layer $BaTiO_3$ films (Sect. 5.3).

5.2 Three-Component $BaTiO_3/SrTiO_3/CaTiO_3$ Superlattices

Similar to binary $BaTiO_3/SrTiO_3$ superlattices described in the previous section, UV Raman spectroscopy was applied to studies of three-component $SrTiO_3/BaTiO_3/CaTiO_3$ short-period superlattices, combining ferroelectric $BaTiO_3$, and non-ferroelectric titanates, $SrTiO_3$ and $CaTiO_3$ [154]. Synthesis of such structures by molecular beam epitaxy [55] and high-precision pulsed laser deposition [25] was recently reported. Use of $CaTiO_3$ (bulk pseudocubic lattice constant $a = 3.83$ Å) and $BaTiO_3$ ($a = 3.9920$ Å) in combination with $SrTiO_3$ can result in a superlattice growing commensurately on $SrTiO_3$ substrate ($a = 3.905$ Å). In such a superlattice, the constituent $BaTiO_3$ and $CaTiO_3$ layers are subject to significant epitaxial strain of opposite signs. Due to the strong coupling between strain and polarization in ferroelectrics, this can result in substantial enhancement of the polarization relative to that of the bulk constituents, as has been observed experimentally [25] and demonstrated theoretically [26, 27].

UV Raman experiments were performed on a series of three-component $SrTiO_3$/$BaTiO_3$/$CaTiO_3$ superlattices, focusing on the temperature evolution of the vibrational spectra and the phase transition behavior. The superlattices with atomically smooth interfaces were grown on $SrTiO_3$ (001)-oriented substrates with $SrRuO_3$ buffer layers by pulsed laser deposition with atomic-layer precision by Lee et al. [25] As in the case of MBE growth of binary superlattices (previous section), the deposition process was monitored by reflection high energy electron diffraction. The samples were characterized by atomic force microscopy (AFM), high resolution scanning transmission electron microscopy (STEM), XRD, and reciprocal space mapping (RSM) [25]. AFM and STEM data demonstrated high structural quality of both the superlattices and underlying $SrRuO_3$ layers: the single-monolayer terrace steps on both the $SrRuO_3$ surface and on the top SL surface after the growth, and abrupt interfaces between the layers. The θ–2θ XRD scans show excellent diffraction patterns with all the superlattice peaks present, thus confirming the designed periodicity of the SLs. The SLs $(SrTiO_3)_l/(BaTiO_3)_m/(CaTiO_3)_n$ SLs are labeled S*l*B*m*C*n* (*l*, *m*, and *n* are thicknesses of $SrTiO_3$, $BaTiO_3$, and $CaTiO_3$ layers, respectively (in unit cells)).

RSM data were used to examine the strain state of the SLs and $SrRuO_3$ buffer layers, and demonstrated that the $SrRuO_3$ layers in the samples studied were commensurate with $SrTiO_3$ substrates. Also, RSM data showed that the SLs with $BaTiO_3$ layer thickness not exceeding the combined thickness of $SrTiO_3$ and $CaTiO_3$ layers were coherently strained; the SLs with thicker $BaTiO_3$ layers were partially relaxed [25, 154]. While the S2B6C2 SL is still nearly fully strained (strain in $BaTiO_3$ layers is still 2.15%), the S2B8C2 structure was significantly relaxed (strain decreased by 1.21%) [154]. All the SLs studied had the total thickness of 200 nm, which was more than enough to completely block the UV light used (325 nm) from penetrating into the $SrRuO_3$ layers and the substrate, so only the superlattice signal was observed in Raman spectra. A 200-nm-thick $BaTiO_3$ film grown using the same technique was also studied for comparison. This film was fully relaxed due to its large thickness.

Figure 21.6a shows the low temperature (30 K) UV Raman spectra of $SrTiO_3$/$BaTiO_3$/$CaTiO_3$ SLs and the 200-nm-thick $BaTiO_3$ film. All the SLs studied are ferroelectric at low temperatures, and the first-order Raman peaks of superlattice phonons are seen in the spectra. The peaks of optical phonons are identified and labeled similarly to the case of binary superlattices $BaTiO_3$/$SrTiO_3$. The peaks at about 290 cm^{-1} in the spectra of fully or nearly fully strained SLs (top five spectra in Fig. 21.6a) have similar positions and shapes to the TO_2 modes of A_1 symmetry of $BaTiO_3$ in the tetragonal phase, although there is probably a contribution of TO_3 and LO_2 modes observed in the spectra of bulk $BaTiO_3$ [85] at 304–308 cm^{-1}. Similar features are characteristic for the spectra of bulk and thin film $BaTiO_3$ [85, 100]. Therefore, these peaks are assigned to the $BaTiO_3$ phonons (note that this feature is almost absent in the spectra of the sample S1B1C1 having the smallest $BaTiO_3$ content). One can conclude then that $BaTiO_3$ layers are likely tetragonal in these SLs, which is reasonable because $BaTiO_3$ layers are compressively strained, like in the case of binary $BaTiO_3$/$SrTiO_3$ SLs grown on $SrTiO_3$ substrates.

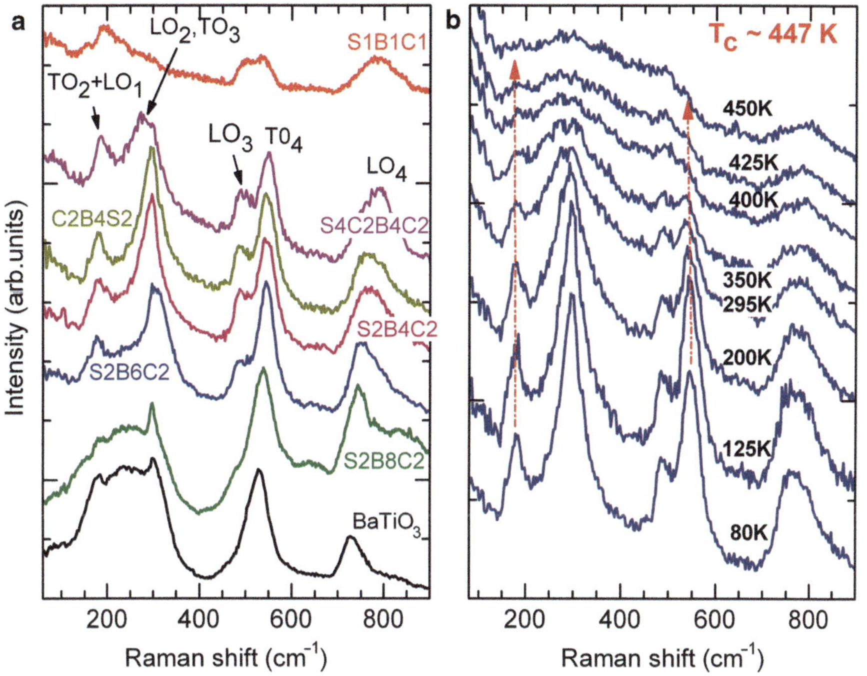

Fig. 21.6 (**a**) Low-temperature (30 K) UV Raman ($\lambda_i = 325$ nm) spectra of $SrTiO_3/BaTiO_3/CaTiO_3$ superlattices and a 200-nm-thick $BaTiO_3$ film. (**b**) Temperature evolution of Raman spectra for $(SrTiO_3)_2/(BaTiO_3)_4/(CaTiO_3)_2$ superlattice. *Arrows* indicate Raman peaks used for T_c determination (After Tenne et al. [154])

The line at ~180 cm^{-1} in the SL spectra, also observed in $BaTiO_3/SrTiO_3$ SLs, is similarly attributed to the first-order Raman scattering by $SrTiO_3$-like TO_2 phonons in SLs (see previous section). $BaTiO_3$ also has a TO_1 mode of A_1 symmetry at about the same position (177 cm^{-1}), and $CaTiO_3$ has phonons in this frequency range, too [155]. Therefore, this phonon is likely not localized within thin $SrTiO_3$, $BaTiO_3$ or $CaTiO_3$ layers. These vibrations likely involve the atoms of the neighboring layers of other materials, so they belong to the whole superlattice. This is also the case for the LO_3 mode, observed at about 485 cm^{-1}, which has nearly the same frequency in these materials, and the peak of TO_4 phonon observed at about 540–545 cm^{-1}. The frequencies of the TO_4 phonons in bulk $BaTiO_3$, $SrTiO_3$, and $CaTiO_3$ are also close (522, 545 and 547 cm^{-1}, respectively) and can be even closer in the superlattices studied because of strain.

The broad feature assigned to the LO_4 phonons was observed in the range 700–800 cm^{-1}. In bulk $BaTiO_3$, this feature is seen at about 720 cm^{-1}, while in $SrTiO_3$ [98] and $CaTiO_3$ [155] its frequency is significantly higher, ~795 cm^{-1}. Therefore, this phonon can be considered as localized either in $BaTiO_3$ or in $SrTiO_3$ and $CaTiO_3$ layers. The individual contributions of $BaTiO_3$ and $SrTiO_3$ + $CaTiO_3$ layers are hard to distinguish because the spectral line is broad. However, one can

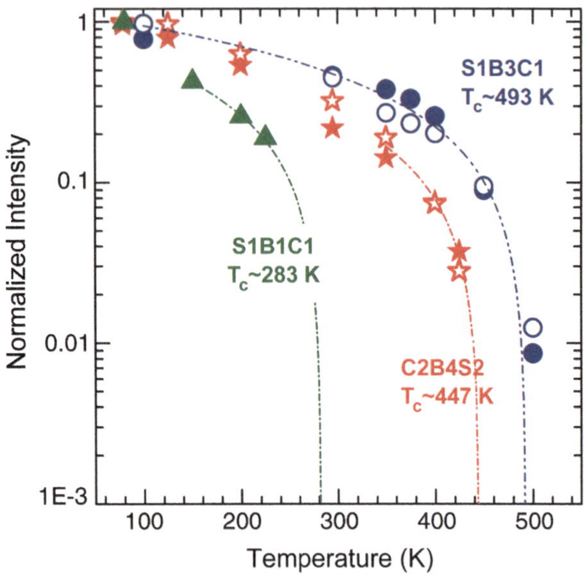

Fig. 21.7 Temperature dependencies of normalized Raman intensities of TO_2 (*solid symbols*) and TO_4 (*open symbols*) phonons for three $CaTiO_3/BaTiO_3/SrTiO_3$ superlattices. The *dash-dotted lines* are fits to a linear temperature dependence (After Tenne et al. [154])

see that the peak shifts toward lower frequencies in SLs with larger $BaTiO_3$ content. As the $BaTiO_3$ layer thickness increases, the structures become partially relaxed, and the spectra closely resemble that of single $BaTiO_3$ film (compare two bottom spectra in Fig. 21.6a).

Raman spectra as a function of temperature are shown in Fig. 21.6b for the C2B4S2 SL. Other superlattices exhibit similar temperature evolution of Raman spectra. These data were used to determine T_c using the same approach as described in the previous section, based on the fact that cubic centrosymmetric perovskite-type crystals have no first-order Raman active modes in the paraelectric phase. The temperature evolution of Raman spectra has indicated that all SLs remain in the tetragonal ferroelectric phase with out-of-plane polarization in the entire temperature range below T_c. The *Tc* determination is illustrated in Fig. 21.7 for three of the SLs studied: S1B1C1, S2B4C2, and S1B3C1. Again, the normalized intensities of the TO_2 and TO_4 phonon peaks (marked by arrows in Fig. 21.6b) were used. In the three-component SLs studied, a structural asymmetry is introduced by the presence of the three different layers, $BaTiO_3$, $SrTiO_3$, and $CaTiO_3$, in each period. Therefore, the phonon peaks should not disappear from the spectra completely upon transition to the paraelectric phase at T_c. Raman intensity should rather drop to some small but non-zero value. However, this inversion symmetry breakdown appears to have a small effect in terms of atomic displacement patterns associated with phonons, and this residual above-*Tc* Raman intensity appears too small to be detected. Therefore, the observed temperature evolution of Raman intensities shows a behavior similar to that of symmetric two-component superlattices.

Figure 21.8 summarizes the results of the *Tc* determination for all the SLs plotted as a function of the thickness ratio $R = \frac{d_{BT}}{d_{ST}+d_{CT}}$, where d_{BT}, d_{ST}, and d_{CT} are the thicknesses of $BaTiO_3$, $SrTiO_3$, and $CaTiO_3$ layers, respectively. Raman results

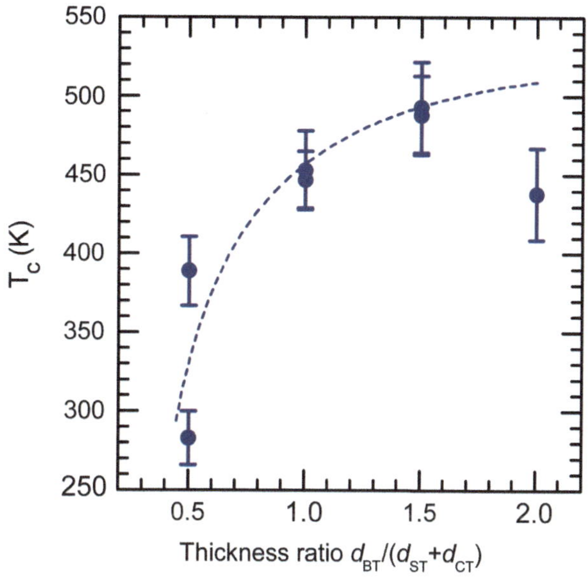

Fig. 21.8 Results of T_c determination from Raman data for three-component $CaTiO_3/BaTiO_3/SrTiO_3$ SLs studied. T_c is plotted as a function of the thickness ratio d_{BT} $(d_{ST} + d_{CT})$. *Dashed line* is guide to an eye (After Tenne et al. [154])

have demonstrated that even the SLs having the ferroelectric $BaTiO_3$ layers as thin as 1 unit cell (S1B1C1) are ferroelectric with T_c near room temperature (283 K, as determined from Raman data). Significant variation in T_c was observed; the SLs with the highest fraction of ferroelectric $BaTiO_3$ layers respective to non-ferroelectric $SrTiO_3$ and $CaTiO_3$ layers have the highest T_c, provided that $BaTiO_3$ layers remain fully (or nearly fully) strained. Strain relaxation causes a drastic decrease in the ferroelectric phase transition temperature, as demonstrated by the rightmost point in Fig. 21.8. It corresponds to the sample with the largest $BaTiO_3$ fraction (S2B8C2), but it is significantly relaxed, hence T_c decreases. These results correlate with the data on ferroelectric polarization in these SLs. As reported by Lee et al. [25], the strongest polarization enhancement was also observed in fully strained SLs with the largest ratio *R*.

5.3 Size Effect on Phase Transitions in Strained Ultrathin $BaTiO_3$ Films

One of the remarkable demonstrations of the capabilities of ultraviolet Raman spectroscopy to probe extremely thin ferroelectric oxide layers reported so far has been its application for studies of ultrathin $BaTiO_3$ films [48]. In order to investigate the size effect on the ferroelectric phase transitions, variable temperature UV Raman spectroscopy was applied to studies of a series of $BaTiO_3$ films with layer thicknesses varied from 1.6 to 10 nm (4–25 unit cells).

Strained $BaTiO_3$ films were grown by molecular-beam epitaxy on TiO_2-terminated (001) $SrTiO_3$ substrates, monitoring the growth by reflection high

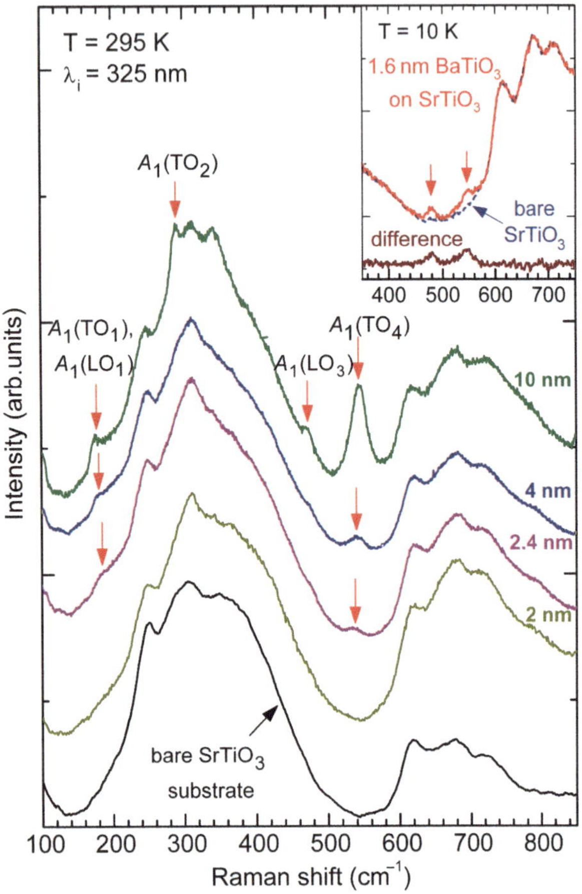

Fig. 21.9 Room temperature Raman spectra of four $BaTiO_3$ ultrathin films on $SrTiO_3$ substrates compared to the spectrum of a bare substrate. Arrows mark the phonon peaks of the $BaTiO_3$ films. The *inset* shows Raman spectra of 1.6-nm-thick $BaTiO_3$ film and $SrTiO_3$ substrate measured at 10 K, and the difference between the two spectra (After Tenne et al. [48])

energy electron diffraction. Both uncapped $BaTiO_3$ films and the same thickness films capped with 10 nm of $SrTiO_3$ were studied. The samples have been studied by synchrotron X-ray scattering using the Advanced Photon Source at Argonne National Laboratory. Synchrotron X-ray data showed that all but the 10-nm-thick samples were commensurate to the $SrTiO_3$ substrates, which implies 2.2% biaxial compressive strain in the $BaTiO_3$ films. In the 10-nm films, slight strain relaxation (less than 0.5% of the $BaTiO_3$ volume) occurred.

Raman spectra were measured in a broad temperature range (10–950 K). 325-nm laser line, used for excitation, provided short enough penetration depth and no thermal background at high temperatures. Room temperature spectra of four $BaTiO_3$ films and a bare $SrTiO_3$ substrate are shown in Fig. 21.9. The $BaTiO_3$ films studied were too thin to absorb the UV light completely, and all spectra contain broad second-order Raman features of $SrTiO_3$ substrates in the ranges 200–500 and 600–750 cm^{-1} [86]. However, in the 10, 5, and 2.4 nm-thick films, the first-order Raman peaks of $BaTiO_3$ are clearly seen, indicating that the films are

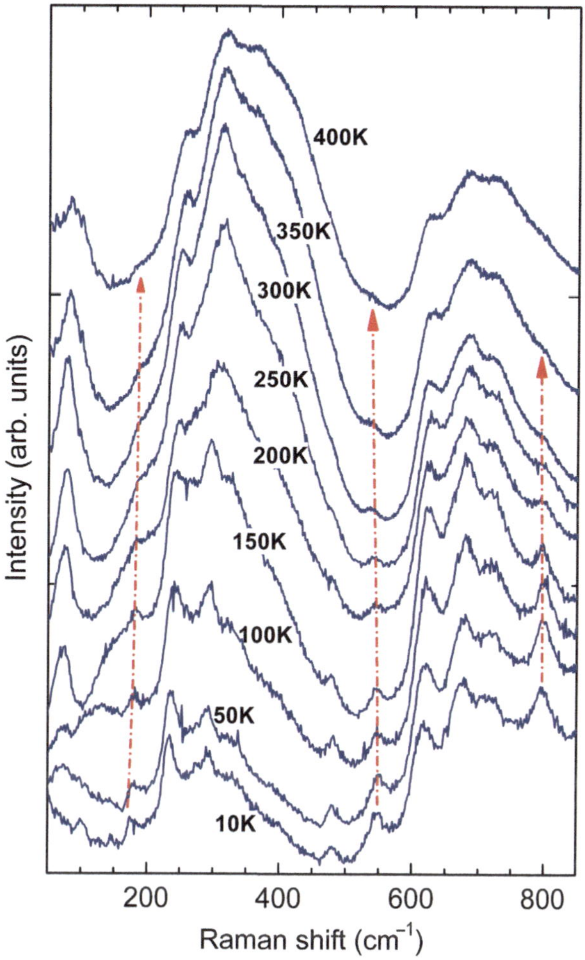

Fig. 21.10 Temperature evolution of Raman spectra of a 2.4-nm-thick $BaTiO_3$ film on a $SrTiO_3$ substrate. *Arrows* mark the phonon peaks of the film used for T_c determination (After Tenne et al. [48])

polar at room temperature. Thinner, 2 and 1.6-nm-thick $BaTiO_3$ films are still paraelectric and do not exhibit the first-order Raman scattering at room temperature (the 295 K spectrum of the 2-nm film shown in Fig. 21.9 contains the second-order peaks only). However, even these thinnest films become ferroelectric at low temperatures, around 60–70 K. The inset to Fig. 21.9 shows the spectra measured at 10 K from the 1.6-nm-thick $BaTiO_3$ film, bare substrate, and the difference spectrum, clearly indicating the presence of $BaTiO_3$ peaks. Positions of the observed phonon peaks (labeled in Fig. 21.9) are characteristic to the spectra of tetragonal $BaTiO_3$, indicating that $BaTiO_3$ layers are tetragonal in these strained films.

Variable-temperature Raman spectra for the 2.4-nm-thick $BaTiO_3$ film is shown in Fig. 21.10. Other films exhibit similar temperature evolution. As can be seen, the peak positions and lineshapes remain nearly unchanged with increasing temperature. Hence, the films remain in the single ferroelectric phase (tetragonal) and the low-temperature orthorhombic and rhombohedral phases characteristic for bulk

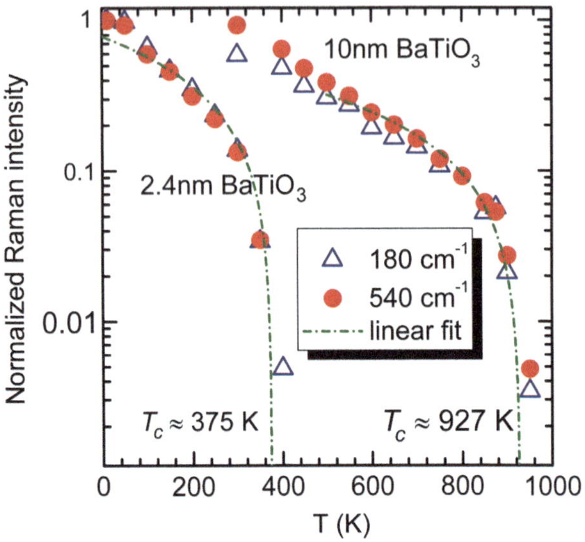

Fig. 21.11 Temperature dependencies of normalized Raman intensities of the peaks at ~180 and 540 cm^{-1} for 2.4- and 10-nm-thick $BaTiO_3$ films capped with 10-nm $SrTiO_3$. The *dash-dotted lines* are linear fits (After Tenne et al. [48])

$BaTiO_3$ are suppressed by biaxial compressive strain, which stabilizes the tetragonal *c* phase. Similar behavior was also observed in compressively strained $BaTiO_3/SrTiO_3$ superlattices described in Sect. 4.1.

Raman spectra measured as a function of temperature allow ferroelectric phase transition T_c to be determined, by plotting the first-order Raman intensity as a function of temperature as described in Sect. 4.1. This is illustrated in Fig. 21.11 for two of the films studied. In this case, the most distinct lines due to the phonons of $BaTiO_3$ films were observed at about 175–185, 475, and 540 cm^{-1}, and attributed to $TO_2 + LO_1$, LO_3, and TO_4 modes of A_1 symmetry, respectively. These peaks can be used for T_c determination; the normalized intensities of the first two peaks as a function of temperature are shown in Fig. 21.11.

Figure 21.12 summarizes the results of the T_c determination for all the samples as a function of $BaTiO_3$ film thickness. Even the films containing only four monolayers of $BaTiO_3$ are ferroelectric with T_c ~70 K. Raman data demonstrate the dramatic effect of $BaTiO_3$ film thickness on T_c. The latter varies over a very broad range, from 70 to about 925 K (more than 500 K above the bulk $BaTiO_3$ value), as the film thickness increases from 1.6 to 10 nm, provided that the films remain fully (or nearly fully) strained. This is the case even for the 10-nm-thick films (less than 0.5% of the film volume is relaxed, as follows from synchrotron X-ray scattering data).

Synchrotron X-ray scattering results indicated the presence of 180° domains in the 2.4, 4.0, and 10-nm-thick $BaTiO_3$ films capped with 10-nm $SrTiO_3$. Similar to the results obtained for $PbTiO_3$ films [38, 156], diffuse intensity induced by the periodic nature of the ferroelectric 180° domains appeared in the scattering around the $BaTiO_3$ Bragg peaks [48]. In particular, for the 10-nm-thick capped film the diffuse intensity was absent at 950 K, but it was present at 870 K and below, yielding a *Tc* value consistent with Raman results. Similar diffuse scattering was observed for the 4.0- and 2.4-nm-thick capped samples at room temperature, while

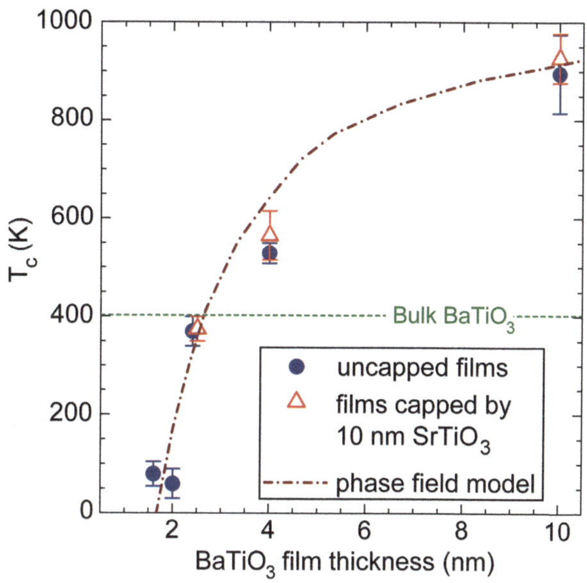

Fig. 21.12 Ferroelectric phase transition temperature, T_c in strained $BaTiO_3$ films on $SrTiO_3$ substrates as a function of the film thickness, as determined from Raman data for all films studied (*symbols*). The *dashed-dotted line* is a result of the phase-field model calculation with open circuit boundary conditions. *Dotted line* shows T_c in bulk $BaTiO_3$ (After Tenne et al. [48])

the data for 2.0 and 1.6 nm films provided no evidence for diffuse scattering, suggesting that T_c was below room temperature for these films, also consistent with T_cs obtained from Raman data. While the satellites were not clearly observed for the uncapped $BaTiO_3$ samples, this may be due to the existence of nonperiodic domains or domain periods larger than observable by the experimental resolution (~80 nm). Raman spectra and their temperature evolution for uncapped and capped $BaTiO_3$ films show almost no difference in terms of the shape and relative intensity of phonon peaks. This indicates that both for capped and uncapped films have similar ferroelectric properties. The T_c values are pretty close for capped and uncapped of the same $BaTiO_3$ thickness, being slightly higher for capped films.

Observed T_c for the thickest, 10-nm films (925 K) approaches the value obtained from the phase diagram calculated without considering the finite thickness effect [22] at the relevant strain, −2.2% (~1000 K). For thinner films the size effect becomes significant and causes a dramatic decrease in T_c. Figure 21.12 also shows the result of the phase-field model calculation of T_c in $BaTiO_3$ films clamped to $SrTiO_3$ substrates as a function of film thickness. The time-dependent Ginzburg-Landau equations [22] have been used, considering open circuit electrostatic boundary conditions, corresponding to the case of the films with no electrodes, studied here. The result agrees very well with Raman spectroscopy data, as one can see in Fig. 21.12. Comparing Raman results with other calculations of the phase diagrams of ultrathin $BaTiO_3$ films for ideal short-circuit electrical boundary conditions [157, 158], films with imperfect-screening metal electrodes [156], and under open-circuit conditions [159], a good agreement with the latter calculations was found. (T_c value for 2.4-nm $BaTiO_3$ films is 370 K from Raman data presented here and ~380 K from Fig. 21.3 of Ref. [159], extrapolated to −2.2% strain).

5.4 Ferroelectricity in Strain-Free $SrTiO_3$ Films: Effect of Non-stoichiometry

Pure $SrTiO_3$ is a well-known incipient ferroelectric (quantum paraelectric) material, and in bulk remains paraelectric down to 0.3 K, where quantum fluctuations of atomic positions suppress ferroelectric ordering [142]. Slight perturbations in the lattice structure can, however, break the delicate balance of forces and lead to the appearance of ferroelectric polarization. Ferroelectricity induced by strain [29, 30], doping [161, 162], or oxygen isotope substitution [163] was reported. Recent study [164] by Jang et al. reported a study of strained and strain-free $SrTiO_3$ films by dielectric, ferroelectric, optical, and nanoscale piezoelectric property measurements. (001) $SrTiO_3$ films were grown on (110) $NdGaO_3$ and (001) $SrTiO_3$ substrates using pulsed-laser deposition (PLD) with in situ growth monitoring by reflection high-energy electron diffraction. The lattice mismatch between pseudocubic in-plane lattice parameters of the $NdGaO_3$ substrate and $SrTiO_3$ induced a biaxial compressive strain of $\sim$1:18% in a fully commensurate $SrTiO_3$ film on this substrate, while the films grown on $SrTiO_3$ were strain-free. For electrical measurements, 60-nm-thick lattice-matched $Sr_{0.2}Ca_{0.8}RuO_3$ electrodes were used with the samples grown on $NdGaO_3$ substrates; strain-free $SrTiO_3$ films were deposited over 50-nm-thick $SrRuO_3$ bottom electrodes.

By a direct comparison of the strained and strain-free $SrTiO_3$ films using temperature-dependent capacitance, ferroelectric polarization, Raman, second harmonic generation, and piezoresponse force microscopy measurements, it was demonstrated that both strained and strain-free $SrTiO_3$ films are relaxor ferroelectrics, and the role of strain is to stabilize longer-range correlation of preexisting polar nanoregions. UV Raman results provided one of the key points of the study, shedding light on the likely origin of these nanoregions. Figure 21.13a shows Raman spectra of five $SrTiO_3$-based materials: bulk single crystals, nearly stoichiometric and intentionally Sr-deficient 100-nm-thick homoepitaxial films grown on SrTiOs substrates by molecular beam epitaxy, and two strain-free nominally stoichiometric films (50- and 1000-nm-thick) grown on $SrRuO_3/SrTiO_3$ by pulsed laser deposition. Figure 21.13b shows the temperature evolution of Raman spectra for the 50-nm-thick $SrTiO_3$ film on $SrRuO_3/SrTiO_3$. The shape and temperature evolution of Raman spectra of the MBE-grown strontium-deficient films are remarkably similar to the spectra of PLD-grown nominally stoichiometric films that were proved to be ferroelectric by a combination of techniques including the polarization hysteresis loops measured by piezoresponse force microscopy [164].

The key result is the appearance of first order Raman scattering (forbidden in centrosymmetric paraelectric phase) by optical phonon modes (TO_2, LO_3, TO_4, and LO_4) in all the film systems. These phonon peaks are weak in the nearly stoichiometric MBE-grown film, and are the strongest in the non-stoichiometric ($Sr_{0.9}TiO_{3-x}$) film with intentionally introduced large Sr deficiency. The nominally stoichiometric PLD-grown $SrTiO_3$ films also exhibit strong first order Raman peaks; nearly identical behavior was observed in both 50- and 1000-nm-thick films. The temperature evolution of Raman spectra (Fig. 21.13b) clearly indicates

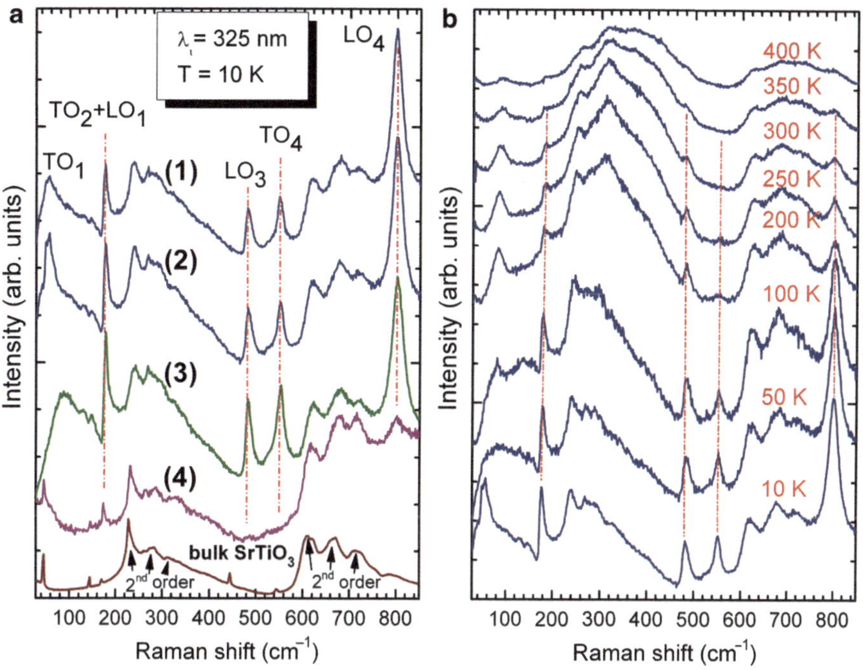

Fig. 21.13 (**a**) UV Raman spectra of four $SrTiO_3$ films and a bulk single crystal at 10 K. (*1*) and (*2*): nominally stoichiometric 1000-nm and 50-nm-thick films grown on $SrRuO_3/SrTiO_3$ by pulsed laser deposition; an intentionally Sr-deficient $Sr_{0.9}TiO_{3-x}$ film (*3*) and a near-stoichiometric $SrTiO_3$ film (*4*) grown on $SrTiO_3$ by molecular beam epitaxy. (**b**) Variable-temperature Raman spectra of the film (*2*): 50-nm-thick $SrTiO_3$ film on $SrRuO_3/SrTiO_3$ (After Jang et al. [164])

that the films remain polar up to ~350 K, which is consistent with the optical second harmonic generation and piezoresponse force microscopy observations [164]. Chemical composition analysis of the 1000-nm-thick $SrTiO_3$ film by wavelength dispersive X-ray spectroscopy indicated that the film had a Sr/Ti ratio of ~0.97, being stoichiometric within experimental error. These findings indicate that the most likely reason for the observed ferroelectricity is a small amount of Sr-deficiency in our nominally stoichiometric $SrTiO_3$ samples, which can lead to the formation of polar nanoregions.

6 Summary and Future Perspective

Ultraviolet Raman spectroscopy has been proven to be an effective technique for probing lattice dynamics and studying phase transformations in nanoscale ferroelectric superlattices and ultrathin films. Advantages of ultraviolet excitation, including shorter penetration depth, enhanced scattering intensity, and absence of thermal radiation background at high temperatures, helped to obtain results discussed in this chapter, which contributed to improved understanding nanoscale ferroelectricity. UV Raman spectroscopy demonstrated that coherently strained $BaTiO_3/SrTiO_3$ and $CaTiO_3/BaTiO_3/SrTiO_3$ superlattices can be ferroelectric even when the thickness of ferroelectric $BaTiO_3$ layer is only one unit cell (4 Å), and the ferroelectric phase transition T_c in superlattices varies largely depending on layer thicknesses and strain.

In single-layer trained $BaTiO_3$ films, using UV excitation allowed observation of phonons in films as thin as 1.6 nm (4 unit cells), indicating that even such ultrathin films become ferroelectric at low temperatures ~70 K, if subject to sufficient strain. Variable-temperature Raman spectroscopy showed Tc in these strained films to vary in a very broad range as a function of layer thickness, being as high as ~930 K for the 10-nm-thick films, that is, exceeding the bulk $BaTiO_3$ value by over 500°. This demonstration of the size effect on the Curie temperature in $BaTiO_3$ films, the ability to adjust, measure, and understand tuning of the phase transition temperature by nearly 900 K using strain and thickness is an example of the great utility of ultraviolet Raman spectroscopy for studying nanoscale ferroelectrics and showing an example of the dramatic modification of material properties in artificially engineered epitaxial heterostructures.

UV Raman studies of ferroelectricity in strain-free non-stoichiometric and nominally stoichiometric $SrTiO_3$ films, in combination with dielectric, ferroelectric, nonlinear optical and nanoscale piezoelectric property measurements highlighted the sensitive role of stoichiometry when exploring strain and epitaxy-induced electronic phenomena in oxide films, heterostructures, and interfaces.

Nevertheless, investigation of lattice dynamics in ferroelectric nanostructures is still at its very beginning, and the ultraviolet Raman spectroscopy technique has just begun to show its effectiveness in the study of nanoscale ferroelectricity. New materials and structures are to be studied, and many questions related to their vibrational spectra, effects of strain, temperature, size, stoichiometry on their behavior are to be answered. Generally, ultraviolet Raman spectroscopy is still a relatively new technique, and its applications to dielectric and ferroelectric materials systems are still rare. We believe the results presented here, such as observation of Raman scattering by an oxide film as thin as 1.6 nm, will attract interest for researchers working not only in the ferroelectrics field, but also those dealing with thin films and nanostructures of wide bandgap semiconductors and high-k dielectrics. These materials are in the focus of many fundamental and applied studies in the field of nanoscience and nanotechnology. Therefore, we believe the applications of ultraviolet Raman spectroscopy for characterization of nanomaterials will continue to expand.

Acknowledgments The author gratefully acknowledges his colleagues and collaborators, without whose invaluable contributions the results presented in this chapter would not have been possible to obtain. In particular, the author thanks S. H. Baek, N. Balke, C. W. Bark, M. Biegalski, M. Bernhagen, C. M. Brooks, A. Bruchhausen, A Cantarero, L. Q. Chen, K. J. Choi, A. Cros, S. Denev, J. A. Eastman, C. B. Eom, A. Fainstein, A. K. Farrar, C. M. Folkman, D. Fong, P. H. Fuoss, V. Gopalan, S. V. Kalinin, J. H. Haeni, T. Heeg, H. W. Jang, Q. X. Jia, R. S. Katiyar, D. M. Kim, A. Kumar, N. D. Lanzillotti-Kimura, H. N. Lee, Y. L. Li, P. Maksymovych, S. M. Nakhmanson, X. Q. Pan, K. M. Rabe, P. Reiche, D. G. Schlom, J. D. Schmidt, J. Schubert, A. Soukiassian,

G. B. Stephenson, S. K. Streiffer, H. P. Sun, W. Tian, C. Thompson, S. Trolier-McKinstry, P. Turner, R. Uecker, V. Vaithyanathan, X. X. Xi, and S. Y. Xu.
The financial support from the following agencies is gratefully acknowledged: the National Science Foundation under Grant Nos. DMR-0705127 and DMR-1006136, the US Department of Energy EPSCoR program under Grant No. DE-FG02-04ER46142, and Research Corporation for Science Advancement (Grant No. 7134).

References

1. Lines ME, Glass AM (1977) Principles and applications of ferroelectrics and related materials. Clarendon, Oxford
2. Ramesh R (ed) (1997) Thin film ferroelectric materials and devices. Kluwer, Boston
3. Ramesh R, Aggarwal S, Auciello O (2001) Science and technology of ferroelectric films and heterostructures for non-volatile ferroelectric memories. Mater Sci Eng R 32:191
4. Dawber M, Rabe KM, Scott JF (2005) Physics of thin-film ferroelectric oxides. Rev Mod Phys 77:1083
5. Scott JF (2007) Applications of modern ferroelectrics. Science 315:954
6. Rabe KM, Ahn CH, Triscone J-M (eds) (2007) Physics of ferroelectrics: a modern perspective. Springer, Berlin
7. Tagantsev AK, Sherman VO, Astafiev KF, Venkatesh J, Setter N (2003) Ferroelectric materials for microwave tunable applications. J Electroceram 11:5
8. Schlom DG, Haeni JH, Lettieri J, Theis CD, Tian W, Jiang JC, Pan XQ (2001) Oxide nano-engineering using MBE. Mater Sci Eng B 87:282
9. Posadas A-B, Lippmaa M, Walker FJ, Dawber M, Ahn CH, Triscone J-M (2007) Growth and novel applications of epitaxial oxide thin films. In: Rabe KM, Ahn CH, Triscone J-M (eds) Physics of ferroelectrics: a modern perspective. Springer, Berlin, p 219
10. Schlom DG, Chen L-Q, Pan XQ, Schmehl A, Zurbuchen MA (2008) A thin film approach to engineering functionality into oxides. J Am Ceram Soc 91:2429–2454
11. Ahn CH, Rabe KM, Triscone J-M (2004) Ferroelectricity at the nanoscale: local polarization in oxide thin films and heterostructures. Science 303:488
12. Hong S (ed) (2004) Nanoscale phenomena in ferroelectric thin films. Springer, Berlin
13. Rüdiger A, Schneller T, Roelofs A, Tiedke S, Schmitz T, Waser R (2005) Nanosize ferroelectric oxides – tracking down the superparaelectric limit. Appl Phys A 80:1247
14. Dawber M, Stucki N, Lichtensteiger C, Gariglio S, Ghosez P, Triscone J-M (2007) Tailoring the properties of artificially layered ferroelectric superlattices. Adv Mater 19:4153
15. Schlom DG, Chen L-Q, Eom C-B, Rabe KM, Streiffer SK, Triscone J-M (2007) Strain tuning of ferroelectric thin films. Annu Rev Mater Res 37:589, and references therein
16. Pertsev NA, Zembilgotov AG, Tagantsev AK (1998) Effect of mechanical boundary conditions on phase diagrams of epitaxial ferroelectric thin films. Phys Rev Lett 80:1988
17. Rabe KM, Waghmare UV (1996) Strain coupling in the $PbTiO_3$ ferroelectric transition. Philos Trans R Soc Lond A354:2897
18. Roytburd AL (1998) Thermodynamics of polydomain heterostructures. I. Effect of macrostresses. J Appl Phys 83:228
19. Alpay SP, Roytburd AL (1998) Thermodynamics of polydomain heterostructures. III. Domain stability map. J Appl Phys 83:4714
20. Tenne DA, Xi XX, Li YL, Chen LQ, Soukiassian A, Zhu MH, James AR, Lettieri J, Schlom DG, Tian W, Pan XQ (2004) Absence of low-temperature phase transitions in epitaxial $BaTiO_3$ thin films. Phys Rev B 69:174101
21. Li YL, Hu SY, Liu ZK, Chen LQ (2001) Phase-field model of domain structures in ferroelectric thin films. Appl Phys Lett 78:3878
22. Li YL, Chen LQ (2006) Temperature-strain phase diagram for $BaTiO_3$ thin films. Appl Phys Lett 88:072905

23. Choi KJ, Biegalski M, Li YL, Sharan A, Schubert J, Uecker R, Reiche P, Chen YB, Pan XQ, Gopalan V, Chen L-Q, Schlom DG, Eom CB (2004) Enhancement of ferroelectricity in strained $BaTiO_3$ thin films. Science 306:1005
24. Shimuta T, Nakagawara O, Makino T, Arai S, Tabata H, Kawai T (2002) Enhancement of remanent polarization in epitaxial $BaTiO_3/SrTiO_3$ superlattices with asymmetric structure. J Appl Phys 91:2290
25. Lee HN, Christen HM, Chisholm MF, Rouleau CM, Lowndes DH (2005) Strong polarization enhancement in asymmetric three-component ferroelectric superlattices. Nature 433:395
26. Neaton JB, Rabe KM (2003) Theory of polarization enhancement in epitaxial $BaTiO_3/SrTiO_3$ superlattices. Appl Phys Lett 82:1586
27. Nakhmanson SM, Rabe KM, Vanderbilt D (2006) Predicting polarization enhancement in multicomponent ferroelectric superlattices. Phys Rev B 73:060101(R)
28. Uwe H, Sakudo T (1976) Stress-induced ferroelectricity and soft phonon modes in $SrTiO_3$. Phys Rev B 13:271
29. Pertsev NA, Tagantsev AK, Setter N (2000) Phase transition and strain-induced ferroelectricity in $SrTiO_3$ epitaxial thin films. Phys Rev B 61:R825
30. Haeni JH, Irvin P, Chang W, Uecker R, Reiche P, Li YL, Choudhury S, Tian W, Hawley ME, Craigo B, Tagantsev AK, Pan XQ, Streiffer SK, Chen LQ, Kirchoefer SW, Levy J, Schlom DG (2004) Room-temperature ferroelectricity in strained $SrTiO_3$. Nature (Lond) 430:758
31. Biegalski MD, Jia Y, Schlom DG, Trolier-McKinstry S, Streiffer SK, Sherman V, Uecker R, Reiche P (2006) Relaxor ferroelectricity in strained epitaxial $SrTiO_3$ thin films on $DyScO_3$ substrates. Appl Phys Lett 88:192907
32. Warusawithana MP, Cen C, Sleasman CR, Woicik JC, Li YL, Fitting Kourkoutis L, Klug JA, Li H, Ryan P, Wang L-P, Bedzyk M, Muller DA, Chen L-Q, Levy J, Schlom DG (2009) A ferroelectric oxide made directly on silicon. Science 324:367
33. Tyunina M, Narkilahti J, Plekh M, Oja R, Nieminen RM, Dejneka A, Trepakov V (2010) Evidence for strain-induced ferroelectric order in epitaxial thin-film $KTaO_3$. Phys Rev Lett 104:227601
34. Bousquet E, Spaldin NA, Ghosez P (2010) Strain-induced ferroelectricity in simple rocksalt binary oxides. Phys Rev Lett 104:037601
35. Spaldin NA (2004) Fundamental size limits in ferroelectricity. Science 304:1606
36. Junquera J, Ghosez P (2003) Critical thickness for ferroelectricity in perovskite ultrathin films. Nature 422:506
37. Sai N, Kolpak AM, Rappe AM (2005) Ferroelectricity in ultrathin perovskite films. Phys Rev B 72:020101(R)
38. Fong DD, Stephenson GB, Streiffer SK, Eastman JA, Auciello O, Fuoss PH, Thompson C (2004) Ferroelectricity in ultrathin perovskite films. Science 304:1650
39. Lichtensteiger C, Triscone J-M, Junquera J, Ghosez P (2005) Ferroelectricity and tetragonality in ultrathin $PbtiO_3$ films. Phys Rev Lett 94:047603
40. Despont L, Koitzsch C, Clerc F, Garnier MG, Aebi P, Lichtensteiger C, Triscone J-M, Garcia de Abajo FJ, Bousquet E, Ghosez Ph (2006) Direct evidence for ferroelectric polar distortion in ultrathin lead titanate perovskite films. Phys Rev B 73:094110
41. Fong DD, Kolpak AM, Eastman JA, Streiffer SK, Fuoss PH, Stephenson GB, Carol Thompson DM, Kim KJ, Choi CB, Eom IG, Rappe AM (2006) Stabilization of monodomain polarization in ultrathin $PbTiO_3$ films. Phys Rev Lett 96:127601
42. Dawber M, Lichtensteiger C, Cantoni M, Veithen M, Ghosez P, Johnston K, Rabe KM, Triscone J-M (2005) Unusual behavior of the ferroelectric polarization in $PbTiO_3/SrTiO_3$ superlattices. Phys Rev Lett 95:177601
43. Tenne DA, Bruchhausen A, Lanzillotti-Kimura ND, Fainstein A, Katiyar RS, Soukiassian A, Vaithyanathan V, Haeni JH, Tian W, Schlom DG, Choi KJ, Kim DM, Eom CB, Sun HP, Pan XQ, Li YL, Chen LQ, Jia QX, Nakhmanson SM, Rabe KM, Xi XX (2006) Probing nanoscale ferroelectricity by ultraviolet raman spectroscopy. Science 313:1614

44. Bousquet E, Dawber M, Stucki N, Lichtensteiger C, Hermet P, Gariglio S, Triscone J-M, Ghosez P (2008) Improper ferroelectricity in perovskite oxide artificial superlattices. Nature (Lond) 452:732
45. Dawber M, Lichtensteiger C, Triscone J-M (2008) Phase transitions in ultra-thin ferroelectric films and fine period multilayers. Phase Transit 81:623
46. Dawber M, Stucki N, Lichtensteiger C, Gariglio S, Triscone J-M (2008) New phenomena at the interfaces of very thin ferroelectric oxides. J Phys Condens Matter 20:264015
47. Garcia V, Fusil S, Bouzehouane K, Enouz-Vedrenne S, Mathur ND, Barthélémy A, Bibes M (2009) Giant tunnel electroresistance for non-destructive readout of ferroelectric states. Nature 460:81
48. Tenne DA, Turner P, Schmidt JD, Biegalski M, Li YL, Chen LQ, Soukiassian A, Trolier-McKinstry S, Schlom DG, Xi XX, Fong DD, Fuoss PH, Eastman JA, Stephenson GB, Thompson C, Streiffer SK (2009) Ferroelectricity in ultrathin strained $BaTiO_3$ films: probing the size effect by ultraviolet Raman spectroscopy. Phys Rev Lett 103:177601
49. Tabata H, Tanaka H, Kawai T (1994) Formation of artificial $BaTiO_3/SrTiO_3$ superlattices using pulsed laser deposition and their dielectric properties. Appl Phys Lett 65:1970
50. Shimuta T, Nakagawara O, Makino T, Arai S, Tabata H, Kawai T (2002) Enhancement of remanent polarization in epitaxial $BaTiO_3/SrTiO_3$ superlattices with asymmetric structure. J Appl Phys 91:2290
51. Le Marrec F, Farhi R, El Marssi M, Dellis JL, Karkut MG, Ariosa D (2000) Ferroelectric $PbTiO_3/BaTiO_3$ superlattices: growth anomalies and confined modes. Phys Rev B 61: R6447
52. Jiang JC, Pan XQ, Tian W, Theis CD, Schlom DG (1999) Abrupt $PbTiO_3/BaTiO_3$ superlattices grown by reactive molecular beam epitaxy. Appl Phys Lett 74:2851
53. Sigman J, Norton DP, Christen HM, Fleming PH, Boatner LA (2002) Antiferroelectric behavior in symmetric $KNbO_3/KTaO_3$ superlattices. Phys Rev Lett 88:097601
54. Christen HM, Specht ED, Silliman SS, Harshavardhan KS (2003) Ferroelectric and antiferroelectric coupling in superlattices of paraelectric perovskites at room temperature. Phys Rev B 68:020101(R)
55. Warusawithana MP, Colla EV, Eckstein JN, Weissman MB (2003) Artificial dielectric superlattices with broken inversion symmetry. Phys Rev Lett 90:036802
56. Soukiassian A, Tian W, Vaithyanathan V, Haeni JH, Chen LQ, Xi XX, Schlom DG, Tenne DA, Sun HP, Pan XQ, Choi KJ, Eom CB, Li YL, Jia QX, Constantin C, Feenstra RM, Bernhagen M, Reiche P, Uecker R (2008) Growth of nanoscale $BaTiO_3/SrTiO_3$ superlattices by molecular-beam epitaxy. J Mater Res 23:1417–1432
57. Sai N, Meyer B, Vanderbilt D (2000) Compositional inversion symmetry breaking in ferroelectric perovskites. Phys Rev Lett 84:5636
58. Neaton JB, Rabe KM (2003) Theory of polarization enhancement in epitaxial $BaTiO_3/SrTiO_3$ superlattices. Appl Phys Lett 82:1586
59. Bungaro C, Rabe KM (2002) Lattice instabilities of $PbZrO_3/PbTiO_3$ [1:1] superlattices from first principles. Phys Rev B 65:224106
60. Johnston K, Huang X, Neaton JB, Rabe KM (2005) First-principles study of symmetry lowering and polarization in $BaTiO_3/SrTiO_3$ superlattices with in-plane expansion. Phys Rev B 71:100103(R)
61. Nakhmanson SM, Rabe KM, Vanderbilt D (2005) Polarization enhancement in two- and three-component ferroelectric superlattices. Appl Phys Lett 87:102906
62. Nakhmanson SM, Rabe KM, Vanderbilt D (2006) Predicting polarization enhancement in multicomponent ferroelectric superlattices. Phys Rev B 73:060101(R)
63. Lisenkov S, Bellaiche L (2007) Phase diagrams of $BaTiO_3/SrTiO_3$ superlattices from first principles. Phys Rev B 76:020102(R)
64. Pertsev NA, Janolin P-E, Kiat J-M, Uesu Y (2010) Enhancing permittivity of ferroelectric superlattices via composition tuning. Phys Rev B 81:144118
65. Cochran W (1960) Crystal stability and the theory of ferroelectricity. Adv Phys 9:387

66. Worlock JM, Fleury PA (1967) Electric field dependence of optical-phonon frequencies. Phys Rev Lett 19:1176
67. Fleury PA, Worlock JM (1968) Electric-field-induced Raman scattering in $SrTiO_3$ and $KTaO_3$. Phys Rev 174:613
68. Rupprecht G, Bell RO, Silverman BD (1961) Nonlinearity and microwave losses in cubic strontium-titanate. Phys Rev 123:97–98
69. Gurevich VL, Tagantsev AK (1991) Intrinsic dielectric loss in crystals. Adv Phys 40: 719–767
70. Tagantsev AK (1993) Phonon mechanisms of intrinsic dielectric loss in crystals. In: Setter N, Colla EL (eds) Ferroelectric ceramics. Birkhäuser, Basel, p 127
71. Axe JD, Harada J, Shirane G (1970) Anomalous acoustic dispersion in centrosymmetric crystals with soft optic phonons. Phys Rev B 1:1227
72. Stirling WG (1972) Neutron inelastic scattering study of the lattice dynamics of strontium titanate: harmonic models. J Phys C 5:2711
73. Shirane G, Axe JD, Harada J, Linz A (1970) Inelastic neutron scattering from single-domain $BaTiO_3$. Phys Rev B 2:3651
74. Shirane G, Axe JD, Harada J, Remeika JP (1970) Soft ferroelectric modes in lead titanate. Phys Rev B 2:155
75. Tomeno I, Ishii Y, Tsunoda Y, Oka K (2006) Lattice dynamics of tetragonal $PbTiO_3$. Phys Rev B 73:064116
76. Hlinka J, Kempa M, Kulda J, Bourges P, Kania A, Petzelt J (2006) Lattice dynamics of ferroelectric $PbTiO_3$ by inelastic neutron scattering. Phys Rev B 73:140101(R)
77. Egami T (2007) Local structure of ferroelectric materials. Annu Rev Mater Res 37:297, and references therein
78. Last JT (1957) Infrared-absorption studies on barium titanate and related materials. Phys Rev 105:1740
79. Spitzer WG, Miller RC, Kleinman DA, Howarth LE (1962) Far infrared dielectric dispersion in $BaTiO_3$, $SrTiO_3$, and TiO_2, Phys Rev 126:1710
80. Servoin JL, Luspin Y, Gervais F (1980) Infrared dispersion in $SrTiO_3$ at high temperature. Phys Rev B 22:5501
81. Servoin JL, Gervais F, Quittet AM, Luspin Y (1980) Infrared and Raman responses in ferroelectric perovskite crystals: apparent inconsistencies. Phys Rev B 21:2038
82. Salje EKH, Bismayer U (1997) Hard mode spectroscopy: the concept and applications. Phase Transitions 63:1
83. Petzelt J, Ostapchuk T (2001) Far-infrared ferroelectric soft mode spectroscopy on thin films. Ferroelectrics 249:81
84. DiDomenico M Jr, Wemple SH, Porto SPS, Bauman RP (1968) Raman spectrum of single-domain $BaTiO_3$. Phys Rev 174:522
85. Scalabrin A, Chaves AS, Shim DS, Porto SPS (1977) Temperature-dependence of A_1 and E optical phonons in $BaTiO_3$. Phys Status Solidi B 79:731
86. Nilsen WG, Skinner JG (1968) Raman spectrum of strontium titanate. J Chem Phys 48:2240
87. Fleury PA, Worlock JM (1968) Electric-field-induced Raman scattering in $SrTiO_3$ and $KTaO_3$. Phys Rev 174:613
88. Burns G, Scott BA (1973) Lattice modes in ferroelectric perovskites: $PbTiO_3$. Phys Rev B 7:3088
89. Taguchi I, Pignolet A, Wang L, Proctor M, Levy F, Schmid PE (1993) Raman scattering from $PbTiO_3$ thin films prepared on silicon substrates by radio frequency sputtering and thermal treatment. J Appl Phys 73:394
90. Zhang H, Leppãvuori S, Karjalainen P (1995) Raman spectra in laser ablated lead zirconate titanate thin films near the morphotropic phase boundary. J Appl Phys 77:2691
91. Ching-Prado E, Cordero J, Katiyar RS, Bhalla AS (1996) Temperature-dependent Raman scattering in PT and PMN-PT thin films. J Vac Sci Technol A 14:762

92. Feng ZC, Kwak BS, Erbil A, Boatner LA (1993) Difference Raman spectra of $PbTiO_3$ thin films grown by metalorganic chemical vapor deposition. Appl Phys Lett 62:349
93. Robins LH, Kaiser DL, Rotter LD, Schenck PK, Stauf GT, Rytz D (1994) Investigation of the structure of barium titanate thin films by Raman spectroscopy. J Appl Phys 76:7487
94. Yuzyuk YI, Alyoshin VA, Zakharchenko IN, Sviridov EV, Almeida A, Chaves MR (2002) Polarization-dependent Raman spectra of heteroepitaxial (Ba, Sr)TiO_3/MgO thin films. Phys Rev B 65:134107
95. Yuzyuk YI, Katiyar RS, Alyoshin VA, Zakharchenko IN, Markov DA, Sviridov EV (2003) Stress relaxation in heteroepitaxial (Ba, Sr)TiO_3/(001)MgO thin film studied by micro-Raman spectroscopy. Phys Rev B 68:104104
96. Merkulov VI, Fox JR, Li H-C, Si W, Sirenko AA, Xi XX (1998) Metal-oxide bilayer Raman scattering in $SrTiO_3$ thin films. Appl Phys Lett 72:3291
97. Sirenko AA, Akimov IA, Fox JR, Clark AM, Li H-C, Si W, Xi XX (1999) Observation of the first-order Raman scattering in $SrTiO_3$ thin films. Phys Rev Lett 82:4500
98. Akimov IA, Sirenko AA, Clark AM, Hao J-H, Xi XX (2000) Electric-field induced soft-mode hardening in $SrTiO_3$ films. Phys Rev Lett 84:4625
99. Tenne DA, Clark AM, James AR, Chen K, Xi XX (2001) Soft phonon modes in $Ba_{0.5}Sr_{0.5}TiO_3$ thin films studied by Raman spectroscopy. Appl Phys Lett 79:3836–3838
100. Tenne DA, Xi XX, Li YL, Chen LQ, Soukiassian A, Zhu MH, James AR, Lettieri J, Schlom DG, Tian W, Pan XQ (2004) Absence of low-temperature phase transitions in epitaxial $BaTiO_3$ thin films. Phys Rev B 69:174101
101. Tenne DA, Soukiassian A, Zhu MH, Clark AM, Xi XX, Choosuwan H, He Q, Guo R, Bhalla AS (2003) Raman study of $Ba_xSr_{1-x}TiO_3$ films: evidence for the existence of polar nanoregions. Phys Rev B 67:012302
102. Tenne DA, Soukiassian A, Xi XX, Choosuwan H, Guo R, Bhalla AS (2004) Lattice dynamics in $Ba_xSr_{1-x}TiO_3$ thin films studied by Raman spectroscopy. J Appl Phys 96:6597
103. Sirenko AA, Bernhard C, Golnik A, Clark AM, Hao J, Si W, Xi XX (2000) Soft-mode hardening in $SrTiO_3$ thin films. Nature (Lond) 404:373
104. Tenne DA, Soukiassian A, Xi XX, Taylor TR, Hansen PJ, York RA, Speck JS (2004) Effect of thermal strain on the ferroelectric phase transition in polycrystalline $Ba_{0.5}S_{r0.5}TiO_3$ thin films studied by Raman spectroscopy. Appl Phys Lett 85:4124
105. Petzelt J, Ostapchuk T (2002) Soft-mode spectroscopy in $SrTiO_3$, $BaTiO_3$, and BST films and ceramics. Ferroelectrics 267:93
106. Ostapchuk T, Petzelt J, Zelezny V, Pashkin A, Pokorny J, Drbohlav I, Kuzel R, Rafaja D, Gorshunov BP, Dressel M, Ohly Ch, Hoffmann-Eifert S, Waser R (2002) Origin of soft-mode stiffening and reduced dielectric response in $SrTiO_3$ thin films. Phys Rev B 66:235406
107. Tenne DA, Xi XX (2008) Raman spectroscopy of ferroelectric thin films and superlattices. J Am Ceram Soc 91:1820
108. Weber WH, Merlin R (eds) (2000) Raman scattering in materials science. Springer, Berlin
109. Cardona M (ed) (1975/1983) Light scattering in solids, 1st edn; Light scattering in solids I: introductory concepts, 2nd edn, vol 8, Topics in applied physics. Springer, Berlin
110. Cardona M, Güntherodt G (eds) (1982) Light scattering in solids II: basic concepts and instrumentation, vol 50, Topics in applied physics. Springer, Berlin
111. Cardona M, Güntherodt G (eds) (1982) Light scattering in solids III: recent results, vol 51, Topics in applied physics. Springer, Berlin
112. Cardona M, Güntherodt G (eds) (1984) Light scattering in solids IV: electronic scattering, spin effects, SERS, and morphic effects, vol 54, Topics in applied physics. Springer, Berlin
113. Cardona M, Güntherodt G (eds) (1989) Light scattering in solids V: superlattices and other microstructures, vol 66, Topics in applied physics. Springer, Berlin
114. Cardona M, Güntherodt G (eds) (1991) Light scattering in solids VI: recent results, including high-T_c superconductivity, vol 68, Topics in applied physics. Springer, Berlin

115. Cardona M, Güntherodt G (eds) (2000) Light scattering in solids VII: crystal field and magnetic excitations, vol 75, Topics in applied physics. Springer, Berlin
116. Cardona M, Güntherodt G (eds) (2000) Light scattering in solids VIII: fullerenes, semiconductor surfaces, coherent phonons, vol 76, Topics in applied physics. Springer, Berlin
117. Cardona M, Merlin R (eds) (2007) Light scattering in solids IX: novel materials and techniques, vol 108, Topics in applied physics. Springer, Berlin
118. Hayes W, Loudon R (1978) Scattering of light by crystals. Wiley, New York
119. Asher SA, Johnson CR, Murtaugh J (1983) Development of a new UV resonance Raman spectrometer for the 217–400-nm spectral region. Rev Sci Instrum 54:1657
120. Pajcini V, Munro CH, Bormett RW, Witkowski RE, Asher SA (1997) UV Raman microspectroscopy: spectral and spatial selectivity with sensitivity and simplicity. Appl Spectrosc 51:81
121. Sparrow MC, Jackovitz JF, Munro CH, Hug WF, Asher SA (2001) New 224 nm hollow cathode laser–UV Raman spectrometer. Appl Spectrosc 55:66
122. Sands HS, Demangeot F, Bonera E, Webster S, Bennett R, Hayward IP, Marchi F, Smith DA, Batchelder DN (2002) Development of a combined confocal and scanning near-field Raman microscope for deep UV laser excitation. J Raman Spectrosc 33:730
123. Asher SA (2002) Ultraviolet Raman spectrometry. In: Chalmers JM, Griffiths PR (eds) Handbook of vibrational spectroscopy, vol 1. Wiley, Chichester, p 557
124. Yashima M, Kakihana M, Shimidzu R, Fujimori H, Yoshimura M (1997) Ultraviolet 363.8-nm Raman spectroscopic system for *in situ* measurements at high temperatures. Appl Spectrosc 51:1224
125. Fujimori H, Yashima M, Kakihana M, Yoshimura M (2000) Phase transition and soft phonon modes in $SrZrO_3$ around 1200°C by ultraviolet laser Raman spectroscopy. Phys Rev B 61:3971
126. Nakashima S-I, Okumura H, Yamamoto T, Shimidzu R (2004) Deep-ultraviolet Raman microspectroscopy: characterization of wide-gap semiconductors. Appl Spectrosc 58:224
127. Feng ZC, Wang W, Chua SJ, Zhang PX, Williams KPJ, Pitt GD (2001) Raman scattering properties of GaN thin films grown on sapphire under visible and ultraviolet excitation. J Raman Spectrosc 32:840
128. Alexson D, Bergman L, Nemanich RJ, Dutta M, Stroscio MA, Parker CA, Bedair SM, El-Masry NA, Adar F (2001) Ultraviolet Raman study of A_1(LO) and E_2 phonons in $In_xGa_{1-x}N$ alloys. J Appl Phys 89:798
129. Bormett RW, Asher SA, Witowski RE, Partlow WD, Lizewski R, Pettit F (1995) Ultraviolet Raman spectroscopy characterizes chemical vapor deposition diamond film growth and oxidation. J Appl Phys 77:5916
130. Merkulov VI, Lannin JS, Munro CH, Asher SA, Veerasamy VS, Milne WI (1997) UV studies of tetrahedral bonding in diamondlike amorphous carbon. Phys Rev Lett 78:4869
131. Prawer S, Nemanich RJ (2004) Raman spectroscopy of diamond and doped diamond. Philos Trans Math Phys Eng Sci 362:2537
132. Ravindran TR, Badding JV (2002) Ultraviolet Raman analysis of the formation of diamond from C_{60}. Sol St Commun 121:391
133. Ravindran TR, Badding JV (2006) UV Raman studies on carbon nitride structures. J Mater Sci 41:7145
134. Liu R, Canonico M (2003) Applications of UV-Raman spectroscopy to microelectronic materials and devices. AIP Conf Proc 683:738
135. Tisinger LH, Liu R, Kulik J, Zhang X, Ramdani J, Demkov AA (2003) Ultraviolet-Raman studies of $SrTiO_3$ ultrathin films on Si. J Vac Sci Technol B 21:53
136. Shiozaki Y, Nakamura E, Mitsui T (eds) (2001) Landolt-Börnstein numerical data and functional relationships in science and technology, new series, Group III, vol 36, subvol A1. Springer, Berlin
137. Fujimori H, Komatsu H, Ioku K, Goto S, Yoshimura M (2002) Anharmonic lattice mode of Ca_2SiO_4: ultraviolet laser Raman spectroscopy at high temperatures. Phys Rev B 66:064306

138. Fujimori H, Komatsu H, Ioku K, Goto S, Watanabe T (2005) Vibrational spectra of Ca_3SiO_5: ultraviolet laser Raman spectroscopy at high temperatures. J Am Ceram Soc 88:1995
139. Fujimori H, Yashima M, Kakihana M, Yoshimura M (2001) In situ ultraviolet Raman study on the phase transition of hafnia up to 2085 K. J Am Ceram Soc 84:663
140. Moutanabbir O, Reiche M, Hähnel A, Erfurth W, Motohashi M, Tarun A, Hayazawa N, Kawata S (2010) UV-Raman imaging of the in-plane strain in single ultrathin strained silicon-on-insulator patterned structure. Appl Phys Lett 96:233105
141. See supporting online material for Tenne et al. [43]
142. Müller KA, Burkard H (1979) $SrTiO_3$: an intrinsic quantum paraelectric below 4 K. Phys Rev B 19:3593
143. B. Jusserand and M. Cardona, "Raman Spectroscopy of Vibrations in Superlattices," in [113]; p. 49.
144. Colvard C, Gant TA, Klein MV, Merlin R, Fischer R, Morkoc H, Gossard AC (1985) Folded acoustic and quantized optic phonons in (GaAl)As superlattices. Phys Rev B 31:2080
145. Soukiassian A, Tian W, Tenne DA, Xi XX, Schlom DG, Lanzillotti-Kimura ND, Bruchhausen A, Fainstein A, Sun HP, Pan XQ, Cros A, Cantarero A (2007) Acoustic Bragg mirrors and cavities made using piezoelectric oxides. Appl Phys Lett 90:042909
146. Bruchhausen A, Fainstein A, Soukiassian A, Schlom DG, Xi XX, Bernhagen M, Reiche P, Uecker R (2008) Ferroelectricity-induced coupling between light and terahertz-frequency acoustic phonons in $BaTiO_3/SrTiO_3$ superlattices. Phys Rev Lett 101:197402
147. Lanzillotti-Kimura ND, Fainstein A, Perrin B, Jusserand B, Soukiassian A, Xi XX, Schlom DG (2010) Enhancement and inhibition of coherent phonon emission of a Ni film in a $BaTiO_3/SrTiO_3$ cavity. Phys Rev Lett 104:187402
148. Chen L-Q (2008) Phase-field method of phase transitions/domain structures in ferroelectric thin films: a review. J Am Ceram Soc 91:1835
149. Chen L-Q, Shen J (1998) Applications of semi-implicit Fourier-spectral method to phase field equations. Comput Phys Commun 108:147
150. Li YL, Hu SY, Tenne D, Soukiassian A, Schlom DG, Xi XX, Choi KJ, Eom CB, Saxena A, Lookman T, Jia QX, Chen LQ (2007) Prediction of ferroelectricity in $BaTiO_3/SrTiO_3$ superlattices with domains. Appl Phys Lett 91:112914
151. Li YL, Hu SY, Tenne D, Soukiassian A, Schlom DG, Chen LQ, Xi XX, Choi KJ, Eom CB, Saxena A, Lookman T, Jia QX (2007) Interfacial coherency and ferroelectricity of $BaTiO_3/SrTiO_3$ superlattice films. Appl Phys Lett 91:252904
152. Uecker R, Wilke H, Schlom DG, Velickov B, Reiche P, Polity A, Bernhagen M, Rossberg M (2006) Spiral formation during czochralski growth of rare-earth scandates. J Cryst Growth 295:84
153. Uecker R, Velickov B, Klimm D, Bertram R, Bernhagen M, Rabe M, Albrecht M, Fornari R, Schlom DG (2008) Properties of rare-earth scandate single crystals (Re = Nd–Dy). J Cryst Growth 310:2649
154. Tenne DA, Lee HN, Katiyar RS, Xi XX (2009) Ferroelectric phase transitions in three-component short-period superlattices studied by ultraviolet Raman spectroscopy. J Appl Phys 105:054106
155. Železný V, Cockayne E, Petzelt J, Limonov MF, Usvyat DE, Lemanov VV, Volkov AA (2002) Temperature dependence of infrared-active phonons in $CaTiO_3$: a combined spectroscopic and first-principles study. Phys Rev B 66:224303
156. Streiffer SK, Eastman JA, Fong DD, Thompson C, Munkholm A, Ramana Murty MV, Auciello O, Bai GR, Stephenson GB (2002) Observation of nanoscale 180° stripe domains in ferroelectric $PbTiO_3$ thin films. Phys Rev Lett 89:067601
157. Lai B-K, Kornev IA, Bellaiche L, Salamo GJ (2005) Phase diagrams of epitaxial $BaTiO_3$ ultrathin films from first principles. Appl Phys Lett 86:132904
158. Paul J, Nishimatsu T, Kawazoe Y, Waghmare UV (2007) Ferroelectric phase transitions in ultrathin films of $BaTiO_3$. Phys Rev Lett 99:077601

159. Yu J, Wu ZQ, Liu ZR, Yan QM, Wu J, Duan WH (2008) Phase diagram of ferroelectric $BaTiO_3$ ultrathin films under open-circuit conditions. J Phys Condens Matter 20:135203
160. Mitsui T, Westphal WB (1961) Dielectric and X-ray studies of $Ca_xBa_{1-x}TiO_3$ and $Ca_xSr_{1-x}TiO_3$. Phys Rev 124:1354
161. Bednorz JG, Müller KA (1984) $Sr_{1-x}Ca_xTiO_3$: an XY quantum ferroelectric with transition to randomness. Phys Rev Lett 52:2289
162. Bianchi U, Kleemann W, Bednorz JC (1994) Raman scattering of ferroelectric $Sr_{1-x}Ca_xTiO_3$, $x = 0.007$. J Phys Condens Matter 6:1229
163. Itoh M, Wang R, Inaguma Y, Yamaguchi T, Shan YJ, Nakamura T (1999) Ferroelectricity induced by oxygen isotope exchange in strontium titanate perovskite. Phys Rev Lett 82:3540
164. Jang HW, Kumar A, Denev S, Biegalski MD, Maksymovych P, Bark CW, Nelson CT, Folkman CM, Baek SH, Balke N, Brooks CM, Tenne DA, Schlom DG, Chen LQ, Pan XQ, Kalinin SV, Gopalan V, Eom CB (2010) Ferroelectricity in strain-free $SrTiO_3$ thin films. Phys Rev Lett 104:197601

Appendix

A

Aberration Aberration is a deviation from the normal.

Amorphization It is the process of transforming a metallic or nonmetallic material from the crystalline state into the amorphous or glassy state.

Anharmonicity Anharmonicity is the deviation of a system from being in a simple harmonic motion.

Anharmonicity effects in nanocrystals Materials' properties, especially the physical properties, are dependent on temperature. A change in the lattice parameters of crystalline materials is expected when population of the different levels for each normal mode is influenced by variation in temperatures. Therefore, any change of the lattice parameters with temperature is attributed to the anharmonicity of the lattice potential. Raman spectroscopy is a great tool to investigate these effects. The Raman spectra of various nanocrystals as well as other amorphous or crystalline materials show changes in line position and bandwidth with temperature. These changes manifest in shift of line position and a change in line width and intensity.

Anisotropy While isotropy is homogeneity of a property (absorbance, refractive index, density, etc.) in all directions, anisotropy is the property of being directionally dependent.

Anti-Stokes The atoms or molecules in excited state show special lines called anti-Stokes in the Raman spectra.

B

Backscattering In physics, backscatter (or backscattering) is the reflection of waves, particles, or signals back to the direction they came from. It is a diffuse reflection due to scattering, as opposed to specular reflection like a mirror. Backscattering has important applications in astronomy, photography, and medical ultrasonography.

Bimetallic nanomaterials Currently used monometallic SERS tags such as gold and silver NPs have a main drawback in that the metal surface is unprotected. The assays are therefore unreliable as components of the analyte can adsorb on the metal surface, leading to the possibility of replacing the label species. One way to overcome this problem is to encapsulate the metal particle/label molecule in a protective shell. Bimetallic nanostructures, with two different metals in a single particle, such as core-shell-type nanomaterials are gaining importance as Raman tags [1].

Bioanalysis Surface-enhanced Raman scattering (SERS) is seen in molecules that are in close proximity to nanostructured metal surfaces, primarily silver and gold, that are capable of supporting plasmon resonances in the visible spectral region where Raman scattering is excited. The phenomenon provides the basis for a powerful analytical technique offering both quantitative and qualitative molecular information about biological molecules. In contrast to IR spectroscopy, the Raman scattering cross section of water molecules is small, allowing vibrational information to be obtained from biological molecules in their native aqueous environment and their efficient detection and discrimination from the background. Low detection limits, narrow spectral bandwidths, the ability to quench fluorescence, and the capacity to be used with or without optical labels make SERS a good choice for DNA or RNA analysis, genetics and proteomics, medical diagnostics, and the detection of chemical warfare agents.

Biosensor SERS possesses many desirable characteristics as a tool for biosensing with high specificity, attomole to high zeptomole mass sensitivity, and to picomolar concentration sensitivity. Recently, SERS has been used in the quantitative detection and analysis of bio-related molecules [2].

Blackbody radiation Blackbody radiation, coming from all objects simply because of their temperature is greater than absolute zero, is a potential source of spectrally broad background signal.

C

Cancer diagnosis Now it is well established that Raman spectroscopy has great potential in detection of tissue abnormalities and therefore is a promising new tool for noninvasive cancer diagnosis. Raman spectroscopy offers detailed information about tissue biochemistry (including conformations and concentrations of constituents). It particularly provides molecular specific information about tissues, which is necessary for cancer diagnosis. Surface-enhanced Raman spectroscopy (SERS) is an excellent technique that can detect molecular signatures in trace amounts. Similarly, near-infrared (NIR) dispersive Raman spectroscopy, in which NIR excitation minimizes fluorescence and absorption by tissue, has also proved to be a sensitive technique. Raman spectroscopy is particularly suited for diagnosing cancer because of its sensitivity in detecting small molecular changes that are associated with cancer, such as an increased nucleus-to-cytoplasm ratio, disordered chromatin, higher metabolic activity, and changes in lipid and protein levels. Thus, many researchers have applied NIR Raman spectroscopy in vitro, ex vivo, as well as in vivo for the diagnosis of cancer with varying degrees of success [4].

Cell-based biosensors Sensing of toxic chemicals and biological toxins at relevant concentrations and in real time is the need of the day. Using living cells as the sensor elements offers a key advantage that they are not engineered to respond specifically to a single toxic agent but are free to react to many biologically active compounds. Of the many techniques to develop cell-based sensors, which can measure changes in cell behavior such as, cell-cell and cell-substrate contact, metabolism, or induction of cell death following exposure of cells to toxic agents, optical method offers a significant advantage as it is a noninvasive tool. This is truer in the case of Raman spectroscopy as it has already been well established as a powerful analytical technique that can provide useful biochemical information regarding live cells, which can be related to the interaction with toxic agents or drugs, disease, cell death, and differentiation. [3] Several different variants of Raman spectroscopy such as resonant Raman (RS) spectroscopy, surface-enhanced Raman spectroscopy (SERS), and coherent anti-Stokes Raman spectroscopy (CARS), in addition to conventional nonresonant Raman spectroscopy have been used for studying live cells and for developing sensors and biosensors. Of these, conventional nonresonant Raman spectroscopy has a great potential for interrogation of cells. Since different toxic chemicals have different effects on living cells and induce specific time-dependent biochemical changes related to cell death mechanisms, one can obtain comprehensive information of the overall biochemical composition of the cell using the Raman spectrum. It is expected that different toxic agents that initiate different cellular responses and biochemical changes should produce distinct changes in the Raman spectra. In addition, time-resolved Raman spectroscopy of living cells under real-time conditions can be obtained noninvasively. The information obtained will be purely based on intrinsic molecular composition of the cell without any alterations due to the use of labels or other contrast agents as is the case in other types of cell biosensors. The detection of time-dependent biochemical changes of cells has the potential to provide the additional level of information needed for quantification and discrimination of a wider range of toxic agents.

Cellular imaging Raman spectroscopy can be used for imaging of cells at the resolution of conventional microscopy based on the spectral signatures of cell's components; Raman imaging of cellular organelles such as nucleus, chromatin, mitochondria or lipid bodies has been demonstrated. Particularly advantageous is the possibility to obtain the associated chemical information noninvaisively. Raman imaging is also proving to be a useful tool to image uptake and distribution patterns of several drug delivery carriers and nanoparticle-based drug delivery systems.

Charge-transfer resonance Raman (CT-RR) scattering There is a chemical enhancement (CE) because of an increase in the static polarizability of a molecule due to adsorption on the metal and light-induced charge transfer between the molecule and the metal surface resulting in enhancement in the Raman cross section similar to molecular resonance Raman scattering. This effect is sometimes called charge-transfer resonance Raman (CT-RR) scattering.

Chemical enhancement (CE) The interaction of molecules with metallic surfaces may be behind an additional increase in the Raman scattering and this is called chemical enhancement (CE).

Coherent anti-Stokes Raman spectroscopy (CARS) Unlike spontaneous Raman emission, the magnitude of signal in CARS is number of orders stronger as multiple photons are employed to address the molecular vibrations, resulting in production of a signal in which the emitted waves are coherent with one another. In this technique, which is a third-order nonlinear optical process, at least two different laser beams interact with the sample and generate a coherent optical signal at the anti-Stokes frequency. The Raman signal relies on a spontaneous transition, competes with other fluorescent processes and is detected on the red side of the incoming radiation. On the other hand, CARS signal relies on a coherently driven transition and is detected on the blue side, which is free from fluorescence. However, it comes with a nonresonant contribution.

Coherent Raman spectroscopic imaging Coherent anti-Stokes Raman scattering (CARS) microscopy/microspectroscopy has become a powerful technique for three-dimensional vibrational imaging of chemical and biological systems. Recently, significant progress has been made in multiplex CARS micro-spectroscopy where a CARS spectrum covering 3,500 cm^{-1} region was obtainable with the use of a white light laser source.

Coherent Raman spectroscopy Coherent Raman spectroscopy is a term that refers to a series of closely related nonlinear Raman techniques in which the scattered Raman radiation emerges from the sample as a coherent beam – coherent meaning that the photons are all in phase with one another. The coherent techniques include Stimulated Raman Scattering (SRS), Coherent anti-Stokes Raman Spectroscopy (CARS), Coharent Stokes Raman Spectroscopy (CSRS), and Stimulated Raman Gain Spectroscopy (SRGS). Although most of the nonlinear Raman techniques are also coherent techniques, there is one incoherent nonlinear Raman process called Hyper Raman.

Confocal microscopy It is an optical imaging technique to increase optical resolution and contrast. While in a conventional microscope, the entire specimen is flooded evenly in light from a light source resulting in detection of a large unfocused background part, the confocal microscope uses point illumination and a pinhole in an optically conjugate plane in front of the detector to eliminate out-of-focus signal – the name "confocal" stems from this configuration.

D

Dark-field resonant Rayleigh scattering spectroscopy and imaging One approach to obtain excellent signal-to-noise ratio for both imaging and spectroscopy is by utilizing a dark field configuration. In this, a sample is obliquely illuminated through a dark-field condenser that partially blocks the incident light thereby resulting in the scattered light to be collected while providing a dark background. With this low background, it is now possible to carry out spectroscopy on a single nanoparticle or a nanostructure with the ability to make non-ensemble measurements so that individual LSPR spectra can be correlated either with nanostructure topography or surface-enhanced spectroscopic intensity depending on the secondary measurement method used.

Dark-field spectroscopy (DFS) Dark-field spectroscopy, also known as resonant Rayleigh scattering spectroscopy (RRSS), is a technique that correlates the LSPR of single nanoparticles to enhancements in fluorescence and Raman scattering.

Deep-UV tip-enhanced Raman scattering In order to analyze molecules using Raman spectroscopy with a nanoscale resolution beyond the diffraction limit, the technique of choice is tip-enhanced Raman spectroscopy (TERS). It has been successfully applied in the field of nanoscience and nanotechnology for the analysis and imaging of carbon nanotubes, organic dye molecules, DNA bases, various cellular components, and so on. In this technique, the localized surface plasmon resonance (LSPR) supported by a sharp metallic tip provides a strong and highly confined electromagnetic field at the tip apex, which is used to locally excite and enhance the Raman scattering of molecules under investigation. In TERS, the metal tips (e.g., in gold and silver) have an optical property that has negative real and minimal imaginary part in the dielectric function at the excitation wavelength. On the other hand, for deep ultraviolet (DUV) wavelengths there are some metal tips such as aluminum that show dielectric function with small imaginary part and negative real part. Therefore, the technique that utilizes tip-enhancement in the DUV region is called deep-UV tip-enhanced Raman scattering. This technique is proving to be an excellent one for imaging of proteins and nucleic acids as they exhibit electronic resonance in the DUV region. It is to be anticipated that DUV resonance Raman scattering technique along with the tip-enhancement technique will be invaluable for gathering information in the nanoscale for a number of biomedical applications [5].

Degenerate four-wave mixing (DFWM) Degenerate four-wave mixing (DFWM) is another well-known nonlinear technique based on third-order nonlinear susceptibility similar to CARS

Density of states (DOS) In statistical and condensed matter physics, the density of states (DOS) of a system is used to describe the number of states at each energy level that are available to be occupied. A high DOS at a specific energy level means that there are many states available for occupation. A DOS of zero means that no states can be occupied at that energy level.

Dispersive Raman spectroscopy This spectroscopy technique uses grating and/or prism dispersing elements.

Double-resonant Raman scattering In a double-resonant process, different wave vectors of phonons corresponding to different incident phonon energies scatter the excited electron across the band minimum. For phonons with dispersion, the different *k*-vectors correspond to different phonon energies, and hence the excitation-energy dependence of the phonon peak follows naturally. In order to fulfill momentum conservation (both incident and scattered photon momentums are very small), the electron has to be scattered back near to where it was excited. This process is usually ascribed to an elastically scattering defect (*D*-mode and high-energy mode) or a second phonon (second order mode scattering). Of course, in a full calculation of the Raman intensity in double resonance the single resonant process (only the photon is resonant) is automatically included; however, its contribution to the total signal is small. The process is shown schematically in Fig. 1. On the left, one can see double-resonant process for a metallic carbon nanotube with different incident phonon energies implying different phonon wave vectors in the double resonance. On the right, different

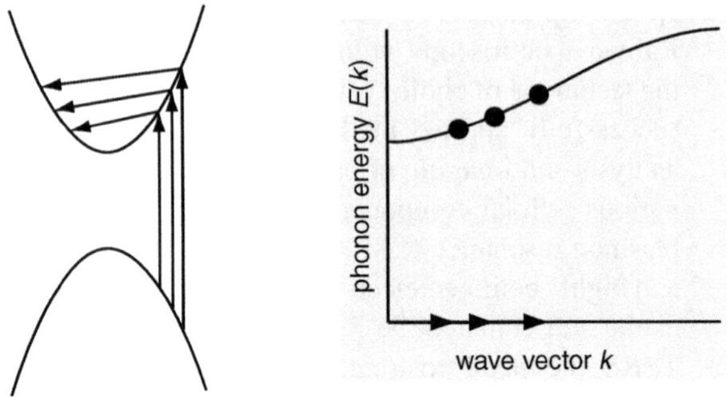

Fig. 1 A schematic representation of process in double-resonant Raman scattering [6]

scattering vectors corresponding to different phonon frequencies are shown.One very well established process for describing a number of features of the Raman spectra of *sp*2-bonded carbon compounds such as graphite or carbon nanotubes is the double-resonant Raman scattering [6].

E

ESM The elastic sphere model [7].

F

Far-from-resonance (FFR) limit The Raman scattering gets simplified dramatically where the exciting laser radiation is far from the lowest allowed excited electronic state of a molecule and this is called as Far-from-resonance (FFR) limit in Raman Spectroscopy. In this limit, the interaction of light with molecules is approximately same for both incident and scattered radiation.

Femtosecond stimulated Raman spectroscopy Femtosecond stimulated Raman spectroscopy (FSRS) is a tool for investigating real-time structural measurements of a chemical change through recording of vibrational structural information with high temporal (50-fs) and spectral (10-cm^{-1}) resolution. It, therefore, enables studies of chemical and biochemical reaction dynamics, giving previously unattainable insight into the structural dynamics of reactively evolving systems. In FSRS, the simultaneous interaction of a narrow-bandwidth picoseconds Raman pulse and a broadband, femtosecond continuum probe pulse leads to the appearance of sharp vibrational gain features on top of the probe envelope [8]. Figure 2 shows a schematic of broadband vibrational probing employed in femtosecond stimulated Raman spectroscopy. In this figure, the Raman pulse is a narrow-bandwidth, picosecond pulse (*green*), whereas the probe is a broadband femtosecond continuum pulse (*purple*). When both pulses are overlapped spatially and temporally in a Raman-active medium, photons are transferred from the high-intensity Raman pulse to the weak probe pulse at the vibrational resonances of a sample. A typical spectrum obtained in a single laser shot for cyclohexane is depicted in Fig. 2. Division of probe spectra obtained in the presence (*black*) and absence (*purple*) of the Raman pulse produces a vibrational spectrum in the expanded trace (*blue*). Conceptually it is similar to femtosecond

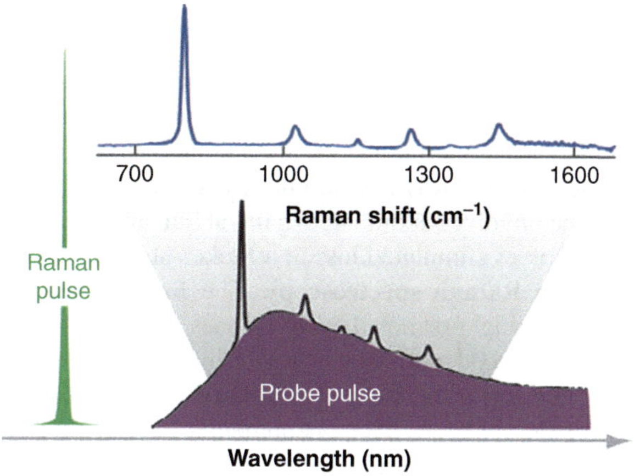

Fig. 2 Illustration of broadband vibrational probing employed in femtosecond stimulated Raman spectroscopy [8]

dynamic electronic absorption spectroscopy where similar disentanglement of time and energy resolution is exploited. However, the difference is that FSRS drives vibrational rather than electronic coherence. Because vibrational dephasing times are much longer (up to picoseconds) than their electronic counterparts (<50 fs), the corresponding energy resolution is excellent (<10 cm^{-1}) [8].

FRET: Forster resonance energy transfer Förster resonance energy transfer, named after German scientist Theodor Förster, is also known as fluorescence resonance energy transfer, resonance energy transfer (RET), or electronic energy transfer (EET). It is a process that describes energy transfer through nonradiative dipole–dipole coupling between two chromophores: a donor chromophore and an acceptor chromophore. When both chromophores are fluorescent, the term "fluorescence resonance energy transfer" is often used in scientific literature.

H

Hyper Raman Hyper Raman is a nonlinear effect in which the vibrational modes interact with the second harmonic of the excitation beam, thereby, allowing observation of vibrational modes that are normally "silent." It frequently relies on SERS-type enhancement to boost the sensitivity.

I

Impulsive stimulated Raman scattering (ISRS) Impulsive stimulated Raman scattering (ISRS) is the creation of coherent ground-state nuclear motion through an impulsive force caused by the interaction of a Raman-active medium with an ultrashort light pulse.

Interferometric Raman spectroscopy Interferometric Raman Spectroscopy is a measurement technique that utilizes time-domain or space-domain measurements of electromagnetic radiation or other type of radiation for collecting Raman spectra based on the coherence of a radiative source. An example is a Fourier transform (FT) Raman spectrometer.

Inverse Raman scattering Inverse Raman scattering (IRS) is a coherent process involving stimulated loss at an anti-Stokes-shifted frequency. The term inverse Raman refers to the fact that, at resonance, the probe radiation is attenuated. In spontaneous Raman spectroscopy, on the other hand radiation at Raman-active frequencies would he generated in the course of the experiment. Inverse Raman scattering (IRS) and stimulated Raman gain (SRG) are closely related. While one involves stimulated gain at an anti-Stokes-shifted frequency, the other involves stimulated loss at a Stokes-shifted frequency.

Inverse Raman spectroscopy The Inverse Raman effect is a form of Raman scattering, first noted by W.J. Jones and B.P. Stoicheff, wherein stokes scattering can exceed anti-Stokes scattering resulting in an absorption line (a dip in intensity) at the sum of irradiated monochromatic light and Raman frequency of the material. This phenomenon is referred to as the *inverse Raman Effect*, application of the phenomenon is referred to as *inverse Raman spectroscopy*, and a record of the continuum is referred to as an *inverse Raman spectrum*.

Inverse spatially offset Raman spectroscopy (SORS) Inverse SORS is a useful sub-variant of SORS that improves certain measurements such as analysis of tissue in vivo. Rather than use spot collection geometry and a circular spot for illumination, the constant offset can be maintained by illuminating the sample with a ring of light centered on the collection region. This has several advantages, including lowering the total power density and allowing simple manipulation of offset distance.

L

Lattice vibrations (Phonons) The vibrations of a crystal are classically described in terms of collective motions in the form of waves called lattice vibrations.

Localized surface plasmon (LSP) The surface plasmon (SP) cannot propagate on the surface of metallic nanoparticles and therefore, is localized and hence known as "localized surface plasmon (LSP)." The LSP resonance of gold and silver NPs occurs in the visible range of the spectrum, which makes these two metals particularly useful for a number of applications ranging from ultrasensitive diagnostic tools to biosensing devices.

M

Michelson interferometer The Michelson interferometer was invented by Albert Abraham Michelson. It is the most common configuration for optical interferometry in which an interference pattern is produced by splitting a beam of light into two paths, bouncing the beams back and recombining them. It is possible that different paths may be of different lengths or be composed of different materials to create alternating interference fringes on a back detector.

Microspectroscopy Raman spectroscopy is a scattering technique and hence specimens do not need to be fixed or sectioned resulting in collecting spectra from a very small volume (<1 μm in diameter) leading to identification of species present in that volume. Raman spectroscopy, therefore, offers several advantages for microscopic analysis, particularly, suitable for microscopic examination of minerals, materials such as polymers and ceramics, cells, and proteins. There are a number of approaches for Raman microspectroscopy. For example, it is possible to record the distribution of a molecule within a cell by selecting a wavenumber characteristic for that molecule from the scattering over a small range of wavenumbers (Raman shifts). This is called as direct imaging. In *hyperspectral imaging* or *chemical imaging*, one could see the distribution of any biomolecule ignoring the presence of water, culture media, buffers, and others that may interfere. Particularly, Raman confocal microscopy has a very high spatial resolution with the lateral and depth resolutions around 250 nm and 1.7 μm, respectively. It is also possible to obtain in vivo time- and space-resolved Raman spectra of microscopic regions of samples such as proteins, cells and organs. In addition, it has also been used in microscopic examination of inorganic specimens, such as rocks and ceramics and polymers.

Multiplexing Multiplexing is the ability to send multiple signals at the same time using a single carrier in the form of a single complex signal and then recovering the separate signals at the receiving end.

N

Nanoparticle -enhanced backscattering Raman imaging

Nanotags – Raman spectroscopy Surface-enhanced Raman spectroscopy (SERS) is gaining in its popularity due to several unique advantages. However, current bottleneck in its utility is the lack of signal reproducibility due to variation in nanoparticle size and shape or aggregation and quantification. Overcoming this is especially important in the case of in vitro or in vivo imaging applications. While aggregation substantially enhances SERS, it is undesirable from an imaging or sensing application standpoint of view since aggregation cannot be controlled and hence results in SERS signal fluctuation. An alternate approach is to encapsulate organic dyes as signature reporter dyes between metallic nanoparticles and a layer of silica or polyethylene glycol (PEG) which prevents agglomeration of the nanoparticles. These are called nanotags, each with a unique Raman spectrum (color). They can be utilized as beacons for imaging with target ligands attached to the PEG or silica surface with well-established bioconjugation chemistries [9].

Near-infrared surface-enhanced Raman spectroscopy Some of the major irritants in Raman measurements are sample fluorescence and photochemistry. However, with the help of Fourier transform (FT) Raman instruments, near-infrared (near-IR) Raman spectroscopy has become an excellent technique for eliminating sample fluorescence and photochemistry in Raman measurements. As demonstrated recently, the range of near-IR Raman techniques can be extended to include near-IR SERS. Near-IR SERS reduces the magnitude of the fluorescence problem because near-IR excitation eliminates most sources of luminescence. Potential applications of near-IR SERS are in environmental monitoring and ultrasensitive detection of highly luminescent molecules [11].

Nonlinear Raman spectroscopy The nonlinear techniques include stimulated Raman scattering (SRS), hyper Raman, stimulated Raman gain (SRG), inverse

Raman scattering (IRS), coherent anti-Stokes Raman spectroscopy (CARS), and coherent Stokes Raman spectroscopy (CSRS). Figure 3 shows Quantum diagrams of the nonlinear Raman processes, with spontaneous Raman included as a reference [10]. The Levels shown are virtual states (——) and ground and excited vibrational levels (-) in the ground electronic state of a molecule. Ground vibrational levels are marked "G." For each Raman process shown, arrows entering at left are incident photons, up arrows represent photon absorption, down arrows represent photon generation, and arrows at right are output photons. Note the conservation in the number of photons created and lost in each process (each photon is represented as an arrow); vI = laser frequency, ω_s = Stokes frequency, Y. = anti-Stokes frequency, Y, = vibrational frequency.

Nyquist frequency The Nyquist frequency is also sometimes known as the folding frequency of a sampling system. Originally discovered by the Swedish-American engineer Harry Nyquist, it is defined as half the sampling frequency of a discrete signal processing system.

P

Partial least squares (PLS) Partial Least Squares (PLS) is a chemo metric technique that enables identification and utilization of the regions of complex overlapping Raman spectra.

Photoacoustic Raman spectroscopy (PARS) Photoacoustic Raman spectroscopy (PARS) is again a nonlinear spectroscopic technique. In this technique, selective population of a given energy state of a system (transitions must involve change in polarizability) is amplified using coherent Raman amplification (also known as stimulated Raman scattering). In this process, it is also important that the frequency difference of the two incident laser beams must be adjusted to equal the frequency of Raman-active transition.

Polarizability Polarizability is defined as the measure of the ability to move the electron cloud in a chemical bond using an external electric field. Since electron cloud in a chemical bond changes with the change in the position of atoms held together by the bond, measurement of polarizability, especially the polarization of Raman scattered light, is important. For a particular molecular vibration, it will be in the same direction as the changing polarizability of the electron cloud.

Polarized analysis There is useful spectral information arising from the analysis of polarization of Raman scattered light. This, typically called as polarized analysis, provides an insight into molecular orientation, molecular shape, and vibrational symmetry. One can also calculate the depolarization ratio. Overall, this technique enables correlation between group theory, symmetry, Raman activity, and peaks in the corresponding Raman spectra. It has been applied successful for solving problems in synthetic chemistry; understanding macromolecular orientation in crystal lattices, liquid crystals or polymer samples and in polymorph analysis.

Polarized Raman spectroscopy Polarized Raman spectroscopy is utilized to investigate molecular orientation and has special significance in the study of polymer systems.

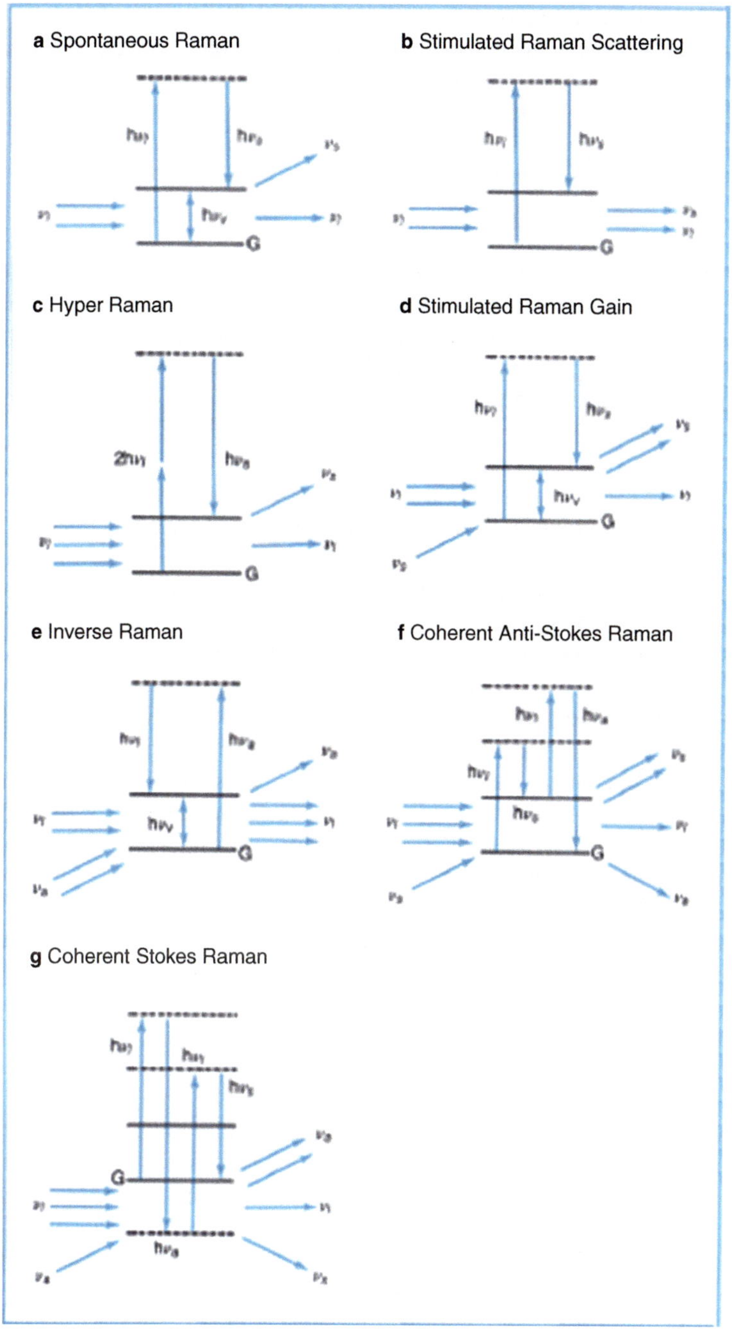

Fig. 3 Quantum diagrams of different nonlinear Raman processes [10]

PRET: Plasmon resonance energy transfer Similar to SERS, plasmon resonance energy transfer (PRET) between biomolecules and the plasmonic nanoparticles that they are adsorbed to allows for detection of biomolecules without appending a fluorophore. This approach yields a considerably higher sensitivity than previously known techniques in addition to spatial resolution amenable for in vivo assays.

Q

Quantitative Raman analysis The intensity of Raman band of an analyte is linearly proportional to the analyte concentration. A plot of analyte band area (integrated intensity) vs, analyte concentration is used to create an equation that predicts analyte concentration from Raman band area. This is an example of quantitative Raman analysis.

R

Raman effect The change in the wavelength of light that occurs when a light beam is deflected by molecules is called Raman Effect. When a beam of light traverses dust-free through a transparent sample of a chemical compound, a small fraction of the light emerges in directions other than that of the incident (incoming) beam. Most of this scattered light is of unchanged wavelength. A small part, however, has wavelengths different from that of the incident light; its presence is a result of the Raman Effect. The phenomenon, discovered in 1929 by Chandrasekhara Venkata Raman, was named after the discoverer.

Raman intensity invariants Rayleigh scattering is an elastic scattering of light or other electromagnetic radiation by particles much smaller than the wavelength of the light, which may be individual atoms or molecules. Named after the British physicist Lord Rayleigh, it can occur when light travels in transparent solids and liquids, but is most prominently seen in gases. It is a function of the electric polarizability of the particles.

Raman near-field scanning optical microscopy (RNSOM) Tip-enhanced Raman spectroscopy (TERS) came originally as an implementation of Raman near-field scanning optical microscopy (RNSOM). In RNSOM, the optical field is confined to a small aperture at the tip of a metal-coated optical fiber brought extremely close to the sample. This allowed overcoming resolution-limiting diffraction phenomena but only a faint signal could be collected owing to the fiber cutoff. In TERS, the optical fiber is replaced with an apertureless metallic tip, which favors surface enhancement of the Raman signal (the so-called SERS effect).

Raman optical activity (ROA) Due to molecular chirality there is a difference in the intensity of Raman scattered right and left circularly polarized light. Raman optical activity (ROA) is a vibrational spectroscopic technique that is reliant on this difference and the spectrum of intensity differences recorded over a range of wavenumbers reveals information about chiral centers within a sample molecule. It is a useful probe to study biomolecular structures and their behavior in aqueous solution especially those of proteins, nucleic acids, carbohydrates, and viruses. The information obtained is in realistic conditions complementing the data from crystallographic approaches in static conditions. Depending on the polarization of the incident and the scattered light, ROA can be observed in a number of forms. When the incident light is linearly polarized and the differences in circular polarization of the scattered light are measured, it is called as the scattered circular polarization (SCP). However, when both the incident and the scattered light are circularly polarized, either in phase (DCPI) or out of phase (DCPII), it is called the dual circular polarization (DCP).

Raman tensors The Raman scattering of a molecule is generated by the interaction of its electrons with an incident light. The electric vector of the scattered light is related to the electric vector of the incident light through a characteristic Raman tensor. A unique Raman tensor exists for each Raman-active molecular vibrational mode [12].

Raman-induced Kerr effect spectroscopy (RIKES) There is an intensity-dependent birefringence induced in a material through its interaction with applied laser fields. This effect is called as the Raman-induced Kerr effect (RIKE). Generally, a media containing ellipsoidal molecules which are free to reorient themselves under the influence of an external electric field show a large optical Kerr effect. The field tends to orient the molecules so that their polarizabilities are largest in the direction of the field. RIKES offers a relatively easy method for suppressing the nonlinear background signal when there are several Raman modes in addition to a nonresonant Kerr susceptibility [13].

S

Scattering The scattering of light is defined as the redirection of light that takes place when an incident light encounters a scattering material. The light can be an electromagnetic (EM) wave. When the light or EM wave interacts with molecules, it results in perturbation of the electron cloud leading to a periodic separation of charge within the molecules, which is called an induced dipole moment. The oscillating induced dipole moment manifests as a source of EM radiation, thereby, resulting in scattered light. The scattering is of two types: elastic and inelastic. In elastic scattering, majority of the light scattered is emitted at the identical frequency of the incident light. However, in inelastic scattering, additional light is scattered at different frequencies and Raman scattering is one such example of inelastic scattering. If the scattered frequency corresponds to the incident frequency, it is elastic scattering (e.g., Mie or Rayleigh). If the frequencies are shifted to lower or higher frequencies, these are therefore inelastic processes with the down-shifted frequency (longer wavelength) referred to as Stokes scattering, and the up-shifted frequency (shorter wavelength) referred to as anti-Stokes scattering.

Scattering cross section The scattering cross section, σ_{scat}, is a hypothetical area which describes the likelihood of light (or other radiation) being scattered by a particle. In general, the scattering cross section is different from the geometrical cross section of a particle, and it depends on the wavelength of the light and the permittivity, shape, and size of the particle. The total amount of scattering in a sparse medium is determined by the product of scattering cross section and the number of particles present.

Shell-isolated nanoparticle-enhanced Raman spectroscopy (SHINERS) Traditionally, tip-enhanced Raman spectroscopy (TERS) studies have been largely limited to molecules having large Raman cross sections. The reason is that the total Raman scattering signal from the tip area (about 20–50 nm in diameter) is rather weak. Adding to this is its high cost and complexity. Therefore, TERS is impractical for many applications. It has been recently demonstrated that if the Au tip is replaced by a monolayer of Au nanoparticles, each coated with an ultrathin shell of silica or alumina (here denoted Au/SiO_2 or Au/Al_2O_3 nanoparticles), with each nanoparticle acting as an Au tip in the TERS system, it is possible to simultaneously bring thousands of TERS tips to the substrate surface to be probed. With this approach one can obtain combined enhanced Raman signal contributed by all of these nanoparticles; which is a two to three orders of magnitude higher than that obtained with a single TERS tip. This new technique is called as shell-isolated nanoparticle-enhanced Raman spectroscopy (SHINERS). Since a chemically inert shell coating is around the Au nanoparticle, it protects the SERS-active nanostructure from contact with whatever is being probed. The main virtue of such a shell-isolated mode is its much higher detection sensitivity and vast number of practical applications utilizing various materials with diverse morphologies [14].

Shifted excitation Raman difference spectroscopy (SERDS) One way to remove the fluorescent background in traditional Raman Spectroscopy is to take advantage of the shift response of the Raman Effect to excitation wavelength shifts. In SERDS, two spectra of a sample are acquired with slightly different excitation wavelengths, and are then subtracted to estimate the Raman spectrum of a sample. This difference will impact the Raman spectra where the entire spectrum will shift in energy by the amount of excitation shift [16].

SIERA: Surface-enhanced infrared absorption As in the case of surface-enhanced Raman scattering (SERS), molecules adsorbed on metal island films or particles exhibit intense infrared absorption several folds higher than what one would expect from conventional measurements without the metal. This effect is referred to as surface-enhanced infrared absorption (SEIRA).

Single-electronic-state (SES) limit The Raman spectroscopy in SES limit is where the incident photon energy is very close to, or falls within, the absorption band of an excited electronic state of a molecule, and the resulting Resonance Raman (RR) scattering is dominated by the properties of this resonant electronic state.

Spatially offset Raman spectroscopy (SORS) Conventional Raman Spectroscopy is limited to the near-surface of diffusely scattering objects and to the first few hundred micrometers depth of surface material. Spatially Offset Raman Spectroscopy (SORS) is a variant of Raman Spectroscopy that allows highly accurate chemical analysis of objects beneath obscuring surfaces. This is done by making at least two Raman measurements; one at the surface and one at an offset position of typically a few millimeters away. To do this without using an offset measurement would be severely restricted by photon shot noise generated by Raman and fluorescence signals originating from the surface layer. It is possible to extend this to multilevel systems requiring multivariate analysis. Accurate chemical analysis of bone beneath skin, tablets inside plastic bottles, explosives inside containers, and counterfeit tablets inside blister packs have been previously carried out. It is also used to discover counterfeit drugs without opening their internal packaging, and for noninvasive monitoring of biological tissue. In inverse SORS, rather than use spot collection geometry and a circular spot for illumination, the constant offset can be maintained by illuminating the sample with a ring of light centered on the collection region.

Stimulated Raman gain Stimulated Raman gain (SRG) and inverse Raman scattering (IRS) are closely related. While one involves stimulated gain at a Stokes-shifted frequency, the other involves stimulated loss at an anti-Stokes-shifted frequency. SRG can be viewed as an induced emission process at the Stokes frequency. Both SRG and IRS are coherent processes.

Stimulated Raman spectroscopy Conventional Raman Spectroscopy is insensitive to be useful in sub cellular imaging [10, 15]. An extension of that technique is Coherent anti-Stokes Raman scattering (CARS) in which two lasers (a pump beam and a Stokes beam) interact with a sample, causing identical molecules to vibrate in phase, generating an enhanced anti-Stokes signal. Thus CARS increases the sensitivity of detection several folds over conventional Raman microscopy. However, the problem with CARS is that the resulting spectrum's strong background makes assigning all but well-isolated bond frequencies very difficult. One approach to eliminate the strong background is to amplify the Raman signal only when the difference in laser frequencies matches a particular molecular frequency and generate no signal if the difference frequencies do not match any molecular frequency. This new approach called stimulated Raman spectroscopy (SRS) trains a pump beam and a Stokes beam on the sample; when the difference in laser frequencies matches a particular molecular frequency, the Raman signal is amplified. When the difference frequency does not match any molecular frequency, no signal is generated, thereby eliminating the spectral background. Thus the SRS spectrum is essentially identical to a conventional Raman spectrum, making spectral assignment much easier than in CARS. Also, SRS microscopy allows three-dimensional imaging of specific molecular species to a depth of about 0.3 mm making it particularly useful for imaging lipids and small molecules such as drugs. It is to be noted that stimulated Raman scattering, unlike stimulated Raman gain (SRG) is a nonlinear technique in which the spontaneous Raman radiation is amplified and the amplified radiation emerges as a coherent beam coincident with the direction of the incident laser radiation.

Stokes and anti-Stokes scattering There are two types of Raman scattering, Stokes scattering and anti-Stokes scattering. In Stokes scattering, molecules absorb energy and the resulting photon of lower energy generates a Stokes line on the red side of the incident spectrum. On the other hand, in anti-Stokes scattering, molecules lose energy because the incident photons are shifted to the blue side of the spectrum, thus generating an anti-Stokes line. Since lower energy states will have more molecules in them than in the case of higher (excited) energy states, the Stokes spectrum will be more intense than the anti-Stokes spectrum.

Surface plasmons Surface plasmons (SPs) are collective excitations of the electrons within the conduction band of a metal.

T

Thermo-Raman spectroscopy Raman spectroscopy is a useful technique to extract information during dynamic thermal processes and this specific application is termed as thermo-Raman spectroscopy (TRS). It is possible to investigate thermally induced changes in Raman band positions, band intensities, and bandwidths and correlate with corresponding structural changes in samples. TRS can also provide quantitative information related to the dynamics thermal processes. Unlike techniques such as thermogravimetric analysis (TGA) and differential thermal analysis (DTA) which can only provide bulk information associated with thermal properties of a solid sample, TRS can be used to study thermally induced structural transformation in solids [17].

Time-resolved Raman spectroscopy (TRRAS) Time-resolved Raman spectroscopy (TRRAS) is used to analyze low energy excitation on ultrafast timescale. While it is not directly sensitive with respect to the atomic positions, it is possible to extract information related to structural, magnetic, and electronic properties. In this technique, by monitoring the Stokesto anti-Stokes ratio, one can obtain information about the thermodynamics of different excitations measuring directly the temperature of the excitation investigated, and thereby to disentangle. A unique aspect of TRRAS is that one can measure the "strength" of the crystalline potential in real time. Since it is inexpensive to carry out TRRAS, this is one of the most sought after tools for studying electronic dynamics. It is also a power tool for investigating structural and magnetic dynamics.

Tomography-Raman Even though the Raman Effect is weaker than fluorescence, algorithms developed for fluorescence imaging can be applied to Raman signals. Using the waves of energy one can image section by section of an object, thereby creating what is known as tomography. Since Raman signals are used for imaging, the tomography is called Raman Tomography.

Transmission Raman spectroscopy (TRS) Raman spectroscopy has several variants and Transmission Raman Spectroscopy (TRS) is one such. In a conventional back-scattering Raman spectroscopy the signal tends to be a representative of the surface and near-surface composition. In TRS Raman photons can be created at all points such that the light passing through the total scrambled Raman signal can be measured on the opposite face of an object. Therefore, Transmission Raman Spectroscopy (TRS) is highly representative of the bulk of the material and a large thickness can be measured in the absence of photon absorption. This produces an analysis representative of the entire mixture and is typically insensitive to coatings, or thin containers. The technique is also somewhat similar to Spatially Offset Raman Spectroscopy, where the light in a diffusely scattering sample spreads through the object randomly. Therefore,

TRS can be regarded as an extreme example of SORS. Most importantly, Transmission Raman is rapid as it requires no sample preparation and involves no phase change.

W

Waveguide Raman spectroscopy (WRS) Waveguide Raman spectroscopy (WRS), in which the pump laser beam in traditional Raman spectroscopy is coupled into a planar dielectric waveguide in order to probe the Raman-active vibrations of the waveguiding material, is used to study thin films, surfaces, and interfaces. Since micrometer dimensions of a waveguide can generate large electric field strengths resulting in high intensities, WRS signals are three to four orders of magnitude higher compared to those from conventional Raman scattering [18].

References

1. Shen A, Chen L, Xie W, Hu J, Zeng A, Richards R, Hu J (2010) Triplex Au–Ag–C core–shell nanoparticles as a novel Raman label. Adv Funct Mater 20:969–975
2. Zhang X, Yonzon CR, Young MA, Stuart DA, Van Duyne RP (2005) Surface-enhanced Raman spectroscopy biosensors: excitation spectroscopy for optimisation of substrates fabricated by nanosphere lithography. IEE Proc Nanobiotechnol 152(6):195–206
3. Notingher I (2007) Raman spectroscopy cell-based biosensors. Sensors 7:1343–1358
4. Keller MD, Kanter EM, Mahadevan-Jansen A (2006) Raman spectroscopy for cancer diagnosis. Spectroscopy 21(11):33–41
5. Taguchi A, Hayazawa N, Furusawa K, Ishitobi H, Kawata S (2009) Deep-UV tip-enhanced Raman scattering. J Raman Spectrosc 40:1324–1330
6. Thomsen C, Maultzsch J, Reich S (2003) Double-resonant Raman scattering in an individual carbon nanotube. AIP Conf Proc 685:225–229. http://www.physik.fu-berlin.de/einrichtungen/ag/ag-reich/docs/Thomsen2003.pdf
7. (a) Lamb H (1882) On the vibrations of an elastic sphere. Proc Lond Math Soc 13:189–212; (b) Duval E (1992) Far-infrared and Raman vibrational transitions of a solid sphere: selection rules. Phys Rev B 46:5795–5797; (c) Verma P, Cordts W, Irmer G, Monecke J (1999) Acoustic vibrations of semiconductor nanocrystals in doped glasses. Phys Rev B 60: 5778–5785
8. Kukura P, McCamant DW, Mathies RA (2007) Femtosecond stimulated Raman spectroscopy. Annu Rev Phys Chem 58:461–488
9. Xiao M, Nyagilo J, Arora V, Kulkarni P, Xu D, Sun X, Davé DP (2010) Gold nanotags for combined multi-colored Raman spectroscopy and x-ray computed tomography. Nanotechnology 21(3):035101
10. Borman SA (1982) Nonlinear Raman spectroscopy. Anal Chem 54(9):1021A
11. Angel SM, Myrick ML (1989) Near-infrared surface-enhanced Raman spectroscopy using a diode laser. Anal Chem 61:1648–1652
12. Tsuboi M, Benevides JM, Thomas GJ Jr (2009) Raman tensors and their application in structural studies of biological systems. Proc Jpn Acad Ser B 85:83–87
13. Borysow J, Taylor RH, Keto JW (1989) Raman-induced Kerr effect spectroscopy of solutions. J Raman Spectrosc 20(4):203–208
14. Li JF, Huang YF, Ding Y, Yang ZL, Li SB, Zhou XS, Fan FR, Zhang W, Zhou ZY, Wu DY, Ren B, Wang ZL, Tian ZQ (2010) Shell-isolated nanoparticle-enhanced Raman spectroscopy. Nature 464:392–395

15. Freudiger CW et al (2008) Label-free biomedical imaging with high sensitivity by stimulated Raman scattering microscopy. Science 322:1857–1861
16. McCain MT, Willett RM, Brady DJ (2008) Multi-excitation Raman spectroscopy technique for fluorescence rejection. Opt Express 16(15):10975–10991
17. Widjaja E, Chong HH, Tjahjono M (2010) Use of thermo-Raman spectroscopy and chemometric analysis to identify dehydration steps of hydrated inorganic samples – application to copper sulfate pentahydrate. J Raman Spectrosc 41:181–186
18. Kanger JS, Otto C, Slotboom M, Greve J (1996) Waveguide Raman spectroscopy of thin polymer layers and monolayers of biomolecules using high refractive index waveguides. J Phys Chem 100:3288–3292

Index

GPSR Compliance

The European Union's (EU) General Product Safety Regulation (GPSR) is a set of rules that requires consumer products to be safe and our obligations to ensure this.

If you have any concerns about our products, you can contact us on ProductSafety@springernature.com

In case Publisher is established outside the EU, the EU authorized representative is:

Springer Nature Customer Service Center GmbH
Europaplatz 3
69115 Heidelberg, Germany

Batch number: 10755310

Printed by Printforce, the Netherlands